K. Tanaka · E.W. Davie (Eds.)
Y. Ikeda · S. Iwanaga · H. Saito · K. Sueishi (Associate Eds.)
Recent Advances in Thrombosis and Hemostasis 2008

K. Tanaka · E.W. Davie (Eds.)
Y. Ikeda · S. Iwanaga · H. Saito · K. Sueishi
(Associate Eds.)

Recent Advances in Thrombosis and Hemostasis 2008

 Springer

Editors

Kenzo Tanaka, M.D., Ph.D., Professor Emeritus
Department of Pathology, Kyushu University
3-1-1 Maidashi, Higashi-ku, Fukuoka 812-8582, Japan

Earl W. Davie, Ph.D., Professor
Department of Biochemistry, University of Washington
J317 Health Science Center, Box 357350, Seattle WA 98195, USA

Associate editors

Yasuo Ikeda, M.D., Ph.D., Professor
Keio University, School of Medicine, Internal Medicine
35 Shinanomachi, Shinjuku-ku, Tokyo 160-8582, Japan

Sadaaki Iwanaga, Ph.D., Professor Emeritus
Department of Biology, Kyushu University
6-10-1 Hakozaki, Higashi-ku, Fukuoka 812-8581, Japan

Hidehiko Saito, M.D., Ph.D., Professor Emeritus
Department of Medicine, Nagoya University
26 Tsurumai-cho, Showa-ku, Nagoya 466-8550, Japan

Katsuo Sueishi, M.D., Ph.D., Professor
Pathophysiological and Experimental Division, Department of Pathology
Kyushu University
3-1-1 Maidashi, Higashi-ku, Fukuoka 812-8582, Japan

Library of Congress Control Number: 2008923271

ISBN 978-4-431-78846-1   e-ISBN 978-4-431-78847-8

Springer is a part of Springer Science+Business Media
springer.com

© Springer 2008
Printed in Japan

Typesetting: SNP Best-set Typesetter Ltd., Hong Kong
Printing and binding: Kato Bunmeisha, Japan

Printed on acid-free paper

Preface

Now more than ever, thrombotic and thromboembolic disorders as well as related diseases such as malignancies, arteriosclerosis, diabetes mellitus, hypertension, and obesity are the leading causes of morbidity and mortality. They have become urgent medical problems with serious economic consequences in industrialized and developing countries alike. At the same time, the impact of molecular biology and genetics on our understanding of thrombosis and hemostasis is rapidly growing stronger as well as our knowledge of regeneration and development of specific tissues, organs, and embryos. Researchers are also constantly learning more about cardiovascular diseases as well as regulatory mechanisms for various intrinsic and extrinsic stimuli in viable tissues. In this volume, our intention has been to present the latest relevant information in molecular biology and genetics as well as the clinical implications of a better understanding of pathophysiology, novel diagnostic methodologies, and therapeutic applications for new methods of prevention in thrombosis/hemostasis and related disorders, including atherosclerosis.

The dramatic advances in knowledge of thrombosis/hemostasis and vascular biology since the first publication of *Recent Advances in Thrombosis and Fibrinolysis*, edited with Japanese colleagues, in 1991, have required extensive revision in order to highlight and review recent progress in the field. The editors also gratefully welcome the seven distinguished non-Japanese authors, who, with their valuable contributions on subjects beyond the coverage by Japanese authors, have made this new edition truly international.

We hope that this book will prove useful to readers who wish to expand their knowledge of the biology and medicine of thrombosis and hemostasis, thus justifying the effort and hard work of all those involved in its publication. We also note that the XXIIIrd International Congress of the International Society of Thrombosis and Haemostasis (ISTH) will be held in Kyoto, Japan, in July 2011. We earnestly hope that the endeavors that went into the publishing of this book will help to promote the scientific excitement that we anticipate for the 2011 ISTH Congress in Kyoto and will contribute to its greater success.

March 2008
Kenzo Tanaka and Earl W. Davie, Editors
Yasuo Ikeda, Sadaaki Iwanaga, Hidehiko Saito,
and Katsuo Sueishi, Associate Editors

Acknowledgments

A special acknowledgment is due to the following colleagues who undertook the major task of coordinating various chapters and providing helpful critiques: Ikuro Maruyama, Osamu Matsuo, Yukio Ozaki, Yoichi Sakata, Koji Suzuki, and Akira Yoshioka (all from the Japanese Society of Thrombosis and Hemostasis). Our deepest appreciation also goes to the various contributors for their professional expertise and cooperation.

Contents

Contributors

Adams, Ty E.
Department of Haematology, Division of Structural Medicine, Thrombosis Research Unit, Cambridge Institute for Medical Research, University of Cambridge
Wellcome Trust/MRC, Cambridge CB2 2XY, UK

Amengual, Olga
Department of Medicine II, Hokkaido University Graduate School of Medicine
N-15 W-7, Kita-ku, Sapporo 060-8638, Japan

Asada, Yujiro
Department of Pathology, Faculty of Medicine, University of Miyazaki
5200 Kihara, Kiyotake, Miyazaki-gun, Miyazaki 889-1692, Japan

Atsumi, Tatsuya
Department of Medicine II, Hokkaido University Graduate School of Medicine
N-15 W-7, Kita-ku, Sapporo 060-8638, Japan

Banno, Fumiaki
Department of Vascular Physiology, National Cardiovascular Center Research Institute
5-7-1 Fujishiro-dai, Suita, Osaka 565-8565, Japan

Bolotova, Tatiana
Age Dimension Research Center, National Institute of Advanced Industrial Science and Technology (AIST)
AIST Tsukuba Center 6-13, 1-1-1 Higashi, Tsukuba, Ibaraki 305-8566, Japan

Callaghan, Michael U.
Fellow Department of Pediatrics, Division of Hematology/Oncology, Children's Hospital of Michigan
3901 Beaubien Blvd., Detroit, MI 48201, USA

Everse, Stephen J.
Department of Biochemistry, University of Vermont
89 Beaumont Avenue, Given B418A, Burlington, VT 05405-0068, USA

Fujikawa, Kazuo
Department of Biochemistry, University of Washington
Box 357350, Seattle, Washington 98195-7350, USA

Fujimura, Yoshihiro
Department of Blood Transfusion Medicine, Nara Medical University Hospital
840 Shijo-cho, Kashihara, Nara 634-8522, Japan

Fukudome, Kenji
Department of Biomolecular Sciences, Division of Immunology, Saga Medical School
Room 2158, 5-1-1 Nabeshima, Saga 849-8501, Japan

Gabazza, Esteban
Department of Immunology, Mie University Graduate School of Medicine
2-174 Edobashi, Tsu, Mie 514-8507, Japan

Gokudan, Soutaro
The Chemo-Sero-Therapeutic Research Institute (KAKETSUKEN)
1-6-1 Okubo, Kumamoto 813-0041, Japan

Goto, Shinya
Department of Medicine, Tokai University School of Medicine
143 Shimokasuya, Isehara, Kanagawa 259-1143, Japan

Hamada, Toshiyuki
Age Dimension Research Center, National Institute of Advanced Industrial Science and Technology (AIST)
AIST Tsukuba Center 6-13, 1-1-1 Higashi, Tsukuba, Ibaraki 305-8566, Japan

Hamako, Jiharu
Fujita Health University College
1-98 Dengakugakubo, Kutsukake-cho,Toyoake, Aichi 470-1192, Japan

Hamasaki, Naotaka
Faculty of Pharmaceutical Science, Nagasaki International University
2825-7 Huis Ten Bosch, Sasebo, Nagasaki 859-3598, Japan

Harada, Naoaki
Department of Translational Medical Science Research, Nagoya City University Graduate School of Medical Sciences
1 Kawasumi, Mizuho-cho, Mizuho-ku, Nagoya 467-8601, Japan

Haruta, Shoji
Department of Obstetrics and Gynecology, Nara Medical University
840 Shijo-cho, Kashihara, Nara 634-8522, Japan

Hashiguchi, Teruto
Laboratory and Vascular Medicine, Cardiovascular and Respiratory Disorders, Advanced Therapeutics Course, Kagoshima University Graduate School of Medical and Dental Sciences
8-35-1 Sakuragaoka, Kagoshima 890-8544, Japan

Hayashi, Tatsuya
Department of Immunology, Mie University Graduate School of Medicine
2-174 Edobashi, Tsu, Mie 514-8507, Japan

Hebling, Christine M.
Department of Chemistry, University of North Carolina at Chapel Hill
A008 Kenan Labs, Chapel Hill, NC 27599-3290, USA

Huntington, James A.
Department of Haematology, University of Cambridge, Division of Structural Medicine, Thrombosis Research Unit, Cambridge Institute for Medical Research
Wellcome Trust/MRC Building, Hills Road, Cambridge CB2 2XY, UK

Ichinose, Akitada
Department of Molecular Patho-Biochemistry and Patho-Biology on Blood and Circulation, School of Medicine, Yamagata University Faculty of Medicine
2-2-2 Iida-Nishi, Yamagata, Yamagata 990-9585, Japan

Ikeda, Yasuo
Division of Hematology, Department of Internal Medicine, Keio University School of Medicine
35 Shinanomachi, Shinjuku-ku, Tokyo 160-8582, Japan

Isaka, Naoki
Division of Clinical Medicine and Biomedical Science, Mie University School of Medicine
2-174 Edobashi, Tsu, Mie 514-8507, Japan

Ito, Masaaki
Division of Clinical Medicine and Biomedical Science, Mie University School of Medicine
2-174 Edobashi, Tsu, Mie 514-8507, Japan

Ito, Takashi
Laboratory and Vascular Medicine, Cardiovascular and Respiratory Disorders, Advanced Therapeutics Course, Kagoshima University Graduate School of Medical and Dental Sciences
8-35-1 Sakuragaoka, Kagoshima 890-8544, Japan

Iwanaga, Sadaaki
The Chemo-Sero-Therapeutic Research Institute (KAKETSUKEN)
2-34-16 Mizutani, Higashi-ku, Fukuoka 813-0041, Japan

Jung, Stephanie M.
Division of Protein Biochemistry, Institue of Life Sciences, Kurume University
1-1 Hyakunen-Koen, Kurume, Fukuoka 839-0842, Japan

Kanaji, Taisuke
Division of Hematology, Department of Medicine, Kurume University School of Medicine
67 Asahimachi, Kurume, Fukuoka 830-0011, Japan

Kanayama, Seiji
Department of Obstetrics and Gynecology, Nara Medical University
840 Shijo-cho, Kashihara, Nara 634-8522, Japan

Kashiwagi, Hirokazu
Department of Hematology and Oncology, Graduate School of Medicine, Osaka University
2-2 Yamadaoka, Suita, Osaka 565-0871, Japan

Kato, Hisao
National Cardiovascular Center Research Institute
5-7-1 Fujishirodai, Suita, Osaka 565-8565, Japan

Kaufman, Randal J.
Departments of Biological Chemistry and Internal Medicine, MSRBII 4554, University of Michigan Medical Center, Howard Hughes Medical Institute
Ann Arbor, MI 48109-0650, USA

Kawaguchi, Ryuji
Department of Obstetrics and Gynecology, Nara Medical University
840 Shijo-cho, Kashihara, Nara 634-8522, Japan

Kawahara, Ko-ichi
Laboratory and Vascular Medicine, Cardiovascular and Respiratory Disorders, Advanced Therapeutics Course, Kagoshima University Graduate School of Medical and Dental Sciences
8-35-1 Sakuragaoka, Kagoshima 890-8544, Japan

Key, Nigel S.
932 Mary Ellen Jones Bldg, CB# 7035, University of North Carolina
Chapel Hill, NC 27599-7035, USA

Kimura, Yumi
Department of Neurology, Tokyo Women's Medical University
8-1 Kawada-cho, Shinjuku-ku, Tokyo 162-8666, Japan

Kobayashi, Hiroshi
Department of Obstetrics and Gynecology, Nara Medical University
840 Shijo-cho, Kashihara, Nara 634-8522, Japan

Koide, Takehiko
Graduate School of Life Science, University of Hyogo
3-2-1 Kouto, Kamigori-cho, Ako-gun, Hyogo 678-1297, Japan

Koike, Takao
Department of Medicine II, Hokkaido University Graduate School of
Medicine
N-15 W-7, Kita-ku, Sapporo 060-8638, Japan

Kojima, Soichi
Molecular Cellular Pathology Research Unit, Discovery Research Institute,
RIKEN
2-1 Hirosawa, Wako, Saitama 351-0198, Japan

Kojima, Tetsuhito
Department of Medical Technology, Nagoya University School of Health
Sciences
1-1-20 Daiko-Minami, Higashi-ku, Nagoya 461-8673, Japan

Kurachi, Kotoku
Age Dimension Research Center, National Institute of Advanced Industrial
Science and Technology (AIST)
AIST Tsukuba Center 6-13, 1-1-1 Higashi, Tsukuba, Ibaraki 305-8566, Japan

Kurachi, Sumiko
Age Dimension Research Center, National Institute of Advanced Industrial
Science and Technology (AIST)
AIST Tsukuba Center 6-13, 1-1-1 Higashi, Tsukuba, Ibaraki 305-8566, Japan

Mann, Kenneth G.
Department of Biochemistry, University of Vermont
208 South Park Drive Suite 2, Room T227, Colchester, VT 05446, USA

Maruyama, Ikuro
Laboratory and Vascular Medicine, Cardiovascular and Respiratory Disorders, Advanced Therapeutics Course, Kagoshima University Graduate School of Medical and Dental Sciences
8-35-1 Sakuragaoka, Kagoshima 890-8544, Japan

Matsubara, Yumiko
Division of Hematology, Department of Internal Medicine, Keio University School of Medicine
35 Shinanomachi, Shinjuku-ku, Tokyo 160-8582, Japan

Matsui, Taei
Department of Biology, Fujita Health University, School of Health Sciences
1-98 Dengakugakubo, Kutsukake-cho, Toyoake, Aichi 470-1192, Japan

Matsumoto, Masanori
Department of Blood Transfusion Medicine, Nara Medical University
840 Shijo-cho, Kashihara, Nara 634-8522, Japan

Matsuo, Osamu
Department of Physiology, Kinki University School of Medicine
377-2 Ohnohigashi, Osakasayama, Osaka 589-8511, Japan

Miyahara, Masatoshi
Division of Clinical Medicine and Biomedical Science, Mie University School of Medicine
2-174 Edobashi, Tsu, Mie 514-8507, Japan

Miyata, Toshiyuki
Department of Etiology and Pathogenesis, National Cardiovascular Center Research Institute
5-7-1 Fujishiro-dai, Suita, Osaka 565-8565, Japan

Mizuguchi, Jun
The Chemo-Sero-Therapeutic Research Institute (KAKETSUKEN)
1-6-1 Okubo, Kumamoto, Kumamoto 813-0041, Japan

Monroe, Dougald M.
932 Mary Ellen Jones Bldg, CB# 7035, University of North Carolina
Chapel Hill, NC 27599-7035, USA

Morita, Takashi
Department of Biochemistry, Meiji Pharmaceutical University
2-522-1 Noshio, Kiyose, Tokyo 204-8588, Japan

Moroi, Masaaki
Division of Protein Biochemistry, Institute of Life Sciences, Kurume University
1-1 Hyakunen-Koen, Kurume, Fukuoka 839-0842, Japan

Mosesson, Michael W.
Blood Research Institute, Blood Center of Wisconsin
PO Box 2178, Milwaukee, WI 53201-2178, USA

Murata, Mitsuru
Division of Hematology, Department of Internal Medicine, Keio University School of Medicine
35 Shinanomachi, Shinjuku-ku, Tokyo 160-8582, Japan

Nagai, Nobuo
Department of Physiology, Kinki University School of Medicine
377-2 Ohnohigashi, Osakasayama, Osaka 598-8511, Japan

Nakamura, Mashio
Division of Clinical Medicine and Biomedical Science, Mie University School of Medicine
2-174 Edobashi, Tsu, Mie 514-8507, Japan

Nakamura, Tomomi
Department of Neurology, Tokyo Women's Medical University
8-1 Kawada-cho, Shinjuku-ku, Tokyo 162-8666, Japan

Nakano, Takeshi
Division of Clinical Medicine and Biomedical Science, Mie University School of Medicine
2-174 Edobashi, Tsu, Mie 514-8507, Japan

Nakashima, Yutaka
Division of Pathology, Fukuoka Red Cross Hospital
3-1-1 Ogusu, Minami-ku, Fukuoka 815-8555, Japan

Nishino, Masato
Nara Prefectural Mimuro Hospital
1-14-16 Mimuro, Sango-cho, Ikoma-gun, Nara 636-0802, Japan

Oi, Hidekazu
Department of Obstetrics and Gynecology, Nara Medical University
840 Shijo-cho, Kashihara, Nara 634-8522, Japan

Okada, Kiyotaka
Department of Physiology, Kinki University School of Medicine
377-2 Ohnohigashi, Osakasayama, Osaka 589-8511, Japan

Okajima, Kenji
Department of Translational Medical Science Research, Nagoya City University, Graduate School of Medical Sciences
1 Kawasumi, Mizuho-cho, Mizuho-ku, Nagoya 467-8601, Japan

Onimaru, Mitsuho
Pathophsiological and Experimental Division, Department of Pathology, Graduate School of Medical Sciences, Kyushu University
3-1-1 Maidashi, Higashi-ku, Fukuoka 812-8582, Japan

Ota, Satoshi
Division of Clinical Medicine and Biomedical Science, Mie University School of Medicine
2-174 Edobashi, Tsu, Mie 514-8507, Japan

Ozaki, Yukio
Department of Clinical Laboratory Medicine, University of Yamanashi
1110 Shimogato, Chuo, Yamanashi 409-3898, Japan

Roberts, Harold R.
932 Mary Ellen Jones Bldg, CB# 7035, University of North Carolina
Chapel Hill, NC 27599-7035, USA

Saito, Hidehiko
Nagoya Central Hospital
3-7-7 Taiko, Nakamura-ku, Nagoya 453-0801, Japan

Sakata, Mariko
Department of Obstetrics and Gynecology, Nara Medical University
840 Shijo-cho, Kashihara, Nara 634-8522, Japan

Sakata, Yoichi
Division of Cell and Molecular Medicine, Center for Molecular Medicine, Jichi Medical University
3311-1 Yakushiji, Shimotsuke, Tochigi 329-0498, Japan

Sejima, Takayuki
Department of Otolaryngology-Head & Neck Surgery, Jichi Medical University
3311-1 Yakushiji, Shimotsuke, Tochigi 329-0498, Japan

Shima, Midori
Department of Pediatrics, Nara Medical University
840 Shijo-cho, Kashihara, Nara 634-8522, Japan

Shiraga, Masamichi
Department of Hematology and Oncology, Graduate School of Medicine, Osaka University
2-2 Yamadaoka, Suita, Osaka 565-0871, Japan

Solovieva, Elena
Age Dimension Research Center, National Institute of Advanced Industrial Science and Technology (AIST)
AIST Tsukuba Center 6-13, 1-1-1 Higashi, Tsukuba, Ibaraki 305-8566, Japan

Stafford, Darrel W.
The University of North Carolina at Chapel Hill, Department of Biology
442 Wilson Hall, Chapel Hill, NC 27599-3280, USA

Sueishi, Katsuo
Pathophsiological and Experimental Division, Department of Pathology, Graduate School of Medical Sciences, Kyushu University
3-1-1 Maidashi, Higashi-ku, Fukuoka 812-8582, Japan

Suenaga, Emi
Age Dimension Research Center, National Institute of Advanced Industrial Science and Technology (AIST)
AIST Tsukuba Center 6-13, 1-1-1 Higashi, Tsukuba, Ibaraki 305-8566, Japan

Suzuki, Koji
Department of Molecular Pathobiology, Mie University Graduate School of Medicine
2-174 Edobashi, Tsu, Mie 514-8507, Japan

Taguchi, Osamu
Department of Pulmonary and Critical Care Medicine, Mie University Graduate School of Medicine
2-174 Edobashi, Tsu, Mie 514-8507, Japan

Takeya, Hiroyuki
Division of Pathological Biochemistry, Department of Biomedical Sciences, School of Life Science, Faculty of Medicine Tottori University
86 Nishimachi, Yonago, Tottori 683-8503, Japan

Titani, Koiti
Department of Biology, Fujita Health University, School of Health Sciences
1-98 Dengakugakubo, Kutsukake-cho, Toyoake, Aichi 470-1192, Japan

Tomiyama, Yoshiaki
Department of Blood Transfusion, Osaka University Hospital
2-15 Yamadaoka, Suita, Osaka 565-0871, Japan
Department of Hematology and Oncology, Graduate School of Medicine, Osaka University
2-2 Yamadaoka, Suita, Osaka 565-0871, Japan

Tsuji, Yoriko
Department of Obstetrics and Gynecology, Nara Medical University
840 Shijo-cho, Kashihara, Nara 634-8522, Japan

Uchiyama, Shinichiro
Department of Neurology, Tokyo Women's Medical University
8-1 Kawada-cho, Shinjuku-ku, Tokyo 162-8666, Japan

Ueshima, Shigeru
Department of Physiology, Kinki University School of Medicine
377-2 Ohnohigashi, Osakasayama, Osaka 589-8511, Japan
Department of Food Science and Nutrition, Kinki University School of Agriculture
3327-204 Nakamachi, Nara, Nara 631-8505, Japan

Urano, Tetsumei
Department of Physiology, Hamamatsu University School of Medicine
1-20-1 Handa-yama, Higashi-ku, Hamamatsu, Shizuoka 431-3192, Japan

Wada, Hideo
Department of Molecular and Laboratory Medicine, Mie University Graduate School of Medicine
2-174 Edobashi, Tsu, Mie 514-8507, Japan

Yagi, Hideo
Department of Hematology, Nara Hospital, Kinki University School of Medicine
1248-1 Otoda-cho, Ikoma, Nara 630-0293, Japan

Yamada, Norikazu
Division of Clinical Medicine and Biomedical Science, Mie University School of Medicine
2-174 Edobashi, Tsu, Mie 514-8507, Japan

Yamada, Yoshihiko
Department of Obstetrics and Gynecology, Nara Medical University
840 Shijo-cho, Kashihara, Nara 634-8522, Japan

Yamashita, Atsushi
Department of Pathology, Faculty of Medicine, University of Miyazaki
5200 Kihara, Kiyotake, Miyazaki-gun, Miyazaki 889-1692, Japan

Yamashita, Jun K.
Laboratory of Stem Cell Differentiation, Stem Cell Research Center, Institute
for Frontier Medical Sciences, Kyoto University
53 Shogoin, Kawahara-cho, Sakyo-ku, Kyoto 606-8507, Japan

Yamazaki, Masako
Department of Neurology, Tokyo Women's Medical University
8-1 Kawada-cho, Shinjuku-ku, Tokyo 162-8666, Japan

Yamazaki, Yasuo
Department of Biochemistry, Meiji Pharmaceutical University
2-522-1 Noshio, Kiyose, Tokyo 204-8588, Japan

Yatomi, Yutaka
Department of Laboratory Medicine, Graduate School of Medicine and Faculty
of Medicine, The University of Tokyo
7-3-1 Hongo, Bunkyo-ku, Tokyo 113-8655, Japan

Yoshioka, Akira
Department of Pediatric, Nara Medical University
840 Shijo-cho, Kashihara, Nara 634-8522, Japan

Part 1 Coagulation Mechanism

Structure and Functions of Fibrinogen and Fibrin

Michael W. Mosesson

Summary. Fibrinogen molecules consist of three pairs of disulfide-bridged chains joined together in the amino-terminal central E domain, which is connected by coiled coils to its outer D domains. Thrombin cleavage of fibrinopeptide A (FPA) from fibrinogen to form fibrin results in double-stranded fibrin fibrils through end-to-middle D:E associations. Lateral fibril associations form fibers and also diverge to form bilateral branches. Formation of equilateral branch junctions completes the basic clot network structures. Intermolecular transversely aligned C-terminal γ chains in fibrils become crosslinked by factor XIII to form γ-dimers. Concomitantly, inter-molecular α chain crosslinking results in α-polymers. Transversely crosslinked γ chains in fibrin account for the complete elastic recovery of maximally stretched fibrin after maximum clot deformation. Fibrin(ogen) participates in other biological functions including: (1) molecular and cellular interactions of fibrin β15-42, which binds to heparin and also mediates platelet and endothelial cell spreading, as well as capillary tube formation via VE-cadherin, an endothelial cell receptor; (2) leukocyte binding to fibrin(ogen) via integrin $\alpha_M\beta_2$ (Mac-1), a receptor on stimulated monocytes and neutrophils; (3) enhanced extracellular matrix interactions by binding to fibronectin; (4) binding to the platelet $\alpha_{IIb}\beta_3$ receptor, which facilitates platelet incorporation into a thrombus; (5) enhanced plasminogen activation resulting from ternary tPA–plasminogen–fibrin complex formation; (6) binding of inhibitors such as α_2-antiplasmin, plasminogen activator inhibitor-2, lipoprotein(a), and histidine-rich glycoprotein, impairing fibrinolysis; (7) down-regulation of factor XIII-mediated crosslinking activity via factor XIII A_2B_2 complex binding to fibrin; (8) antithrombin I, a fibrin activity that inhibits thrombin generation in plasma by sequestering thrombin in the clot and by reducing the catalytic potential of fibrin-bound thrombin.

Key words. Fibrin · Fibrinogen · Fibrinolysis · Thrombin · Leukocytes · Platelets

Introduction

Fibrinogen and fibrin play overlapping roles in blood clotting, fibrinolysis, cellular and matrix interactions, the inflammatory response, wound healing, and neoplasia. These general functions are regulated by specific sites on fibrin(ogen) molecules, some of which are masked in fibrinogen. Commonly, they evolve as a consequence of

3

fibrin formation or fibrinogen-surface interactions. This chapter relates the structural features of fibrin(ogen) to its numerous biological functions, including thrombin binding (antithrombin I), fibrinolysis, regulation of factor XIII activity, growth factor binding, and cellular interactions, including platelets, leukocytes, fibroblasts, and endothelial cells.

Fibrinogen Conversion to Fibrin

Fibrinogen molecules are elongated 45-nm structures that consist of two outer D domains, each connected by a coiled-coil segment to the central E domain (Fig. 1). The molecule is comprised of two sets of three polypeptide chains termed Aα, Bβ, and γ that are joined together in the N-terminal E domain by five symmetrical disulfide bridges [1–4]. Other nonsymmetrical disulfide bridges in this domain form a disulfide ring [1, 3]. Aα chains consists of 610 residues, Bβ chains 461 residues, and the major γ chain form, γA, 411 residues [5]. A minor γ chain variant termed γ′, arises through alternative processing of the primary mRNA transcript [6], resulting in substitution of the C-terminal γ$_A$408-411V sequence with a unique anionic 20-amino-acid sequence (γ′408-427L) containing two sulfated tyrosines [7, 8]. Gamma prime chains account for ~8% of the total fibrinogen γ chain population and are mainly found in

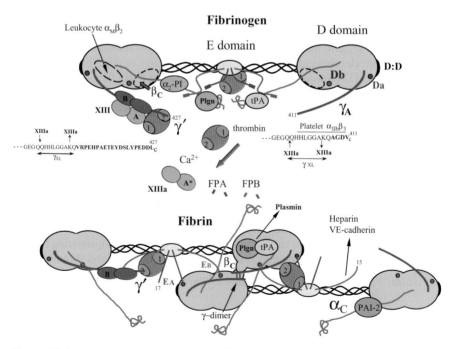

Fig. 1. Fibrinogen structure, its conversion to fibrin, and the thrombin-mediated conversion of native factor XIII to XIIIa. Binding sites for proteins, enzymes, receptors, and other molecules that participate in fibrin(ogen) functions are illustrated (From Mosesson [212], with permission)

heterodimeric fibrinogen molecules that account for ~15% of plasma fibrinogen molecules [9]. Homodimeric γ'/γ' molecules amount to less than 1% of the fibrinogen molecules in blood [10].

The Aα chain contains an amino-terminal fibrinopeptide A (FPA) sequence, the cleavage of which by thrombin initiates fibrin assembly [11–13] by exposing a polymerization site termed E_A. One portion of E_A is at the N-terminus of the fibrin α chain comprising residues 17–20 (GPRV) [14], and another portion is in the fibrin β chain between residues 15 and 42 [15–18]. Each E_A site combines with a constitutive complementary binding pocket (Da) in the D domain of neighboring molecules that is located between γ337 and γ379 [18–20]. The initial E_A:Da associations cause molecules to align in a staggered overlapping end-to-middle domain arrangement to form double-stranded twisting fibrils [21–24] (Fig. 2). Fibrils also undergo lateral associations to create multistranded fibers [25, 26].

There are two types of branch junctions in fibrin networks [27]. The first occurs when a double-stranded fibril converges with another such fibril to form a so-called bilateral junction. Lateral convergence of fibrils results in multistranded versions of this structure. A second type of branch junction, now termed equilateral, is formed by convergent noncovalent interactions among three fibrin molecules that give rise to three fibrils of equal widths. Equilateral junctions form with greater frequency

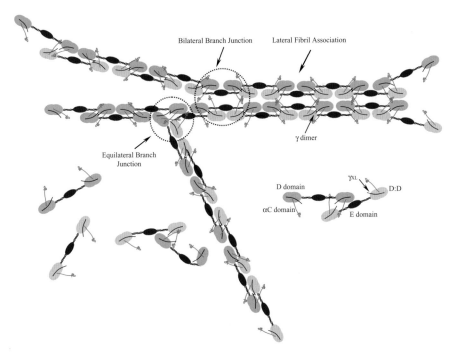

FIG. 2. Fibrin assembly, branching, lateral fibril association, and γ chain crosslinking. Fibrin molecules are represented in two color schemes for ease of recognition. Crosslinked γ chains are positioned "transversely" between fibril strands, as discussed in the text (From Mosesson [213], with permission)

when FPA cleavage is slow [28]; and under such conditions, the networks are more branched and the matrix is "tighter" (i.e., less porous) than those formed at high levels of thrombin [29].

Fibrinopeptide B (FPB, Bβ1-14) release takes place more slowly than release of FPA [11–13] and exposes an independent polymerization site, E_B [30], beginning with β15-18 (GHRP) [14], that interacts with a constitutive complementary Db site in the β chain segment of the D domain [20, 31]. This interaction contributes to lateral association by inducing rearrangements in the $β_C$ region of the D domain that promote intermolecular $β_C:β_C$ contacts [32], as illustrated in Fig. 1. Polymerization of des-BB fibrin results in the same type of fibril structure as occurs with des-AA fibrin [28], but the clot strength is lower than that of des-AA fibrin [30].

The αC' domain originates at residue 220 in the D domain, not far from where it emerges from the D domain, and terminates at Aα610 [33]. Fibrin clots formed from fibrinogen catabolite fractions I-6 to I-9, which lack C-terminal portions of αC domains, display prolonged thrombin times, reduced turbidity, and generate thinner fibers [34–36]. In fibrinogen, αC domains tend to be noncovalently tethered to the E domain [37–39] but dissociate from it following FPB cleavage [38, 39]. This event evidently makes αC domains available for interaction with other αC domains, thereby promoting lateral fibril associations and more extensive network assembly.

There are two constitutive self-association sites in the γ chain region of each D domain (the γ-module) that participate in fibrin or fibrinogen assembly and/or cross-linking, namely $γ_{XL}$ and D:D [40, 41]. The $γ_{XL}$ site overlaps the γ chain crosslinking site (Fig. 1). Intermolecular association between two $γ_{XL}$ sites promotes alignment of crosslinking regions for factor XIII- or XIIIa-mediated transglutamination [40, 42, 43]. Each D:D site is situated at the outer portion of a fibrin(ogen) D domain between residues 275 and 300 of the γ-module [44]. These sites are necessary for proper end-to-end alignment of fibrinogen or fibrin molecules in assembling polymers. Congenital dysfibrinogenemic molecules such as fibrinogen Tokyo II (γR275C) [41], which have defective D:D site interactions, are characterized by networks displaying increased fiber branching, which evidently results from slowed fibrin assembly, plus inaccurate end-to-end positioning of assembling fibrin monomers.

Fibrin Crosslinking and Its Effects on Fibrin Viscoelasticity

The C-terminal region of each fibrinogen or fibrin γ chain contains a crosslinking site at which factor XIII or XIIIa catalyzes the formation of γ-dimers [40, 42, 45, 46] by introducing reciprocal intermolecular ε-(γ-glutamyl)lysine covalent bonds between the γ406 lysine of one γ chain and a glutamine at g398/399 of another (cf. Fig. 1) [47–49]. The same type of intermolecular ε-(γ-glutamyl)lysine bridging occurs more slowly among amine donor and lysine acceptor sites in Aα or α chains [50, 51], thereby creating α-oligomers and larger α-polymers [42, 45, 52]. Crosslinking also occurs among α and γ chains [25, 53], and intramolecular crosslinked α-γ chain heterodimers have been identified in fibrinogen molecules [54].

Da:E_A interactions drive fibrin assembly and facilitate intermolecular antiparallel alignment of γ chain pairs at $γ_{XL}$ sites, thereby accelerating the crosslinking rate [40,

43, 46, 55]. At physiological concentrations, alignment of γ chains in fibrinogen is slower in the presence of factor XIII than it is in fibrin, and therefore crosslinking takes place more slowly in fibrinogen. However, because the rate of fibrinogen crosslinking is concentration dependent, its crosslinking rate can be accelerated even beyond that of fibrin simply by raising fibrinogen concentration sufficiently [46].

The exact location of crosslinked γ chains in assembled fibrin fibrils has been controversial. Whether they are positioned transversely between fibril strands or longitudinally along each strand of a fibril has been formally debated by John Weisel and me [56–59]. Those wishing to gain a detailed perspective of the basis for these opposing views should consult the debate articles themselves. In this chapter I have assumed transverse positioning of crosslinked γ chains (i.e., between D domains of opposing fibril strands). One very good reason for this representation, apart from abundant previously summarized experimental evidence [56], is that *only* this crosslinking arrangement can account for an important viscoelastic property of fibrin—that after being stretched maximally to 1.8 times its original length, crosslinked fibrin films can recover nearly 100% of their original form [60]. Non-crosslinked fibrin clots cannot achieve this elastic recovery [61] nor could longitudinally crosslinked fibrin fibrils if they existed. Figure 3 illustrates how transversely positioned crosslinked γ chains confer this viscoelastic property. Other evidence bearing on this crosslinking arrangement in fibrin is contained in the following section.

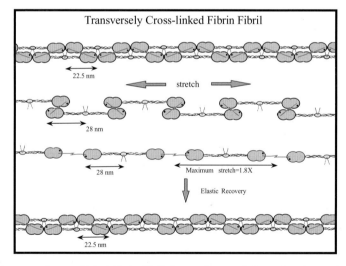

FIG. 3. Transversely crosslinked fibrin fibril that undergoes deformation due to the stress of stretching and its elastic recovery after relaxing the stress. When a double-stranded fibril is maximally stretched 1.8 times [60], it becomes single-stranded, and fibril constituents remain "connected" through the covalently crosslinked γ chains. After stress relaxation, the fibrin film recovers its original form. Only crosslinked γ chains that are positioned transversely between fibril strands, as shown, can account for this viscoelastic behavior (Adapted from Mosesson [56], with permission)

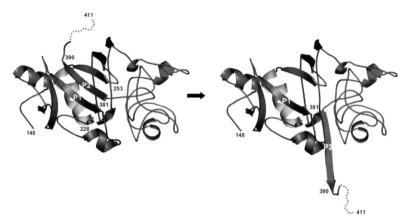

Fig. 4. Known γ-module crystal structure (residues γ148-411) (*left*), and the hypothetical structure resulting from "pull-out" of the γ381-390 β-strand insert (P2) (*right*) (Adapted from Yakovlev et al. [68], with permission)

Leukocyte Integrin Receptor $\alpha_M\beta_2$

The leukocyte integrin, $\alpha_M\beta_2$ (Mac-1), is a high-affinity receptor for fibrin(ogen) on stimulated monocytes and neutrophils [62, 63]; it is important for fibrin(ogen)–leukocyte interactions contributing to the inflammatory response, as lucidly discussed by Flick et al. [64]. The $\alpha_M\beta_2$ binding site within the fibrinogen D domain is situated between two peptide sequences, γ190-202 and γ377-395, now designated P1 and P2, which form adjacent antiparallel β strands [65]. P1 is an integral part of the γ-module, whereas P2 is inserted into the γ-module and forms an antiparallel β-strand with P1, as demonstrated in the X-ray structures of this region [44, 66] (Fig. 4). In further experiments concerned with expression of the $\alpha_M\beta_2$ binding site in fibrin(ogen), Ugarova et al. [67] found that deletion of the minimal recognition sequence, γ390-395, was not sufficient to ablate $\alpha_M\beta_2$ binding since mutants were still effective in supporting adhesion of $\alpha_M\beta_2$-expressing cells. Yakovlev et al. [68] extended these observations by identifying a second sequence in the P1 segment between γ228-253 that contributes to the binding activity.

Exposure of the $\alpha_M\beta_2$ Binding Site and Its Relation to Chain Crosslinking

Fibrin or immobilized fibrinogen bind to $\alpha_M\beta_2$ with great avidity, whereas soluble fibrinogen is a relatively poor ligand [69, 70]. This indicates that the sites for $\alpha_M\beta_2$ are cryptic in fibrinogen but become exposed when fibrinogen is immobilized on plastic or converted to fibrin, clearly suggesting that significant conformational changes occur in this region of the D domain. Other evidence indicating conformational changes in this region involves the exposure of a receptor-induced binding site (RIBS-1), that has been localized to γ373-385 [71, 72]. The γ373-385 sequence is recognized only after the platelet fibrinogen receptor $\alpha_{IIb}\beta_3$ binds to fibrinogen.

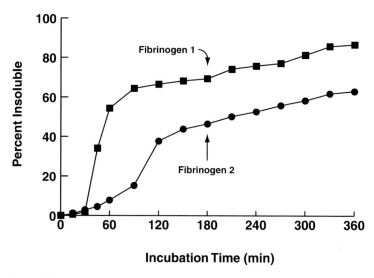

FIG. 5. Rates of fibrinogens 1 and 2 crosslinking mediated by plasma factor XIII at equivalent concentrations. The introduction of crosslinks renders fibrinogen insoluble in acetic acid. Measuring this property provides a quantitative measure of the crosslinking rate. (From Siebenlist et al. [42], with permission)

Based on X-ray structures of the D domain and its γ-module [44, 66] Doolittle inferred that the folding of β-strands in this region is immutable [73]. This presumption has provided a shaky rationale for dismissing the possibility of transverse crosslinking in that the folding revealed in X-ray structures would not permit the emerging C-terminal segment of the γ chain to extend far enough from the D domain to become engaged in transverse crosslinking. Such reasoning abnegates the possibility of transverse crosslinking without seriously considering the large body of biochemical and biophysical evidence that strongly favors it [56, 57].

The question of β-strand folding has been examined by Yakovlev et al. [74], who showed that the P2 segment containing the γ381-390 sequence could be displaced without disrupting the compact structure of the γ-module. This finding led to a so-called pull-out hypothesis (Fig. 5), which suggested that unfolding of the P2 β-strand insert could enable γ chain crosslinking sites to extend sufficiently to become aligned for transverse crosslinking to take place, thereby allaying the conceptual objections raised above. Evidence recited earlier concerning exposure of $\alpha_M\beta_2$ binding sites as well as the RIBS-1 epitope, bolsters the pull-out hypothesis because unfolding of P2 could readily account for their exposure.

Fibrinolysis

Exposure of tPA-Binding and Plasminogen-Binding Sites in Fibrin

Tissue-type plasminogen activator (tPA) is synthesized by vascular endothelial cells [75] and circulates in blood [76]. tPA-mediated plasminogen activation is accelerated in the presence of fibrin, but there is little or no stimulatory effect with fibrinogen [77,

78]. Nevertheless, there is a high-affinity plasminogen-binding site in fibrinogen [79] that is located in the distal portion of each αC domain; there also is a tPA binding site in this same region [80]. Studies of fibrinogen Dusart (AαR554C-albumin), which showed defective plasminogen binding, and fibrinogen Marburg, which lacks substantial portions of the αC domain, showed impaired fibrin-stimulated plasminogen activation by tPA [81–84]. These results suggest that plasminogen binding in the αC domain may regulate fibrinolysis by making bound plasminogen readily available for ternary complex formation in a fibrin system.

tPA-stimulated plasminogen activation is strongly promoted by fibrin polymers and also by crosslinked fibrinogen polymers [85]. Plasminogen activation occurs through tPA binding to fibrin followed by plasminogen addition to form a ternary complex [77]. Subsequent proteolytic cleavage of fibrin by plasmin creates additional lysine-binding sites [86, 87], thereby enhancing fibrinolysis by increasing plasminogen accumulation. Two sites in fibrin are involved in enhanced plasminogen activation by tPA: Aα148-160 and γ312-324 [88–90]. Both sites are cryptic in fibrinogen but become exposed during fibrin assembly, primarily as a consequence of intermolecular D:E interactions that induce conformational changes in the D region and result in exposure of tPA- and plasminogen-binding sites. The exposure is reversed after the complex dissociates [91]. Aα148-160 binds tPA or plasminogen with similar affinity [92–94]. However, plasminogen is preferentially bound at this site in vivo because the circulating plasminogen concentration is much higher than that of tPA [91, 95]. γ312-324 interacts exclusively with tPA [96, 97].

Proteins that Bind to Fibrin(ogen) and Affect Fibrinolysis

α_2-Antiplasmin (α_2AP) can be crosslinked to fibrin α chains at Aα303 [98, 99], and its presence enhances resistance to fibrinolysis [100]. It was recognized many years ago that there was a potent antiplasmin activity associated with fibrinogen [101], but its identification as α_2AP itself remained uncertain until years later with the demonstration that α_2AP was covalently bound to plasma fibrinogen [102]. As discussed earlier, native factor XIII is capable of crosslinking fibrinogen [42], and it is also capable of incorporating α_2AP into a potent inhibitor of plasmin, fibrinogen (unpublished experiments). In addition, PAI-2, an inhibitor of plasmin generation by urokinase or tPA, can be crosslinked to fibrin at several sites in the αC domain that are remote from the Aα303 site for α_2AP [103, 104], thus amplifying the resistance of fibrin to lysis.

Lipoprotein(a) [LP(a)] is an atherogenic lipoprotein complex formed from apolipoprotein(a) that is disulfide bound to the apolipoprotein B-100 moiety of low density lipoprotein (LDL). LP(a) binds to plasmin-degraded fibrin and fibrinogen [105] and to the αC domain by a lysine-independent mechanism [106], competes with plasminogen for binding sites in fibrin(ogen) [107, 108], and becomes crosslinked to fibrinogen in the presence of XIIIa [109]. Its presence on fibrin(ogen) has an inhibitory effect on fibrinolysis [107, 108] and provides a mechanism for depositing LP(a) at places of fibrin deposition, such as injured blood vessels or atheroscleotic lesions.

Histidine-rich glycoprotein (HRGP) is a plasma and platelet protein that binds specifically to fibrinogen and fibrin [110]. A high proportion of HRGP circulates as a complex bound to the lysine-binding site of plasminogen, thereby reducing the effective plasminogen concentration, inhibiting binding of plasminogen to fibrin(ogen), and retarding fibrinolysis in vitro [111]. Although the physiological relevance of the HRGP effect has been questioned [112], the fact remains that both high plasma levels of HRGP [113, 114] and HRGP deficiencies [115–117] are associated with thrombophilia.

Molecular and Cellular Interactions of Fibrin(ogen)

Heparin and Endothelial Cell Binding

In addition to its role in mediating fibrin assembly, the fibrin β15-42 sequence binds heparin [118–120], thereby participating in cell–matrix interactions. β15-42 mediates platelet spreading [121], fibroblast proliferation [122], endothelial cell spreading, proliferation and capillary tube formation [119, 122–124], and release of von Willebrand factor [125, 126]. Furthermore, a dimeric heparin-binding fragment containing the β15-57 sequence (β15-66)$_2$ was shown to mimic the high-affinity heparin-binding characteristics of fibrin or the fibrin E fragment [120]. The presence of the fibrinogen Bβ1-14 fibrinopeptide segment interferes with heparin binding.

Interaction between the fibrin β15-42 sequence and the endothelial cell receptor, VE-cadherin, stimulates capillary tube formation and promotes angiogenesis [124, 127]. A dimeric β chain peptide (β15-66)$_2$ has a much higher affinity [128], indicating that the VE-cadherin site in fibrin involves two fibrin β chain segments.

Integrin Binding to Platelets

Many cellular interactions with fibrinogen and fibrin occur through integrin binding to RGD sequences at Aα572–575 (RGDS) and possibly at Aα95-98 (RGDF) [129–133]. Crosslinking of αC domains promotes integrin-dependent cell adhesion and signaling [134]. In the case of platelets, RGD sites compete with the fibrinogen $γ_A$400-411 sequence [135] (Fig. 1) for binding to platelet integrin $α_{IIb}β_3$ [136–138]. The subject of fibrinogen–platelet interactions has been adequately reviewed [139].

Proteins, Growth Factors, and Cytokines that Bind to Fibrin(ogen)

In addition to the many molecular-fibrin(ogen) interactions covered in this chapter, I list below some other examples of proteins that bind to fibrin(ogen) and affect its biological behavior. *Plasma fibronectin* binds to the Aα chain of fibrinogen in its C-terminal region as fibrinogen molecules lacking this part of the molecule do not interact with fibronectin [140]. Covalent crosslinking of fibronectin to fibrin [98, 141] takes place mainly between Gln 3 of fibronectin [142] and the C-terminal region of the Aα chain of fibrin [143]. Such interactions would be advantageous for incorporating fibrin(ogen) molecules into the extracellular matrix.

Fibroblast growth factor-2 (FGF-2, bFGF) and *vascular endothelial growth factor* (VEGF) bind to fibrinogen and are able to potentiate endothelial cell proliferation when bound [144–146]. The cytokine, interleukin-1 (IL-1), is a participant in the inflammatory response. IL-1β but not IL-1α bound to fibrin(ogen) displaced bound FGF-2 and displayed enhanced stimulatory activity of endothelial cells in the bound form [147]. These additional examples of fibrin(ogen) binding interactions are not intended to be exhaustive in scope; rather, they further illustrate how fibrin(ogen) participates in regulating its own physiological destiny.

Factor XIII Activity in Plasma

Plasma factor XIII circulates as an A_2B_2 tetramer that is bound by its B subunits to γ′ chain-containing fibrinogen molecules (fibrinogen 2) [148] (Fig. 1). Fibrinogen 2 serves in this capacity as the carrier protein for circulating factor XIII. In addition to the transglutaminase activity possessed by thrombin-activated factor XIII, it is now known that "native" factor XIII displays constitutive enzymatic (transglutaminase) activity and can catalyze crosslinking of preferred substrates, such as fibrinogen or fibrin, without requiring prior proteolytic activation by thrombin [42]. The observed fibrin crosslinking rates attainable with factor XIII approach those observed with factor XIIIa. This then raises the question as to how such potentially potent factor XIII activity is regulated in the circulation. For example, under normal circumstances only small amounts of crosslinked fibrin(ogen) are detectable in plasma, although large amounts of crosslinked fibrin(ogen) products have been reported under pathological circumstances, such as in patients with familial Mediterranean fever [149].

Part of the answer to this question lies in the following: (1) γ′ Chain-containing fibrinogen 2 (γ′/γA) becomes crosslinked much more slowly than γA-containing fibrinogen 1 (γA/γA) (Fig. 5), suggesting that binding to fibrinogen 2 provides one mechanism for suppressing factor XIII crosslinking activity. (2) Excess B subunits in plasma provide another suppressive element as these subunits prevent thrombin-independent activation of cellular factor XIII (A_2) [150], and their presence in the A_2B_2 tetramer causes a lag in the onset of fibrinogen crosslinking by plasma XIII [42]. Nevertheless, other as yet undefined factor XIII regulatory mechanisms might exist.

Evidence for suppression of constitutive factor XIII activity lies in the observation that very little crosslinked fibrinogen or fibrin is found in normal plasma [149]. Nevertheless, another important substrate for factor XIII, α_2-antiplasmin (α2AP), which is the potent plasmin inhibitor, has been identified in plasma fibrinogen molecules [102], suggesting that this reflects the footprint of native factor XIII activity in circulating blood, as further discussed below. Although it is well recognized that α2AP becomes crosslinked by thrombin-activated factor XIIIa to fibrin [98, 151, 152] or to fibrinogen [99, 153] specifically at position 303 of the Aα chain [99, 103], it is not yet widely appreciated that α2AP can also be crosslinked to fibrinogen in plasma by native factor XIII (unpublished experiments). Substantial amounts of α2AP have been found covalently complexed to normal plasma fibrinogen as well as to a dysfibrinogen, fibrinogen Cedar Rapids [102], and more recently in all fibrinogen specimens thus far isolated from plasma (unpublished experiments). This suggests that although intermolecular crosslinking of fibrinogen in plasma is well suppressed that may not be the case for α2AP.

Antithrombin I

Thrombin binds to its substrate, fibrinogen, through an anion-binding region termed exosite 1 [154, 155]. Substrate binding results in cleavage and release of fibrinopeptides A (FPA) and eventually fibrinopeptide B (FPB). But concomitantly, the resulting fibrin exhibits residual nonsubstrate thrombin-binding potential. Howell recognized nearly a century ago that the fibrin clot itself exhibits significant thrombin-binding potential [156], and the thrombin binding associated with fibrin formation in plasma was termed antithrombin I by Seegers and colleagues more than six decades ago [157–159]. In recent years, recognition of the functional importance of antithrombin I for down-regulation of thrombin generation in plasma has brought a new perspective on its physiological role [160, 161]. This section provides an update on the constituents in fibrin that comprise antithrombin I, their mechanisms of action, and their physiological roles.

Antithrombin I activity is defined by two classes of nonsubstrate thrombin-binding sites in fibrin [162, 163]: one of relatively low affinity in the E domain (~two sites per molecule), and the other of higher affinity in D domains of fibrin(ogen) molecules containing a γ' chain variant (γ'^{427L}) [163] (Fig. 1). Altogether, γ' chains comprise ~8% of the total γ chain population [7, 9]. Virtually all γ' chains in plasma fibrinogen are found in a chromatographic subfraction termed fibrinogen 2, each molecule of which also contains a platelet-binding γ_A chain. Fibrinogen 1 is homodimeric with respect to its γ_A chain population and accounts for ~85% of human plasma fibrinogen.

Low-affinity thrombin-binding activity reflects thrombin exosite 1 binding in E domain of fibrin, as recently detailed by analyses of thrombin-fibrin fragment E crystals by Pechik et al. [164] (Fig. 6). In contrast, high-affinity thrombin binding to γ' chains takes place through exosite 2 [165–167] (Fig. 7). The γ' chain thrombin-binding site is situated between residues 414 and 427, and tyrosine sulfation at $\gamma'418$ and $\gamma'422$ increases thrombin-binding potential [8]. The binding affinity of thrombin for γ'-containing fibrin molecules is increased by concomitant fibrin binding to thrombin exosite 1 [168] (cf. Fig. 1).

Several studies of fibrin-bound thrombin have focused on the prothrombotic potential of thrombin bound to fibrin or to fibrin degradation products [169–174], and such observations are of course valid. Nevertheless, it seems less well appreciated that thrombin binding to fibrin in clotting blood significantly suppresses thrombin activity in terms of thrombin generation. Although an antithrombin I deficiency or defect per se has in the past not been considered to be a thrombotic risk factor, this certainly does appear to be the case, as summarized below.

The effect of γ' chain binding to thrombin exosite 2 is more complex than binding to thrombin exosite 1 at the fibrin E domain. The γ' chain-thrombin interaction takes place at exosite 2 (Fig. 7) and induces noncompetitive allosteric down-regulation of amidolytic activity at the thrombin catalytic site and consequently slowed release of fibrinopeptide A among other effects [175]. This effect is independent of the slow–fast transition induced by Na^+ binding [176] or the effect that is reflected in the slow cleavage of fibrinogen induced by thrombin binding to thrombomodulin. [177]. Because of delayed fibrinopeptide cleavage in γ' chain-containing fibrinogen 2, the fibrin that is produced has finer network fibers and contains more branches than does fibrin 1

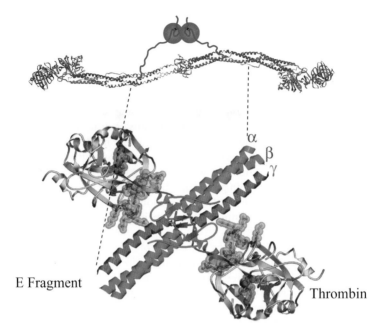

FIG. 6. Three-dimensional structure of two thrombin molecules bound to a fibrin E fragment that is projected from a ribbon diagram of a fibrin molecule. The thrombin–fibrin complex is drawn as a ribbon diagram along a twofold symmetry axis perpendicular to the plane of the page. Aα, Bβ, and γ chain fragments are blue, green, and red, respectively. Thrombin molecules are beige, and the residues included in exosite 1 are orange. The PPACK inhibitor bound to the active site is magenta. (Adapted from Pechik et al. [164], with permission)

[175, 178]. This structural modification of matrix structure also results in delayed fibrinolysis [175]. The down-regulating effect of the γ′ peptide sequence on catalytic site activity is similar to that induced in thrombin by other exosite 2-binding proteins such as GP1bα or prothrombin fragment 2 [179–182], a monoclonal antibody directed against an epitope in thrombin exosite 2 [183], and DNA aptamer HD-22 [184]. This suggests that thrombin exosite 2 binding interactions (e.g., with GP1bα or with γ′ chains) play a role in vivo in regulating thrombin generation. In addition to the effects of γ′ chain binding on fibrin formation and lysis, fibrin-mediated enhancement of factor XIII activation [185–188] was slower in the presence of fibrinogen 2 than with fibrinogen 1 [175]. (Fig. 5)

In addition to the full length γ′ chains, a shortened version of this chain, γ'^{423P}, is present in most plasmas [189–192]. We believe that γ'^{423P} chains arise by posttranslational proteolytic processing of intact γ'^{427L} fibrinogen chains, but to date we have not been able to identify the basis for this occurrence. Because the ultimate C-terminal γ′ 424-427 sequence is required for thrombin binding at the γ′ site [8], γ'^{423P} chains lack thrombin-binding potential, and their formation would effectively reduce antithrombin I activity at the expense of the γ'^{427L} chains.

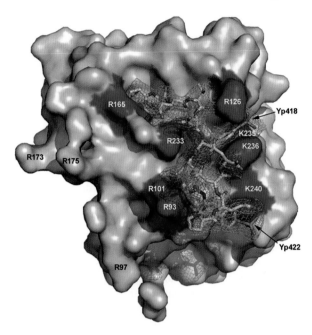

FIG. 7. Crystal structure of the γ′ peptide (γ′408-427) making contact with basic residues (blue) in exosite 2 of thrombin. The γ′ peptide interactions closely reproduce heparin binding at this site [214], thereby explaining why binding of the γ′ peptide and heparin are mutually exclusive and why thrombin bound to fibrin is resistant to inactivation by antithrombin III and heparin cofactor II [165, 215]. (Adapted from Pineda et al. [167], with permission)

Antithrombin I and Its Relation to Thrombotic Disease

Antithrombin I is an important regulator of thrombin activity in clotting blood [161], a concept based on a number of prior observations and reports. (1) Fibrin from certain congenital dysfibrinogens (e.g., fibrinogen New York I [162] and fibrinogen Naples I [168, 193, 194] exhibit reduced thrombin binding capacity and are associated with marked venous or arterial thromboembolism). (2) Paradoxically, severe thromboembolic disease, both venous and arterial, occurs in afibrinogenemia and hypofibrinogenemia [195–203] often in association with the infusion of fibrinogen. (3) Increased levels of prothrombin activation fragment F_{1+2} [204, 205] or TAT complexes [203, 204] are found in afibrinogenemic plasma (i.e., congenital antithrombin I deficiency), and these abnormal levels can be normalized by fibrinogen infusions [203, 205], further suggesting that an underlying hypercoagulable state exists in this condition. (4) The report that an afibrinogenemic subject developed occlusive peripheral arterial thrombosis in the absence of a fibrinogen infusion [203] seems to be analogous to studies in ferric chloride-injured afibrinogenemic mice, which developed abundant intravascular thrombi at the site of injury that characteristically embolized downstream [206]. 5) The demonstration by Dupuy et al. [203] that increased thrombin generation in their patient's plasma was normalized by addition of fibrinogen underscored the thrombin inhibitory role of plasma fibrinogen. Demonstrating that fibrinogen 2 (γA/γ') had a more profound effect in normalizing thrombin generation in afibrinogenemic plasma than did fibrinogen 1 (γA/γA) [160] emphasized the dominant role of γ' chains in thrombin binding and inhibition by fibrin. Overall, these considerations indicate that antithrombin I is a major thrombin inhibitor that deserves a place alongside other established thrombin inhibitors such as antithrombin III.

In addition to evidence discussed above, more recent reports have suggested that the content of γ'-containing fibrinogen in plasma has a relation to the incidence of thrombotic disease [207–209]. Uitte de Willige et al. [209] investigated the effect of γ'-fibrinogen/total fibrinogen ratios on the risk of venous thrombosis in the Leiden Thrombophilia Study [210]. They demonstrated that reduced γ'-fibrinogen/total fibrinogen ratios were associated with an increased thrombosis risk and were correlated with a particular γ chain gene haplotype termed FGG-H2. The potentially relevant single nucleotide polymorphisms (SNPs) of that haplotype are located in intron 9 (9615 C/T) and just downstream from the polyadenylation site of exon 10 (10034 C/T). These may individually or collectively result in reduced production of γ' chain transcripts, although other explanations may exist.

On the other hand, Drouet et al. [207] suggested that subjects with elevated γ'-fibrinogen/total fibrinogen ratios correlated with a higher incidence of arterial thrombosis, and Lovely et al. [208] reported an association between elevated levels of γ' chains and coronary artery disease, but this association did not hold with respect to the γ'-fibrinogen/total fibrinogen ratios. Significant elevations in γ' chain concentration were also reported by Manilla et al. [211] in patients with myocardial infarction, although the differences were rather small (~10%), and there were no differences in the γ'-fibrinogen/total fibrinogen ratios. More studies are required to place these several observations in the proper physiological context and to develop a coherent mechanistic understanding of the findings.

Finally, thrombotic microangiopathy (TMA) is a life-threatening syndrome with major forms that include thrombotic thrombocytopenic purpura (TTP) and hemolytic uremic syndrome (HUS). The syndrome is characterized by microangiopathic hemolytic anemia, thrombocytopenia, and microvascular thrombosis accompanied by varying degrees of tissue ischemia and infarction. We investigated a group of TMA patients and found that there was an association between the syndrome and a lowered plasma γ' chain content, suggesting that a low content of antithrombin I activity contributes to the microvascular thrombosis that characterizes TMA [192].

Acknowledgments. Work from my laboratory has been supported most recently by NIH grant R01 HL-70627.

References

1. Blombäck B, Hessel B, Hogg D (1976) Disulfide bridges in NH$_2$-terminal part of human fibrinogen. Thromb Res 8:639–658
2. Huang S, Cao Z, Davie EW (1993) The role of amino-terminal disulfide bonds in the structure and assembly of human fibrinogen. Biochem Biophys Res Commun 190:488–495
3. Zhang J-Z, Redman CM (1992) Identification of Bβ chain domains involved in human fibrinogen assembly. J Biol Chem 267:21727–21732

4. Hoeprich PD Jr, Doolittle RF (1983) Dimeric half-molecules of human fibrinogen are joined through disulfide bonds in an antiparallel orientation. Biochemistry 22:2049–2055

5. Henschen A, Lottspeich F, Kehl M, et al (1983) Covalent structure of fibrinogen. Ann N Y Acad Sci 408:28–43

6. Chung DW, Davie EW (1984) γ and γ′ chains of human fibrinogen are produced by alternative mRNA processing. Biochemistry 23:4232–4236

7. Wolfenstein-Todel C, Mosesson MW (1981) Carboxy-terminal amino acid sequence of a human fibrinogen g chain variant (γ′). Biochemistry 20:6146–6149

8. Meh DA, Siebenlist KR, Brennan SO, et al (2001) The amino acid sequences in fibrin responsible for high affinity thrombin binding. Thromb Haemost 85:470–474

9. Mosesson MW, Finlayson JS, Umfleet RA (1972) Human fibrinogen heterogeneities. III. Identification of γ chain variants. J Biol Chem 247:5223–5227

10. Wolfenstein-Todel C, Mosesson MW (1980) Human plasma fibrinogen heterogeneity: evidence for an extended carboxyl-terminal sequence in a normal gamma chain variant (γ′). Proc Natl Acad Sci U S A 77:5069–5073

11. Scheraga HA, Laskowski M Jr (1957) The fibrinogen-fibrin conversion. Adv Prot Chem 12:1–131

12. Blombäck B (1958) Studies on the action of thrombotic enzymes on bovine fibrinogen as measured by N-terminal analysis. Arkiv Kemi 12:321–335

13. Blombäck B, Hessel B, Hogg D, et al (1978) A two-step fibrinogen-fibrin transition in blood coagulation. Nature 275:501–505

14. Laudano AP, Doolittle RF (1978) Studies on synthetic peptides that bind to fibrinogen and prevent fibrin polymerization. Proc Natl Acad Sci U S A 75:3085–3089

15. Liu CY, Koehn JA, Morgan FJ (1985) Characterization of fibrinogen New York 1. J Biol Chem 260:4390–4396

16. Pandya BV, Cierniewski CS, Budzynski AZ (1985) Conservation of human fibrinogen conformation after cleavage of the Bβ chain NH_2-terminus. J Biol Chem 260: 2994–3000

17. Siebenlist KR, DiOrio JP, Budzynski AZ, et al (1990) The polymerization and thrombin-binding properties of des-(Bβ1–42)-fibrin. J Biol Chem 265:18650–18655

18. Shimizu A, Nagel GM, Doolittle RF (1992) Photoaffinity labeling of the primary fibrin polymerization site: isolation of a CNBr fragment corresponding to γ 337-379. Proc Natl Acad Sci U S A 89:2888–2892

19. Pratt KP, Côté HCF, Chung DW, et al (1997) The primary fibrin polymerization pocket: three-dimensional structure of a 30-kDa C-terminal γ chain fragment complexed with the peptide gly-pro-arg-pro. Proc Natl Acad Sci U S A 94:7176–7181

20. Everse SJ, Spraggon G, Veerapandian L, et al (1998) Crystal structure of fragment double-D from human fibrin with two different bound ligands. Biochemistry 37:8637–8642

21. Ferry JD (1952) The mechanism of polymerization of fibrinogen. Proc Natl Acad Sci U S A 38:566–569

22. Krakow W, Endres GF, Siegel BM, et al (1972) An electron microscopic investigation of the polymerization of bovine fibrin monomer. J Mol Biol 71:95–103

23. Fowler WE, Hantgan RR, Hermans J, et al (1981) Structure of the fibrin protofibril. Proc Natl Acad Sci U S A 78:4872–4876

24. Müller MF, Ris HA, Ferry JD (1984) Electron microscopy of fine fibrin clots and fine and coarse fibrin films. J Mol Biol 174:369–384

25. Mosesson MW, Siebenlist KR, Amrani DL, et al (1989) Identification of covalently linked trimeric and tetrameric D domains in crosslinked fibrin. Proc Natl Acad Sci U S A 86:1113–1117

26. Hewat EA, Tranqui L, Wade RH (1983) Electron microscope structural study of modified fibrin and a related modified fibrinogen aggregate. J Mol Biol 170:203–222

27. Mosesson MW, DiOrio JP, Siebenlist KR, et al (1993) Evidence for a second type of fibril branch point in fibrin polymer networks, the trimolecular junction. Blood 82:1517–1521
28. Mosesson MW, DiOrio JP, Muller MF, et al (1987) Studies on the ultrastructure of fibrin lacking fibrinopeptide B (β-fibrin). Blood 69:1073–1081
29. Blomback B, Carlsson K, Fatah K, et al (1994) Fibrin in human plasma: gel architectures governed by rate and nature of fibrinogen activation. Thromb Res 75:521–538
30. Shainoff JR, Dardik BN (1983) Fibrinopeptide B in fibrin assembly and metabolism: physiologic significance in delayed release of the peptide. Ann N Y Acad Sci 408: 254–267
31. Medved LV, Litvinovich SV, Ugarova TP, et al (1993) Localization of a fibrin polymerization site complimentary to Gly-His-Arg sequence. FEBS Lett 320:239–242
32. Yang Z, Mochalkin I, Doolittle RF (2000) A model of fibrin formation based on crystal structures of fibrinogen and fibrin fragments complexed with synthetic peptides. Proc Natl Acad Sci U S A 97:14156–14161
33. Weisel JW, Medved LV (2001) The structure and function of the αC domains of fibrinogen. Ann NY Acad Sci 936:312–327
34. Mosesson MW, Sherry S (1966) The preparation and properties of human fibrinogen of relatively high solubility. Biochemistry 5:2829–2835
35. Mosesson MW (1983) Fibrinogen heterogeneity. Ann NY Acad Sci 408:97–113
36. Hasegawa N, Sasaki S (1990) Location of the binding site "b" for lateral polymerization of fibrin. Thromb Res 57:183–195
37. Mosesson MW, Hainfeld JF, Haschemeyer RH, et al (1981) Identification and mass analysis of human fibrinogen molecules and their domains by scanning transmission electron microscopy. J Mol Biol 153:695–718
38. Veklich YI, Gorkun OV, Medved LV, et al (1993) Carboxyl-terminal portions of the a chains of fibrinogen and fibrin. J Biol Chem 268:13577–13585
39. Gorkun OV, Veklich YI, Medved LV, et al (1994) Role of the αC domains of fibrin in clot formation. Biochemistry 33:6986–6997
40. Mosesson MW, Siebenlist KR, Hainfeld JF, et al (1995) The covalent structure of factor XIIIa crosslinked fibrinogen fibrils. J Struct Biol 115:88–101
41. Mosesson MW, Siebenlist KR, DiOrio JP, et al (1995) The role of fibrinogen D domain intermolecular association sites in the polymerization of fibrin and fibrinogen Tokyo II (γ275 Arg®Cys). J Clin Invest 96:1053–1058
42. Siebenlist KR, Meh D, Mosesson MW (2001) Protransglutaminase (factor XIII) mediated crosslinking of fibrinogen and fibrin. Thromb Haemost 86:1221–1228
43. Mosesson MW, Siebenlist KR, Hernandez I, et al (2002) Fibrinogen assembly and crosslinking on a fibrin fragment E template. Thromb Haemost 87:651–658
44. Spraggon G, Everse SJ, Doolittle RF (1997) Crystal structures of fragment D from human fibrinogen and its crosslinked counterpart from fibrin. Nature 389:455–462
45. McKee PA, Mattock P, Hill RL (1970) Subunit structure of human fibrinogen, soluble fibrin, and crosslinked insoluble fibrin. Proc Natl Acad Sci U S A 66:738–744
46. Kanaide H, Shainoff JR (1975) Crosslinking of fibrinogen and fibrin by fibrin-stabilizing factor (factor XIIIa). J Lab Clin Med 85:574–597
47. Doolittle RF, Chen R, Lau F (1971) Hybrid fibrin: Proof of the intermolecular nature of γ-γ crosslinking units. Biochem Biophys Res Commun 44:94–100
48. Chen R, Doolittle RF (1971) γ-γ Crosslinking sites in human and bovine fibrin. Biochemistry 10:4486–4491
49. Purves LR, Purves M, Brandt W (1987) Cleavage of fibrin-derived D-dimer into monomers by endopeptidase from puff adder venom (Bitis arietans) acting at crosslinked sites of the γ chain: sequence of carboxy-terminal cyanogen bromide γ-chain fragments. Biochemistry 26:4640–4646

50. Sobel JH, Gawinowicz MA (1996) Identification of the a chain lysine donor sites involved in factor XIII$_a$ fibrin crosslinking. J Biol Chem 271:19288–19297

51. Matsuka YV, Medved LV, Migliorini MM, et al (1996) Factor XIIIa-catalyzed crosslinking of recombinant αC fragments of human fibrinogen. Biochemistry 35:5810–5816

52. Folk JE, Finlayson JS (1977) The ε-(γ-glutamyl)lysine crosslink and the catalytic role of transglutaminases. Adv Prot Chem 31:1–133

53. Shainoff JR, Urbanic DA, DiBello PM (1991) Immunoelectrophoretic characterizations of the crosslinking of fibrinogen and fibrin by factor XIIIa and tissue transglutaminase: identification of a rapid mode of hybrid alpha-/gamma-chain crosslinking that is promoted by the gamma-chain crosslinking. J Biol Chem 266:6429–6437

54. Siebenlist KR, Mosesson MW (1996) Evidence for intramolecular crosslinked Aα γ chain heterodimers in plasma fibrinogen. Biochemistry 35:5817–5821

55. Samokhin GP, Lorand L (1995) Contact with the N termini in the central E domain enhances the reactivities of the distal D domains of fibrin to factor XIIIa. J Biol Chem 270:21827–21832

56. Mosesson MW (2004) The fibrin crosslinking debate: crosslinked gamma-chains in fibrin fibrils bridge "transversely" between strands: yes. J Thromb Haemost 2:388–393

57. Mosesson MW (2004) Crosslinked gamma-chains in a fibrin fibril are situated transversely between its strands. J Thromb Haemost 2:1469–1471

58. Weisel JW (2004) Crosslinked gamma-chains in fibrin fibrils bridge transversely between strands: no. J Thromb Haemost 2:394–399

59. Weisel JW (2004) Crosslinked gamma-chains in a fibrin fibril are situated transversely between its strands. J Thromb Haemost 2:1467–1469

60. Roska FJ, Ferry JD (1982) Studies of fibrin film. I. Stress relaxation and birefringence. Biopolymers 21:1811–1832

61. Ferry JD (1988) Structure and rheology of fibrin networks. In: Kramer O (ed) Biological and synthetic polymer networks. Elsevier Applied Science, Amsterdam, pp 41–55

62. Altieri DC, Mannucci PM, Capitanio AM (1986) Binding of fibrinogen to human monocytes. J Clin Invest 78:968–976

63. Altieri DC, Bader R, Mannucci PM, et al (1988) Oligospecificity of the cellular adhesion receptor Mac-1 encompasses an inducible recognition specificity for fibrinogen. J Cell Biol 107:1893–1900

64. Flick MJ, Du X, Witte DP, et al (2004) Leukocyte engagement of fibrin(ogen) via the integrin receptor alphaMbeta2/Mac-1 is critical for host inflammatory response in vivo. J Clin Invest 113:1596–1606

65. Ugarova TP, Solovjov DA, Zhang L, et al (1998) Identification of a novel recognition sequence for integrin $α_M β_2$ within the γ-chain of fibrinogen. J Biol Chem 273:22519–22527

66. Yee VC, Pratt KP, Cote HC, et al (1997) Crystal structure of a 30 kDa C-terminal fragment from the gamma chain of human fibrinogen. Structure 5:125–138

67. Ugarova TP, Lishko VK, Podolnikova NP, et al (2003) Sequence gamma 377-395(P2), but not gamma 190-202(P1), is the binding site for the alpha MI-domain of integrin alpha M beta 2 in the gamma C-domain of fibrinogen. Biochemistry 42:9365–9373

68. Yakovlev S, Zhang L, Ugarova T, et al (2005) Interaction of fibrin(ogen) with leukocyte receptor alpha M beta 2 (Mac-1): further characterization and identification of a novel binding region within the central domain of the fibrinogen gamma-module. Biochemistry 44:617–626

69. Loike JD, Silverstein R, Wright SD, et al (1992) The role of protected extracellular compartments in interactions between leukocytes, and platelets, and fibrin/fibrinogen matrices. Ann N Y Acad Sci 667:163–172

70. Lishko VK, Kudryk B, Yakubenko VP, et al (2002) Regulated unmasking of the cryptic binding site for integrin alpha M beta 2 in the gamma C-domain of fibrinogen. Biochemistry 41:12942–12951
71. Zamarron C, Ginsberg MH, Plow EF (1990) Monoclonal antibodies specific for a conformationally altered state of fibrinogen. Thromb Haemost 64:41–46
72. Zamarron C, Ginsberg MH, Plow EF (1991) A receptor-induced binding site in fibrinogen elicited by its interaction with platelet membrane glycoprotein IIb-IIIa. J Biol Chem 266:16193–16199
73. Doolittle RF (2003) X-ray crystallographic studies on fibrinogen and fibrin. J Thromb Haemost 1:1559–1565
74. Yakovlev S, Litvinovich S, Loukinov D, et al (2000) Role of the beta-strand insert in the central domain of the fibrinogen gamma-module. Biochemistry 39: 15721–15729
75. Levin E (1983) Latent tissue plasminogen activator produced by human endothelial cells in culture: evidence for an enzyme-inhibitor complex. Proc Natl Acad Sci U S A 80:6804–6808
76. Collen D (1980) On the regulation and control of fibrinolysis. Thromb Haemost 43:77–89
77. Hoylaerts M, Rijken DC, Lijnen HR, et al (1982) Kinetics of the activation of plasminogen by human tissue plasminogen. J Biol Chem 257:2912–2919
78. Rånby M (1982) Studies on the kinetics of plasminogen activation by tissue plasminogen activator. Biochim Biophys Acta 704:461–469
79. Bok RA, Mangel WF (1985) Quantitative characterization of the binding of plasminogen to intact fibrin clots, lysine-sepharose, and fibrin cleaved by plasmin. Biochemistry 24:3279–3286
80. Tsurupa G, Medved L (2001) Identification and characterization of novel tPA- and plasminogen-binding sites within fibrin(ogen) alpha C-domains. Biochemistry 40:801–808
81. Soria J, Soria C, Caen P (1983) A new type of congenital dysfibrinogenaemia with defective fibrin lysis-Dusard syndrome: possible relation to thrombosis. Br J Haematol 53:575–586
82. Lijnen HR, Soria J, Soria C, et al (1984) Dysfibrinogenemia (fibrinogen Dusard) associated with impaired fibrin-enhanced plasminogen activation. Thromb Haemost 51:108–109
83. Mosesson MW, Siebenlist KR, Hainfeld JF, et al (1996) The relationship between the fibrinogen D domain self-association/crosslinking site (γ_{XL}) and the fibrinogen Dusart abnormality (Aα R554C-albumin). J Clin Invest 97:2342–2350
84. Sugo T, Nakamikawa C, Takebe M, et al (1998) Factor XIIIa crosslinking of the Marburg fibrin: formation of alpha and gamma-heteromultimers and the alpha-chain-linked albumin gamma complex, and disturbed protofibril assembly resulting in acquisition of plasmin resistance relevant to thrombophilia. Blood 91:3282–3288
85. Mosesson MW, Siebenlist KR, Voskuilen M, et al (1998) Evaluation of the factors contributing to fibrin-dependent plasminogen activation. Thromb Haemost 79:796–801
86. Suenson E, Lützen O, Thorsen S (1984) Initial plasmin-degradation of fibrin as the basis of a positive feed-back mechanism in fibrinolysis. Eur J Biochem 140:513–522
87. Harpel PC, Chang TS, Verderber E (1985) Tissue plasminogen activator and urokinase mediate the binding of Glu-plasminogen to plasma fibrin I. Evidence for new binding sites in plasmin-degraded fibrin I. J Biol Chem 260:4432–4440
88. Schielen WJ, Voskuilen M, Tesser GI, et al (1989) The sequence A alpha-(148-160) in fibrin, but not in fibrinogen, ia accessible to monoclonal antibodies. Proc Natl Acad Sci U S A 86:8951–8954
89. Schielen WJ, Adams HP, van Leuven K, et al (1991) The sequence gamma-(312-324) is a fibrin-specific epitope. Blood 77:2169–2173

90. Schielen WJG, Adams HPHM, Voskuilen M, et al (1991) The sequence Aα-(154-159) of fibrinogen is capable of accelerating the tPA catalyzed activation of plasminogen. Blood Coag Fibrinolys 2:465–470

91. Yakovlev S, Makogonenko E, Kurochkina N, et al (2000) Conversion of fibrinogen to fibrin: mechanism of exposure of tPA- and plasminogen-binding sites. Biochemistry 39:15730–15741

92. Lezhen TI, Kudinov SA, Medved' LV (1986) Plasminogen-binding site of the thermostable region of fibrinogen fragment D. FEBS Lett 197:59–62

93. Bosma PJ, Rijken DC, Nieuwenhuizen W (1988) Binding of tissue-type plasminogen activator to fibrinogen fragments. Eur J Biochem 172:399–404

94. de Munk GA, Caspers MP, Chang GT, et al (1989) Binding of tissue-type plasminogen activator to lysine, lysine analogues, and fibrin fragments. Biochemistry 28: 7318–7325

95. Nieuwenhuizen W (2001) Fibrin-mediated plasminogen activation. Ann NY Acad Sci 936:237–246

96. Yonekawa O, Voskuilen M, Nieuwenhuizen W (1992) Localization in the fibrinogen gamma-chain of a new site that is involved in the acceleration of the tissue-type plasminogen activator-catalysed activation of plasminogen. Biochem J 283: 187–191

97. Grailhe P, Nieuwenhuizen W, Angles-Cano E (1994) Study of tissue-type plasminogen activator binding sites on fibrin using distinct fragments of fibrinogen. Eur J Biochem 219:961–967

98. Tamaki T, Aoki H (1981) Crosslinking of α₂-plasmin inhibitor and fibronectin to fibrin by fibrin-stabilizing factor. Biochim Biophys Acta 661:280–286

99. Kimura S, Aoki N (1986) Crosslinking site in fibrinogen for alpha 2-plasmin inhibitor. J Biol Chem 261:15591–15595

100. Sakata Y, Aoki H (1982) Significance of crosslinking of α₂-plasmin inhibitor to fibrin in inhibition of fibrinolysis and in hemostasis. J Clin Invest 69:536–542

101. Mosesson MW, Finlayson JS (1963) Biochemical and chromatographic studies of certain activities associated with human fibrinogen preparations. J Clin Invest 42:747–755

102. Siebenlist KR, Mosesson MW, Meh DA, et al (2000) Coexisting dysfibrinogenemia (gammaR275C) and factor V Leiden deficiency associated with thromboembolic disease (fibrinogen Cedar Rapids). Blood Coagul Fibrinolysis 11:293–304

103. Ritchie H, Lawrie LC, Crombie PW, et al (2000) Crosslinking of plasminogen activator inhibitor 2 and α2- antiplasmin to fibrin(ogen). J Biol Chem 275:24915–24920

104. Ritchie H, Robbie LA, Kinghorn S, et al (1999) Monocyte plasminogen activator inhibitor 2 (PAI-2) inhibits u-PA-mediated fibrin clot lysis and is crosslinked to to fibrin. Thromb Haemost 81:96–103

105. Harpel PC, Gordon BR, Parker TS (1989) Plasmin catalyzes binding of lipoprotein(a) to immobilized fibrinogen and fibrin. Proc Natl Acad Sci U S A 86:3847–3851

106. Tsurupa G, Ho-Tin-Noe B, Angles-Cano E, et al (2003) Identification and characterization of novel lysine-independent apolipoprotein(a)-binding sites in fibrin(ogen) alphaC-domains. J Biol Chem 278:37154–37159

107. Loscalzo J, Weinfeld M, Fless GM, et al (1990) Lipoprotein(a), fibrin binding, and plasminogen activation. Arteriosclerosis 10:240–245

108. Hervio L, Durlach V, Girard-Globa A, et al (1995) Multiple binding with identical linkage: a mechanism that explains the effect of lipoprotein(a) on fibrinolysis. Biochemistry 34:13353–13358

109. Romanic AM, Arleth AJ, Willette RN, et al (1998) Factor XIIIa crosslinks lipoprotein(a) with fibrinogen and is present in human atherosclerotic lesions. Circ Res 83:264–269

110. Leung LLK (1986) Interaction of histidine-rich glycoprotein with fibrinogen and fibrin. J Clin Invest 77:1305–1311

111. Lijnen HR, Hoylaerts M, Collen D (1980) Isolation and characterization of a human plasma protein with affinity for the lysine binding sites in plasminogen: role in the regulation of fibrinolysis and identification as histidine-rich glyocoprotein. J Biol Chem 255:10214–10222

112. Anglés-Cano E, Rouy D, Lijnen HR (1992) Plasminogen binding by alpha 2-antiplasmin and histidine-rich glycoprotein does not inhibit plasminogen activation at the surface of fibrin. Biochim Biophys Acta 1156:34–42

113. Anglés-Cano E, Gris JC, Schved JF (1992) Familial association of high levels of histidine-rich glycoprotein and plasminogen activator inhibitor-1 with venous thromboembolism. J Lab Clin Med 121:646–653

114. Castaman G, Ruggeri M, Burei F, et al (1993) High levels of histidine-rich glycoprotein and thrombotic diathesis. Thromb Res 69:297–305

115. Souto JC, Garí M, Falkon L, et al (1996) A new case of hereditary histidine-rich glycoprotein deficiency with familial thrombophilia. Thromb Haemost 75:374–375

116. Shigekiyo T (1998) [Congenital histidine-rich glycoprotein deficiency.] Ryoikibetsu Shokogun Shirizu 491–493

117. Shigekiyo T, Yoshida H, Matsumoto K, et al (1998) HRG Tokushima: molecular and cellular characterization of histidine-rich glycoprotein (HRG) deficiency. Blood 91:128–133

118. Odrljin TM, Shainoff JR, Lawrence SO, et al (1996) Thrombin cleavage enhances exposure of a heparin binding domain in the N-terminus of the fibrin beta chain. Blood 88:2050–2061

119. Odrljin TM, Francis CW, Sporn LA, et al (1996) Heparin-binding domain of fibrin mediates its binding to endothelial cells. Arterioscler Thromb Vasc Biol 16:1544–1551

120. Yakovlev S, Gorlatov S, Ingham K, et al (2003) Interaction of fibrin(ogen) with heparin: further characterization and localization of the heparin-binding site. Biochemistry 42:7709–7716

121. Hamaguchi M, Bunce LA, Sporn LA, et al (1993) Spreading of platelets on fibrin is mediated by the amino terminus of the β chain including peptide β 15-42. Blood 81:2348–2356

122. Sporn LA, Bunce LA, Francis CW (1995) Cell proliferation on fibrin: modulation by fibrinopeptide cleavage. Blood 86:1801–1810

123. Chalupowicz DG, Chowdhury ZA, Bach TL, et al (1995) Fibrin II induces endothelial cell capillary tube formation. J Cell Biol 130:207–215

124. Bach TL, Barsigian C, Yaen CH, et al (1998) Endothelial cell VE-cadherin functions as a receptor for the β15-42 sequence of fibrin. J Biol Chem 273:30719–30728

125. Ribes JA, Bunce LA, Francis CW (1989) Mediation of fibrin-induced release of von Willebrand factor from cultured endothelial cells by the fibrin β chain. J Clin Invest 84:435–441

126. Francis CW, Bunce LA, Sporn LA (1993) Endothelial cell responses to fibrin mediated by FPB cleavage and the amino terminus of the β chain. Blood Cells 19:291–307

127. Martinez J, Ferber A, Bach TL, et al (2001) Interaction of fibrin with VE-cadherin. Ann N Y Acad Sci 936:386–405

128. Gorlatov S, Medved L (2002) Interaction of fibrin(ogen) with the endothelial cell receptor VE-cadherin: mapping of the receptor-binding site in the NH_2-terminal portions of the fibrin beta chains. Biochemistry 41:4107–4116

129. Cheresh DA (1987) Human endothelial cells synthesize and express an Arg-Gly-Asp-directed adhesion receptor involved in attachment to fibrinogen and von Willebrand factor. Proc Natl Acad Sci U S A 84:6471–6475

130. Felding-Habermann B, Ruggeri ZM, Cheresh DA (1992) Distinct biological consequences of integrin alpha v beta 3-mediated melanoma cell adhesion to fibrinogen and its plasmic fragments. J Biol Chem 267:5070–5077

131. Gailit J, Clarke C, Newman D, et al (1997) Human fibroblasts bind directly to fibrinogen at RGD sites through integrin alpha(v)beta3. Exp Cell Res 232:118–126
132. Asakura S, Niwa K, Tomozawa T, et al (1997) Fibroblasts spread on immobilized fibrin monomer by mobilizing a β_1-class integrin, together with a vitronectin receptor $\alpha v \beta 3$ on their surface. J Biol Chem 272:8824–8829
133. Suehiro K, Gailit J, Plow EF (1997) Fibrinogen is a ligand for integrin alpha5beta1 on endothelial cells. J Biol Chem 272:5360–5366
134. Belkin AM, Tsurupa G, Zemskov E, et al (2005) Transglutaminase-mediated oligomerization of the fibrin(ogen) αC-domains promotes integrin-dependent cell adhesion and signaling. Blood 105:3561–3568
135. Kloczewiak M, Timmons S, Lukas TJ, et al (1984) Platelet receptor recognition site on human fibrinogen: synthesis and structure-function relationship of peptides corresponding to the C-terminal segment of the γ chain. Biochemistry 23: 1767–1774
136. Andrieux A, Hudry-Clergeon G, Ryckwaert J-J (1989) Amino acid sequences in fibrinogen mediating its interaction with its platelet receptor, GP IIbIIIa. J Biol Chem 264:9258–9265
137. Lam SCT, Plow EF, Smith MA, et al (1987) Evidence that arginyl-glycyl-aspartate peptides and fibrinogen γ chain peptides share a common binding site on platelets. J Biol Chem 262:947–950
138. Santoro SA, Lawing WJ Jr (1987) Competition for related but nonidentical binding sites on the glycoprotein IIb-IIIa complex by peptides derived from platelet adhesive proteins. Cell 48:867–873
139. Bennett JS (2001) Platelet-fibrinogen interactions. Ann NY Acad Sci 936:340–354
140. Stathakis NE, Mosesson MW (1977) Interactions among heparin, cold-insoluble globulin, and fibrinogen in formation of the heparin precipitable fraction of plasma. J Clin Invest 60:855–865
141. Mosher DF (1975) Crosslinking of cold-insoluble globulin by fibrin-stabilizing factor. J Biol Chem 250:6614–6621
142. McDonagh RP, McDonagh J, Petersen TE, et al (1981) Amino acid sequence of the factor XIIIa acceptor site in bovine plasma fibronectin. FEBS Lett 127:174–178
143. Matsuka YV, Migliorini MM, Ingham KC (1997) Crosslinking of fibronectin to C-terminal fragments of the fibrinogen alpha-chain by factor XIIIa. J Prot Chem 16:739–745
144. Sahni A, Odrljin T, Francis CW (1998) Binding of basic fibroblast growth factors to fibrinogen and fibrin. J Biol Chem 273:7554–7559
145. Sahni A, Sporn LA, Francis CW (1999) Potentiation of endothelial cell proliferation by fibrin(ogen)-bound fibroblast growth factor-2. J Biol Chem 274:14936–14941
146. Sahni A, Francis CW (2000) Vascular endothelial growth factor binds to fibrinogen and fibrin and stimulates endothelial cell proliferation. Blood 96:3772–3778
147. Sahni A, Guo M, Sahni SK, et al (2004) Interleukin-1beta but not IL-1alpha binds to fibrinogen and fibrin and has enhanced activity in the bound form. Blood 104:409–414
148. Siebenlist KR, Meh DA, Mosesson MW (1996) Plasma factor XIII binds specifically to fibrinogen molecules containing γ' chains. Biochemistry 35:10448–10453
149. Mosesson MW, Wautier JL, Amrani DL, et al (1982) Evidence for circulating fibrin in familial Mediterranean fever. J Lab Clin Med 99:559–567
150. Polgár J, Hidasi V, Muszbek L (1990) Non-proteolytic activation of cellular protransglutaminase (placenta macrophage factor XIII). Biochem J 267:557–560
151. Sakata Y, Aoki N (1980) Crosslinking of alpha 2-plasmin inhibitor to fibrin by fibrin-stabilizing factor. J Clin Invest 65:290–297
152. Tamaki T, Aoki N (1982) Crosslinking of alpha 2-plasmin inhibitor to fibrin catalyzed by activated fibrin-stabilizing factor. J Biol Chem 257:14767–14772

153. Ichinose A, Aoki N (1982) Reversible crosslinking of alpha 2-plasmin inhibitor to fibrinogen by fibrin-stabilizing factor. Biochim Biophys Acta 706:158–164
154. Fenton JW II, Olson TA, Zabinski MP, et al (1988) Anion-binding exosite of human α-thrombin and fibrin(ogen) recognition. Biochemistry 27:7106–7112
155. Stubbs MT, Bode W (1993) A player of many parts: the spotlight falls on thrombin's structure. Thromb Res 69:1–58
156. Howell WH (1910) The preparation and properties of thrombin, together with observations on antithrombin and prothrombin. Am J Physiol 26:453–473
157. Seegers WH, Nieft M, Loomis EC (1945) Note on the adsorption of thrombin on fibrin. Science 101:520–521
158. Seegers WH (1947) Multiple protein interactions as exhibited by the blood-clotting mechanism. J Phys Colloid Chem 51:198–206
159. Seegers WH, Johnson JF, Fell C (1954) An antithrombin reaction related to prothrombin activation. Am J Physiol 176:97–103
160. De Bosch NB, Mosesson MW, Ruiz-Sáez A, et al (2002) Inhibition of thrombin generation in plasma by fibrin formation (antithrombin I). Thromb Haemost 88:253–258
161. Mosesson MW (2003) Antithrombin I: inhibition of thrombin generation in plasma by fibrin formation. Thromb Haemost 89:912
162. Liu CY, Nossel HL, Kaplan KL (1979) Defective thrombin binding by abnormal fibrin associated with recurrent thrombosis. Thromb Haemost 42:79 (abstract)
163. Meh DA, Siebenlist KR, Mosesson MW (1996) Identification and characterization of the thrombin binding sites on fibrin. J Biol Chem 271:23121–23125
164. Pechik I, Madrazo J, Mosesson MW, et al (2004) Crystal structure of the complex between thrombin and the central "E" region of fibrin. Proc Natl Acad Sci U S A 101:2718–2723
165. Pospisil CH, Stafford AR, Fredenburgh JC, et al (2003) Evidence that both exosites on thrombin participate in its high affinity interaction with fibrin. J Biol Chem 278:21584–21591
166. Lovely RS, Moaddel M, Farrell DH (2003) Fibrinogen γ′ chain binds thrombin exosite II. J Thromb Haemost 1:124–131
167. Pineda AO, Chen ZW, Marino F, et al (2007) Crystal structure of thrombin in complex with fibrinogen gamma' peptide. Biophys Chem 125:556–559
168. Meh DA, Mosesson MW, Siebenlist KR, et al (2001) Fibrinogen Naples I (Bβ A68T) non-substrate thrombin binding capacities. Thromb Res 103:6373
169. Kumar R, Béguin S, Hemker C (1994) The influence of fibrinogen and fibrin on thrombin generation: evidence for feedback activation of the clotting system by clot bound thrombin. Thromb Haemost 72:713–721
170. Nossel HL, Ti M, Kaplan KL, et al (1976) The generation of fibrinopeptide A in clinical blood samples: evidence for thrombin activity. J Clin Invest 58:1136–1144
171. Francis CW, Markham RE, Barlow GH, et al (1983) Thrombin activity of fibrin thrombi and soluble plasmic derivatives. J Lab Clin Med 102:220–230
172. Owen J, Friedman KD, Grossman BA, et al (1988) Thrombolytic therapy with tissue plasminogen activator or streptokinase induces transient thrombin activity. Blood 72:616–620
173. Weitz JI, Hudoba M, Massel D, et al (1990) Clot-bound thrombin is protected from inhibition by heparin-antithrombin III but is susceptible to inactivation by antithrombin III-independent inhibitors. J Clin Invest 86:385–391
174. Mutch NJ, Robbie LA, Booth NA (2001) Human thrombi contain an abundance of active thrombin. Thromb Haemost 86:1028–1034
175. Siebenlist KR, Mosesson MW, Hernandez I, et al (2005) Studies on the basis for the properties of fibrin produced from fibrinogen containing γ′ chains. Blood 106:2730–2736
176. Di Cera E, Dang QD, Ayala YM (1997) Molecular mechanisms of thrombin function. Cell Mol Life Sci 53:701–730

177. Esmon CT (1989) The roles of protein C and thrombomodulin in the regulation of blood coagulation. J Biol Chem 264:4743–4746
178. Cooper AV, Standeven KF, Ariens RA (2003) Fibrinogen gamma-chain splice variant gamma' alters fibrin formation and structure. Blood 102:535–540
179. Fredenburgh JC, Stafford AR, Weitz JI (1997) Evidence for allosteric linkage between exosites 1 and 2 of thrombin. J Biol Chem 272:25493–25499
180. Bock LC, Griffin LC, Latham JA, et al (1992) Selection of single-stranded DNA molecules that bind and inhibit human thrombin. Nature 355:564–566
181. Liaw PC, Fredenburgh JC, Stafford AR, et al (1998) Localization of the thrombin-binding domain on prothrombin fragment 2. J Biol Chem 273:8932–8939
182. Li CQ, Vindigni A, Sadler JE, et al (2001) Platelet glycoprotein Ib alpha binds to thrombin anion-binding exosite II inducing allosteric changes in the activity of thrombin. J Biol Chem 276:6161–6168
183. Colwell NS, Blinder MA, Tsiang M, et al (1998) Allosteric effects of a monoclonal antibody against thrombin exosite II. Biochemistry 37:15057–15065
184. Tasset DM, Kubik MF, Steiner W (1997) Oligonucleotide inhibitors of human thrombin that bind distinct epitopes. J Mol Biol 272:688–698
185. Credo RB, Curtis CG, Lorand L (1978) Ca^{2+}-related regulatory function of fibrinogen. Proc Natl Acad Sci U S A 75:4234–4237
186. Janus TJ, Lewis SD, Lorand L, et al (1983) Promotion of thrombin-catalyzed activation of factor XIII by fibrinogen. Biochemistry 22:6268–6272
187. Greenberg CS, Miraglia CC, Rickles FR, et al (1985) Cleavage of blood coagulation factor XIII and fibrinogen by thrombin during in vitro clotting. J Clin Invest 75:1463–1470
188. Naski MC, Lorand L, Shafer JA (1991) Characterization of the kinetic pathway for fibrin promotion of alpha-thrombin-catalyzed activation of plasma factor XIII. Biochemistry 30:934–941
189. Francis CW, Kraus DH, Marder VJ (1983) Structural and chromatographic heterogeneity of normal plasma fibrinogen associated with the presence of three gamma-chain types with distinct molecular weights. Biochim Biophys Acta 744:155–164
190. Francis CW, Keele EM, Marder VJ (1984) Purification of three gamma-chains with different molecular weights from normal human plasma fibrinogen. Biochim Biophys Acta 797:328–335
191. Francis CW, Muller E, Henschen A, et al (1988) Carboxy-terminal amino acid sequences of two large variant forms of the human plasma fibrinogen g chain. Proc Natl Acad Sci U S A 85:3358–3362
192. Mosesson MW, Hernandez I, Raife TJ, et al (2007) Plasma fibrinogen gamma' chain content in the thrombotic microangiopathy syndrome. J Thromb Haemost 5:62–69
193. Koopman J, Haverkate F, Lord ST, et al (1992) Molecular basis of fibrinogen Naples associated with defective thrombin binding and thrombophilia: homozygous substitution of B beta 68 Ala → Thr. J Clin Invest 90:238–244
194. Di Minno G, Martinez J, Cirillo F, et al (1991) A role for platelets and thrombin in the juvenile stroke of two siblings with defective thrombin-absorbing capacity of fibrin(ogen). Arterioscler Thromb 11:785–796
195. Caen J, Faur Y, Inceman S, et al (1964) Nécrose ischémique bilatérale dans un cas de grande hypofibrinogénémie congénitale. Nouv Rev Fr Hematol 4:321–326
196. Marchal G, Duhamel G, Samama M, et al (1964) Thrombose massive des vaisseaux d'un membre au cours d'une hypofibrinémie congénitale. Hemostase 4:81–89
197. Nilsson IM, Niléhn J-E, Cronberg S, et al (1966) Hypofibrinogenemia and massive thrombosis. Acta Med Scand 180:65–76
198. Ingram GI, McBrien DJ, Spencer H (1966) Fatal pulmonary embolism in congenital fibrinopenia. Acta Haematol 35:56–62

199. Mackinnon HH, Fekete JF (1971) Congenital afibrinogenemia: vascular changes and multiple thrombosis induced by fibrinogen infusions and contraceptive medication. Can Med Assoc J 140:597–599

200. Cronin C, Fitzpatrick D, Temperly I (1988) Multiple pulmonary emboli in a patient with afibrinogenaemia. Acta Haematol 7:53–54

201. Drai E, Taillan B, Schneider S, et al (1992) Thrombose portale révélatrice d'une afibrinogénémie congénitale. La Presse Med 21:1820–1821

202. Chafa O, Chellali T, Sternberg C, et al (1995) Severe hypofibrinogenemia associated with bilateral ischemic necrosis of toes and fingers. Blood Coagul Fibrinolysis 6:549–552

203. Dupuy E, Soria C, Molho P, et al (2001) Embolized ischemic lesions of toes in an afibrinogenemic patient: possible relevance to in vivo circulating thrombin. Thromb Res 102:211–219

204. De Bosch N, Sáez A, Soria C, et al (1997) Coagulation profile in afibrinogenemia. Thromb Haemost (Suppl):625 (abstract)

205. Korte W, Feldges A (1994) Increased prothrombin activation in a patient with congenital afibrinogenemia is reversible by fibrinogen substitution. Clin Invest 72:396–398

206. Ni H, Denis CV, Subbarao S, et al (2000) Persistence of platelet thrombus formation in arterioles of mice lacking both von Willebrand factor and fibrinogen. J Clin Invest 106:385–392

207. Drouet L, Paolucci F, Pasqualini N, et al (1999) Plasma gamma'/gamma fibrinogen ratio, a marker of arterial thrombotic activity: a new potential cardiovascular risk factor? Blood Coagul Fibrinolysis 10(Suppl 1):S35–S39

208. Lovely RS, Falls LA, Al Mondhiry HA, et al (2002) Association of γA/γ' fibrinogen levels and coronary artery disease. Thromb Haemost 88:26–31

209. Uitte de Willige S, de Visser MC, Houwing-Duistermaat JJ, et al (2005) Genetic variation in the fibrinogen gamma gene increases the risk of deep venous thrombosis by reducing plasma fibrinogen γ' levels. Blood 106:4176–4183

210. Van der Meer FJ, Koster T, Vandenbroucke JP, et al (1997) The Leiden Thrombophilia Study (LETS). Thromb Haemost 78:631–635

211. Manilla MN, Lovely RS, Kazmierczak SC, et al (2007) Elevated plasma fibrinogen γ' concentration is associated with myocardial infarction: effects of variation in fibrinogen genes and environmental factors. J Thromb Haemost 5:766–733

212. Mosesson MW (2005) Fibrinogen and fibrin structure and functions. J Thromb Haemost 3:1894–1904

213. Mosesson MW (2003) Fibrinogen gamma chain functions. J Thromb Haemost 1:231–238

214. Carter WJ, Cama E, Huntington JA (2005) Crystal structure of thrombin bound to heparin. J Biol Chem 280:2745–2749

215. Becker DL, Fredenburgh JC, Stafford AR, et al (1999) Exosites 1 and 2 are essential for protection of fibrin-bound thrombin from heparin-catalyzed inhibition by antithrombin and heparin cofactor II. J Biol Chem 274:6226–6233

Vitamin K Cycles and γ-Carboxylation of Coagulation Factors

Darrel W. Stafford and Christine M. Hebling

Summary. The vitamin K cycle consists of two enzymes, vitamin K oxido reductase (VKOR) and γ-glutamyl carboxylase (VKGC). VKGC acts on vitamin K-dependent (VKD) substrates using co-substrates: reduced vitamin K (KH_2), O_2, and CO_2 to convert glutamic acid (Glu) to γ-carboxyglutamic acid (Gla). The posttranslational modification of Glu to Gla is required for the activity of VKD proteins. VKD proteins participate in a broad range of physiologies including proteins involved in hemostasis, calcification, and cell growth. Concomitant with the carboxylation reaction, KH_2 is converted to vitamin K epoxide (KO). After carboxylation, reduced VKOR converts KO to vitamin K and regenerates KH_2, allowing the cycle to continue. Presumably, a third enzyme, which reduces a single disulfide bond in VKOR after each reaction cycle is also required. The details of these reactions are the subject of this chapter.

Key words. Vitamin K · γ-Glutamyl carboxylase · VKOR · VKD protein · Blood coagulation

Introduction

Vitamin K was first discovered in 1929 when Henrik Dam was studying the synthesis of cholesterol in chickens [1]. Dam noted that chickens being fed a lipid-free diet developed a bleeding diathesis. This observation later led to the discovery of the fat-soluble nutrient vitamin K, named after the Scandinavian spelling of the word *koagulation* for its crucial role in blood coagulation [2]. After the initial discovery of vitamin K, Edward Doisy determined that the main source of vitamin K was through the diet, particularly from green vegetables and fish meal [3].

Further characterization of the chemical structure classified vitamin K in a family of methylated naphthoquinone ring structures occupying variable aliphatic side chains at position 3 [3]. Vitamin K_1, also called phylloquinone, has a side chain consisting of four isoprenoid units, one of which is unsaturated. Vitamin K_2, also called menaquinone, contains an unsaturated side chain of repeating isoprenoid groups that vary in length from 4 to 13 units and is found in milk, cheese, and fermented soy products. K_2 analogues are typically called menaquinone-*n*, or MK-*n*, where *n* is the number of isoprenoid groups. There is a third synthetic form of vitamin K, called vitamin K_3 or menadione (MK0), which has a methyl group present at the second position on the naphthoquinone ring. It is primarily used in animal feed and is converted by the body to menaquinone for biological function (Davidson RT, Foley AL,

Engelke JA, and Suttie JW (1998) Conversion of dietary phylloquinone to tissue mena-quinone-4 in rats is not dependent on gut bacteria. J Nutrition 128: 220–223). Structural representations of vitamin K analogs are shown in Fig. 1.

Shortly after the discovery of vitamin K, another important discovery in blood coagulation was made. During the early 1940s, cows were bleeding to death after eating moldy sweet clover hay [4]. Karl Link discovered that these deaths could be attributed to a fungal vitamin K antagonist, dicumarol. Further studies led to the synthesis of a series of derivatives of dicumarol. One of these derivatives, warfarin (Wisconsin Alumni Research Foundation) was initially used as a rat poison before being used as an anticoagulant. Despite the potential harmful side effects, warfarin has continued to be regarded as the preferred choice of anticlotting agent.

The actual role of vitamin K in blood coagulation did not become apparent until 1974 when Steflo, Nelsestuen, and Magnusson independently discovered the novel amino acid γ-carboxyglutamic acid (Gla) [5–7]. As represented in Fig. 2, a Gla residue is a modified glutamic acid (Glu) residue with the addition of a carboxyl group at

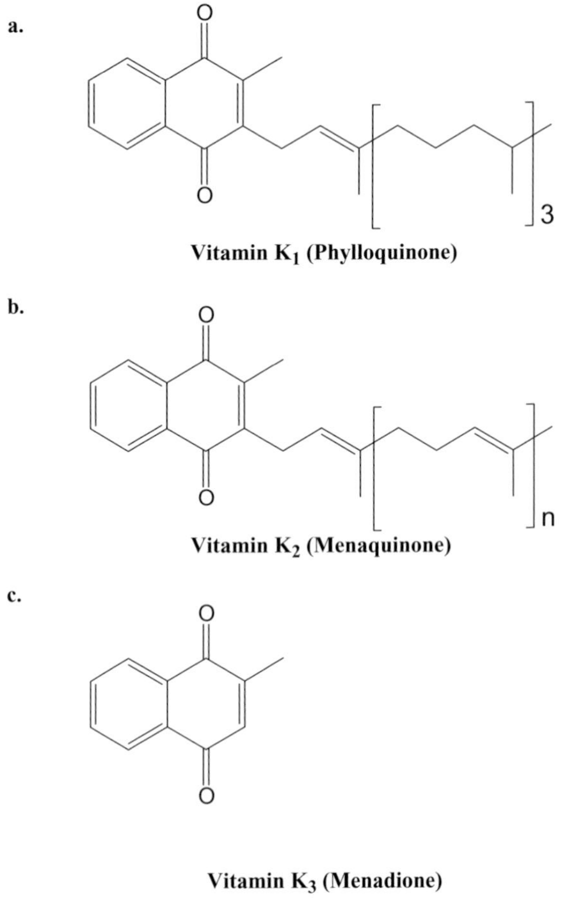

a.

Vitamin K$_1$ (Phylloquinone)

b.

Vitamin K$_2$ (Menaquinone)

c.

Vitamin K$_3$ (Menadione)

FIG. 1. Vitamin K homologues. a Vitamin K$_1$. b Vitamin K$_2$. c Vitamin K$_3$

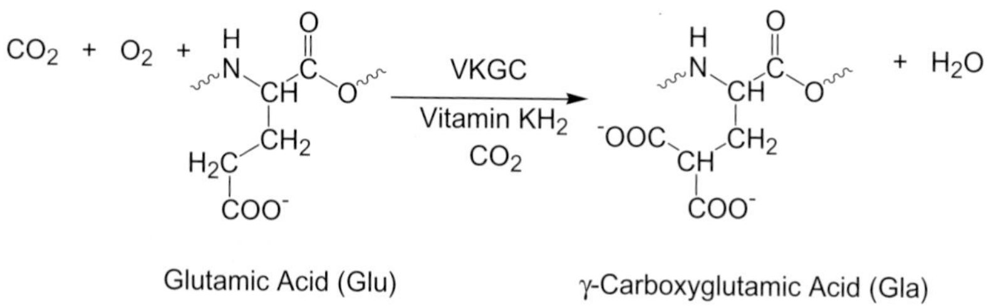

FIG. 2. Conversion of glutamic acid (*Glu*) to γ-carboxyglutamic acid (*Gla*) in the presence of γ-glutamyl carboxylase (*VKGC*), vitamin K hydroquinone (*KH₂*), oxygen (*O₂*), and carbon dioxide (*CO₂*)

position 4. In this study, it was observed that Gla residues were absent from bovine prothrombin following vitamin K-deficient diets or treatment with the anticoagulant dicumarol, a vitamin K antagonist. It was then hypothesized that vitamin K serves as an obligatory cofactor for carboxylation of protein-bound Glu to Gla in prothrombin and other vitamin K-dependent proteins.

Furhther work investigating the role of vitamin K, allowed identification of a family of Gla proteins referred to as vitamin K-dependent (VKD) proteins. Most of these proteins play a role in blood coagulation, including those with procoagulant functions (prothrombin, factor VII, factor IX, factor X) and those with anticoagulant function (proteins C, S, and Z). Outside of the coagulation cascade, VKD proteins are also involved in calcification and cell growth. One such example was demonstrated with the observation that a knockout of matrix Gla protein (MGP) leads to complete calcification of arteries within two months of age [8]. This condition may be reversed, however, if MGP is locally expressed in smooth muscle cells, demonstrating the importance and specificity of this reaction [9]. Transgenic expression of MGP with two or four of its Gla residues modified to aspartate prevents carboxylation and renders MGP incapable of rescuing the calcification of arteries in the MGP knockout mouse. Thus, the carboxylation of Glu residues is found necessary for the function of MCP in bone metabolism. Other VKD proteins include osteocalcin (involved in calcification), the growth arrest protein Gas 6 (activates the Axl tryrosine kinase receptor), and four proline-rich proteins of unknown function: PRRG1, PRRG2, PRRG3, and PRRG4 [10].

Several factors influence the impact of vitamin K during the carboxylation process including source, location, and side chain additions. The amount of vitamin K in the diet is often a limiting factor in the carboxylation reaction. For example, osteocalcin, or bone Gla protein, is partially carboxylated in normal healthy volunteers, however, becomes fully carboxylated only when vitamin K is supplemented into the diet [11]. Likewise, it has been commonly assumed that vitamin K produced by enteric bacteria can be absorbed. If coprophagy is prevented, however, rats fed a vitamin K-free diet develop severe bleeding problems within weeks [12, 13]. In addition to vitamin K source, the location and half-lives of vitamin K analogs also play an important role. A recent observation concluded that K_1 appears to be taken up primarily in the liver, whereas K_2 appears to preferentially accumulate in arteries and extrahepatic locations [13, 14]. The half-lives of vitamin K derivatives were also seen to vary depending on side-chain additions. Increasing the hydrophobicity of vitamin K_2 analogues MK7 and

MK9 has been shown to contribute to longer half-lives in plasma compared to short-chain analogues such as MK4 [15, 16].

Further investigation of vitamin analogues disproved the former assumption that vitamin K functions solely as a co-substrate for carboxylation of VKD proteins. Although it was known that K_2 promoted bone formation, it was assumed that its function was exclusively through its action on the Gla proteins osteocalcin and MGP [17]. Recently, however, it has been reported that K_2 can directly stimulate mRNA production of osteoblast mRNA markers as well as function as a transcriptional regulator of bone-specific genes [18]. Such findings not only expand current knowledge of the overall physiological importance of vitamin K, but specifically provide additional information on vitamin K involvement in carboxylation.

The enzyme that modifies vitamin K-dependent proteins is gamma-glutamyl carboxylase (VKGC). Glu residues in VKD proteins are carboxylated to form Gla amino acids (Fig. 3, I). Concomitant with Gla modification, reduced vitamin K (KH_2) is converted to vitamin K 2,3-epoxide (KO). Before KH_2 can be reused in the carboxylation mechanism, it must be converted from KO to vitamin K by vitamin K epoxide reductase (VKOR) (Fig. 3, II). During this recycling, cysteine residues in VKOR at positions 132 and 135 are oxidized to a disulfide bond. The highly conserved redox center in VKOR plays a crucial role not only in vitamin K formation but has also been shown to be the target site for the anticoagulant warfarin. Despite the known functionality of this center, the identity of electron donation for VKOR remains unknown. As KH_2 is regenerated (Fig. 3, III), the disulfide bond in VKOR is reduced back to its thiol form and the cycle continues [19, 20]. As a result of the interdependence of proteins in the vitamin K cycle, warfarin treatment not only results in the inhibition of VKOR activity, but subsequently results in a dramatic depletion of KH_2, reduced carboxylation, and a decrease in blood clotting. The vitamin K cycle is depicted in Fig. 3.

It is clear that VKOR itself can catalyze both the conversion of KO to Vitamin K and Vitamin K to KH_2 [21, 22]. Nevertheless, patients poisoned with warfarin or brodifacoum (superwarfarin) can be maintained with vitamin K treatment indicating that enzymes other than VKOR may also be involved in the conversion of vitamin K to KH_2. For example, it has been shown that DT diaphorase (NQO1) and (NQO2) can function to convert K to KH_2. Reports studying this quinone reductase, have concluded that DT diaphorase acts as the antidotal enzyme [23, 24]. With the recent availability of knockout mice NQO1, NQO2, and their combination, [25–27] future experimentation will determine if one (or both) of these enzymes are responsible for the conversion of K to KH_2 in warfarin poisoned patients.

Carboxylation

Vitamin K dependent γ-glutamyl carboxyase was first discovered in 1975 but was not purified and cloned until 1991 [28–30]. VKGC is a 758-amino-acid integral membrane protein with five transmembrane domains and one disulfide bond between residues 99 and 450 [31, 32]. Topological studies demonstrated that each transmembrane domain passes through the endoplasmic reticulum (ER), with the N-terminal in the cytoplasm and the C-terminal in the lumen. The membrane topology of the enzyme is shown in Fig. 4.

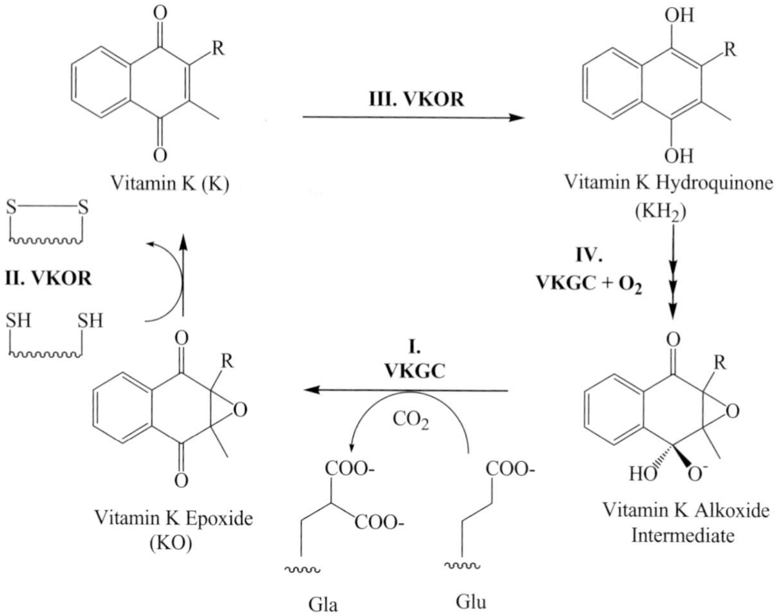

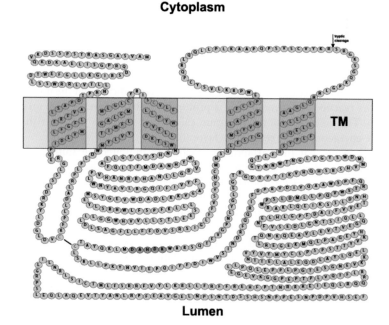

FIG. 3. Roles of γ-glutamyl carboxylase (*VKGC*) and vitamin K oxido reductase (*VKOR*) in the vitamin K cycle

Cytoplasm

Lumen

FIG. 4. Proposed topology of vitamin K γ-carboxylase with five transmembrane domains (*TM*) and a disulfide bond between cysteine residues 99 and 450

VKCG recognizes VKD protein substrates through an 18-amino-acid N-terminal propeptide domain that serves to bind the substrate to an exosite on VKGC and anchor the substrate while multiple Glu residues in the substrate protein are modified. Although propeptide domains have similar sequences, their affinities are seen to vary more than 100-fold [33]. Unlike typical VKD proteins, the propeptide domain in bone proteins is slightly modified. In the case of MGP, the propeptide domain is found internally and remains after carboxylation of Glu residues [34]. The propeptide domain of osteocalcin, on the other hand, has very poor affinity for carboxyase; however, binding is found to occur more strongly in the remaining protein sequence [33].

In most VKD proteins, a highly conserved Gla domain follows the propeptide. This approximately 45-amino-acid residue region contains the Glu residues that are converted to Gla. In a typical VKD protein, approximately 9–13 Glu residues are carboxylated [35–37]. Exceptions to this are osteocalcin and MGP, which do not have typical Gla domains, carboxylating only 3–5 Glu residues. Once carboxylation converts the selected Glu residues in the clotting cascade proteins to Gla residues, the propeptide is cleaved, and the mature protein is secreted from the cell. When the VKD protein is in the presence of calcium, the Gla domain undergoes a conformational change, allowing association with phospholipids on the membrane surface near damaged vascular tissue.

An elegant theory explaining a chemical model for carboxylation was proposed by Dowd and colleagues [38]. According to this model, described as base amplification, KH_2 is transformed to a strong base by VKGC and O_2. This strong base then abstracts a proton from the γ-carbon at position 4 of selected glutamates in vitamin K-dependent proteins and, through nucleophilic attack by CO_2, forms the Gla product. Dowd initially proposed two possible structures for the strong base: a dialkoxide and an alkoxide of vitamin K.

Recently Davis et al. investigated quantum mechanical methods to evaluate Dowd's hypothesis [39]. The geometries of the proposed model intermediates were energetically optimized; however, they were unable to isolate a stable dialkoxide. Thus, it was concluded that the alkoxide of vitamin K (Fig. 3, IV) is most likely the strong base that abstracts a proton from the γ-carbon of glutamic acid to create a carbanion and react with carbon dioxide. Results from this study also indicate that the initial, crucial step in the reaction is the removal of a proton from the hydroxyl group on position 1 of KH_2. The prediction of the quantum calculation supports that after this initial event the energetics of the reaction is favorable.

After showing that the carboxylase reaction is inhibited by sulfhydryl-specific reagents such as N-ethylmaleimide, it was initially assumed that cysteines were the catalytic residue on VKGC responsible for KH_2 deprotonation. In keeping with Dowd's theory, Pudota et al. reported that modification of cysteine residues 99 and 450 in VKGC resulted in the loss of carboxylation, indicating that these two cysteine residues are involved as active site residues (Pudota BN, Miyagi M, Hallgren KW, et al (2000) Identification of the vitamin K-dependent carboxylase active site: Cys-99 and Cys-250 are required for both epoxidation are carboxylation. PNAS 97(24): 13033–13038). In 2003, Tie et al. provided evidence that cysteines are not involved in the active glutamate recognition site and further revealed that C99 and C450 were involved in disulfide linkage crucial for carboxylase activity [31]. Confirmation of disulfide bond formation was shown by Wu et al. and Tie et al. where purified carboxylase was cleaved at residues 349 and 351 under modified trypsin conditions [31, 40]. Before reduction of any disulfide bond, this proteolyzed, purified two-chain carboxylase migrates as a single band on sodium dodecyl sulfate polyacrylamine gel electrophoresis (SDS-PAGE) and exhibits normal carboxylase activity. After reduction, the purified, trypsin-digested protein migrates as two bands [40]. Moreover, the mutation of C598 and C700 (but not residue C450) after limited trypsin cleavage results in a mutant enzyme without loss of activity. This mutated enzyme is still shown to be held together by a disulfide bond based on gel analyses. Because C450 is the only remaining cysteine in the mutated 351–758 tryptic fragment, this residue must participate in disulfide bond formation.

As a result of these findings, other residues in carboxylase were investigated. Rishavy et al. demonstrated that when carboxylase was incubated with diethyl pyrocarbonate, a histidine-specific reagent, carboxylase activity was decreased. Thus, the authors concluded that a histidine was likely the active site residue of VKGC [41]. Further studies by the same laboratory reported that lysine 218 was the catalytic base to deprotonate KH_2. This work looked at the ability of exogenous amines to restore activity of the mutated K218A and concluded that recovery of activity was not only feasible but dependent on the basic form of the amine. Although no further experimentation has been completed on residue 218, it had previously been demonstrated by Shimizu et al. that a mutation of lysine 218 resulted in lack of carboxylase activity [42]. Moreover, this scenario is consistent with quantum calculations [43, 44].

Our current theory on how carboxylation occurs is that the substrate binds to VKGC primarily through its propeptide. The binding is relatively tight, and all substrate carboxylations (that are to occur for a given substrate) occur during a single binding event [45–47]. Thus, when a kinetic time course experiment monitoring the incorporation of radioactive $^{14}CO_2$ is measured in the presence of a large excess of substrate including the propeptide and Gla domain of human factor IX, a burst phase followed by a prolonged steady state is observed (Fig 5a). The slower steady-state rate constant corresponds to the release of the fully carboxylated K-dependent substrate from VKGC. This is indicated by the observation that the steady-state rate constant of $4.5 \times 10^{-4}\,s^{-1}$ is nearly identical to the dissociation rate constant of a fully carboxylated substrate, $3.7 \times 10^{-4}\,s^{-1}$. The burst phase represents the incorporation of $^{14}CO_2$ during one binding event (for the factor IX substrate, 12 $^{14}CO_2$ molecules) and is fast compared to the release of the carboxylated substrate. The experimental compliance between the extent of the burst, the intercept of the extrapolated steady-state rate, and the concentration of active enzyme calculated independently is remarkably agreeable. The rate of incorporation during the burst phase is calculated to be $0.013\,s^{-1}$. This agrees well, as it should, with the rate constant calculated for an enzyme excess experiment, $0.014\,s^{-1}$ (Fig. 5b). These results are slightly different from those previously reported ($0.016\,s^{-1}$ and $0.02\,s^{-1}$); however, the data is identical, with the only difference being that all points were fit simultaneously instead of analyzing each experiment separately. The results shown in Fig. 5b reveal that the enzyme-excess rate constant is linear up to ten Gla residues per molecule and then slows for the addition of the last two $^{14}CO_2$ molecules. Incidentally, only ten factor IX Glas are required for full activity [48]. Based on these in vitro experiments, it was concluded that a VKD substrate binds VKGC through its propeptide domain and while bound modifies all necessary Glu residues. The extent of carboxylation is not only determined by the relative rates of carboxylation itself but is also dependent on the level of vitamin K available, the dissociation rate of the substrate from VKGC, and the addition or modification of co-substrates.

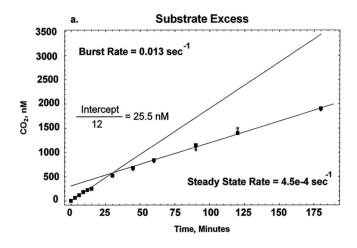

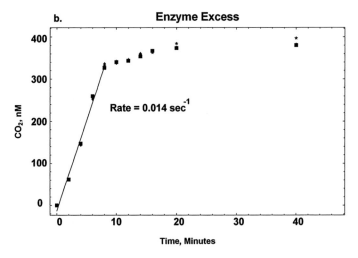

FIG. 5. Time course of carboxylation of FIXproGla41. **a** Substrate-excess conditions at 17°C with 25 nM vitamin K-expoxide reductase, and 1500 nM FIXproGla. The initial burst phase is followed by a slower steady-state phase. The lines represent linear regressions. **b** Enzyme-excess conditions. The concentration of enzyme was 250 nM, and the concentration of FIXproGla41 was 50 mM. The rate is linear up to 10 Glas and is considerably slower for the last two $^{14}CO_2$ points. Each time point represents three sets of data points that were individually fitted to a regression equation

Although, the observation that carboxylation is processive has been confirmed by Stenina et al. [47], it was speculated that the propeptide is bound to a second site. Additional experimentation, however, provided further evidence that the stoichiometry of the propeptide binding to VKGC is 1:1 [49], discounting the possibility of a second propeptide binding site. Comparing in vitro rate constants ($0.013\,s^{-1}$ for the burst phase of a substrate-excess time course and $0.014\,s^{-1}$ for an enzyme excess study) to an in vivo rate constant for carboxylation ($0.047\,s^{-1}$), it may be noted that comparable values were obtained when accounting for experimental assumptions and measurement error [50].

Additional Components of the Vitamin K Cycle

The Wallin laboratory attempted to identify VKOR by cross-linking rat microsomal proteins to a radioactive azide derivative of warfarin. From this study, three spots on a two-dimensional gel were identified to be radioactive. One of the spots was identified as cytochrome b_5, one remained unidentified, and the third was calumenin. Because cytochrome b_5 was unlikely to have VKOR activity, it was concluded that calumenin must be the warfarin-sensitive subunit of VKOR. Indeed when further studies were conducted looking at warfarin-resistant rats, calumenin was found to be in high concentrations in the liver. Thus, it was concluded that calumenin interacts with VKOR and is responsible for warfarin resistance [51]. Based on Wallin's data, two groups suggested the possibility of a relation between calumenin and the vitamin K cycle and began by looking for polymorphisms associated with warfarin dosage [52, 53]. Kimura et al. [53] found no warfarin-resistant patients associated with calumenin polymorphisms, and Vecsler et al. [52] reported one patient with a calumenin polymorphism that required very high warfarin doses to achieve a therapeutic INR (International Normalization Ratio). Although there was no mention of the sequence of the VKOR gene from that patient, it was recently learned (Eva Gak, 2007, personal communication), that this particular patient suffers from a mutation in VKOR that renders it warfarin resistant, thus discounting a relation between warfarin resistance and calumenin influence in humans. Wallin's laboratory also reported that the percentage of active factor IX in a cell line increases when calumenin production is inhibited with siRNA [54]. Data for the importance of calumenin in carboxylation is accumulating, and further investigation is warranted.

Correlation of In Vitro and In Vivo Data

Our conclusion drawn from in vitro experiments that substrate release is mechanistically important is consistent with several observations drawn from experimental studies conducted in cells and from patients. In the first example, a transfected line of HEK 293 cells exhibits only 30% carboxylation of the factor X (FX) vitamin K-dependent protein, leaving 70% uncarboxylated. It was postulated that because FX has the tightest binding propeptide of any known Gla protein, reducing the affinity of the propeptide should increase the turnover rate and lead to more complete carboxylation. To test this hypothesis, the FX propeptide was replaced with a prothrombin propeptide exhibiting affinity for VKGC about 50 times less than that of FX. The study confirmed that the modified substrate resulted in an increase to 90%–95% carboxylation of the FX Gla domain [55].

A second study focused on a patient with a factor IX propeptide mutation that affected carboxylation. This patient demonstrated no bleeding diathesis until placed on warfarin therapy, where he was found to develop a hemophilia B phenotype [56]. Upon sequencing this patient's factor IX gene, a propeptide mutation (A-10T) was found to result in an 11-fold decrease in substrate affinity for VKGC [33]. This can be rationalized to fit kinetic data by assuming that the rate of carboxylation is reduced because of low vitamin K concentrations during warfarin therapy. Under this assumption, and given the reduced affinity of the mutated propeptide domain, the factor IX substrate is thought to dissociate from VKGC before carboxylation is complete.

A third example looks at osteocalcin from ostensibly normal volunteers that is partially carboxylated unless vitamin K supplements are introduced, allowing the substrate to become fully carboxylated. The propeptide of osteocalcin has the lowest affinity of any of the known VKD proteins. Although osteocalcin's Gla domain has a relatively high affinity for VKGC, it is assumed that the low affinity of the propeptide combined with a reduced vitamin K concentration results in the dissociation of osteocalcin prior to complete carboxylation [11]. Additional supplementation of vitamin K, however, was found to stimulate further the rate of carboxylation.

The behavior of vitamin K-dependent proteins with VKGC varies extensively depending on protein specificity. Of the VKD proteins expressed (FVIII, FIX, FX), it is apparent that the protein is secreted from the cell regardless of carboxylation status. Numerous reports support this conclusion [57, 58]; and other than that of Hallgren et al. [59], no evidence has been shown that overexpressed VKGC inhibits secretion of the VKD protein. In contrast, undercarboxylated protein C was seen not to be secreted efficiently [60]. In support of this realization, a study conducted in Suttie's laboratory reported undercarboxylated prothrombin was either degraded intracellularly, maintained in an intracellular pool, or secreted depending on whether it was rat or human [61].

VKOR

The enzyme responsible for the regeneration of vitamin K from vitamin K epoxide is VKOR. This activity was first reported in 1970 [62], but it proved refractory to purification for more than 30 years until 2004, when the gene that codes for VKOR was successfully identified [63, 64]. The epoxide reductase is encoded in about 5 kb of DNA on human chromosome 16. Several alternative splice variants for VKOR have been identified, but the cDNA encoding the enzymatic activity is MGC11276.

The VKOR enzyme, 163 amino acids in length, is an integral membrane protein with three transmembrane domains (Fig. 6) [65]. The amino-terminus is located in the lumen of the ER, and the carboxy-terminus is exposed to the cytoplasm. There is a CXXC motif that is characteristic of the active site of many redox proteins [66]. In the case of VKOR, the CXXC motif resides within a transmembrane helix motif. Mutation of cysteine residue 132 or 135 results in the loss of enzymatic activity, indicating the likelihood of active site recognition [67].

Wallin et al. reported the purification of VKOR consisting of two components: microsomal epoxide hydrolase and glutathione S-transferase [68, 69]. Using this multisubunit model, Rost et al. [64] named the gene encoding VKOR, VKORC1, for catalytic subunit 1. Creation of a knockout mouse for microsomal epoxide hydrolase, however, exhibited no abnormal phenotype [70]. Moreover, there is no further evidence that glutathione S-transferase is a component of VKOR. To date, all known mutations that affect vitamin K-dependent carboxylation can be assigned to either VKOR or VKGC [64].

Another reason for evoking additional components of a VKOR complex is that enzyme activity is lost when solubilized with detergents. One explanation for this

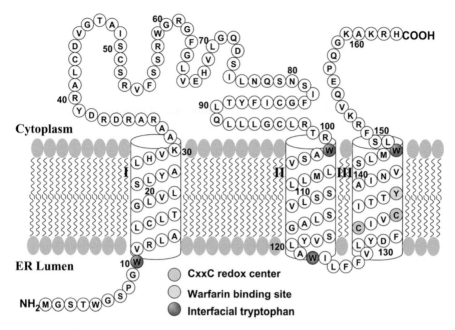

FIG. 6. Topology of VKOR. The amino-terminus resides in the endoplasmic reticulum (*ER*) lumen, and the carboxy-terminus resides in the cytoplasm

occurrence is that the C*XX*C active site motif lies within a transmembrane helix requiring a membrane for proper structure and activity. Although to date there is no definitive evidence that the VKOR enzyme is part of a higher-order complex, the possibility of complex formation is still open to debate. One such compelling argument is that there must at least be a temporary complex that occurs when oxidized VKOR is reduced by an in vivo reductant, similar to the thioredoxin reductase that reduces thioredoxin in *Escherichia coli*. In in vitro microsomal preparations, lipoamide reductase [71], reduced thioredoxin [72], and protein disulfide isomerase [73] are seen to support VKOR activity and may be targets of interest. Recently, Wallin's laboratory reported that protein disulfide isomerase (PDI) is the reductant that reduces the disulfide bond in VKOR created during each K cycle [74]. Wallin's laboratory demonstrated the ability to immunoprecipitate VKOR together with PDI when using an anti-PDI antibody tag. It was concluded from this study that treating the cells with either RNAi against PDI or treatment with bacitracin decreased the activity of VKOR in microsomes. Although sufficient PDI is found in the ER lumen, providing concrete evidence that PDI is responsible for reduction of the disulfide bond in VKOR will be difficult to achieve in vivo. The question remains of interest and further study is encouraged.

VKOR Purification

The purification of active VKOR from insect cells was reported by Chu et al. in 2006 [21]. The most critical aspect of the purification was to find conditions that allowed reconstitution of the enzyme after purification. Since VKOR must maintain membrane structure for activity, the addition of detergents to microsomes containing recombinant VKOR in insect cells results in a loss of activity. Thus, the activity measured must be for the portion of the microsome that is not completely solubilized. Depending on the amount of phospholipids in the microsomes, variability in the amount of residual activity makes it difficult to compare VKOR studies. When looking at endogenous VKOR in bovine microsomes, however, some detergent must be added for optimal activity. Although VKOR has encountered problems with solubility during purification, understanding such solubility issues may begin to open the door towards understanding the complex nature of VKOR.

It was later determined that the purified enzymes, VKOR, could be reconstituted into pure dioleoyl-phosphatidylcholine by a long dialysis to remove the deoxycholate detergent and by reduction with dithiothreitol (DTT). During this procedure, the active site cysteines appear to become oxidized during purification and must be reduced by DTT to regain epoxide activity. Using this protocol, the purity of VKOR was estimated to be greater than 93% with minimal interference from contaminants [21]. In addition, the turnover number of the purified protein was greater than that of the starting microsomes, confirming that activity was not lost during purification. The state of the membrane provided by dioleoyl-phosphatidylcholine is still under study. Presumably, micelles become present during VKOR purification, which may change with storage lifetime. If activity is lost during storage, it may be regained by adding back detergent and repeating the dialysis step.

Except for the difficulty of solubilization, the purification of VKOR is straightforward and reproducible. As an example of the reproducibility of VKOR purification, Jin et al. [22] purified nine VKOR mutations. In this study, each of the seven cysteine residues in VKOR were mutated. Of the seven cysteine residues in VKOR, four (C43, C51, C132, and C135) are found to be conserved across a broad range of species [66]. The functional importance of these residues were investigated through mutational analysis of each cysteine. In addition, VKOR was made with both C43 and C51 mutated to alanine, as well as VKOR with residues C43–C51 deleted. Similarly, Rost et al. [75] also mutated each of the cysteines in VKOR and measured VKOR activity in whole cell extracts. Similar results were achieved in both experiments confirming that C132 and C135 of the CXXC motif participate as the redox active site in VKOR. One exception, however, was found in studying the residue C51. Jin et al. [22] concluded full activity was maintained for VKOR C51A, whereas, Rost et al. [75] determined complete loss of activity with the mutation C51S. Although the evolutionary argument for the importance of C43 and C51 is pretty persuasive, removing cysteines 43 and 51 together with the intervening amino acids was found to persist without significant loss of activity [22]. Even though these residues were found from molecular modeling to be on the opposite side of the membrane from the presumed active site cysteine residues 132 and 135, it is too soon to conclude that C43 and C51 have no functional importance in vivo.

Practical Consequences of VKOR Identification

In the United States alone, warfarin was prescribed to more than 19 million patients in 2006 [76]. Because the dose of warfarin required for a therapeutic level of anticoagulation varies greatly among patients, the utilization of warfarin is accompanied by a significant risk of side effects. For example, a study was conducted investigating 565 patients starting outpatient therapy with warfarin. The results from this study concluded that major bleeding occured in 12% of these patients and death resulted in two percent of the patients observed [77]. Despite these statistics, it has been estimated that warfarin use can prevent approximately 20 strokes per induced bleeding episode [78]. Thus, there has been considerable interest in developing new oral anticoagulant agents that exhibit the positive benefits of warfarin's function without the potential harmful side effects.

Current warfarin therapy utilizes warfarin as a racemic mixture, with S-warfarin being the potent enantiomer. It was determined, however, that S-warfarin is degraded by cytochrome P450 2C9. Further investigation indicated that single nucleotide polymorphisms (SNPs) of cytochrome P450 2C9 can be correlated with the dose of warfarin required for effective anticoagulation [79, 80]. An additional point of interest is that much of the variation in warfarin resistance can be attributed to polymorphisms in VKOR. Several closely linked polymorphisms have been described; and as of this writing, at least 35 articles describing an association between warfarin dose and VKOR polymorphisms have been published. In the Asian and Caucasian populations alone, more than 95% of patients examined are homozygous for a low-dose, intermediate-dose, or high-dose haplotype group. The concordance between results obtained by different laboratories is remarkable. One such study, conducted by Li et al. [81], concluded that the low-dose, intermediate-dose, and high-dose groups required 3.8, 5.1 and 6.7 mg warfarin per day, respectively, to achieve the desired INR. D'Andrea et al. [82] reported low and high warfarin doses corresponding to the same genotypes as 3.5 and 6.2 mg/day. In addition, Rieder et al. [83] reported warfarin doses of 2.7, 4.9, and 6.2 mg/day for low-, intermediate-, and high-dose warfarin requirements. Thus, the genotyping of patients prior to warfarin therapy is likely to reduce the incidence of severe bleeding. Other factors that affect warfarin dosage may be attributed to the variability in the coding region of the gene for VKOR itself. For example, Rost et al. [64] found a patient with the VKOR mutation R98W to exhibit reduced VKOR activity. It was also determined that mutations R58G, V29L, V45A, and L128R resulted in warfarin resistance in humans and the Y139C mutation in rats.

How might one improve current therapy? One way is to screen for the polymorphisms that affect warfarin dosage and adjust the dosage appropriately. This should help with some of the extreme levels of dosage and would be particularly helpful at the onset of treatment. In addition, part of the variability in a single patient's response to warfarin over time probably arises from variation in the dietary intake of vitamin K. It seems logical, therefore, that dosing patients with warfarin and vitamin K simultaneously should buffer this variation. The problem of using this model with K_1 as the buffer species, according to Price et al., was that it stimulated arterial calcification in rats [84]. As further confirmation, Vermeer's laboratory repeated Price's experiments with warfarin and K_1; and in contrast, when rats were fed warfarin and K_2 no arterial calcification was observed [14]. To improve warfarin therapy, it is proposed that combining vitamin K_2 with warfarin will provide both a buffer to reduce fluctuations in the INR resulting from variation in dietary control as well as reduce calcification that may accompany warfarin treatment [85].

Conclusions

Recent developments in the mechanics of VKGC and VKOR have had a considerable impact on understanding the technicalities of the vitamin K cycle. Despite the progress made with VKGC and VKOR, our understanding of carboxylation remains rudimentary. The recent success of Schmidt-Krey et al. in obtaining crystals suitable for electron diffraction is encouraging [86], providing the possibility of structural information that will allow a quantum leap forward in our understanding of VKGC protein dynamics. The same level of desired understanding holds true for VKOR; however, it is likely that obtaining structural information on VKOR will be even more challenging. Additional points of interest focus on expanding in vitro studies on carboxylation to substrates other than factor IX. It has been previously mentioned that VKD substrates exhibit affinities that vary by more than two orders of magnitude. The understanding on how these substrates that are made within a single cell and can compete for the enzyme continues to remain a question of interest.

Due to their close association, one might have expected VKOR and VKGC to be involved in a higher-order complex formation. Convincing evidence for such a multiplexed unit, however, has not yet been provided. Focusing on VKOR alone, another important quest is to identify unequivocally the enzyme that reduces VKOR after each reaction cycle. One potential approach would be to look at mutations of the active site residues in VKOR to trap intermediates between VKOR and its natural reductant. In addition, the roles of the alternatively spliced forms of VKOR also need to be examined in detail. Finally, utilizing VKOR SNPs to predict warfarin dosage together with vitamin K_2 can potentially produce warfarin or warfarin-like anticoagulants to be developed with greater efficacy and fewer undesirable side effects.

References

1. Dam H (1929) Cholesterine metabolism in hens and chickens. Biochim Z 215:475–492
2. Dam H (1935) The antihaemorrhagic vitamin of the chick. Biochem J 29:1273–1285
3. Binkley S, MacCorquodale D, Thayer S, et al (1939) The isolation of vitamin K J Biol Chem 130:219–234
4. Link K (1959) The discovery of dicumarol and its sequels. Circulation 19:97–107
5. Nelsestuen G, Zytkovic T, Howard J (1974) The mode of action of vitamin K: identification of γ-carboxyglutamic acid as a component of prothrombin. J Biol Chem 249:6347–6350
6. Magnusson S, Sottrup-Jensen L, Petersen TE, et al (1974) Primary structure of the vitamin K-dependent part of prothrombin. FEBS Lett 44:189–193
7. Stenflo J, Fernlund P, Egan W, et al (1974) Vitamin-K dependent modifications of glutamic-acid residues in prothrombin. Proc Natl Acad Sci U S A 71:2730–2733
8. Luo GB, Ducy P, McKee MD, et al (1997) Spontaneous calcification of arteries and cartilage in mice lacking matrix GLA protein. Nature 386:78–81
9. Murshed M, Schinke T, McKee MD, et al (2004) Extracellular matrix mineralization is regulated locally; different roles of two gla-containing proteins. J Cell Biol 165:625–630
10. Kulman JD, Harris JE, Xie L, et al (2001) Identification of two novel transmembrane gamma-carboxyglutamic acid proteins expressed broadly in fetal and adult tissues. Proc Natl Acad Sci U S A 98:1370–1375

11. Binkley NC, Krueger DC, Kawahara TN, et al (2002) A high phylloquinone intake is required to achieve maximal osteocalcin gamma-carboxylation. Am J Clin Nutr 76:1055–1060
12. Groenen-van Dooren MM, Ronden JE, Soute BA, et al (1995) Bioavailability of phylloquinone and menaquinones after oral and colorectal administration in vitamin-K-deficient rats. Biochem Pharmacol 50:797–801
13. Davidson RT, Foley AL, Engelke JA, et al (1998) Conversion of dietary phylloquinone to tissue menaquinone-4 in rats is not dependent on gut bacteria. J Nutr 128:220–223
14. Spronk HM, Soute BA., Schurgers LJ, et al (2003) Tissue-specific utilization of menaquinone-4 results in the prevention of arterial calcification in warfarin-treated rats. J Vasc Res 40:531–537
15. Schurgers LJ, Vermeer C (2002) Differential lipoprotein transport pathways of K-vitamins in healthy subjects. Biochim Biophys Acta 1570:27–32
16. Schurgers LJ, Teunissen KJ, Hamulyak K, et al (2007) Vitamin K-containing dietary supplements: comparison of synthetic vitamin K_1 and natto-derived menaquinone-7. Blood 109:3279–3283
17. Akedo Y, Hosoi T, Inoue S, Ikegami A, et al (1992) Vitamin K_2 modulates proliferation and function of osteoblastic cells in vitro. Biochem Biophys Res Commun 187:814–820
18. Tabb MM, Sun A, Zhou C, et al (2003) Vitamin K_2 regulation of bone homeostasis is mediated by the steroid and xenobiotic receptor SXR. J Biol Chem 278: 43919–43927
19. Silverman RB (1981) Chemical model studies for the mechanism of vitamin K epoxide reductase. J Am Chem Soc 103:5939–5941
20. Davis CH II, Deerfield D, Wymore T, et al (2006) A quantum chemical study of the mechanism of action of vitamin K carboxylase (VKC). III. Intermediates and transition states. J Mol Graph Model 26:409–414
21. Chu PH, Huang TY, Williams J, et al (2006) Purified vitamin K epoxide reductase alone is sufficient for conversion of vitamin K epoxide to vitamin K and vitamin K to vitamin KH_2. Proc Natl Acad Sci U S A 103:19308–19313
22. Jin DY, Tie JK, Stafford DW (2007) The conversion of vitamin K epoxide to vitamin K quinone and vitamin K quinone to vitamin K hydroquinone uses the same active site cysteines. Unpublished data
23. Wallin R, Cain D, Sane DC (1999) Matrix gla protein synthesis and gamma-carboxylation in the aortic vessel wall and proliferating vascular smooth muscle cells: a cell system which resembles the system in bone cells. Thromb Haemost 82: 1764–1767
24. Schurgers LJ, Spronk HM, Soute BA, et al (2006) Regression of warfarin-induced medial elastocalcinosis by high intake of vitamin K in rats. Blood 109:2823–2831
25. Iskander K, Li J, Han S, et al (2006) NQO1 and NQO2 regulation of humoral immunity and autoimmunity. J Biol Chem 281:30917–30924
26. Radjendirane V, Joseph P, Lee YH, et al (1998) Disruption of the DT diaphorase (NQO1) gene in mice leads to increased menadione toxicity. J Biol Chem 273:7382–7389
27. Long DJ, Gaikwad A, Multani A, et al (2002) Disruption of the NAD(P)H:quinone oxidoreductase 1 (NQO1) gene in mice causes myelogenous hyperplasia. Cancer Res 62:3030–3036
28. Esmon CT, Sadowski JA, Suttie JW (1975) A new carboxylation reaction: the vitamin K-dependent incorporation of H-$^{14}CO_3^-$ into prothrombin. J Biol Chem 250:4744–4748
29. Wu SM, Cheung WF, Frazier D, et al (1991) Cloning and expression of the cDNA for human gamma-glutamyl carboxylase. Science 254:1634–1636
30. Wu SM, Morris DP, Stafford DW (1991) Identification and purification to near homogeneity of the vitamin K-dependent carboxylase. Proc Natl Acad Sci U S A 88: 2236–2240

31. Tie JK, Mutucumarana VP, Straight DL, et al (2003) Determination of disulfide bond assignment of human vitamin K-dependent gamma-glutamyl carboxylase by matrix-assisted laser desorption/ionization time-of-flight mass spectrometry. J Biol Chem 278:45468–454475

32. Tie JK, Wu SM, Jin DY, et al (2000) A topological study of the human gamma-glutamyl carboxylase. Blood 96:973–978

33. Stanley TB, Jin DY, Lin PJ, et al (1999) The propeptides of the vitamin K-dependent proteins possess different affinities for the vitamin K-dependent carboxylase. J Biol Chem 274:16940–16944

34. Price PA, Fraser JD and Metz-Virca G (1987) Molecular cloning of matrix Gla protein: implications for substrate recognition by the vitamin K-dependent gamma-carboxylase. Proc Natl Acad Sci U S A 84:8335–8339

35. Katayama K, Ericsson LH, Enfield DL, et al (1979) Comparison of amino acid sequence of bovine coagulation factor IX (Christmas factor) with that of other vitamin K-dependent plasma proteins. Proc Natl Acad Sci U S A 76:4990–4994

36. Suttie JW, Geweke LO, Martin SL, et al (1980) Vitamin K epoxidase: dependence of epoxidase activity on substrates of the vitamin K-dependent carboxylation reaction. FEBS Lett 109:267–270.

37. Stenflo J, Suttie JW (1977) Vitamin K-dependent formation of gamma-carboxyglutamic acid. Annu Rev Biochem 46:157–172.

38. Dowd P, Hershline R, Ham SW, et al (1995) Vitamin K and energy transduction: a base strength amplification mechanism. Science 269:1684–1691.

39. Davis CH, II Deerfield D, Wymore T, et al (2006) A quantum chemical study of the mechanism of action of vitamin K epoxide reductase (VKOR). II. Transition states. J Mol Graph Model 26:401–408

40. Wu SM, Mutucumarana VP, Geromanos S, et al (1997) The propeptide binding site of the bovine gamma-glutamyl carboxylase. J Biol Chem 272:11718–11722

41. Rishavy MA, Pudota BN, Hallgren KW, et al (2004) A new model for vitamin K-dependent carboxylation: the catalytic base that deprotonates vitamin K hydroquinone is not Cys but an activated amine. Proc Natl Acad Sci U S A 101:13732–13737.

42. Shimizu A, Sugiura I, Matsushita T, et al (1998) Identification of the five hydrophilic residues (Lys-217, Lys-218, Arg-359, His-360, and Arg-513) essential for the structure and activity of vitamin K-dependent carboxylase. Biochem Biophys Res Commun 251:22–26.

43. Deerfield D II, Davis CH, Wymore T, et al (2006) Quantum chemical study of the mechanism of action of vitamin K epoxide reductase (VKOR). Int J Quantum Chem 106:2944–2952

44. Davis CH II, Deerfield, D, Stafford DW, et al (2007) Quantum chemical study of the mechanism of action of vitamin K carboxylase (VKC). IV. Intermediates and transition states. J Phys Chem A 111:7257–7261

45. Morris DP, Stevens RD, Wright DJ, et al (1995) Processive post-translational modification: vitamin K-dependent carboxylation of a peptide substrate. J Biol Chem 270:30491–30498

46. Lin PJ, Straight DL, Stafford DW (2004) Binding of the factor IX gamma-carboxyglutamic acid domain to the vitamin K-dependent gamma-glutamyl carboxylase active site induces an allosteric effect that may ensure processive carboxylation and regulate the release of carboxylated product. J Biol Chem 279:6560–6566

47. Stenina O, Pudota BN, McNally BA, et al (2001) Tethered processivity of the vitamin K-dependent carboxylase: factor IX is efficiently modified in a mechanism which distinguishes Gla's from Glu's and which accounts for comprehensive carboxylation in vivo. Biochemistry 40:10301–10309

48. Bond M, Jankowski M, Patel H, et al (1998) Biochemical characterization of recombinant factor IX. Semin Hematol 35:11–17

49. Presnell SR, Tripathy A, Lentz BR, et al (2001) A novel fluorescence assay to study propeptide interaction with gamma-glutamyl carboxylase. Biochemistry 40:11723–11733

50. Hallgren KW, Qian W, Yakubenko AV, et al (2006) r-VKORC1 expression in factor IX BHK cells increases the extent of factor IX carboxylation but is limited by saturation of another carboxylation component or by a shift in the rate-limiting step. Biochemistry 45:5587–5598

51. Wallin R, Hutson SM, Cain D, et al (2001) A molecular mechanism for genetic warfarin resistance in the rat. FASEB J 15:2542–2544

52. Vecsler ML, Loebstein R, Almog S, et al (2006) Combined genetic profiles of components and regulators of the vitamin K-dependent gamma-carboxylation system affect individual sensitivity to warfarin. Thromb Haemost 95:205–211

53. Kimura R, Miyashita K, Kokubo Y, et al (2007) Genotypes of vitamin K epoxide reductase, gamma-glutamyl carboxylase, and cytochrome P450 2C9 as determinants of daily warfarin dose in Japanese patients. Thromb Res 120:181–186

54. Wajih N, Hutson SM, Wallin R (2006) siRNA silencing of calumenin enhances functional factor IX production. Blood 108:3757–3760

55. Camire RM, Larson PJ, Stafford DW, et al (2000) Enhanced gamma-carboxylation of recombinant factor X using a chimeric construct containing the prothrombin propeptide. Biochemistry 39:14322–14329

56. Chu K, Wu SM, Stanley T, et al (1996) A mutation in the propeptide of factor IX leads to warfarin sensitivity by a novel mechanism. J Clin Invest 98:1619–1625

57. De la Salle H, Altenburger W, Elkaim R, et al (1985) Active gamma-carboxylated human factor IX expressed using recombinant DNA techniques. Nature 316:268–270

58. Busby S, Kumar A, Joseph M, et al (1985) Expression of active human factor IX in transfected cells. Nature 316:271–273

59. Hallgren KW, Hommema EL, McNally BA, et al (2002) Carboxylase overexpression effects full carboxylation but poor release and secretion of factor IX: implications for the release of vitamin K-dependent proteins. Biochemistry 41:15045–15055

60. Tokunaga F, Wakabayashi S, Koide T (1995) Warfarin causes the degradation of protein C precursor in the endoplasmic reticulum. Biochemistry 34:1163–1170

61. Wu W, Bancroft JD, Suttie JW (1996) Differential effects of warfarin on the intracellular processing of vitamin K-dependent proteins. J Thromb Haemost 76:46–52

62. Bell RG, Matschiner JT (1970) Vitamin K activity of phylloquinone oxide. Arch Biochem Biophys 141:473–476

63. Li T, Chang CY, Jin DY, et al (2004) Identification of the gene for vitamin K epoxide reductase. Nature 427:541–544

64. Rost S, Fregin A, Ivaskevicius V, et al (2004) Mutations in VKORC1 cause warfarin resistance and multiple coagulation factor deficiency type 2. Nature 427:537–541

65. Tie JK, Nicchitta C, von Heijne G, et al (2005) Membrane topology mapping of vitamin K epoxide reductase by in vitro translation/cotranslocation. J Biol Chem 280:16410–16416

66. Goodstadt L, Ponting CP (2004) Vitamin K epoxide reductase: homology, active site and catalytic mechanism. Trends Biochem Sci 29:289–292

67. Wajih N, Sane DC, Hutson SM, et al (2005) Engineering of a recombinant vitamin K-dependent gamma-carboxylation system with enhanced gamma-carboxyglutamic acid forming capacity: evidence for a functional CXXC redox center in the system. J Biol Chem 280:10540–10547

68. Guenthner TM, Cai D, Wallin R (1998) Co-purification of microsomal epoxide hydrolase with the warfarin-sensitive vitamin K_1 oxide reductase of the vitamin K cycle. Biochem Pharmacol 55:169–175

69. Cain D, Hutson SM, Wallin R (1997) Assembly of the warfarin-sensitive vitamin K 2,3-epoxide reductase enzyme complex in the endoplasmic reticulum membrane. J Biol Chem 272:29068–29075

70. Miyata M, Kudo G, Lee YH, et al (1999) Targeted disruption of the microsomal epoxide hydrolase gene: microsomal epoxide hydrolase is required for the carcinogenic activity of 7,12-dimethylbenz[a]anthracene. J Biol Chem 274:23963–23968

71. Thijssen HH, Janssen YP, Vervoort LT (1994) Microsomal lipoamide reductase provides vitamin K epoxide reductase with reducing equivalents. Biochem J 297:277–280

72. Silverman RB, Nandi DL (1988) Reduced thioredoxin: a possible physiological cofactor for vitamin K epoxide reductase: further support for an active site disulfide. Biochem Biophys Res Commun 155:1248–1254

73. Soute BA, Groenen-van Dooren MM, Holmgren A, et al (1992) Stimulation of the dithiol-dependent reductases in the vitamin K cycle by the thioredoxin system: strong synergistic effects with protein disulphide-isomerase. Biochem J 281:255–259

74. Wajih N, Hutson SM, Wallin R (2007) Disulfide-dependent protein folding is linked to operation of the vitamin K cycle in the endoplasmic reticulum: a protein disulfide isomerase-VKORC1 redox enzyme complex appears to be responsible for vitamin K1 2,3-epoxide reduction. J Biol Chem 282:2626–2635

75. Rost S, Fregin A, Hunerberg M, et al (2005) Site-directed mutagenesis of coumarin-type anticoagulant-sensitive VKORC1: evidence that highly conserved amino acids define structural requirements for enzymatic activity and inhibition by warfarin. Thromb Haemost 94:780–786

76. Drug Topics (2007) Top 200 generic drugs by units in 2006. Drug Topics Online

77. Landefeld CS, Goldman L (1989) Major bleeding in outpatients treated with warfarin: incidence and prediction by factors known at the start of outpatient therapy. Am J Med 87:144–152

78. Horton JD, Bushwick BM (1999) Warfarin therapy: evolving strategies in anticoagulation. Am Fam Physician 59:635–646

79. Higashi MK, Veenstra DL, Kondo LM, et al (2002) Association between CYP2C9 genetic variants and anticoagulation-related outcomes during warfarin therapy. JAMA 287:1690–1698

80. Yasar U, Eliasson E, Dahl ML, et al (1999) Validation of methods for CYP2C9 genotyping: frequencies of mutant alleles in a Swedish population. Biochem Biophys Res Commun 254:628–631

81. Li T, Lange LA, Li X, et al (2006) Polymorphisms in the VKORC1 gene are strongly associated with warfarin dosage requirements in patients receiving anticoagulation. J Med Genet 43:740–744

82. D'Andrea G, D'Ambrosio RL, Di Perna P, et al (2005) A polymorphism in the VKORC1 gene is associated with an interindividual variability in the dose-anticoagulant effect of warfarin. Blood 105:645–649

83. Rieder MJ, Reiner AP, Gage, BF, et al (2005) Effect of VKORC1 haplotypes on transcriptional regulation and warfarin dose. N Engl J Med 352:2285–2293

84. Price PA, Faus SA, Williamson MK (1998) Warfarin causes rapid calcification of the elastic lamellae in rat arteries and heart valves. Arterioscler Thromb Vasc Biol 18:1400–1407

85. Stafford DW, Roberts HR, Vermeer C (2007) Vitamin K supplementation during oral anticoagulation: concerns. Blood 109:3607; author reply 3607–3608

86. Schmidt-Krey I, Haase W, Mutucumarana V, et al (2007) Two-dimensional crystallization of human vitamin K-dependent gamma-glutamyl carboxylase. J Struct Biol 157:437–442

Synthesis and Secretion of Coagulation Factor VIII

Michael U. Callaghan and Randal J. Kaufman

Summary. Coagulation factor VIII (FVIII) is a complex glycoprotein that is deficient in the X chromosome-linked bleeding disorder hemophilia A. FVIII has a domain structure of A1-A2-B-A3-C1-C2. Upon synthesis, FVIII is translocated into the lumen of the endoplasmic reticulum (ER), where it undergoes extensive processing including cleavage of a signal peptide and glycosylation at 25 asparagine residues. In the ER lumen FVIII interacts with the protein chaperones calnexin, calreticulin, and immunoglobulin-binding protein prior to trafficking to the Golgi compartment. Trafficking from the ER to the Golgi compartment is facilitated by the ER Golgi intermediate compartment protein 53 and multiple combined factor deficiency protein chaperone complex. Upon secretion from the cell, FVIII is cleaved at two sites in the B-domain to form a heterodimer consisting of the heavy chain containing the A1-A2-B domains in a metal ion-dependent complex with the light chain consisting of the A3-C1-C2 domains. In the plasma, FVIII is stabilized through interaction with von Willebrand factor. Upon damage to blood vessel walls, thrombin cleaves FVIII and releases the B-domain to form an active heterotrimer that binds activated coagulation factor IX on the surface of platelet phospholipid to form the active factor Xase complex. This complex efficiently cleaves factor X to its active form, which activates prothrombin and leads to the formation of a stable fibrin clot.

Key words. Factor VIII · Hemophilia · Secretion · Endoplasmic reticulum · Structure

Introduction

Coagulation factor VIII (FVIII) is an essential cofactor for coagulation factor IXa (FIXa) in the intrinsic clotting cascade. The presence of activated FVIII increases the catalytic efficiency of FIXa-mediated activation of factor X (FX) by several orders of magnitude. Deficiency or dysfunction of FVIII results in the bleeding disorder hemophilia A, which is the most common severe congenital bleeding diathesis, affecting 1/5000 males. Hemophilia was described in rabbinical writings and the Talmud almost 2000 years ago [1]. Deficiency of FVIII results in a severe defect in secondary hemostasis and frequently bleeds into muscles and joints caused by minimal or no trauma. Affected patients also have severe and prolonged bleeding after trauma or surgery.

Diagnosis and treatment of hemophilia has improved drastically since whole blood transfusion was first described as a treatment by Lane in 1840 [2]. In 1936, Patek and Stetson determined that the absence of a factor that is present in the plasma of normal individuals leads to hemophilia, and they later isolated and named antihemophilic factor [3, 4]. Plasma products were first developed for the treatment of hemophilia during the late 1920s [5]. Although the requirement for FVIII in coagulation was known, its specific role as a cofactor for FIX was not determined until the late 1960s [6, 7]. During the late 1940s Cohn isolated fractions of plasma containing FVIII in large quantities, providing the opportunity for home treatment and effective treatment of bleeds encountered by patients with hemophilia, drastically altering the prognosis of the disease and the quality of life of the patients [8, 9]. Unfortunately, these products required large pools of donors, and during the late 1970s and early 1980s many patients acquired viral hepatitis and acquired immunodeficiency syndrome (AIDS). Monoclonal antibody purification has significantly reduced this risk, although fear of blood-borne illness is still a major deterrent to the use of these products [10].

In 1984, the FVIII gene was isolated [11, 12], paving the way for the development in just a few short years of recombinant FVIII, produced in mammalian cells, thereby diminishing the risk of spread of blood borne infections. Current treatment strategies employ monoclonal antibody purified concentrated human plasma-derived FVIII or recombinant-derived FVIII given by intravenous infusion [13, 14]. Unfortunately, recombinant products are difficult to manufacture, and supplies are unreliable and expensive [13, 15]. In 2002, the cost of therapy for hemophilia A was estimated at more than $1 billion (US$) [5]. A major problem with current treatment strategies is the development of FVIII inhibitors in 20% of hemophilia A patients; these antibodies neutralize the effects of the infused factor, leading to the need for more frequent and higher doses of FVIII or other therapies. Recombinant activated factor VII and activated prothrombin complexes, which are the mainstay of treatment in hemophilia A patients with inhibitors, act by bypassing the intrinsic coagulation cascade. Unfortunately, these products are costly. Furthermore, the former requires an onerous dosing schedule owing to a short half-life, and the latter carries the risk of blood-borne pathogens and thrombosis. Refinement in therapy for hemophilia A has generally followed scientific discoveries describing the characteristics of FVIII. Approximately 300 000 of the world's 400 000 people living with hemophilia receive little or no treatment. Improved understanding of FVIII synthesis and secretion is the key to new therapies to address these and other issues.

Factor VIII Genetics

The gene for FVIII, *F8c*, is located on the X chromosome at Xq28, and therefore hemophilia A displays an X-linked recessive inheritance pattern. It is closely linked to the genes for color blindness and glucose-6-phosphate dehydrogenase (G6PD), both of which have associated X-linked disorders that can be co-inherited with hemophilia [16]. Indeed, polymorphisms in the G6PD gene were historically exploited for hemophilia carrier testing [17]. The *F8c* gene consists of 186 kilobases encoding

26 exons and 25 introns with a coding sequence of 7053 nucleotides. Prior to the therapeutic advances of factor replacement therapies, hemophilia A was usually fatal before the attainment of reproductive viability, and therefore one-third of *F8c* mutations were de novo, found in patients with no family history of the disorder, as is predicted by the Haldane hypothesis. As of June 2007, the Human Gene Mutation Database listed 885 unique mutations, including missense, nonsense, frame-shift, inversion, and large deletion mutations. Approximately 5%–10% of patients with hemophilia A have no identifiable mutation in the *F8c* gene. Mutations cause varying degrees of deficiency and/or dysfunction of FVIII and are categorized by the resulting cofactor activity in clotting assays as severe (<1% of normal activity), moderate (1%–5%), or mild (5%–40%). The type of mutation can also help predict inhibitor risk, with large deletion generally carrying the greatest risk and point mutations the smallest [18–21]. Approximately 25% of patients with hemophilia A and 40% of those with severe hemophilia A harbor an intron 22 inversion mutation resulting from an aberrant recombination event in the gamete of the maternal grandfather [22, 23].

Factor VIII Gene Expression

Factor VIII is produced in the liver, and liver transplantation has been shown to cure hemophilia [24]. However, transplantation of hemophiliac dog livers into wild-type dogs did not cause hemophilia, suggesting extrahepatic sites of FVIII production [25]. FVIII mRNA is expressed in many tissues, but some of this is likely illegitimate transcription and does not result in any substantial quantity of FVIII protein [26]. Although the site of FVIII synthesis has not been absolutely defined, most data support the notion that FVIII is synthesized in the hepatic reticuloendothelial system [24, 27–31], hepatic sinusoidal endothelial cells [32], and endothelial cells, megakaryocytes [29, 33, 34], and possibly Kupffer cells [35]. Recently, human lung has also been shown to produce FVIII [36]. The quantity of FVIII mRNA and protein endogenously produced naturally is too low to study at a biochemical level; therefore, most of our knowledge of FVIII biosynthesis is derived from cultured cells engineered to express FVIII at a high level. In mammalian cells, FVIII expression is significantly lower than that of similar proteins, such as factor V (FV) [37, 38]. Deletion of the B-domain and other modifications to the FVIII gene can result in higher levels of expression [38, 39].

Factor VIII Structure and Function

Factor VIII is a large glycoprotein synthesized as a single polypeptide chain of 2351 amino acids (Fig. 1). A 19-amino-acid signal sequence is cleaved upon translocation into the lumen of the endoplasmic reticulum (ER). FVIII is homologous to FV, both having a domain structure A1-A2-B-A3-C1-C2 [11, 12, 40–44]. In FVIII, three regions are rich in acidic amino acids and are designated a1, a2, and a3. In the Golgi compartment, FVIII is cleaved at two sites in the B-domain to form a heterodimer consisting of a heavy chain (made up of the A1, A2, and B domains) in a divalent metal ion-dependent interaction with the light chain (made up of the A3, C1, and C2 domains).

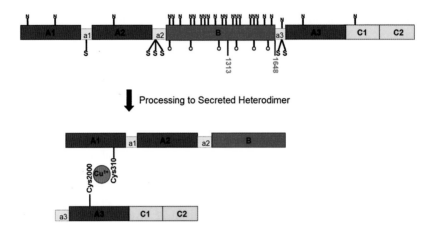

FIG. 1. Intracellular factor VIII (FVIII) processing. FVIII is processed to its circulating copper ion-dependent heterodimer form through cleavage at two sites in the B-domain. Tyrosine sulfation (*S*) is at amino acids 346, 718, 719, 723, 1664, and 1680 in the acid-rich a1, a2, and a3 regions. The potential sites of N-linked glycosylation are represented by *N*. O-linked glycosylation is represented by *O*, although the actual sites of O-linked glycosylation are unknown

In plasma, FVIII is present in a tight association with von Willibrand factor (vWF) [45–49].

A Domains

The FVIII A domains are homologous to the copper-binding protein ceruloplasmin, and FVIII contains a single type I copper ion that enhances the interaction of the light and heavy chains [11, 12, 50–52]. This copper ion is lost upon dissociation of the heavy and light chains [53]. Potential sites of copper binding have been identified in both the A1 and A3 domains, although the exact binding site remains elusive [54]. Cys310 in the A1 domain and Cys2000 in the A3 domain both appear to play important roles in the copper ion-dependent interchain association [50, 52]. The A1 domain also contains a calcium/manganese binding site at residues 110–126 that is important for FVIII cofactor activity [55]. Residues in the A1 and A3 domains interact through hydrophobic and electrostatic forces that stabilize the heterodimer [56, 57].

B Domain

The B domain is released during the cleavage and activation of FVIII [58]. Although the B domains of FV and FVIII have extensively diverged in their amino acid sequence [42], they both contain a large number of N-linked oligosaccharides [59, 60]. The B domains of different species differ considerably but have conserved a large number of N-linked oligosaccharides [61]. The N-linked oligosaccharides are likely important for secretion, as complete deletion of the B domain reduces the efficiency of secretion. These N-linked oligosaccharides are also recognized by the asialoglycoprotein receptor that may be involved in FVIII clearance from the plasma [62].

C Domains

The C domains resemble milk fat globule protein and galactose oxidase, members of the discoidin family [63–66]. The crystal structure of the C2 domain has been solved and demonstrates four hydrophobic surface residues and four positively charge residues involved in phospholipid and vWF binding. Numerous hemophilia-causing mutations and epitopes for inhibitory antibodies are localized to these sites [63, 67, 68]. The interaction of vWF with phospholipid requires different residues at these sites [67]. Mutations at R2307G, Q2311P, W2313, L2210F, W2229C, and C2326S cause a protein secretion defect, likely leading to protein misfolding and degradation in the ER [67, 69].

von Willebrand Factor Interaction

Patients with type 3, severe von Willebrand disease (vWD) have very little circulating vWF and a parallel deficiency of FVIII; in contrast, patients with mild vWD have slightly reduced FVIII levels and patients with vWD type 2 Normandy, in which vWF is unable to bind FVIII, have very low FVIII levels. These diseases highlight the important role of vWF in protecting FVIII in the circulation and extending its half-life. The C2 domain of FVIII interacts with vWF at Arg2307 and Trp2229; and Thr2154, Gln2100, and Arg2150 are important to induce a C1-C2 conformation compatible for vWF binding [70]. A group of amino acids (Ser2119, Arg2116, Tyr2105) in the C1 domain also appear to play a role in the interaction between FVIII and vWF [70].

Phospholipid Interaction

Upon activation of platelets, large areas of phosphatidylserine-rich phospholipid are exposed to fluid phase coagulation factors and facilitate the action of these proteins [71]. These negatively charged lipids are required for activation of FX [72–74]. FVIII binds to phospholipid through both electrostatic and hydrophobic interactions [75, 76]. Upon thrombin cleavage, FVIII binds phospholipid with a much higher affinity than vWF, thus releasing FVIII from vWF [77–80]. Phospholipid binding is largely mediated through interactions of surface hydrophobic residues in the C2 domain [67, 81].

Factor IX Interaction

Factor IX is serine protease consisting of 415 amino acids that undergo vitamin K-dependent γ-carboxylation, a modification that improves FIX binding to phospholipid. FIXa forms the catalytic site of the Xase complex and cleaves FX to its active form in the presence of Ca^{2+}, phospholipid, FVIII, and Na^+ [82]. FVIII interacts with FIXa largely through the light chain, with FVIII residues 1778–1840 playing an important role [83]. The helix at residues 330–338 of activated FIX binds to FVIII at residues 558–565 in the A2 domain and protects FVIII from proteolytic cleavage by activated protein C [84–86]. These interaction sites appear to be important for FVIII cofactor activity [87] as mutations at residues 558–565 in FVIII cause hemophilia A [87] and mutations at residues 330–338 in FIX cause hemophilia B [88]. Mutation at residue 225 in FIX may increase flexibility of the 330 helix and decrease interaction with FVIII

[82]. Residues 708–715 of the FVIII A2 domain may also contribute to the interaction between FVIII and FIXa through a salt bridge between Asp712 on FVIII and Lys294 [89] in FIXa. Arg527 is another site in FVIII that may be involved in interaction with FIXa [90].

Factor X Interaction

Factor X (FX) is a serine protease that is activated by FIXa on phospholipid surfaces in the presence of FVIIIa and Ca^{2+}. FXa participates in the conversion of prothrombin to thrombin in the common portion of the coagulation cascade. FVIII residues 484–509 in the A2 domain form an important site for interaction with FX in the Xase complex [91]. The A1 domain residues 337–372 also interact with FX, and mutation at Arg336 decreases binding to FX [92–94]. The interaction between FVIIIa and FX appears to depend in large part on the presence of FIXa [95]. However, the primary role of FVIIIa in the FXase complex is to change conformation of the FIXa binding site to promote binding and cleavage of FX. FXa can inactivate FVIIIa through cleavage at Arg336 in the A1 domain and release of the A2 domain [96].

Thrombin Interaction

Thrombin cleaves FVIII after Arg372 at the junction of the a1–A2 domains to expose the FIXa binding site [97]. Thrombin also cleaves after Arg740 at the a2–B domain border and releases the B domain (Fig. 2). Cleavage by thrombin at Arg1689 at the

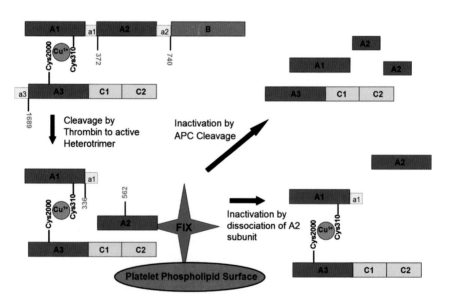

Fig. 2. Factor VIII activation, inactivation, and interactions with the components of coagulation. Thrombin cleaves FVIII at three sites to form the active heterotrimer of the Xase complex. Activated protein C (APC) cleaves FVIII at two sites to its inactive products. The A2 domain spontaneously dissociates and inactivates FVIII. Cleavage sites are labeled by amino acid number and are red

a3–A3 junction promotes dissociation from vWF [77]. After cleavage of FVIII by thrombin, FVIIIa is generated that is comprised of a heterotrimer of the A1, A2, and A3-C1-C2 subunits. The A2 domain interacts weakly through ionic interactions with the A1/A3-C1-C2 heterodimer and can dissociate and inactivate FVIIIa at physiological concentrations and pH. Several point mutations have been described that weaken this interaction, thereby reducing FVIII activity, resulting in hemophilia in which plasma FVIII displays a discrepancy in the measure of FVIII where the one-stage assay gives approximately two- to sevenfold higher values than the two-stage assay [98–102].

Activated Protein C Interaction

Activated protein C (APC) is a serine protease anticoagulant that, in the presence of its cofactor protein S, inactivates both FVa and FVIIIa by proteolytic cleavage. A common polymorphism of FV, FV Leiden, is known to slow inactivation of FV and predispose affected individuals to thrombosis [103–105]. Patients with deficiency of APC are also predisposed to thrombosis. APC cleaves FVIII at Arg562, a site of FIXa interaction, and causes FVIII inactivation [94, 106]. APC also cleaves FVIII after Arg336 within the A1 domain that may cause A2 domain dissociation and result in inactivation [96, 106].

LRP/LDL Receptor Interaction

The low density lipoprotein (LDL) receptor-related protein (LRP) is a member of the LDL receptor family responsible for cellular uptake of lipoproteins and may participate in the clearance of FVIII from plasma [107–109]. LRP is able to bind FVIII through interactions in the C2 domain [107], residues 1804–1834 [110] in the A3 domain, and residues 484–509 in the A2 domain [108]. However, vWF binding to FVIII prevents C2 domain interaction with LRP [107]. Mice lacking LRP do have modestly increased FVIII levels, although the mechanism for this increase remains in question [111, 112].

Factor VIII Folding and Trafficking Within the Cell

Proteins undergo posttranslational modifications and processing in the ER and take on secondary, tertiary, and quaternary structures that are essential for secretion and proper function. The ER also has a "quality control" function that determines whether a protein has been properly translated, modified, and folded [113]. Proteins that are improperly folded are unfolded and refolded, sequestered in the ER, or degraded by the proteasome [114].

Our knowledge of FVIII expression is derived from studies of FVIII biosynthesis in transfected mammalian cells, as there are no natural cell lines that produce FVIII at measurable levels. In the transfected cell lines, FVIII protein is secreted at significantly lower levels than similar proteins, such as FV [37]. One reason for the low level of FVIII secretion is inefficient intracellular trafficking from the ER to the Golgi compartment. It is well established that the rate-limiting step in protein secretion is the rate at which proteins attain their properly folded conformation in the ER. The requirements for folding of FVIII into its final conformation are much more onerous than the requirements for folding of the homologous FV protein [115]. FVIII biosynthesis studied in Chinese hamster ovary cells, the cells used for recombinant FVIII production, demonstrated that FVIII behaves as a classic secreted protein. Upon translation, FVIII is translocated to the lumen of the ER, where the 19-amino-acid signal sequence is removed and the addition of high-mannose core oligosaccharides occurs. FVIII polypeptide folding is facilitated with the aid of two chaperone systems in the ER: immunoglobulin binding protein (BiP) and calnexin/calreticulin (CNX/CRT) (Fig. 3). When folded into its correct conformation, FVIII binds to the ER Golgi intermediate compartment protein 53/multiple combined factor deficiency protein 2 (ERGIC-53/MCFD-2) heterodimeric cargo trafficking complex and is transported to the Golgi in coating protein II (COPII)-coated vesicles. Improperly folded or defective FVIII is degraded through ER-associated degradation by retrotranslocation into the cytosol and degradation by the 26S proteasome [116–118]. In the Golgi compartment, FVIII is cleaved after residues 1313 and 1648 in the B domain to form a heterodimer. N-linked oligosaccharides are modified to their complex form, O-linked oligosaccharides are added to serine and threonine residues within the B domain, and six tyrosine residues are sulfated in the acidic amino acid-rich regions [119]. FVIII is then secreted from the cell and stabilized by binding to vWF.

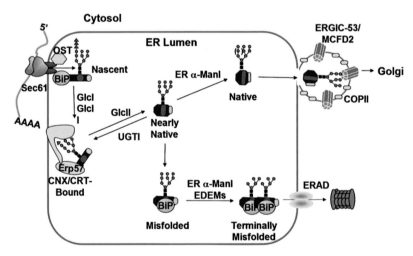

Fig. 3. Factor VIII trafficking in the endoplasmic reticulum (*ER*). Nascent chains enter the ER lumen through a proteinaceous channel (*Sec 61*) translocon complex. Most of them are modified by covalent addition of one or more oligosaccharides. Native proteins are released into the secretory pathway. Terminally misfolded proteins are retranslocated into the cytosol. *Triangles*, glucose; *circles*, mannose residues. Polypeptides that fail to acquire their native form are removed from the ER through ER-associated degradation (*ERAD*). *EDEMs*, degradation-enhancing mannosidase-like proteins; *BiP*, immunoglobulin-binding protein; *ERGIC-53*, Golgi intermediate compartment protein 53; *MCFD2*, multiple combined factor deficiency protein 2; *CNX*, calnexin; *CRT*, calreticulin; *OST*, oligosaccharyl transferase; *Glcl*, glucosidase I; *Erp57*, thiol-disulfide isomerase ERp57; *α-Manl*, α-mannosidase-like protein; *COPII*, coating protein complex II

BiP Interaction with FVIII

BiP/GRP78 is an ER luminal chaperone protein of the heat shock protein 70 (hsp70) family that is induced in response to diverse cellular stresses (e.g., glucose deprivation or inhibition of N-linked glycosylation) that cause accumulation of unfolded proteins in the ER lumen. High levels of FVIII expression induce transcription of the BiP gene [120]. Upon translocation to the ER, BiP interacts with FVIII transiently and retains FVIII in the ER lumen. The level of BiP inversely correlates with FVIII secretion efficiency. Overexpression of wild-type BiP or BiP mutants with defective ATPase function resulted in accumulation of BiP and FVIII complexes in the ER with decreased FVIII secretion [121]. It was proposed that BiP binds to hydrophobic β-sheet in the A1 domain. Mutation of F309S of this β-sheet reduced BiP interaction and improved FVIII secretion efficiency [44, 122]. Disassociation of wild-type BiP from FVIII requires much higher levels of ATP than dissociation from the F309S mutant [44, 123].

Calnexin and Calreticulin Interaction with FVIII

Calnexin (CNX) is an ER transmembrane protein chaperone with homology to the ER luminal protein chaperone calreticulin (CRT) [124, 125]. Both proteins bind Ca^{2+} with high affinity and are integral to the folding of nascent glycoproteins into their final conformation. These two proteins form a quality control system, preventing unfolded or misfolded proteins from continuing along the secretory pathway [126–130]. Interaction with CNX/CRT has been shown to improve disulfide bond formation, possibly by increasing interaction with thiol-disulfide isomerase ERp57 [131]. Upon entrance into the ER, a preassembled oligosaccharide core of 14 residues is added to selective residues having the consensus sequence of Asn-Xxx-Ser/Thr, where Xxx is any amino acid except proline. This core oligosaccharide structure is subsequently trimmed by the sequential action of glucosidase I and glucosidase II. CNX and CRT bind to the monoglucosylated intermediate during the trimming process. Release from CNX/CRT is coupled with cleavage of the terminal glucose residue by glucosidase II. Uridine diphosphate (UDP)-glucose:glycoprotein glucosyltransferase (UGT1) can recognize misfolded glycoproteins released from CNX/CRT and reglucosylate the N-linked core, thereby promoting another round of interaction with CNX/CRT for continued attempts at proper folding [113, 132, 133]. FVIII interacts with both CNX and CRT in the ER through the B domain, and this interaction is likely important for both FVIII secretion and FVIII degradation through the 26S proteasome through ER-associated protein degradation [134].

Intracellular FVIII Aggregation

Factor VIII, along with other proteins—major histocompatibility complex class II (MHC II), procollagen, thyroglobulin—form aggregates when critical amounts of the unfolded protein accumulate in the ER [135–141]. Upon translocation into the ER lumen, FVIII transiently forms high-molecular -weight non-disulfide-linked aggregates [122]. A portion of these aggregates are degraded through ER-associated degradation (ERAD), but others may represent a rate-limiting folding intermediate, as they have the ability to disaggregate and be secreted in an ATP-dependent fashion [122]. Aggregation is increased under conditions of ATP depletion, increased FVIII expression, and inhibition of N-linked glycosylation with tunicamycin [122]. These aggregates can bind to BiP, an ER chaperone protein that uses energy from its ATPase function during the folding of proteins [44, 120–122, 134, 141, 142].

FVIII Misfolding and ERAD

A significant portion of FVIII remains misfolded and is shunted into the ERAD pathway and degraded through the 26S proteasome. Inhibition of the 26S proteasome with lactacystin results in accumulation of FVIII intracellularly but does not increase FVIII secretion [122, 134]. ER degradation-enhancing mannosidase-like protein (EDEM) has been implicated in directing glycoproteins to ERAD [116–118].

FVIII and UPR

The ER can adjust its protein-folding capacity to meet the protein-folding demand through a collection of intracellular signaling pathways termed the unfolded protein response (UPR). The UPR decreases the amount of unfolded protein in the ER lumen and increases the ER protein-folding capacity. These pathways act to decrease mRNA translation, increase proteosome-mediated destruction of unfolded proteins, and upregulate the nuclear transcription of genes encoding functions required for proper protein folding. A deficiency of cellular ATP is one of a number of triggers for the UPR. The ATP-dependent function of BiP is required for proper folding of FVIII [44]. FV does not require nearly as much ATP for proper folding despite its close homology to FVIII. Electron microscopy of mammalian cells transfected with FVIII genes demonstrate a marked increase in the number mitochondria surrounding the ER, suggesting an increased energy requirement for FVIII secretion. Furthermore, the same cells placed in an ATP-depleting environment develop enlarged ER with aggregates of unfolded FVIII. In contrast, similar depletion of ATP from cells did not affect secretion of vWF or FV. These unfolded proteins are toxic to the cells and lead to programmed cell death. One of the three pathways of the unfolded protein response, PERK-directed phosphorylation of eIF2α, leads to apoptosis when unfolded proteins persist in the ER [143].

The UPR could have major implications for gene therapy for hemophilia as the increased production of FVIII in the host cell could lead to accumulation of unfolded FVIII in the ER lumen, programmed cell death, and dysfunction of the organ in which the cells were transfected. The UPR is an important process in the pathology of numerous diseases associated with protein misfolding, such as type II diabetes, amyloidosis, Alzheimer's disease, prion disease, and multiple myeloma [144–148].

Factor VIII Transport from the ER to the Golgi Compartment and Combined Deficiency of Factors V and VIII

Proteins destined for the cell surface enter the secretory pathway by translocation across the ER membrane in an unfolded state. In the ER lumen, secretory proteins undergo folding and posttranslational modification prior to transit to the Golgi compartment. Properly folded proteins from the ER are packaged into COPII-coated vesicles and directed to the Golgi [149].

Patients with a bleeding disorder characterized by deficiency of both FV and FVIII have been described since the 1950s [150]. This disease is inherited in an autosomal recessive manner although the F8 gene is located on X chromosome and FV on chromosome 1. Most patients have levels of FV and FVIII that are 5%–30% of normal [151]. The disorder was linked to mutations in a protein of the ER/Golgi complex, ERGIC-53 [152]. Patients with combined FV and FVIII deficiency have a complete absence of ERGIC-53 [153, 154]. It was also noted that 30% of patients with combined FV and FVIII deficiency did not harbor mutations in the ERGIC-53 gene, and these patients were subsequently found to have mutations in MCFD2 [153–155].

ERGIC-53 was originally identified as a protein that localized to the ER/Golgi intermediate compartment. ERGIC-53 is a transmembrane lectin with a selective mannose-binding property [156, 157]. MCFD2 forms a calcium-dependent interaction with ERGIC-53 in the secretory pathway and interacts with N-linked oligosaccharides in the B-domains of FV and FVIII. MCFD2 can crosslink to both FVIII and ERGIC-53, and the three proteins are co-localized in the secretory pathway [158]. Furthermore, ERGIC-53 is required for the retention of MCFD2 in the secretory pathway, where deficiency of ERGIC-53 leads to MCFD2 secretion [158]. Interestingly, although ERGIC-53 and MCFD2 are thought to be specific chaperones for FV and FVIII, cathepsin C, and cathepsin Z [159, 160], they are up-regulated in cells in response to ER stress [158, 161].

Future Perspectives

Bioengineered FVIII with Improved Activities

As knowledge of FVIII structure and secretion has grown, logic driven site-directed DNA mutagenesis has yielded FVIII molecules with improved properties (Table 1). Swaroop et al. demonstrated that Phe309Ser mutation within the A1 domain improved FVIII secretion by threefold [44]. Miao et al. made modifications and truncations to the B domain and showed that a shortened B domain of 226 amino acids containing six N-linked oligosaccharides (226/N6) improved secretion approximately eightfold compared to wild-type FVIII or full B domain-deleted FVIII [39]. The combination of the 226/N6 with the F309S mutation was additive, resulting in 15- to 25-fold higher secretion [39]. Other modifications have increased mRNA expression [38, 58, 162], improved secretion [163, 164], slowed inactivation and increased half-lives [165, 166], increased activation [167], reduced antigenicity [168], and reduced immunogenicity [169].

Strategies to Reduce Inhibitor Antibodies

One of the most feared complications of hemophilia is the development of an inhibitory antibody to infused FVIII, which occurs in approximately 30% of patients [170, 171]. These immunoglobulin G (IgG) antibodies have been mapped to numerous epitopes in the FVIII molecule. These antibodies can inhibit interaction with vWF, decrease the half-life of FVIII in plasma, or inhibit interaction of FVIII with

TABLE 1. Bioengineered factor VIII

Bioengineered factor	Manipulation	Possible benefit	Current status
BDD FVIII [38, 192]	Deletion of B-domain	Improved expression properties for manufacturing smaller cDNA for viral vectors in gene therapy [192]	Commercially available [195]
226 N6 FVIII [196]	Partial B-domain deletion	Improved properties for secretion	Preclinical
F309S FVIII [44]	Point mutation decreasing interaction with BiP	Improved secretion	Preclinical
A2 Epitope-free FVIII [169]	R484A/R489A/P492A	Decreased immunogenicity	Preclinical
Recombinant porcine FVIII [197]	FVIII with porcine FVIII structure	Decreased immunogenicity for use in inhibitor patients	In clinical trials
LRP-resistant FVIII [196]	Mutations at residues 484–509	Improved half-life	Preclinical
Disulfide bond stabilized FVIII [198, 199]	Disulfide bridge C664–C1826 and C662–C1828	Improved half-life	Preclinical
IR8 [166]	R336I, R562K, R740A	Improved half-life	Preclinical

BDD, B-domain deleted; FVIII, factor VIII; LRP, low density lipoprotein receptor-related protein

phospholipid or FIXa. Current strategies for treating patients with these inhibitory antibodies include the use of bypassing agents such as activated FVII or activated prothrombin complexes [172–174]. Eradication of the antibody using immune tolerance induction protocols employing high doses of FVIII given over long durations are often successful [175]. Recently, anti-CD20 monoclonal antibodies have been employed to eradicate inhibitors and have met with some success [176–187]. Another approach to inducing immune tolerance using gene therapy has been proposed [188, 189]. The use of porcine FVIII, which lacks many of the epitopes targeted by inhibitor antibodies, is another strategy that has been attempted [190, 191]. Attempts to create bioengineered FVIII with reduced immunogenicity have been reported [169]. Improved knowledge of FVIII structure and secretion along with better understanding of the immunology of inhibitor formation are likely to improve therapy for patients with inhibitors.

Gene Therapy for Hemophilia A

Gene therapy offers immense promise as a cure for hemophilia. If gene transfer could provoke even 2%–5% of normal FVIII production in hemophilia A patients, it would drastically alter the course of the disease. Gene transfer has been attempted and, to date, has produced only modest transient increases in FVIII levels [192, 193]. Many factors may contribute to the poor clinical response. FVIII mRNA expression is low, and modifications such as deletion of the B domain offer promise in reducing this problem [38]. FVIII is inefficiently folded and accumulates in transfected cells, leading to cellular stress and death [194]. New bioengineered FVIII proteins that are more efficiently secreted offer promise in resolving this issue [39]. There is fear that an immune response to the vectors, transfected cells, or protein products may limit gene therapy; and numerous efforts are ongoing to address these problems.

Acknowledgments. R.J.K. is an Investigator of the Howard Hughes Medical Institute and is supported by NIH grants PO1-HL057346 and RO1-HL052173. M.U.C. is a National Hemophilia Foundation/Baxter Clinical Fellow.

References

1. Rosner F (1969) Hemophilia in the Talmud and rabbinic writings. Ann Intern Med 70:833–837
2. Lane S (1840) Haemorrhagic diathesis: successful transfusion of blood. Lancet 1:185
3. Patek AJ, Stetson RP (1936) Hemophilia. I. The abnormal coagulation of the blood and its relation to the blood platelets. J Clin Invest 15:531–542
4. Patek AJ, Taylor FH (1937) Hemophilia. I. Some properties of a substance obtained from normal human plasma effective in accelerating the coagulation of hemophilic blood. J Clin Invest 16:113–124
5. Kingdon HS, Lundblad RL (2002) An adventure in biotechnology: the development of haemophilia A therapeutics—from whole-blood transfusion to recombinant DNA to gene therapy. Biotechnol Appl Biochem 35:141–148
6. Hougie C, Denson KW, Biggs R (1967) A study of the reaction product of factor 8 and factor IX by gel filtration. Thromb Diath Haemorrh 18:211–222
7. Osterud B, Rapaport SI (1970) Synthesis of intrinsic factor X activator: inhibition of the function of formed activator by antibodies to factor VIII and to factor IX. Biochemistry 9:1854–1861
8. Alexander B, Landwehr G (1948) Studies of hemophilia. I. The assay of the antihemophilic clot-promoting principle in normal human plasma with some observations on the relative potency of certain plasma fractions. J Clin Invest 27:98–105
9. McMillan CW, Diamond LK, Surgenor DM (1961) Treatment of classic hemophilia: the use of fibrinogen rich in factor VIII for hemorrhage and for surgery. N Engl J Med 265:277–283
10. Tabor E (1999) The epidemiology of virus transmission by plasma derivatives: clinical studies verifying the lack of transmission of hepatitis B and C viruses and HIV type 1. Transfusion 39:1160–1168
11. Toole JJ, Knopf JL, Wozney JM, et al (1984) Molecular cloning of a cDNA encoding human antihaemophilic factor. Nature 312:342–327
12. Vehar GA, Keyt B, Eaton D, et al (1984) Structure of human factor VIII. Nature 312:337–342
13. Mannucci PM (2003) Hemophilia: treatment options in the twenty-first century. J Thromb Haemost 1:1349–355
14. Tusell J, Perez-Bianco R (2002) Prophylaxis in developed and in emerging countries. Haemophilia 8:183–188
15. Rogoff EG, Guirguis HS, Lipton RA, et al (2002) The upward spiral of drug costs: a time series analysis of drugs used in the treatment of hemophilia. Thromb Haemost 88:545–553

16. Bell J, Haldane JB (1986) The linkage between the genes for colour-blindness and haemophilia in man: by Julia Bell and J.B.S. Haldane, 1937. Ann Hum Genet 50:3–34
17. Edgell CJ, Kirkman HN, Clemons E, et al (1978) Prenatal diagnosis by linkage: hemophilia A and polymorphic glucose-6-phosphate dehydrogenase. Am J Hum Genet 30:80–84
18. Schwaab R, Brackmann HH, Meyer C, et al (1995) Haemophilia A: mutation type determines risk of inhibitor formation. Thromb Haemost 74:1402–1406
19. Oldenburg J, Schroder J, Brackmann HH, et al (2004) Environmental and genetic factors influencing inhibitor development. Semin Hematol 41:82–88
20. Gill JC (1999) The role of genetics in inhibitor formation. Thromb Haemost 82: 500–504
21. Astermark J, Berntorp E, White GC, et al (2001) The Malmo International Brother Study (MIBS): further support for genetic predisposition to inhibitor development in hemophilia patients. Haemophilia 7:267–272
22. Lakich D, Kazazian HH Jr, Antonarakis SE, et al (1993) Inversions disrupting the factor VIII gene are a common cause of severe haemophilia A. Nat Genet 5:236–241
23. Rossiter JP, Young M, Kimberland ML, et al (1994) Factor VIII gene inversions causing severe hemophilia A originate almost exclusively in male germ cells. Hum Mol Genet 3:1035–1039
24. Bontempo FA, Lewis JH, Gorenc TJ, et al (1987) Liver transplantation in hemophilia A. Blood 69:1721–1724
25. Webster WP, Zukoski CF, Hutchin P, et al (1971) Plasma factor VIII synthesis and control as revealed by canine organ transplantation. Am J Physiol 220:1147–1154
26. Chelly J, Concordet JP, Kaplan JC, et al (1989) Illegitimate transcription: transcription of any gene in any cell type. Proc Natl Acad Sci U S A 86:2617–2621
27. Kelly DA, Summerfield JA, Tuddenham EG (1984) Localization of factor VIIIC: antigen in guinea-pig tissues and isolated liver cell fractions. Br J Haematol 56:535–543
28. Lewis JH, Bontempo FA, Spero JA, et al (1985) Liver transplantation in a hemophiliac. N Engl J Med 312:1189–1190
29. Schick PK, Walker J, Profeta B, et al (1997) Synthesis and secretion of von Willebrand factor and fibronectin in megakaryocytes at different phases of maturation. Arterioscler Thromb Vasc Biol 17:797–801
30. Wion KL, Kelly D, Summerfield JA, et al (1985) Distribution of factor VIII mRNA and antigen in human liver and other tissues. Nature 317:726–729
31. Zelechowska MG, van Mourik JA, Brodniewicz-Proba T (1985) Ultrastructural localization of factor VIII procoagulant antigen in human liver hepatocytes. Nature 317:729–730
32. Do H, Healey JF, Waller EK, et al (1999) Expression of factor VIII by murine liver sinusoidal endothelial cells. J Biol Chem 274:19587–19592
33. Chiu HC, Schick PK, Colman RW (1985) Biosynthesis of factor V in isolated guinea pig megakaryocytes. J Clin Invest 75:339–346
34. Galbusera M, Zoja C, Donadelli R, et al (1997) Fluid shear stress modulates von Willebrand factor release from human vascular endothelium. Blood 90:1558–1564
35. Hollestelle MJ, Thinnes T, Crain K, et al (2001) Tissue distribution of factor VIII gene expression in vivo—a closer look. Thromb Haemost 86:855–861
36. Jacquemin M, Neyrinck A, Hermanns MI, et al (2006) FVIII production by human lung microvascular endothelial cells. Blood 108:515–517
37. Lynch CM, Israel DI, Kaufman RJ, et al (1993) Sequences in the coding region of clotting factor VIII act as dominant inhibitors of RNA accumulation and protein production. Hum Gene Ther 4:259–272
38. Pittman DD, Alderman EM, Tomkinson KN, et al (1993) Biochemical, immunological, and in vivo functional characterization of B-domain-deleted factor VIII. Blood 81:2925–2935

39. Miao HZ, Sirachainan N, Palmer L, et al (2004) Bioengineering of coagulation factor VIII for improved secretion. Blood 103:3412–3419

40. Cripe LD, Moore KD, Kane WH (1992) Structure of the gene for human coagulation factor V. Biochemistry 31:3777–3785

41. Jenny RJ, Pittman DD, Toole JJ, et al (1987) Complete cDNA and derived amino acid sequence of human factor V. Proc Natl Acad Sci U S A 84:4846–4850

42. Kane WH, Davie EW (1988) Blood coagulation factors V and VIII: structural and functional similarities and their relationship to hemorrhagic and thrombotic disorders. Blood 71:539–555

43. Mann KG (1999) Biochemistry and physiology of blood coagulation. Thromb Haemost 82:165–174

44. Swaroop M, Moussalli M, Pipe SW, et al (1997) Mutagenesis of a potential immunoglobulin-binding protein-binding site enhances secretion of coagulation factor VIII. J Biol Chem 272:24121–24124

45. Foster PA, Fulcher CA, Houghten RA, et al (1988) An immunogenic region within residues Val1670-Glu1684 of the factor VIII light chain induces antibodies which inhibit binding of factor VIII to von Willebrand factor. J Biol Chem 263:5230–5234

46. Foster PA, Fulcher CA, Marti T, et al (1987) A major factor VIII binding domain resides within the amino-terminal 272 amino acid residues of von Willebrand factor. J Biol Chem 262:8443–8446

47. Kaufman RJ, Wasley LC, Davies MV, et al (1989) Effect of von Willebrand factor coexpression on the synthesis and secretion of factor VIII in Chinese hamster ovary cells. Mol Cell Biol 9:1233–1242

48. Kaufman RJ, Wasley LC, Dorner AJ (1988) Synthesis, processing, and secretion of recombinant human factor VIII expressed in mammalian cells. J Biol Chem 263:6352–6362

49. Takahashi Y, Kalafatis M, Girma JP, et al (1987) Localization of a factor VIII binding domain on a 34 kilodalton fragment of the N-terminal portion of von Willebrand factor. Blood 70:1679–1682

50. Wakabayashi H, Koszelak ME, Mastri M, et al (2001) Metal ion-independent association of factor VIII subunits and the roles of calcium and copper ions for cofactor activity and inter-subunit affinity. Biochemistry 40:10293–10300

51. Koschinsky ML, Funk WD, van Oost BA, et al (1986) Complete cDNA sequence of human preceruloplasmin. Proc Natl Acad Sci U S A 83:5086–5090

52. Tagliavacca L, Moon N, Dunham WR, et al (1997) Identification and functional requirement of Cu(I) and its ligands within coagulation factor VIII. J Biol Chem 272:27428–27434

53. Bihoreau N, Pin S, de Kersabiec AM, et al (1994) Copper-atom identification in the active and inactive forms of plasma-derived FVIII and recombinant FVIII-delta II. Eur J Biochem 222:41–48

54. Pemberton S, Lindley P, Zaitsev V, et al (1997) A molecular model for the triplicated A domains of human factor VIII based on the crystal structure of human ceruloplasmin. Blood 89:2413–2421

55. Wakabayashi H, Schmidt KM, Fay PJ (2002) Ca(2−) binding to both the heavy and light chains of factor VIII is required for cofactor activity. Biochemistry 41:8485–8492

56. Sudhakar K, Fay PJ (1996) Exposed hydrophobic sites in factor VIII and isolated subunits. J Biol Chem 271:23015–23021

57. Fay PJ (1988) Reconstitution of human factor VIII from isolated subunits. Arch Biochem Biophys 262:525–531

58. Toole JJ, Pittman DD, Orr EC, et al (1986) A large region (approximately equal to 95 kDa) of human factor VIII is dispensable for in vitro procoagulant activity. Proc Natl Acad Sci U S A 83:5939–5942

59. Hironaka T, Furukawa K, Esmon PC, et al (1992) Comparative study of the sugar chains of factor VIII purified from human plasma and from the culture media of recombinant baby hamster kidney cells. J Biol Chem 267:8012–8020
60. Kumar HP, Hague C, Haley T, et al (1996) Elucidation of N-linked oligosaccharide structures of recombinant human factor VIII using fluorophore-assisted carbohydrate electrophoresis. Biotechnol Appl Biochem 24(Pt 3):207–216
61. Elder B, Lakich D, Gitschier J (1993) Sequence of the murine factor VIII cDNA. Genomics 16:374–379
62. Bovenschen N, Rijken DC, Havekes LM, et al (2005) The B domain of coagulation factor VIII interacts with the asialoglycoprotein receptor. J Thromb Haemost 3:1257–1265
63. Pratt KP, Shen BW, Takeshima K, et al (1999) Structure of the C2 domain of human factor VIII at 1.5 A resolution. Nature 402:439–442
64. Poole S, Firtel RA, Lamar E, et al (1981) Sequence and expression of the discoidin I gene family in *Dictyostelium discoideum*. J Mol Biol 153:273–289
65. Stubbs JD, Lekutis C, Singer KL, et al (1990) cDNA cloning of a mouse mammary epithelial cell surface protein reveals the existence of epidermal growth factor-like domains linked to factor VIII-like sequences. Proc Natl Acad Sci U S A 87:8417–8421
66. Pellequer JL, Gale AJ, Griffin JH, et al (1998) Homology models of the C domains of blood coagulation factors V and VIII: a proposed membrane binding mode for FV and FVIII C2 domains. Blood Cells Mol Dis 24:448–461
67. Spiegel PC, Murphy P, Stoddard BL (2004) Surface-exposed hemophilic mutations across the factor VIII C2 domain have variable effects on stability and binding activities. J Biol Chem 279:53691–53698
68. Spiegel PC Jr, Jacquemin M, Saint-Remy JM, et al (2001) Structure of a factor VIII C2 domain-immunoglobulin G4kappa Fab complex: identification of an inhibitory antibody epitope on the surface of factor VIII. Blood 98:13–19
69. Pipe SW, Kaufman RJ (1996) Factor VIII C2 domain missense mutations exhibit defective trafficking of biologically functional proteins. J Biol Chem 271:25671–25676
70. Liu ML, Shen BW, Nakaya S, et al (2000) Hemophilic factor VIII C1- and C2-domain missense mutations and their modeling to the 1.5-angstrom human C2-domain crystal structure. Blood 96:979–987
71. Bevers EM, Comfurius P, Zwaal RF (1983) Changes in membrane phospholipid distribution during platelet activation. Biochim Biophys Acta 736:57–66
72. Van Dieijen G, Tans G, Rosing J, et al (1981) The role of phospholipid and factor VIIIa in the activation of bovine factor X. J Biol Chem 256:3433–3442
73. Gilbert GE, Furie BC, Furie B (1990) Binding of human factor VIII to phospholipid vesicles. J Biol Chem 265:815–822
74. Kemball-Cook G, Barrowcliffe TW (1992) Interaction of factor VIII with phospholipids: role of composition and negative charge. Thromb Res 67:57–71
75. Andersson LO, Thuy LP, Brown JE (1981) Affinity chromatography of coagulation factors II, VIII, IX and X on matrix-bound phospholipid vesicles. Thromb Res 23:481–489
76. Atkins JS, Ganz PR (1992) The association of human coagulation factors VIII, IXa and X with phospholipid vesicles involves both electrostatic and hydrophobic interactions. Mol Cell Biochem 112:61–71
77. Lollar P, Hill-Eubanks DC, Parker CG (1988) Association of the factor VIII light chain with von Willebrand factor. J Biol Chem 263:10451–10455
78. Andersson LO, Brown JE (1981) Interaction of factor VIII-von Willebrand factor with phospholipid vesicles. Biochem J 200:161–167
79. Lajmanovich A, Hudry-Clergeon G, Freyssinet JM, et al (1981) Human factor VIII procoagulant activity and phospholipid interaction. Biochim Biophys Acta 678:132–136

80. Saenko EL, Scandella D, Yakhyaev AV, et al (1998) Activation of factor VIII by thrombin increases its affinity for binding to synthetic phospholipid membranes and activated platelets. J Biol Chem 273:27918–27926

81. Arai M, Scandella D, Hoyer LW (1989) Molecular basis of factor VIII inhibition by human antibodies: antibodies that bind to the factor VIII light chain prevent the interaction of factor VIII with phospholipid. J Clin Invest 83:1978–1984

82. Schmidt AE, Stewart JE, Mathur A, et al (2005) Na$^+$ site in blood coagulation factor IXa: effect on catalysis and factor VIIIa binding. J Mol Biol 350:78–91

83. Lenting PJ, Donath MJ, van Mourik JA, et al (1994) Identification of a binding site for blood coagulation factor IXa on the light chain of human factor VIII. J Biol Chem 269:7150–7155

84. Regan LM, Lamphear BJ, Huggins CF, et al (1994) Factor IXa protects factor VIIIa from activated protein C: factor IXa inhibits activated protein C-catalyzed cleavage of factor VIIIa at Arg562. J Biol Chem 269:9445–9452

85. Fay PJ, Beattie T, Huggins CF, et al (1994) Factor VIIIa A2 subunit residues 558–565 represent a factor IXa interactive site. J Biol Chem 269:20522–20527

86. Amano K, Sarkar R, Pemberton S, et al (1998) The molecular basis for cross-reacting material-positive hemophilia A due to missense mutations within the A2-domain of factor VIII. Blood 91:538–548

87. Jenkins PV, Freas J, Schmidt KM, et al (2002) Mutations associated with hemophilia A in the 558-565 loop of the factor VIIIa A2 subunit alter the catalytic activity of the factor Xase complex. Blood 100:501–508

88. Mathur A, Bajaj SP (1999) Protease and EGF1 domains of factor IXa play distinct roles in binding to factor VIIIa: importance of helix 330 (helix 162 in chymotrypsin) of protease domain of factor IXa in its interaction with factor VIIIa. J Biol Chem 274:18477–18486

89. Jenkins PV, Dill JL, Zhou Q, et al (2004) Contribution of factor VIIIa A2 and A3-C1-C2 subunits to the affinity for factor IXa in factor Xase. Biochemistry 43: 5094–5101

90. Mertens K, van Wijngaarden A, Bertina RM, et al (1985) The functional defect of factor VIII Leiden, a genetic variant of coagulation factor VIII. Thromb Haemost 54:650–653

91. Jenkins PV, Dill JL, Zhou Q, et al (2004) Clustered basic residues within segment 484-510 of the factor VIIIa A2 subunit contribute to the catalytic efficiency for factor Xa generation. J Thromb Haemost 2:452–458

92. Lapan KA, Fay PJ (1997) Localization of a factor X interactive site in the A1 subunit of factor VIIIa. J Biol Chem 272:2082–2088

93. Lapan KA, Fay PJ (1998) Interaction of the A1 subunit of factor VIIIa and the serine protease domain of factor X identified by zero-length cross-linking. Thromb Haemost 80:418–422

94. Regan LM, O'Brien LM, Beattie TL, et al (1996) Activated protein C-catalyzed proteolysis of factor VIIIa alters its interactions within factor Xase. J Biol Chem 271:3982–3987

95. Mathur A, Zhong D, Sabharwal AK, et al (1997) Interaction of factor IXa with factor VIIIa: effects of protease domain Ca^{2+} binding site, proteolysis in the autolysis loop, phospholipid, and factor X. J Biol Chem 272:23418–23426

96. Eaton D, Rodriguez H, Vehar GA (1986) Proteolytic processing of human factor VIII: correlation of specific cleavages by thrombin, factor Xa, and activated protein C with activation and inactivation of factor VIII coagulant activity. Biochemistry 25:505–512

97. Fay PJ, Mastri M, Koszelak ME, et al (2001) Cleavage of factor VIII heavy chain is required for the functional interaction of a2 subunit with factor IXA. J Biol Chem 276:12434–12439

98. Pipe SW, Eickhorst AN, McKinley SH, et al (1999) Mild hemophilia A caused by increased rate of factor VIII A2 subunit dissociation: evidence for nonproteolytic inactivation of factor VIIIa in vivo. Blood 93:176–183
99. Pipe SW, Saenko EL, Eickhorst AN, et al (2001) Hemophilia A mutations associated with 1-stage/2-stage activity discrepancy disrupt protein-protein interactions within the triplicated A domains of thrombin-activated factor VIIIa. Blood 97:685–691
100. Hakeos WH, Miao H, Sirachainan N, et al (2002) Hemophilia A mutations within the factor VIII A2-A3 subunit interface destabilize factor VIIIa and cause one-stage/two-stage activity discrepancy. Thromb Haemost 88:781–787
101. Duncan EM, Duncan BM, Tunbridge LJ, et al (1994) Familial discrepancy between the one-stage and two-stage factor VIII methods in a subgroup of patients with haemophilia A. Br J Haematol 87:846–848
102. Rudzki Z, Duncan EM, Casey GJ, et al (1996) Mutations in a subgroup of patients with mild haemophilia A and a familial discrepancy between the one-stage and two-stage factor VIII:C methods. Br J Haematol 94:400–406
103. Dahlback B, Hildebrand B (1994) Inherited resistance to activated protein C is corrected by anticoagulant cofactor activity found to be a property of factor V. Proc Natl Acad Sci U S A 91:1396–1400
104. Dahlback B (1997) Resistance to activated protein C as risk factor for thrombosis: molecular mechanisms, laboratory investigation, and clinical management. Semin Hematol 34:217–234
105. Bertina RM, Koeleman BP, Koster T, et al (1994) Mutation in blood coagulation factor V associated with resistance to activated protein C. Nature 369:64–67
106. Fay PJ, Smudzin TM, Walker FJ (1991) Activated protein C-catalyzed inactivation of human factor VIII and factor VIIIa: identification of cleavage sites and correlation of proteolysis with cofactor activity. J Biol Chem 266:20139–20145
107. Lenting PJ, Neels JG, van den Berg BM, et al (1999) The light chain of factor VIII comprises a binding site for low density lipoprotein receptor-related protein. J Biol Chem 274:23734–23739
108. Saenko EL, Yakhyaev AV, Mikhailenko I, et al (1999) Role of the low density lipoprotein-related protein receptor in mediation of factor VIII catabolism. J Biol Chem 274:37685–37692
109. Turecek PL, Schwarz HP, Binder BR (2000) In vivo inhibition of low density lipoprotein receptor-related protein improves survival of factor VIII in the absence of von Willebrand factor. Blood 95:3637–3638
110. Bovenschen N, Boertjes RC, van Stempvoort G, et al (2003) Low density lipoprotein receptor-related protein and factor IXa share structural requirements for binding to the A3 domain of coagulation factor VIII. J Biol Chem 278:9370–9377
111. Espirito Santo SM, Pires NM, Boesten LS, et al (2004) Hepatic low-density lipoprotein receptor-related protein deficiency in mice increases atherosclerosis independent of plasma cholesterol. Blood 103:3777–3782
112. Bovenschen N, Herz J, Grimbergen JM, et al (2003) Elevated plasma factor VIII in a mouse model of low-density lipoprotein receptor-related protein deficiency. Blood 101:3933–3939
113. Sousa MC, Ferrero-Garcia MA, Parodi AJ (1992) Recognition of the oligosaccharide and protein moieties of glycoproteins by the UDP-Glc:glycoprotein glucosyltransferase. Biochemistry 31:97–105
114. Kaufman RJ (1999) Stress signaling from the lumen of the endoplasmic reticulum: coordination of gene transcriptional and translational controls. Genes Dev 13:1211–1233
115. Pittman DD, Tomkinson KN, Kaufman RJ (1994) Post-translational requirements for functional factor V and factor VIII secretion in mammalian cells. J Biol Chem 269:17329–17337

116. Eriksson KK, Vago R, Calanca V, et al (2004) EDEM contributes to maintenance of protein folding efficiency and secretory capacity. J Biol Chem 279:44600–44605
117. Molinari M, Calanca V, Galli C, et al (2003) Role of EDEM in the release of misfolded glycoproteins from the calnexin cycle. Science 299:1397–1400
118. Oda Y, Hosokawa N, Wada I, et al (2003) EDEM as an acceptor of terminally misfolded glycoproteins released from calnexin. Science 299:1394–1397
119. Kaufman RJ (1998) Post-translational modifications required for coagulation factor secretion and function. Thromb Haemost 79:1068–1079
120. Dorner AJ, Wasley LC, Kaufman RJ (1989) Increased synthesis of secreted proteins induces expression of glucose-regulated proteins in butyrate-treated Chinese hamster ovary cells. J Biol Chem 264:20602–20607
121. Morris JA, Dorner AJ, Edwards CA, et al (1997) Immunoglobulin binding protein (BiP) function is required to protect cells from endoplasmic reticulum stress but is not required for the secretion of selective proteins. J Biol Chem 272:4327–4334
122. Tagliavacca L, Wang Q, Kaufman RJ (2000) ATP-dependent dissociation of non-disulfide-linked aggregates of coagulation factor VIII is a rate-limiting step for secretion. Biochemistry 39:1973–1981
123. Dorner AJ, Wasley LC, Kaufman RJ (1990) Protein dissociation from GRP78 and secretion are blocked by depletion of cellular ATP levels. Proc Natl Acad Sci U S A 87:7429–7432
124. Bergeron JJ, Brenner MB, Thomas DY, et al (1994) Calnexin: a membrane-bound chaperone of the endoplasmic reticulum. Trends Biochem Sci 19:124–128
125. Michalak M, Milner RE, Burns K, et al (1992) Calreticulin. Biochem J 285(Pt 3):681–692
126. Hebert DN, Foellmer B, Helenius A (1996) Calnexin and calreticulin promote folding, delay oligomerization and suppress degradation of influenza hemagglutinin in microsomes. EMBO J 15:2961–2968
127. Arunachalam B, Cresswell P (1995) Molecular requirements for the interaction of class II major histocompatibility complex molecules and invariant chain with calnexin. J Biol Chem 270:2784–2790
128. Pind S, Riordan JR, Williams DB (1994) Participation of the endoplasmic reticulum chaperone calnexin (p88, IP90) in the biogenesis of the cystic fibrosis transmembrane conductance regulator. J Biol Chem 269:12784–12788
129. Gelman MS, Chang W, Thomas DY, et al (1995) Role of the endoplasmic reticulum chaperone calnexin in subunit folding and assembly of nicotinic acetylcholine receptors. J Biol Chem 270:15085–15092
130. Kim PS, Arvan P (1995) Calnexin and BiP act as sequential molecular chaperones during thyroglobulin folding in the endoplasmic reticulum. J Cell Biol 128:29–38
131. Zapun A, Darby NJ, Tessier DC, et al (1998) Enhanced catalysis of ribonuclease B folding by the interaction of calnexin or calreticulin with ERp57. J Biol Chem 273:6009–6012
132. Trombetta SE, Ganan SA, Parodi AJ (1991) The UDP-Glc:glycoprotein glucosyltransferase is a soluble protein of the endoplasmic reticulum. Glycobiology 1:155–161
133. Sousa M, Parodi AJ (1995) The molecular basis for the recognition of misfolded glycoproteins by the UDP-Glc:glycoprotein glucosyltransferase. EMBO J 14:4196–4203
134. Pipe SW, Morris JA, Shah J, et al (1998) Differential interaction of coagulation factor VIII and factor V with protein chaperones calnexin and calreticulin. J Biol Chem 273:8537–8544
135. Hurtley SM, Helenius A (1989) Protein oligomerization in the endoplasmic reticulum. Annu Rev Cell Biol 5:277–307
136. Bonnerot C, Marks MS, Cosson P, et al (1994) Association with BiP and aggregation of class II MHC molecules synthesized in the absence of invariant chain. EMBO J 13:934–944

137. Cotner T, Pious D (1995) HLA-DR beta chains enter into an aggregated complex containing GRP-78/BiP prior to their degradation by the pre-Golgi degradative pathway. J Biol Chem 270:2379–2386

138. Kellokumpu S, Suokas M, Risteli L, et al (1997) Protein disulfide isomerase and newly synthesized procollagen chains form higher-order structures in the lumen of the endoplasmic reticulum. J Biol Chem 272:2770–2777

139. Kim PS, Bole D, Arvan P (1992) Transient aggregation of nascent thyroglobulin in the endoplasmic reticulum: relationship to the molecular chaperone, BiP. J Cell Biol 118:541–549

140. Marks MS, Germain RN, Bonifacino JS (1995) Transient aggregation of major histocompatibility complex class II chains during assembly in normal spleen cells. J Biol Chem 270:10475–10481

141. Marquette KA, Pittman DD, Kaufman RJ (1995) A 110-amino acid region within the A1-domain of coagulation factor VIII inhibits secretion from mammalian cells. J Biol Chem 270:10297–10303

142. Dorner AJ, Wasley LC, Kaufman RJ (1992) Overexpression of GRP78 mitigates stress induction of glucose regulated proteins and blocks secretion of selective proteins in Chinese hamster ovary cells. EMBO J 11:1563–1571

143. Moussalli M, Pipe SW, Hauri HP, et al (1999) Mannose-dependent endoplasmic reticulum (ER)-Golgi intermediate compartment-53-mediated ER to Golgi trafficking of coagulation factors V and VIII. J Biol Chem 274:32539–32542

144. Brancaccio D, Ghiggeri GM, Braidotti P, et al (1995) Deposition of kappa and lambda light chains in amyloid filaments of dialysis-related amyloidosis. J Am Soc Nephrol 6:1262–270

145. Davis PD, Raffen R, Dul LJ, et al (2000) Inhibition of amyloid fiber assembly by both BiP and its target peptide. Immunity 13:433–442

146. Deng HX, Tainer JA, Mitsumoto H, et al (1995) Two novel SOD1 mutations in patients with familial amyotrophic lateral sclerosis. Hum Mol Genet 4:1113–1116

147. Dobson CM (2004) Principles of protein folding, misfolding and aggregation. Semin Cell Dev Biol 15:3–16

148. Kelly JW, Colon W, Lai Z, et al (1997) Transthyretin quaternary and tertiary structural changes facilitate misassembly into amyloid. Adv Protein Chem 50:161–181

149. Schekman R, Orci L (1996) Coat proteins and vesicle budding. Science 271:1526–1533

150. Oeri J, Matter M, Isenschmid H, et al (1954) [Congenital factor V deficiency (parahemophilia) with true hemophilia in two brothers.] Bibl Paediatr 58:575–588

151. Zhang B, Ginsburg D (2004) Familial multiple coagulation factor deficiencies: new biologic insight from rare genetic bleeding disorders. J Thromb Haemost 2:1564–1572

152. Nichols WC, Seligsohn U, Zivelin A, et al (1998) Mutations in the ER-Golgi intermediate compartment protein ERGIC-53 cause combined deficiency of coagulation factors V and VIII. Cell 93:61–70

153. Nichols WC, Terry VH, Wheatley MA, et al (1999) ERGIC-53 gene structure and mutation analysis in 19 combined factors V and VIII deficiency families. Blood 93:2261–2266

154. Neerman-Arbez M, Johnson KM, Morris MA, et al (1999) Molecular analysis of the ERGIC-53 gene in 35 families with combined factor V-factor VIII deficiency. Blood 93:2253–2260

155. Zhang B, Cunningham MA, Nichols WC, et al (2003) Bleeding due to disruption of a cargo-specific ER-to-Golgi transport complex. Nat Genet 34:220–225

156. Arar C, Carpentier V, Le Caer JP, et al (1995) ERGIC-53, a membrane protein of the endoplasmic reticulum-Golgi intermediate compartment, is identical to MR60,

an intracellular mannose-specific lectin of myelomonocytic cells. J Biol Chem 270:3551–3553

157. Itin C, Roche AC, Monsigny M, et al (1996) ERGIC-53 is a functional mannose-selective and calcium-dependent human homologue of leguminous lectins. Mol Biol Cell 7:483–493

158. Nyfeler B, Zhang B, Ginsburg D, et al (2006) Cargo selectivity of the ERGIC-53/MCFD2 transport receptor complex. Traffic 7:1473–14781

159. Appenzeller C, Andersson H, Kappeler F, et al (1999) The lectin ERGIC-53 is a cargo transport receptor for glycoproteins. Nat Cell Biol 1:330–334

160. Vollenweider F, Kappeler F, Itin C, et al (1998) Mistargeting of the lectin ERGIC-53 to the endoplasmic reticulum of HeLa cells impairs the secretion of a lysosomal enzyme. J Cell Biol 142:377–389

161. Nyfeler B, Nufer O, Matsui T, et al (2003) The cargo receptor ERGIC-53 is a target of the unfolded protein response. Biochem Biophys Res Commun 304:599–604

162. Plantier JL, Rodriguez MH, Enjolras N, et al (2001) A factor VIII minigene comprising the truncated intron I of factor IX highly improves the in vitro production of factor VIII. Thromb Haemost 86:596–603

163. Doering CB, Healey JF, Parker ET, et al (2002) High level expression of recombinant porcine coagulation factor VIII. J Biol Chem 277:38345–38349

164. Doering CB, Healey JF, Parker ET, et al (2004) Identification of porcine coagulation factor VIII domains responsible for high level expression via enhanced secretion. J Biol Chem 279:6546–6552

165. Amano K, Michnick DA, Moussalli M, et al (1998) Mutation at either Arg336 or Arg562 in factor VIII is insufficient for complete resistance to activated protein C (APC)-mediated inactivation: implications for the APC resistance test. Thromb Haemost 79:557–563

166. Pipe SW, Kaufman RJ (1997) Characterization of a genetically engineered inactivation-resistant coagulation factor VIIIa. Proc Natl Acad Sci U S A 94:11851–11856

167. Voorberg J, van Stempvoort G, Bos JM, et al (1996) Enhanced thrombin sensitivity of a factor VIII-heparin cofactor II hybrid. J Biol Chem 271:20985–20988

168. Barrow RT, Healey JF, Gailani D, et al (2000) Reduction of the antigenicity of factor VIII toward complex inhibitory antibody plasmas using multiply-substituted hybrid human/porcine factor VIII molecules. Blood 95:564–568

169. Parker ET, Healey JF, Barrow RT, et al (2004) Reduction of the inhibitory antibody response to human factor VIII in hemophilia A mice by mutagenesis of the A2 domain B-cell epitope. Blood 104:704–710

170. Ehrenforth S, Kreuz W, Scharrer I, et al (1992) Incidence of development of factor VIII and factor IX inhibitors in haemophiliacs. Lancet 339:594–598

171. Kreuz W, Escuriola-Ettingshausen C, Martinez-Saguer I, et al (1996) Epidemiology of inhibitors in haemophilia A. Vox Sang 70(Suppl 1):2–8

172. Hay CR, Brown S, Collins PW, et al (2006) The diagnosis and management of factor VIII and IX inhibitors: a guideline from the United Kingdom Haemophilia Centre Doctors Organisation. Br J Haematol 133:591–605

173. Lusher JM, Shapiro SS, Palascak JE, et al (1980) Efficacy of prothrombin-complex concentrates in hemophiliacs with antibodies to factor VIII: a multicenter therapeutic trial. N Engl J Med 303:421–425

174. Lusher J, Ingerslev J, Roberts H, et al (1998) Clinical experience with recombinant factor VIIa. Blood Coagul Fibrinolysis 9:119–128

175. Brackmann HH, Gormsen J (1977) Massive factor-VIII infusion in haemophiliac with factor-VIII inhibitor, high responder. Lancet 2:933

176. Carcao M, St Louis J, Poon MC, et al (2006) Rituximab for congenital haemophiliacs with inhibitors: a Canadian experience. Haemophilia 12:7–18

177. Chuansumrit A, Husapadol S, Wongwerawattanakoon P, et al (2007) Rituximab as an adjuvant therapy to immune tolerance in a haemophilia A boy with high inhibitor titre. Haemophilia 13:108–110

178. Collins PW (2006) Novel therapies for immune tolerance in haemophilia A. Haemophilia 12(Suppl 6):94–100; discussion 100–101

179. Field JJ, Fenske TS, Blinder MA (2007) Rituximab for the treatment of patients with very high-titre acquired factor VIII inhibitors refractory to conventional chemotherapy. Haemophilia 13:46–50

180. Fox RA, Neufeld EJ, Bennett CM (2006) Rituximab for adolescents with haemophilia and high titre inhibitors. Haemophilia 12:218–222

181. Hay C, Recht M, Carcao M, et al (2006) Current and future approaches to inhibitor management and aversion. Semin Thromb Hemost 32(Suppl 2):15–21

182. Jy W, Gagliano-DeCesare T, Kett DH, et al (2003) Life-threatening bleeding from refractory acquired FVIII inhibitor successfully treated with rituximab. Acta Haematol 109:206–208

183. Lillicrap D (2006) The role of immunomodulation in the management of factor VIII inhibitors. Hematology 2006:421–425

184. Mateo J, Badell I, Forner R, et al (2006) Successful suppression using Rituximab of a factor VIII inhibitor in a boy with severe congenital haemophilia: an example of a significant decrease of treatment costs. Thromb Haemost 95:386–387

185. Mathias M, Khair K, Hann I, et al (2004) Rituximab in the treatment of alloimmune factor VIII and IX antibodies in two children with severe haemophilia. Br J Haematol 125:366–368

186. Moschovi M, Aronis S, Trimis G, et al (2006) Rituximab in the treatment of high responding inhibitors in severe haemophilia A. Haemophilia 12:95–99

187. Stasi R, Brunetti M, Stipa E, et al (2004) Selective B-cell depletion with rituximab for the treatment of patients with acquired hemophilia. Blood 103:4424–4428

188. Dobrzynski E, Herzog RW (2005) Tolerance induction by viral in vivo gene transfer. Clin Med Res 3:234–240

189. Lei TC, Scott DW (2005) Induction of tolerance to factor VIII inhibitors by gene therapy with immunodominant A2 and C2 domains presented by B cells as Ig fusion proteins. Blood 105:4865–4870

190. Brettler DB, Forsberg AD, Levine PH, et al (1989) The use of porcine factor VIII concentrate (Hyate:C) in the treatment of patients with inhibitor antibodies to factor VIII: a multicenter US experience. Arch Intern Med 149:1381–1385

191. Hay CR, Laurian Y, Verroust F, et al (1990) Induction of immune tolerance in patients with hemophilia A and inhibitors treated with porcine VIIIC by home therapy. Blood 76:882–886

192. Powell JS, Ragni MV, White GC 2nd, et al (2003) Phase 1 trial of FVIII gene transfer for severe hemophilia A using a retroviral construct administered by peripheral intravenous infusion. Blood 102:2038–2045

193. Roth DA, Tawa NE Jr, O'Brien JM, et al (2001) Nonviral transfer of the gene encoding coagulation factor VIII in patients with severe hemophilia A. N Engl J Med 344:1735–1742

194. Kaufman RJ, Pipe SW, Tagliavacca L, et al (1997) Biosynthesis, assembly and secretion of coagulation factor VIII. Blood Coagul Fibrinolysis 8(Suppl 2):S3–S14

195. Sandberg H, Almstedt A, Brandt J, et al (2001) Structural and functional characteristics of the B-domain-deleted recombinant factor VIII protein, r-VIII SQ. Thromb Haemost 85:93–100

196. Pipe SW (2005) The promise and challenges of bioengineered recombinant clotting factors. J Thromb Haemost 3:1692–1701

197. Barrow RT, Lollar P (2006) Neutralization of antifactor VIII inhibitors by recombinant porcine factor VIII. J Thromb Haemost 4:2223–2229

198. Radtke KP, Griffin JH, Riceberg J, et al (2007) Disulfide bond-stabilized factor VIII has prolonged factor VIIIa activity and improved potency in whole blood clotting assays. J Thromb Haemost 5:102–108
199. Gale AJ, Radtke KP, Cunningham MA, et al (2006) Intrinsic stability and functional properties of disulfide bond-stabilized coagulation factor VIIIa variants. J Thromb Haemost 4:1315–1322

Structure and Biological Function of Factor XI

Kazuo Fujikawa

Summary. Factor XI was discovered in 1953 as a missing factor in three family members who had a history of mild hemorrhage especially after tooth extraction [1]. Its disorder is inherited as an autosomal recessive trait. Symptoms are variable, mild to severe, and associated with injury or surgery [2]. A half century from the discovery, its purification, amino acid sequence, and crystal structure have been established. Factor XI is the zymogen of a homodimeric serine protease circulating in blood, 5 µg/ml, as a complex with high-molecular-weight kininogen. The amino-terminal half of the molecule is composed of four unique structures called the apple domain. These domains provide exosites to activators, substrates, or cofactors to modulate its biological activity. The carboxy-terminal half of the molecule is a catalytic domain with a structure typical of serine proteases. Factor XI could play a role in factor XII-independent intrinsic pathway to maintain thrombin generation following the initial stage of coagulation, which is triggered by exposed tissue factor. A number of mutated genes were identified in factor XI deficiency, and their recombinant molecules were studied in relation to their impaired biological functions. Some of these studies provided useful information for functional properties of the molecule. For review see ref. [3].

Key words. Factor XI · Factor IX · Platelets · Tissue factor · Apple domain

Introduction

Factor XI is the precursor of a disulfide-linked homodimer of a serine protease. The monomer is a single glycosylated polypeptide composed of 507 amino acid residues. It migrates on sodium dodecyl sulfate polyacrylamide gel electrophoresis (SDS-PAGE) as a 140- to 150-kDa band without reduction or a 75-kDa band with reduction [4, 5]. The N-terminal portion of the monomer consists of four tandem repeats, with each repeat having 90 or 91 amino acid residues with six Cys similarly paired by three internal disulfide bonds. These repeats are called the apple domain (A domain), with A1, A2, A3, and A4 arranged in order from the N-terminus. A1 and A4 have one extra Cys residue. The extra Cys_2 in A1 domain is conjugated with free cysteine, and the extra Cys_{321} in A4 domain is engaged in the dimer formation. Factor XIa is a dimer composed of two each of heavy (Glu_1-Arg_{369}) and light (Ile_{370}-Val_{607}) chains connected by disulfide bonds. Cys_{362} present in a short peptide chain, Met_{358}-Arg_{369}, extended from the A4 domain is engaged in an interchain disulfide bond with Cys_{482} in the light chain. Factor XI contains five potential N-linked glycosylation sites: three in the heavy chains and two in the light chains [6, 7] (Fig. 1).

Factor XI has high sequence identity (58%) with plasma prekallikrein (PK), which also has four apple domains [8]. It circulates as a monomer, forming a complex with high-molecular-weight kininogen (HMWK). Factor XI and PK bind to the light chain of HMWK with binding constants of $K_d = 69 \pm 4\,nM$ and $K_d = 18 \pm 5\,nM$, respectively [9]. Both factor XI and PK bind to the near C-terminal region of the light chain of HMWK.

The mode of three disulfide bond pairings formed in four A domains had not been found in other proteins, and a unique structure of A domain was expected. The

FIG. 1. Amino acid sequence and location of disulfide bonds of factor XI monomer. Four apple domains are labeled A_1, A_2, A_3, and A_4. Cys_{321} in A4 (*asterisk*) is engaged in the interchain disulfide bond to form the dimer. The cleavage site, Arg_{369}-Ile_{370}, by thrombin or factor XIIa is shown by an *arrow*. *Circled* residues—His_{413}, Asp_{462}, Ser_{557}—are members of the catalytic triad. Residues marked by a *diamond* are potential sites of N-linked carbohydrate chain attachment. Residues marked by a *solid diamond* were found by sequence analysis to have a carbohydrate chain attached. (From McMullen et al. [7], with permission)

light chain has a sequence typical of serine proteases, and a structure similar to chymotrypsin [10] or thermolysin [11] was highly predicted. Recently, a crystal structure of factor XI zymogen was determined at 2.9 Å resolution [12]. As expected, four A domains form independent structures, each of which is composed of seven β-strands and one α-helix with seven loops present between the β-strands and the α-helix. Four A domains are arrange in tandem: A1/A2 tandem is facing A3/A4 tandem at 180°. This arrangement results in the formation of a unique flat saucer-like structure of the heavy chain with the dimension of 60 × 60 × 20 Å. A large hole formed in the center of four A domains is surrounded predominantly by basic and aromatic residues (Fig. 2a). Two heavy chains are assembled at 70° to form an inverted V-shaped structure of the dimer held by A4 domains. The disulfide bond between the A4 domains is located at the top of the structure (Fig. 2b). Two heavy chains hold reversibly each other; one side of the β-sheets and the β4-β5 loops participate in mutual

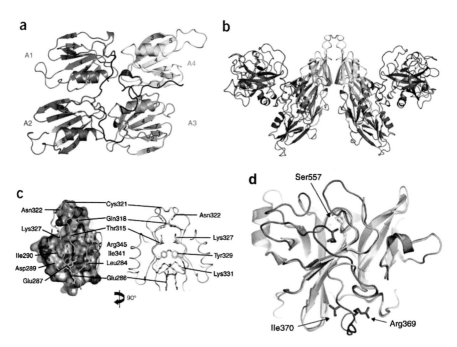

FIG. 2. Structure of factor XI. **a** Topology of four apple domains. Connecting loops between A domains are shown in black, and the connecting loop between A4 and the catalytic domain is dark blue. **b** Topology of four apple domains together with a protease domain, which is red. An interchain disulfide bond (orange) is located at the top of the structure. **c** *Left* Surface view of interacting charge residues of A4 in one of the monomers, with a side chain of A4 in the other monomer. Positively charged residues are blue, negatively charged residues are red, and side chains are yellow. *Right* Ribbon diagram shows the interaction of two A4 domains with hydrogen bonds and electrostatic interactions (orange). **d** Ribbon diagram showing the conformational changes before and after factor XI activation. The activated form is green, and the inactive form is yellow. Newly formed N-terminal Ile$_{370}$ is purple. (From Papagrigoriou et al. [12], with permission)

binding of A4 domains (Fig. 2c). This structure is in accordance with an earlier observation that the A4 domain itself is capable of forming the dimer without disulfide linkage [13], as detailed below. A2 domains are apart each other by a distance of 50 Å, and the A1 and A3 domains are located close to each other, separated by 5 Å. The heavy chain connects with the light chain by a short peptide (Met_{358}-Arg_{369}) that extends from the A4 domain. This peptide chain bends 90° at Cys_{362} and connects to the light chain.

The protease domain, a light chain, resides away from the A domain assembly and the catalytic site, Ser_{557}, is located at the outer side of the protease domain. It has a structure typical of trypsin-type serine proteases: two β-barrels connected by a central loop. The crystal structure of the protease domain of factor XIa has been determined [14]. When the two forms of the protease domains—one from zymogen and the other from the activated form—are compared, their structures are not largely different except in one region. During activation, the newly formed N-terminal Ile_{370} residue moves toward the catalytic site, Ser_{557}, together with an accompanying activation loop, Val_{371}-Ser_{376}. Ile_{370} relocates 20 Å distant from the original position, where a liberated NH_2 group of Ile_{370} forms a salt bridge with the carboxylate group of Asp_{556} to yield the catalytic machinery (Fig. 2d).

It was shown that the dimer is stable in the absence of the interchain disulfide bond between A4 domains. A chimera, a tissue plasminogen activator linked to the A4 domain, in which Cys_{321} is replaced by Ser to break the interchain disulfide bond, elutes as a dimer from a gel filtration column but migrates as a monomer on SDS-PAGE. This result shows that the A4 domains are capable of holding each other without an interchain disulfide bond [13].

Dimeric structure is unique in serine proteases, and it is interesting to determine if the dimeric form of factor XI is requisite for biological function. A chimera, factor XI/PK4 (A4 of factor XI is replaced by A4 of PK), was expressed in fibroblasts as the monomer and XI/PK4-Gly_{326} (Cys_{326} in PK4 is replaced by Gly to force Cys_{321} to form an interchain disulfide bond) as the dimer. XI/PK4, XI/PK4-Gly_{326}, and factor XI bind to activated platelets by similar affinities and had similar plasma clotting activities in a kaolin-induced clotting assay using purified phospholipids. These proteins were then activated and their clotting activities tested in factor XI-deficient plasma using activated platelets. XI/PK4-Gly_{326} had 72% clotting activity, whereas XI/PK4 had <0.05% of normal factor XIa clotting activity. XI/PK4-Gly_{326} and factor XI cleaved factor IX similarly, but XI/PK4 poorly cleaved factor IX. These results showed that the dimeric form of factor XI is required for activation of factor IX under physiological conditions [15].

Biological Functions of Apple Domains

Factor XI was shown to bind to stimulated platelets in the presence of HMWK and Zn^{2+}. The binding is specific, reversible, and saturable. Scatchard analysis showed that HMWK binds at 1500 sites/cell with $K_d = 10$ nM. Because factor XI binds to HMWK, it is not clear either factor XI binds platelets as the complex with HMWK or indirectly via HMWK [16]. To find the HMWK-binding site on factor XI, chimeras, in which each A domain of factor XI was linked to tissue plasminogen activator, were expressed in eukaryotic cells and tested for their inhibitory activities against HMWK binding to factor XI. The chimeras with A2 and A4 had higher inhibitory activities than those with A1 and A3. Monoclonal antibody specific to A2 had stronger inhibitory activity than antibodies to A1 or A4. These data indicate that major HMWK-binding site is located in A2 domain, and other domains are involved for full binding activity to HMWK [17].

To study the binding site of factor XI to activated platelets, recombinant (r)-A3 domain was expressed in prokaryotic cells. [^{125}I]-Factor XI binds to activated platelets in the presence of HMWK and Zn^{2+} at 1035 sites/cell, and it was inhibited by r-A3 with the same potency as unlabeled factor XI with $K_i = 5–6 \times 10^{-9}$ M. These results indicate that platelet-binding sites reside solely in the A3 domain. A synthetic peptide, corresponding to Thr_{249}-Phe_{260} in the C-terminal region of A3 domain, inhibited factor XI binding to activated platelets. The A3 domain of PK, PK3, had no binding activity, whereas a molecule in which the C-terminal sequence of PK3 was replaced by the corresponding factor XI sequence, Thr_{249}-Val_{271}, restored the binding activity. In this region, residues Arg_{250}, Lys_{255}, Phe_{260}, and Gln_{263} were found to be important for platelet binding by alanine scanning mutagenic analysis [18]. Thrombin activation of factor XI is enhanced by heparin, which directly binds to factor XI and factor XIa with $K_d = 1.1 \times 10^{-7}$ M and 0.7×10^{-9} M, respectively. The r-A3 domain was found to have heparin-binding affinity similar to that of factor XI, indicating that full heparin-binding activity is contained in A3. A peptide, Thr_{249}-Phe_{260}, in A3 domain, which contains a consensus heparin-binding sequence, Leu_{251}-Lys-Lys-Ser-Lys_{254}, inhibited heparin binding of factor XI. Alanine scanning mutagenetic analysis of r-A3 showed that residues Lys_{252} and Lys_{253} are important for heparin binding [19]. r-Factor XI mutants with alanine substitution for basic residues in the consensus heparin-binding sequence were expressed in fibroblasts and tested for heparin-binding activity. The results showed that the substitution of Lys_{252}, Lys_{253}, and Lys_{255} weakened heparin-binding activity, with replacement of Lys_{253} being most effective [20]. These results together showed that the region Thr_{249}-Phe_{260} shares the binding sites between platelets and heparin.

The binding site of factor IX on factor XI was studied using chimeras expressed in kidney fibroblasts, in which each factor XI A domain was replaced by the corresponding A domains of PK. A chimera, factor XI/PK3, had <1% of coagulant activity of wild-type factor XI, whereas chimeras, XI/PK1, XI/PK2, and XI/PK4 had 68%–140% activity. The weak activity of XI/PK3 is caused by a lower binding affinity to factor IX with $K_m = 12.7\,\mu M$ comparing with $0.11–0.37\,\mu M$ for factor XI and two other chimeras. Two antibodies specific to the A3 domain inhibited factor IX activation by factor XIa. These results showed that the binding site to factor IX locates in A3 domain [21]. r-Factor XI mutants with alanine substitution in N-terminal two regions (Ile_{183}-Val_{191} and Ser_{195}-Ile_{197}) and one C-terminal region (Ser_{258}-Ser_{264}) in A3 domain had markedly reduced factor IX-activating activity of factor XI. A chimera, XI/PK3, has weak factor IX activating activity caused by poor substrate-binding affinity, 35-fold larger K_m, than wild-type factor XI. This weak activity was restored in a mutant chimera, factorXI/PK3, in which the above three regions in PK3 were replaced by the corresponding regions of factor XI A3 domain. These results show that the region Ile_{183}-Val_{191} is highly important for factor IX activation and additional regions (Ser_{195}-Ile_{197} and Ser_{258}-Ser_{264}) are required for full activity [22].

Activation and Enzyme Activity

Factor XI is activated to factor XIa by factor XIIa, and HMWK with Zn^{2+} enhances its activation [4, 5]. It is also activated by thrombin (see below). Resulting factor XIa activates factor IX in the presence of Ca^{2+} [23, 24], leading to thrombin generation through the coagulation pathway [25]. Factor XIa cleaves chromogenic or fluorogenic peptide substrates—Glu-Pro-Arg-pNA or Boc-Glu(OBzl)-Ala-Arg-MCA—which are used for quantitative determination of enzyme activity. Factor XIa is inhibited by serine protease inhibitors (e.g., p-aminobenzamidine, diisopropylfluorophosphate, aprotinin). It is also completely inhibited by antithrombin III in the presence of heparin at a 1:2 molar ratio, indicating that each monomer of factor XIa has independent catalytic activity [5]. The inhibition of factor IX activation by factor XIa by leupeptin or aprotinin shows mixed-type, noncompetitive inhibition [26]. A similar pattern was obtained for the inhibition of factor XIa catalysis of peptidic substrates by factor IX. However, the isolated catalytic domain of factor XIa was competitively inhibited by p-aminobenzamidine and not inhibited by factor IX [27]. These results indicate that factor IX interacts with exosite in the heavy chain of factor XIa in addition to interaction with the catalytic site, consistent with the existence of a factor IX-binding site in the A3 domain of factor XI [21, 22].

Role in Hemostasis

Deficiency of both factor XII and HMWK does not cause a bleeding disorder. Thus, neither is factor XIIa the activator of factor XI, nor is HMWK a cofactor in physiological conditions. The amino acid sequence of factor XI deduced from its cDNA revealed that the cleavage site by factor XIIa is the Arg_{369}-Ile_{370} bond in the Pro-Arg-Ile sequence [6]. The presence of Pro in the P2' position in the cleavage site indicated that thrombin might activate factor XI. Such a sequence is present in the cleavage sites of factor VIII, factor V, and other procoagulant zymozens, which are activated by thrombin under physiological conditions.

Factor XI was found to be activated by thrombin in the presence of dextran sulfate, sulfatide, or heparin. In the presence of these negatively charged macromolecules, factor XI can be autoactivated [28, 29]. Other negatively charged molecules, such as dermatan sulfate and chondroitin sulfate, can support factor XI activation by thrombin [30]. This system, called thrombin-mediated factor XI activation or factor XII-independent factor XI activation, may be playing a physiological role in the coagulation system. When tissue factor (TF) is exposed at trauma sites, extrinsic coagulation is triggered; exposed TF binds to factor VII/VIIa, and the complex of TF/factor VIIa activates factor IX. The resulting factor IXa activates factor X, leading to thrombin generation. However, once factor Xa is formed, circulating tissue factor pathway inhibitor (TFPI) instantaneously inhibits TF/factor VIIa activity to terminate the extrinsic pathway. To maintain coagulation, another system is required to generate thrombin activity. The thrombin-mediated factor XI activation system may serve this need; a trace amount of thrombin initially generated by extrinsic pathway can activate factor XI to factor XIa, which reciprocally generates thrombin activity via the intrinsic pathway (Fig. 3).

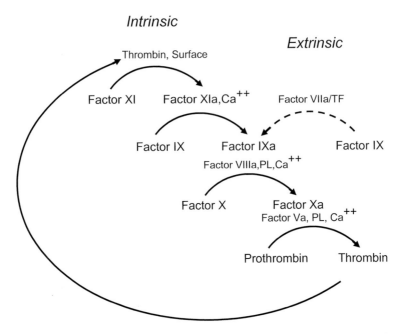

FIG. 3. Intrinsic (thrombin-mediated factor XI activation pathway) and extrinsic pathways. Exposure of the tissue factor (*TF*) at trauma sites and following factor IX activation are probably the initial reactions in coagulation, which is shown by the *broken line* curve. (From Fujikawa [3], with permission)

A few years later, several groups showed that thrombin-mediated factor XI activation takes place in plasma or reconstituted systems in the absence of nonphysiological materials, such as dextran sulfate. In these experiments, activation of factor XI by factor XIIa was prevented by using factor XII-deficient plasma or adding factor XIIa inhibitors to the test samples. Clot formation and lysis time induced by tissue-type plasminogen activator were measured by turbidity, after TF was added to the recalcified plasma. The clot formation was greater (>40%) in the presence of factor XI than in the absence of factor XI, when coagulation was triggered by a small amount of TF. Clot lysis time was more prolonged in the presence of factor XI than its absence. However, when larger amounts of TF were used to trigger the reactions, the presence of factor XI had no effect on either the clot formation rate or its lysis time. The amount of factor XIa generated was a trace, in a range of 0.01% [32]. The delayed lysis time was found to be associated with activation of thrombin activatable fibrinolysis inhibitor (TAFI). These results first showed that thrombin-mediated factor XI activation plays a role in coagulation as well as fibrinolysis [33].

Thrombin-mediated factor XI activation during the initial phase of coagulation was studied in a reconstituted system. When coagulation was induced by a small amount of TF (1.25 pM) relipidated with phosphatidylserine (PS)/phosphatidylcholine (PC), thrombin activity increased followed by factor XIa activity, which corresponded to ~0.1% of initial factor XI. The increase in factor XIa activity was blocked by the addition of hirudin, showing that thrombin is a sole activator of factor XI. However, the increase of thrombin activity was not influenced by the presence of factor XI, indicating that thrombin was generated via an extrinsic rather than an intrinsic pathway. These data showed that the concentration of factor XIa activity generated during the initial phase is not enough to activate factor IX, leading to thrombin generation [34]. A cell-based model system was prepared to be as close to in vivo conditions as possible [35]. In this system, monocytes were used as a source of TF, and platelets were included along with plasma levels of all other coagulation factors (except factor XII), TAFI, and antithrombin-III. Among 12 platelet preparations from different donors, thrombin generation was completely dependent on factor XI in some preparations and partially dependent in others. Prior activation of platelets by a thrombin receptor activation peptide, SFLLRN, did not have any effect on the thrombin generation. HMWK with Zn^{2+}/Ca^{2+} had no effect on the thrombin generation in this system. Only <0.02% of the plasma level of factor XI was converted to factor XIa, which was precipitated with platelets [36].

Another study also showed that slow thrombin-mediated factor XI activation proceeds in normal platelet-rich plasma (PRP) or platelet-poor plasma (PPP), triggered by r-TF (7.5 pg/ml) relipidated with PC/PS (normal concentration of TF is ~10 ng/ml). In these studies, polybrene was used throughout the experiments to avoid factor XI activation by factor XIIa. Thrombin generation was undetectable in factor XI-deficient plasma, but it was completely restored by the addition of a small volume (10%) of normal plasma. Thrombin generation was about twofold higher in PRP than in PPP [37]. It is interesting to note that a small increase in the factor XI level in plasma also improves the conditions of factor XI deficiency: Patients with a type II homozygote mutation and a factor XI level of ~1% have more serious bleeding problem than patients with the type III homozygote mutation and ~9% factor XI [38]. Thrombin-mediated factor XI activation was studied in PRP without adding TF. In these experiments, corn trypsin inhibitor, a specific factor XIIa inhibitor, was added to prevent factor XI activation by factor XIIa. The addition of small amounts of thrombin to normal PRP increased thrombin generation dose-dependently, whereas thrombin generation was not detectable in factor XI deficient PRP or normal PPP. The thrombin generation in normal PRP was not influenced by the addition of anti-TF antibody, excluding the possibility that thrombin generated via the extrinsic pathway triggered by TF released from platelet α-granules. These results showed that thrombin and factor XIa are reciprocally activated through a platelet-associated intrinsic pathway [39]. Lastly, in TF-coated microtiter plates, factor XI was not required for small clot formation (100 μl plasma) in recalcified plasma but was required for a larger clot formation (300 μl plasma), indicating that clots are initially formed on the surface coated with TF independently of factor XI, but factor XI is required to keep the clot growing [40].

The requirement of factor XI in coagulation was demonstrated in animal models. In an $FeCl_3$-induced vena cava thrombosis model, thrombus formation in factor XI-deficient mice is less sensitive to $FeCl_3$ than in normal mice; and the tail bleeding time is longer in factor XI-deficient mice than in normal mice [41]. Intravenous administration of monoclonal antibody to factor XI/ XIa (3 mg/kg) markedly reduced arterial thrombus growth under repeated balloon injury; the activated partial thromboplastin time was prolonged, whereas the prothrombin time, bleeding time, and platelet aggregation were not affected by antibody administration [42]. Although it is not clear if factor XI was activated by factor XIIa in these animal experiments, thrombin-mediated factor XI activation could contribute to thrombus formation in these injury-related coagulation states.

All of these experiments showed that thrombin-mediated factor XI activation proceeds slowly in plasma or in reconstitution systems mimicking the plasma milieu when TF is absent or present in very low concentrations either by restricted supply or the diminution of VIIa/TF activity by TFPI. Some of experiments were carried out without platelets. It is possible that plasma preparations contain microparticles that may serve as a surface. Other studies were performed in the presence of platelets, but the activities were not significantly enhanced. Accordingly, it is not clear whether thrombin-mediated factor XI activation requires a cell surface (i.e., platelets or endothelial cells). It is also unclear if slow thrombin-mediated factor XI activation can account for a part or all of the missing thrombogenic activity in factor XI deficiency.

Mutation

Factor XI deficiency is rare: one case per million in the general population, high in Ashkenazi Jews (12%), and relatively high in several European communities. To date, more than 100 mutations have been found in factor XI deficiency. A database for these mutations was created, and genetic and phenotypic data are available (www.factorXI. org) [43]. Missense and nonsense mutations and insertion, deletion, or splicing in the factor XI gene causes the deficiency. Antigens are not detectable in most cases owing to the secretion problem of misfolded molecules. In an analysis of Ashkenazi Jews, three types of mutation—splicing (type I), nonsense (type II), missense (type III)—were found in six unrelated patients with bleeding disorders. Three genotypes—I/II, II/III, III/III—were found in these patients [38]. In a general Ashkenazi population, II/III and III/III are prevalent [44]. Deficiency with antigen-positivity is rare, with only a few cases reported. Mutants (Q226R and S248N in the A3 domain) were found in an African American family with a history of excessive bleeding. r-Factor XI mutant of the former has a fivefold higher K_m for factor IX, showing that the A3 domain has an effect on factor XI affinity to factor IX, consistent with the presence of the factor IX binding site in the A3 domain [18]. r-Factor XI of the latter has fivefold lower affinity for activated platelets, which is also consistent with the platelet-binding site being in the A3 domain [45]. A patient with G555E homozygote mutation in the protease domain has <1% activity with a normal level of antigen in plasma. r-E555 factor XI is normally activated by factor XIIa or thrombin, and its active form cleaves a peptide substrate similar to wild-type factor XI. However, r-E555 activates factor IX 400-fold more slowly than the wild-type and is resistant to antithrombin III. Gly_{555} locates two residues prior the active site Ser residue. The side chain of Glu residue in this position probably interferes with the interaction between the factor XIa S2' site and the P2' site of factor IX and causes weak catalytic activity [46]. A patient with T475I mutation in the protease domain has factor XI activity equivalent to that in normal plasma but a lower level of antigen (FXI:C, 39 U/dl; Ag 27 U/dl). This mutant loses the N-glycosylation at Asn_{473} because replacement of Thr_{475} by Ile ruins the consensus N-glycosylation sequence, Asn_{473}-X-Thr_{475}. The possibility that loss of one carbohydrate chain causes poor secretion was tested. r-Mutant factor XI, A473, which loses one of carbohydrate chain at the same position, was expressed in Chinese hamster ovary (CHO) cells. This variant secreted with a high level of antigen. These results suggest that poor secretion is not caused by losing glycosylation but by replacement of this highly conserved Thr_{475} by Ile [47].

Note

Walsh and his colleagues have studied the factor XI activation mechanism on stimulated platelet membranes involving the glycoprotein Ib-IX-V complex and published several articles on this subject. Recently, they retracted two entire articles and several data in other papers because of the difficulty of reproducing the experiments. Therefore, their studies related to this subject are not included in this article (see Walsh PN and his colleagues. *Biochemistry* 46: 12886–7, J Biol Chem 282: 29067, Blood 110: 4164 [2007]).

References

1. Rosenthal RL, Dreskin OH, Rosenthal N (1953) New hemophilia-like disease caused by deficiency of a third thromboplastin factor. Proc Soc Exp Biol Med 82:171–174
2. O'Connell NM (2004) Factor XI deficiency. Semin Hematol 41:76–81
3. Fujikawa K (2005) Historical perspective of factor XI. Thromb Res 115:441–450
4. Bouma BN, Griffin JH (1977) Human blood coagulation factor XI: purification, properties, and mechanism of activation by activated factor XII. J Biol Chem 252:6432–6437
5. Kurachi K, Davie EW (1977) Activation of human factor XI (plasma thromboplastin antecedent) by factor XIIa (activated Hageman factor). Biochemistry16:5831–5839
6. Fujikawa K, Chung DW, Hendrickson LE, et al (1986) Amino acid sequence of human factor XI, a blood coagulation factor with four tandem repeats that are highly homologous with plasma prekallikrein. Biochemistry 25:2417–2424
7. McMullen BA, Fujikawa K, Davie EW (1991) Location of the disulfide bonds in human coagulation factor XI: the presence of tandem apple domains. Biochemistry 30:2056–2060
8. Chung DW, Fujikawa K, McMullen BA, et al (1986) Human plasma prekallikrein, a zymogen to a serine protease that contains four tandem repeats. Biochemistry 25:2410–2417
9. Tait JF, Fujikawa K (1987) Primary structure requirements for the binding of human high molecular weight kininogen to plasma prekallikrein and factor XI. J Biol Chem 262:11651–11656
10. Freer ST, Kraut J, Robertus JD, et al (1970) Chymotrypsinogen: 2.5-Ångstrom crystal structure, comparison with alpha-chymotrypsin, and implications for zymogen activation. Biochemistry 9:1997–2009
11. Holmes MA, Mathews BW (1982) Structure of thermolysin refined at 1.6 A resolution. J Mol Biol 160:623–639
12. Papagrigoriou E, McEwan PA, Walsh PN, et al (2006) Crystal structure of the factor XI zymogen reveals a pathway for transactivation. Nat Struct Mol Biol 13:557–558
13. Meijers JC, Mulvihill ER, Davie EW, et al (1992) Apple four in human blood coagulation factor XI mediates dimer formation. Biochemistry 31:4680–4684
14. Navaneetham D, Jin L, Pandey P, et al (2005) Structural and mutational analyses of the molecular interactions between the catalytic domain of factor XIa and the Kunitz protease inhibitor domain of protease nexin. J Biol Chem 280:36165–36167

15. Gailani D, Ho D, Sun MF, et al (2001) Model for a factor IX activation complex on blood platelets: dimeric conformation of factor XIa is essential. Blood 97:3117–122

16. Greengard JS, Heeb MJ, Ersdal E, et al (1986) Binding of coagulation factor XI to washed human platelets. Biochemistry 25:3884–3890

17. Renne T, Gailani D, Meijers JC, et al (2002) Characterization of the H-kininogen-binding site on factor XI: a comparison of factor XI and plasma prekallikrein. J Biol Chem 277:4892–4899

18. Ho DH, Baglia FA, Walsh PN (2000) Factor XI binding to activated platelets is mediated by residues R(250), K(255), F(260), and Q(263) within the apple 3 domain. Biochemistry 39:316–323

19. Ho DH, Badellio K, Baglia FA, et al (1998) A binding site for heparin in the apple 3 domain of factor XI. J Biol Chem 273:16382–16390

20. Zhao M, Abdel-Razek T, Sun MF, et al (1998) Characterization of a heparin binding site on the heavy chain of factor XI. J Biol Chem 273:31153–31159

21. Sun Y, Gailani D (1996) Identification of a factor IX binding site on the third apple domain of activated factor X. J Biol Chem 271:29023–29028

22. Sun MF, ZhaoM, Gailani D (1999) Identification of amino acids in the factor XI apple 3 domain required for activation of factor IX. J Biol Chem 274:36373–36378

23. Osterud B, Bouma B, Griffin JH (1978) Human blood coagulation factor IX: purification, properties, and mechanism of activation by activated factor XI. J Biol Chem 253:5946–5951

24. Di Scipio RG, Kurachi K, Davie EW (1978) Activation of human factor IX (Christmas factor). J Clin Invest 61:1528–1538

25. Davie EW, Fujikawa K, Kisiel W (1991) The coagulation cascade: initiation, maintenance, and regulation. Biochemistry 30:10363–10370

26. Pedicord DL, Seiffert D, Blat Y (2004) Substrate-dependent modulation of the mechanism of factor XIa inhibition. Biochemistry 43:11883–11888

27. Ogawa T, Verhamme IM, Sun MF, et al (2005) Exosite-mediated substrate recognition of factor IX by factor XIa: the factor XIa heavy chain is required for initial recognition of factor IX. J Biol Chem 280:23523–235230

28. Naito K, Fujikawa K (1991) Activation of human blood coagulation factor XI independent of factor XII: factor XI is activated by thrombin and factor XIa in the presence of negatively charged surfaces. J Biol Chem 266:7353–7358

29. Gailani D, Broze GJ (1991) Factor XI activation in a revised model of blood coagulation. Science 253:909–912

30. Gailani D, Broze GJ (1993) Effects of glycosaminoglycans on factor XI activation by thrombin. Blood Coagul Fibrinolysis 4:5–20

31. Broze GJ (1992) The role of tissue factor pathway inhibitor in a revised coagulation cascade. Semin Hematol 29:159–169

32. Von dem Borne PA, Meijers JC, Bouma BN (1995) Feedback activation of factor XI by thrombin in plasma results in additional formation of thrombin that protects fibrin clots from fibrinolysis. Blood 86:3035–3042

33. Von dem Borne PA, Bajzar L, Meijers JC, et al (1997) Thrombin-mediated activation of factor XI results in a thrombin-activatable fibrinolysis inhibitor-dependent inhibition of fibrinolysis. J Clin Invest 99:2323–2327

34. Butenas S, Veer Cvt, Mann KG (1997) Evaluation of the initiation phase of blood coagulation using ultrasensitive assays for serine proteases. J Biol Chem 272:21527–1533

35. Roberts HR, Monroe DM, Oliver JA, et al (1998) Newer concepts of blood coagulation. Haemophilia 4:331–334

36. Oliver JA, Monroe D, Roberts HR (1999) Thrombin activates factor XI on activated platelets in the absence of factor XII. Arterioscler Thromb Vasc Biol 19:170–177

37. Keularts IM, Zivelin A, Seligsohn U, et al (2001) The role of factor XI in thrombin generation induced by low concentrations of tissue factor. Thromb Haemost 85:1060–1065

38. Asakai R, Chung DW, Davie EW, et al (1991) Factor XI deficiency in Ashkenazi Jews in Israel. N Engl J Med 325:153–158
39. Wielders SJ, Beguin S, Hemker HC, et al (2004) Factor XI-dependent reciprocal thrombin generation consolidates blood coagulation when tissue factor is not available. Arterioscler Thromb Vasc Biol 24:1138–1142
40. Von dem Borne PA, Cox L, Bouma BN (2006) Factor XI enhances fibrin generation and inhibits fibrinolysis in a coagulation model initiated by surface-coated tissue factor. Blood Coagul Fibrinolysis 17:251–257
41. Wang X, Smith PL, Hsu MY, et al (2006) Effects of factor XI deficiency on ferric chloride-induced vena cava thrombosis in mice. J Thromb Haemost 4:1982–1988
42. Yamashita A, Nishihara K, Kitazawa T, et al (2006) Factor XI contributes to thrombus propagation on injured neointima of the rabbit iliac artery. J Thromb Haemost 4:1496–1501
43. Saunders RE, O'Connell NM, Lee CA, et al (2005) Factor XI deficiency database: an interactive web database of mutations, phenotypes, and structural analysis tools. Hum Mutat 26:192–198
44. Salomon, O Zivelin A, Livant T (2003) Prevalence, causes, and characterization of factor XI inhibitors in patients with inherited factor XI deficiency. Blood 101:4783–4788
45. Sun MF, Baglia FA, Ho D (2001) Defective binding of factor XI-N248 to activated human platelets. Blood 98:125–129
46. Zivelin A, Ogawa T, Bulvik S, et al (2004) Severe factor XI deficiency caused by a Gly555 to Glu mutation (factor XI-Glu555): a cross-reactive material positive variant defective in factor IX activation. J Thromb Haemost 2:1782–1789
47. McVey JH, Lal K, Imanaka Y, et al (2005) Characterisation of blood coagulation factor XI T475I. Thromb Haemost 93:1082–1088

Structural Insights into the Life History of Thrombin

JAMES A. HUNTINGTON

Summary. Thrombin is the ultimate coagulation factor. Not only is it the final protease generated by the blood coagulation cascade, it has more than 12 substrates and 5 cofactors. How thrombin specificity is directed during the four stages of hemostasis is of great interest to the medical community, as insufficient thrombin activity leads to bleeding and excessive activity results in thrombosis. Over the last three decades we have learned a great deal about how thrombin is generated and how it recognizes its several cofactors, substrates, and inhibitors. Although much has been inferred from biochemical studies, our current understanding is primarily based on numerous crystallographic structures of thrombin complexes. In this chapter I provide an overview of the multiple roles thrombin plays in the initiation, amplification, propagation, and attenuation phases of hemostasis, and describe how the special structural features of thrombin are exploited to achieve regulation and substrate selectivity.

Key words. Thrombin · Substrate · Recognition · Crystal structure · Exosite

Introduction

It is difficult to say exactly who "discovered" thrombin. In the mid-19th century Alexander Buchanan observed that the addition of a clot into serum initiated rapid clotting. The presence of a "fibrinoplastic" substance that converted fibrinogen to fibrin was later inferred by the Estonian physiologist Alexander Schmidt, who subsequently purified the material by alcohol precipitation. Schmidt's work culminated in the formulation of a theory close to what we today hold as true—that thrombin is formed from an inactive zymogen by cellular factors and subsequently converts fibrinogen into a fibrin clot [1]. Although Buchanan is often credited with the discovery of thrombin, this is due in large part to a posthumous campaign by an interested party [2]. It is more accurate to credit Schmidt with the initial description of thrombin, and his work over three decades (1860–1895) laid the experimental and theoretical foundation of modern hemostasis.

Multiple Functions of Thrombin in Hemostasis

Thrombin was initially described for its ability to stimulate the formation of a fibrin clot in blood ex vivo. Indeed, the widely accepted definition of thrombin is "a proteolytic enzyme that . . . facilitates the clotting of blood by catalyzing conversion of fibrinogen to fibrin" (Encyclopaedia Britannica, www.britannica.com). There is, however, a big difference between a blood clot formed in the test tube and one formed in blood vessels. Although thrombin activity is limited to generating fibrin in blood plasma ex vivo, it has multiple roles during the evolution of a blood clot in vivo. The best way to conceptualize the multiple roles of thrombin is to consider the four stages of hemostasis and the three cell surfaces involved (excellent recent reviews include [3–6]) (Fig. 1).

Initiation

Hemostasis is normally initiated by damage to the endothelial cell layer of a blood vessel. The exposed components of the extracellular matrix [primarily collagen and tissue factor (TF)] bind to circulating platelets and coagulation factors. Resting platelets adhere to the exposed collagen via platelet receptors GpVI and GpI (via von Willebrand factor) and become partially activated, resulting in limited degranulation and secretion of several factors (f), including fVa. At the same time, fibrinogen accumulates at the site of injury through interaction with platelet receptors. The proteolytic cascade, formally known as the extrinsic pathway, begins with the binding of circulating fVIIa to the exposed TF. This complex begins to convert fX to fXa (also fVII to fVIIa and fIX to fIXa), which then assembles with the secreted fVa to form the initial prothrombinase complex (Fig. 1a). Only small amounts of thrombin can be generated before the fVIIa–TF–fXa complex is inhibited by tissue factor pathway inhibitor (TFPI), and fXa alone is inhibited by antithrombin (AT, formerly known as ATIII), but some of the generated thrombin diffuses away from the subendothelial surface to adhered platelets. This thrombin binds to the platelet receptor GpIα and then cleaves protease-activated receptor 1 (PAR1) (Fig. 1a, right). The combined stimulation provided by the collagen receptor interactions and cleavage of PAR1 results in full activation of the platelets, leading to exposure of a negatively charged phosphatidylserine (PS)-rich surface, morphological changes, and full degranulation. Thrombin produced during the initiation phase also cleaves some of the accumulated fibrinogen to fibrin (Fig. 1b). The result is an initial hemostatic plug composed of fibrin and activated platelets. The PS-rich outer leaf of activated platelets is the membrane that supports the subsequent amplification and propagation phases of hemostasis.

Amplification

A small amount of thrombin is now associated with the surface of platelets on the nascent clot. It is critical to generate more thrombin to stabilize the clot; this is the role of the arm of the cascade previously called the intrinsic pathway. Cofactors V and VIII are now littering the surface of the platelet and are activated by thrombin to fVa and fVIIIa (Fig. 1c, right). Factor VIIIa binds to the small amount of fIXa (made during the initiation stage) to form the intrinsic tenase complex, which subsequently activates

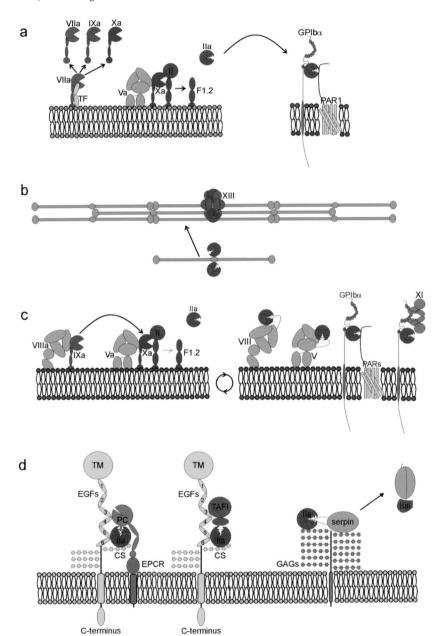

FIG. 1. Thrombin formation and activities throughout the hemostatic response. **a** Initiation takes place on two membrane surfaces, the subendothelium (*left*) and the resting platelet (*right*). (Membranes are shown as lipid bilayers with head groups colored according to type: subendothelium is pink; resting platelets have a green phosphatidylcholine-rich outer surface and an red phosphatidylserine-rich inner surface; activated platelets have a red outer leaf; and the intact endothelium is gray). Rupture of the endothelium exposes the tissue factor (*TF*)-bearing subendothelium. TF binds to circulating factor VIIa (fVIIa) to generate more fVIIa, fIXa, and fXa. Factor Xa then assembles with fVa to form the initial prothrombinase complex, which converts prothrombin (*II*) to thrombin (*IIa*). Some of the formed thrombin binds to platelets via receptor GPIbα and cleaves protease-activated receptor1 (*PAR1*), resulting in platelet activation. **b** Some of the thrombin generated during the initiation phase converts fibrinogen (pink rods) to fibrin; however, most fibrin formation occurs during the latter stages of the hemostatic response. The fibrin mesh is stabilized by the action of fXIIIa (green), which is formed by fibrin-assisted thrombin cleavage. **c** Amplification and propagation phases both take place on the activated platelet surface. Factors Va and VIIIa bind to the activated platelets to allow formation of the intrinsic tenase complex (fVIIIa/fIXa) and the prothrombinase complex. Thrombin generated at the amplification phase can then activate more fV and fVIII to up-regulate further thrombin formation. During the propagation phase, thrombin also cleaves substrates requiring higher thrombin levels, PAR4 and fXI, with the aid of the cofactor GPIbα. **d** Attenuation phase of the hemostatic response takes place on the intact endothelium adjacent to the clot. Thrombin binds to EGF domains 5 and 6 and sometimes the chondroitin sulfate moiety (*CS*) of thrombomodulin (*TM*, yellow). This event blocks recognition of prothrombotic factors and enhances recognition of protein C (green, *PC*). Thrombin-activatable fibrinolysis inhibitor (*TAFI*, pink) is also activated by the thrombin–TM complex. Finally, thrombin is irreversibly inhibited by members of the serpin family (*right*) when bound to the glycosaminoglycans (*GAGs*) lining the endothelium

fX on the PS-rich platelet surface (Fig. 1c, left). The resulting fXa binds to the thrombin-activated fVa to form the prothrombinase complex, which begins the large-scale conversion of prothrombin to thrombin. Thrombin has thus played a central role in setting the stage for its own mass production during the subsequent propagation phase.

Propagation

Nearly 95% of the thrombin that is produced during the course of a normal hemostatic response occurs during the propagation phase [7]. The prothrombinase complex rapidly converts prothrombin to thrombin, so the local concentration of thrombin becomes elevated. Fibrinogen circulates at a high concentration in the plasma and is rapidly converted to fibrin. Thrombin cleaves fibrinogen at two places, leading to linear and lateral polymerization. This mesh is stabilized by covalent crosslinking through the action of fXIII, and fibrinolysis is further inhibited by the action of TAFI, both of which are activated by thrombin (Fig. 1b,d). Thrombin continues to feed back to stimulate its own formation by activation of fV and fVIII, and now it can also cleave fXI (Fig. 1c, right). Factor XIa activates more fIX to reinforce generation of fXa, which then creates more thrombin. All the while, platelets continue to adhere to the mesh as coagulation rages. At some point, however, the procoagulant platelet-rich surface of the clot encounters intact endothelium, and the attenuation phase begins.

Attenuation

Most descriptions of hemostasis do not consider attenuation as a separate phase of the normal coagulation process, but without it we have thrombosis, not hemostasis. The endothelial cell surface has two main features that resist overgrowth of the clot: It expresses the integral membrane protein thrombomodulin (TM), and it is covered by highly sulfated proteoglycans (Fig. 1d). Thrombin binding to TM blocks further procoagulant substrates from being cleaved by thrombin and stimulates the activation of protein C (PC). Activated protein C (APC) attenuates thrombin production by cleavage inactivation of fVa and fVIIIa. The heparin-like glycosaminoglycans (GAGs) stimulate inhibition of thrombin by serpins (Fig. 1d, right), AT and heparin cofactor II (HCII). All of the thrombin generated during the propagation phase is either retained within the clot (bound to fibrin or platelets) or is inhibited by serpins and cleared from the circulation.

Thrombin Structure

Like all coagulation proteases, thrombin is generated from a zymogen form (prothrombin) by the action of another protease. Prothrombin is secreted from the liver as a single-chain glycoprotein of 579 amino acids, composed of a γ-carboxyglutamate (Gla) domain, two consecutive kringle domains, and a serine protease domain (Fig. 2). The Gla domain contains 10 γ-carboxylated glutamate residues and is required for the Ca^{2+}-dependent association of prothrombin to membrane surfaces (especially the PS-rich surface of activated platelets). The anticoagulant effect of the widely used drug warfarin is mainly due to suppression of γ-carboxylation of prothrombin [8], which limits prothrombin association to the prothrombinase complex on the surface of activated platelets. The function of the kringle domains is unknown, but they may play some role in docking to prothrombinase [9]. Thrombin is formed by the sequential cleavage of two peptide bonds by fXa: Cleavage at Arg320 results in the formation of meizothrombin, and the subsequent cleavage at Arg271 produces mature, active thrombin, which dissociates from the Gla and kringle domains (F1.2 fragment) [10].

This is a remarkable event in hemostasis, as thrombin is the only coagulation protease to be "liberated" from its membrane association domain upon activation. The result is a molecule that can diffuse between membranes and within the luminal space to encounter numerous substrates and cofactors. Other forms of thrombin, created by autolysis in vitro, have been described [11]. They include the most common form, α-thrombin, which is shorter on the N-terminus of the light chain by 13 residues and is otherwise indistinguishable from thrombin; and β and γ thrombins, which are cleaved within the heavy chain and have altered substrate recognition properties. However, the products of thrombin autolysis are unlikely to be formed in vivo and are thus not of any physiological relevance.

Mature thrombin is a two-chain protein with the 49-residue light chain associated to the protease domain by a disulfide bond and intimate noncovalent contacts. The light chain aids in the correct folding of thrombin, but has no known functional role [12]. The heavy chain of thrombin (259 residues) is larger than the corresponding sequence of trypsin by 35 residues owing to the presence of several "insertion loops." Of particular relevance to thrombin function is the 60-loop and the 147-loop (also

Prothrombin

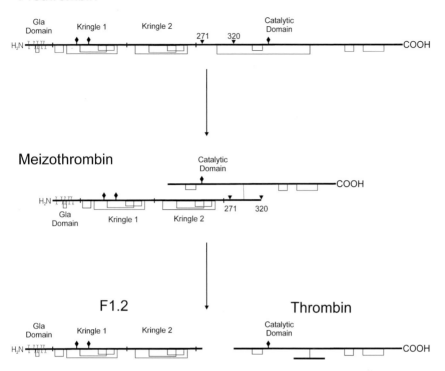

FIG. 2. Formation of thrombin from prothrombin proceeds via a single intermediate. Prothrombin consists of a Gla domain, two kringle domains, and a catalytic domain. *Top* Gla residues are indicated by γ, disulfide bonds by *connecting lines*, glycosylation sites by *diamonds*, and activation cleavage sites by *triangles*. The action of prothrombinase cleaves initially at the 320 site to form the intermediate meizothrombin (*middle*), followed by cleavage at the 271 site to release the final products thrombin and the F1.2 fragment (*bottom*). Thrombin is no longer associated with the membrane surface and is uniquely able to diffuse away to encounter substrates and cofactors at other sites

known as the γ-loop). Thrombin is usually numbered according to one of two schemes: sequential heavy chain numbering from 1 to 259, or, as used in this chapter, chymotrypsin template numbering starting the heavy chain at Ile16 [13] (prothrombin numbering is used when discussing thrombin formation, as in Fig. 2). Because thrombin has more residues than chymotrypsin, the insertion loops are lettered starting from "a." Thus, the 60-loop extends from 60a to 60i, and the γ-loop extends from 147a to 147e. The catalytic residues are conserved with all chymotrypsin family serine proteases and include the catalytic triad of His57, Asp102, and Ser195, as well as the critical Gly193 (forming the oxyanion hole with Ser195) and Asp194, which forms a salt bridge with Ile16 to activate the protease. Thrombin does not bind Ca^{2+} as does trypsin and chymotrypsin, but it does bind specifically to the monovalent cation Na^+ [14].

The first crystallographic structure of thrombin was published in 1989 [13]. Since then, more than 240 structures of thrombin have been deposited in the Protein Data

Bank (www.rcsb.org). It is safe to say that thrombin is one of the most thoroughly studied protein structures, and yet much remains to be discovered about how the features of thrombin are employed to give rise to its manifold functions in hemostasis. The regions on the surface of thrombin, which are normally discussed in relation to substrate and cofactor recognition, are the active site cleft, the Na$^+$ binding site, and the positively charged patches called anion-binding exosites I and II. Traditionally, representations of thrombin are oriented with the active site cleft facing so a bound substrate peptide would have its N-terminus on the left and its C-terminus on the right (Fig. 3). This standard orientation (also sometimes called the classic view or the Bode orientation) allows viewers to identify immediately the functional regions of thrombin [15], which are discussed in the following sections.

Active Site Cleft

Serine proteases of the chymotrypsin family are all composed of two β-barrel domains, which form a cleft at their interface (Fig. 3a). This cleft contains the catalytic machinery of the protease, the catalytic triad and the oxyanion hole and is thus known as the active site cleft. The mechanism of catalysis is understood in great detail and can be summarized as a potentiated nucleophilic attack of the side chain oxygen of Ser195 on the scissile peptide bond of a substrate. This is really the subject of a basic text on biochemistry, so for the purposes of this chapter suffice it to say that the precise orientation of these residues in a substrate-accessible cleft is required for an active protease. Surface representations of thrombin (Fig. 3b,c) reveal thrombin's deep active site cleft, which has even been described as a "canyon" [13]. The depth of the cleft is primarily due to the 60 and γ insertion loops, which lie just above and below the active site, respectively (Fig. 3). It has been demonstrated that the depth of the active site limits the accessibility of macromolecular substrates and inhibitors. An example of this is aprotinin (bovine pancreatic trypsin inhibitor), which binds to thrombin with low affinity. The crystal structure of aprotinin bound to thrombin shows that accommodation of this macromolecular inhibitor requires large-scale movement of the surface loops [16]. In this way the 60 and γ loops contribute to the specificity of thrombin by limiting potential substrates. The position of these loops also allows them to contribute directly to the specificity of interactions with substrates, with the 60-loop binding to residues N-terminal to the scissile bond and the γ-loop interacting with those residues on the C-terminal side. For macromolecular substrates, it is also common for these loops to be making exosite contacts (contacts not involving the substrate peptide and the active site cleft), which further contribute to specificity.

Substrate Sequence Preference

The convention for discussing the interaction between a substrate peptide and the active site of a protease is to number the substrate from the peptide bond that is attacked (the scissile bond) out toward its N- and C-termini [17]. The scissile bond is denoted P1-P1', and residues to either end are numbered consecutively. Normal substrates interact by docking residues stretching from P4 on the N-terminal side to P3' on the C-terminal side (Fig. 4). The pockets within the active site cleft that accommodate these residues are denoted S, so the P1 residue binds in the S1 pocket, and so

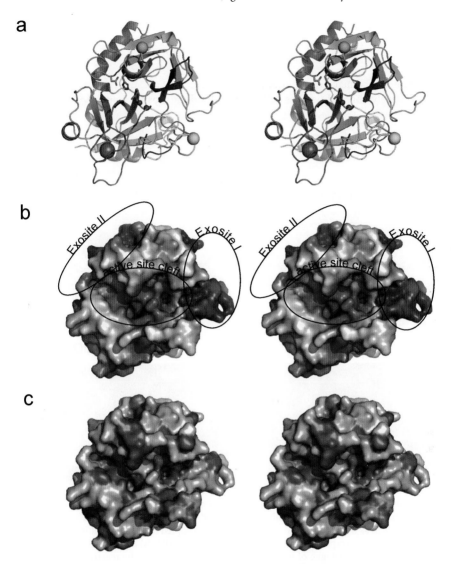

Fig. 3. Stereo views of thrombin structure shown in the standard orientation. **a** Ribbon diagram of thrombin colored from blue to red from the N- to the C-terminus for the heavy chain, with the light chain colored gray. In the standard orientation, the active site cleft is facing so a substrate would bind from left to right from its N-terminus to its C-terminus. The catalytic triad is shown as *rods*, and the oxyanion hole as *blue balls*. The 60 and γ-loops are indicated by *a cyan ball* and *a yellow ball*, respectively, and bound Na$^+$ is shown as a *purple ball*. **b** The surface electrostatic representation reveals the deep acidic active site cleft and basic exosites I and II (red for negative potential, blue for positive potential). **c** The same surface colored according to hydrophobic potential (green) reveals the hydrophobic nature of the active site cleft and exosite I

PPACK (1PPB)   PAR1 (1NRS)

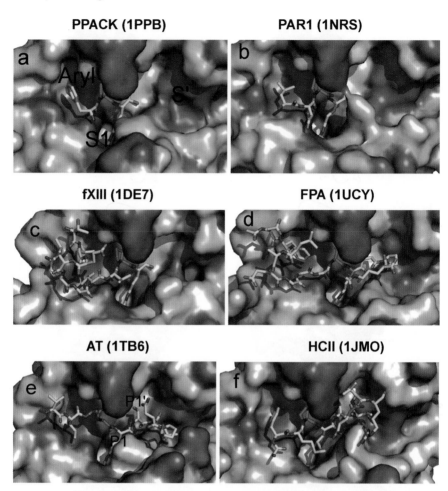

FIG. 4. Surface hydrophobic representations of thrombin's active site interactions with peptides. Crystallographic structures of thrombin bound to PPACK (**a**) (showing the aryl binding and S1 pockets, as well as indicating the S′ side, which is not occupied in most thrombin-peptide structures); a peptide corresponding to the N-terminus of the PAR1 (**b**) and a fXIII-derived peptide (**c**); an uncleavable analogue of fibrinopeptide Aα (FPA) (**d**); and the reactive center loops of serpins AT (**e**) (P1 Arg and P1′ Ser indicated) and HCII (**f**) illustrate the depth and plasticity of thrombin's active site cleft. Protein Data Bank accession codes are given in parentheses for each structure. Note that fXIII and FPA utilize the P9 residue to bind in the aryl-binding pocket, normally occupied by P4. This is achieved by a short helix in this region of the peptide substrate. The AT and HCII complexes are the only crystal structures showing significant P′ side interactions

TABLE 1. Natural substrate sequences of thrombin

Substrate	P4	P3	P2	P1	P1′	P2′	P3′
Fibrinogen (A)	Gly	Gly	Val	Arg	Gly	Pro	Arg
Fibrinogen (B)	Phe	Ser	Ala	Arg	Gly	His	Arg
Factor V (709)	Leu	Gly	Ile	Arg	Ser	Phe	Arg
Factor V (1018)	Leu	Ser	Pro	Arg	Thr	Phe	His
Factor V (1545)	Trp	Tyr	Leu	Arg	Ser	Asn	Asn
Factor VIII (372)	Ile	Gln	Ile	Arg	Ser	Val	Ala
Factor VIII (740)	Ile	Glu	Pro	Arg	Ser	Phe	Ser
Factor VIII (1689)	Gln	Ser	Pro	Arg	Ser	Phe	Gln
Factor XIII	Gly	Val	Pro	Arg	Gly	Val	Asn
PAR1	Leu	Asp	Pro	Arg	Ser	Phe	Leu
PAR4	Pro	Ala	Pro	Arg	Gly	Tyr	Pro
Factor XI	Ile	Lys	Pro	Arg	Ile	Val	Gly
Protein C	Val	Asp	Pro	Arg	Leu	Ile	Asp
TAFI	Val	Ser	Pro	Arg	Ala	Ser	Ala
Antithrombin	Ile	Ala	Gly	Arg	Ser	Leu	Asn
HCII	Phe	Met	Pro	Leu	Ser	Thr	Gln
PCI	Phe	Thr	Phe	Arg	Ser	Ala	Arg
Consensus	L/I/V/F	X	Pro	Arg	S/T/G/A	X	X

on. The properties of the active site cleft of thrombin favor basic residues at the P1 position and hydrophobic residues at P4. The preference for Arg at the P1 position is explained by a hydrophobic S1 pocket with an acidic base. The preference for hydrophobic residues at P4 is similarly explained by the hydrophobic nature of the S4 pocket (also known as the aryl-binding pocket). Many studies have been undertaken to determine the substrate preference of thrombin in the context of small peptides, and it is possible to obtain a consensus "best" substrate for thrombin of Leu-X-Pro-Arg-Ser-X-X [18–20]. However, it is important to note that most of the natural substrates of thrombin do not have this sequence, and that some substrates have sequences that are poorly recognized by thrombin (Table 1). This suggests that specificity is only partially determined by substrate sequence. Indeed, it may be that these poor substrate sequences exist so that cleavage depends on additional factors (cofactors), as is discussed later.

Na+ Binding Site

The active site of thrombin is "formed" after cleavage of the Arg15-Ile16 bond of prothrombin (320 in prothrombin numbering). The new positively charged amino group on the N-terminus of the heavy chain (Ile16) inserts into the body of thrombin and forms a critical salt bridge with Asp194 in the active site loop. This event stabilizes the active conformation of the catalytic residues and forms the oxyanion hole. Another site formed by activation of thrombin is the Na^+ binding site. Na^+ has been shown to coordinate to the main chain oxygens of Arg221a, Lys224, and four water molecules [21] (Fig. 5a). The functional consequences of Na^+ binding have been described in great detail in many articles, principally from the Di Cera group in St. Louis (for review see [22, 23]). The Na^+-free form of thrombin is generally less active

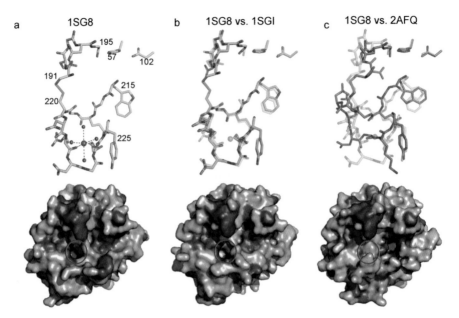

FIG. 5. Structural consequences of Na$^+$ coordination. In each panel, the top portion shows the Na$^+$ binding loop, the active site loop, and the catalytic triad as *rods*; and the bottom portion is a hydrophobic surface representation of thrombin, with a *red circle* indicating the S1 pocket. **a** Thrombin coordinates Na$^+$ (purple) via two main chain oxygens (Arg221a, Lys224) and four conserved waters (*red balls*, bonds indicated by *broken lines*). The Na$^+$ binding loop (Trp215–Tyr225) is linked directly to the active site loop (Cys191–Ser195) by the Cys191–Cys220 disulfide bond (green). Catalytic residues His57 and Asp102 are shown for reference. The structure 1SG8 is one of the many acknowledged to be in the "fast" conformation. The surface representation (below) reveals an open active site cleft with a fully formed S1 pocket (*circled in red*). **b** The same crystal form grown in the absence of Na$^+$ (1SGI) shows no significant change in conformation in the Na$^+$ or active site loops (1SGI in cyan and 1SG8 in semitransparent yellow; the position of Na$^+$ is indicated), and the S1 pocket is still fully formed. This is a putative "slow" form of thrombin, based solely on the fact that the crystal was grown without Na$^+$. **c** Another Na$^+$-free structure of thrombin (2AFQ) reveals large-scale conformational changes in the Na$^+$-binding and active site loops (2AFQ in magenta and 1SG8 in semitransparent yellow). The active site cleft is relatively open, but the S1 pocket is not formed

than the Na$^+$-bound form; these forms have thus become known as the "slow" and "fast" forms, respectively. The slow form is thought to exhibit anticoagulant activities because it is less able to cleave the procoagulant substrates of thrombin while maintaining its ability to cleave PC in the presence of TM. The anticoagulant effect of slow thrombin has been demonstrated in vivo with variants of thrombin engineered to exhibit reduced or abrogated ability to bind Na$^+$ [24–26].

The structural consequences of Na$^+$ binding are currently being debated in the literature. Several crystallographic structures have been solved in the absence of Na$^+$, revealing many possible conformational differences with the Na$^+$-bound structure. They range from a zymogen-like state with a collapsed active site pocket and noncatalytic active site architecture [27, 28] (Fig 5c) to subtle movements of water molecules and the side chain of Ser195 [29, 30] (Fig. 5b). Whatever the conformation of the slow form, its physiological relevance depends on the Na$^+$-free state being significantly populated in the blood. The affinity of thrombin for Na$^+$ has been reported to be highly dependent on temperature; at room temperature, the K$_d$ is ~20 mM, and at 37°C it is >100 mM [31]. Any argument that Na$^+$ is an allosteric cofactor of thrombin depends on such a high K$_d$ because the Na$^+$ concentration is tightly regulated at 143 mM. Another way of conceptualizing this issue is to consider Na$^+$ a ligand necessary for stabilizing the correct, active thrombin fold. Several other hemostatic serine proteases bind to metal ions (including Na$^+$) in their catalytic domains to confer higher activity; however, the metal ion concentrations and K$_d$ values predict that all will be fully coordinated in blood. More work is required to determine whether the slow form of thrombin is more than just an interesting in vitro artefact. The purpose of describing Na$^+$ binding briefly in this chapter is to make the reader aware that thrombin does contain Na$^+$ as part of its structure, and that an active and interesting debate exists about the conformational consequences of Na$^+$ binding and its potential physiological relevance.

Anion Binding Exosites

The basic patches on the surface of thrombin were first designated anion-binding sites I and II by Wolfram Bode and colleagues in their 1992 article detailing the analysis of the refined thrombin structure [15]. They argued that the anionic surfaces of proteins that bind to these regions need not form direct contacts to "experience the effect of this positive field at some distance." This phenomenon, "electrostatic steering" [32], could thus directly influence the rate at which substrates diffuse toward thrombin and also exercise an important indirect effect on which surface loops could access the active site cleft. As with the active site and the Na$^+$ biding site of thrombin, the exosites are also "formed" upon conversion from prothrombin. Exosite I in prothrombin binds to ligands with significantly reduced affinity due to conformational differences with exosite I of active thrombin [33]. Exosite II does not undergo a conformational change upon activation; rather, it is blocked in prothrombin by the F1.2 fragment [34]. Thus, allosteric and steric mechanisms are responsible for the formation of these two critical thrombin exosites.

Exosite I is particularly well positioned to influence binding in the active site and thus thrombin specificity (Fig. 3). It is located just to the right of the active site and is contiguous with the region that binds to the P$'$ side of the substrate. The residues that make up exosite I and that are of principal importance to substrate and cofactor binding can be chosen based on crystal structures of various complexes (TM [35], HCII [36], fibrin E domain [37]). They include Phe34, Lys36, Ser36a, Pro37, Gln38, Leu65, Arg67, Ser72, Thr74, Arg75, Tyr76, Arg77a, Asn78, Lys81, Ile82, Met84, and Lys110. One sees from the residues that make up exosite I that only a fraction is positively charged (5/17). Indeed, most of the binding energy for exosite I interactions is conferred by hydrophobic contacts. Surface representations of exosite I (shown in the top portion of Fig. 6) illustrate its dual electrostatic and hydrophobic nature.

Exosite II is farther away from the active site and is located on the top left-hand side in the standard orientation (Fig. 3); therefore, its participation in substrate

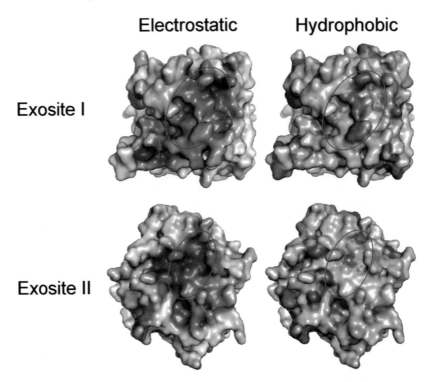

FIG. 6. Exosite properties of thrombin. Thrombin is rotated from the standard orientation to reveal the electrostatic (*left*) and hydrophobic (*right*) properties of exosites I (*top*) and II (*bottom*). As suggested by the ligands, which bind specifically to either site, exosite I is basic and hydrophobic in nature, and exosite II is primarily basic. The principal ligand-binding regions are indicated by the *red ovals*

recognition by thrombin is necessarily indirect. It is the more basic of the two exosites (Fig. 6, bottom) and was originally identified as the putative heparin-binding site. The residues that compose exosite II can be identified by analyzing the crystal structures of thrombin bound to exosite II ligands (GPIbα [38], heparin [39]). They are His91, Arg93, Arg101, Arg126, Arg165, Asn179, His230, Phe232, Arg233, Leu234, Lys235, Lys236, and Lys240. Thus, exosite II interactions primarily involve basic residues (10/13), with only minimal involvement of hydrophobic residues (2/13). There is convincing structural and biochemical evidence that anionic ligands bind with exclusivity to either exosite I or II, never to both sites. Apparent exceptions are usually due to artifactual crystal contacts. Recently, a general rule for determining which site will interact with a particular peptide was derived based on known interactions [40]. As one would predict, the ratio of basic-to-hydrophobic residues in the ligand sequence determines the binding exosite, with hydrophobic content corresponding to exosite I binding. All of the thrombin activities relevant to hemostasis require the involvement of at least one of the anion-binding exosites; most require both. Thus, determining the exosite interactions is crucial to understanding how thrombin activity is directed throughout the hemostatic response.

Structural Basis of Thrombin Activities

Much of what is known about how thrombin interacts with cofactors and substrates has been deduced from biochemical characterization of libraries of thrombin variants (e.g., see [41]). In addition to this inferred interaction data, several important crystal structures of thrombin in complex with cofactors, substrates, and inhibitors have been solved (for review see [40]). To date, there is no structure of thrombin bound to a physiological protein substrate in its recognition complex (Michaelis complex), with the exception of thrombin bound to inhibitors AT and HCII. However, all thrombin–cofactor complexes have been determined by X-ray crystallography (for review see [42]). Just how the active site and exosites I and II are involved in recognizing substrates in the initiation, amplification, propagation, and attenuations phases is discussed next.

Initiation

When thrombin is activated on the subendothelial surface, it finds itself surrounded by fibrinogen (Fig. 7) and platelets, which have accumulated at the site of injury. Each fibrinogen monomer is cleaved by thrombin at four sites (fibrinogen is a dimer containing two Aα and two Bβ cleavage sites) in a cofactor-independent manner. All cleavage sites are accessible to the two thrombin molecules docked to the central E-domain of fibrinogen via extensive exosite I interactions (Fig. 7). Indeed, exosite I was formerly known as the fibrinogen recognition exosite. The recently solved structure of two thrombin monomers bound to the E-domain of fibrinogen [37] suggests a plausible model of how cleavage at all four positions can occur without thrombin disengaging from its exosite I contacts [43].

The thrombin generated during the initiation phase also finds its way to the platelet surface, where it encounters its tight-binding cofactor glycoprotein Ibα (GPIbα). GPIbα is thought to serve as a cofactor for all of the substrate cleavage reactions of thrombin that take place on the platelet surface. The interaction between thrombin and GPIbα has been investigated by mutagenesis and competition studies [44, 45] and has been shown to involve exosite II of thrombin and residues 271–284 of GPIbα (Fig. 8b). Two crystal structures of the thrombin-GPIbα fragment complex revealed quite different exosite II interactions [38, 46]. Based on the number of salt bridges, it is likely that the structure described by Dumas and colleagues [38] best represents the binding mode in solution. It is important that binding is exclusively through exosite II as it leaves exosite I available to interact directly with substrates. The first substrate that thrombin cleaves on the platelet surface is PAR1. The activation peptide of PAR1 contains a negatively charged P′ region that binds to exosite I [47] (Fig. 8a). Although there is no single structure showing the active site and exosite I interactions, a model has been made from two partial structures [48] and suggests that the active site interactions are contiguous with those in exosite I [40]. The role of the exosite II interaction with cofactor GPIbα is to localize thrombin to the platelet surface, and the engagement of exosite I by the C-terminal part of the activation peptide serves to

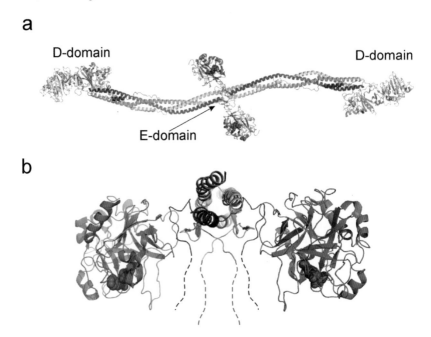

a

D-domain   D-domain

E-domain

b

FIG. 7. Fibrinogen recognition by thrombin. **a** Fibrinogen is a dimer of trimers (α, β, and γ chains colored blue, green, and yellow, respectively), forming a central E-domain and two terminal D-domains. Thrombin (magenta, with bound PPACK in red space-filling to indicate the position of the active site) binds to the central E-domain of fibrinogen via exosite I, with the stoichiometry indicated (two thrombin monomers per fibrinogen monomer). **b** Close-up view of the interaction between the E-domain and two thrombin molecules is rotated to reveal the proximity of the Aα and Bβ cleavage sites (*dashed lines*) to the active sites of thrombin

FIG. 8. Structures of thrombin complexes illustrate how substrates and cofactors compete for exosites I and II during different phases of hemostasis. Exosite I and II orientations correspond precisely to the electrostatic and hydrophobic surfaces shown in Fig. 6. The surface representation of thrombin is colored yellow to indicate the regions within 4.5 Å of the ligand. **a** Cleavage activation of PAR1 (magenta rods) depends on the extension of the active site interactions onto exosite I. **b** This event is supported by the localization of thrombin to the platelet surface via exosite II interactions with a portion of GPIbα (green rods). **c** Fibrinogen cleavage requires exosite I docking of thrombin to the fibrin E-domain (ribbons colored as in Fig. 7). **d** Thrombin is also capable of binding with high affinity to the fibrinogen γ chain when the γ′ sequence is found (on about 15% of circulating fibrinogen) via exosite II. **e** Binding of thrombin to the endothelial cell surface receptor TM (semitransparent cyan ribbons) is primarily hydrophobic in nature and is thus mediated via exosite I. It should also be noted that TM, like many exosite I ligands, encroaches on the S′ region of the active site cleft, thus influencing substrate specificity. **f** Interaction between thrombin and heparin (cyan rods) is mediated via exosite II

Exosite I

Exosite II

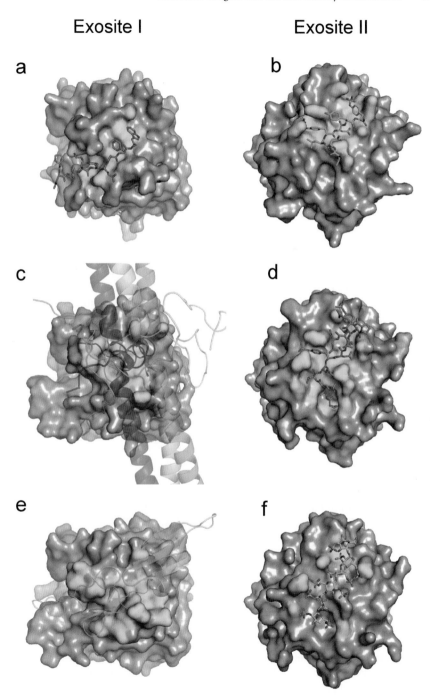

a

b

c

d

e

f

orient the scissile bond of PAR1 in the active site cleft of thrombin. Both recognition events help small amounts of thrombin to cleave PAR1 efficiently at the initiation of hemostasis, and the localization of thrombin via its cofactor interaction ensures that multiple rounds of receptor cleavage are possible before thrombin can diffuse away.

Amplification

The surface of activated platelets in the initial hemostatic plug are decorated with procofactors V and VIII, and progression of hemostasis depends on their activation by thrombin. Factors V and VIII share a homologous structure composed of domains A1-A2-B-A3-C1-C2 and are each cleaved in three positions by thrombin to produce the active cofactors (Table 1). Although the domain structures are similar, there are differences in the sites cleaved and how thrombin recognizes the individual procofactors. Thrombin exosite I is crucial for recognition and cleavage of all three sites on fV and fVIII, as demonstrated using exosite I competitors and altered or mutated [49] thrombins [49–55]. Binding to exosite I may be mediated by the A1, A2, or C2 domains of the procofactors [56–58]. The role of exosite II in procofactor conversion is less well established. Exosite II may contribute to enhancing the rate of one of the three fV cleavage events and probably all three steps of fVIII activation [52], but activation does not appear to depend critically on these interactions. There is still much to learn about how thrombin recognizes fV and fVIII. To date, there is no crystal structure of either procofactor (or active cofactor), and we are likely to require several crystal structures of thrombin–procofactor complexes before we understand how the structural features of thrombin are employed to confer specificity for all of the cleavage steps involved in procofactor activation. What we can say with some certainty is that exosite I is critical to activate both procofactors and that exosite II is more important for activation of fVIII than it is for fV.

Propagation

At this stage the platelet surface is mediating the production of large amounts of thrombin. Generated thrombin continues to cleave fibrinogen into fibrin, and it is here that most fibrin is formed. In addition to the exosite I interaction between thrombin and the E-domain of fibrin(ogen), thrombin is also capable of binding to the γ-chain of fibrin when the alternatively spliced sequence γ' is found (in about 15% of chains) [59]. The γ' peptide binds to thrombin via exosite II [60] with a dissociation constant in the nanomolar range [61]. The fibrin meshwork provides the cofactor for the thrombin activity responsible for stabilizing the fibrin polymer itself. Factor XIIIa is a transglutaminase that covalently crosslinks Lys and Gln residues from different fibrin monomers at multiple positions (for a thorough review see [62]). This reaction takes place on the surface of fibrin, with thrombin bound via exosite I to the E-domain, as discussed earlier (Figs. 1b, 7, 8c). Exosite II does not appear to be involved in recognition of fXIII [63]. Fibrinolysis is further inhibited by the action of thrombin through cleavage activation of TAFI [64]. TAFIa is a metalloprotease that removes the C-terminal Lys residues from fibrin, thereby preventing the association of fibrinolytic proteases tPA and plasmin. Somewhat paradoxically, TAFI activation is significantly dependent on the cofactor TM [65]. TM binds to exosite I of thrombin (discussed below in greater detail) via epidermal growth factor (EGF) domains 5 and 6 (Fig. 8e), and it has been demonstrated that EGF domain 3 of TM directly interacts with TAFI to stimulate activation by thrombin [66] (Fig. 1d). No other exosite interactions are expected; but again, no crystal structure of TAFI or its complex with thrombin–TM have been solved.

Thrombin cleavage of PAR1 continues on the surface of platelets aided by the exosite II-mediated association of thrombin with the platelet surface via GPIbα. As mentioned above, PAR1 contains an acidic C-terminal region that binds to exosite I of thrombin. The other PAR on platelets is PAR4 [67]. PAR4 cleavage is not aided by an exosite I interaction with thrombin [68], and thus cleavage does not occur until high concentrations of thrombin are obtained [69]. Another substrate dependent on high thrombin concentrations is fXI [70], an unusual dimeric serine protease that generates fIXa. Activation of fXI takes place on the platelet surface, where it along with thrombin binds to the cofactor GPIbα [71] (Fig. 1c). Because of the involvement of GPIbα, exosite II is involved indirectly in fXI activation, but exosite I residues are also involved through direct interaction with fXI (likely mediated by the A1 domain of fXI) [41, 72]. Although there is a crystal structure of the zymogen form of fXI [73] (and of the isolated the protease domain [74]), there is little information regarding how thrombin recognizes fXI or of the role played by the cofactor GPIbα. It is even unclear if both of the protease domains in dimeric fXI require activation for proper functioning. (Note: since the writing of this chapter, several papers including references 41 and 71 have been partially retracted. It is thus unlikely that GPIbα is actually a cofactor for thrombin activation of fXI.)

Attenuation

Normal hemostasis requires that clotting be limited to the site of vascular damage and that the clot does not grow so large as to occlude the blood vessel. The ramped-up thrombin generation experienced during the propagation phase must eventually be stopped, and the thrombin that has already formed must be inhibited and cleared from the circulation. These events primarily occur on the surface of the adjacent intact endothelium, where two cofactors lie in wait for any active thrombin: TM and GAGs (Fig. 1d). TM is an integral membrane protein composed of an N-terminal lectin-like domain, six consecutive EGF-like domains, a glycosylated Ser/Thr-rich region, a single transmembrane helix, and a small C-terminal cytoplasmic tail (for reviews see [75–77]). Thrombin binds to TM with very high affinity (dissociation constant ~30 nM) via EGF domains 5 and 6 [78]. In about 30% of TM molecules a polysulfated chondroitin sulfate (CS) moiety is attached to the Ser/Thr region [79], increasing its affinity for thrombin by ~10-fold [80]. TM is found throughout the vasculature; but its density and fraction containing the CS moiety is dependent on the type of blood vessel [79]. The structure of TM (EGF domains 4–6) bound to thrombin, published in 2000 [35], shows a typical exosite I interaction composed primarily of hydrophobic contacts and a single salt bridge (Fig. 8e). This tight interaction with exosite I effectively blocks the prothrombotic activity of thrombin, as it can no longer cleave fibrinogen, PAR1, fV, fVIII, or any of the other substrates of thrombin that rely on exosite I interactions. The substrate specificity of thrombin is thus effectively altered by TM so it preferentially cleaves PC. APC is a serine protease that, with some help from protein S, inactivates fVa and fVIIIa, effectively shutting down the propagation phase [76]. TM aids in thrombin recognition of PC mainly by co-localization and substrate presentation. PC binds to an endothelial cell surface receptor (EPCR) [81, 82] so it is bound to the same endothelial cell surface as TM-bound thrombin [83]. PC also binds to the fourth EGF domain of TM [84] and effectively presents the activation loop to the active site of thrombin bound to EGF domains 5 and 6 (Fig. 1d). It is unlikely that any direct exosite contacts between PC and thrombin contribute to recognition, as exosite I is always tightly bound to EGF domains 5 and 6, and exosite II is occasionally occupied by the CS moiety of TM. Mutagenesis studies support this conclusion [85].

The other cofactor for the attenuation phase of hemostasis is found on the proteoglycans lining the surface of endothelial cells. Of particular relevance to thrombin inhibition are the GAGs heparan sulfate (HS, a close cousin of heparin) and dermatan sulfate (DS) (for recent review see [86] and [87]). The crystal structure of thrombin bound to a heparin fragment [39] confirmed that exosite II is the binding site for HS and DS (and also the CS moiety on TM) (Fig. 8d). The interaction is primarily electrostatic in nature, with some additional hydrogen bonds. HS is a cofactor for the recognition of AT by thrombin, and DS is a cofactor for the recognition of HCII by thrombin [87]. How recognition is conferred differs for the two inhibitors, but it is thought that both thrombin and the serpin must bind to the same GAG chain, although it may be possible that they bind to adjacent chains in the physiological context. Crystal structures of both serpin-thrombin recognition complexes have been solved [36, 88] (Fig. 9). AT makes contacts with the 60-, γ-, and Na$^+$-binding loops of thrombin; and exosite II is involved in binding to heparin (Fig. 9b). Exosite I, however, is not involved in the recognition of thrombin by AT. This contrasts with the mechanism used by HCII. HCII has an 80-residue N-terminal loop that contains an anionic region with two sulfated tyrosines. This "hirudin-like" region is released from contacts with the body of HCII upon DS binding and subsequently recruits thrombin through interaction with exosite I (Fig. 9c,d). This interaction is critical for thrombin recognition of HCII because the P1 residue of HCII is the nonconsensus amino acid Leu (Table 1). Thus, thrombin recognition of HCII utilizes exosite I to bind to the hirudin-like domain and exosite II for interaction with DS (although the DS interaction is not seen in the published structure).

Another circulating serpin capable of inhibiting thrombin is protein C inhibitor (PCI) [89]. As the name suggests, PCI inhibits APC, but it is also the physiological inhibitor of thrombin bound to TM. TM accelerates inhibition of thrombin by PCI by two orders of magnitude, probably by making direct contacts to PCI via EGF domains 4 and/or 5 [90].

Inhibition and Clearance

The thrombin produced during the hemostatic response must eventually be removed from the circulation. It is thought that a fraction of this thrombin is sequestered in the clot itself [91] and therefore inaccessible to physiological inhibitors. Eventually, however, this material must also be inhibited, as the clot dissolves during wound healing. The inhibitors of thrombin (AT, HCII, and PCI, discussed above) are all members of the serpin family of serine protease inhibitors (see [92–94] and other reviews). The conserved serpin mechanism of inhibition involves dramatic conformational change and the partial unfolding of the protease in the final complex [95] (Fig. 10).

Several good review articles deal with the serpin mechanism in detail (e.g., [96]), but it can be described briefly as a two-step reaction (Fig. 10a). The first step is the

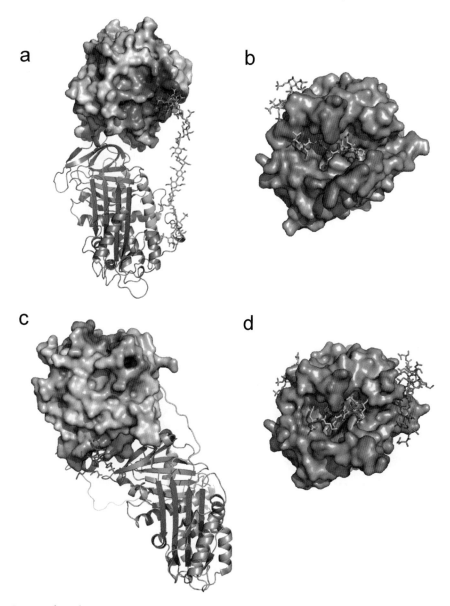

FIG. 9. Thrombin recognition by serpins antithrombin (AT) and heparin cofactor II (HCII) relies on extensive active site and exosite contacts. **a** The structure of the complex between AT (ribbon diagram, with β-sheet A in red; reactive center loop, orRCL, in yellow; heparin-binding helix D in cyan), thrombin (electrostatic surface representation), and heparin (green rods) illustrates how a single heparin chain bridges AT to thrombin to form the Michaelis complex. **b** Thrombin in the standard orientation is colored orange to illustrate regions within 4.5 Å of AT or heparin. The surface representation reveals extensive interactions in and around the active site clef (the RCL of AT in yellow rods) and the remote binding to heparin (cyan rods) via exosite II. **c** Thrombin (electrostatic surface representation) docks in a different position on HCII (acidic tail colored magenta). **d** Thrombin (shown as in b) binds to the RCL of HCII in a manner similar to that of AT but with significantly less contact surface. Specificity is largely determined by the exosite I interaction with the acidic tail (magenta rods). Because a glycosaminoglycan is likely to be involved in the physiological setting, heparin is shown binding to exosite II, although it was not seen in the crystal structure

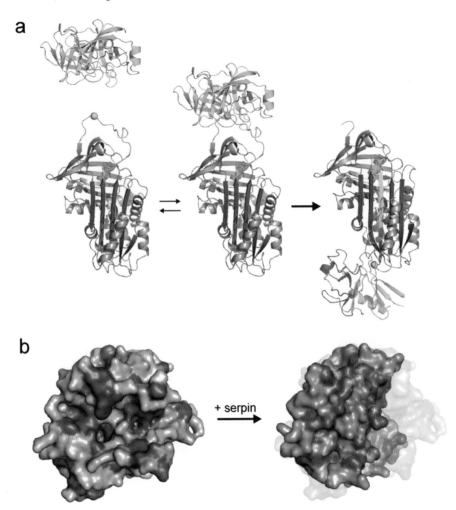

FIG. 10. Serpin mechanism of protease inhibition and its effect on the structure of thrombin. **a** Serpins inhibit serine proteases via a minimal two-step mechanism. In the first step (*left*) the protease (cyan) recognizes its cognate sequence within the RCL (yellow) of the serpin (β-sheet A is red). Of principal importance for this recognition event is the P1 residue (depicted as a yellow ball). The rate of formation and stability of this recognition complex (also known as the Michaelis complex, *center*) determines whether proteolysis of the RCL can progress. At the acyl-enzyme intermediate step, the RCL is rapidly incorporated into the β-sheet A (*right*). This event stabilizes the serpin and prevents deacylation by rearranging the catalytic loop of the protease. As a consequence, the protease is destabilized (colored according to B-factor, from blue to red at high values), and about 40% of the structure becomes disordered. **b** The effect of thrombin inhibition by serpins. *Left* Hydrophobic surface representation of native thrombin (as in Fig. 3c). *Right* A model of thrombin in complex with a serpin (same orientation). The active site and the Na⁺-binding site are altered, and exosite I has been unfolded. A semitransparent surface representation of native thrombin is overlaid with the surface of crushed thrombin to illustrate the amount of structure lost by serpin complexation

recognition of a solvent-exposed loop on the serpin by a target protease. This involves typical substrate-like docking of a sequence containing the scissile P1-P1' bond into the active site of the protease (e.g., Fig. 4e,f) but, as described above, can also involve exosite contacts (e.g. Fig. 9b,d). The first step, formation of the Michaelis complex, typically determines the rate of complex formation and thus specificity [97]. The second step begins when the protease cleaves the scissile bond and proceeds to the acyl-enzyme intermediate, where the P1' has detached from the P1 residue and an ester bond exists between the catalytic Ser195 and the carbonyl carbon of the P1 on the serpin. Before deacylation can occur, the serpin incorporates the N-terminal portion of the loop containing the P1 residue into the center of a large β-sheet. This conformational change flings the protease from one pole of the serpin to the other and distorts the structure of the protease to prevent deacylation from occurring. The structure of the serpin–protease complex [95] suggests that about 40% of the heavy chain of thrombin would become reordered or unfolded when inhibited by serpins, including residues 16–41, 60–84, 94–98, 110–120, 139–156, 186–199, and 213–224. Thus, inhibition of thrombin by a serpin would effectively destroy the active site, the Na^+ binding site, and exosite I, whereas exosite II would remain intact. Biochemical studies support this prediction by demonstrating a concomitant loss of functional active site with loss of ability to bind to ligands via exosite I but retention of the ability to bind exosite II ligands [98, 99]. The effect on thrombin is nicely illustrated by comparing the surface representation of native thrombin to that of a model of thrombin inhibited by a serpin (Fig. 10b). The functional consequence of the destruction of exosite I is that thrombin is released from its cofactor- and substrate-binding sites; this is of particular relevance to fibrin and TM-bound thrombin. Once released from the cell surface, the serpin–thrombin complex is rapidly cleared from the circulation primarily by liver-based receptors of the LRP family [96]. The conformational change in thrombin induced by complexation with a serpin appears to be critical for recognition by low density lipoprotein receptor-related protein (LRP) because the complex is cleared much more rapidly than either of the individual components [100–102].

Conclusion

Thrombin has special structural features that are used by several substrates and cofactors in a cell surface-dependent fashion. The active site, the Na^+ binding site, and exosites I and II are cryptic in prothrombin and are formed or revealed by the action of the prothrombinase complex. The limited thrombin produced during the initiation phase is able to active some platelets to begin the amplification phase. At this stage, thrombin begins to feed back to stimulate its own formation, leading to the propagation phase where most of the thrombin is generated. Here thrombin cleaves multiple substrates to extend and reinforce the clot. When the clot encounters the intact vascular bed, thrombin actively participates in the attenuation of coagulation and eventually its own inhibition by serpins. Complexation by serpins destroys exosite I, the active site, and the Na^+ binding site, and reveals a cryptic LRP recognition site to allow rapid clearance from the circulation. Thrombin thus has a complex life history with several chapters, multiple characters, and conflicting subplots, culminating in a dramatic and violent ending.

References

1. Owen, CAJ (2001) Prothrombin. In: Nichols WL, Bowie EJ (eds) A history of blood coagulation. Mayo Foundation for Medical Education and Research, Rochester, MN, pp 27–35
2. Marcum, JA (1998) Defending the priority of "remarkable researches": the discovery of fibrin ferment. Hist Philos Life Sci 20:51–76
3. Davie EW, Kulman JD (2006) An overview of the structure and function of thrombin. Semin Thromb Hemost 32(Suppl 1):3–15
4. Lane DA, Philippou H, Huntington JA (2005) Directing thrombin. Blood 106:2605–2612
5. Monroe DM, Hoffman M (2006) What does it take to make the perfect clot? Arterioscler Thromb Vasc Biol 26:41–48
6. Jesty J, Beltrami E (2005) Positive feedbacks of coagulation: their role in threshold regulation. Arterioscler Thromb Vasc Biol 25:2463–2469
7. Mann KG, Brummel K, Butenas S (2003) What is all that thrombin for? J Thromb Haemost 1:1504–1514
8. Furie B, Liebman HA, Blanchard RA, et al (1984) Comparison of the native prothrombin antigen and the prothrombin time for monitoring oral anticoagulant therapy. Blood 64:445–451
9. Deguchi H, Takeya H, Gabazza EC, et al (1997) Prothrombin kringle 1 domain interacts with factor Va during the assembly of prothrombinase complex. Biochem J 321 (Pt 3):729–735
10. Bianchini EP, Orcutt SJ, Panizzi P, et al (2005) Ratcheting of the substrate from the zymogen to proteinase conformations directs the sequential cleavage of prothrombin by prothrombinase. Proc Natl Acad Sci U S A 102:10099–10104
11. Boissel JP, Le Bonniec B, Rabiet MJ, et al (1984) Covalent structures of beta and gamma autolytic derivatives of human alpha-thrombin. J Biol Chem 259:5691–5697
12. De Cristofaro R, Akhavan S, Altomare C, et al (2004) A natural prothrombin mutant reveals an unexpected influence of A-chain structure on the activity of human alpha-thrombin. J Biol Chem 279:13035–13043
13. Bode W, Mayr I, Baumann U, et al (1989) The refined 1.9 A crystal structure of human alpha-thrombin: interaction with D-Phe-Pro-Arg chloromethylketone and significance of the Tyr-Pro-Pro-Trp insertion segment. EMBO J 8:3467–3475
14. Di Cera E, Guinto ER, Vindigni A, et al (1995) The Na$^+$ binding site of thrombin. J Biol Chem 270:22089–22092
15. Bode W, Turk D, Karshikov A (1992) The refined 1.9-A X-ray crystal structure of D-Phe-Pro-Arg chloromethylketone-inhibited human alpha-thrombin: structure analysis, overall structure, electrostatic properties, detailed active-site geometry, and structure-function relationships. Protein Sci 1:426–471
16. Van de Locht A, Bode W, Huber R, et al (1997) The thrombin E192Q-BPTI complex reveals gross structural rearrangements: implications for the interaction with antithrombin and thrombomodulin. EMBO J 16:2977–2984
17. Schechter I, Berger A (1967) On the size of the active site in proteases. I. Papain. Biochem Biophys Res Commun 27:157–162
18. Harris JL, Backes BJ, Leonetti F, et al (2000) Rapid and general profiling of protease specificity by using combinatorial fluorogenic substrate libraries. Proc Natl Acad Sci U S A 97:7754–7759
19. Petrassi HM, Williams JA, Li J, et al (2005) A strategy to profile prime and non-prime proteolytic substrate specificity. Bioorg Med Chem Lett 15:3162–3166
20. Ohkubo S, Miyadera K, Sugimoto Y, et al (2001) Substrate phage as a tool to identify novel substrate sequences of proteases. Comb Chem High Throughput Screen 4:573–583

21. Zhang E, Tulinsky A (1997) The molecular environment of the Na$^+$ binding site of thrombin. Biophys Chem 63:185–200
22. Page MJ, Di Cera E (2006) Is Na$^+$ a coagulation factor? Thromb Hemost 95:920–921
23. Di Cera E, Page MJ, Bah A, et al (2007) Thrombin allostery. Phys Chem Chem Phys 9:1291–1306
24. Tsiang M, Paborsky LR, Li WX, et al (1996) Protein engineering thrombin for optimal specificity and potency of anticoagulant activity in vivo. Biochemistry 35:16449–16457
25. Cantwell AM, Di Cera E (2000) Rational design of a potent anticoagulant thrombin. J Biol Chem 275:39827–39830
26. Gruber A, Marzec UM, Bush L, et al (2007) Relative antithrombotic and antihemostatic effects of protein C activator versus low-molecular-weight heparin in primates. Blood 109:3733–3740
27. Johnson DJ, Adams TE, Li W, et al (2005) Crystal structure of wild-type human thrombin in the Na$^+$-free state. Biochem J 392:21–28
28. Pineda AO, Chen ZW, Bah A, et al (2006) Crystal structure of thrombin in a self-inhibited conformation. J Biol Chem 281:32922–32928
29. Pineda AO, Carrell CJ, Bush LA, et al (2004) Molecular dissection of Na$^+$ binding to thrombin. J Biol Chem 279:31842–31853
30. Pineda AO, Savvides S, Waksman G, et al (2002) Crystal structure of the anticoagulant slow form of thrombin. J Biol Chem 277:40177–40180
31. Wells CM, Di Cera E (1992) Thrombin is a Na(+)-activated enzyme. Biochemistry 31:11721–11730
32. Karshikov A, Bode W, Tulinsky A, et al (1992) Electrostatic interactions in the association of proteins: an analysis of the thrombin-hirudin complex. Protein Sci 1:727–735
33. Anderson PJ, Nesset A, Dharmawardana KR, et al (2000) Characterization of pro-exosite I on prothrombin. J Biol Chem 275:16428–16434
34. Arni RK, Padmanabhan K, Padmanabhan KP, et al (1994) Structure of the non-covalent complex of prothrombin kringle 2 with PPACK-thrombin. Chem Phys Lipids 67-68:59–66
35. Fuentes-Prior P, Iwanaga Y, Huber R, et al (2000) Structural basis for the anticoagulant activity of the thrombin-thrombomodulin complex. Nature 404:518–525
36. Baglin TP, Carrell RW, Church FC, et al (2002) Crystal structures of native and thrombin-complexed heparin cofactor II reveal a multistep allosteric mechanism. Proc Natl Acad Sci U S A 99:11079–11084
37. Pechik I, Madrazo J, Mosesson MW, et al (2004) Crystal structure of the complex between thrombin and the central "E" region of fibrin. Proc Natl Acad Sci U S A 101:2718–2723
38. Dumas JJ, Kumar R, Seehra J, et al (2003) Crystal structure of the GpIbalpha-thrombin complex essential for platelet aggregation. Science 301:222–226
39. Carter WJ, Cama E, Huntington JA (2004) Crystal structure of thrombin bound to heparin. J Biol Chem 280:2745–2749
40. Huntington JA (2005) Molecular recognition mechanisms of thrombin. J Thromb Haemost 3:1861–1872
41. Yun TH, Baglia FA, Myles T, et al (2003) Thrombin activation of factor XI on activated platelets requires the interaction of factor XI and platelet glycoprotein Ib alpha with thrombin anion-binding exosites I and II, respectively. J Biol Chem 278:48112–48119
42. Adams TE, Huntington JA (2006) Thrombin-cofactor interactions: structural insights into regulatory mechanisms. Arterioscler Thromb Vasc Biol 26:1738–1745
43. Pechik I, Yakovlev S, Mosesson MW, et al (2006) Structural basis for sequential cleavage of fibrinopeptides upon fibrin assembly. Biochemistry 45:3588–3597

44. Li CQ, Vindigni A, Sadler JE, et al (2001) Platelet glycoprotein Ib alpha binds to thrombin anion-binding exosite II inducing allosteric changes in the activity of thrombin. J Biol Chem 276:6161–6168

45. De Candia E, Hall SW, Rutella S, et al (2001) Binding of thrombin to glycoprotein Ib accelerates the hydrolysis of Par-1 on intact platelets. J Biol Chem 276:4692–4698

46. Celikel R, McClintock RA, Roberts JR, et al (2003) Modulation of alpha-thrombin function by distinct interactions with platelet glycoprotein Ibalpha. Science 301:218–221

47. Vu TK, Wheaton VI, Hung DT, et al (1991) Domains specifying thrombin-receptor interaction. Nature 353:674–677

48. Mathews II, Padmanabhan KP, Ganesh V, et al (1994) Crystallographic structures of thrombin complexed with thrombin receptor peptides: existence of expected and novel binding modes. Biochemistry 33:3266–3279

49. Myles T, Yun TH, Leung LL (2002) Structural requirements for the activation of human factor VIII by thrombin. Blood 100:2820–2826

50. Naski MC, Fenton JW, Maraganore JM, et al (1990) The COOH-terminal domain of hirudin: an exosite-directed competitive inhibitor of the action of alpha-thrombin on fibrinogen. J Biol Chem 265:13484–13489

51. Bukys MA, Orban T, Kim PY, et al (2006) The structural integrity of anion binding exosite I of thrombin is required and sufficient for timely cleavage and activation of factor V and factor VIII. J Biol Chem 281:18569–18580

52. Esmon CT, Lollar P (1996) Involvement of thrombin anion-binding exosites 1 and 2 in the activation of factor V and factor VIII. J Biol Chem 271:13882–13887

53. Myles T, Yun TH, Hall SW, et al (2001) An extensive interaction interface between thrombin and factor V is required for factor V activation. J Biol Chem 276:25143–25149

54. Dharmawardana KR, Olson ST, Bock PE (1999) Role of regulatory exosite I in binding of thrombin to human factor V, factor Va, factor Va subunits, and activation fragments. J Biol Chem 274:18635–18643

55. Arocas V, Lemaire C, Bouton MC, et al (1998) Inhibition of thrombin-catalyzed factor V activation by bothrojaracin. Thromb Haemost 79:1157–1161

56. Nogami K, Shima M, Hosokawa K, et al (2000) Factor VIII C2 domain contains the thrombin-binding site responsible for thrombin-catalyzed cleavage at Arg1689. J Biol Chem 275:25774–25780

57. Suzuki H, Shima M, Nogami K, et al (2006) Factor V C2 domain contains a major thrombin-binding site responsible for thrombin-catalyzed factor V activation. J Thromb Haemost 4:1354–1360

58. Edwards C, Armstrong P, Goode G, et al (1907) Cross-talking between calcium and histamine in the expression of MAPKs in hypertensive vascular smooth muscle cells. Cell Mol Biol (Noisy-le-grand) 53:61–66

59. Chung DW, Davie EW (1984) gamma and gamma' chains of human fibrinogen are produced by alternative mRNA processing. Biochemistry 23:4232–4236

60. Pineda AO, Chen ZW, Marino F, et al (2007) Crystal structure of thrombin in complex with fibrinogen gamma' peptide. Biophys Chem 125:556–559

61. Meh DA, Siebenlist KR, Brennan SO, et al (2001) The amino acid sequence in fibrin responsible for high affinity thrombin binding. Thromb Haemost 85:470–474

62. Lorand L (1907) Factor XIII: structure, activation, and interactions with fibrinogen and fibrin. Ann N Y Acad Sci 936:291–311

63. Philippou H, Rance J, Myles T, et al (2003) Roles of low specificity and cofactor interaction sites on thrombin during factor XIII activation: competition for cofactor sites on thrombin determines its fate. J Biol Chem 278:32020–32026

64. Bouma BN, Meijers JC (2003) Thrombin-activatable fibrinolysis inhibitor (TAFI, plasma procarboxypeptidase B, procarboxypeptidase R, procarboxypeptidase U). J Thromb Haemost 1:1566–1574

65. Bajzar L, Morser J, Nesheim M (1996) TAFI, or plasma procarboxypeptidase B, couples the coagulation and fibrinolytic cascades through the thrombin-thrombomodulin complex. J Biol Chem 271:16603–16608

66. Kokame K, Zheng X, Sadler JE (1998) Activation of thrombin-activable fibrinolysis inhibitor requires epidermal growth factor-like domain 3 of thrombomodulin and is inhibited competitively by protein C. J Biol Chem 273:12135–12139

67. Kahn ML, Zheng YW, Huang W, et al (1998) A dual thrombin receptor system for platelet activation. Nature 394:690–694

68. Jacques SL, Kuliopulos A (2003) Protease-activated receptor-4 uses dual prolines and an anionic retention motif for thrombin recognition and cleavage. Biochem J 376:733–740

69. Wu CC, Teng CM (2006) Comparison of the effects of PAR1 antagonists, PAR4 antagonists, and their combinations on thrombin-induced human platelet activation. Eur J Pharmacol 546:142–147

70. Butenas S, Dee JD, Mann KG (2003) The function of factor XI in tissue factor-initiated thrombin generation. J Thromb Haemost 1:2103–2111

71. Baglia FA, Badellino KO, Li CQ, et al (2002) Factor XI binding to the platelet glycoprotein Ib-IX-V complex promotes factor XI activation by thrombin. J Biol Chem 277:1662–1668

72. Baglia FA, Walsh PN (1996) A binding site for thrombin in the apple 1 domain of factor XI. J Biol Chem 271:3652–3658

73. Papagrigoriou E, McEwan PA, Walsh PN, et al (2006) Crystal structure of the factor XI zymogen reveals a pathway for transactivation. Nat Struct Mol Biol 13:557–558

74. Jin L, Pandey P, Babine,RE, et al (2005) Crystal structures of the FXIa catalytic domain in complex with ecotin mutants reveal substrate-like interactions. J Biol Chem 280:4704–4712

75. Esmon CT (2003) The protein C pathway. Chest 124:26S–32S

76. Dahlback B, Villoutreix BO (2003) Molecular recognition in the protein C anticoagulant pathway. J Thromb Haemost 1:1525–1534

77. Weiler H, Isermann BH (2003) Thrombomodulin. J Thromb Haemost 1:1515–1524

78. Vindigni A, White CE, Komives EA, et al (1997) Energetics of thrombin-thrombomodulin interaction. Biochemistry 36:6674–6681

79. Lin JH, McLean K, Morser J, et al (1994) Modulation of glycosaminoglycan addition in naturally expressed and recombinant human thrombomodulin. J Biol Chem 269:25021–25030

80. Ye J, Rezaie AR, Esmon CT (1994) Glycosaminoglycan contributions to both protein C activation and thrombin inhibition involve a common arginine-rich site in thrombin that includes residues arginine 93, 97, and 101. J Biol Chem 269:17965–17970

81. Oganesyan V, Oganesyan N, Terzyan S, et al (2002) The crystal structure of the endothelial protein C receptor and a bound phospholipid. J Biol Chem 277:24851–24854

82. Esmon CT (2000) The endothelial cell protein C receptor. Thromb Haemost 83:639–643

83. Stearns-Kurosawa DJ, Kurosawa S, Mollica JS, et al (1996) The endothelial cell protein C receptor augments protein C activation by the thrombin-thrombomodulin complex. Proc Natl Acad Sci U S A 93:10212–10216

84. Yang L, Rezaie AR (2003) The fourth epidermal growth factor-like domain of thrombomodulin interacts with the basic exosite of protein C. J Biol Chem 278:10484–10490

85. Hall SW, Nagashima M, Zhao L, et al (1999) Thrombin interacts with thrombomodulin, protein C, and thrombin-activatable fibrinolysis inhibitor via specific and distinct domains. J Biol Chem 274:25510–25516

86. Sasisekharan R, Raman R, Prabhakar V (1907) Glycomics approach to structure-function relationships of glycosaminoglycans. Annu Rev Biomed Eng 8:181–231

87. Huntington JA (2005) Heparin activation of serpins. In: Garg HG, Linhardt RJ, Hales CA (eds) Chemistry and biology of heparin and heparan sulfate. Elsevier, Oxford, pp 367–398
88. Li W, Johnson DJ, Esmon CT, et al (2004) Structure of the antithrombin-thrombin-heparin ternary complex reveals the antithrombotic mechanism of heparin. Nat Struct Mol Biol 11:857–862
89. Geiger M (2007) Protein C inhibitor, a serpin with functions in- and outside vascular biology. Thromb Haemost 97:343–347
90. Rezaie AR, Cooper ST, Church FC, et al (1995) Protein C inhibitor is a potent inhibitor of the thrombin-thrombomodulin complex. J Biol Chem 270:25336–25339
91. Meddahi S, Bara L, Fessi H, et al (1907) Standard measurement of clot-bound thrombin by using a chromogenic substrate for thrombin. Thromb Res 114:51–56
92. Huntington JA (2006) Shape-shifting serpins: advantages of a mobile mechanism. Trends Biochem Sci 31:450–455
93. Silverman GA, Bird PI, Carrell RW, et al (2001) The serpins are an expanding superfamily of structurally similar but functionally diverse proteins: evolution, mechanism of inhibition, novel functions, and a revised nomenclature. J Biol Chem 276:33293–33296
94. Law RH, Zhang Q, McGowan S, et al (1907) An overview of the serpin superfamily. Genome Biol 7:216
95. Huntington JA, Read RJ, Carrell RW (2000) Structure of a serpin-protease complex shows inhibition by deformation. Nature 407:923–926
96. Gettins PG (2002) Serpin structure, mechanism, and function. Chem Rev 102:4751–4804
97. Olson ST, Swanson R, Day D, et al (2001) Resolution of Michaelis complex, acylation, and conformational change steps in the reactions of the serpin, plasminogen activator inhibitor-1, with tissue plasminogen activator and trypsin. Biochemistry 40: 11742–11756
98. Bock PE, Olson ST, Bjork I (1997) Inactivation of thrombin by antithrombin is accompanied by inactivation of regulatory exosite I. J Biol Chem 272:19837–19845
99. Fredenburgh JC, Stafford AR, Weitz JI (2001) Conformational changes in thrombin when complexed by serpins. J Biol Chem 276:44828–44834
100. Long GL, Kjellberg M, Villoutreix BO, et al (2003) Probing plasma clearance of the thrombin-antithrombin complex with a monoclonal antibody against the putative serpin-enzyme complex receptor-binding site. Eur J Biochem 270:4059–4069
101. Kounnas MZ, Church FC, Argraves WS, et al (1996) Cellular internalization and degradation of antithrombin III-thrombin, heparin cofactor II-thrombin, and alpha 1-antitrypsin-trypsin complexes is mediated by the low density lipoprotein receptor-related protein. J Biol Chem 271:6523–6529
102. Corral J, Rivera J, Guerrero JA, et al (2007) Latent and polymeric antithrombin: clearance and potential thrombotic risk. Exp Biol Med (Maywood) 232:219–226

A Molecular Model of the Human Prothrombinase Complex

Stephen J. Everse, Ty E. Adams, and Kenneth G. Mann

Summary. Within the prothrombinase complex, the cofactor protein factor Va increases the catalytic efficiency of the serine protease factor Xa in the presence of calcium ions on a phospholipid membrane surface. The precise details leading to the alteration in catalytic activity of factor Xa have been the center of debate for decades. Our recent crystal structure of activated protein C-inactivated bovine factor Va (factor Va$_i$) revealed a domain organization unlike any previous model of either factor Va or factor VIIIa. Based on this structure, several homology models of human factor Va have been built along with models of human factor Xa (based on the porcine factor IXa structure). It is obvious that a deeper understanding of the interaction of the various components in the prothrombinase complex will eventually lead to elucidation of the precise mechanism of cofactor function.

Introduction

The successive activation of zymogen components in the coagulation cascade is dependent on the formation of enzymatic complexes. The only physiologically relevant generator of thrombin for blood coagulation is the prothrombinase complex, which is composed of the cofactor protein factor Va, the serine protease factor Xa, and calcium ions assembled on a phospholipid membrane surface [1, 2]. Assembly of the complex increases the catalytic efficiency of factor Xa toward its substrate, prothrombin, by approximately 300 000-fold when compared to factor Xa alone [3, 4]. The principle function of factor Va is to recruit factor Xa to an appropriate membrane surface (provided physiologically by the activated platelet), where it acts as part of the receptor for factor Xa and forms the prothrombinase complex [5, 6]. Accumulated data suggests that the binding of factor Va to factor Xa does not directly affect the catalytic site; rather, it induces a conformational change in factor Xa that exposes a prothrombin-binding site. Upon exposure of this binding site, the prothrombinase complex can bind prothrombin on the membrane surface, leading to a huge burst of thrombin, the signature enzyme of blood coagulation. Thus, assembly of the prothrombinase complex is essential for the generation of thrombin at levels required to maintain hemostasis [7]. In this review we examine efforts to model this complex and relate structure to function.

Factor X

Factor X (55 kDa) circulates in plasma as a two-chain zymogen. As a vitamin K-dependent coagulation protein, the light chain (16.5 kDa) contains an N-terminal Gla domain with 11 Gla residues, which is believed to be the point of contact with the membrane, and two EGF domains in tandem [8] (Fig. 1A). An activation peptide and the serine protease domain form the heavy chain (30 kDa), which is linked to the light chain by a single disulfide bond. Factor X is activated to factor Xa with the release of the activation peptide by either the extrinsic (tissue factor/factor VIIa) or the intrinsic (factor VIIIa/factor IXa) tenase complexes.

Structure of des(Gla) Factor Xa

The structure of des(Gla) factor Xa has provided a new target for the development of antithrombotic agents as well as the first crystallographic view of an epidermal growth factor (EGF) domain [9–11]. As expected, the differences between the nuclear magnetic resonance (NMR) structures and this EGF domain were minimal, and the overall fold of the protease domain is similar to thrombin, factor VIIa, and factor IXa, with the differences arising in the insertion loops. The positively charged fibrinogen-binding exosite in thrombin is completely obliterated in factor Xa as that region is primarily populated by negatively charged amino acids.

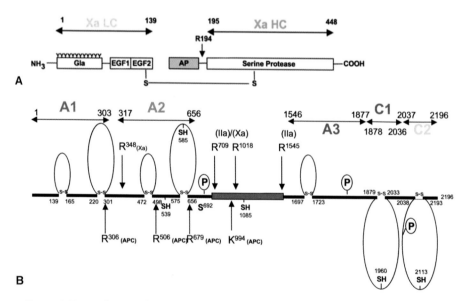

FIG. 1. **A** Human factor X domain structure. *Y*, Gla residues. Domains are labeled, and the cleavage site indicated with an *arrow*. **B** Human factor V domain structure. Domains are labeled above, cleavage sites are indicated (*arrows*) for IIa, Xa, and APC, as are disulfides (*S-S*), free cysteines (*SH*), and sulfation (*S*) sites. (Modified from Kalafatis et al. [159], with permission)

Gla Domain: Insights into Membrane Interaction

In the ultracentrifuge, factor Xa behaves like an elongated molecule ($\sim$100 Å), a finding confirmed by the X-ray structure [9, 12]. Fluorescence energy transfer measurements place the distance between the active site and the phospholipid surface between 61Å and 69 Å regardless of whether factor Va was present [13]. Based on these observations it is believed that the domains of factor Xa are projecting perpendicularly to the membrane surface in the prothrombinase complex and that the EGF domains serve as spacers to place the protease domain at the correct height to interact with factor Va and to cleave substrate (prothrombin).

Factor V

Factor V[1] circulates in human plasma at a concentration of $\sim$20 nM as a large single-chain species (330 kDa, 2196 amino acids) devoid of coagulant activity [14, 15] with $\sim$20% of it stored within platelet α-granules [16]. Factor V is a modular glycoprotein (domains A1-A2-B-A3-C1-C2) containing copper and calcium ions, each of which plays a structural role in the function of the cofactor (Fig. 1B) [17, 18]. Before factor V is secreted from the hepatocyte, numerous posttranslational modifications occur, including both N- and O-linked glycosylation, sulfation of tyrosines, and phosphorylation [19–25].

In addition, factor V shares 30%–40% identity with factor VIII (another coagulation cofactor) in the A and C domains; the B domain, however, is poorly conserved even within species. The A domains also share $\sim$30% identity with the plasma copper-binding protein ceruloplasmin [18].

Human factor V is physiologically activated through selective proteolysis by thrombin at Arg^{1545}, Arg^{1018}, and Arg^{709} [26], thus removing the B domain in two fragments. The resulting heterodimer, factor Va, is composed of a heavy chain (HC) (A1-A2, 105 kDa) and a light chain (LC) (A3-C1-C2, 74 kDa) that associate in a calcium-dependent manner (A1-A2·A3-C1-C2) [27].

Factor Va is proteolytically inactivated by the anticoagulant-activated protein C (APC), which is one of the key reactions in the regulation of thrombin formation [28–35]. APC and factor Xa compete for binding to the LC of factor Va [36]. Once APC is bound, cleavage first occurs at Arg^{506}, partially inactivating the cofactor, followed by cleavage at Arg^{306}, allowing spontaneous release of the A2 domain and complete inactivation of factor Va [37]. An additional cleavage occurs at Arg^{679} [37–41]. Recent data demonstrate that even when these sites are removed in factor Va it can still be inhibited by APC; thus, APC must also cleave at other residues in the HC [42].

C Domains: Insights into Membrane Interaction

The C domains show a weak homology ($\sim$20% identity) to lectin-binding proteins, the discoidins, from *Dictyostelium discoideum* [43]. Interestingly, discoidin I has been found to bind negatively charged phospholipids, one of the functions ascribed to the C domains in the coagulation cofactors [44]. At the core of each domain is an eight-stranded β-sandwich fold that was found to be homologous to the discoidin proteins

---

[1] Numbering is for the human molecule.

[45, 46]. In addition, two extensive regions of exposed hydrophobic surface were observed. One occurs at the top of the β-sandwich, and the other is on the bottom forming a pocket that is defined by three β-hairpin loops called "spikes." The interior of the pocket is lined with polar side chains that appear ideally arranged for interactions with the head group of phosphatidylserine (PS). Mutagenesis studies have indicated that phospholipid binding is primarily due to spike C2-1 (Ser^{2058}-Trp^{2068}) in factor V [47–50]. In fact, Trp^{2063} and Trp^{2064} in this spike have been implicated in high-affinity binding to PS membranes and regulating the formation of a productive prothrombinase complex [51].

In the structure of the factor V C2 domain, two crystal forms were observed—one with the spike pocket "open," exposing maximum hydrophobic character, and the other "closed," reducing its hydrophobic surface by more than one-third [45]. The latter form was predicted to be favored in the aqueous environment of the circulating procofactor, and this prediction has been supported by antibody-binding experiments [49]. Recently, these structures have been used to evaluate the mechanism, dynamics, and energetics of membrane binding [52]. The data supported the "closed" form as the circulating procofactor but surprisingly found that both the "open" and "closed" forms are energetically compatible with membrane binding.

Structure of Factor V, Va, and Va$_i$

Available structural information from ultracentrifugation experiments show that factor V is a highly asymmetrical (rod-like) single-chain protein (330 kDa) with a Stokes radius of 93 Å [53]. Electron microscopy (EM) of two-dimensional (2D) crystals confirmed these results. Factor V appears as an irregular ellipsoid, 100–200 Å long, that displays no major structural rearrangement upon activation with the release of the B region. This implies that the HC and LC from factor Va form a globular structure with the B domain, extending from the ellipsoid as a rod-like structure that can be removed proteolytically without disturbing the complex [54–56]. Thus, the factor Xa binding site might be present in factor V but is masked by the B region. This idea was elegantly tested by Toso and Camire [57], who deleted various portions of the B region and discovered that some of the new constructs had the same affinity for factor Xa as the activated cofactor Va. Interestingly these constructs functioned in a manner equivalent to that of factor Va even though they remained as single-chain, unactivated molecules.

The 2.8-Å crystal structure of activated protein C-inactivated bovine factor Va (factor Va$_i$), which includes the A1 domain from the HC and the entire LC, indicates a domain organization quite different from that of previous models [46]. Most notably, the factor Va$_i$ structure places the C1 and C2 domains in a "side-by-side" orientation in which both C domains could potentially interact with the phospholipid surface (Fig. 2A) versus earlier models in which the C1 domain was stacked upon the C2 domain (Fig. 2B). Support for this C domain arrangement has come from alanine-scanning site-directed mutagenesis data [58], which demonstrated that residues in the C1 domain are also essential for both phospholipid binding and prothrombinase activity.

Factor Va Models

Several models of the factor Va "A" domain trimer, as well as models of the homologous factor VIIIa domains, have been based on the known homology of the "A"

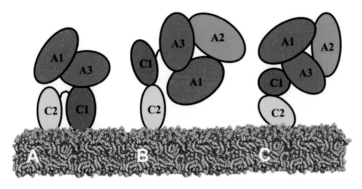

FIG. 2. Domain organization of our factor Va$_i$ structure (A) [46], the factor VIIIa electron crystallographic model (B) [160], and the factor Va model (C) [62]

domains to the plasma copper-binding protein ceruloplasmin [59, 60]. The crystal structures of the C2 domains in 1999 from both factors Va and VIIIa allowed generation of homology models of factor Va and were used in an 8-Å EM model of factor VIIIa [45, 61–63] (Fig. 2C). Gale et al. [64] used the factor Va$_i$ structure and the A domains from Pellequer et al. [62] to generate an updated factor Va model, whereas Orban et al. [65] used a similar construct to predict the nonhomologous COOH-termini of the heavy chain (Ser^{664}-Arg^{709}), which has been implicated in prothrombin binding [66, 67].

Prothrombinase Complex Assembly and Function

Prothrombinase complex formation proceeds through a mechanism whereby factor Va and factor Xa interact with a 1:1 stoichiometry on a negatively charged phospholipid membrane in the presence of calcium [1]. Binding of factor Va to the membrane is mediated by the LC (Table 1) in a Ca^{2+}-independent manner that is in marked contrast to the vitamin K (Gla)-dependent proteins [68–72]. Using competitive assays with synthetic peptides representing regions of factor Va, residues Ile^{311}-Phe^{325} and Gly^{493}-Arg^{506} in the HC have been shown to be important for factor Xa binding [73, 74]. Further studies with synthetic peptides identified residues Glu^{323}-Val^{331}—specifically Glu^{323}, Tyr^{324}, Glu^{330}, and Val^{331}—as essential for the expression of full cofactor activity [75, 76]. Not surprisingly, APC inactivation, which ultimately removes the A2 domain, abolishes the interaction between the cofactor and factor Xa [37]. Finally, the creation of novel glycosylation sites also identified the region surrounding residues Glu^{467}, Ala^{511}, Arg^{652}, and His^{1683} in the A2 and A3 domains as important for interaction with factor Xa [77].

Conversely, binding analyses using factor Va peptides identified residues Leu$^{211(32)}$-*Gly$^{222(43)}$, Glu$^{254(74)}$-Phe$^{274(94)}$, and Ala$^{404(220)}$-Phe$^{418(234)}$ on factor Xa as potentially

---

*Positions of residues in the proteinase domain of factor Xa are given and the topographically equivalent residues in chymotrypsinogen are in parentheses.

TABLE 1. Summary of structural observations regarding plasma factor Va and its multiple associations

Association and observation	Reference(s)
Va·membrane	
Binding to PS/PC membranes is mediated by the Va·LC (A3 and C2) and is Ca^{2+} independent	[47, 48, 70, 71, 95, 97, 136–138]
PS binds to a single site in Va·C2	[47, 49, 71, 97, 139]
Va·HC and Va·LC each has two PS binding sites in the presence of Ca^{2+} and in the absence they each have one	[140]
Hydrophobic contributions to membrane binding by Va·HC	[141]
94–345 (Va·HC) and 1828–2196 (Va·LC) required for membrane binding	[142]
Presence of carbohydrate at Asn^{2181} decreases the Va·LC affinity for membranes and perhaps Xa	[143, 144]
Va·Trp2063 and Trp2064 required for high-affinity lipid binding	[51]
Residues in Va·C1 necessary for membrane binding	[58]
Cooperative roles of Va and PS membranes as cofactors in IIa activation	[145]
Va·Xa	
Va·HC and Va·LC both interact	[6, 84, 92, 93, 146, 147]
Va·506 has a lessened affinity for Xa	[31, 37]
Membrane-bound Va·LC does not interact with Xa	[95]
Va·311–325 forms an Xa and/or II binding site	[148]
Va·307–348 critical for function/interactions with Xa and/or II	[149]
Va·323–331 contains Xa binding site	[76]
Va·493–505 forms an Xa binding site	[73]
Va·499–506 define an extended Xa binding site	[64]
Removal of Va·HC 683–709 inhibits Xa and/or II binding	[85]
Carbohydrates at 467, 511, 652 in the Va·HC and 1683 in Va·LC display attenuated Xa binding	[77]
Presence of carbohydrate at Asn2181 decreases the Va·LC affinity for membranes and perhaps Xa	[143, 144]
Removal of Va·B domain not needed for cofactor function	[57]
Only Xa EGF2 and protease domain involved in Va binding	[121, 122, 150]
Xa·211–222 and 254–274 involved in Va interaction	[78]
Xa·347, 351, and 414 play key roles in Va and/or II recognition	[80]
Xa·404–418 forms Va binding site	[79]
Va interacts with 368–373 loop of Xa	[83]
Va·Xa·II	
Interaction promoted by Va·HC	[151, 152]
Xa·270 used in recognizing II	[128]
Removal of Va·HC 683–709 inhibits Xa and/or II binding	[85]
Removal of Va·HC 697–709 inhibits II binding	[66]
Va·HC 695–706 do not form a productive binding site for II	[129]
Va·311–325 forms a Xa and/or II binding site	[148]
Va·314–348 critical for function/interactions with Xa and/or II	[149]
Xa·347, 351, and 414 play key roles in Va and/or II recognition	[80]
Active site independent recognition of substrate by prothrombinase	[117]
Preexosite I prethrombin-1 mutants are a poor substrate	[116]
II·67 mutants are a poor substrate for prothrombinase	[153]
Va·HC binds exosite 1 of IIa	[154, 155]
II may interacts with 262–271 & 282–296 of Xa	[88]
II·557–571 and Xa·415–429 mediate interactions in the prothrombinase	[79]
II·473–487 form Va binding site in prothrombinase	[156]
Xa interacts with II·205–220	[157]
APC·Va·Xa	
APC binds Va·LC cooperatively	[36, 38, 39, 41]
Comparison of thrombomodulin binding on APC with Va binding site	[158]

important cofactor interaction sites [78, 79]. Mutation of several key basic residues, including $Arg^{347(165)}$, $Lys^{351(169)}$, and $Lys^{414(230)}$, have been shown to reduce factor Xa incorporation into the prothrombinase complex, specifically reducing factor Va binding [80–82]. It has also been suggested that the $Asp^{368(185)}$-$Asp^{373(188)}$ loop may be critical for Na^+ binding as well as factor Va interaction [83]. Binding and inhibition studies have also identified regions on both the cofactor and the proteinase that are important for prothrombin binding [84–88]. Finally, several groups have identified interacting residues in the analogous enzyme–cofactor coagulation complex, the intrinsic tenase, including the 330 helix of factor IXa, and residues 558–565 in the A2 domain of factor VIIIa [89–91].

In the absence of membrane, the factor Va–factor Xa interaction is relatively weak ($K_d \sim 0.8$ mM) [92]; however, when coupled with membrane binding, a tight association ($K_d \sim 1$ nM) results [6, 93, 94]. Thus, efficient binding involves not only electrostatic and hydrophobic interactions but also penetration into the lipid bilayer [45, 95, 96]. Several studies have now indicated that the preferred lipid membrane for prothrombinase formation is rich in phosphatidylserine [47, 51, 97, 98].

The fundamental contribution of factor Va to the prothrombinase complex may be retention of factor Xa on the membrane surface. Factor Xa has a weak affinity for membranes ($K_d \sim 0.1$ μM), which implies that retention of factor Xa on the membrane is short-lived [99]. It has been postulated that membrane-bound factor Va acts as a receptor on the platelet surface and effectively increases the affinity of factor Xa for the membrane [5, 6]. Formation of the prothrombinase complex increases the catalytic efficiency of the enzyme by five orders of magnitude compared to factor Xa alone [7].

Enzyme–Substrate–Cofactor Complex

Prothrombin (72 kDa) is the most abundant of the vitamin K-dependent coagulation factors, with circulating levels ranging between 1 and 2 μM in human plasma [100]. It is composed of four structural elements: a Gla domain, two kringle domains in tandem, and a serine protease domain. Physiologically, the prothrombinase-catalyzed activation of prothrombin to thrombin is the result of two cleavages. The first cleavage occurs after Arg323, thereby forming the active intermediate meizothrombin, which remains membrane bound because of the attached Gla domain [101]. A second cleavage at Arg274 releases thrombin and fragment 1·2.

Based on kinetic data, factor Va appears to interact with factor Xa and prothrombin to present one to the other effectively in the formation of a ternary enzyme–substrate–cofactor complex. If meizothrombin accurately reflects these molecular interactions, factor Va "induces" a recognition site in prothrombin for factor Xa or modifies factor Xa such that it readily catalyzes the cleavage of prothrombin, or both [102]. Evidence for a factor Va-induced change in the active site of factor Xa is provided by the change in extrinsic fluorescence of DEGR–Xa observed in the presence and absence of PL [5, 13, 103].

Explanations for the substrate specificity of factor Xa, especially in the prothrombinase complex, have been based on its active site geometry (from the crystal structures of factor Xa) [9, 10], by mutagenesis of the residues surrounding the active site [104], and studies using synthetic peptide substrates. Although these approaches describe the properties of factor Xa in solution, they provide no information regarding the substrate specificity of factor Xa incorporated into prothrombinase [105]. The affinity of the complex for its substrate is not believed to be determined by prothrombin binding to the active site of factor Xa but, rather, by a complex set of interactions involving extended recognition sites (exosites) removed from the active site and substrate regions that are distant from the region surrounding the scissile bond [106].

Prothrombinase Models

Autin et al. [107] recently generated the first published models of the human prothrombinase complex. Their factor Va model was based on their previously generated A domain trimer [60, 108] combined with the C1–C2 domains of bovine factor Va$_i$ structure [46]. Their factor Xa model was based on the porcine factor IXa structure [109]. They used three docking methodologies and applied the following initial criteria: (1) The protease domain of factor Xa must make contact with the A2 and/or A3 domains of factor Va. (2) Factor Xa residues Val$^{415(231)}$-Thr$^{428(244)}$ and Tyr$^{344(162)}$-Leu$^{352(170)}$ should bind directly to factor Va. (3) These factor Xa residues should be within 15 Å of Ala^{511} of factor Va. From thousands of initial models, they were able to distill five distinct models that bury between 1046 Å^2 and 2151 Å^2 of surface area and included Gly^{493}-Arg^{506} [73] and Arg^{501}/Arg^{510}/Ala^{511} as part of the binding site. Their best model (M3) placed the active site of factor Xa approximately 75 Å from the membrane surface (Fig. 3), which best matches known biochemical data.

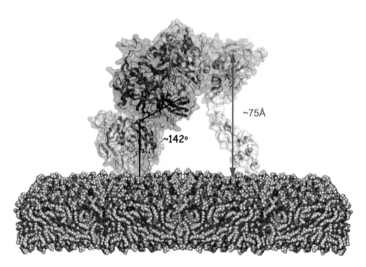

FIG. 3. Three-dimensional ribbon diagram and surface projection of the human prothrombinase model M3 from Autin et al. [107]. Their factor Va model (A1 domain is shown in red, A2 in orange, A3 in blue, and C1 in green; the C2 domain is depicted in yellow) is on the *left*; and their factor Xa (proteinase domain is shown in bright green and the Gla-EGF1-EGF2 domains in light yellow) on the *right* are depicted on a membrane surface. In this model, the factor Xa active site is around 75 Å above the membrane. All structure-based images were created using PyMOL [161]

Our aim in this chapter is to review the current literature and present a theoretical model of the prothrombinase complex using homology models of factor Va and factor Xa based on the crystal structures of bovine factor Va$_i$ and porcine factor IXa. Our model is consistent with known biochemical data and may provide insight into the mechanism for the extraordinary increase in catalytic efficiency of the prothrombinase complex observed during blood coagulation.

New Prothrombinase Model

Models of human factor Va and human factor Xa were built as described in the Supplemental Material section at the end of the chapter. Suffice it to say, the model of human factor Va built using the structures of bovine factor Va$_i$ (1SDD) and human ceruloplasmin (1KCW) and the model of factor Xa was based on porcine factor IXa (1PFX). These models were then subjected to protein docking protocols and the resulting complexes evaluated with regard to the available biochemical data. From the initial 9240 models, one was selected as the best and is described below.

Structural Mapping of Biochemical Data

Studies on the binding of factor Va to factor Xa have identified several potential regions of interaction (Table 1) (i.e., Ile^{311}-Phe^{325}, Glu^{323}-Val^{331}, and Gly^{493}-Arg^{506}), and these residues are all localized on one face of the model, on the under side of the A domain trimer (Fig. 4A). Additionally, in our factor Va model, residues Lys^{499}-Arg^{506} are found in a solvent exposed loop, and residues Glu^{323}-Val^{331} reside on top of the A2 domain. Similarly, the APC cleavage site Arg^{306} is found in a flexible linker between the A1 and A2 domains. Taken together, these regions of factor Va represent a substantial potential interaction surface mostly comprising the A2 domain.

Most mutational analyses and peptide-binding studies have clustered the potential factor Va-binding residues of factor Xa to the proteinase domain, concentrated around two regions that correspond to the anion-binding exosites of thrombin (Fig. 4B). Exosite I is positively charged in thrombin, but in factor Xa this presumed site contains a high concentration of negatively charge amino acids, including three glutamic acid residues between Ile$^{213(34)}$-Gly$^{219(40)}$ and five within residues Glu$^{254(74)}$-Glu$^{266(86)}$. In contrast, the putative factor Xa exosite II contains positively charged patches that include residues Arg$^{347(165)}$, Lys$^{351(169)}$, and Lys$^{414(230)}$ shown to be important for binding heparin and prothrombinase complex assembly [80]. Interestingly, recent biochemical data now suggest that exosite I is not critical for either binding of the enzyme to factor Va or recognition of its substrate prothrombin [110].

Model of the Prothrombinase Complex

Once the homology models were created and energy minimized, they were subjected to protein–protein docking studies using factor Va as the receptor molecule and factor Xa as the pseudo-ligand. Docking experiments generated 9240 rotational and translational results, which were filtered in accordance with specific biochemical data as detailed in the Supplemental Material. The list was further filtered to

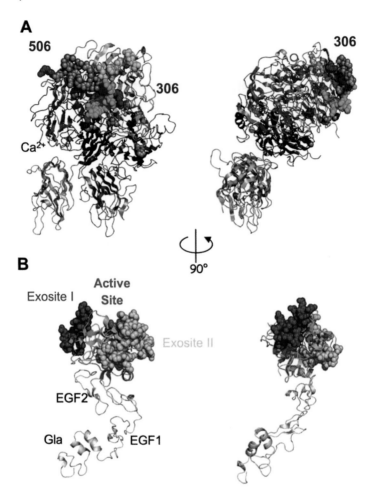

FIG. 4. **A** Potential binding surfaces based on biochemical information mapped onto the factor Va homology model shown in two orientations. Residues identified as important in prothrombinase complex assembly are shown as colored spheres; residues Ile[311]–Trp[322] are shown in cyan, residues Glu[323]–Val[331] in yellow-green, residues Gly[493]–Arg[506] in blue-green, residues Glu[467], Ala[511], Phe[651], and His[1683] in gray, and APC cleavage sites in bright blue. **B** Potential binding surfaces based on biochemical information mapped onto the factor Xa homology model shown in two orientations. Residues identified as important in prothrombinase complex assembly are shown as colored spheres: Exosite I is purple-blue, exosite II is aquamarine, Arg[347] is magenta

eliminate complexes in which both proteins could not simultaneously interact with the membrane surface. The top 10 models were visually inspected, and a single model was chosen for description here. Our model shows an extensive binding interface between the cofactor and proteinase, involving both the catalytic and EGF2 domains (Fig. 5). Binding occurs primarily through interactions centered around exosite II of factor Xa. There are also weak interactions near the NH_2-terminal of the EGF2 domain

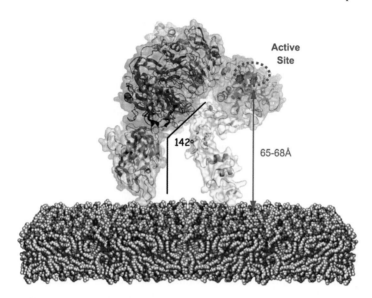

FIG. 5. Best fit solution of our prothrombinase complex from protein–protein docking simulations. The three-dimensional diagram with factor Va shown as a surface representation colored as previously described. (Note that the C1 domain is directly behind the C2 domain.) The factor Xa model (colored as previously described) was rotated and translated resulting in a protein–protein docking solution based on surface complementarity. Both were placed onto a hypothetical membrane

that aid in the orientation of the factor Xa light chain for optimal positioning of the active site and substrate-binding exosites for prothrombin activation.

In our model, factor Xa binds to the lateral face of factor Va at a site principally involving the A2 domain and select residues of the A3 domain. These residues are localized at or near the sites identified previously by peptide competition and site-directed glycosylation experiments (Table 1). The proteinase domain of factor Xa fits into a groove created at the interface of the two plastocyanin-like motifs in the A2 domain. This groove extends toward the center of the threefold axis created by the factor Va A domains. The NH_2-terminal portion of the EGF2 domain of factor Xa and the linker to the EGF1 domain follow this groove with only limited contacts with factor Va. Upon assembly of the prothrombinase complex, approximately $1080\,\text{Å}^2$ of surface area of the two molecules is buried at the interface.

The entire complex can be oriented so that the C domains of factor Va bury approximately 8 Å into the membrane (as predicted by Mollica et al. [52]), and the Gla domain of factor Xa binds to the membrane without significant insertion (Fig. 5). To allow the Gla domain of factor Xa to bind the membrane, factor Va aligns the C domains perpendicular to the surface, which tilts the A domain trimer 38° toward the lipid membrane. This lowers the overall height of factor Va to approximately 100 Å and brings the binding region of factor Va into contact with exosite II of factor Xa, which would allow maximal binding energy between the membrane, factor Va, and factor Xa.

The structure of factor Xa in the complex is similar to the unbound factor Xa model, where the Gla-EGF1-EGF2 domains of factor Xa extend from the potential membrane at a 30° incline. This serves to position the active site of factor Xa at 65–68 Å above the membrane surface, consistent with previous reports using fluorescence-labeled phospholipid vesicles in conjunction with labeled active site inhibited factor Xa [13]. The active site of factor Xa in our complex faces away from both factor Va and the potential membrane surface; this orientation would more than likely ensure that the Arg^{320} cleavage occurs first in the reaction mechanism for prothrombin activation. No major conformational changes are predicted in the proteinase domain at or near the active site of the enzyme, which is supported by biochemical evidence, indicating that the increase in catalytic efficiency observed by factor Xa in the prothrombinase complex is not due to specific rearrangements in the active site of the enzyme [111].

Prothrombinase-Binding Interface

The importance of many residues previously identified by mutagenesis studies is explained by a detailed examination of the binding interface (Fig. 6). The extensive interface between factor Va and factor Xa involves both hydrophobic and electrostatic interactions. For instance, in our model, Arg$^{347Xa(165)}$ makes a critical salt bridge with Asp^{504Va}. When mutated to either Asn or Ala, factor Xa affinity for factor Va is reduced 10- to 25-fold, suggesting the importance of this salt bridge in the assembly of the prothrombinase complex [80, 81]. Additionally Lys$^{351Xa(169)}$ interacts with Asn^{398Va} and the backbone carbonyl of Asp^{504Va}. Arg^{506} of factor Va forms a hydrogen bond to the amide oxygen of Gln$^{360Xa(178)}$, and Arg^{505Va} interacts with the backbone carbonyl of Lys$^{351Xa(169)}$. On factor Va, Arg^{510} intercalates between Lys$^{414Xa(230)}$ and Glu$^{310Xa(129)}$, forming a triplet where either Arg^{510Va} or Lys$^{414Xa(230)}$ may exist in a partially charged state. Two hydrogen bonds exist between the factor Va A3 domain and the EGF2 domain of factor Xa, and involve the backbone carbonyl of Pro^{1693Va} to the hydroxyl of Thr^{85Xa} and the amide oxygen of Gln^{1828Va} to the side-chain amino of Arg^{86Xa}. These are the only interactions not between the proteinase domain of factor Xa and the A2 domain of factor Va.

A principal mechanism for shutting down the coagulation cascade is factor Va inactivation through cleavage at two arginine residues, Arg^{306} and Arg^{506}, by the anticoagulant APC. Incorporation of factor Xa into the prothrombinase complex has been shown to prevent inactivation of factor Va by protecting it from cleavage at Arg^{506} [36]. In our model of the prothrombinase complex, Arg^{506} of the cofactor is protected by protein–protein contacts with factor Xa, thereby excluding cleavage at this site by APC. This arginine makes a hydrogen bond to the amide of Gln$^{360(178)}$ of factor Xa, preventing recognition by the active site of APC.

A detailed comparison of the interface between our prothrombinase model and the M3 model of Autin et al. [107] uncovered many similarities and a few differences with highlights presented below: root mean square differences (RMSD) between factor Va's, 1.45 Å; between factor Xa's, 0.65 Å; and between complexes, 10.3 Å. This large difference observed between complexes is principally the result of the proteinase domain

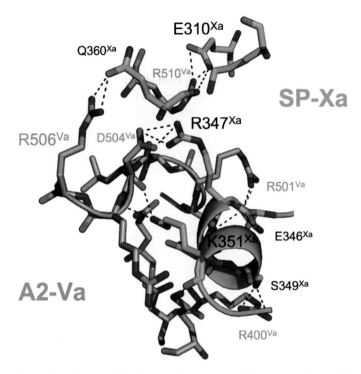

Fig. 6. Hydrogen bonding interface between factor Va and factor Xa in the prothrombinase complex. The proteinase domain of factor Xa (bright green) and the A2 domain of factor Va (orange) are both overlayed with standard CPK coloring schemes. Hydrogen bonding potentials are represented by *black dashed lines*. Potential hydrogen bonds were selected based on stereochemistry, partial formal charge, and distance restraints of each residue with respect to surrounding residues

being placed farther up the A2 domain in the M3 model and a shift in the orientation of Gla-EGF1 domains in the factor Xa models. This can best be observed by mapping the interaction surfaces of both factor Va and factor Xa (Fig. 7). Although both models present similar faces toward each other and occlude access to the region surrounding Arg[506], in our model this region makes direct contacts with factor Xa (Fig. 6) whereas in the M3 model this region has not yet been refined to maximize contacts. Additionally, in our model both the A1 and A3 domains closely approach factor Xa, whereas in the M3 model these domains are minimally or not involved. In our factor Xa model EGF2 approaches but does not contact the A3 domain of factor Va, whereas in the M3 model it approaches the A2 domain (Fig. 7). Factor Xa interaction with both the heavy and light chains of factor Va is supported by sedimentation equilibrium data [92].

Proposed Model   **M3**

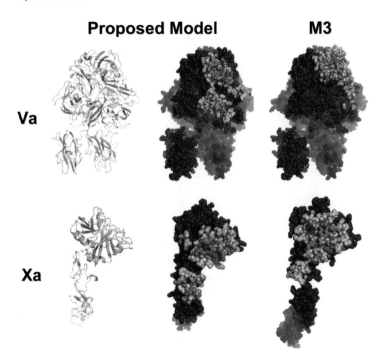

FIG. 7. Binding interfaces of our prothrombinase complex model and the M3 model of Autin et al. [107]. Interatomic distances were calculated between all atoms of factor Va and factor Va and those less than 10 Å colored according to their distance from the other factor, red being the shortest and greater distances shaded toward green. The orientations presented are the same for both molecules after a least-squares fit (only the proteinase domain was used for the factor Xa). Domain colors for the ribbon diagrams are described in Fig. 3

Discussion

Several of the steps in the coagulation process involve activation of a zymogen catalyzed by complexes composed of a chymotrypsin-like serine protease bound to a membrane-anchored cofactor protein [112, 113]. Some of these serine proteases, such as thrombin and factor Xa, contain sites of interaction with substrates and modifiers away from the active site, termed exosites. Initially thought of as merely stabilizing active site interactions between enzymes and substrates, exosite contacts have now emerged as dominant contributors to enzyme/substrate specificity; this is exemplified in the case of the prothrombinase complex [88, 105, 106, 114–118]. In a similar fashion, exosite interactions have also been shown to play a role in the assembly of additional reactive complexes that are essential for efficient coagulation [80].

Analogous to the interaction between factor IXa and factor VIIIa, where the EGF2 and the proteinase domains are found to mediate binding [119, 120], our model of the prothrombinase complex shows binding between the cofactor and enzyme occurring primarily through the A2 domain of factor Va and exosite II of factor Xa (Figs. 4, 7). This region on factor Xa resembles that of thrombin anion-binding exosite II and involves residues $Arg^{306(126)}$-$Lys^{317(134)}$, $Asp^{346(164)}$-$Gln^{360(178)}$, and, potentially, $Lys^{414(230)}$ in the proteinase domain. Minor interactions in the EGF2 domain at Thr^{85} and Arg^{86} and the A3 domain of factor Va may also play a role in orienting the enzyme in the complex. Our model shows, as do most of the models of Autin et al. [107], no interactions with the putative factor Xa cation binding exosite I; however, conflicting results about the importance of these residues for the assembly of the prothrombinase complex have been reported [78, 110]. In our model, neither the Gla or EGF1 domains display any contacts with factor Va. This is consistent with previous experiments using factor IX/factor X chimera proteins, where the Gla and EGF1 domains of factor X were exchanged with their factor IX equivalents, which showed that the EGF2 and protease domains of factor Xa were sufficient for factor Va binding [121, 122]. The Gla-EGF1 region of the enzyme seem to be more important for zymogen activation by the tissue factor–factor VIIa complex.

Charge complementarity combined with the extensive hydrogen bonding network observed at the interface of the complex certainly would contribute to the tight affinity ($K_d \sim 1\,nM$) [123] of the two components seen on a phospholipid membrane. Because factor Xa alone has a binding affinity for membrane of $\sim 0.1\,\mu M$, the presence of factor Va increases the ability of factor Xa to bind to lipid by nearly 5000-fold [69]. For this reason, factor Va is typically thought of as a cell surface receptor for factor Xa, increasing the local concentration of the enzyme at the membrane to levels needed to drive coagulation. Although factor V and factor Va can bind to activated platelets with equal affinity, factor V has no appreciable binding capacity for factor Xa. Removal of the B domain, which occurs during activation of factor V to factor Va, has been shown to be necessary for the full expression of cofactor activity and assembly of the prothrombinase complex [57, 124]. A recent study reported that cleavage at Arg^{1545} in factor V was sufficient for the realization of cofactor function [66]. In our model of the prothrombinase complex, the footprint of factor Xa on factor Va suggests that the binding site on factor V would be occluded by the presence of the B domain, in particular the C-terminal portion near the Arg^{1545} thrombin cleavage site.

Assembly of the prothrombinase complex also leads to 10^4-fold enhancement in the catalytic rate (k_{cat}) of factor Xa activation of prothrombin [125]. Initial hypotheses involved a cofactor-dependent active site rearrangement of factor Xa producing a more catalytically efficient enzyme [126]. However, experiments using small chromogenic substrates show no difference in observed factor Xa activity in the presence or absence of factor Va [125, 127], and recent data demonstrating equivalent binding of substrates and product to free and active site inhibited prothrombinase suggest that the active site residues do not undergo substantial conformational change upon complex assembly and contribute negligibly to substrate specificity [117]. Current opinion now leans toward the binding of prothrombin exosites to complementary exosites on prothrombinase, leading to efficient presentation of cleavage sequences to the active site of factor Xa [118]. On factor Xa, residues $Lys^{270(96)}$, $Val^{262(82)}$-$His^{271(91)}$, $Asp^{282(102)}$-$Met^{296(116)}$, and $Val^{415(241)}$-$Arg^{429(252)}$ were identified as potential prothrombin contact sites [79, 88, 128]. These residues are nicely clustered together and suggest interaction with the A2 domain of factor Va. On the A2 domain, it has been suggested that the C-terminal of the heavy chain (Ser^{664}-Arg^{709}) may provide a binding site for prothrombin, but recent data suggest otherwise [129]. Regardless, it would be difficult to imagine prothrombin not interacting with factor Va considering its proximity.

Extensive interaction between prothrombin and the prothrombinase complex has been suggested to involve every domain of the zymogen substrate. Fluorescence measurements have revealed that the Gla domain, residues 1–46, can directly interact with factor Va [130]. Both kringle domains potentially interact with the prothrombinase complex. In the presence of factor Va, kringles 1 and 2 can inhibit prothrombinase-dependent activation of prothrombin, although only prothrombin lacking kringle 2 displays significant reductions in the K_M and k_{cat} reaction constants, suggesting an important function of this domain in prothrombin activation [131–134]. In addition, only kringle 2 inhibits factor Xa cleavage of prothrombin in the absence of factor Va [134]. This implies that kringle 2 may bind simultaneously to both factor Xa and factor Va in the prothrombinase complex. Several studies have focused on the contribution of regions in the protease domain on the binding and activation of prothrombin and its intermediates. Sites remote from the substrate cleavage sites have been shown to be important for recognition by the prothrombinase complex [88, 105, 106, 114]. Specifically, Bock and colleagues discovered interactions between proexosite I of the substrate and factor Va [87, 135]. These data suggest that binding of prothrombin involves interactions primarily with the cofactor, cementing its importance in coagulation as an activatable cell surface receptor for both enzyme and substrate. Binding to factor Va effectively raises the local concentration of each, resulting in increased interaction as well as orientating the substrate for optimal presentation of the activation peptide scissile bond for efficient cleavage.

Conclusion

As fragments of, and homologous structures to, factor V have become available, new and better models have been produced through the years. With each new model, biochemical experiments were proposed to validate or refute it. In a few cases the model did not survive, and in others it did. Ultimately, it will be a crystal structure of factor Va that reveals the intricate details of its A2 domain and binding sites. However, the current models of both factor Va and factor Xa are now robust enough to be useful in docking simulations, which have matured greatly during the last few years, generating prothrombinase complexes. Two papers describing new prothrombinase models have appeared, see: Stoilova-McPhie et al. [169] and Lee et al. [170]. Although these models will never reveal all of the secrets of the prothrombinase complex, multiple models generated by different algorithms and constraints will be one of our best tools to guide and understand newly generated biochemical data of the prothrombinase complex.

Acknowledgments. We thank Bruno Villoutreix for sharing the Autin et al. [107] M3 model with us. This work was supported by grants from the National Institute of Health (5R01 HL064891, 5R01 HL 034575). T.E.A. was the recipient of a DOE EPSCoR structural biology graduate fellowship.

References

1. Nesheim ME, Taswell JB, Mann KG (1979) The contribution of bovine factor V and factor Va to the activity of prothrombinase. J Biol Chem 254:10952–10962

2. Rosing J, Tans G, Grovers Riemslag JW, et al (1980) The role of phospholipids and factor Va in the prothrombinase complex. J Biol Chem 255:274
3. Mann KG (1984) Membrane-bound enzyme complexes in blood coagulation. Prog Haemost Thromb 7:1–23
4. Kane W, Davie EW (1988) Blood coagulation factors V and VIII: structural and functional similarities and their relationships to hemorrhagic and thrombotic disorders. Blood 71:539–555
5. Krishnaswamy S, Jones KC, Mann KG (1988) Prothrombinase complex assembly: kinetic mechanism of enzyme assembly on phospholipid vesicles. J Biol Chem 263:3823–3834
6. Krishnaswamy S (1990) Prothrombinase complex assembly: contributions of protein-protein and protein-membrane interactions toward complex formation. J Biol Chem 265:3708–3718
7. Mann KG, Kalafatis M (2003) Factor V: a combination of Dr. Jekyll and Mr. Hyde. Blood 101:20–30
8. Nelsestuen GL, Ostrowski BG (1999) Membrane association with multiple calcium ions: vitamin-K-dependent proteins, annexins and pentraxins. Curr Opin Struct Biol 9:433–437
9. Padmanabhan K, Padmanabhan KP, Tulinsky A, et al (1993) Structure of human Des (1-45) factor Xa at 2.2 a resolution. J Mol Biol 232:947–966
10. Brandstetter H, Kuhne A, Bode W, et al (1996) X-ray structure of active site-inhibited clotting factor Xa. implications for drug design and substrate recognition. J Biol Chem 271:29988–29992
11. Kamata K, Kawamoto H, Honma T, et al (1998) Structural basis for chemical inhibition of human blood coagulation factor Xa. Proc Natl Acad Sci U S A 95:6630–6635
12. Lim TK, Bloomfield VA, Nelsestuen GL (1977) Structure of the prothrombin- and blood clotting factor X-membrane complexes. Biochemistry 16:4177–4181
13. Husten EJ, Esmon CT, Johnson AE (1987) The active site of blood coagulation factor Xa its distance from the phospholipid surface and its conformational sensitivity to components of the prothrombinase complex. J Biol Chem 262:12953–12961
14. Nesheim ME, Myrmel KH, Hibbard LS, et al (1979) Isolation and characterization of single chain bovine factor V. J Biol Chem 254:508–517
15. Dahlback B (1980) Human coagulation factor V purification and thrombin-catalyzed activation. J Clin Invest 66:583–591
16. Tracy PB, Eide LL, Bowie EJ, et al (1982) Radioimmunoassay of factor V in human plasma and platelets. Blood 60:59–63
17. Esmon CT (1979) The subunit structure of thrombin-activated factor V isolation of activated factor V, separation of subunits, and reconstitution of biological activity. J Bioll Chem 254:964–973
18. Mann KG, Lawler CM, Vehar GA, et al (1984) Coagulation factor V contains copper ion. J Biol Chem 259:12949–12951
19. Kane WH, Davie EW (1986) Cloning of a cDNA coding for human factor V, a blood coagulation factor homologous to factor VIII and ceruloplasmin. Proc Natl Acad Sci U S A 83:6800–6804
20. Jenny RJ, Pittman DD, Toole JJ, et al (1987) Complete cDNA and derived amino acid sequence of human factor V. Proc Natl Acad Sciences U S A 84:4846–4850
21. Hortin GL (1990) Sulfation of tyrosine residues in coagulation factor V. Blood 76:946–952
22. Cripe LD, Moore KD, Kane WH (1992) Structure of the gene for human coagulation factor V. Biochemistry 31:3777–3785
23. Guinto ER, Esmon CT, Mann KG, et al (1992) The complete cDNA sequence of bovine coagulation factor V. J Biol Chem 267:2971–2978
24. Kalafatis M, Rand MD, Jenny RJ, et al (1993) Phosphorylation of factor Va and factor VIIIa by activated platelets. Blood 81:704–719

25. Rand MD, Kalafatis M, Mann KG (1994) Platelet coagulation factor Va: the major secretory platelet phosphoprotein. Blood 83:2180–2190
26. Nesheim ME, Foster WB, Hewick R, et al (1984) Characterization of factor V activation intermediates. J Biol Chem 259:3187–3196
27. Krishnaswamy S, Russell GD, Mann KG (1989) The reassociation of factor Va from its isolated subunits. J Biol Chem 264:3160–3168
28. Kisiel W, Canfield WM, Ericsson LH (1977) Anticoagulant properties of bovine plasma protein C following activation by thrombin. Biochemistry 16:5824–5831
29. Kalafatis M, Rand MD, Mann KG (1994) The mechanism of inactivation of human factor V and human factor Va by activated protein C. J Biol Chem 269:31869–31880
30. Kalafatis M, Bertina RM, Rand MD, et al (1995) Characterization of the molecular defect in factor VR506Q. J Biol Chem 270:4053–4057
31. Nicolaes GAF, Tans G, Thomassen MCLGD, et al (1995) Peptide bond cleavages and loss of functional activity during inactivation of factor Va and factor Va R506Q by activated protein C. J Biol Chem 270:21158–21166
32. Rosing J, Hoekema L, Nicolaes GAF (1995) Effects of protein S and factor Xa on peptide bond cleavages during inactivation of factor Va and factor Va R506Q by activated protein C. J Biol Chem 270:27852–27858
33. Egan JO, Kalafatis M, Mann KG (1997) The effect of Arg306→Ala and Arg506→Gln substitutions in the inactivation of recombinant human factor Va by activated protein C and protein S. Protein Sci 6:2016–2027
34. Hockin MF, Kalafatis M, Shatos M, et al (1997) Protein C activation and factor Va inactivation on human umbilical vein endothelial cells. Arterioscler Thromb Vasc Biol 17:2765–2775
35. Dahlback B, Villoutreix BO (2003) Molecular recognition in the protein C anticoagulant pathway. J Thromb Haemost 1:1525–1534
36. Nesheim ME, Canfield WM, Kisiel W (1982) Studies of the capacity of factor Xa to protect factor Va from inactivation by activated protein C. J Biol Chem 257:1443–1447
37. Mann KG, Hockin M, Begin KJ, et al (1997) Activated protein C cleavage of factor Va leads to dissociation of the A2 domain. J Biol Chem 272:20678–20683
38. Krishnaswamy S, Williams EB, Mann KG (1986) The binding of activated protein C to factors V and Va. J Biol Chem 261:9684–9693
39. Solymoss S, Tucker MM, Tracy PB (1988) Kinetics of inactivation of membrane-bound factor Va by activated protein C: protein S modulates factor Xa protection. J Biol Chem 263:14884–14890
40. Kalafatis M, Mann KG (1993) Role of the membrane in the inactivation of factor Va by activated protein C. J Biol Chem 268:27246–27257
41. Hockin MF, Cawthern KM, Kalafatis M (1999) A model describing the inactivation of factor Va by APC: bond cleavage, fragment dissociation, and product inhibition. Biochemistry 38:6918–6934
42. Van der Neut Kolfschoten M, Dirven RJ, Vos HL, et al (2004) Factor Va is inactivated by activated protein C in the absence of cleavage sites at Arg-306, Arg-506, and Arg-679. J Biol Chem 279:6567–6575
43. Poole S, Firtel R, Lamar E, et al (1981) Sequence and expression of the discoidin I gene family in *Dictyostelium discoideum*. J Mol Biol 153:273–289
44. Bartles JR, Galvin NJ, Frazier WA (1982) Discoidin. I. Membrane interactions. II. Discoidin I binds to and agglutinates negatively charged phospholipid vesicles. Biochim Biophys Acta 1982:4189
45. Macedo-Ribeiro S, Bode W, Huber R, et al (1999) Crystal structures of the membrane-binding C2 domain of human coagulation factor V. Nature 402:434–439
46. Adams TE, Hockin MF, Mann KG, et al (2004) The crystal structure of activated protein C-inactivated bovine factor Va: implications for cofactor function. Proc Natl Acad Sci U S A 101:8918–8923

47. Kim SW, Quinn-Allen MA, Camp JT, et al (2000) Identification of functionally important amino acid residues within the C2-domain of human factor V using alanine-scanning mutagenesis. Biochemistry 39:1951–1958
48. Nicolaes GA, Villoutreix BO, Dahlback B (2000) Mutations in a potential phospholipid binding loop in the C2 domain of factor V affecting the assembly of the prothrombinase complex. Blood Coagul Fibrinolysis 11:89–100
49. Izumi T, Kim SW, Greist A, et al (2001) Fine mapping of inhibitory anti-factor V antibodies using factor V C2 domain mutants: identification of two antigenic epitopes involved in phospholipid binding. Thromb Haemost 85:1048–1054
50. Gilbert GE, Kaufman RJ, Arena AA (2002) Four hydrophobic amino acids of the factor VIII C2 domain are constituents of both the membrane-binding and von Willebrand factor-binding motifs. J Biol Chem 277:6374–6381
51. Peng W, Quinn-Allen MA, Kim SW, et al (2004) Trp2063 and Trp2064 in the factor Va C2 domain are required for high-affinity binding to phospholipid membranes but not for assembly of the prothrombinase complex. Biochemistry 43:4385–4393
52. Mollica L, Fraternali F, Musco G (2006) Interactions of the C2 domain of human factor V with a model membrane. Proteins 64:363–375
53. Mann KG, Nesheim ME, Tracy PB (1981) Molecular weight of undegraded plasma factor V. Biochemistry 20:28–33
54. Fowler WE, Fay PJ, Arvan DS, et al (1990) Electron microscopy of human factor V and factor VIII: correlation of morphology with domain structure and localization of factor V activation fragments. Proc Natl Acad Sci U S A 87:7648–7652
55. Mosesson MW, Church WR, DiOrio JP, et al (1990) Structural model of factors V and Va based on scanning transmission electron microscope images and mass analysis. J Biol Chem 265:8863–8868
56. Stoylova S, Mann KG, Brisson A (1994) Structure of membrane-bound human factor Va. FEBS Lett 351:330–334
57. Toso R, Camire RM (2004) Removal of B-domain sequences from factor V rather than specific proteolysis underlies the mechanism by which cofactor function is realized. J Biol Chem 279:21643–21650
58. Saleh M, Peng W, Quinn-Allen MA, et al (2004) The factor V C1 domain is involved in membrane binding: identification of functionally important amino acid residues within the C1 domain of factor V using alanine scanning mutagenesis. Thromb Haemost 91:16–27
59. Pemberton S, Lindley P, Zaitsev V, et al (1997) A molecular model for the triplicated A domains of human factor VIII based on the crystal structure of human ceruloplasmin. Blood 89:2413–2421
60. Villoutreix BO, Dahlback B (1998) Structural investigation of the A domains of human blood coagulation factor V by molecular modeling. Protein Sci 7:1317–1325
61. Pratt KP, Shen BW, Takeshima K, et al (1999) Structure of the C2 domain of human factor VIII at 1.5 Å resolution. Nature 402:439–442
62. Pellequer JL, Gale AJ, Getzoff ED (2000) Three-dimensional model of coagulation factor Va bound to activated protein C. Thromb Haemost 84:849–857
63. Stoilova-McPhie S, Villoutreix BO, Mertens K, et al (2002) 3-Dimensional structure of membrane-bound coagulation factor VIII: modeling of the factor VIII heterodimer within a 3-dimensional density map derived by electron crystallography. Blood 99:1215–1223
64. Gale AJ, Yegneswaren S, Xu X, et al (2007) Characterization of a factor Xa binding site on factor Va near Arg506 APC cleavage site. J Biol Chem 282:21848–21855
65. Orban T, Kalafatis M, Gogonea V (2005) Completed three-dimensional model of human coagulation factor Va: molecular dynamics simulations and structural analyses. Biochemistry 44:13082–13090
66. Kalafatis M, Beck DO, Mann KG (2003) Structural requirements for expression of factor Va activity. J Biol Chem 278:33550–33561

67. Beck DO, Bukys MA, Singh LS, et al (2004) The contribution of amino acid region ASP695-TYR698 of factor V to procofactor activation and factor Va function. J Biol Chem 279:3084–3095
68. Higgins DL, Mann KG (1983) The interaction of bovine factor V and factor V derived peptides wiht phospholipid vesicles. J Biol Chem 258:6503–6508
69. Krishnaswamy S, Mann KG (1988) The binding of factor Va to phospholipid vesicles. J Biol Chem 263:5714–5723
70. Kalafatis M, Rand MD, Mann KG (1994) Factor Va–membrane interaction is mediated by two regions located on the light chain of the cofactor. Biochemistry 33:486–493
71. Ortel TL, Quinn-Allen MA, Keller FG, et al (1994) Localization of functionally important epitopes within the second C-type domain of coagulation factor V using recombinant chimeras. J Biol Chem 269:15898–15905
72. McDonald JF, Shah AM, Schwalbe RA, et al (1997) Comparison of naturally occurring vitamin K-dependent proteins: correlation of amino acid sequences and membrane binding properties suggests a membrane contact site. Biochemistry 36:5120–5127
73. Heeb MJ, Kojima Y, Hackeng TM, et al (1996) Binding sites for blood coagulation factor Xa and protein S involving residues 493-506 in factor Va. Protein Sci 5:1883–1889
74. Kojima Y, Heeb MJ, Gale AJ, et al (1998) Binding site for blood coagulation factor Xa involving residues 311-325 in factor Va. J Biol Chem 273:14900–14905
75. Kalafatis M, Beck DO (2002) Identification of a binding site for blood coagulation factor Xa on the heavy chain of factor Va: amino acid residues 323-331 of factor V represent an interactive site for activated factor X. Biochemistry 41:12715–12728
76. Singh LS, Bukys MA, Beck DO, et al (2003) Amino acids Glu323, Tyr324, Glu330, and Val331 of factor Va heavy chain are essential for expression of cofactor activity. J Biol Chem 278:28335–28345
77. Steen M, Villoutreix BO, Norstrom EA, et al (2002) Defining the factor Xa-binding site on factor Va by site-directed glycosylation. J Biol Chem 277:50022–50029
78. Chattopadhyay G, James HL, Fair DS (1992) Molecular recognition sites on factor Xa which participate in the prothrombin complex. J Biol Chem 267:12323–12329
79. Yegneswaran S, Mesters RM, Griffin JH (2003) Identification of distinct sequences in human blood coagulation factor Xa and prothrombin essential for substrate and cofactor recognition in the prothrombinase complex. J Biol Chem 278:33312–33318
80. Rezaie AR (2000) Identification of basic residues in the heparin-binding exosite of factor Xa critical for heparin and factor Va binding. J Biol Chem 275:3320–3327
81. Rudolph AE, Porche-Sorbet R, Miletich JP (2000) Substitution of asparagine for arginine 347 of recombinant factor Xa markedly reduces factor Va binding. Biochemistry 39:2861–2867
82. Rudolph AE, Porche-Sorbet R, Miletich JP (2001) Definition of a factor Va binding site in factor Xa. J Biol Chem 276:5123–5128
83. Rezaie AR, Kittur FS (2004) The critical role of the 185-189-loop in the factor Xa interaction with Na⁺ and factor Va in the prothrombinase complex. J Biol Chem 279:48262–48269
84. Guinto ER, Esmon CT (1984) Loss of prothrombin and of factor Xa–factor Va interactions upon inactivation of factor Va. J Biol Chem 259:13986–13992
85. Bakker HM, Tans G, Thomassen MC, et al (1994) Functional properties of human factor Va lacking the Asp683-Arg709 domain of the heavy chain. J Biol Chem 269:20662–20667
86. Dharmawardana KR, Bock PE (1998) Demonstration of exosite I-dependent interactions of thrombin with human factor V and factor Va involving the factor Va heavy chain: analysis by affinity chromatography employing a novel method for active- site-selective immobilization of serine proteinases. Biochemistry 37:13143–13152
87. Anderson PJ, Nesset A, Dharmawardana KR (2000) Role of proexosite I in factor Va-dependent substrate interactions of prothrombin activation. J Biol Chem 275:16435–16442
88. Wilkens M, Krishnaswamy S (2002) The contribution of factor Xa to exosite-dependent substrate recognition by prothrombinase. J Biol Chem 277:9366–9374

89. Bajaj SP (1999) Region of factor IXa protease domain that interacts with factor VIIIa: analysis of select hemophilia B mutants. Thromb Haemost 82:218–225

90. Mathur A, Bajaj SP (1999) Protease and EGF1 domains of factor IXa play distinct roles in binding to factor VIIIa: importance of helix 330 (helix 162 in chymotrypsin) of protease domain of factor IXa in its interaction with factor VIIIa. J Biol Chem 274:18477–18486

91. Bajaj SP, Schmidt AE, Mathur A, et al (2001) Factor IXa:factor VIIIa interaction: helix 330-338 of factor IXa interacts with residues 558-565 and spatially adjacent regions of the a2 subunit of factor VIIIa. J Biol Chem 276:16302–16309

92. Pryzdial ELG, Mann KG (1991) The association of coagulation factor Xa and factor Va. J Biol Chem 266:8969–8977

93. Krishnaswamy S, Mann KG, Nesheim ME (1986) The prothrombinase-catalyzed activation of prothrombin proceeds through the intermediate meziothrombin in an ordered, sequential reaction. J Biol Chem 261:8977–8984

94. Kalafatis M, Xue J, Lawler CM, et al (1994) Contribution of the heavy and light chains of factor Va to the interaction with factor Xa. Biochemistry 33:6538–6545

95. Pusey ML, Nelsestuen GL (1984) Membrane binding properties of blood coagulation factor V and derived peptides. Biochemistry 23:6202–6210

96. Lecompte MF, Bouix G, Mann KG (1994) Electrostatic and hydrophobic interactions are involved in factor Va binding to membranes containing acidic phospholipids. J Biol Chem 269:1905–1910

97. Ortel TL, Devore-Carter D, Quinn-Allen MA, et al (1992) Deletion analysis of recombinant human factor V: evidence for a phosphatidylserine binding site in the second C-type domain. J Biol Chem 267:4189–4198

98. Comfurius P, Smeets EF, Willems GM, et al (1994) Assembly of the prothrombinase complex on lipid vesicles depends on the stereochemical configuration of the polar headgroup of phosphatidylserine. Biochemistry 33:10319–10324

99. Nelsestuen GL, Broderius M (1977) Interaction of prothrombin and blood clotting factor X with membranes of varying composition. Biochemistry 16:4172–4177

100. McDuffie FC, Giffin C, Niedringhaus R, et al (1979) Prothrombin, thrombin and prothrombin fragments in plasma of normal individuals and of patients with laboratory evidence of disseminated intravascular coagulation. Thromb Res 16:759–773

101. Krishnaswamy S, Church WR, Nesheim ME, et al (1987) Activation of human prothrombin by human prothrombinase: influence of factor Va on the reaction mechanism. J Biol Chem 262:3291–3299

102. Boskovic DS, Giles AR, Nesheim ME (1990) Studies of the role of factor Va in the factor Xa-catalyzed activation of prothrombin, fragment 1.2-prethrombin-2, and dansyl-L-glutamyl-glycyl-L-arginine-meizothrombin in the absence of phospholipid. J Biol Chem 265:10497–10505

103. Nesheim M, Kettner C, Shaw E, et al (1981) Cofactor dependence of factor Xa incorporation into the prothrombinase complex. J Biol Chem 256:6537–6540

104. Rezaie AR, Esmon CT (1995) Contribution of residue 192 in factor Xa to enzyme specificity and function. J Biol Chem 270:16176–16181

105. Krishnaswamy S, Betz A (1997) Exosites determine macromolecular substrate recognition by prothrombinase. Biochemistry 36:12080–12086

106. Betz A, Krishnaswamy S (1998) Regions remote from the site of cleavage determine macromolecular substrate recognition by the prothrombinase complex. J Biol Chem 273:10709–10718

107. Autin L, Steen M, Dahlback B, et al (2006) Proposed structural models of the prothrombinase (FXa–FVa) complex. Proteins 63:440–450

108. Pellequer JL, Gale AJ, Griffin JH, et al (1998) Homology models of the C domains of blood coagulation factors V and VIII: a proposed membrane binding mode for FV and FVIII C2 domains. Blood Cells Mol Dis 24:448–461

109. Brandstetter H, Bauer M, Huber R, et al (1995) X-ray structure of clotting factor IXa: active site and module structure related to Xase activity and hemophilia B. Proc Natl Acad Sci U S A 92:9796–9800
110. Bianchini EP, Pike RN, Le Bonniec BF (2004) The elusive role of the potential factor X cation-binding exosite-1 in substrate and inhibitor interactions. J Biol Chem 279:3671–3679
111. Walker RK, Krishnaswamy S (1993) The influence of factor Va on the active site of factor Xa. J Biol Chem 268:13920–13929
112. Mann KG, Jenny RJ, Krishnaswamy S (1988) Cofactor proteins in the assembly and expression of blood clotting enzyme complexes. Annu Rev Biochem 57:915–956
113. Davie EW, Fujikawa K, Kisiel W (1991) The coagulation cascade: initiation, maintenance and regulation. Biochemistry 30:10363–10370
114. Baugh RJ, Dickinson CD, Ruf W, et al (2000) Exosite interactions determine the affinity of factor X for the extrinsic Xase complex. J Biol Chem 275:28826–28833
115. Boskovic DS, Krishnaswamy S (2000) Exosite binding tethers the macromolecular substrate to the prothrombinase complex and directs cleavage at two spatially distinct sites. J Biol Chem 275:38561–38570
116. Chen L, Yang L, Rezaie AR (2003) Proexosite-1 on prothrombin is a factor Va-dependent recognition site for the prothrombinase complex. J Biol Chem 278:27564–27569
117. Boskovic DS, Troxler T, Krishnaswamy S (2004) Active site-independent recognition of substrates and product by bovine prothrombinase: a fluorescence resonance energy transfer study. J Biol Chem 279:20786–20793
118. Krishnaswamy S (2005) Exosite-driven substrate specificity and function in coagulation. J Thromb Haemost 3:54–67
119. Wong MY, Gurr JA, Walsh PN (1999) The second epidermal growth factor-like domain of human factor IXa mediates factor IXa binding to platelets and assembly of the factor X activating complex. Biochemistry 38:8948–8960
120. Wilkinson FH, Ahmad SS, Walsh PN (2002) The factor IXa second epidermal growth factor (EGF2) domain mediates platelet binding and assembly of the factor X activating complex. J Biol Chem 277:5734–5741
121. Hertzberg MS, Ben-Tal O, Furie B, et al (1992) Construction, expression, and characterization of a chimera of factor IX and factor X: the role of the second epidermal growth factor domain and serine protease domain in factor Va binding. J Biol Chem 267:14759–14766
122. Thiec F, Cherel G, Christophe OD (2003) Role of the Gla and first epidermal growth factor-like domains of factor X in the prothrombinase and tissue factor-factor VIIa complexes. J Biol Chem 278:10393–10399
123. Krishnaswamy S (1990) Prothrombinase complex assembly: contributions of protein-protein and protein-membrane interactions toward complex formation. J Biol Chem 265:3708–3718
124. Steen M, Dahlback B (2002) Thrombin-mediated proteolysis of factor V resulting in gradual B-domain release and exposure of the factor Xa-binding site. J Biol Chem 277:38424–38430
125. Brufatto N, Nesheim ME (2003) Analysis of the kinetics of prothrombin activation and evidence that two equilibrating forms of prothrombinase are involved in the process. J Biol Chem 278:6755–6764
126. Mann KG, Nesheim ME, Church WR, et al (1990) Surface-dependent reactions of the vitamin K-dependent enzyme complexes. Blood 76:1–16
127. Nesheim M, Eid S, Mann KG (1981) Assembly of the prothrombinase complex in the absence of prothrombin. J Biol Chem 29:9874–9882
128. Manithody C, Rezaie AR (2005) Functional mapping of charged residues of the 82-116 sequence in factor Xa: evidence that lysine 96 is a factor Va independent recognition site for prothrombin in the prothrombinase complex. Biochemistry 44:10063–10070

129. Toso R, Camire RM (2006) Role of hirudin-like factor Va heavy chain sequences in prothrombinase function. J Biol Chem 281:8773–8779
130. Blostein MD, Rigby AC, Jacobs M, et al (2000) The Gla domain of human prothrombin has a binding site for factor Va. J Biol Chem 275:38120–38126
131. Esmon CT, Jackson CM (1974) The conversion of prothrombin to thrombin. IV. The function of the fragment 2 region during activation in the presence of factor. J Biol Chem 25:7791–7797
132. Bajaj SP, Butkowski RJ, Mann KG (1975) Prothrombin fragments: Ca^{2+} binding and activation kinetics. J Biol Chem 250:2150–2156
133. Kotkow KJ, Deitcher SR, Furie B, et al (1995) The second kringle domain of prothrombin promotes factor Va mediated prothrombin activation by prothrombinase. J Biol Chem 270:4551–4557
134. Deguchi H, Takeya H, Gabazza EC, et al (1997) Prothrombin kringle 1 domain interacts with factor Va during the assembly of prothrombinase complex. Biochem J 321(Pt 3):729–735
135. Anderson PJ, Nesset A, Dharmawardana KR, et al (2000) Characterization of pro-exosite I on prothrombin. J Biol Chem 275:16428–16434
136. Lecompte MF, Krishnaswamy S, Mann KG, et al (1987) Membrane penetration of bovine factor V and Va detected by labeling with 5-iodonaphthalene-1-azide. J Biol Chem 262:1935
137. Krishnaswamy S, Mann KG (1988) The binding of factor Va to phospholipid vesicles. J Biol Chem 263:57154–5723
138. Kalafatis M, Jenny RJ, Mann KG (1990) Identification and characterization of a phospholipid-binding site of bovine factor Va. J Biol Chem 265:21580–21589
139. Srivastava A, Quinn-Allen MA, Kim SW, et al (2001) Soluble phosphatidylserine binds to a single identified site in the C2 domain of human factor Va. Biochemistry 40:8246–8255
140. Zhai X, Srivastava A, Drummond DC, et al (2002) Phosphatidylserine binding alters the conformation and specifically enhances the cofactor activity of bovine factor Va. Biochemistry 41:5675–5684
141. Koppaka V, Talbot WF, Zhai X, et al (1997) Roles of factor Va heavy and light chains in protein and lipid rearrangements associated with the formation of a bovine factor Va–membrane complex. Biophys J 73:2638–2652
142. Zeibdawi AR, Pryzdial EL (2001) Mechanism of factor Va inactivation by plasmin: loss of A2 and A3 domains from a Ca^{2+}-dependent complex of fragments bound to phospholipid. J Biol Chem 276:19929–19936
143. Rosing J, Bakker HM, Thomassen MCLGD, et al (1993) Characterization of two forms of human factor Va with different cofactor activities. J Biol Chem 268:21130–21136
144. Kim SW, Ortel TL, Quinn-Allen MA, et al (1999) Partial glycosylation at asparagine-2181 of the second C-type domain of human factor V modulates assembly of the prothrombinase complex. Biochemistry 38:11448–11454
145. Weinreb GE, Mukhopadhyay K, Majumder R, et al (2003) Cooperative roles of factor V(a) and phosphatidylserine-containing membranes as cofactors in prothrombin activation. J Biol Chem 278:5679–5684
146. Annamalai AE, Rao AK, Chiu HC, et al (1987) Epitope mapping of functional domains of human factor Va with human and murine monoclonal antibodies: evidence for the interaction of heavy chain with factor Xa and calcium. Blood 70:139–146
147. Kalafatis M, Xue J, Lawler CM, et al (1994) Contribution of the heavy and light chains of factor Va to the interaction with factor Xa. Biochemistry 33:6538–6545
148. Kojima Y, Heeb MJ, Gale AJ, et al (1998) Binding site for blood coagulation factor Xa involving residues 311-325 in factor Va. J Biol Chem 273:14900–14905
149. Kalafatis M, Mann KG (2001) The role of the membrane in the inactivation of factor Va by plasmin: amino acid region 307-348 of factor V plays a critical role in factor Va cofactor function. J Biol Chem 276:18614–18623

150. Kittur FS, Manithody C, Rezaie AR (2004) Role of the N-terminal epidermal growth factor-like domain of factor X/Xa. J Biol Chem 279:24189–24196
151. Guinto ER, Esmon CT (1984) Loss of prothrombin and of factor Xa–factor Va interactions upon inactivation of factor Va by activated protein C. J Biol Chem 259:13986–13992
152. Luckow EA, Lyons DA, Ridgeway TM, et al (1989) Interaction of clotting factor V heavy chain with prothrombin and prethrombin 1 and role of activated protein C in regulating this interaction: analysis by analytical ultracentrifugation. Biochemistry 28:2348–2354
153. Akhavan S, De Cristofaro R, Peyvandi F, et al (2002) Molecular and functional characterization of a natural homozygous Arg67His mutation in the prothrombin gene of a patient with a severe procoagulant defect contrasting with a mild hemorrhagic phenotype. Blood 100:1347–1353
154. Dharmawardana KR, Olson ST, Bock PE (1999) Role of regulatory exosite I in binding of thrombin to human factor V, factor Va, factor Va subunits, and activation fragments. J Biol Chem 274:18635–18643
155. Dharmawardana KR, Bock PE (1998) Demonstration of exosite I-dependent interactions of thrombin with human factor V and factor Va involving the factor Va heavy chain: analysis by affinity chromatography employing a novel method for active-site-selective immobilization of serine proteinases. Biochemistry 37:13143–13152
156. Yegneswaran S, Mesters RM, Fernandez JA, et al (2004) Prothrombin residues 473-487 contribute to factor Va binding in the prothrombinase complex. J Biol Chem 279:49019–49025
157. Taneda H, Andoh K, Nishioka J, et al (1994) Blood coagulation factor Xa interacts with a linear sequence of the kringle 2 domain of prothrombin. J Biochem (Tokyo) 116:589–597
158. Gale AJ, Griffin JH (2004) Characterization of a thrombomodulin binding site on protein C and its comparison to an activated protein C binding site for factor Va. Proteins 54:433–441
159. Kalafatis M, Egan JO, van't Veer C, et al (1997) The regulation of clotting factors. Crit Rev Eukaryot Gene Expr 7:241–280
160. Stoylova SS, Lenting PJ, Kemball-Cook G, et al (1999) Electron crystallography of human blood coagulation factor VIII bound to phospholipid monolayers. J Biol Chem 274:36573–36578
161. The PyMOL Molecular Graphics System [http://www.pymol.org]
162. Cuff JA, Clamp ME, Siddiqui AS, et al (1998) JPred: a consensus secondary structure prediction server. Bioinformatics 14:892–893
163. Van Gunsteren WF, Mark AE (1992) On the interpretation of biochemical data by molecular dynamics computer simulation. Eur J Biochem 204:947–961
164. Brunger AT, Adams PD, Clore GM, et al (1998) Crystallography & NMR system: a new software suite for macromolecular structure determination. Acta Crystallogr D Biol Crystallogr 4:905–921
165. Brooks B, Bruccoleri R, Olafson B, et al (1983) A program for macromolecular energy, minimization and dynamics calculations. J Comp Chem 4:187–217
166. Katchalski-Katzir E, Shariv I, Eisenstein M, et al (1992) Molecular surface recognition: determination of geometric fit between proteins and their ligands by correlation techniques. Proc Natl Acad Sci U S A 89:2195–2199
167. Gabb HA, Jackson RM, Sternberg MJ (1997) Modelling protein docking using shape complementarity, electrostatics and biochemical information. J Mol Biol 272:106–120
168. Jackson RM, Gabb HA, Sternberg M (1998) Rapid refinement of protein interfaces incorporating solvation: application to the docking problem. J Mol Biol 276:265–285
169. Stoilova-McPhie S, Parmenter CD, Segers K, et al (2008) Defining the structure of membrane-bound human blood coagulation factor Va. J Thromb Haemost 6:76–82
170. Lee CJ, Lin P, Chandrasekaran V, et al (2008) Proposed structural models of human factor Va and prothrombinase. J Thromb Haemost 6:83–89

Supplemental Material: Detailed Materials and Methods

Sequence Alignment

Human Factor Va

A ClustalW alignment, consistent with other alignments of both factors V and VIII with ceruloplasmin, indicated an overall sequence identity of approximately 33.5% [60, 62]. The A2 domain (Ile^{296}-Arg^{709}) of human factor V shared the most sequence identity with the A2 domain (Val^{327}-Gly^{752}) of human ceruloplasmin. Additional alignments between bovine and human factor Va sequences were performed using residues Ala^{1}-Ile^{296} and Ser^{1546}-Tyr^{2196} from the human molecule aligned with residues Ala^{1}-Ile^{296} and Ser^{1537}-Tyr^{2183} from the bovine molecule.

Human Factor Xa

A sequence alignment was generated for the light-chain (residues 1–139) fragment of human factor X encompassing the Gla, EGF1, and EGF2 domains, with the corresponding residues of porcine factor IX using ClustalW.

Homology Modeling

The crystal structure of bovine factor Va$_{i}$ (PDB accession code 1SDD) was used as the template for the A1 domain and light chain. The A2 domain of ceruloplasmin (PDB accession code 1KCW) was used as a template to build the missing A2 domain of factor Va$_{i}$. Owing to extremely low sequence homology, residues Asp^{661}-Arg^{709} of the A2 domain of factor Va were not modeled. The light chain of factor Xa was generated using the crystal structure of the homologous coagulation serine protease porcine factor IXa (PDB accession code 1PFX) as the template to provide the orientation of the Gla-EGF1-EGF2 domains with respect to the protease domain of human factor Xa (PDB accession code 1FAX) [10, 109].

Initial models were subject to a single round of manual refinement and an additional round of side-chain energy minimization. For factor Va, missing loops in the A1 and A3 domain were built automatically using ProModII by scanning through a database of structural fragments derived from the PDB. Side-chain orientations were chosen from a rotamer library and adjusted using a van der Waals exclusion test. Three loops within the A3 domain (residues Gln^{1570}-Ile^{1581}, Tyr^{1649}-Asp^{1669}, Lys^{1758}-Lys^{1772}) were manually altered based on predictions of the secondary structure program JPRED [162]. Each loop fragment was modeled independently and manually inserted. The rebuilt models for full-length factor Va and factor Xa were subject to a single round of energy minimization in SWISS-MODEL using the GROMOS96 force-field matrix [163]. Both calcium and copper ions were added to the factor Va, and the models were energy-minimized a final time using CNS [164] with the CHARMM22 force field [165].

The overall domain structure of the three A domains in the current factor Va model is consistent with previously described models of factor Va and with the crystal structure of ceruloplasmin [993 residues, root-mean-square deviation (RMSD) 1.68 Å] and bovine factor Va$_{i}$ (550 residues, RMSD 0.87 Å). The C1-C2 dimer closely resembles that of the factor Va$_i$ structure, arranged in a "side-by-side" orientation along the long edge of each domain and positioned such that both domains could potentially interact with a membrane surface simultaneously.

Protein–Protein Docking

The final models of factors Va and Xa were docked together using the 3D-Dock Suite of programs [166, 167]. Both protein models are input independently and placed on orthogonal grids. For our experiments, factor Va was held static while factor Xa was rotated and translated in three-dimensional space, with each complex being scored on surface complementarities using the program FTDock version 2.0 [167]. The complexes were then ranked based on their residue level pair potentials using RPScore. The resulting 9240 potential complexes were filtered using known biochemical information for important residues involved in binding and residues necessary for prothrombinase function. The following criteria were utilized in progressive filtering: (1) residues Gly^{493}-Arg^{506} of factor Va are within 10 Å of factor Xa; (2) residues Ile^{311}-Val^{331} of factor Va are within 15 Å of factor Xa; and (3) Arg$^{347(165)}$ of factor Xa is within 5 Å of factor Va. The resulting complexes were visually inspected, and a single complex was identified where both enzyme and cofactor were oriented appropriately to interact with a cellular membrane concurrently. The resulting complex was then subjected to a single round of energy minimization and residue side-chain refinement using the program Multi-Dock version 1.0 [168].

Several models of factor Xa were generated based on the factor VIIa (PDB accession code 1DAN) and factor IXa (PDB accession code 1PFX) crystal structures. These two homologous proteins differ in the orientations of the Gla and EGF1 domains with respect to the protease domain. Protein docking simulations using the factor VIIa-based factor Xa model resulted in complexes that differed significantly from known information about potential factor Va–factor Xa binding interfaces and were therefore excluded. Our current factor Xa model orients the light chain (Gla-EGF1-EGF2) to the proteinase domain, as observed in the factor IXa structure.

Lipid Mediators and Tissue Factor Expression

Hiroyuki Takeya and Koji Suzuki

Summary. Tissue factor (TF) is the key initiator of the coagulation cascade. Increased TF expression in circulating and vascular cells, which can be induced in response to exogenous and endogenous stimuli, is responsible for the thrombotic complications associated with acute and chronic inflammation. Lipid mediators play key roles in controlling inflammation. Among them, eicosanoids, platelet-activating factor, and lysophospholipids are synthesized in and released from circulating and vascular cells at sites of inflammation and vascular injury, and they participate in a range of physiological and pathological processes including hemostasis, thrombosis, wound healing, angiogenesis, and atherosclerosis. These lipid mediators appear to have important roles in the cross-talk between inflammation and the coagulation protease cascade by either facilitating or suppressing TF expression. This chapter summarizes recent findings that have uncovered lipid mediators as potent regulators of TF expression and activity. These findings provide an opportunity to explore new therapeutic approaches for thrombotic complications associated with inflammation.

Key words. Tissue factor · Eicosanoid · Platelet-activating factor · Lysophosphatidic acid · Sphingosine-1-phosphate

Introduction

Tissue factor (TF) is the principal cellular activator of the extrinsic pathway of the coagulation cascade [1]. This membrane-anchored glycoprotein functions as a high-affinity receptor for factor VII/VIIa, with which it forms a complex that activates factors IX and X, eventually leading to thrombin generation (see Fig. 5). Thrombin generated at sites of vascular injury plays a pivotal role in hemostasis by mediating fibrin formation and platelet activation. In addition, thrombin and other upstream coagulation proteases transduce cell signaling via activation of protease-activated receptors (PARs), which are G protein-coupled receptors (GPCRs) [see the chapter by S. R. Coughlin, this volume]. These actions, together with the cell activation induced by the platelet-derived secretory and surface molecules, facilitate the completion of hemostasis and the wound healing process.

Consistent with a protective role of TF in the hemostatic response, it is constitutively expressed in several extravascular cell types surrounding blood vessels and at boundaries of organs. TF expression can be induced de novo in many cell types including monocytes, endothelial cells (ECs), and smooth muscle cells (SMCs) in response to various exogenous and endogenous stimuli, such as endotoxin, inflammatory cytokines, growth factors, oxidized low density lipoprotein (oxidized LDL), thrombin, and other coagulation proteases [2, 3]. Increased TF activity of these cells is a feature of several pathophysiological conditions, in particular acute and chronic inflammation and atherosclerosis. The uncontrolled TF expression is responsible for the development of disseminated intravascular coagulation during sepsis [see the chapter by H. Wada, this volume]. TF is also abundantly expressed by macrophages and SMCs in atherosclerotic plaques, where TF is a critical determinant of thrombogenicity [2]. On the other hand, the TF-triggered coagulation cascade exaggerates inflammation through PARs, which if not attenuated might lead to endothelial dysfunction, atherosclerosis lesion initiation, or progression to advanced plaques.

Lipid mediators are generated from phospholipid membranes and regulate a variety of physiological and pathological processes, particularly related to inflammation, immunity, vasculogenesis, or angiogenesis [4, 5]. Examples include eicosanoids, plate-

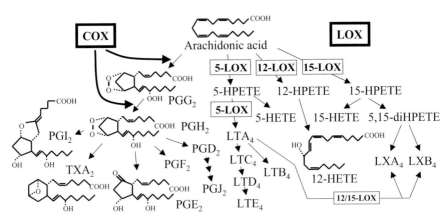

FIG. 1. Arachidonic acid metabolism and eicosanoid generation. Eicosanoids are compounds derived from eicosa (20-C)-polyenoic fatty acids and play a major role in the host defense. Prostanoids contain a cyclic ring produced via the action of cyclooxygenase (*COX*) and can be further grouped as thromboxanes (TXs), which are named after their first identification in platelets, and prostaglandins (*PG*), which are initially detected in the seminal fluid. COX consists of two isoenzymes; COX-1 is expressed constitutively, and COX-2 is generated by ligand-coupled stimulation. Tissue-specific isomerases and synthases convert PGH$_2$ into distinct prostanoids. Leukotrienes (*LT*), which are named after their initial identification in leukocytes, are generated via the 5-lipoxygenase (5-*LOX*) pathway. LTs are either produced by leukocytes at sites of inflammation or formed through transcellular metabolism after uptake and metabolism of leukocyte-derived LTA$_4$ by downstream enzymes of the 5-LOX pathway (LTA$_4$ hydrolase and LTC$_4$ synthase) in cells that normally do not express 5-LOX, such as endothelial cells (ECs). LTC$_2$, LTD$_2$, and LTE$_2$ are also called cysteinyl leukotrienes. 12-LOX, which inserts oxygen into position 12 of arachidonic acid, can be differentiated into several subtypes (platelet-type 12-LOX, leukocyte-type 12-LOX, and epidermis-type 12-LOX), and each is encoded for by a separate gene. Lipoxins (*LX*) are synthesized via dual oxygenation by 5-LOX together with 15-LOX or 12-LOX. Aspirin-acetylated COX-2 might also contribute to the synthesis of lipoxines by converting arachidonic acid into 15-epimeric hydroxy-eicosatetraenoic acid (*HETE*), which is further metabolized by 5-LOX into 15-epimeric LXs

let-activating factor (PAF), and lysophospholipids. Eicosanoids include the cyclooxy-genase (COX) and lipoxygenase (LOX) products of arachidonic acid (Fig. 1). In plasma membrane phospholipids, arachidonic acid is exclusively located at the sn-2 position of glycerol moiety, and its release is carefully regulated by phospholipase A_2. The pre-cursor of PAF, 1-O-alkyl-lysophosphatidylcholine (alkyl-LPC; lyso PAF), is also a product of phospholipase A_2 reaction and is subsequently acetylated by acetyl-CoA: lysoPAF acetyltransferase to yield PAF (Fig. 2) [6]. The best studied examples of lyso-phospholipid mediators are lysophosphatidic acid (LPA) and sphingosine 1-phos-phate (S1P), having a glycerol or sphingoid backbone to which a single carbon chain and a polar head group are attached. Most of the receptors for the lipid mediators belong to the GPCR superfamily.

Lipid mediators are important for vascular homeostasis, and some of them play direct roles in hemostasis and thrombosis. For example, thromboxane A_2 (TXA_2) is a potent platelet activator, vasoconstrictor, and SMC mitogen; and aspirin affords car-dioprotection through inhibition of TXA_2 synthesis by platelet COX-1. In addition, lipid mediators are increasingly recognized as important regulators of the TF-medi-ated positive feedback loop between the prothrombotic and proinflammatory signal-ing effects. Here, we describe the current status of the studies on the role of eicosanoids, PAF, LPA, S1P, and other lipid mediators in TF expression and activity.

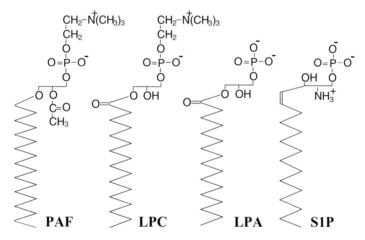

FIG. 2. Platelet-activating factor (*PAF*) and lysophospholipid mediators. Lysophosphatidylcho-line (*LPC*) is formed by phospholipase A_2-mediated hydrolysis of phosphatidylcholine. LPC concentration of 200–300 µM are found in the plasma in part due to catalysis by lecithine:cho-lesterol acyltransferase that transfers fatty acids on the *sn*-2 position of phosphatidylcholine to free cholesterol in plasma, forming cholesterol esters and LPC. LPC is also a component of oxi-dized low density lipoprotein (LDL), which is involved in the pathogenesis of atherosclerosis and inflammation. During LDL oxidation, ~40% of phosphatidylcholine contained in the LDL molecule is converted to LPC. See text for platelet-activating factor (PAF) and lysophosphatidic acid (*LPA*) generation and Fig. 4 for sphingosine-1-phosphate (*S1P*) generation

Eicosanoids

Platelets have been demonstrated to augment leukocyte TF expression and activity [7]. In platelets, arachidonic acid is converted into mainly TXA_2 by the COX-1 pathway and into 12-hydroxyeicosatetraenoic acid (12-HETE) by the platelet–12-LOX pathway. Platelet 12-HETE, but not TXA_2, is involved in platelet-mediated enhancement of leukocyte TF expression [8]. Exogenous 12-HETE up-regulates monocyte TF synthesis induced by endotoxin [8, 9] or P-selectin [10], but it has no effect in resting monocytes. LOX pathway products have been implicated as mediators of early inflammatory events in atherogenesis [11], in which 12-HETE as well as cysteinyl leukotrienes (LTs) such as LTD_4 among other LOX products [12] may participate by driving the TF-mediated inflammatory circuit.

Exogenous arachidonic acid (20:4; n-6) (Fig. 3) has been reported to induce TF production in mononuclear cells in a platelet-independent manner [13], whereas it has no effect on the basal and stimulated levels of TF activity in ECs [14, 15]. COX-1 metabolites such as TXA_2 or endoperoxides (PGG_2 or PGH_2) play a critical role in the TF induction by arachidonic acid [13]. This finding is in stark contrast to the insignificant involvement of the COX-1 pathway in the platelet-mediated enhancement of monocyte TF expression [8]. Interestingly, exogenous arachidonic acid induces increased TXA_2 synthesis and TF expression in monocytes isolated from type 2 diabetic patients compared to controls [16]. Consumption of marine n-3 polyunsaturated fatty acids (n-3 PUFAs) (Fig. 3), such as eicosapentaenoic acid (EPA; 20:5, n-3) and docosahexaenoic acid (DHA; 22:6, n-3), is suggested to protect against cardiovascular disease [17] by reducing serum triglycerides, arrhythmias, inflammation, and thrombotic tendency and improving EC function [18]. It has been demonstrated that admin-

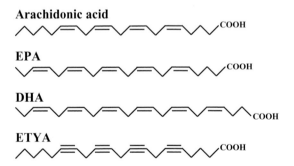

Arachidonic acid

EPA

DHA

ETYA

Fig. 3. Polyunsaturated fatty acids (PUFAs). Arachidonic acid is an n-6 PUFA, where the first double bond is at carbon atom number 6 counted from the methyl end of the carbon chain backbone. Eicosapentaenoic acid (*EPA*) and docosahexaenoic acid (*DHA*) found in fatty fish and fish oil belong to n-3 PUFA, where the first double bond is at carbon atom number 3. Because mammals cannot desaturate the fatty acids fully, mammals have to acquire "essential" PUFAs as a part of their diet. In addition, mammals cannot convert n-6 into n-3 PUFAs. Eicosatetraynoic acid (*ETYA*) is the nonmetabolizable acetylenic analogue of arachidonic acid and is a ligand of PPARα

istration of a diet rich in these n-3 PUFAs to animals or humans results in a reduction of TF activity by monocytes [19–22]. In addition, EPA and DHA have been reported to inhibit TF activity in stimulated monocytic cell lines [23, 24]. However, n-3 and n-6 PUFAs have complex effects on TF expression [14, 25, 26] in part due to a variety of their metabolites that can positively and negatively regulate TF expression (see below). Further studies are required to confirm the real effects of PUFAs on TF expression. Eicosatetraynoic acid (ETYA) (Fig. 3), which is a selective peroxisome proliferators-activated receptor-α (PPAR-α) activator, inhibits endotoxin-induced TF expression in monocyte lineage cells in a PPAR-α-dependent manner [27].

Exposure of monocytes to endotoxin induces COX-2 expression, which is associated with enhanced levels of TXA$_2$. This COX-2-derived TXA$_2$ is suggested to mediate partly the endotoxin-stimulated TF synthesis in monocytes and whole blood through the interaction with its GPCR [28, 29]. In ECs, PGI$_2$ (prostacyclin) (Fig. 1) is produced as the main COX-2 product in inflammatory states. PGI$_2$ has opposing biological properties to TXA$_2$, promoting vasodilation and inhibiting platelet aggregation via the specific GPCR on vascular SMCs and platelets; and, indeed, disruption of this GPCR in mice increases the risk of thrombosis [30]. Stable PGI$_2$ analogues such as iloprost, which are used to treat peripheral vascular disease, inhibit TF synthesis induced in monocytes and monocytic THP-1 cells in response to various stimuli including endotoxin, tumor necrosis factor-α (TNFα) and interleukin-1β (IL-1β) [31, 32], despite no effect on endothelial TF synthesis [33]. The PGI$_2$ receptor is predominantly coupled to the Gs family of heterotrimeric G proteins, leading to the accumulation of cAMP. As with other agents that elevate intracellular levels of cAMP (e.g., pentoxifylline, forskolin, isobutylmethylxanthine, and dibutyryl cAMP [34]), it is suggested that the accumulation of cAMP by iloprost treatment is related to its inhibitory effects on TF synthesis [35]. Although PGE$_2$ is also reported to down-regulate TF and procoagulant activity induced by various stimuli [31, 36], a recent study has shown that PGE$_2$ may have a prothrombotic role in vivo [37]. Disruption of both the GPCRs for TXA$_2$ and PGI$_2$ has provided evidence that PGI$_2$ modulates cardiovascular homeostasis by regulating the response to TXA$_2$ [38]. Thus, it is envisioned in vivo that the net effect on TF expression in a variety of physiological and pathological settings reflects a balance of the opposing functions of these eicosanoids.

Nonconventional prostanoids such as PGJ$_2$ and its derivatives, collectively categorized as cyclopentenone prostaglandins, are known to evoke cell-protective and antiinflammatory responses. The data regarding their effects on TF expression are conflicting. Eligini et al. indicated that 15-deoxy-D^{12,14}-PGJ$_2$ (15d-PGJ$_2$) inhibits TF expression in macrophages and ECs induced by endotoxin and TNFα [39], but others have described the opposite or no effect [27, 40]. These prostanoids are thought to be ligands of PPAR-γ, a nuclear receptor that targets gene expression regulating glucose and lipoprotein metabolism, inflammation, and cellular proliferation and differentiation. However, in contrast to PPAR-α [27], it has been consistently suggested that a role of PPAR-γ activation in the regulation of TF expression is excluded [27, 39, 40]. Lipoxin A$_4$ (LXA$_4$) is a LOX-derived eicosanoid (Fig. 1) that also displays potent antiinflammatory actions. Intriguingly, LXA$_4$ has been shown to trigger TF expression in monocytes, ECs, and ECV304 cells (nonendothelial parenchymal cells) [41].

Platelet-Activating Factor

Platelet-activating factor (Fig. 2), which was originally described as a platelet aggregation mediator released from basophils during antigen-induced anaphylaxis [42], is now known as a potent proinflammatory and vasoactive lipid mediator. Upon stimulation, PAF is produced by and activates a variety of cells including platelets, neutrophils, eosinophils, monocytes, and ECs [43]. The effects of PAF are mediated through stimulation of a specific GPCR, which is the first lipid mediator receptor to be cloned [44]. The PAF receptor is mainly coupled to Gq/11, Go, and Gi [43].

PAF enhances endotoxin-induced TF activity in macrophages, although PAF alone has minimal effects [45–47]. Activation of macrophages by PAF augments inositol phosphate generation and calcium mobilization [43]. As a result of this initial activation, macrophages are "primed" for enhanced activation by endotoxin [45, 46]. The effect of PAF on TF expression is stimulus dependent as PAF does not enhance TF expression in macrophages induced by IL-1β [48]. Unlike macrophages, circulating monocytes cannot be primed with PAF [49, 50], and both granulocytes and platelets are required for PAF-mediated potentiation of TF induction in monocytes [50, 51]. It has been proposed that PAF acts as a juxtacrine or autocrine mediator of the augmentation of TF expression in ECs in some situations, including hypoxia [48], and lymphocyte adhesion to ECs [52, 53]. PAF mediates TF production after treatment of interferon-γ-stimulated ECs with antibodies to E-selectin and intracellular adhesion molecule-1 (ICAM-1) together [54]. In cultured SMCs, PAF alone can induce TF activity without any other stimuli [55, 56]. This result, which is contrast to the priming effect of PAF on TF induction in macrophages and ECs, should be interpreted with caution because isolated and cultured SMCs change from a contractile to a synthetic active phenotype. Nevertheless, because a similar phenotype modulation of SMCs characterized by active proliferation appears to be an early event in the pathogenesis of atherosclerosis, PAF per se might be able to induce TF in atherogenic situations.

PAF may control TF expression in neutrophils and eosinophils and may have a role in the release of so-called blood-borne TF from these cells [57, 58]. Several research groups have reported the presence of biologically active TF circulating in blood, although the origin, concentration, and regulation of activity (e.g., soluble or membrane-bound, cryptic or exposed) of blood-borne TF remain topics of debate [59]. Monocytes/macrophages are thought to be cellular sources, although ECs, platelets, and other blood cells [60] may contribute to the pool of blood-borne TF. In a rabbit model of acute obstructive cholangitis, one of the most virulent causes of sepsis, TF expression is detected in infiltrating neutrophils in the rabbit liver [60]. It was demonstrated using PAF antagonist in this in vivo animal model that PAF plays an important role in neutrophil TF expression [57]. Recently, immunoelectron microscopy has revealed that eosinophils harbor TF [58]. PAF, and more pronounced in combination with granulocyte-macrophage colony-stimulating factor (GM-CSF) induces TF activity on eosinophils by translocating TF, which is preformed and located in the specific granules in resting eosinophils, to the cell membrane. PAF/GM-CSF treatment also increases the TF transcript levels in eosinophils [58]. It remains to be elucidated whether circulating TF actually arises from neutrophils and eosinophils in response to PAF in vivo.

Lysophosphatidic Acid and Lysophosphatidylcholine

In 1996 and 1998, two related GPCRs named ventricular zone gene-1 (vzg-1) and endothelial differentiation gene-1 (EDG-1) were shown to be the receptors for LPA and S1P, respectively [61–63]. Since then, there has been an enormous increase in knowledge about these two lysophospholipids (see below for S1P). LPA (Fig. 2) elicits cell proliferation, migration and survival, neurite retraction, smooth muscle contraction, and platelet aggregation by acting on its specific GPCRs; and it is implicated in a variety of physiological and pathological responses including brain development, wound healing, obesity, and cancer [64]. Thrombogenic and atherogenic activities of LPA are suggested by the finding that LPA accumulates in oxidized LDL and atherosclerotic plaques, where it mediates platelet activation, SMC dedifferentiation, and neointima formation [65]. In addition, it was recently suggested that LPA exerts thrombogenic activity through procoagulant phosphatidylserine exposure and microvesicle formation in erythrocytes [66]. LPA is mainly produced extracellularly via cleavage of LPC by the lysophospholipid-specific phospholipase D that is identical with the autocrine cell-motility factor autotaxin [64]. LPC (Fig. 2) is the major component of oxidized LDL and can mimic some of the atherogenic effects of oxidized LDL [67, 68]. LPC acts as a chemotactic factor for monocytes [69], growth factor for SMCs [70], inhibitor of endothelium-dependent arterial relaxation [71], and inducer of gene expression of vascular cell adhesion molecule-I (VCAM-I) and ICAM-I in ECs [72], and of plasminogen activator inhibitor-1 (PAI-1) and monocyte chemoattractant protein-1 (MCP-1) in SMCs [73, 74].

Oxidized LDL, especially minimally oxidized LDL, induces TF expression at the level of transcription in SMCs [75] and ECs [76–78]. LPA has been reported to be the active component of oxidized LDL responsible for the TF induction in SMCs [79]. The effect of LPC on TF induction in SMCs has not been evaluated. No definite effect of LPA and LPC on TF induction has been observed in ECs [80]. Instead, oxidized phospholipids such as oxidized 1-palmitoyl-2-arachidonoyl-sn-glycero-3-phosphoryl-choline are suggested to be the major active components of oxidized LDL that induce TF expression in ECs [81]. In monocytes/macrophages, the effects of oxidized LDL and its components on TF expression are less clear. Several reports have noted that oxidized LDL per se induces TF activity in monocyte lineage cells [82–84]. However, results have not been consistent among publications; oxidized LDL alone does not induce TF expression in these cells [85, 86]. The observed discrepancy might be related to differences in the intensity of LDL oxidation, the amount of oxidized LDL, and actual state of differentiation or activation of cells that were used [86]. Furthermore, accumulating data suggest that components of oxidized LDL, including LPC and oxidized phospholipids, although promoting several inflammatory, atherogenic, and thrombotic cellular responses can stimulate antiinflammatory, tissue-protective, and antithrombotic mechanisms [87–89]. In this context, LPC has been shown to suppress endotoxin-induced TF expression in monocytes [90]. Further studies are needed to examine the role of LPA, LPC, oxidized phospholipids, and other compounds of oxidized LDL in the regulation of TF expression in different cell types.

Sphingosine-1-Phosphate and Other Sphingolipids

Sphingolipids comprise a complex class of lipids named for their enigmatic "sphinx-like" properties [91]. Once thought to be useful only as membrane-forming molecules, sphingolipids have been found to be highly bioactive and regulate diverse cellular functions. Sphingomyelin metabolites (Fig. 4) act as first and second messengers in a variety of signaling pathways. S1P mainly works extracellularly as a lipid mediator, and ceramide acts primarily as an intracellular signaling lipid. The so-called sphingolipid rheostat proposes that the relative levels of S1P and ceramide are important determinants of cell fate: S1P promotes proliferation, survival, and inhibition of apoptosis, whereas ceramide is a growth-inhibiting lipid implicated in differentiation and apoptosis. In addition to being messengers, sphingolipids are involved in dynamic plasma membrane functions—e.g., sphingolipid- and cholesterol-enriched microdomains (lipid rafts and caveolae) as platforms for the local assembly of signaling complexes. With regard to the regulation of TF activity, TF is localized in the sphingolipid-cholesterol microdomains [92–97]; and disruption of the microdomains alters the coagulant activity of TF, shedding of TF-bearing microvesicles [98], and cell signaling induced by TF–factor VIIa [99].

Sphingosine-1-phosphate is thought to be abundantly stored in platelets and released upon stimulation at sites of vascular injury. It induces robust activation of ECs and SMCs through its GPCRs encoded by endothelial differentiation genes [see the chapter by Y. Yatomi, this volume]. S1P regulates physiological and pathological processes similar to those regulated by thrombin, which include embryogenesis, wound healing, tissue remodeling, vasculogenesis, and angiogenesis. This platelet-derived lipid mediator strongly potentiates TF expression in ECs induced by throm-

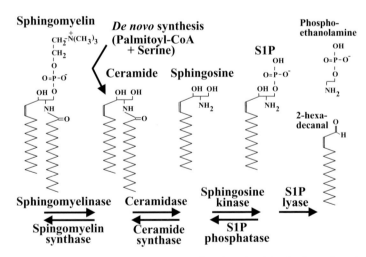

Fig. 4. Sphingolipid metabolism. Sphingosine-1-phosphate (*S1P*) is a biologically active sphingolipid produced in many cells by the phosphorylation of sphingosine and irreversibly degraded by S1P lyase. The sphingosine precursor ceramide can be formed by breakdown of the plasma membrane component sphingomyelin or de novo synthesis

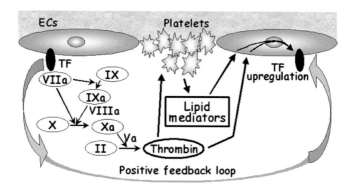

FIG. 5. Tissue factor (TF)-mediated amplified loop of thrombotic and inflammatory events. Thrombin is known to be a potent stimulator of TF expression in ECs, suggesting the TF-mediated positive feedback loop: TF up-regulation increases thrombin, which in turn stimulates ECs to express TF. Several lipid mediators are known to be synthesized in, and released from, circulating and vascular cells at sites of inflammation and vascular injury and modulate EC function. It has been shown in vitro that S1P, which is released from activated platelets, up-regulates TF induction in ECs, suggesting that S1P might participate in the positive feedback loop

bin, although S1P alone does not affect TF expression [36]. The enhancement of TF expression by S1P likely promotes further thrombin and S1P generation, suggesting that S1P might play a critical role in the TF-mediated positive feedback loop between coagulation activation reactions and inflammation at wound sites (Fig. 5). On the other hand, several S1P-induced EC responses are thought to be antiatherogenic [see the chapter by Y. Yatomi, this volume]. The pathophysiological impact of S1P-promoted TF up-regulation remains to be examined.

Conclusions

The actions of lipid mediators on TF expression are varied and complex when compared to those of protein mediators, mainly due to multiple metabolic and nonenzymatic oxidative pathways that generate a number of bioactive lipid molecules from a single molecular lipid species. However, it is becoming increasingly clear that lipid mediators play important roles in controlling TF expression and activity, particularly during interactions of circulating and vascular cells at sites of inflammation and vascular injury. Further research is required to understand the delicate balance between their "good" and "bad" actions on TF expression, how they are regulated, and how to manipulate their actions therapeutically in the prevention or treatment of thrombotic complications.

References

1. Monroe DM, Key NS (2007) The tissue factor-factor VIIa complex: procoagulant activity, regulation, and multitasking. J Thromb Haemost 5:1097–1105
2. Steffel J, Luscher TF, Tanner FC (2006) Tissue factor in cardiovascular diseases: molecular mechanisms and clinical implications. Circulation 113:722–731

3. Deguchi H, Takeya H, Wada H, et al (1997) Dilazep, an antiplatelet agent, inhibits tissue factor expression in endothelial cells and monocytes. Blood 90:2345–2356

4. Hla T (2005) Genomic insights into mediator lipidomics. Prostaglandins Other Lipid Mediat 77:197–209

5. Eyster KM (2007) The membrane and lipids as integral participants in signal transduction: lipid signal transduction for the non-lipid biochemist. Adv Physiol Educ 31:5–16

6. Shindou H, Hishikawa D, Nakanishi H, et al (2007) A single enzyme catalyzes both platelet-activating factor production and membrane biogenesis of inflammatory cells: cloning and characterization of acetyl-CoA:LYSO-PAF acetyltransferase. J Biol Chem 282:6532–6539

7. Niemetz J, Marcus AJ (1974) The stimulatory effect of platelets and platelet membranes on the procoagulant activity of leukocytes. J Clin Invest 54:1437–1443

8. Lorenzet R, Niemetz J, Marcus AJ, et al (1986) Enhancement of mononuclear procoagulant activity by platelet 12-hydroxyeicosatetraenoic acid. J Clin Invest 78:418–423

9. Eilertsen KE, Olsen JO, Osterud B (2003) Ex-vivo regulation of endotoxin-induced tissue factor in whole blood by eicosanoids. Blood Coagul Fibrinolysis 14:41–48

10. Pellegrini G, Malandra R, Celi A, et al (1998) 12-Hydroxyeicosatetraenoic acid upregulates P-selectin-induced tissue factor activity on monocytes. FEBS Lett 441:463–466

11. Funk CD (2006) Lipoxygenase pathways as mediators of early inflammatory events in atherosclerosis. Arterioscler Thromb Vasc Biol 26:1204–1206

12. Uzonyi B, Lotzer K, Jahn S, et al (2006) Cysteinyl leukotriene 2 receptor and protease-activated receptor 1 activate strongly correlated early genes in human endothelial cells. Proc Natl Acad Sci U S A 103:6326–6331

13. Cadroy Y, Dupouy D, Boneu B (1998) Arachidonic acid enhances the tissue factor expression of mononuclear cells by the cyclo-oxygenase-1 pathway: beneficial effect of n-3 fatty acids. J Immunol 160:6145–6150

14. Tardy B, Bordet JC, Berruyer M, et al (1992) Priming effect of adrenic acid (22:4(n-6)) on tissue factor activity expressed by thrombin-stimulated endothelial cells. Atherosclerosis 95:51–58

15. Del Turco S, Basta G, Lazzerini G, et al (2007) Parallel decrease of tissue factor surface exposure and increase of tissue factor microparticle release by the n-3 fatty acid docosahexaenoate in endothelial cells. Thromb Haemost 98:210–219

16. Konieczkowski M, Skrinska VA (2001) Increased synthesis of thromboxane A_2 and expression of procoagulant activity by monocytes in response to arachidonic acid in diabetes mellitus. Prostaglandins Leukot Essent Fatty Acids 65:133–138

17. He K, Song Y, Daviglus ML, et al (2004) Accumulated evidence on fish consumption and coronary heart disease mortality: a meta-analysis of cohort studies. Circulation 109:2705–2711

18. Connor WE (2000) Importance of n-3 fatty acids in health and disease. Am J Clin Nutr 71:171S–175S

19. Hansen JB, Olsen JO, Wilsgard L, et al (1989) Effects of dietary supplementation with cod liver oil on monocyte thromboplastin synthesis, coagulation and fibrinolysis. J Intern Med Suppl 731:133–139

20. Harker LA, Kelly AB, Hanson SR, et al (1993) Interruption of vascular thrombus formation and vascular lesion formation by dietary n-3 fatty acids in fish oil in nonhuman primates. Circulation 87:1017–1029

21. Tremoli E, Eligini S, Colli S, et al (1994) n-3 Fatty acid ethyl ester administration to healthy subjects and to hypertriglyceridemic patients reduces tissue factor activity in adherent monocytes. Arterioscler Thromb 14:1600–1608

22. Chu AJ, Moore J, Sampsell RN (1997) Daily supplementation with MaxEPA suppresses endotoxin-inducible monocytic procoagulation in dogs. J Surg Res 71:93–99

23. Chu AJ, Moore J (1991) Differential effects of unsaturated fatty acids on modulation of endotoxin-induced tissue factor activation in cultured human leukemia U937 cells. Cell Biochem Funct 9:231–238
24. Chu AJ, Walton MA, Prasad JK, et al (1999) Blockade by polyunsaturated n-3 fatty acids of endotoxin-induced monocytic tissue factor activation is mediated by the depressed receptor expression in THP-1cells. J Surg Res 87:217–224
25. Crutchley DJ (1984) Effects of inhibitors of arachidonic acid metabolism on thromboplastin activity in human monocytes. Biochem Biophys Res Commun 19:179–184
26. Bates EJ, Ferrante A, Poulos A, et al (1995) Inhibitory effects of arachidonic acid (20:4,n-6) and its monohydroperoxy- and hydroxy-metabolites on procoagulant activity in endothelial cells. Atherosclerosis 116:125–133
27. Marx N, Mackman N, Schonbeck U, et al (2001) PPARα activators inhibit tissue factor expression and activity in human monocytes. Circulation 103:213–219
28. Eilertsen KE, Osterud B (2002) The central role of thromboxane and platelet activating factor receptors in ex vivo regulation of endotoxin-induced monocyte tissue factor activity in human whole blood. J Endotoxin Res 8:285–293
29. Eligini S, Violi F, Banfi C, et al (2006) Indobufen inhibits tissue factor in human monocytes through a thromboxane-mediated mechanism. Cardiovasc Res 69:218–226
30. Murata T, Ushikubi F, Matsuoka T, et al (1997) Altered pain perception and inflammatory response in mice lacking prostacyclin receptor. Nature 388:678–682
31. Crutchley DJ, Hirsh MJ (1991) The stable prostacyclin analog, iloprost, and prostaglandin E$_1$ inhibit monocyte procoagulant activity in vitro. Blood 78:382–386
32. Crutchley DJ, Conanan LB, Que BG (1994) Effects of prostacyclin analogs on the synthesis of tissue factor, tumor necrosis factor-alpha and interleukin-1 beta in human monocytic THP-1 cells. J Pharmacol Exp Ther 271:446–51
33. Crutchley DJ, Conanan LB, Toledo AW et al (1993) Effects of prostacyclin analogues on human endothelial cell tissue factor expression. Arterioscler Thromb 13:1082–1089
34. Parry GC, Mackman N (1997) Role of cyclic AMP response element-binding protein in cyclic AMP inhibition of NF-κB-mediated transcription. J Immunol 159:5450–5456
35. Crutchley DJ, Solomon DE, Conanan LB (1992) Prostacyclin analogues inhibit tissue factor expression in the human monocytic cell line THP-1 via a cyclic AMP-dependent mechanism. Arterioscler Thromb 12:664–670
36. Abecassis M, Falk JA, Makowka L, et al (1987) 16,16-Dimethyl prostaglandin E$_2$ prevents the development of fulminant hepatitis and blocks the induction of monocyte/macrophage procoagulant activity after murine hepatitis virus strain 3 infection. J Clin Invest 80:881–889
37. Gross S, Tilly P, Hentsch D, et al (2007) Vascular wall-produced prostaglandin E$_2$ exacerbates arterial thrombosis and atherothrombosis through platelet EP3 receptors. J Exp Med 204:311–320
38. Cheng Y, Austin SC, Rocca B, et al (2002) Role of prostacyclin in the cardiovascular response to thromboxane A$_2$. Science 296:539–541
39. Eligini S, Banfi C, Brambilla M, et al (2002) 15-Deoxy-delta12,14-prostaglandin J$_2$ inhibits tissue factor expression in human macrophages and endothelial cells: evidence for ERK1/2 signaling pathway blockade. Thromb Haemost 88:524–532
40. Xie S, O'Regan DJ, Kakkar VV, et al (2002) 15-Deoxy-delta12,14 PGJ$_2$ induces procoagulant activity in cultured human endothelial cells. Thromb Haemost 87:523–529
41. Maderna P, Godson C, Hannify G, et al (2000) Influence of lipoxin A(4) and other lipoxygenase-derived eicosanoids on tissue factor expression. Am J Physiol Cell Physiol 279:C945–C953
42. Benveniste J, Henson PM, Cochrane CG (1972) Leukocyte-dependent histamine release from rabbit platelets: the role of IgE, basophils, and a platelet-activating factor. J Exp Med 136:1356–1377

43. Ishii S, Shimizu T (2000) Platelet-activating factor (PAF) receptor and genetically engineered PAF receptor mutant mice. Prog Lipid Res 39:41–82
44. Honda Z, Nakamura M, Miki I, et al (1991) Cloning by functional expression of platelet-activating factor receptor from guinea-pig lung. Nature 349:342–346
45. Kucey DS, Kubicki EI, Rotstein OD (1991) Platelet-activating factor primes endotoxin-stimulated macrophage procoagulant activity. J Surg Res 50:436–441
46. Kucey DS, Cheung PY, Rotstein OD (1992) Platelet-activating factor modulates endotoxin-induced macrophage procoagulant activity by a protein kinase C-dependent mechanism. Infect Immun 60:944–950
47. Maier RV, Hahnel GB, Fletcher JR (1992) Platelet-activating factor augments tumor necrosis factor and procoagulant activity. J Surg Res 52:258–264
48. Herbert JM, Corseaux D, Lale A, et al (1996) Hypoxia primes endotoxin-induced tissue factor expression in human monocytes and endothelial cells by a PAF-dependent mechanism. J Cell Physiol 169:290–299
49. Cuschieri J, Gourlay D, Bulger E, et al (2002) Platelet-activating factor priming of inflammatory cell activity requires cellular adherence. Surgery 132:157–166
50. Osterud B (1992) Platelet activating factor enhancement of lipopolysaccharide-induced tissue factor activity in monocytes: requirement of platelets and granulocytes. J Leukoc Biol 51:462–465
51. Halvorsen H, Olsen JO, Osterud B (1993) Granulocytes enhance LPS-induced tissue factor activity in monocytes via an interaction with platelets. J Leukoc Biol 54:275–282
52. Schmid E, Muller TH, Budzinski RM, et al (1995) Lymphocyte adhesion to human endothelial cells induces tissue factor expression via a juxtacrine pathway. Thromb Haemost 73:421–428
53. Reverdiau-Moalic P, Watier H, et al (1998) Human allogeneic lymphocytes trigger endothelial cell tissue factor expression by a tumor necrosis factor-dependent pathway. J Lab Clin Med 132:530–540
54. Schmid E, Muller TH, Budzinski RM, et al (1995) Signaling by E-selectin and ICAM-1 induces endothelial tissue factor production via autocrine secretion of platelet-activating factor and tumor necrosis factor α. J Interferon Cytokine Res 15:819–825
55. D'Andrea D, Ravera M, Golino P, et al (2003) Induction of tissue factor in the arterial wall during recurrent thrombus formation. Arterioscler Thromb Vasc Biol 23:1684–1689
56. Cirillo P, Golino P, Calabro P, et al (2003) Activated platelets stimulate tissue factor expression in smooth muscle cells. Thromb Res 112:51–57
57. Todoroki H, Higure A, Okamoto K, et al (1998) Possible role of platelet-activating factor in the in vivo expression of tissue factor in neutrophils. J Surg Res 80:149–155
58. Moosbauer C, Morgenstern E, Cuvelier SL, et al (2007) Eosinophils are a major intra-vascular location for tissue factor storage and exposure. Blood 109:995–1002
59. Butenas S, Bouchard BA, Brummel-Ziedins KE, et al (2005) Tissue factor activity in whole blood. Blood 105:2764–2770
60. Nakamura S, Imamura T, Okamoto K (2004) Tissue factor in neutrophils: yes. J Thromb Haemost 2:214–217
61. Hecht JH, Weiner JA, Post SR, et al (1996) Ventricular zone gene-1 (vzg-1) encodes a lysophosphatidic acid receptor expressed in neurogenic regions of the developing cerebral cortex. J Cell Biol 135:1071–1083
62. Lee MJ, Van Brocklyn JR, Thangada S, et al (1998) Sphingosine-1-phosphate as a ligand for the G protein-coupled receptor EDG-1. Science 279:1552–1555
63. Okamoto H, Takuwa N, Gonda K, et al (1998) EDG1 is a functional sphingosine-1-phosphate receptor that is linked via a Gi/o to multiple signaling pathways, including phospholipase C activation, Ca^{2+} mobilization, Ras-mitogen-activated protein kinase activation, and adenylate cyclase inhibition. J Biol Chem 273:27104–27110
64. Van Meeteren LA, Moolenaar WH (2007) Regulation and biological activities of the autotaxin-LPA axis. Prog Lipid Res 46:145–160

65. Siess W, Tigyi G (2004) Thrombogenic and atherogenic activities of lysophosphatidic acid. J Cell Biochem 92:1086–1094
66. Chung SM, Bae ON, Lim KM, et al (2007) Lysophosphatidic acid induces thrombogenic activity through phosphatidylserine exposure and procoagulant microvesicle generation in human erythrocytes. Arterioscler Thromb Vasc Biol 27:414–421
67. Parks BW, Lusis AJ, Kabarowski JH (2006) Loss of the lysophosphatidylcholine effector, G2A, ameliorates aortic atherosclerosis in low-density lipoprotein receptor knockout mice. Arterioscler Thromb Vasc Biol 26:2703–2709
68. Lavi S, McConnell JP, Rihal CS, et al (2007) Local production of lipoprotein-associated phospholipase A$_2$ and lysophosphatidylcholine in the coronary circulation: association with early coronary atherosclerosis and endothelial dysfunction in humans. Circulation 115:2715–2721
69. Quinn MT, Parthasarathy S, Steinberg D (1988) Lysophosphatidylcholine: a chemotactic factor for human monocytes and its potential role in atherogenesis. Proc Natl Acad Sci U S A 85:2805–2809
70. Chai YC, Howe PH, DiCorleto PE, et al (1996) Oxidized low density lipoprotein and lysophosphatidylcholine stimulate cell cycle entry in vascular smooth muscle cells: evidence for release of fibroblast growth factor-2. J Biol Chem 271:17791–17797
71. Eizawa H, Yui Y, Inoue R, et al (1995) Lysophosphatidylcholine inhibits endothelium-dependent hyperpolarization and N-omega-nitro-L-arginine/indomethacin-resistant endothelium-dependent relaxation in the porcine coronary artery. Circulation 92:3520–3526
72. Kume N, Cybulsky MI, Gimbrone MA Jr (1992) Lysophosphatidylcholine, a component of atherogenic lipoproteins, induces mononuclear leukocyte adhesion molecules in cultured human and rabbit arterial endothelial cells. J Clin Invest 90:1138–1144
73. Dichtl W, Stiko A, Eriksson P, et al (1999) Oxidized LDL and lysophosphatidylcholine stimulate plasminogen activator inhibitor-1 expression in vascular smooth muscle cells. Arterioscler Thromb Vasc Biol 19:3025–3032
74. Rong JX, Berman JW, Taubman MB, et al (2002) Lysophosphatidylcholine stimulates monocyte chemoattractant protein-1 gene expression in rat aortic smooth muscle cells. Arterioscler Thromb Vasc Biol 22:1617–1623
75. Penn MS, Cui MZ, Winokur AL, et al (2000) Smooth muscle cell surface tissue factor pathway activation by oxidized low-density lipoprotein requires cellular lipid peroxidation. Blood 96:3056–3063
76. Drake TA, Hannani K, Fei HH, et al (1991) Minimally oxidized low-density lipoprotein induces tissue factor expression in cultured human endothelial cells. Am J Pathol 138:601–607
77. Weis JR, Pitas RE, Wilson BD, et al (1991) Oxidized low-density lipoprotein increases cultured human endothelial cell tissue factor activity and reduces protein C activation. FASEB J 5:2459–2465
78. Fei H, Berliner JA, Parhami F, et al (1993) Regulation of endothelial cell tissue factor expression by minimally oxidized LDL and lipopolysaccharide. Arterioscler Thromb 13:1711–1717
79. Cui MZ, Zhao G, Winokur AL, et al (2003) Lysophosphatidic acid induction of tissue factor expression in aortic smooth muscle cells. Arterioscler Thromb Vasc Biol 23:224–230
80. Takeya H, Gabazza EC, Aoki S, et al (2003) Synergistic effect of sphingosine 1-phosphate on thrombin-induced tissue factor expression in endothelial cells. Blood 102:1693–1700
81. Bochkov VN, Mechtcheriakova D, Lucerna M, et al (2002) Oxidized phospholipids stimulate tissue factor expression in human endothelial cells via activation of ERK/EGR-1 and Ca^{++}/NFAT. Blood 99:199–206
82. Wada H, Kaneko T, Wakita Y, et al (1994) Effect of lipoproteins on tissue factor activity and PAI–II antigen in human monocytes and macrophages. Int J Cardiol 47: S21–S25

83. Lewis JC, Bennett-Cain AL, DeMars CS, et al (1995) Procoagulant activity after exposure of monocyte-derived macrophages to minimally oxidized low density lipoprotein: co-localization of tissue factor antigen and nascent fibrin fibers at the cell surface. Am J Pathol 147:1029–1040

84. Ohsawa M, Koyama T, Yamamoto K, et al (2000) 1α,25-dihydroxyvitamin D_3 and its potent synthetic analogs downregulate tissue factor and upregulate thrombomodulin expression in monocytic cells, counteracting the effects of tumor necrosis factor and oxidized LDL. Circulation 102:2867–2872

85. Brand K, Banka CL, Mackman N, et al (1994) Oxidized LDL enhances lipopolysaccharide-induced tissue factor expression in human adherent monocytes. Arterioscler Thromb 14:790–797

86. Van den Eijnden MM, van Noort JT, Hollaar L, et al (1999) Cholesterol or triglyceride loading of human monocyte-derived macrophages by incubation with modified lipoproteins does not induce tissue factor expression. Arterioscler Thromb Vasc Biol 19:384–392

87. Bochkov VN, Kadl A, Huber J, et al (2002) Protective role of phospholipid oxidation products in endotoxin-induced tissue damage. Nature 419:77–81

88. Golodne DM, Monteiro RQ, Graca-Souza AV, et al (2003) Lysophosphatidylcholine acts as an anti-hemostatic molecule in the saliva of the blood-sucking bug *Rhodnius prolixus*. J Biol Chem 278:27766–27771

89. Yan JJ, Jung JS, Lee JE, et al (2004) Therapeutic effects of lysophosphatidylcholine in experimental sepsis. Nat Med 10:161–167

90. Engelmann B, Zieseniss S, Brand K, et al (1999) Tissue factor expression of human monocytes is suppressed by lysophosphatidylcholine. Arterioscler Thromb Vasc Biol 19:47–53

91. Goñi FM, Alonso A, Gómez-Muñoz A (2006) Special issue on sphingolipids. Biochim Biophys Acta 1758:1863–2156

92. Mulder AB, Smit JW, Bom VJ, et al (1996) Association of smooth muscle cell tissue factor with caveolae. Blood 88:1306–1313

93. Mulder AB, Smit JW, Bom VJ, et al (1996) Association of endothelial tissue factor and thrombomodulin with caveolae. Blood 88:3667–3670

94. Sevinsky JR, Rao LV, Ruf W (1996) Ligand-induced protease receptor translocation into caveolae: a mechanism for regulating cell surface proteolysis of the tissue factor-dependent coagulation pathway. J Cell Biol 133:293–304

95. Dietzen DJ, Page KL, Tetzloff TA (2004) Lipid rafts are necessary for tonic inhibition of cellular tissue factor procoagulant activity. Blood 103:3038–3044

96. Mandal SK, Iakhiaev A, Pendurthi UR, et al (2005) Acute cholesterol depletion impairs functional expression of tissue factor in fibroblasts: modulation of tissue factor activity by membrane cholesterol. Blood 105:153–160

97. Dietzen DJ, Page KL, Tetzloff TA, et al (2007) Inhibition of 3-hydroxy-3-methylglutaryl coenzyme A (HMG CoA) reductase blunts factor VIIa/tissue factor and prothrombinase activities via effects on membrane phosphatidylserine. Arterioscler Thromb Vasc Biol 27:690–696

98. Del Conde I, Shrimpton CN, Thiagarajan P, et al (2005) Tissue-factor-bearing microvesicles arise from lipid rafts and fuse with activated platelets to initiate coagulation. Blood 106:1604–1611

99. Awasthi V, Mandal SK, Papanna V, et al (2007) Modulation of tissue factor-factor VIIa signaling by lipid rafts and caveolae. Arterioscler Thromb Vasc Biol 27:1447–1455

Tissue Factor Pathway Inhibitor: Structure and Function

Hisao Kato

Summary. Tissue factor pathway inhibitor (TFPI) is a protease inhibitor with three tandem Kunitz-type inhibitor domains that inhibits the tissue factor (TF)-initiated blood coagulation cascade reactions by forming a complex with factor VIIa/TF via the first Kunitz-type inhibitor domain and with factor Xa via the second Kunitz-type inhibitor domain. The third Kunitz-type inhibitor domain of TFPI has no ability to inhibit protease but plays an important role in the functions of TFPI. After the discovery of TFPI, another inhibitor with three tandem Kunitz-type inhibitor domains was isolated and named as TFPI-2. The first TFPI was then named as TFPI-1. However, TFPI-2 is not the inhibitor for the initiation of the tissue factor pathway. TFPI-2 is now thought most likely to have functions clearly different from those of TFPI-1.

In this review, TFPI-1 is simply written as TFPI. Referring to the most recent literature, the structure and function of TFPI has been reviewed. Two functions of TFPI (i.e., inhibition of the TF-initiated blood coagulation pathway and inhibition of proliferation of vascular wall cells) are discussed. The relation of the plasma TFPI level to various diseases and the prospects for clinical application of TFPI are also discussed. The structure and function of TFPI-2 are briefly summarized.

Key words. Tissue factor pathway inhibitor (TFPI) · Tissue factor (TF) · Blood coagulation · Cardiovascular diseases · GPI anchor

Introduction

Tissue factor pathway inhibitor (TFPI) is a protease inhibitor with three tandem Kunitz-type inhibitor domains. It inhibits the tissue factor (TF)-initiated blood coagulation cascade reactions by forming a complex with factor VIIa/TF via the first Kunitz-type inhibitor domain and with factor Xa via the second Kunitz-type inhibitor domain. Although the presence of the inhibitor at the initiation of the extrinsic blood coagulation pathway in plasma has long been known, the isolation and the determination of the amino acid sequence of the unique inhibitor was finally reported only in 1988 by Broze and his group, who designated the inhibitor a lipoprotein-associated coagulation inhibitor (LACI). Rapaport and his group have published many articles on the inhibitor, which was named an extrinsic pathway inhibitor (EPI). To resolve the confusing situation in which reports from many laboratories used two different names for the same entity, a meeting at the International Society on Thrombosis and

Haemostasis held at Amsterdam in 1991 recommended the name tissue factor pathway inhibitor (TFPI). The history of TFPI has been reviewed by Broze and many other investigators [1, 2].

The typical Kunitz-type inhibitor from bovine pancreas (BPTI) consists of 58 amino acid residues with three disulfide bridges. There are many proteins with the Kunitz-type inhibitor domains in the animal and vegetable world that consist of one, two, or three and multiple domains, homologous to BPTI. A family of these proteins has been assembled as a consequence of exon shuffling through incorporation of genetic elements or domains from different evolutionary precursors [3]. Therefore, the proteins with Kunitz-type inhibitor domains are not necessarily protease inhibitors. After the discovery of TFPI, another inhibitor with three tandem Kunitz-type inhibitor domains was isolated and named TFPI-2. The first TFPI was then renamed TFPI-1. However, TFPI-2 is not an inhibitor of the initiation of the tissue factor pathway. Currently, TFPI-2 is thought most likely to have completely different functions from those of TFPI-1. The properties of TFPI-1 and TFPI-2 are compared in Table 1.

TABLE 1. Comparison of the properties of TFPI-1 and TFPI-2

Parameter	TFPI-1	TFPI-2
Gene		
Chromosome	2p31-32.1	7q22
Size	85 kb	kb
Exons	10	5
Introns	9	4
Molecular weight	36 kDa (α form)	35 kDa
	28 kDa (β form)	
Amino acid residues	276 (α form)	213
	193 (β form)	
Synthesis	Endothelial cells	Endothelial cells
	Monocytes	Keratinocytes
	Platelets	Dermal fibroblasts
	Fibroblast	Placenta, liver
	Smooth muscle cells	Heart, kidney
Plasma conc.	14–20 ng/ml (free α form)	0.4–0.5 ng/ml
	38–150 ng/ml (total α form)	
	Increases about 5- to 10-fold after heparin treatment	
Inhibition[a]	TF/VIIa (Kunitz domain 1)	Plasmin, kallikrein
	Xa (Kunitz domain 2)	Trypsin, chymotrypsin
Other functions	Inhibition of proliferation of vascular wall cells	ECM remodeling
		Inhibition of intravasation and extravasation of tumor cells

TFPI, tissue factor pathway inhibitor

[a]Rate constants for inhibition by TFPI-1
  Inhibition of factor Xa by second Kunitz-type inhibitor domain:
    $K_i = 8.3$ nM
      $= 0.5$ nM (in the presence of protein S and phospholipids)
  Second order rate constant
      9.6×10^8 M^{-1}s^{-1} for the inhibition of Xa by TFPI
      0.2×10^8 M^{-1}s^{-1} for the inhibition of TF/VIIa by TFPI-Xa

In this review, TFPI-1 is simply written as TFPI. The structure and function of TFPI is reviewed based on the most recent literature. The structure and function of TFPI-2 is briefly summarized as well. A previous review of TFPI by the author appeared in 2002 [4]. The reviews on TFPI by other investigators are listed as Review References after the reference section to help the readers understand chronologically the various points of view on TFPI.

Structure and Function of TFPI

Structure of TFPI

Human TFPI gene consists of 10 exons as shown in Fig. 1. TFPIα with signal peptide is expressed from exons 3, 4, 5, 6, 7, 9, and 10. Mature TFPIα consists of 276 amino acid residues with three tandem Kunitz-type inhibitor domains. TFPIβ is an alternatively spliced form of TFPI and is expressed from exons 3, 4, 5, 6, 7, and 8 in which the third Kunitz-type inhibitor domain and the C-terminal region of TFPIα are

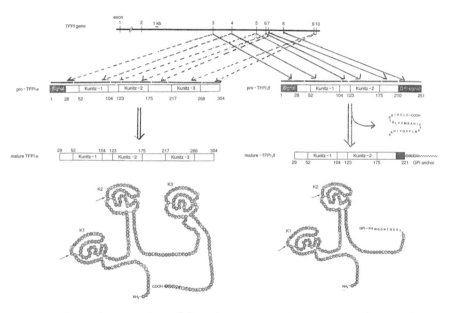

FIG. 1. Synthesis of TFPIα and TFPIβ from the *TFPI* gene. The *TFPI* gene is shown at the top of the figure. Originally, the *TFPI* gene was reported to consist of nine exons. Because a new exon specific for TFPIβ was found between exon 7 and exon 8, the new exon was designated exon 8 in this figure. Therefore, the original exons 8 and 9 were designated exons 9 and 10, respectively. Expression of proforms and mature forms of TFPIα and TFPIβ from the *TFPI* gene is shown. Asn221 of TFPIβ was predicted to be the attachment site of the GPI anchor from the homology of the GPI anchor signal [6]. *Signal*, signal peptide; *GPI signal*, protein portion susceptible for cleavage and attachment of the GPI anchor. Amino acid sequences of TFPIα and TFPIβ are shown at the lower part of the figure. *Arrows* indicate the reactive sites of Kunitz-type inhibitor domains

replaced with an unrelated C-terminal region that directs the attachment of a glyco-sylphosphatidylinositol (GPI) anchor. After cleavage of a peptide bond in the GPI signal at the C-terminal region of TFPIβ, the GPI anchor is attached to the newly formed C-terminus. The homologous amino acid sequences to human TFPIα have been found in monkeys, rabbits, dogs, rats, and mice. The presence of TFPIβ was first found in the mouse gene [5] and then confirmed in the human gene [6]. Mature TFPIα has an acidic N-terminal region, three tandem Kunitz inhibitor domains, and a C-terminal basic region, whereas TFPIβ deletes the third Kunitz-type inhibitor domain and the C-terminal basic region and, instead, associates with the GPI anchor at the C-terminus.

Localization and Distribution of TFPI

Although TFPI is synthesized mainly by vascular endothelial cells, the in vivo distribution of TFPI is highly complicated. Free-form and lipoprotein-associated TFPI are present in plasma. Some, especially the lipoprotein-associated form, are truncated at the C-terminal region. Oxidized low density lipoprotein (LDL) is reported to reduce TFPI activity [7]. TFPI is also present in platelets and synthesized by monocytes. Previously, platelet TFPI was reported to be stored in α granules and released upon activation of platelets. However, a recent report suggests that TFPI is expressed on the surface of platelets after activation [8]. Nevertheless, most of the TFPI is bound to the vascular wall. Both TFPIα and TFPIβ are bound to the cell surface, mostly in a GPI-dependent manner. TFPIβ is supposed to be bound to endothelial cells directly through the GPI anchor at its C-terminus, whereas TFPIα is apparently bound to syndecan and glypican or not yet identified GPI-linked proteins. Treatment with phosphatidylinositol-specific phospholipase C, which cleaves GPI membrane anchors, releases about 80% of total surface TFPI, and the remaining TFPI is released by subsequent heparin treatment [6]. TFPIα is supposed to bind to heparan sulfate and other related glycosaminoglycans on endothelial cells via the third Kunitz-type inhibitor domain and the C-terminal basic region, as these regions are shown to bind with heparin [9–11]. The third Kunitz-type inhibitor domain of TFPIα plays an important role in cell surface binding [12].

TFPI is localized in caveolae on endothelial cells. Caveolin-1 regulates the distribution and function of TFPI [13]. The presence of a GPI-anchored co-receptor for TFPI that controls the trafficking and cell surface expression is also suggested [14]. Figure 2 shows a scheme of the localization of TFPI on the surface of endothelial cells. It has been known that the in vivo administration of heparin causes a prompt release of TFPI into the circulation. The function of heparin was supposed to release TFPI bound to glycosaminoglycans on the cell surface or to release from the intracellular storage. More complicated, a recent report suggests that heparin up-regulates TFPI mRNA expression of endothelial cells in culture [15].

Although the functional difference between TFPIα and TFPIβ remains to be established, Piro and Broze [16] suggested their different roles. Stable clones of ECV304 cells that express reduced levels of TFPIα, TFPIβ, or both were produced using a plasmid-based small-interfering RNA technique. Although TFPIα comprises 80% of the surface TFPI, TFPIβ was responsible for the bulk of the cellular FVIIa/TF inhibitory activity, suggesting a potential alternative role for cell surface TFPIα.

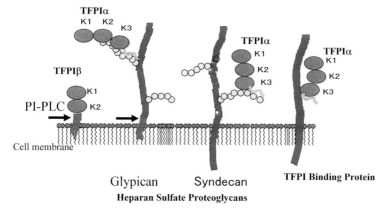

FIG. 2. Localization of TFPIα and TFPIβ on the surface of endothelial cells. *PI-PLC*, phosphatidylinositol-specific phospholipase C; *K1, K2, K3*, first, second, and third Kunitz-type inhibitor domains. ●, GPI anchor; ∞, carbohydrate chains of proteoglycans; ○, C-terminal basic region of TFPI

Inhibition of TF-Initiated Activation of Factors IX and X

The extrinsic blood coagulation is regulated mainly by three mechanisms: (1) inhibition of the activated enzymes by antithrombin (antithrombin III); (2) the protein C pathway, which inactivates cofactors, factor VIIIa, and factor Va; and (3) inhibition of TF-initiated activation of factor X and factor IX by TFPI. TFPI forms a complex with factor Xa via the second Kunitz-type inhibitor domain, and the complex associates with TF/VIIa complex formed on the phospholipids surface via the first Kunitz-type inhibitor domain, leading to the inhibition of further activation of factors IX and X by TF/factor VIIa. In the absence of TFPI, factors X and IX are activated by TF/factor VIIa, as antithrombin weakly inhibits TF/factor VIIa. Therefore, in vivo, factors X and IX are consumed, leading to a bleeding tendency. These mechanisms explain the bleeding tendency of TFPI knockout mice. Although the third Kunitz-type inhibitor domain of TFPI has no ability to bind with protease, the domain is essential for the full inhibitory activity of TFPI [17]. Many studies on kinetic analysis of the inhibitory activities of TFPI have been published. Typical kinetic constants for the inhibition are summarized in Table 1. In plasma, the antithrombin concentration is about 1000-fold higher than that of TFPI. Straightforward data showing inhibition of TF/VIIa by antithrombin and TFPI was shown by Broze et al., as shown in Fig. 3 [18]. They tested the inhibition of factor X by TF/VIIa in the presence of TFPI and antithrombin. In the absence of heparin, antithrombin had no apparent effect on activation, whereas in the presence of heparin it modestly reduced factor X generation. In contrast, the activation of factor X by TF/VII was dramatically delayed in the presence of heparin and antithrombin. These results confirm that TFPI inhibits mainly the activation of factor X by TF/VIIa, whereas antithrombin inhibits the activities of factor Xa at physiological concentrations.

It has been shown that protein S enhances the activity of activated protein C to degrade factors VIIIa and Va, which is one of the important regulatory mechanism of

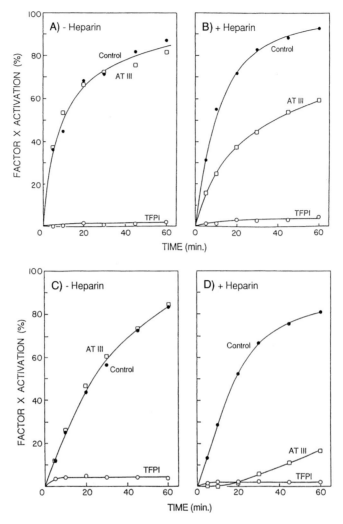

FIG. 3. Effect of antithrombin (antithrombin III) and TFPI on factor X activation by factor VIIa/thromboplastin (TF) (**A, B**) or by factor VII/TF (**C, D**) [18]. **A, B** Reactions contained factor VIIa (4 nmol/l), human brain thromboplastin (TF) (0.5% v/v), CaCl$_2$ (4 nmol/l), ^{3}H-factor X (80 nmol/l), and either antithrombin (5 μmol/l) (*squares*), TFPI (2 nmol/l) (*open circles*), or no other additions (*filled circles*). **C, D** The contents of the reaction were the same as in **A** and **B** except for the presence of zymogen factor VII (10 nml/l). **A, C** No heparin. **B, D** In the presence of heparin (0.5 U/ml). Factor X activation was determined by ^{3}H-factor X activation peptide release. (With permission)

blood coagulation. Recently, Hackeng et al found that protein S stimulates the inhibition of the tissue factor pathway by TFPI and thus demonstrated that protein S and TFPI act in concert in the inhibition of TF activity [19].

Antiangiogenic Activity of TFPI

Antiangiogenic activity of TFPI was first found in studies using cultured endothelial cells and smooth muscle cells [20, 21]. Regarding the inhibitory activity of TFPI, the third Kunitz-type inhibitor domain and the C-terminal basic region play an important role. Although the mechanism on the inhibitory activity of TFPI toward proliferation of these cells remains to be established, binding of TFPI with VLDL receptor is reported to be essential [22–24].

On the other hand, it has been reported that signaling of the TF-initiated coagulation pathway plays important role in angiogenesis and cancer [25]. Ahamed and Ruf suggested that TFPI controls TF-mediated signaling through protease-activated receptors (PARs) [26]. These functions of TFPI suggest a promising therapeutic application of TFPI—not only for thrombotic diseases but also for atherosclerosis and cancer.

Physiological and Pathological Roles of TFPI

Many articles have been published on the relation between plasma TFPI levels and various physiological and pathological states. These results indicate that the TFPI level is reduced in thrombotic diseases in which blood coagulation pathways are activated, such as acute myocardial infarction and disseminated intravascular coagulation. On the other hand, TFPI level in plasma increases, possibly by the stimulation of TFPI synthesis in vascular wall cells, in parallel with the reduction of antithrombotic activity of vascular wall cells such as atherosclerotic diseases. However, these results should be carefully reviewed because the specificity of the methods for measuring TFPI in plasma is limited because of the complexity of the distribution of TFPI, as shown in Fig. 4.

TFPI activity can be measured by the inhibition of factor X activation by TF/factor VIIa using synthetic substrate for factor Xa. This method is influenced by other inhibitors in plasma and does not distinguish free-form TFPI from the lipoprotein-associated forms and truncated forms of TFPI. On the other hand, the EIA method using TFPI antibody is convenient and widely used, but its specificity is dependent on the epitopes the antibody recognizes. Particularly, these methods do not measure the bulk of endothelial cell-associated TFPI. Therefore, the variation of TFPI level in various states should be explained, taking into consideration what molecular entities of TFPI were measured by the method used. It should be kept in mind that TFPI assay method is still developing [27].

It is clear that TFPI plays an important role in the development of atherosclerosis and in acute coronary syndromes and stroke [28–30]. Studies on TFPI knockout mice clearly demonstrate an important role of TFPI [31–36]. However, many other diseases are reported to be associated with TFPI or anti-TFPI antibody, such as antiphospholipid syndrome [37–40] and Fabry's disease [41]. It has been also reported that TFPI is related to circadian rhythm [42–44], fertility [45], and levels of lipoprotein(a) [46] and adrenomedullin [47, 48]. The variable TFPI level in various

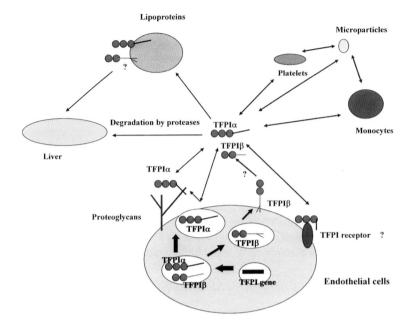

FIG. 4. In vivo distribution of TFPI

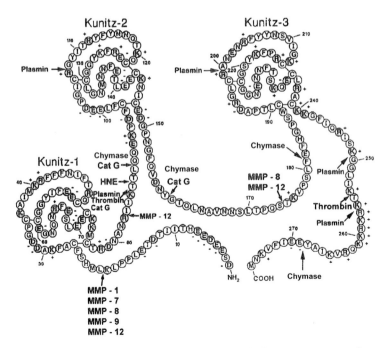

FIG. 5. Sites in TFPI susceptible to proteolytic degradation [51]. *MMP*, matrix metalloprotease; *CatG*, cathepsin G; *HNE*, human neutrophil elastase. (With permission) Cleavage sites by chymase were added, referring to [75]

physiological and pathological states are in part due to the polymorphisms of the upstream and exons/introns of the TFPI gene [49, 50]. Until now, six polymorphisms have been reported: -287T/C, -399C/T, Val264Met(874G/A), Pro151Leu(536C/T), 384T/ C, Int7–33T/C. Although many reports on the relation between the polymorphism of TFPI and the plasma TFPI level have been published, their relation remains to be established.

In summary, the TFPI level in plasma is influenced by many factors. First, TFPI in plasma is cleaved by many proteases, as shown in Fig. 5 [51]. The presence of anti-TFPI antibody in plasma reduces the TFPI level. The plasma TFPI level may be definitively influenced by the clearance rate of TFPI by the liver and release of TFPI from the surface of endothelial cells, platelets, and monocytes or smooth muscle cells. Regulation of TFPI expression of course causes a variation of the plasma TFPI level, as shown in Fig. 4. It is highly probable that elevation or reduction of the plasma TFPI level also reflects the higher or lower concentration of TFPI in vascular wall cells or circulating microparticles [52].

Prospects for Clinical Application of TFPI

Because the major function of TFPI is inhibition of the TF-initiated blood coagulation pathway, the use of recombinant TFPI is supposed to be promising for therapy and prevention of thrombotic diseases [53]. In fact, the efficacy has been shown in in vitro or animal experiments. However, a Phase III randomized, double-blind, placebo-controlled clinical study on severe sepsis could not demonstrate efficacy of intravenous administration of recombinant TFPI from *Escherichia coli* [54]. The reason the clinical study by Chiron Corp/Pharmacia Corp. was unsuccessful is not known. It may be due to the recombinant TFPI lacking carbohydrate chains because the carbohydrate chains of TFPI are supposed to be important for the in vivo functions of TFPI. Otherwise, the application of TFPI to other diseases related to atherosclerosis should be considered. In the studies using recombinant TFPI, the immobilized TFPI may be also promising [55].

On the other hand, since the discovery of another function of TFPI (i.e., inhibition of vascular wall cell proliferation) there have been many reports on gene transfer of TFPI in animal experiments not only for treatment of thrombosis but also for prevention of restenosis of the coronary artery after PTCA [56–59]. These studies showed that gene transfer of TFPI is effective for preventing restenosis and for other diseases. Further clinical studies are needed, however.

TFPI-2

Human TFPI-2 was originally isolated from placental tissue, which was called placental protein 5 (PP5). In 1994, Sprecher and coworkers isolated a cDNA homologous to TFPI-2. Since the rediscovery in 1994, TFPI-2 has been investigated extensively to reveal its physiological and pathological roles [60]. As summarized in Table 1, human TFPI-2 gene maps to chromosome 7 at the 7q22 region, which consists of five exons and four introns. The mature TFPI-2 protein has a short acidic amino-terminal region, three tandem Kunitz-type inhibitor domains, and a C-terminal basic amino acid region. Recombinant TFPI-2 exhibited a strong inhibitory effect toward the amidolytic activities of trypsin, chymotrypsin, plasmin, plasma kallikrein, and factor XIa.

The TFPI-2 gene is expressed in most human tissues. Despite the abundant presence of TFPI-2 in the placenta, its function during pregnancy is not fully understood. Ogawa et al. found that the serum TFPI-2 level was significantly elevated in patients with preeclampsia and suggested that a decrease of glypican-3 in the placenta caused the increase [61].

Interestingly, tumors arising from various tissues have reduced or undetectable expression of TFPI-2. The tumor-related diminished expression of TFPI-2 has been attributed to promoter hypermethylation. The down-regulation of TFPI-2 synthesis during tumorigenesis may play a crucial role in the intravasation and extravasation of the tumor cell during the metastatic process. The additional mechanism for tumor cells to down-regulate TFPI-2 expression was recently reported in the finding of an up-regulation of a novel, aberrantly spliced TFPI-2 transcript in tumor cells [62]. These reduced functions of TFPI-2 are supposed to cause enhancement of the degradation of ECM by plasmin and the activation of metalloproteinases (MMPs) by plasmin [63]. Intraviteral injection of recombinant TFPI-2 rescued sodium iodate-induced retinal degeneration in rabbits and naturally occurring retinal degeneration in RCS rats. These results suggest that TFPI-2 can be utilized in the treatment of retinal degeneration [64].

TFPI-2 has been shown to function as a mitogen for vascular smooth muscle cells and retinal pigment epithelial cells [65, 66]. Proteomic and transcriptomic analyses of retinal pigment epithelial cells exposed to TFPI-2 revealed that TFPI-2 augmented c-myc synthesis [67].

On the other hand, TFPI-2 was shown to have an antiangiogenic role with direct inhibitory effects on endothelial cells and therefore may be a novel therapeutic target for angiogenic disease processes [68]. In fact, adenoviral-mediated TFPI-2 gene transfer inhibited the growth of a human glioblastoma cell line [69] and of laryngeal squamous cell carcinoma in a nude mice model [70].

Future Outlook

As described in this review, TFPI has mainly two functions: inhibition of TF-initiated blood coagulation reactions and inhibition of the proliferation of vascular wall cells. Although the mechanism for inhibiting blood coagulation by TFPI has been well understood, the function of TFPI toward vascular wall cells remains to be established. Nevertheless, clinical use of TFPI as an anticoagulant drug and an antiangiogenic drug is expected in near future. A similar clinical application has also been attempted on other Kunitz-type inhibitors, such as bikunin from inter- α trypsin inhibitor and Ixolaris from a tick salivary gland, which has two Kunitz-type inhibitor domains [71, 72]. An anticoagulant similar to Ixolaris, Penthalaris was also found in tick saliva and has five Kunitz-type inhibitor domains [73]. The recombinant proteins or genes of these inhibitors have become an attractive target for the development of novel anticoagulant and novel anticancer drug. Low-molecular-weight substances that can be administered orally should be also developed for thrombotic diseases and cancer. Selective and orally available inhibitors for the TF–VIIa complex have been chemically synthesized [74]. Clinical application of these inhibitors for TF/VIIa activity will give an insight on the role of TF/VIIa in blood coagulation and angiogenesis.

On the basic aspect of TFPI, the focus should be on the function and localization of TFPIα and TFPIβ. Furthermore, the specific assay methods for two forms of TFPI should be developed from these basic studies on TFPI. Using novel assay methods, the plasma TFPI level can be established as a new risk factor in the various diseases related to atherosclerosis.

References

1. Broze GJ Jr (2003) The rediscovery and isolation of TFPI. J Thromb Haemost 1:1671–1675
2. Hjort PF (2004) Tissue factor pathway inhibitor revisited. J Thromb Haemost 2:2241–2249
3. Ikeo K, Takahashi K, Gojobori T (1992) Evolutionary origin of a Kunitz-type trypsin inhibitor domain inserted in the amyloid beta precursor protein of Alzheimer's disease. J Mol Evol 34:536–543
4. Kato H (2002) Regulation of functions of vascular wall cells by tissue factor pathway inhibitor: basic and clinical aspects. Arterioscler Thromb Vasc Biol 22:539–548
5. Chang JY, Monroe DM, Oliver JA, et al (1999) TFPIβ, a second product from the mouse tissue factor pathway inhibitor (TFPI) gene. Thromb Haemostas 81:45–49
6. Zhang J, Piro O, Lu L, et al (2003) Glycosyl phosphatidylinositol anchorage of tissue factor pathway inhibitor. Circulation 108:623–627
7. Ohkura N, Hiraishi S, Itabe H, et al (2004) Oxidized phospholipids in oxidized low-density lipoprotein reduce the activity of tissue factor pathway inhibitor through association with its carboxy-terminal region. Antioxid Redox Signal 6:705–712
8. Maroney SA, Haberichter SL, Friese P, et al (2007) Active tissue factor pathway inhibitor is expressed on the surface of coated platelets. Blood 109:1931–1937
9. Enjyoji K, Miyata T, Kamikubo Y, et al (1995) Effect of heparin on the inhibition of factor Xa by tissue factor pathway inhibitor: a segment, Gly212-Phe243, of the third Kunitz domain is a heparin-binding site. Biochemistry 34:5725–5735
10. Mine S, Yamazaki T, Miyata T, et al (2002) Structural mechanism for heparin-binding of the third Kunitz domain of human tissue factor pathway inhibitor. Biochemistry 41:78–85
11. Ye Z, Takano R, Hayashi K, et al (1998) Structural requirements of human tissue factor pathway inhibitor (TFPI) and heparin for TFPI–heparin interaction. Thromb Res 89:263–270
12. Piro O, Broze GJ Jr (2004) Role for the Kunitz-3 domain of tissue factor pathway inhibitor alpha in cell surface binding. Circulation 110:3567–3572
13. Lupu C, Hu X, Lupu F (2005) Caveolin-1 enhances tissue factor pathway inhibitor exposure and function on the cell surface. J Biol Chem 280:22308–22317
14. Maroney SA, Cunningham AC, Ferrel J, et al (2006) A GPI-anchored co-receptor for tissue factor pathway inhibitor controls its intracellular trafficking and cell surface expression. J Thromb Haemost 4:1114–1124
15. Thyzel E, Kohli S, Siegling S, et al (2007) Relative quantification of glycosaminoglycan-induced upregulation of TFPI-mRNA expression in vitro. Thromb Res 119:785–791
16. Piro O, Broze GJ Jr (2005) Comparison of cell-surface TFPIalpha and beta. J Thromb Haemost 3:2677–2683
17. Lockett JM, Mast AE (2002) Contribution of regions distal to glycine-160 to the anti-coagulant activity of tissue factor pathway inhibitor. Biochemistry 41:4989–4997

18. Broze GJ Jr, Likert K, Higuchi D (1993) Inhibition of factor VIIa/tissue factor by anti-thrombin III and tissue factor pathway inhibitor. Blood 82:1679–1680

19. Hackeng TM, Sere KM, Tans G, et al (2006) Protein S stimulates inhibition of the tissue factor pathway by tissue factor pathway inhibitor. Proc Natl Acad Sci U S A 103:3106–3111

20. Hamuro T, Kamikubo Y, Nakahara Y, et al (1998) Human recombinant tissue factor pathway inhibitor induces apoptosis in cultured human endothelial cells. FEBS Lett 421:197–202

21. Kamikubo Y, Nakahara Y, Takemoto S, et al (1997) Human recombinant tissue-factor pathway inhibitor prevents the proliferation of cultured human neonatal aortic smooth muscle cells. FEBS Lett 407:116–120

22. Shirotani-Ikejima H, Kokame K, Hamuro T, et al (2002) Tissue factor pathway inhibitor induces expression of JUNB and GADD45B mRNAs. Biochem Biophys Res Commun 299:847–852

23. Hembrough TA, Ruiz JF, Swerdlow BM, et al (2004) Identification and characterization of a very low density lipoprotein-binding peptide from tissue factor pathway inhibitor that has antitumor and antiangiogenic activity. Blood 103:3374–3380

24. Hembrough TA, Ruiz JF, Papathanassiu AE, et al (2001) Tissue factor pathway inhibitor inhibits endothelial cell proliferation via association with the very low density lipo-protein receptor. J Biol Chem 276:12241–12248

25. Belting M, Ahamed J, Ruf W (2005) Signaling of the tissue factor coagulation pathway in angiogenesis and cancer. Arterioscler Thromb Vasc Biol 25:1545–1550

26. Ahamed J, Belting M, Ruf W (2005) Regulation of tissue factor-induced signaling by endogenous and recombinant tissue factor pathway inhibitor-1. Blood 105:2384–2391

27. Dahm AEA, Andersen TO, Rosendaal F, et al (2005). A novel anticoagulant activity assay of tissue factor pathway inhibitor I (TFPI). J Thromb Haemost 3:651–658

28. Morange PE, Blankenberg S, Alessi MC, et al (2007) Prognostic value of plasma tissue factor and tissue factor pathway inhibitor for cardiovascular death in patients with coronary artery disease: the AtheroGene study. J Thromb Haemost 5:475–482

29. Adams MJ, Thom J, Hankey GJ, et al (2006) The tissue factor pathway in ischemic stroke. Blood Coagul Fibrinolysis 17:527–532

30. Sakata T, Mannami T, Baba S, et al (2004) Potential of free-form TFPI and PAI-1 to be useful markers of early atherosclerosis in a Japanese general population (the Suita Study): association with the intimal-medial thickness of carotid arteries. Atherosclerosis 176:355–360

31. Huang ZF, Higuchi D, Lasky N, et al (1997) Tissue factor pathway inhibitor gene disruption produces intrauterine lethality in mice. Blood 90:944–951

32. Chen D, Weber M, Shiels PG, et al (2006). Postinjury vascular intimal hyperplasia in mice is completely inhibited by CD34+ bone marrow-derived progenitor cells expressing membrane-tethered anticoagulant fusion proteins. J Thromb Haemost 4:2191–2198

33. Chen D, Weber M, McVey JH, et al (2004) Complete inhibition of acute humoral rejection using regulated expression of membrane-tethered anticoagulants on xenograft endothelium. Am J Transplant 4:1958–1963

34. Pan S, Kleppe LS, Witt TA, et al (2004) The effect of vascular smooth muscle cell-targeted expression of tissue factor pathway inhibitor in a murine model of arterial thrombosis. Thromb Haemost 92:495–502

35. Pedersen B, Holscheler T, Sato Y, et al (2005) A balance between tissue factor and tissue factor pathway inhibitor is required for embryonic development and hemostasis in adult mice. Blood 105:2777–2782

36. Singh R, Pan S, Mueske CH, et al (2003) Tissue factor pathway inhibitor deficiency enhances neointimal proliferation and formation in a murine model of vascular remodeling. Thromb Haemost 89:747–751

37. Martinuzzo M, Iglesias Varela ML, et al (2005) Antiphospholipid antibodies and antibodies to tissue factor pathway inhibitor in women with implantation failures or a early and late pregnancy losses. J Thromb Haemost 3:2587–2589
38. Liestol S, Sandset PM, Jacobsen EM, et al (2007) Decreased anticoagulant response to tissue factor pathway inhibitor type 1 in plasmas from patients with lupus anticoagulants. Br J Haematol 136:131–137
39. Lean SY, Ellery P, Ivey L, et al (2006) The effects of tissue factor pathway inhibitor and anti-beta2 glycoprotein I IgG on thrombin generation. Haematologica 91:1360–1366
40. Gardiner C, Cohen H, Austin SK, et al (2006) Pregnancy loss, tissue factor pathway inhibitor deficiency and resistance to activated protein C. J Thromb Haemost 4:2724–2726
41. Fedi S, Gensini F, Gori AM, et al (2005) Homocysteine and tissue factor pathway inhibitor levels in patients with Fabry's disease. J Thromb Haemost 3:2117–2119
42. Dahm A, Andersen TO, Rosendaal F, et al (2006) Opposite circadian rhythms in melatonin and tissue factor pathway inhibitor type 1: does daylight affect coagulation? J Thromb Haemost 4:1840–1842
43. Dahm AEA, Iversen PO, Hjeltnes N, et al (2006) Differences in circadian variations of tissue factor pathway inhibitor type 1 between able-bodies and spinal cord injured. Thromb Res 118:281–287
44. Pinotti M, Bertolucci C, Portaluppi F, et al (2005) Daily and circadian rhythms of tissue factor pathway inhibitor and factor VII activity. Arterioscler Thromb Vasc Biol 25:646–649
45. Lwaleed BA, Greenfield RS, Birch BR, et al (2005) Does human semen contain a functional haemostatic system? A possible role for tissue factor pathway inhibitor in fertility through semen liquefaction. Thromb Haemost 93:847–852
46. Nisio MD, Ten Wolde M, Meijers JCM, et al (2005) Effects of high plasma lipoprotein (a) levels on tissue factor pathway inhibitor and the protein C pathway. J Thromb Haemost 3:2123–2125
47. Marutsuka K, Hatakeyama K, Yamashita A, et al (2003) Adrenomedullin augments the release and production of tissue factor pathway inhibitor in human aortic endothelial cells. Cardiovasc Res 57:232–237
48. Liu W, Zhu ZQ, Wang W, et al (2007) Crucial roles of GATA-2 and SP-1 in adrenomedullin-affected expression of tissue factor pathway inhibitor in human umbilical vein endothelial cells exposed to lipopolysaccharide. Thromb Haemost 97:839–846
49. Hoppe B, Tolou F, Dorner T, et al (2006) Gene polymorphisms implicated in influencing susceptibility to venous and arterial thromboembolism: frequency distribution in a healthy German population. Thromb Haemost 96:465–470
50. Almasy L, Soria JC, Souto DM, et al (2005) A locus on chromosome 2 influences levels of tissue factor pathway inhibitor: results from the GAIT study. Arterioscler Thromb Vasc Biol 25:1489–1492
51. Cunningham AC, Hasty KA, Enghild J, et al (2002) Structural and functional characterization of tissue factor pathway inhibitor following degradation by matrix metalloproteinase-8. Biochem J 367:451–458
52. Steppich B, Mattisek C, Sobczyk D, et al (2005) Tissue factor pathway inhibitor on circulating microparticles in acute myocardial infarction. Thromb Haemost 93:35–39
53. Laterre PF, Wittebole X, Collienne C (2006) Pharmacological inhibition of tissue factor. Semin Thromb Haemost 32:71–76
54. Abraham E, Reinhart K, Opal S, et al (2003) Efficacy and safety of Tifacogin (recombinant tissue factor pathway inhibitor) in severe sepsis. JAMA 290:238–247
55. Chandiwal A, Zaman FS, Mast AE, et al (2006) Factor Xa inhibition by immobilized recombinant tissue factor pathway inhibitor. J Biomater Sci Polym Ed 17:1025–1037
56. Chen HH, Vicente CP, Tollefsen DM, et al (2005) Fusion proteins comprising annexin V and Kunitz protease inhibitors are highly potent thrombogenic site-directed anticoagulants. Blood 105:3902–3909

57. Kijiyama N, Ueno H, Sugimoto I, et al (2006) Intratracheal gene transfer of tissue factor pathway inhibitor attenuates pulmonary fibrosis. Biochem Biophys Res Commun 339:1113–1119

58. Chen D, Giannopoulos K, Shiels PG, et al (2004) Inhibition of intravascular thrombosis in murine endotoxemia by targeted expression of hirudin and tissue factor pathway inhibitor analogs to activated endothelium. Blood 104:1344–1349

59. Kopp CW, Holzenbein T, Steiner S, et al (2004) Inhibition of restenosis by tissue factor pathway inhibitor: in vivo and in vitro evidence for suppressed monocyte chemoattraction and reduced gelatinolytic activity. Blood 103:1653–1661

60. Chand HS, Foster DC, Kisiel W (2005) Structure, function and biology of tissue factor pathway inhibitor-2. Thromb Haemost 94:1122–1130

61. Ogawa M, Yanoma S, Nagashima Y, et al (2007) Paradoxical discrepancy between the serum level and the placental intensity of PP5/TFPI-2 in preeclampsia and/or intrauterine growth restriction: Possible interaction and correlation with glypican-3 hold the key. Placenta 28:224–232

62. Kempaiah P, Chand HS, Kisiel W (2007) Identification of a human TFPI-2 splice variant that is upregulated in human tumor tissues. Mol Cancer 6:20

63. Rollin J, Regina S, Vourc'h P, et al (2007) Influence of MMP-2 and MMP-9 promoter polymorphisms on gene expression and clinical outcome of non-small cell lung cancer. Lung Cancer 56:273–280

64. Obata R, Yanagi Y, Tamaki Y, et al (2005) Retinal degeneration is delayed by tissue factor pathway inhibitor-2 in RCS rats and a sodium-iodate-induced model in rabbits. Eye 19:464–468

65. Shinoda E, Yui Y, Hattori R, et al (1999) Tissue factor pathway inhibitor-2 is a novel mitogen for vascular smooth muscle cell. J Biol Chem 274:5379–5384

66. Tanaka Y, Utsumi J, Matsui M, et al (2004) Purification, molecular cloning and expression of a novel growth promoting factor for retinal pigment epithelial cell, REF-1/TFPI-2. Invest Ophthalmol Vis Sci 45:245–252

67. Shibuya M, Okamoto H, Nozawa T, et al (2007) Proteomic and transcriptomic analyses of retinal pigment epithelial cells exposed to REF-1/TFPI-2. Invest Ophthalmol Vis Sci 48:516–521

68. Xu Z, Maiti D, Kisiel W, et al (2006) Tissue factor pathway inhibitor-2 is upregulated by vascular endothelial growth factor and suppresses growth factor-induced proliferation of endothelial cells. Arterioscler Thromb Vasc Biol 26:2819–2825

69. Yanamandra N, Kondraganti S, Gondi CS, et al (2005) Recombinant adeno-associated virusexpressing TFPI-2 inhibits invasion, angiogenesis and tumor growth in a human glioblastoma cell line. Int J Cancer 115:998–1005

70. Sun Y, Xie M, Liu M, et al (2006) Growth suppression of human laryngeal squamous cell carcinoma by adenovirus-mediated tissue factor pathway inhibitor gene 2. Laryngoscope 116:596–601

71. Monteiro RQ, Rezaie AR, Ribeiro JMC, et al (2005) Ixolaris: a factor X heparin-binding exocite inhibitor. Biochem J 387:871–877

72. Nazareth RA, Tomaz LS, Ortiz-Costa S, et al (2006) Antithrombotic properties of Ixolaris, a potent inhibitor of the extrinsic pathway of the coagulation cascade. Thromb Haemost 96:7–13

73. Fancischetti IM, Mather TN, Ribeiro JM (2004) Penthalaris, a novel recombinant five-Kunitz tissue factor pathway inhibitor (TFPI) from the salivary gland of the tick vector of Lyme disease, *Ixodes scapularis*. Thromb Haemost 91:886–898

74. Miura M, Seki N, Koike T, et al (2007) Design, synthesis and biological activity of selective and orally available TF/VIIa complex inhibitors containing non-amidine P1 ligands. Bioorg Med Chem 15:160–173

75. Hamuro T, Kido H, Asada Y, et al (2007) Tissue factor pathway inhibitor is highly susceptible to chymase-mediated proteolysis. FEBS J 274:3065–3077

Review References

1. Lwaleed BA, Bass PS (2006) Tissue factor pathway inhibitor: structure, biology and involvement in disease. J Pathol 208:327–339
2. Sajadi S, Ezekowitz MD, Dhond A, et al (2003) Tissue factor pathway inhibitors as a novel approach to antithrombotic therapy. Drug News Perspect 16:363–369
3. Golino P, Ragni M, Cimmino G, et al (2002) Role of tissue factor pathway inhibitor in the regulation of tissue factor-dependent blood coagulation. Cardiovasc Drug Rev 20:67–80
4. Schwartz AL, Broze GJ Jr (1997) Tissue factor pathway inhibitor endocytosis. Trends Cardiovasc Med 7:234–239
5. Bajaj MS, Bajaj SP (1997) Tissue factor pathway inhibitor: potential therapeutic applications. Thromb Haemost 78:471–477
6. Lindahl AK (1997) Tissue factor pathway inhibitor: from unknown coagulation inhibitor to major antithrombotic principle. Cardiovasc Res 33:286–291
7. Sandset PM, Bendz B (1997) Tissue factor pathway inhibitor: clinical deficiency states. Thromb Haemost 78:467–470
8. Sandset PM (1996) Tissue factor pathway inhibitor (TFPI)—an update. Haemostasis 26(Suppl 4):154–165
9. Broze GJ Jr (1995) Tissue factor pathway inhibitor. Thromb Haemost 74:90–93
10. Lindahl AK (1995) Tissue factor pathway inhibitor in health and disease. Trends Cardiovasc Med 5:167–171
11. Broze GJ Jr (1995) Tissue factor pathway inhibitor and the revised theory of coagulation. Annu Rev Med 46:103–112
12. Wun TC (1995) Tissue Factor Pathway Inhibitor. In: Robert HR, High KA (eds) Molecular biology of thrombosis and hemostasis. Marcel Dekker, New York, pp 331–353
13. Petersen LC, Valentin S, Hedner U (1995) Regulation of the extrinsic pathway system in health and disease: the role of factor VIIa and tissue factor pathway inhibitor. Thromb Res 79:1–47
14. Novotny WF (1994) Tissue factor pathway inhibitor. Semin Thromb Hemost 20:101–108
15. Broze GJ Jr, Tollefsen DM (1994) Regulation of blood coagulation by protease inhibitors. In: Stamatoyanpoulos G, Nenhus AW, Majerus PW, et al (eds) The molecular basis of blood diseases, 2nd edn. Saunders, Philadelphia, pp 629–656
16. Rapaport SI, Rao LVM (1992) Initiation and regulation of tissue factor-dependent blood coagulation. Arterioscler Thromb 12:1111–1121
17. Broze GJ Jr, Girard TJ, Novotny WF (1990) Regulation of coagulation by a multivalent Kunitz-type inhibitor. Biochemistry 29:7539–7546

Biology of an Antithrombotic Factor—ADAMTS13

Fumiaki Banno and Toshiyuki Miyata

Summary. ADAMTS13 is a newly identified plasma protease that limits platelet thrombus formation through cleavage of von Willebrand factor (VWF) multimers. Both congenital and acquired defects in ADAMTS13 lead to a potentially fatal syndrome, thrombotic thrombocytopenic purpura (TTP). Here, we describe the structural and functional features of ADAMTS13 and discuss the association between ADAMTS13 deficiency and TTP pathogenesis. ADAMTS13 is mainly synthesized in hepatic stellate cells and is composed of multidomains in which the proximal C-terminal domains, especially the spacer domain, play a critical role in the cleavage of VWF. Conformational changes of VWF, which can be induced by flow shear force and ligand binding, are necessary to expose the ADAMTS13 binding and cleavage sites on VWF. A peptide of 73 amino acid residues around the ADAMTS13 cleavage site in VWF has been identified as the minimal substrate for ADAMTS13 and was named VWF73. ADAMTS13 can bind via its spacer domain to a core sequence within the C-terminal portion of VWF73 and specifically cleave this substrate. The VWF73-based assays for measuring ADAMTS13 activity have been established and applied for the clinical diagnosis of TTP. Accumulating clinical information from TTP patients and their families suggests that ADAMTS13 deficiency alone is not sufficient to cause TTP. No TTP-like phenotype in ADAMTS13-deficient mice supports this suggestion, and the deficient mice may be useful for elucidating the genetic and/or environmental triggers for TTP.

Key words. ADAMTS13 · von Willebrand factor · Thrombotic thrombocytopenic purpura · Microangiopathy · Platelet

Introduction

ADAMTS13 is a plasma protease that specifically cleaves von Willebrand factor (VWF) [1–4]. VWF has a binding site with platelet GPIbα and subendothelial collagen, and plays a critical role in platelet adhesion and aggregation on vascular lesions. VWF is mainly synthesized in endothelial cells and secreted into the circulating blood in the form of large homomultimers. Larger VWF multimers have stronger thrombotic activity; in particular, ultra-large VWF (UL-VWF) multimers, exceeding 20 000 kDa, can induce excessive platelet aggregation under shear stress [5, 6]. UL-VWF multimers are normally cleaved by ADAMTS13 to smaller forms, thus restraining platelet thrombus formation. The lack of ADAMTS13 activity allows UL-VWF multimers to persist in the circulation and leads to the development of thrombotic thrombocytopenic purpura (TTP) (Fig. 1) [7–10].

Thrombotic thrombocytopenic purpura is a serious disorder with thrombocytopenia and hemolytic anemia as prominent symptoms; it often accompanies renal failure, neurological dysfunction, and fever. The first description of this disease was published in 1924; the case involved an individual who died of high fever and multiple organ failure [11]. Autopsies of affected patients have revealed widespread hyaline thrombi in small vessels [12]. Systemic symptoms are caused by microvascular thrombosis as follows: The consumption of platelets in thrombi causes thrombocytopenia; the fragmentation of red blood cells, which occurs at the thrombotic angiostenosis site, causes anemia; and thrombotic occlusion of small vessels causes renal and cerebral failure and is also associated with other organ damage. Without treatment, the mortality rate

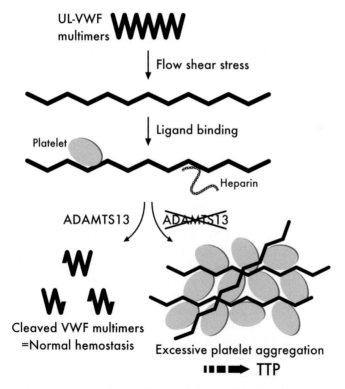

FIG. 1. Role of ADAMTS13 in the regulation of platelet thrombus formation. Under flow shear stress, ADAMTS13 cleaves ultra-large von Willebrand factor (*UL-VWF*) multimers into smaller forms and supports normal hemostasis. The absence of ADAMTS13 permits UL-VWF-mediated platelet aggregation, which leads to an attack of thrombotic thrombocytopenic purpura (*TTP*)

of affected patients exceeds 90%, but plasma exchange reduces the death rate to approximately 20% [13].

In 2001, ADAMTS13 was identified as VWF-cleaving protease [3, 4, 10]. This discovery dramatically increased our understanding of TTP pathophysiology [see also the chapter by Y. Fujimura, this volume]. Congenital TTP (Upshaw-Schulman syndrome) is recognized to be associated with the *ADAMTS13* gene mutations. Some forms of acquired TTP (e.g., idiopathic TTP, ticlopidine-induced TTP) are known to be associated with the production of autoantibodies against ADAMTS13. In this chapter, we summarize the progress that has been made in elucidating the biology of ADAMTS13, focusing on structure and synthesis, domain functions, enzymatic properties, assay methods, *ADAMTS13* gene mutations, and ADAMTS13-deficient mice.

Primary Structure and Biosynthesis of ADAMTS13

ADAMTS13 is a zinc-dependent metalloprotease that belongs to the ADAMTS (a disintegrin-like and metalloprotease with thrombospondin type 1 motifs) family. There are 19 *ADAMTS* genes (*ADAMTS1–20*, with 5 the same as 11) in the human genome. ADAMTS family proteins are thought to contribute to the construction or resolution of the extracellular matrix and inhibition of angiogenesis, reproduction, and organogenesis [14], but the precise function of each member has not been fully elucidated. ADAMTS13 is one of the members that has been intensively studied.

Human ADAMTS13 contains 1427 amino acid residues and has common domains in the ADAMTS family, including a signal peptide, a propeptide, a metalloprotease domain, a disintegrin-like domain, a thrombospondin type 1 (Tsp1) motif, a cysteine (Cys)-rich domain, and a spacer domain in order from the N-terminus (Fig. 2, Table 1). Many of the ADAMTS proteins include additional Tsp1 motifs, and ADAMTS13 has seven more Tsp1 motifs. In addition, ADAMTS13 has two C-terminal CUB (complement components C1r/C1s, urchin epidermal growth factor, and bone morphogenic protein-1) domains that are not found in other members. A calculated molecular mass of the polypeptide is 145 kDa, whereas the apparent molecular mass of plasma ADAMTS13 is approximately 190 kDa. The difference is largely due to glycosylation [15]. ADAMTS13 has 10 potential *N*-glycosylation sites and seven potential *O*-glycosylation sites, at least six of which are modified with an *O*-fucose disaccharide [16]. This *O*-fucosylation is required for secretion of ADAMTS13.

The human *ADAMTS13* gene is located on chromosome 9q34 and consists of 29 exons (Fig. 2, Table 1). The mRNA is approximately 4.7 kb and is mainly expressed in hepatic stellate cells [17, 18], which suggests that these cells are the major source of plasma ADAMTS13. Indeed, a reduction of plasma ADAMTS13 activity has been observed with liver cirrhosis [19] or hepatectomy [20] in humans, and with hepatic stellate cell damage in rat models [21]. Platelets and endothelial cells also express ADAMTS13 [22, 23], which may contribute to local VWF cleavage. The plasma concentration of ADAMTS13 is approximately 0.5–1.0 µg/ml with a 2- to 4-day half-life in the circulation [24, 25].

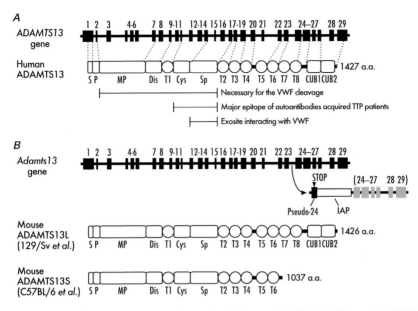

FIG. 2. Structure of ADAMTS13 genes and proteins. **A** Human ADAMTS13. **B** Mouse ADAMTS13. In certain strains of mice (e.g., C57BL/6), an IAP-retrotransposon (*IAP*) is inserted into intron 23 of the *Adamts13* gene (*arrow*) and creates a pseudo-exon 24 including a premature stop codon. ADAMTS13S with a truncated C-terminus is mainly expressed in these strains. *S*, signal peptide; *P*, propeptide; *MP*, metalloprotease domain; *Dis*, disintegrin-like domain; *T* (numbered 1–8), thrombospondin type 1 motif; *Cys*, cysteine-rich domain; *Sp*, spacer domain; *CUB*, complement components C1r/C1s, urchin epidermal growth factor, and bone morphogenic protein-1 domain

TABLE 1. Characteristic of human and mouse ADAMTS13

Characteristic	Human ADAMTS13	Mouse ADAMTS13L	Mouse ADAMTS13S
Genomic location	9q34	2A3	2A3
No. of exons	29	29	24
mRNA (kb)	4.7	4.6	3.6
Amino acid residues			
Precursor	1427	1426	1037
Mature	1353	1350	961
Molecular weight (kDa)			
Predicted	145	147	105
Observed	190	Not determined	Not determined
Production	Hepatic stellate cells, endothelial cells, platelets	Hepatic stellate cells	Hepatic stellate cells
Plasma concentration	0.5–1.0 µg/ml	Not determined	Not determined

Functional Roles of ADAMTS13 Domains

The propeptide of ADAMTS13 contains a recognition sequence [RX(K/R)R] for furin that is localized in the trans-Golgi network. This domain is removed in the secretory process, and the N-terminus of mature ADAMTS13 in plasma starts with the metalloprotease domain [1–3]. In other members of the ADAMTS family, the propeptide usually acts to maintain enzymatic latency or assist in protein folding [14]. However, the propeptide of ADAMTS13 is exceptionally short and does not have such functions [26]. Deletion of the propeptide does not impair the active enzyme secretion. ADAMTS13, with its intact propeptide, is also proteolytically active. The role of propeptide in ADAMTS13 is unknown at present.

Like other ADAMTS members, the metalloprotease domain of ADAMTS13 includes the active site sequence (HEXXHXXGXXHD) that coordinates Zn^{2+}. The conserved methionine residue that supports the active site Zn^{2+} within the sequence (V/I)M(A/S) and the residues that form a predicted Ca^{2+} coordination site are also present in the metalloprotease domain.

The proximal C-terminal domains, including the disintegrin-like domain, the first Tsp1 motif, the Cys-rich domain, and the spacer domain, participate in substrate recognition and are required for the cleavage of VWF (Fig. 2) [27]. Deletion of the spacer domain markedly impairs the enzymatic activity [28, 29]. The target epitope of inhibitory antibodies for ADAMTS13 is concentrated on the Cys-rich and spacer domains (Fig. 2) [28, 30] and thus supports the important function of these domains. The Cys-rich domain contains an RGD sequence, a potential integrin-binding site, but the mutation of RGD to RGE does not affect the secretion or enzymatic activity in vitro [28]. The specific binding between ADAMTS13 and integrins has not been reported.

The functional role of more distal C-terminal domains, including seven Tsp1 motifs and two CUB domains, remains controversial. Recombinant ADAMTS13 mutants lacking these distal C-terminal domains still bind to immobilized VWF with an affinity similar to that of full-length ADAMTS13 [31] and maintain VWF-cleaving activity under static and flow conditions [28, 29, 32]. These results indicate that the distal C-terminal domains are dispensable, at least in vitro. However, removal of the two CUB domains from ADAMTS13 reduces the binding affinity for VWF threefold, whereas additional removal of the eighth Tsp1 motif increases the affinity twofold [31]. Recombinant polypeptides and synthetic peptides derived from the first CUB domain can block VWF cleavage by ADAMTS13 under flow [33]. These data suggest that the distal C-terminal domains influence the substrate recognition. Future in vivo studies may help define the functions of the distal C-terminal domains.

Enzymatic Properties and Assay Methods of ADAMTS13

The only known physiological substrate for ADAMTS13 is VWF, which is cleaved at the Tyr1605-Met1606 bond in the VWF-A2 domain [34]. The activity requires both Zn^{2+} and Ca^{2+} and is cooperatively enhanced by these cations [35]. Ba^{2+} can also enhance the ADAMTS13 activity in citrated plasma and is usually applied to clinical

ADAMTS13 assays. This effect of Ba^{2+} may reflect the release of chelated Ca^{2+} or Zn^{2+} because Ba^{2+} cannot stimulate ADAMTS13 in citrate-free conditions [35].

The cleavage rate of VWF by ADAMTS13 is markedly elevated by flow shear stress [36] or by denaturing reagents [37, 38], suggesting that conformational changes in VWF are required to expose the cleavage site for ADAMTS13. The modeled structure of the VWF-A2 domain supports this concept, indicating that the Tyr1605-Met1606 bond is buried in a core ß-sheet of the native structure [39]. Flow shear stress also promotes the interaction between the VWF-A1 domain and its ligand, platelet GPIbα or heparin, and this interaction is thought to accelerate the VWF cleavage by ADAMTS13 (Fig. 1) [40]. On the other hand, physiological concentrations of NaCl inhibit the VWF cleavage by ADAMTS13 in denaturing conditions [41]. The inhibitory effect has been ascribed to the specific binding of Cl^- to the VWF-A1 domain, which stabilizes the folded conformation of VWF [42]. Some proteases, such as thrombin and plasmin, can degrade ADAMTS13 in vitro [43], but the physiological significance of the degradation has not been fully established.

The assays used to measure ADAMTS13 activity were initially developed by Furlan et al. and Tsai and Lian [37, 38]. In these assays, purified human VWF was incubated with plasma ADAMTS13 in the presence of either urea or guanidine HCl. The reaction products were subjected to sodium dodecyl sulfate (SDS)-agarose or SDS-polyacrylamide gel electrophoresis and Western blot analysis using anti-VWF antibodies. These assays offered a breakthrough in the discovery of ADAMTS13 but are not suitable for routine clinical tests because of technical complications. As alternatives, several assays, including a collagen-binding assay, an immunoradiometric assay and a ristocetin-cofactor assay, were developed [44] but must be improved in terms of their specificity and sensitivity.

In 2004, we identified a 73-amino-acid sequence spanning residues D1596 to R1668 of the VWF A2 domain, VWF73, as the minimum region for ADAMTS13 cleavage [45]. We introduced a chemical modification into VWF73 and developed a fluorogenic substrate, FRETS-VWF73 (Fig. 3) [46]. The FRETS-VWF73 assay is specific and

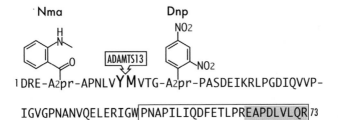

FIG. 3. Structure of FRETS-VWF73. In the VWF73 peptide (D1596-R1668 of VWF), Q1599 and N1610 residues are substituted for 2,3-diaminopropionic residues (*A2pr*) modified with a 2-(*N*-methylamino) benzoyl group (*Nma*) and a 2,4-dinitrophenyl group (*Dnp*), respectively. In the intact FRET-VWF73, when the Nma group is excited at 340 nm, fluorescence is quenched by the neighboring Dnp group. After cleavage of the Y-M bond by ADAMTS13, the emission of fluorescence at 440 nm from Nma is detected. The 24-amino-acid residues with ADAMTS13 inhibitory activity are *boxed*. The proposed exosite for interaction with the ADAMTS13 spacer domain is shown in *gray*

sensitive for ADAMTS13 using a simple procedure. It does not require denaturing conditions; therefore, the incubation time of the substrates and plasma samples is reduced to less than 1 h. Several new assays based on the VWF73 sequence have been developed [47]. Most of the assays are compatible with the 96-well microplates, so they have the potential to be widely used in clinical settings.

Assays for ADAMTS13 activity can be utilized to detect the inhibitory antibodies against ADAMTS13. For this purpose, heat-inactivated plasma samples are mixed with normal plasma, and then the residual ADAMTS13 activity of normal plasma is determined. However, it should be noted that some pathogenic autoantibodies could enhance the clearance of ADAMTS13 from the circulation without blocking the enzymatic activity [48]. Several enzyme-linked immunosorbent assay (ELISA) methods of measuring antibodies against ADAMTS13 and ADAMTS13 antigen levels have recently been developed [47].

Under the physiological pH and ionic strength condition, ADAMTS13 cleaves VWF with a $K_{m\ app}$ of $3.7 \pm 1.4\,\mu g/ml$ (15 nM in VWF subunits) [35], which is comparable to the plasma VWF concentration of $5-10\,\mu g/ml$. Under the same condition, ADAMTS13 cleaves FRETS-VWF73 with a $K_{m\ app}$ of $3.2 \pm 1.1\,\mu M$ [35], indicating that the affinity of ADAMTS13 to FRETS-VWF73 is decreased approximately 200-fold in comparison with its affinity for VWF. However, ADAMTS13 cleaves VWF and FRETS-VWF73 with substantially comparable catalytic efficiencies ($k_{cat}/K_{m\ app}$) of $55\,\mu M^{-1}\,min^{-1}$ and $18\,\mu M^{-1}\,min^{-1}$, respectively, because the $k$cat for FRETS-VWF73 cleavage is approximately 70-fold higher than that for VWF cleavage.

The cleavage of VWF73 by ADAMTS13 is markedly impaired by deletion of the C-terminal nine residues between E1660 and R1668 from VWF73 [45]. The peptide of 24 residues containing P1645–K1668 can competitively inhibit the cleavage of VWF by ADAMTS13 (Fig. 3) [49]. Thus, ADAMTS13 appears to interact with VWF by binding to the region comprising the C-terminus of VWF73. The sequence of E1660–R1668 has recently been shown to be the exosite that interacts with the spacer domain of ADAMTS13 (Figs. 2, 3) [50].

Mutations and Polymorphisms in *ADAMTS13* Gene

A number of causative mutations for congenital TTP have been identified in the *ADAMTS13* gene (Table 2) and are distributed throughout the gene. Some of these mutations have been confirmed to produce deleterious effects on secretion or proteolytic activity by in vitro expression analysis [15, 51, 52]. Congenital TTP is inherited in autosomal recessive fashion, and affected patients are compound heterozygotes or homozygotes for mutant alleles. Most of the mutations have been found in single families, but some have been observed in different, unrelated patients. In particular, the 4143insA mutation has been recurrently identified in northern and central European families, and haplotype analysis suggests that this mutation was derived from a common origin [53]. The 4143insA mutation may be a major cause of congenital TTP in northern and central Europe, and its allele frequency should be determined in a large population.

Several common missense polymorphisms have been reported in the *ADAMTS13* gene (Table 2). Approximately 10% of the Japanese population may be heterozygous

TABLE 2. TTP-causative mutations and missence polymorphisms in *ADAMTS13* gene

Type and region	Nucleotide	Effect	Domain
Missense			
Exon 3	237C>G	179M	Metalloprotease
Exon 3	262G>A	V88M	Metalloprotease
Exon 3	286C>G	H96D	Metalloprotease
Exon 3	304C>T	R102C	Metalloprotease
Exon 6	577C>T	R193W	Metalloprotease
Exon 6	587C>T	T196I	Metalloprotease
Exon 6	607T>C	S203P	Metalloprotease
Exon 7	695T>A	L232Q	Metalloprotease
Exon 7	702C>A	H234Q	Metalloprotease
Exon 7	703G>C	D235H	Metalloprotease
Exon 7	749C>T	A250V	Metalloprotease
Exon 7	788C>G	S263C	Metalloprotease
Exon 7	803G>C	R268P	Metalloprotease
Exon 8	932G>A	C311Y	Disintegrin-like
Exon 9	1039T>A	C347S	Disintegrin-like
Exon 9	1058C>T	P353L	Disintegrin-like
Exon 10	1170G>C	W390C	Tsp1-1
Exon 10	1193G>A	R398H	Tsp1-1
Exon 12	1370C>T	P457L	Cys-rich
Exon 13	1520G>A	R507Q	Cys-rich
Exon 13	1523G>A	C508Y	Cys-rich
Exon 13	1582A>G	R528G	Cys-rich
Exon 16	1787C>T	A596V	Spacer
Exon 17	2012C>T	P671L	Spacer
Exon 17	2017A>T	I673F	Spacer
Exon 17	2074C>T	R692C	Tsp1-2
Exon 19	2272T>C	C758R	Tsp1-3
Exon 21	2723G>A	C908Y	Tsp1-5
Exon 21	2723G>C	C908S	Tsp1-5
Exon 22	2851T>G	C951G	Tsp1-5
Exon 24	3070T>G	C1024G	Tsp1-7
Exon 24	3178C>T	R1060W	Tsp1-7
Exon 25	3367C>T	R1123C	Tsp1-8
Exon 26	3638G>A	C1213Y	CUB-1
Exon 26	3655C>T	R1219W	CUB-1
Exon 27	3716G>T	G1239V	CUB-1
Exon 28	4006C>T	R1336W	CUB-2
Nonsense			
Exon 2	130C>T	Q44X	Propeptide
Exon 10	1169G>A	W390X	Tsp1-1
Exon 12	1345C>T	Q449X	Cys-rich
Exon 21	2728C>T	R910X	Tsp1-5
Exon 24	3047G>A	W1016X	Tsp1-7
Exon 24	3100A>T	R1034X	Tsp1-7
Exon 26	3616C>T	R1206X	CUB-1
Exon 27	3735G>A	W1245X	CUB-1

TABLE 2. *Continued*

Type and region	Nucleotide	Effect	Domain
Insertion/deletion			
Exon 1	83insT	Frameshift	Signal peptide
Exon 4	291del29	Frameshift	Metalloprotease
Exon 7	719del6	G241C242del	Metalloprotease
Exon 10	1095del18	W365-R370del	Disintegrin-like
Exon 15	1783delTT	Frameshift	Spacer
Exon 19	2279delG	Frameshift	Tsp1-3
Exon 19	2376del26	Frameshift	Tsp1-3
Exon 20	2459delAT	Frameshift	Tsp1-4
Exon 23	2930del6	C977W, A978R979del	Tsp1-6
Exon 25	3252delCT	Frameshift	Tsp1-8
Exon 27	3769insT	Frameshift	CUB-1
Exon 29	4143insA	Frameshift	CUB-2
Splice			
Intron 3	330+1G>A	Abnormal splicing	Metalloprotease
Intron 4	414+1G>A	Abnormal splicing	Metalloprotease
Intron 6	686+1G>A	Abnormal splicing	Metalloprotease
Intron 6	687−2A>G	Abnormal splicing	Metalloprotease
Intron 7-Exon 8	del29	Abnormal splicing	Disintegrin-like
Intron 10	1244+2T>G	Abnormal splicing	Tsp1-1
Intron 11	1309−1G>A	Abnormal splicing	Cys-rich
Intron 13	1584+5G>A	Abnormal splicing	Cys-rich
Missense polymorphism			
Exon 1	19C>T	R7W	Signal peptide
Exon 12	1342C>G	Q448E	Cys-rich
Exon 12	1423C>T	P475S	Cys-rich
Exon 16	1852C>G	P618A	Spacer
Exon 16	1874G>A	R625H	Spacer
Exon 18	2195C>T	A732V	Tsp1-2
Exon 21	2699C>T	A900V	Tsp1-5
Exon 21	2708C>T	S903L	Tsp1-5
Exon 24	3097G>A	A1033T	Tsp1-7

for the P475S polymorphism, which reduces ADAMTS13 activity [15]. This polymorphism is also found in Chinese with low frequency but not in Caucasians [54, 55]. Although it has been shown that the P475S is not a genetic risk for deep vein thrombosis [56], the relation to other thrombotic disorders should be examined. It is noteworthy that some missense polymorphisms can interact and modify the effect of rare mutations on ADAMTS13 secretion or activity [57].

Mouse ADAMTS13

The cDNA sequence and the gene structure of mouse *Adamts13* were determined [58]. Unexpectedly, two kinds of *Adamts13* gene were found in a mouse strain-specific manner (Fig. 2, Table 1). The *Adamts13* gene of the 129/Sv, FVB/NJ, and CAST/EiJ strains contains 29 exons, like the human *ADAMTS13* gene, and encodes ADAMTS13L with the same domain constitutions as human ADAMTS13. On the other hand, several strains of mice, including the BALB/c, C3H/He, C57BL/6, and DBA/2 strains, harbor the insertion of an intracisternal A-particle (IAP) retrotransposon into intron 23 of the *Adamts13* gene. The inserted IAP is one of the endogenous transposable elements present at approximately 2000 sites in the mouse genome. Like retroviruses, the IAP contains two long terminal repeats with signals for the initiation and regulation of transcription and for the polyadenylation of transcripts. The IAP insertion into the *Adamts13* gene produces a cryptic splicing site followed by a premature in-frame stop codon and a polyadenylation signal derived from the long terminal repeat. As a result, ADAMTS13S that lacks the C-terminal two Tsp1 motifs and two CUB domains is predominantly expressed in these strains.

Both forms of mouse ADAMTS13 are mainly expressed in the liver and retain a furin-recognition sequence in the propeptide domain, Zn^{2+}-binding and Ca^{2+} coordination sequences in the metalloprotease domain, and an RGD sequence in the Cys-rich domain. They show VWF-cleaving activity in vitro, but the truncated form exhibits considerably lower activity than the full-length form for purified human VWF multimers [59]. The physiological significance of this natural variation for hemostatic function of laboratory mice remains to be elucidated.

ADAMTS13-Deficient Mice

Mouse models of congenital ADAMTS13 deficiency have been developed by gene-targeting approaches [60, 61]. We generated *Adamts13$^{-/-}$* mice on a 129/Sv genetic background using a targeting vector that eliminates exons 3–6 of the *Adamts13* gene, encoding the metalloprotease domain [60]. In the *Adamts13$^{-/-}$* mice, ADAMTS13 mRNA is not detected in the liver. Plasma ADAMTS13 activity is completely absent, indicating our successful targeting strategy. The intercrosses of *Adamts13$^{+/-}$* mice have produced offspring in the expected Mendelian distribution, and *Adamts13$^{-/-}$* mice are viable and fertile. Analysis of plasma VWF multimer patterns detects UL-VWF multimers in *Adamts13$^{-/-}$* mice but not in *Adamts13$^{+/+}$* or *Adamts13$^{+/-}$* mice, suggesting that the deficiency of ADAMTS13 alone can support the generation of plasma UL-VWF multimers. However, no apparent signs of TTP have been observed in *Adamts13$^{-/-}$* mice. Blood platelet counts, plasma haptoglobin levels, and peripheral blood smears are normal in *Adamts13$^{-/-}$* mice, suggesting a lack of thrombocytopenia and hemolytic anemia in *Adamts13$^{-/-}$* mice. Although pregnancy is a triggering event for TTP, renal histology of pregnant *Adamts13$^{-/-}$* mice does not show thrombi deposition or excessive VWF accumulation in microvessels. Further analysis of thrombogenic properties using a parallel plate flow chamber revealed that platelet thrombus formation under shear stress is significantly promoted in *Adamts13$^{-/-}$* mice compared to *Adamts13$^{+/+}$* mice. Consumptive thrombocytopenia is more severely induced in *Adamts13$^{-/-}$* mice than in *Adamts13$^{+/+}$* mice after intravenous injection of a mixture of collagen and epinephrine. Therefore, a complete lack of ADAMTS13 in mice is a prothrombotic state, but it alone is not sufficient to cause TTP. Factors in addition to ADAMTS13 deficiency may be necessary for the development of TTP in mice.

Other *Adamts13$^{-/-}$* mice have been generated with exons 1–6 eliminated [61]. On a mixed C57BL/6J and 129X1/SvJ genetic background, they are viable without any TTP-like phenotypes, similar to our *Adamts13$^{-/-}$* mice on a 129/Sv genetic background. After activation of microvenule endothelium with calcium ionophore A23187, VWF-mediated platelet adhesion is significantly prolonged in *Adamts13$^{-/-}$* mice compared to *Adamts13*[+/+] controls. When arterioles are injured by ferric chloride treatment, platelet adhesion and thrombus formation on the exposed subendothelium are accelerated in *Adamts13*[−/−] mice [62]. Thus, ADAMTS13 down-regulates platelet adhesion and aggregation in vivo, and ADAMTS13 deficiency can provide enhanced thrombus formation at the site of vascular lesions. It is notable that plasma VWF multimer patterns in *Adamts13*[+/+] mice and *Adamts13*[−/−] mice are indistinguishable on the mixed C57BL/6J and 129X1/SvJ genetic background.

Introduction of the genetic background of the CASA/Rk strain, which has elevated plasma VWF levels, onto *Adamts13*[−/−] mice produces the appearance of plasma UL-VWF multimers along with a significant reduction of mean platelet counts [61]. After the challenge with shigatoxin, the mice exhibit TTP-like symptoms, including thrombocytopenia, hemolysis with schistocytes, and VWF-rich and fibrin-poor hyaline thrombosis in microvessels. Moreover, a subset of these *Adamts13*[−/−] mice can spontaneously develop TTP symptoms. Because no correlation has been observed between plasma VWF levels and the severity of symptoms, TTP-modifier genes that are not associated with plasma VWF levels may be delivered from a CASA/Rk genetic background. Shigatoxin is known to induce endothelial dysfunction; therefore, TTP can be triggered in *Adamts13*[−/−] mice by environmental factors causing endothelial activation or damage.

There is a large variation in the phenotypes of TTP patients with ADAMTS13 deficiency. Many TTP patients with congenital ADAMTS13 deficiency have their first acute episode during the neonatal period or early infancy, but late-onset cases and asymptomatic carriers in adulthood have also been reported [63–66]. Patients with identical *ADAMTS13* mutations but different clinical courses have been described [65–67], indicating that ADAMTS13 deficiency brings a serious risk but is not a sufficient condition for TTP. Data from the mouse models with ADAMTS13 deficiency support this view, and ADAMTS13-deficient mice can contribute to the identification of additional genetic and/or environmental TTP triggers.

An attempt has also been made to develop a mouse model for acquired ADAMTS13 deficiency [62]. Infusion of an anti-ADAMTS13 antibody into wild-type mice causes VWF-mediated platelet thrombus formation in activated microvenules similar to that seen in *Adamts13*[−/−] mice. Thus, anti-ADAMTS13 antibodies appear to induce functional ADAMTS13 deficiency and may be utilized for an animal model of acquired TTP. The efficacy of recombinant ADAMTS13 administration has been assessed in mice with or without ADAMTS13 deficiency [62]. Recombinant ADAMTS13 can reduce platelet adhesion and aggregation and promote thrombus dissolution in both wild-type and *Adamts13*[−/−] mice, suggesting that recombinant ADAMTS13 may be used as an antithrombotic agent.

Conclusion

Since the discovery of ADAMTS13 in 2001, marked advances have been made in elucidating the biochemical and physiological functions of this new antithrombotic factor. Its enzymatic characteristics and domain functions involving the recognition and cleavage of VWF have been significantly clarified. Simple, rapid, sensitive assays for measuring ADAMTS13 activity have been developed and are beginning to be utilized for the clinical diagnosis of TTP and its related disorders. The accumulation of clinical information about TTP patients highlights the possibility that ADAMTS13 deficiency is not sufficient for the development of TTP. Additional factors that contribute to the onset or severity of TTP should be explored. Mouse models of ADAMTS13 deficiency have the potential to help identify TTP triggers.

References

1. Fujikawa K, Suzuki H, McMullen B, et al (2001) Purification of human von Willebrand factor-cleaving protease and its identification as a new member of the metalloproteinase family. Blood 98:1662–1666
2. Gerritsen HE, Robles R, Lämmle B, et al (2001) Partial amino acid sequence of purified von Willebrand factor-cleaving protease. Blood 98:1654–1661
3. Soejima K, Mimura N, Hirashima M, et al (2001) A novel human metalloprotease synthesized in the liver and secreted into the blood: possibly, the von Willebrand factor-cleaving protease? J Biochem (Tokyo) 130:475–480
4. Zheng X, Chung D, Takayama TK, et al (2001) Structure of von Willebrand factor-cleaving protease (ADAMTS13), a metalloprotease involved in thrombotic thrombocytopenic purpura. J Biol Chem 276:41059–41063
5. Sporn LA, Marder VJ, Wagner DD (1987) von Willebrand factor released from Weibel-Palade bodies binds more avidly to extracellular matrix than that secreted constitutively. Blood 69:1531–1534
6. Federici AB, Bader R, Pagani S, et al (1989) Binding of von Willebrand factor to glycoproteins Ib and IIb/IIIa complex: affinity is related to multimeric size. Br J Haematol 73:93–99
7. Moake JL, Rudy CK, Troll JH, et al (1982) Unusually large plasma factor VIII: von Willebrand factor multimers in chronic relapsing thrombotic thrombocytopenic purpura. N Engl J Med 307:1432–1435
8. Furlan M, Robles R, Galbusera M, et al (1998) von Willebrand factor-cleaving protease in thrombotic thrombocytopenic purpura and the hemolytic-uremic syndrome. N Engl J Med 339:1578–1584
9. Tsai HM, Lian EC (1998) Antibodies to von Willebrand factor-cleaving protease in acute thrombotic thrombocytopenic purpura. N Engl J Med 339:1585–1594
10. Levy GG, Nichols WC, Lian EC, et al (2001) Mutations in a member of the ADAMTS gene family cause thrombotic thrombocytopenic purpura. Nature 413:488–494
11. Moschcowitz E (1924) Hyaline thrombosis of the terminal arterioles and capillaries; a hitherto undescribed disease. Proc N Y Pathol Soc 24:21–24
12. Asada Y, Sumiyoshi A, Hayashi T, et al (1985) Immunohistochemistry of vascular lesion in thrombotic thrombocytopenic purpura, with special reference to factor VIII related antigen. Thromb Res 38:469–479
13. Rock GA, Shumak KH, Buskard NA, et al (1991) Comparison of plasma exchange with plasma infusion in the treatment of thrombotic thrombocytopenic purpura: Canadian Apheresis Study Group. N Engl J Med 325:393–397
14. Porter S, Clark IM, Kevorkian L, et al (2005) The ADAMTS metalloproteinases. Biochem J 386:15–27
15. Kokame K, Matsumoto M, Soejima K, et al (2002) Mutations and common polymorphisms in ADAMTS13 gene responsible for von Willebrand factor-cleaving protease activity. Proc Natl Acad Sci U S A 99:11902–11907
16. Ricketts LM, Dlugosz M, Luther KB, et al (2007) O-Fucosylation is required for ADAMTS13 secretion. J Biol Chem 282:17014–17023
17. Uemura M, Tatsumi K, Matsumoto M, et al (2005) Localization of ADAMTS13 to the stellate cells of human liver. Blood 106:922–924

18. Zhou W, Inada M, Lee TP, et al (2005) ADAMTS13 is expressed in hepatic stellate cells. Lab Invest 85:780–788
19. Mannucci PM, Canciani MT, Forza I, et al (2001) Changes in health and disease of the metalloprotease that cleaves von Willebrand factor. Blood 98:2730–2735
20. Ko S, Okano E, Kanehiro H, et al (2006) Plasma ADAMTS13 activity may predict early adverse events in living donor liver transplantation: observations in 3 cases. Liver Transpl 12:859–869
21. Kume Y, Ikeda H, Inoue M, et al (2007) Hepatic stellate cell damage may lead to decreased plasma ADAMTS13 activity in rats. FEBS Lett 581:1631–1634
22. Suzuki M, Murata M, Matsubara Y, et al (2004) Detection of von Willebrand factor-cleaving protease (ADAMTS-13) in human platelets. Biochem Biophys Res Commun 313:212–216
23. Turner N, Nolasco L, Tao Z, et al (2006) Human endothelial cells synthesize and release ADAMTS-13. J Thromb Haemost 4:1396–1404
24. Soejima K, Nakamura H, Hirashima M, et al (2006) Analysis on the molecular species and concentration of circulating ADAMTS13 in blood. J Biochem (Tokyo) 139:147–154
25. Furlan M, Robles R, Morselli B, et al (1999) Recovery and half-life of von Willebrand factor-cleaving protease after plasma therapy in patients with thrombotic thrombocytopenic purpura. Thromb Haemost 81:8–13
26. Majerus EM, Zheng X, Tuley EA, et al (2003) Cleavage of the ADAMTS13 propeptide is not required for protease activity. J Biol Chem 278:46643–46648
27. Ai J, Smith P, Wang S, et al (2005) The proximal carboxyl-terminal domains of ADAMTS13 determine substrate specificity and are all required for cleavage of von Willebrand factor. J Biol Chem 280:29428–29434
28. Soejima K, Matsumoto M, Kokame K, et al (2003) ADAMTS-13 cysteine-rich/spacer domains are functionally essential for von Willebrand factor cleavage. Blood 102:3232–3237
29. Zheng X, Nishio K, Majerus EM, et al (2003) Cleavage of von Willebrand factor requires the spacer domain of the metalloprotease ADAMTS13. J Biol Chem 278:30136–30141
30. Klaus C, Plaimauer B, Studt JD, et al (2004) Epitope mapping of ADAMTS13 autoantibodies in acquired thrombotic thrombocytopenic purpura. Blood 103:4514–4519
31. Majerus EM, Anderson PJ, Sadler JE (2005) Binding of ADAMTS13 to von Willebrand factor. J Biol Chem 280:21773–21778
32. Tao Z, Wang Y, Choi H, et al (2005) Cleavage of ultralarge multimers of von Willebrand factor by C-terminal-truncated mutants of ADAMTS-13 under flow. Blood 106:141–143
33. Tao Z, Peng Y, Nolasco L, et al (2005) Recombinant CUB-1 domain polypeptide inhibits the cleavage of ULVWF strings by ADAMTS13 under flow conditions. Blood 106:4139–4145
34. Dent JA, Berkowitz SD, Ware J, et al (1990) Identification of a cleavage site directing the immunochemical detection of molecular abnormalities in type IIA von Willebrand factor. Proc Natl Acad Sci U S A 87:6306–6310
35. Anderson PJ, Kokame K, Sadler JE (2006) Zinc and calcium ions cooperatively modulate ADAMTS13 activity. J Biol Chem 281:850–857
36. Tsai HM, Sussman II, Nagel RL (1994) Shear stress enhances the proteolysis of von Willebrand factor in normal plasma. Blood 83:2171–2179
37. Furlan M, Robles R, Lämmle B (1996) Partial purification and characterization of a protease from human plasma cleaving von Willebrand factor to fragments produced by in vivo proteolysis. Blood 87:4223–4234
38. Tsai HM (1996) Physiologic cleavage of von Willebrand factor by a plasma protease is dependent on its conformation and requires calcium ion. Blood 87:4235–4244

39. Sutherland JJ, O'Brien LA, Lillicrap D, et al (2004) Molecular modeling of the von Willebrand factor A2 domain and the effects of associated type 2A von Willebrand disease mutations. J Mol Model 10:259–270
40. Nishio K, Anderson PJ, Zheng XL, et al (2004) Binding of platelet glycoprotein Ibalpha to von Willebrand factor domain A1 stimulates the cleavage of the adjacent domain A2 by ADAMTS13. Proc Natl Acad Sci U S A 101:10578–10583
41. De Cristofaro R, Peyvandi F, Palla R, et al (2005) Role of chloride ions in modulation of the interaction between von Willebrand factor and ADAMTS-13. J Biol Chem 280:23295–23302
42. De Cristofaro R, Peyvandi F, Baronciani L, et al (2006) Molecular mapping of the chloride-binding site in von Willebrand factor (VWF): energetics and conformational effects on the VWF/ADAMTS-13 interaction. J Biol Chem 281:30400–30411
43. Crawley JT, Lam JK, Rance JB, et al (2005) Proteolytic inactivation of ADAMTS13 by thrombin and plasmin. Blood 105:1085–1093
44. Miyata T, Kokame K, Banno F (2005) Measurement of ADAMTS13 activity and inhibitors. Curr Opin Hematol 12:384–389
45. Kokame K, Matsumoto M, Fujimura Y, et al (2004) VWF73, a region from D1596 to R1668 of von Willebrand factor, provides a minimal substrate for ADAMTS-13. Blood 103:607–612
46. Kokame K, Nobe Y, Kokubo Y, et al (2005) FRETS-VWF73, a first fluorogenic substrate for ADAMTS13 assay. Br J Haematol 129:93–100
47. Miyata T, Kokame K, Banno F, et al (2007) ADAMTS13 assays and ADAMTS13-deficient mice. Curr Opin Hematol 14:277–283
48. Shelat SG, Smith P, Ai J, et al (2006) Inhibitory autoantibodies against ADAMTS-13 in patients with thrombotic thrombocytopenic purpura bind ADAMTS-13 protease and may accelerate its clearance in vivo. J Thromb Haemost 4:1707–1717
49. Wu JJ, Fujikawa K, McMullen BA, et al (2006) Characterization of a core binding site for ADAMTS-13 in the A2 domain of von Willebrand factor. Proc Natl Acad Sci U S A 103:18470–18474
50. Gao W, Anderson PJ, Majerus EM, et al (2006) Exosite interactions contribute to tension-induced cleavage of von Willebrand factor by the antithrombotic ADAMTS13 metalloprotease. Proc Natl Acad Sci U S A 103:19099–19104
51. Matsumoto M, Kokame K, Soejima K, et al (2004) Molecular characterization of ADAMTS13 gene mutations in Japanese patients with Upshaw-Schulman syndrome. Blood 103:1305–1310
52. Kokame K, Miyata T (2004) Genetic defects leading to hereditary thrombotic thrombocytopenic purpura. Semin Hematol 41:34–40
53. Schneppenheim R, Kremer Hovinga JA, Becker T, et al (2006) A common origin of the 4143insA ADAMTS13 mutation. Thromb Haemost 96:3–6
54. Bongers TN, De Maat MP, Dippel DW, et al (2005) Absence of Pro475Ser polymorphism in ADAMTS-13 in Caucasians. J Thromb Haemost 3:805
55. Ruan C, Dai L, Su J, et al (2004) The frequency of P475S polymorphism in von Willebrand factor-cleaving protease in the Chinese population and its relevance to arterial thrombotic disorders. Thromb Haemost 91:1257–1258
56. Kimura R, Honda S, Kawasaki T, et al (2006) Protein S-K196E mutation as a genetic risk factor for deep vein thrombosis in Japanese patients. Blood 107:1737–1738
57. Plaimauer B, Fuhrmann J, Mohr G, et al (2006) Modulation of ADAMTS13 secretion and specific activity by a combination of common amino acid polymorphisms and a missense mutation. Blood 107:118–125
58. Banno F, Kaminaka K, Soejima K, et al (2004) Identification of strain-specific variants of mouse Adamts13 gene encoding von Willebrand factor-cleaving protease. J Biol Chem 279:30896–30903

59. Zhou W, Bouhassira EE, Tsai HM (2007) An IAP retrotransposon in the mouse ADAMTS13 gene creates ADAMTS13 variant proteins that are less effective in cleaving von Willebrand factor multimers. Blood 110:886–893

60. Banno F, Kokame K, Okuda T, et al (2006) Complete deficiency in ADAMTS13 is prothrombotic, but it alone is not sufficient to cause thrombotic thrombocytopenic purpura. Blood 107:3161–3166

61. Motto DG, Chauhan AK, Zhu G, et al (2005) Shigatoxin triggers thrombotic thrombocytopenic purpura in genetically susceptible ADAMTS13-deficient mice. J Clin Invest 115:2752–2761

62. Chauhan AK, Motto DG, Lamb CB, et al (2006) Systemic antithrombotic effects of ADAMTS13. J Exp Med 203:767–776

63. Furlan M, Lämmle B (2001) Aetiology and pathogenesis of thrombotic thrombocytopenic purpura and haemolytic uraemic syndrome: the role of von Willebrand factor-cleaving protease. Best Pract Res Clin Haematol 14:437–454

64. Tao Z, Anthony K, Peng Y, et al (2006) Novel ADAMTS-13 mutations in an adult with delayed onset thrombotic thrombocytopenic purpura. J Thromb Haemost 4:1931–1935

65. Noris M, Bucchioni S, Galbusera M, et al (2005) Complement factor H mutation in familial thrombotic thrombocytopenic purpura with ADAMTS13 deficiency and renal involvement. J Am Soc Nephrol 16:1177–1183

66. Donadelli R, Banterla F, Galbusera M, et al (2006) In-vitro and in-vivo consequences of mutations in the von Willebrand factor cleaving protease ADAMTS13 in thrombotic thrombocytopenic purpura. Thromb Haemost 96:454–464

67. Veyradier A, Lavergne JM, Ribba AS, et al (2004) Ten candidate ADAMTS13 mutations in six French families with congenital thrombotic thrombocytopenic purpura (Upshaw-Schulman syndrome). J Thromb Haemost 2:424–429

Antithrombin and Heparin Cofactor II: Structure and Functions

Takehiko Koide

Summary. Antithrombin (AT) and heparin cofactor II (HCII) are plasma serpins that function as principal regulators of blood coagulation. These serpins inhibit their target proteinases by forming an inactive enzyme–inhibitor complex through an interaction between their reactive center and the active site of the proteinase. Among the coagulation proteinases, AT mainly inhibits factor Xa and thrombin, and HCII exclusively inhibits thrombin. However, both AT and HCII by themselves are poor inhibitors of coagulation proteinases as these reactions proceed at a slow rate and are time-dependent. These serpins also require glycosaminoglycans for their physiological functions. By binding to glycosaminoglycans, a conformational change occurs in the reactive center loop (RCL), which activates these serpins to become a more efficient inhibitor of coagulation proteinases. The physiological cofactor for AT is heparan sulfate proteoglycans on endothelial cells, and that for HCII is dermatan sulfate proteoglycans on vascular smooth muscle cells. The physiological importance of AT as an anticoagulant in the circulation is well supported by a number of thrombotic disorders in patients with AT deficiency. However, the physiological function of HCII as an anticoagulant is unclear, as HCII deficiency is not a significant risk factor for venous or arterial thrombosis. Recent crystallographic analyses of monomeric native AT variant, AT-thrombin, or factor Xa–pentasaccharide complexes, as well as HCII–thrombin complex, have revealed more detailed mechanisms of proteinase inhibition by these serpins and mechanisms of glycosaminoglycan-dependent enhancement of the reaction rates, together with some revisions of our previous knowledge.

Key words. Antithrombin · Heparin · Serpin · Heparin cofactor II · Thrombin

Introduction

Hemostasis is a balance of procoagulant and anticoagulant systems in the circulation of blood in vascular vessels. During the early stage of injury, procoagulant systems, including platelet activation and the blood coagulation cascade, must be predominant to ensure clot formation at the site of injury and to prevent loss of blood. Subsequently, the anticoagulant system regulates the procoagulant system to prevent thrombus formation from expanding beyond the site of injury. One of the main regulatory systems of blood coagulation is the direct inhibition of active coagulation proteinases by plasma serine proteinase inhibitors (serpins). Antithrombin (AT) and heparin cofactor II (HCII) are two serpins that regulate blood coagulation. The physiological importance of AT as a regulator of hemostasis has been well established from a number of thrombotic disorders caused by AT deficiency [1]. However, the significance of HCII as a regulator of blood coagulation is still controversial, as a convincing association between HCII deficiency and venous or arterial thrombosis has not been established [2]. The target coagulation factors are mainly thrombin and factor Xa for AT, and exclusively thrombin for HCII. It is well known that proteinase inhibition rates of AT and HCII are dramatically accelerated in the presence of heparin (heparan sulfate, in vivo) and dermatan sulfate, respectively. Recent crystallographic analyses of native AT and HCII and their complexed form with a target proteinase and pentasaccharide, which represents the essential part of heparin to interact with AT, revealed a detailed mechanism of activation of AT by heparin [3, 4], and, similarly of HCII [5]. In this chapter, recent progress in studies on the structure–function relation of AT and HCII is reviewed, particularly as it relates to the heparin-activation mechanism of these serpins.

Antithrombin

Human antithrombin (AT) is a single-chain glycoprotein with Mr ~58 000; it is composed of 432 amino acid residues and four carbohydrate sites attaching at Asn96, 135, 155, and 192 [6, 7]. The gene for AT is located on chromosome 1q23.1–23.9 with a size of 13 477 bp, consisting of seven exons (1, 2, 3A, 3B, 4, 5 and 6) and six introns [8]. AT is synthesized mainly in the liver and circulates in human plasma at a concentration of ~2.3 μmol/l (~125 mg/l) [9] with a half-life of about 3 days [10]. About 5%–10% of AT (referred to as antithrombin β) lacks carbohydrate at Asn135, which has a higher affinity for heparin [11]. AT has two specific functional sites; one is referred to as the "heparin-binding site" located in the N-terminal region of the polypeptide, and the other as the "reactive center" (also referred to as "the reactive site") for target proteinases located in the C-terminal region. The reactive center is the amino acid that directly interacts with the active site of a target proteinase. The basic tertiary structure of AT is common to all serpins, comprised of nine helices (designated A to I from the N-terminal region) and three β-sheets (designated A, B, and C, composed of six, six, and four strands, respectively). The functionally important reactive center loop (RCL) is exposed on the surface of the molecule having the reactive center (Arg393-Ser394, P1-P1'[1]) at the top extremity of the loop (Fig. 1) [12].

Approximately 75%–80% of the heparin-dependent thrombin inhibitory activity in human plasma is due to AT, and the remaining 20%–25% is due to HCII. The principal role of AT in the regulation of hemostasis has been well demonstrated by a close

---

[1] In this chapter, residues N-terminal to the reactive center bond are numbered from P1 and those C-terminal are numbered from P1', according to the proteolytic substrate numbering scheme of Schechter and Berger [38].

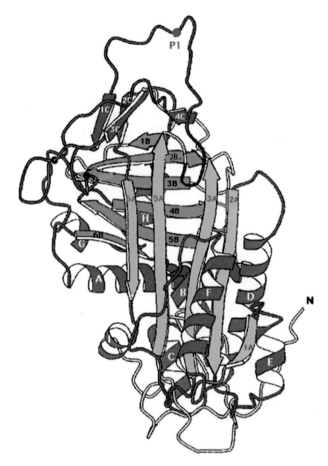

FIG. 1. Structure of antithrombin (AT). Helices (designated *A* to *I* from the N-terminal region) are red, and β-sheets are lime green (sheet A), green (sheet B), and purple (sheet C). The reactive center loop (RCL) is the top line with *P1* (the reactive center Arg393) in the middle part of the loop. (From Stein and Carrell [12], with permission)

association of its deficiency with venous thrombosis, whereas hemorrhagic episodes have been reported at slightly higher plasma levels of AT. Thus, the circulating level of AT is critical for the regulation of hemostasis within a narrow normal range of 80%–120%.

In vitro, AT can inhibit most coagulation proteinases, including thrombin and factors Xa, IXa, XIa, XIIa, and VIIa. Among them, the inhibitions of thrombin and factor Xa are regarded as physiologically significant. In many cases, the protease inhibition rate by AT is greatly enhanced in the presence of heparin (in vitro) or heparan sulfate proteoglycan on the surface of endothelial cells (in vivo) [13], as described in detail below.

Mechanism of Thrombin Inhibition by AT

The basic mechanism of proteinase inhibition by AT is common to those of other serpins (Fig. 2) [14]. AT forms an inactive complex with a protease through covalent bond formation between the reactive center of AT and the catalytic site of the proteinase. This basic inhibition mechanism of AT is common to both thrombin and factor Xa inhibitions. Specifically, the reactive center peptide bond between P1 Arg393 and P1' Ser394 of AT is selectively hydrolyzed by the target proteinase, forming an acyl–enzyme complex. The reaction, however, does not proceed to deacylation, but a dramatic conformational change occurs in the N-terminal portion of the cleaved RCL binding the target proteinase and it forms a new β strand (designated strand 4A) by insertion between strands 3A and 5A. During this dynamic movement of the RCL, the

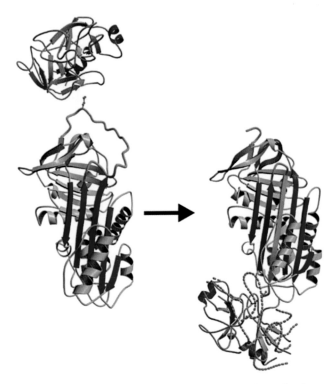

FIG. 2. Proteinase inhibition by a serpin. Trypsin (*left top*, cyan) attacks the reactive center of the serpin α_1-antitrypsin (*left bottom*) and forms an acyl–enzyme complex by hydrolyzing the reactive center peptide bond. Subsequently, the N-terminal portion of the cleaved RCL is inserted into the A-sheet, forming a new β strand. In this striking conformational change, the P1 residue binding the target proteinase moves 70 Å around the α_1-antitrypsin molecule and the structure of the complexed trypsin becomes distorted. (From Huntington et al. [14], with permission)

acyl-bonded proteinase moves as much as 70 Å around the AT molecule and finally settles in a conformationally distorted form of the inactivated proteinase (Fig. 2) [14]. This striking conformational change results in greater thermal stability of the "R" (relaxed) form of the cleaved AT in comparison to the "S" (stressed) form of the native AT.

Heparin-Enhanced Thrombin Inhibition by AT: Heparin Cofactor Activity

Although AT can inhibit many coagulation proteinases in vitro, AT alone is a slow-acting inhibitor and requires a heparin-enhanced instantaneous inhibitory activity (so-called heparin cofactor activity) to play its physiological role as a thrombin or factor Xa inhibitor in vivo. The heparin cofactor activity of AT has been known for more than three decades, and the mode of action of heparin in the inhibition of thrombin has been proposed as "template mechanism." In this mechanism, both AT and thrombin need to bind the same heparin molecule. Therefore, a minimal length of 16–18 monosaccharides are required for the maximal enhancement in the rate of thrombin inhibition by AT. As mentioned above, even in the absence of heparin, thrombin captures the reactive center Arg393 of AT and hydrolyzes its peptide bond, resulting in formation of an acyl–enzyme complex. However, owing to the inside-directed orientation of the side chain of Arg393 in a free form of native AT, the complex formation (i.e., the inhibition of thrombin) proceeds very slowly, taking 20–30 min to complete. By comparison, when AT binds to heparin (or heparan sulfate proteoglycan on endothelial cells) at the heparin-binding site, a drastic conformational change occurs in the AT molecule, and the location and orientation of the reactive center Arg393 becomes easily accessible to the active site Ser195 of thrombin, which also binds to the same heparin molecule at the nearby site. In this case, the inhibition of thrombin is completed almost instantaneously. The second-order rate constants for thrombin inhibition in the absence and presence of heparin are $1.4 \times 10^4 \, M^{-1} s^{-1}$ (37°C) and $1.5\text{-}4 \times 10^7 \, M^{-1} s^{-1}$, respectively. Thus, the thrombin inhibition rate of AT is enhanced 1000-fold in the presence of heparin. The physiological importance of this heparin-dependent inhibition of thrombin or factor Xa by AT has also been well demonstrated by many thrombotic patients whose plasma contains a normal level of AT that is capable of inhibiting thrombin and factor Xa normally in the absence of heparin but reveals no enhancement of the inhibition by heparin, thus lacking heparin-binding ability [1].

Several amino acid residues involved in the binding to heparin have been identified from studies of genetic mutants, site-directed mutagenesis, and chemical modifications [1, 15–23]. They include Arg46 and Arg47 on helix A and Lys114, Lys125, Arg129, Arg132, Lys133, and Lys136 on and near helix D. Those residues involved in the heparin-binding site on AT interact with a specific segment of the heparin molecule, which is typically represented by "pentasaccharide" sequence of heparin (Fig. 3) [24]. The characteristic of this pentasaccharide sequence is the presence of 3-O-sulfate in glucosamine F that is rather rare in heparin.

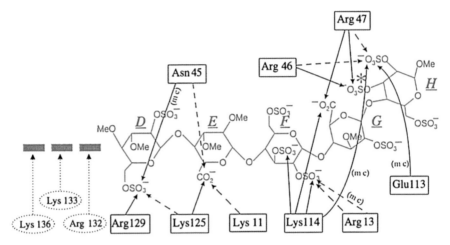

Fɪɢ. 3. Interaction between AT and pentasaccharide. *Solid arrows* indicate likely hydrogen bonds, *dashed arrows* indicate possible bonds, and (*mc*) indicates main-chain bonding. *Dotted arrows* from Arg132, Lys133 and Lys136 indicate interaction with extended oligosaccharide (or heparin). (From Jin et al. [24], with permission)

Mechanisms of AT Activation by Heparin: Heparin-Induced Conformational Change

Precisely what heparin does to AT and thrombin has been a mystery for more than three decades. The mechanism of action of heparin was first suggested in 1997 by X-ray crystallographic analysis of the dimer of intact and latent AT each bound by a synthetic pentasaccharide with the AT-binding sequence [24]. A characteristic feature of AT among serpins is that the N-terminal portion of the RCL was partially inserted between strands 3A and 5A (Figs. 1, 2) [25, 26]. Furthermore, the side chain of the P1 Arg393 is oriented toward the interior of the molecule in contrast to that of α_1-antitrypsin, which is exposed to the outside of the molecule. A mode of heparin action was proposed that postulates that when the pentasaccharide is bound to AT, helices A and D are elongated, with the formation of a new two-turn helix P between the C and D helices, resulting in extrusion of the partially inserted RCL and a more exposed orientation to the side chain of P1 Arg393 [24]. However, a recent crystal structure of monomeric native AT revealed that helix A was already elongated and helix P was already formed in the monomeric native AT in which both V317 on strand 2C and T401 on strand 1C were replaced by C to prevent conversion to a latent form by forming a disulfide bond [3]. Furthermore, the monomeric native AT revealed that the P1 Arg393 of monomeric AT, the side chain of which is shown in green & red color in State N in Fig. 4, has shifted 17 Å toward the left from the P1 Arg393 of the native AT in the heterodimer (located on the top of the red RCL in State N' in Fig. 4), and that the side chain (in green & red) of the P1 Arg393 of monomeric AT is oriented towards the interior of the molecule forms a salt bridge with the side chain of Glu237 and a hydrogen bond with the main chain carbonyl oxygen of Asn233. Thus, the

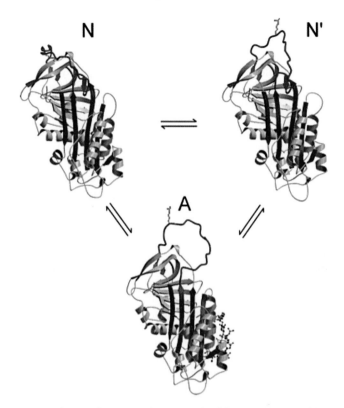

FIG. 4. Comparison of RCLs of native AT from crystals of the monomeric and heterodimer AT, and equilibrium model for the conformational states of AT. State N (*top left*) is the monomeric AT with its RCL held close against the body of AT and the side chain of P1 Arg393 (in green & red) is oriented towards the interior of the molecule. This state is thought to be nonreactive toward proteinases (see the text for details). State N' (*top right*) is the native AT in a heterodimer with latent AT (Protein Data Bank ID code IE04), and the side chain of P1 Arg393 (in green & red) is free to interact with proteinases but the hinge is still partially inserted. State A (*bottom*) is the activated AT in complex with pentasaccharide and S195A factor Xa (factor Xa is not shown) [4]. The hinge region is fully expelled, and the RCL is entirely liberated. (Taken from [3] with a partial modification, with permission)

side chain of Arg393 is buried in a hydrophobic pocket, and these side chain interactions serve to anchor the RCL [3], strongly suggesting that this represents the structural basis for the poor inhibitory activity of native AT without heparin. However, it is also true that AT shows appreciable proteinase inhibition in the absence of heparin. Therefore, it is proposed that the RCL of native AT probably exists in an equilibrium between at least two states: one (state N) with a buried and inaccessible Arg393 side chain and another (state N') with a fully solvent-accessible Arg393 but with the hinge still partially inserted. The hinge region is fully expelled, and the RCL is entirely liberated only after activation by the pentasaccharide (state A) (Fig. 4) [3].

Pentasaccharide-Enhanced Factor Xa Inhibition by AT

Among several factor Xa regulatory systems in the circulating blood, the inhibition of factor Xa by AT would be the most important physiologically. The mechanism of factor Xa inhibition by AT is basically the same as that of thrombin inhibition, described above. However, there is a large difference in that the activation of AT by the pentasaccharide (Fig. 3) is much more prominent for factor Xa inhibition (several hundredfold) than for thrombin (about twofold). For the heparin-dependent activation of AT in thrombin inhibition, both proteins need to bind with the same heparin molecule (template mechanism), whereas a conformational change of AT through the binding with pentasaccharide is sufficiently effective to enhance factor Xa inhibition (allosteric mechanism). The detailed mechanism of Xa inhibition by pentasaccharide-activated AT has been recently demonstrated from X-ray crystallographic analyses of the pentasaccharide-activated AT and S195A factor Xa complex [4]. Amino acid residues on AT involved in pentasaccharide binding are essentially the same as in Fig. 3, mainly including Arg46 and Arg47 on helix A, and Lys114, Lys125, and Arg129 on helix D, among which the principal importance of Lys114 is emphasized [4]. By binding with the pentasaccharide, the partially inserted hinge region of AT is expelled from between β-sheets 3A and 5A (Fig. 1), which shifts the position of the reactive center Arg393 and eventually exposes the side chain of Arg393 toward the outside of the molecule, resulting in the activation of AT (state A in Fig. 4). Thus, activated AT, in addition to the interaction of the reactive center Arg393 with the active site of factor Xa, interacts with two distinct exosites on factor Xa; one is in the autolysis loop (from Arg143 to Gln151) and the other is in the 36-loop (from Glu36 to Phe41). These two exosites on factor Xa contact with the newly formed exosites on activated AT, composed of s4C (Asn233, Arg235, Glu237), s3C (Met251, Tyr253, Glu255), s1B (Arg262), s1C (Thr401), and the C-terminal insert region (Pro397-Asn398-Arg399) of the RCL (Fig. 1). Especially the interaction of Arg150 on the autolysis loop of factor Xa buried in a surface cavity of AT with Asn233 and Glu237 on s4C and the interactions of Glu37 and Glu39 on the 36-loop of factor Xa with Arg262 on s1B and Arg399 on the C-terminal insert, respectively, are important exosite contacts. Mutagenesis studies also revealed that Tyr253 and Glu255 on s3C are key determinants of an exosite made accessible by heparin activation to promote rapid inhibition of factor Xa [27].

AT Deficiency and Its Pathological Implication

The normal range of AT in plasma is 80%–120% of the average level, which is very narrow for plasma proteins. Thus, the plasma level of AT is critical to maintain the hemostatic balance. It is well known that individuals with lower than the normal level of AT repeatedly suffer from thrombotic episodes. Homozygous AT deficiency (except homozygous deficiency of AT with no heparin-binding ability) has not been reported, suggesting that homozygous deficiency of AT is fatal, as shown in KO mice [28]. On the other hand, those with an excess level of AT show a bleeding tendency. Thus, AT deficiency, regardless of whether congenital or acquired, is one of the principal causes for venous thromboembolism. Not only inactive mutants but also those lacking heparin-binding ability result in severe thrombotic disorders in patients [1], demonstrating that the heparin-enhanced rapid inhibitions of thrombin and factor Xa are physiologically essential.

Heparin Cofactor II

Human heparin cofactor II (HCII) is a single-chain glycoprotein with an Mr ~65 600; it is composed of 480 amino acid residues and three carbohydrate chains N-glycosidically linked to Asn30, 169 and 368. The gene for HCII is located on chromosome 22q11.1 with a length of 15 849 bp consisting of five exons and four introns [29]. HCII is mainly synthesized in the liver and circulates in human plasma at a concentration of ~1 µmol/l (~66 mg/l) with a half-life of 2–3 days [30, 31].

HCII also belongs to the serpin superfamily, sharing about 30% sequence homology with AT. Therefore, the tertiary structure of HCII is also quite similar to that of AT, as shown in Fig. 1, including partial insertion of the reactive center loop into the β-sheet A [5]. However, when compared with AT and other plasma serpins, HCII has an extension of about 80 residues in the N-terminal region, where HCII has a unique cluster of acidic amino acids referred to as "acidic region" (also as "acidic tail") with repeated sequences of Glu56-Asp-Asp-Asp-Tyr(SO$_4$)-Leu-Asp62 and Glu69-Asp-Asp-Asp-Tyr(SO$_4$)-Ile-Asp75, similar to the C-terminal sequence of hirudin, a thrombin inhibitor from the medical leech. Another structural characteristic feature of HCII is the heparin/dermatan sulfate-binding site that is essentially the same as AT (see below). It is also unusual that HC has the reactive center Leu444-Ser445 (P1-P1′) (note that P1 is not Arg) for thrombin inhibition. Because of this Leu at the reactive center, HCII inhibits thrombin very slowly despite the normal orientation of the side chain of P1 Leu444, as demonstrated by crystallographic structural analysis [5].

Heparin-Enhanced Thrombin Inhibition by HCII

HCII exclusively inhibits thrombin among coagulation and fibrinolytic proteinases. It also inhibits chymotrypsin and chymase through its reactive center Leu residue. Similar to AT, the thrombin inhibitory activity of HCII is dramatically enhanced (~1000-fold) in the presence of heparin. However, HCII requires about 10 times higher concentration of heparin (about 0.2 U/ml for HCII in contrast to 0.02 U/ml for AT). Because the thrombin inhibitory activity of HCII is enhanced more than 1000-fold in the presence of dermatan sulfate [32], and HCII is the only plasma serpin that binds to dermatan sulfate, the physiological cofactor of HCII is thought to be a dermatan sulfate proteoglycan on smooth muscle cells. Consequently, the thrombin inhibitory activity of HCII is enhanced more than 1000-fold in the presence of dermatan sulfate [32].

From sequence and tertiary structural similarities of the heparin-binding regions of AT and HCII, their heparin-binding properties seem to be common. However, AT has only a heparin-binding property, but no dermatan sulfate-binding property. Therefore, it is a unique feature of HCII that this serpin has both heparin- and dermatan sulfate-binding sites, which are not identical but partially overlapped. Based on studies of genetic mutants and site-directed mutagenesis, amino acid residues involved in the heparin-binding and the dermatan sulfate-binding of HCII have been identified as Lys173 (Lys114), Arg184, and Lys185 (Lys125) and as Arg184, Lys185, Arg189 (Arg129), Arg192 (Arg132), and Arg193 (Lys133), respectively, on or near helix D (corresponding residues in AT are shown in parentheses) [33]. Despite the similarity of these amino acid residues in the heparin-binding site and their three-dimensional locations between HCII and AT [5], HCII does not necessarily require the specific sequence of the high-affinity pentasaccharide for AT shown in Fig. 3 [34].

Mechanism of HCII Activation by Heparin: Heparin-Induced Conformational Change

The mechanism of thrombin inhibition by HCII is similar to AT, as shown in Fig. 2. However, HCII is unique in its ability to be activated not only by heparin but also by dermatan sulfate, a repeating polymer of 2-O-sulfated L-iduronic acid and 4-O- or 6-O-sulfated N-acetyl-D-galactosamine.

A unique model of activation by glycosaminoglycan (GAG—heparin or dermatan sulfate) is proposed for HCII. The sequential steps of GAG-mediated thrombin inhibition by HCI is illustrated in Fig. 5 [5]. In the absence of GAG, the N-terminal acidic domain of native HCII is sequestered by an intramolecular interaction with the main body of HCII (Fig. 5a). Although the counterpart of this interaction has long been proposed to be the aforementioned GAG-binding site [35, 36], it has not yet been confirmed by a crystallographic structural analysis of native HCII. When GAG binds to the GAG-binding site of HCII, some conformational changes that occur in HCII, such as expulsion of the RCL and closure of β-sheet A, and the acidic domain is

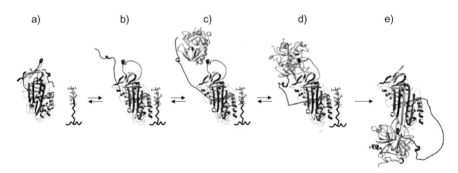

FIG. 5. Model of glycosaminoglycan (GAG)-mediated thrombin inhibition by HCII. **a** Native HCII. Similar to AT, the RCL (yellow) is partially inserted, but the side chain (red ball) of the reactive center P1 Leu 444 is oriented toward the outside of the molecule. The N-terminal acidic domain (purple) is illustrated here to interfere with thrombin attacking (see text for details). **b** GAG-bound HCII releases the N-terminal domain, which binds to the anion-binding exosite I of thrombin and facilitates rapid interaction between the active site of thrombin and the reactive center of HCII **c, d** Acyl–enzyme complex is formed. **e** Concomitant with the cleavage of the reactive center peptide bond, the N-terminal portion of the RCL is inserted into the A-sheet, forming a new strand, 4A. During this conformational change, the tethered thrombin is translocated to the opposite side of the HCII molecule (from top to bottom in **e**) and resulted in a distorted form. (From Baglin et al. [5], with permission)

released (Fig. 5b). Subsequently, HCII binds to the anion-binding exosite I of thrombin (Fig. 5c). This interaction approximates the active site of thrombin and the reactive center P1 Leu444 of HCII (Fig. 5d). When the GAG chain is sufficiently long (>30 monosaccharides), HCII also binds to the anion-binding exosite II of thrombin, forming a bridge between HCII and thrombin. This bridging mechanism of GAG dramatically facilitates the interaction between the reactive center of HCII and the active site of thrombin. Once the reactive center peptide bond is cleaved by thrombin and an acyl–enzyme bond is formed, a dramatic conformational change occurs in the N-terminal portion of the cleaved RCL and it forms a new β strand (shown in lime green in Fig. 5e).

Physiological Function of HCII with Antithrombotic Activity

As already mentioned above, the heparin-dependent thrombin inhibitory activity due to HCII is about 20%–25% in human plasma. Although individuals who completely lack HCII have not been reported and an association between heterozygous HCII deficiency and venous or arterial thrombosis has not been established, it has been suggested that HCII deficiency is not a significant risk factor for thrombotic disease [2]. Furthermore, HCII knockout mice are reported to be indistinguishable from their wild-type littermates in weight and survival at 1 year of age and do not have spontaneous thromboses or other morphological abnormalities [37]. In comparison with wild-type animals, however, HCII knockout mice demonstrate a significantly shorter time to thrombotic occlusion of the carotid artery after photochemically induced endothelial cell injury. This abnormality can be corrected by infusion of purified HCII, suggesting that HCII inhibits arterial thrombosis after endothelial injury [37].

References

1. Lane DA, Bayston T, Olds RJ, et al (1997) Antithrombin mutation database: 2nd (1997) Update. Thromb Haemost 77:197–211
2. Tollefsen DM (2002) Heparin cofactor II deficiency. Arch Pathol Lab Med 126:1394–1400
3. Johnson DJD, Langdown J, Li W, et al (2006) Crystal structure of monomeric native antithrombin reveals a novel reactive center loop conformation. J Biol Chem 281:35478–35486
4. Johnson D, Li W, Adams T, et al (2006) Antithrombin–S195A factor Xa-heparin structure reveals the allosteric mechanism of antithrombin activation EMBO J 25:2029–2037
5. Baglin TP, Carrell RW, Church FC, et al (2002) Crystal structures of native and thrombin-complexed heparin cofactor II reveal a multistep allosteric mechanism. Proc Natl Acad Sci U S A 99:11079–11084
6. Petersen TE, Dudek-Wojciechowska G, Sottrup-Jensen L, et al (1979) Primary structure of antithrombin III (heparin cofactor): partial homology between α_1-antitrypsin and antithrombin III. In: Collen D, Wiman B, Verstraete M (eds) The physiological inhibitors of blood coagulation and fibrinolysis. Elsevier Science Publishers, Amsterdam, pp 43–54

7. Bock SC, Wion KL, Vehar GA, et al (1982) Cloning and expression of the cDNA for human antithrombin III. Nucleic Acids Res 10:8113–8125
8. Olds RJ, Lane DA, Chowdhury V, et al (1993) Complete nucleotide sequence of the antithrombin gene: evidence for homologous recombination causing thrombophilia. Biochemistry 32:4216–4224
9. Conard J, Brosstad F, Larsen ML, et al (1983) Molar antithrombin concentration in normal human plasma. Haemostasis 13:363–368
10. Collen D, Schetz J, de Cock F, et al (1977) Metabolism of antithrombin III (heparin cofactor) in man: effects of venous thrombosis and of heparin administration. Eur J Clin Invest 7:27–35
11. Peterson CB, Blackburn MN (1985) Isolation and characterization of an antithrombin III variant with reduced carbohydrate content and enhanced heparin binding. J Biol Chem 260:610–615
12. Stein PE, Carrell RW (1995) What do dysfunctional serpins tell us about molecular mobility and disease? Nat Struct Biol 2:96–113
13. De Agostini AI, Watkins SC, Slayter HS, et al (1990) Localization of anticoagulantly active heparan sulfate proteoglycans in vascular endothelium: antithrombin binding on cultured endothelial cells and perfused rat aorta. J Cell Biol 111:1293–1304
14. Huntington JA, Read RJ, Carrell RW (2000) Structure of a serpin–protease complex shows inhibition by deformation Nature 407:923–926
15. Peterson CB, Noyes CM, Pecon JM, et al (1987) Identification of a lysyl residue in antithrombin which is essential for heparin binding. J Biol Chem 262:8061–8065
16. Liu CS, Chang JY (1987) The heparin binding site of human antithrombin III: selective chemical modification at Lys114, Lys125, and Lys287 impairs its heparin cofactor activity. J Biol Chem 262:17356–17361
17. Chang JY (1989) Binding of heparin to human antithrombin III activates selective chemical modification at lysine 236: Lys-107, Lys-125 and Lys-136 are situated within the heparin-binding site of antithrombin III. J Biol Chem 264:3111–3115
18. Sun XJ, Chang JY (1990) Evidence that arginine-129 and arginine-145 are located within the heparin binding site of human antithrombin III. Biochemistry 29:8957–8962
19. Koide T, Odani S, Takahashi K, et al (1984) Antithrombin III Toyama: replacement of arginine 47 by cysteine in hereditary abnormal antithrombin III that lacks heparin-binding ability. Proc Natl Acad Sci USA 81:289–293
20. Arocas V, Bock SC, Olson ST, et al (1999) The role of Arg46 and Arg47 of antithrombin in heparin binding. Biochemistry 38:10196–10204
21. Desai U, Swanson R, Bock SC, et al (2000) Role of arginine 129 in heparin binding and activation of antithrombin. J Biol Chem 275:18976–18984
22. Arocas V, Bock SC, Raja S, et al (2001) Lysine 114 of antithrombin is of crucial importance for the affinity and kinetics of heparin pentasaccharide binding. J Biol Chem 276:43809–43817
23. Schedin-Weiss S, Desai UR, Bock SC, et al (2002) Importance of lysine 125 for heparin binding and activation of antithrombin. Biochemistry 41:4779–4788
24. Jin L, Abrahams JP, Skinner R, et al (1997) The anticoagulant activation of antithrombin by heparin. Proc Natl Acad Sci U S A 94:14683–14688
25. Carrell RW, Stein PE, Fermi G, et al (1994) Biological implications of a 3 A structure of dimeric antithrombin. Structure 2:257–270
26. Schreuder HA, de Boer B, Dijkema R, et al (1994) The intact and cleaved human antithrombin III complex as a model for serpin-proteinase interactions. Nature Struct Biol 1:48–54
27. Izaguirre G, Olson ST (2006) Residues Tyr253 and Glu255 in strand 3 of beta-sheet C of antithrombin are key determinants of an exosite made accessible by heparin activation to promote rapid inhibition of factors Xa and IXa. J Biol Chem 281:13424–13432

28. Ishiguro K, Kojima T, Kadomatsu K, et al (2000) Complete antithrombin deficiency in mice results in embryonic lethality. J Clin Invest 106:873–878

29. Herzog R, Lutz S, Blin N, et al (1991) Complete nucleotide sequence of the gene for human heparin cofactor II and mapping to chromosomal band 22q11. Biochemistry 30:1350–1357

30. Jaffe EA, Armellino D, Tollefsen DM (1985) Biosynthesis of functionally active heparin cofactor II by a human hepatoma-derived cell line. Biochem Biophys Res Commun 132:368–374

31. Tollefsen DM, Pestka CA (1985) Heparin cofactor II activity in patients with disseminated intravascular coagulation and hepatic failure. Blood 66:769–774

32. Tollefsen DM, Pestka CA, Monafo WJ (1983) Activation of heparin cofactor II by dermatan sulfate. J Biol Chem 258:6713–6716

33. Tollefsen DM (1997) Heparin cofactor II. Adv Exp Med Biol 425:35–44

34. Maimone MM, Tollefsen DM (1988) Activation of heparin cofactor II by heparin oligosaccharides. Biochem Biophys Res Commun 152:1056–1061

35. Ragg H, Ulshofer T, Gerewitz J (1990) On the activation of human leuserpin-2, a thrombin inhibitor, by glycosaminoglycans. J Biol Chem 265:5211–5218

36. Van Deerlin VMD, Tollefsen DM (1991) The N-terminal acidic domain of heparin cofactor II mediates the inhibition of α-thrombin in the presence of glycosaminoglycans. J Biol Chem 266:20223–20231

37. He L, Vicente CP, Westrick et al (2002) Heparin cofactor II inhibits arterial thrombosis after endothelial injury. J Clin Invest 109:213–219

38. Schechter I, Berger A (1967) On the size of the active site in proteases. I. Papain. Biochem Biophys Res Commun 27:157–162

Part 2 Endothelial Cells and Inflammation

High-Mobility Group Box 1: Missing Link Between Thrombosis and Inflammation?

Takashi Ito, Ko-ichi Kawahara, Teruto Hashiguchi, and Ikuro Maruyama

Summary. High-mobility group box 1 protein (HMGB1) is an abundant nuclear protein with a dual function. Inside the cell, HMGB1 binds to DNA and modulates a variety of intranuclear processes, including transcription. Outside the cell, HMGB1 acts as a signal of tissue damage and can promote inflammation, immune responses, and tissue regeneration. During sepsis and/or disseminated intravascular coagulation, however, massive accumulation of HMGB1 in the systemic circulation can cause multiple organ failure and a subsequent lethal outcome. HMGB1 in the systemic circulation is recognized as a lethal mediator of sepsis and a promising therapeutic target for sepsis. Thrombomodulin (TM), a natural anticoagulant glycoprotein expressed on the surface of endothelial cells, plays an important role in sequestering HMGB1. TM may prevent HMGB1 from reaching remote organs, thereby restricting the range of HMGB1 action to the site of injury. In this chapter, we review recent progress made in defining the physiological and pathological roles of HMGB1 and therapeutic strategies aimed at blocking HMGB1.

Key words. Sepsis · Disseminated intravascular coagulation (DIC) · Thrombomodulin · Alarmin · High-mobility group box 1 (HMGB1)

Introduction

During evolution, multicellular organisms have developed defense mechanisms to counteract life-threatening events, such as infection, tissue injury, and hemorrhage. To initiate and accomplish appropriate protective responses, our defense systems need to recognize potentially life-threatening events. So far, two theories have been proposed to explain how our defense systems detect such threats: the stranger hypothesis and the danger hypothesis (Fig. 1) [1, 2]. Immune cells have molecular microdetectors, called pattern recognition receptors (PRRs) that recognize common molecular patterns of invading pathogens (Fig. 1a) [3, 4]. Such pathogen-associated molecular patterns (PAMPs) include components of bacterial cell walls, flagellar proteins, and viral nucleic acids; and immunocompetent cells can detect PAMPs as life-threatening strangers using PRRs [4–6]. Another strategy to detect life-threatening events is targeted on tissue damage. During trauma or microbial attack, the resulting cellular stress or tissue damage alerts our defense systems. In this case, instead of recognizing

a.   [Stranger hypothesis]

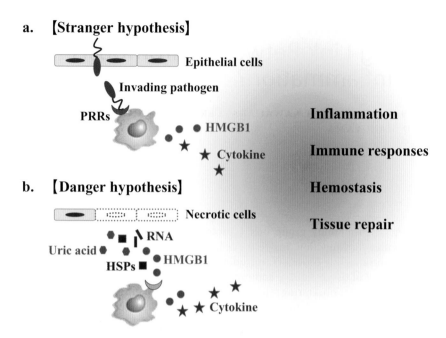

FIG. 1. Stranger hypothesis and danger hypothesis. **a** Using pattern recognition receptors (*PRRs*), immune cells detect invading microorganisms as life-threatening strangers and provoke a series of protective responses: inflammation, immune response, hemostasis, and tissue repair. **b** Our defense systems detect intracellular components as signals of tissue damage and provoke the same or similar protective responses as above. These intracellular molecules include uric acid, RNA, high-mobility group box 1 protein (*HMGB1*), and heat-shock proteins (*HSPs*)

PAMPs, immunocompetent cells recognize molecules that are normally found only inside cells unless released by damage (Fig. 1b). These intracellular molecules include uric acid, RNA, high-mobility group box 1 protein (HMGB1), heat-shock proteins (HSPs), and S100 proteins; and the term damage-associated molecular patterns (DAMPs) or "alarmin" is proposed to categorize such endogenous molecules that signal tissue damage [2, 7–9]. Thus, our defense systems scan for signs of both stranger and danger; and when they are detected, our defense systems initiate protective responses, such as inflammation, the immune response, hemostasis, and tissue repair.

What is HMGB1?

HMGB1 was previously thought to function only as a nuclear factor that enhances transcription. This protein is expressed in almost all eukaryotic cells and is highly conserved through evolution [7, 10]. HMGB1 is highly mobile and can shuttle between the nucleus and the cytoplasm [11]. HMGB1 binds to the minor groove of DNA without sequence specificity, which causes a conformational change in the DNA that promotes its interaction with transcription factors such as nuclear factor κB (NFκB), steroid hormone receptors, and glucocorticoid receptors [12, 13]. HMGB1-deficient mice die shortly after birth because of, at least in part, hypoglycemia caused by deficient glucocorticoid receptor function [14].

In addition to its intracellular role, HMGB1 can be released into the extracellular milieu and plays quite distinct roles. There are two ways in which HMGB1 can be released: passive release from necrotic cells (Fig. 1b) and active secretion from immunocompetent cells (Fig. 1a) [15–17]. HMGB1 is normally bound loosely to chromatin, so if cells become necrotic or leaky the HMGB1 diffuses out into the extracellular milieu [17]. Importantly, apoptotic cells do not release HMGB1 even after undergoing secondary necrosis and partial autolysis [17]. Thus, cells undergoing apoptosis are programmed to withhold the signal, thereby preventing the inflammatory process during apoptosis.

Several cell types, such as inflammatory cells, have the ability to secrete HMGB1 actively and thus produce a danger signal without dying. Active secretion depends on relocalization of HMGB1 from the nucleus to specific cytoplasmic organelles, the secretory lysosomes [16]. Inflammatory stimuli lead to hyperacetylation of HMGB1, which causes an accumulation of HMGB1 in the cytoplasm and blocks reentry to the nuclear compartment [18, 19]. Cytoplasmic HMGB1 is taken up by secretory lysosomes, following which the HMGB1-containing secretory lysosomes are fused with the plasma membrane and are secreted, at least in part, in a specific ATP-binding cassette (ABC) transporter-dependent manner [9, 16].

Once released into the extracellular space, HMGB1 acts as a signal of tissue damage and can promote inflammation (Fig. 1). HMGB1 transduces cellular signals through an advanced glycation end-products receptor (AGER, formerly known as RAGE) and possibly other receptors, such as Toll-like receptor 2 (TLR2) and TLR4 [20–25]. HMGB1 interacts with AGER expressed on endothelial cells, inducing endothelial cell activation, with increased expression of vascular cell adhesion molecule 1 (VCAM-1), intercellular adhesion molecule 1 (ICAM-1), and E-selectin [26, 27]. In addition, HMGB1 activates adhesive and migratory functions of neutrophils and monocytes in a AGER-dependent manner [28, 29]. Engagement of AGER therefore increases adhesion of neutrophils and monocytes to endothelial cells and promotes the recruitment of leukocytes across endothelial barriers. HMGB1 also acts on leukocytes, inducing the expression of proinflammatory cytokines, such as tumor necrosis factor-α (TNFα), interleukin-1 (IL-1), IL-6, and type I interferon (IFN) [15, 30, 31]. In this regard, HMGB1 and PAMPs synergistically activate leukocytes and may contribute to our defense systems against pathogens and to autoimmune pathogenesis [25]. Necrotic cells lacking HMGB1 have a greatly reduced ability to elicit the production of TNFα by neighborhood monocytes, suggesting an important role of HMGB1 in promoting inflammation during necrotic cell death [17].

Extracellular HMGB1 also acts as an endogenous immune adjuvant (Fig. 1). For the clonal expansion of naive T cells, both specific antigen and co-stimulatory signals provided by the same antigen-presenting cell (APC) are required [32]. In bacterial infections, microbial components act as immune adjuvants and induce expression of the co-stimulatory molecules on APCs [33]. However, there are some circumstances where immune responses are generated in the apparent absent of any microbial or other exogenous adjuvants. Such situations include immune responses to tumors, transplants, and possibly certain viruses. It has been postulated that in such conditions APCs can be activated by endogenous signals derived from cells that are stressed, virally infected, or killed necrotically but not by healthy cells [34, 35]. Several actual and putative endogenous adjuvants (e.g., HSPs, uric acid, HMGB1) have been identified, and there are others whose identities are not yet known (Fig. 1b) [36–38].

In addition to its roles in inflammation and immune responses, extracellular HMGB1 has the ability to promote tissue repair. Compared to healthy muscle, dystrophic muscle contains more HMGB1, and exogenously added HMGB1 induces migration and proliferation of vessel-associated stem cells [39]. After myocardial infarction, exogenously added HMGB1 induces proliferation and differentiation of cardiac progenitor cells, thereby promoting cardiac regeneration [40]. Furthermore, cardiac performance in HMGB1-treated mice is significant improved compared to that in control mice, suggesting that HMGB1 can induce functionally relevant myocardial regeneration following myocardial infarction [40]. Although the role of endogenous HMGB1 in tissue regeneration remains to be clarified, exogenously added HMGB1 may be used to promote tissue regeneration. Taken together, HMGB1 released from damaged or activated cells seems to orchestrate postinjury responses: inflammation, immune responses, and subsequent tissue regeneration.

HMGB1 as a Lethal Mediator in Sepsis and/or DIC

Sepsis is a life-threatening disorder that results from systemic inflammatory and coagulatory responses to infection [41]. Hyperactivation of the inflammatory system is the most important feature of sepsis and has been the most common target of therapeutic strategies. So far, diverse therapies directed against proinflammatory mediators such as TNFα and IL-1 have revealed dramatic effects in animal models of sepsis [42, 43]. However, in humans, most of these strategies have not improved survival of septic patients [44, 45] in part because the classic proinflammatory mediators, such as TNFα and IL-1, are released within minutes after PAMP exposure; thus, even a minimal delay in treatment may result in treatment failure.

HMGB1 is a promising therapeutic target for sepsis. In the presence of sepsis and/or disseminated intravascular coagulation (DIC), serum HMGB1 levels are elevated in both humans and animals [46–48]. Accumulation of HMGB1 in the systemic circulation occurs considerably later than that of classically early proinflammatory mediators such as TNFα and IL-1 [46, 49], and this delayed kinetics of HMGB1 makes it an attractive therapeutic target with a wider window of opportunity for the treatment of sepsis [50]. Indeed, blockade of HMGB1, even at later time points after onset of infection, has been shown to rescue animals from lethal sepsis [46, 51–53]. Although data on higher species are not yet available, HMGB1 antagonism provides hope for new approaches that will be therapeutically effective in humans with sepsis.

The mechanisms by which HMGB1 exerts its lethal effect in sepsis are not fully determined and seem to be multifactorial. Proinflammatory activity of HMGB1 is a possible mechanism. Extracellular HMGB1 triggers inflammation, and HMGB1 plays an important role in the pathogenesis of inflammatory diseases, including acute inflammatory lung injury, rheumatoid arthritis, and vasculitis [30, 54–58]. Therefore, massive accumulation of HMGB1 in the systemic circulation will lead to the systemic inflammatory response syndrome (SIRS), a major feature of sepsis. However, it is also reported that the roles of HMGB1 and TNF are distinguishable [7, 50], and the lethal effects of HMGB1 in sepsis cannot be explained by its inflammatory activity alone.

HMGB1 may alter the procoagulant–anticoagulant balance, with an increase in procoagulant factors and a decrease in anticoagulant factors [59]; such alteration of the balance is another important aspect of sepsis [60]. HMGB1 induces tissue factor expression on monocytes and inhibits the anticoagulant protein C pathway mediated by thrombin-thrombomodulin (TM) complexes [59]. In thrombin-induced DIC model rats, HMGB1 accelerates microvascular thrombosis with irreversible renal and respiratory failure, thereby acting as a lethal mediator [59]. In patients with DIC, plasma HMGB1 levels are significantly increased and correlate with the DIC score and sepsis-related organ failure assessment (SOFA) score [48]. Given that DIC and concomitant multiple organ failure are strong predictors of mortality, procoagulant activity of HMGB1 may in part be responsible for its lethal effects.

HMGB1 may act as a lethal mediator by causing epithelial cell barrier dysfunction [61, 62]. Exposure of epithelial cells to HMGB1 causes down-regulation of the expression of cell surface molecules that are responsible for the tight adhesion between adjacent epithelial cells, leading to epithelial hyperpermeability. When injected into animals, HMGB1 increases ileal mucosal permeability and promotes bacterial translocation and hepatocellular injury. HMGB1-specific antibodies confer significant protection against the development of these epithelial abnormalities and subsequent organ dysfunction in animals with endotoxemia or sepsis, indicating that this might be a mechanism for how increased HMGB1 levels cause organ dysfunction [7, 47].

Thrombomodulin as a Security Guard in Blood Vessels

In a recent clinical trial, recombinant human soluble thrombomodulin (rhsTM) significantly alleviated DIC [63]. Endothelial membrane-bound TM forms a high-affinity complex with thrombin, thereby inhibiting thrombin interaction with fibrinogen and protease-activated receptor-1 (PAR-1). In addition, TM enhances thrombin-mediated protein C (PC) activation by more than two orders of magnitude [64, 65]. Like membrane-bound TM, the rhsTM, which is composed of the extracellular domain of TM, also binds to thrombin, thereby promoting PC activation. Activated PC (APC) has antiinflammatory and antiapoptotic activities that involve binding of APC to endothelial protein C receptor (EPCR) and cleavage of PAR-1 [66]. APC exerts an antithrombotic effect by inactivating factors Va and VIIIa, limiting the generation of thrombin. As a result of decreased thrombin levels, the proinflammatory, procoagulant, and antifibrinolytic responses induced by thrombin are reduced [64, 65, 67]. In addition, TM itself has antiinflammatory properties in its N-terminal lectin-like domain [68].

The target molecule of TM is not restricted to thrombin. TM can bind to HMGB1 as well as thrombin, thereby dampening the inflammatory and coagulatory responses [53, 69]. TM also dampens the generation of reactive oxygen species (unpublished observations). Under the conditions in which injured vessel locally loses the integrity

Blood vessel

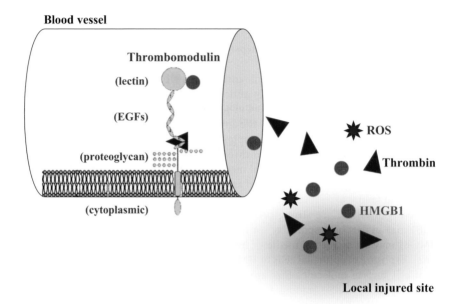

FIG. 2. Thrombomodulin as a security guard in blood vessels. Thrombomodulin-bearing endothelial cells can sequester HMGB1 and thrombin, thereby preventing them from reaching remote
organs

of TM-bearing cells, HMGB1 and thrombin present in the circulation propagate
inflammatory and coagulatory responses [27, 70]. However, once an adjacent portion
of the vessel wall with intact endothelial cells is encountered, TM-bearing cells can
sequester HMGB1 and thrombin, thereby preventing them from reaching remote
organs (Fig. 2). Thus, TM acts as a security guard in blood vessels. Because the expression of endothelial TM is systemically down-regulated during septic conditions
[71], replacement with rhsTM offers therapeutic value in patients with sepsis and/or
DIC.

Conclusion

In the canonical model of sepsis, PAMPs are thought to be responsible for the development of sepsis. However, recent investigations suggest the importance of endogenous
molecules as amplifiers of signals in the pathogenesis of sepsis [72, 73]. Defining their
functions during sepsis in greater detail is necessary for the design of new therapeutic
strategies in clinical sepsis.

Acknowledgments. We thank all of our colleagues who helped us through open discussion to conceive ideas here: S. Yamada, K. Takenouchi, Y. Oyama, K. Kikuchi,
Y. Morimoto, K.K. Biswas, N. Arimura, S. Arimura, N. Miura, B. Shrestha, X. Meng,
S. Tancharoen, Y. Nawa, T. Shimizu, and T. Uchimura. This work was supported by
a Research Grant from the Ministry of Education, Culture, Sports, Science, and

Technology of Japan (Grant-in-Aid 18791341) (T.I.) and by a Health and Labor Sciences Research Grant from the Ministry of Health, Labor, and Welfare (I.M.).

References

1. Heath WR, Carbone FR (2003) Immunology: dangerous liaisons. Nature 425:460–461
2. Bianchi ME (2007) DAMPs, PAMPs and alarmins: all we need to know about danger. J Leukoc Biol 81:1–5
3. Janeway CA Jr (1989) Approaching the asymptote? Evolution and revolution in immunology. Cold Spring Harbor Symp Quant Biol 54(Pt 1):1–13
4. Janeway CA Jr, Medzhitov R (2002) Innate immune recognition. Annu Rev Immunol 20:197–216
5. Takeda K, Kaisho T, Akira S (2003) Toll-like receptors. Annu Rev Immunol 21:335–376
6. Meylan E, Tschopp J, Karin M (2006) Intracellular pattern recognition receptors in the host response. Nature 442:39–44
7. Lotze MT, Tracey KJ (2005) High-mobility group box 1 protein (HMGB1): nuclear weapon in the immune arsenal. Nat Rev Immunol 5:331–342
8. Kannemeier C, Shibamiya A, Nakazawa F, et al (2007) Extracellular RNA constitutes a natural procoagulant cofactor in blood coagulation. Proc Natl Acad Sci U S A
9. Harris HE, Raucci A: Alarmin(g) news about danger (2006) workshop on innate danger signals and HMGB1. EMBO Rep 7:774–778
10. Erlandsson Harris H, Andersson U (2004) Mini-review: the nuclear protein HMGB1 as a proinflammatory mediator. Eur J Immunol 34:1503–1512
11. Falciola L, Spada F, Calogero S, et al (1997) High mobility group 1 protein is not stably associated with the chromosomes of somatic cells. J Cell Biol 137:19–26
12. Agresti A, Lupo R, Bianchi ME, et al (2003) HMGB1 interacts differentially with members of the Rel family of transcription factors. Biochem Biophys Res Commun 302:421–426
13. Boonyaratanakornkit V, Melvin V, Prendergast P, et al (1998) High-mobility group chromatin proteins 1 and 2 functionally interact with steroid hormone receptors to enhance their DNA binding in vitro and transcriptional activity in mammalian cells. Mol Cell Biol 18:4471–4487
14. Calogero S, Grassi F, Aguzzi A, et al (1999) The lack of chromosomal protein Hmg1 does not disrupt cell growth but causes lethal hypoglycaemia in newborn mice. Nat Genet 22:276–280
15. Andersson U, Wang H, Palmblad K, et al (2000) High mobility group 1 protein (HMG-1) stimulates proinflammatory cytokine synthesis in human monocytes. J Exp Med 192:565–570
16. Gardella S, Andrei C, Ferrera D, et al (2002) The nuclear protein HMGB1 is secreted by monocytes via a non-classical, vesicle-mediated secretory pathway. EMBO Rep 3:995–1001
17. Scaffidi P, Misteli T, Bianchi ME (2002) Release of chromatin protein HMGB1 by necrotic cells triggers inflammation. Nature 418:191–195
18. Bonaldi T, Talamo F, Scaffidi P, et al (2003) Monocytic cells hyperacetylate chromatin protein HMGB1 to redirect it towards secretion. EMBO J 22:5551–5560
19. Dumitriu IE, Baruah P, Manfredi AA, et al (2005) HMGB1: guiding immunity from within. Trends Immunol 26:381–387
20. Hori O, Brett J, Slattery T, et al (1995) The receptor for advanced glycation end products (RAGE) is a cellular binding site for amphoterin: mediation of neurite outgrowth and co-expression of rage and amphoterin in the developing nervous system. J Biol Chem 270:25752–25761

21. Taguchi A, Blood DC, del Toro G, et al (2000) Blockade of RAGE-amphoterin signaling suppresses tumour growth and metastases. Nature 405:354–360
22. Schmidt AM, Yan SD, Yan SF, et al (2001) The multiligand receptor RAGE as a progression factor amplifying immune and inflammatory responses. J Clin Invest 108:949–955
23. Park JS, Svetkauskaite D, He Q, et al (2004) Involvement of toll-like receptors 2 and 4 in cellular activation by high mobility group box 1 protein. J Biol Chem 279:7370–7377
24. Park JS, Gamboni-Robertson F, He Q, et al (2006) High mobility group box 1 protein interacts with multiple Toll-like receptors. Am J Physiol Cell Physiol 290:C917–C924
25. Tian J, Avalos AM, Mao SY, et al (2007) Toll-like receptor 9-dependent activation by DNA-containing immune complexes is mediated by HMGB1 and RAGE. Nat Immunol 8:487–496
26. Chavakis T, Bierhaus A, Al-Fakhri N, et al (2003) The pattern recognition receptor (RAGE) is a counterreceptor for leukocyte integrins: a novel pathway for inflammatory cell recruitment. J Exp Med 198:1507–1515
27. Fiuza C, Bustin M, Talwar S, et al (2003) Inflammation-promoting activity of HMGB1 on human microvascular endothelial cells. Blood 101:2652–2660
28. Orlova VV, Choi EY, Xie C, et al (2007) A novel pathway of HMGB1-mediated inflammatory cell recruitment that requires Mac-1-integrin. EMBO J 26:1129–1139
29. Rouhiainen A, Kuja-Panula J, Wilkman E, et al (2004) Regulation of monocyte migration by amphoterin (HMGB1). Blood 104:1174–1182
30. Taniguchi N, Kawahara K, Yone K, et al (2003) High mobility group box chromosomal protein 1 plays a role in the pathogenesis of rheumatoid arthritis as a novel cytokine. Arthritis Rheum 48:971–981
31. Dumitriu IE, Baruah P, Bianchi ME, et al (2005) Requirement of HMGB1 and RAGE for the maturation of human plasmacytoid dendritic cells. Eur J Immunol 35:2184–2190
32. Sharpe AH, Freeman GJ (2002) The B7-CD28 superfamily. Nat Rev Immunol 2:116–126
33. Hunter RL (2002) Overview of vaccine adjuvants: present and future. Vaccine 20(Suppl 3):S7–S12
34. Gallucci S, Lolkema M, Matzinger P (1999) Natural adjuvants: endogenous activators of dendritic cells. Nat Med 5:1249–1255
35. Rock KL, Hearn A, Chen CJ, et al (2005) Natural endogenous adjuvants. Springer Semin Immunopathol 26:231–246
36. Srivastava PK, Maki RG (1991) Stress-induced proteins in immune response to cancer. Curr Top Microbiol Immunol 167:109–123
37. Shi Y, Evans JE, Rock KL (2003) Molecular identification of a danger signal that alerts the immune system to dying cells. Nature 425:516–521
38. Rovere-Querini P, Capobianco A, Scaffidi P, et al (2004) HMGB1 is an endogenous immune adjuvant released by necrotic cells. EMBO Rep 5:825–830
39. Palumbo R, Sampaolesi M, De Marchis F, et al (2004) Extracellular HMGB1, a signal of tissue damage, induces mesoangioblast migration and proliferation. J Cell Biol 164:441–449
40. Limana F, Germani A, Zacheo A, et al (2005) Exogenous high-mobility group box 1 protein induces myocardial regeneration after infarction via enhanced cardiac C-kit+ cell proliferation and differentiation. Circ Res 97:e73–e83
41. Cohen J (2002) The immunopathogenesis of sepsis. Nature 420:885–891
42. Tracey KJ, Fong Y, Hesse DG, et al (1987) Anti-cachectin/TNF monoclonal antibodies prevent septic shock during lethal bacteraemia. Nature 330:662–664
43. Ohlsson K, Bjork P, Bergenfeldt M, et al (1900) Interleukin-1 receptor antagonist reduces mortality from endotoxin shock. Nature 348:550–552

44. Reinhart K, Karzai W (2001) Anti-tumor necrosis factor therapy in sepsis: update on clinical trials and lessons learned. Crit Care Med 29:S121–125
45. Fisher CJ Jr, Dhainaut JF, Opal SM, et al (1994) Recombinant human interleukin 1 receptor antagonist in the treatment of patients with sepsis syndrome: results from a randomized, double-blind, placebo-controlled trial; phase III rhIL-1ra Sepsis Syndrome Study Group. JAMA 271:1836–1843
46. Wang H, Bloom O, Zhang M, et al (1999) HMG-1 as a late mediator of endotoxin lethality in mice. Science 285:248–251
47. Yang H, Ochani M, Li J, et al (2004) Reversing established sepsis with antagonists of endogenous high-mobility group box 1. Proc Natl Acad Sci U S A 101:296–301
48. Hatada T, Wada H, Nobori T, et al (2005) Plasma concentrations and importance of high mobility group box protein in the prognosis of organ failure in patients with disseminated intravascular coagulation. Thromb Haemost 94:975–979
49. Wang H, Yang H, Czura CJ, et al (2001) HMGB1 as a late mediator of lethal systemic inflammation. Am J Respir Crit Care Med 164:1768–1773
50. Wang H, Yang H, Tracey KJ et al (2004) Extracellular role of HMGB1 in inflammation and sepsis. J Intern Med 255:320–331
51. Ulloa L, Ochani M, Yang H, et al (2002) Ethyl pyruvate prevents lethality in mice with established lethal sepsis and systemic inflammation. Proc Natl Acad Sci U S A 99:12351–12356
52. Wang H, Liao H, Ochani M, et al (2004) Cholinergic agonists inhibit HMGB1 release and improve survival in experimental sepsis. Nat Med 10:1216–1221
53. Abeyama K, Stern DM, Ito Y, et al (2005) The N-terminal domain of thrombomodulin sequesters high-mobility group-B1 protein, a novel antiinflammatory mechanism. J Clin Invest 115:1267–1274
54. Abraham E, Arcaroli J, Carmody A, et al (2000) HMG-1 as a mediator of acute lung inflammation. J Immunol 165:2950–2954
55. Kokkola R, Sundberg E, Ulfgren AK, et al (2002) High mobility group box chromosomal protein 1: a novel proinflammatory mediator in synovitis. Arthritis Rheum 46:2598–2603
56. Kokkola R, Li J, Sundberg E, et al (2003) Successful treatment of collagen-induced arthritis in mice and rats by targeting extracellular high mobility group box chromosomal protein 1 activity. Arthritis Rheum 48:2052–2058
57. Jiang W, Pisetsky DS (2007) Mechanisms of disease: the role of high-mobility group protein 1 in the pathogenesis of inflammatory arthritis. Nat Clin Pract Rheumatol 3:52–58
58. Taira T, Matsuyama W, Mitsuyama H, et al (2007) Increased serum high mobility group box-1 level in Churg-Strauss syndrome. Clin Exp Immunol 148:241–247
59. Ito T, Kawahara K, Nakamura T, et al (2007) High-mobility group box 1 protein promotes development of microvascular thrombosis in rats. J Thromb Haemost 5:109–116
60. Russell JA (2006) Management of sepsis. N Engl J Med 355:1699–1713
61. Sappington PL, Yang R, Yang H, et al (2002) HMGB1 B box increases the permeability of Caco-2 enterocytic monolayers and impairs intestinal barrier function in mice. Gastroenterology 123:790–802
62. Sappington PL, Fink ME, Yang R, et al (2003) Ethyl pyruvate provides durable protection against inflammation-induced intestinal epithelial barrier dysfunction. Shock 20:521–528
63. Saito H, Maruyama I, Shimazaki S, et al (2007) Efficacy and safety of recombinant human soluble thrombomodulin (ART-123) in disseminated intravascular coagulation: results of a phase III, randomized, double-blind clinical trial. J Thromb Haemost 5:31–41
64. Weiler H, Isermann BH (2003) Thrombomodulin. J Thromb Haemost 1:1515–1524

65. Esmon CT (2005) The interactions between inflammation and coagulation. Br J Haematol 131:417–430
66. Riewald M, Petrovan RJ, Donner A, et al (2002) Activation of endothelial cell protease activated receptor 1 by the protein C pathway. Science 296:1880–1882
67. Bernard GR, Vincent JL, Laterre PF, et al (2001) Efficacy and safety of recombinant human activated protein C for severe sepsis. N Engl J Med 344:699–709
68. Conway EM, Van de Wouwer M, Pollefeyt S, et al (2002) The lectin-like domain of thrombomodulin confers protection from neutrophil-mediated tissue damage by suppressing adhesion molecule expression via nuclear factor kappaB and mitogen-activated protein kinase pathways. J Exp Med 196:565–577
69. Esmon C (2005) Do-all receptor takes on coagulation, inflammation. Nat Med 11:475–477
70. Levi M, Ten Cate H (1999) Disseminated intravascular coagulation. N Engl J Med 341:586–592
71. Faust SN, Levin M, Harrison OB, et al (2001) Dysfunction of endothelial protein C activation in severe meningococcal sepsis. N Engl J Med 345:408–416
72. Brunn GJ, Platt JL (2006) The etiology of sepsis: turned inside out. Trends Mol Med 12:10–16
73. Sansonetti PJ (2006) The innate signaling of dangers and the dangers of innate signaling. Nat Immunol 7:1237–1242

Coagulation, Inflammation, and Tissue Remodeling

KOJI SUZUKI, TATSUYA HAYASHI, OSAMU TAGUCHI, AND
ESTEBAN GABAZZA

Summary. Tissue remodeling is a repair response to wound injury. Fibrosis is an aberrant type of tissue remodeling characterized by exuberant accumulation of extracellular matrix proteins that often leads to loss of organ function. Recent studies have shown that interaction between the coagulation system and the inflammatory response plays a critical role in wound repair. Soon after tissue injury the coagulation cascade is activated, leading to platelet activation, thrombin generation, fibrin deposition, and release of inflammatory mediators. Activation of the fibrinolytic system and release of growth factors complete the normal repair process. Dysfunction of the fibrinolytic pathway or uncontrollable coagulation activation may be followed by excessive accumulation of matrix proteins, loss of normal tissue architecture, and fibrosis. Inflammatory cytokines may increase activation of the coagulation system by inducing the expression of tissue factor on the surface of injured cells, whereas thrombin, the final enzyme of the coagulation pathway activation, may promote the inflammatory response and collagen deposition by stimulating secretion of cytokines and growth factors. The beneficial effect of inhibitors of thrombin generation in fibrotic disorders supports the biological importance of the coagulation system in tissue remodeling.

Key words. Tissue fibrosis · Coagulation · Protease-activated receptor · Inflammation · Cytokines · Thrombin

Tissue Remodeling

Tissue remodeling is a host response to wound injury characterized by the concomitant processes of extracellular matrix formation and degradation that may lead to normal reconstruction of parenchymal tissue and organ structures or to pathological deposition of matrix proteins and/or excessive proliferation of resident structural cells [1]. During the early reparative stage, damaged and dead cells are replaced by the same types of cell, and the organ parenchyma is reconstructed, conserving the original architecture; however, during the fibroproliferative stage, or fibrosis, excessive accumulation of matrix components leads ultimately to permanent parenchymal scar and sometimes to organ dysfunction [2]. The precise mechanism that governs the fate of organ tissue after injury has not yet been elucidated but it is currently the focus of many investigations. Inflammatory mediators—including cytokines, chemokines, toxins, and procoagulants released by leukocytes or epithelial and vascular endothelial cells during acute or chronic infections or immunological or coagulation disorders—are potent inducers of tissue damage and subsequent fibroproliferative changes [3].

Tissue injury is followed by increased secretion of inflammatory cytokines [interleukin-1β (IL-1β), tumor necrosis factor-α (TNFα)], transforming growth factor-β1 (TGFβ1), and growth factors such as connective tissue growth factor (CTGF) and platelet-derived growth factors (PDGF) [4]. CTGF and PDGF induce increased accumulation and proliferation of fibroblasts and myofibroblasts in the extravascular space. TGFβ1 stimulates secretion of connective components such as collagen and fibronectin from fibroblasts and myofibroblasts. In addition, TGFβ may contribute to the fibroproliferative process by promoting epithelial-to-mesenchymal cell transition. Other cytokines, such as IL-1β and IL-13, have been also described to stimulate the secretion of extracellular matrix proteins from mesenchymal cells [4]. Monocytes, macrophages, and epithelial and endothelial cells may be the source of cytokines and growth factors including IL-1β, TGFβ1, and PDGF [5]. The coagulation system also contributes to the process of tissue scarring during wound healing. Thrombin, the final enzyme formed after activation of the coagulation system, may promote tissue scarring by mediating formation of the fibrin matrix and by stimulating secretion of inflammatory cytokines, growth factors, and other procoagulant factors [6].

Interaction between the Coagulation System and the Inflammatory Response

The coagulation system may be activated by the extrinsic or intrinsic pathway [7]. The extrinsic pathway is activated by tissue factor (TF) bound to activated factor VII (factor VIIa) at sites of tissue injury, forming the factor X-activating complex. The intrinsic pathway is activated when negatively charged components of the extravascular matrix interact with clotting factors, leading to activation of the kallikrein/kinin system [7]. The extrinsic and intrinsic pathways then merge in a common pathway in which activated factor V (factor Va) and activated factor X (factor Xa) bind on negatively charged phospholipids (phosphatidylserine and phosphatidylethanolamine)-rich cell membrane, forming the prothrombinase complex, which then generates thrombin by proteolytic cleavage of prothrombin [7]. Inflammatory cytokines (TNFα, IL-1β, IL-6), oxidants, mechanical stress, proteases, and thrombin itself may increase this cascade of thrombin generation by enhancing the surface expression of TF on a variety of cells including vascular endothelial cells, epithelial cells, macrophages, and neural and smooth muscle cells [8]. The classic function ascribed to thrombin is its role in thrombosis and hemostasis. Thrombin induces clot formation by converting fibrinogen to fibrin. This hemostatic activity of thrombin is fundamental for preserving the integrity of the circulation system and wound repair at sites of tissue damage [7, 8].

In addition to its function in hemostasis, thrombin is also a potent proinflammatory factor. Thrombin promotes the inflammatory response by stimulating the migration, activation, and secretion of cytokines and proteases from leukocytes, macrophages, and endothelial and smooth muscle cells. Platelets can be also activated by thrombin inducing the secretion of various inflammatory mediators such as serotonin, thromboxane A_2, platelet factor-4, prostacyclin, procoagulants, and growth factors [8]. Recent studies have shown that thrombin participates in the inflammatory response by stimulating cells through its protease-activated receptors (PARs) [8]. PARs are 7-transmembrane domain G protein-coupled receptors that are activated by proteolytic cleavage and the unmasking of an amino-terminal receptor sequence, which initiates cell signaling by binding to the second extracellular loop of the receptor. There are four PARs, of which PAR-1, PAR-3, and PAR-4 are thrombin receptors, whereas PAR-2 is a trypsin receptor; PAR-2 is also cleaved by tryptase, factor Xa, and factor VIIa [8]. The PAR-mediated cellular effects have implicated thrombin in several physiological and pathological conditions, including platelet aggregation, cell survival and apoptosis, inflammatory and immune responses, metastasis, tumor growth, and angiogenesis [9]. It is believed that PAR-1 mediates the proinflammatory activity of thrombin [10]; however, recent studies suggest that PAR-1 may also exert inhibitory activity on inflammation. One example is the suppression of cytokine expression after cleavage of PAR-1 by activated protein C (APC) [11].

Coagulation Activation and Tissue Fibrosis

Recent studies have demonstrated that coagulation proteases play pivotal roles in the fibroproliferative response during wound healing and tissue repair. Enhanced activation of the coagulation cascade observed in diseases that ultimately evolve into fibrosis provides evidence for the role of coagulation proteases in the process of tissue scarring. The concentration of thrombin and the TF–factor VIIa complex is elevated in bronchoalveolar lavage from patients with interstitial lung diseases and acute respiratory distress syndrome [12–15]. Patients with hepatitis-associated liver cirrhosis and glomerulonephritis have also shown increased activation of the coagulation system [16, 17]. Deposition of activated coagulation factors has been also detected in vascular walls with atherosclerotic changes [18]. Furthermore, enhanced activation of the coagulation cascade has been found to precede the development of fibrosis in animal models of tissue injury, such as bleomycin- or lipopolysaccharide-induced lung injury and endotoxemia-associated liver injury [16] (Fig. 1).

The process of tissue repair results from an exquisite balance between degradation and formation of extracellular matrix proteins. Following tissue injury, thrombin generated during coagulation activation converts fibrinogen to fibrin, which is then deposited at sites of damaged tissue. At the initial stages, fibrin deposition plays an important role in tissue repair because it serves as a provisional structural matrix for the proliferation of mesenchymal cells and the accumulation of extracellular matrix proteins (e.g., collagen, fibronectin) at sites of injury [16]. Fibrin and its degradation products may also accelerate the reparative process by stimulating the migration of fibroblasts and leukocytes [19–21]. Under physiological conditions, fibrin is totally degraded by plasmin formed by activation of the fibrinolysis system, and it is subsequently replaced by normally structured extracellular matrix proteins. However, continuous activation of the coagulation system or failure of the fibrinolytic pathway may

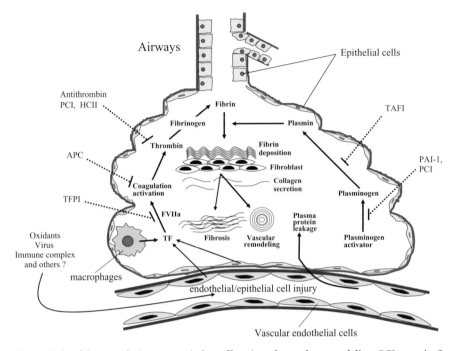

FIG. 1. Role of the coagulation system in lung fibrosis and vascular remodeling. *PCI*, protein C inhibitor; *HCII*, heparin cofactor II; *TFPI*, tissue factor pathway inhibitor; *APC*, activated protein C; *TAFI*, thrombin-activatable fibrinolysis inhibitor; *PAI-1*, plasminogen activator inhibitor-1; →, conversion or stimulation; *dashed lines*, inhibition

cause enhanced deposition of fibrin, leading to a strong fibroproliferative response and destruction of normal organ structures [17]. Persistence or enhanced deposition of fibrin has been associated with the development of fibrosis in several organs including the lung, liver, skin, and kidneys [22, 23] (Fig. 1). It is worth noting that fibrin is not always necessary for tissue fibrosis because pulmonary fibrosis has been shown to develop in mice deficient in fibrinogen [24].

Thrombin may promote fibrogenesis by stimulating the proliferation of collagen-producing cells, the production and secretion of growth factors (PDGF A-A, PDGF-AB, CTGF), the surface expression of PDGF receptors, the recruitment or chemotaxis of fibroblasts, and the secretion of collagen and fibronectin from fibroblasts and smooth muscle and epithelial cells [25]. Most of these cellular effects of thrombin appear to be mediated by PAR-1 [8]. In addition, thrombin may affect tissue remodeling by inducing wound contraction, transformation of fibroblasts to smooth muscle α-actin-positive myofibroblasts, or modulating the production and activation of metalloproteinases [26]. Other coagulation proteases, such as factor Xa and factor VIIa, may also take part in wound repair and tissue fibrosis. Factor Xa and factor VIIa were found to have indirect mitogenic activity on fibroblasts derived from various organs by increasing the secretion of PDGF [26].

Regulation of Tissue Remodeling by Anticoagulants

There are several regulators of the coagulation system activation including antithrombin, heparin cofactor II (HCII), tissue factor pathway inhibitor (TFPI), APC in the protein C pathway with its cofactor protein S, and the multifunctional inhibitor protein C inhibitor (PCI) (Fig. 1). Antithrombin is a natural anticoagulant that inhibits thrombin with more efficiency when it interacts with heparin-like glycosaminoglycans on endothelial cells. Antithrombin was reported to inhibit acute lung injury in animal models; and heparin, which enhances thrombin inhibition by antithrombin, was reported to inhibit bleomycin-induced lung fibrosis [27]. TFPI is a Kunitz-type protease inhibitor of factor VIIa/TF in the presence of factor Xa. Intratracheal gene transfer of TFPI was reported to attenuate the development of pulmonary fibrosis in rats [28]. Inhibitors of fibrinolysis, such as thrombin-activatable fibrinolysis inhibitor (TAFI), and plasminogen activator inhibitor may also accelerate the development of tissue fibrosis [23]. Plasmin, which is the effector enzyme of the fibrinolysis system, plays a critical role in the degradation of fibrin formed at sites of tissue injury. Deficient plasmin generation has been associated with increased tissue deposition of collagen in animal models of lung fibrosis [22]. PCI is another important serine protease inhibitor with multiple functions. We have recently shown that PCI blocks vascular wall remodeling in a mouse model of pulmonary arterial hypertension by inhibiting thrombin [29].

The anticoagulant protein C pathway is activated when thrombin bound to thrombomodulin cleaves the zymogen protein C to APC, which blocks coagulation activation by inhibiting both factor Va and activated factor VIII [23]. Protein S is a vitamin K-dependent plasma glycoprotein that circulates in a free form or in complex with C4b-binding protein [30]. The free form of protein S acts as a cofactor of APC, increasing severalfold its anticoagulant activity. Protein S may also directly inhibit coagulation activation by binding to factors Va and Xa [30]. Decreased activity of the anticoagulant protein C system with increased thrombin generation has been described in patients with inflammatory diseases frequently associated with tissue fibrosis, such as interstitial lung diseases, bronchial asthma, and hepatitis [31–33]. The mechanism of decreased APC generation in fibrotic diseases is believed to occur owing to poor availability of factors required for efficient protein C activation, such as thrombomodulin, endothelial protein C receptor (EPCR), or protein S. For example, thrombomodulin and EPCR are reduced on the endothelial and epithelial cell surface in acute lung injury, and liver cells from rats with endotoxemia express low levels of protein S [34–36]. The role of coagulation cascade activation in the pathogenesis of fibrosis has been substantiated by studies in animal models showing that the administration of APC or direct inhibitors of thrombin blocks the development of fibrosis in organs such as lung and liver [12, 37] (Fig. 1). The significant inverse correlation between decreased APC generation and collagen deposition in patients with lung fibrosis further supports the role of coagulation activation in aberrant tissue remodeling [38]. Quite recently, we found that the multifunctional protease inhibitor PCI directly and markedly inhibits a factor XIIa-like protease, hepatocyte growth factor activator (HGFA), which results in decreased hepatocyte growth factor precursor activation; this in turn causes impaired liver regeneration after partial hepatectomy

[39, 40]. This finding indicates that PCI also participates in regulation of wound healing and tissue remodeling, such as liver regeneration.

Conclusion

Several lines of evidence suggest that interaction of the coagulation system with the inflammatory response plays a fundamental role in normal wound healing as well as in the pathogenesis of tissue fibrosis. In vitro studies have shown that PAR-1, PAR-3, and PAR-4 mediate the proinflammatory effects of thrombin; but whether these PARs exert similar effects in vivo needs further investigation. New insights into the mechanism of thrombin-mediated inflammation may provide clues for the development of novel therapies for several intractable chronic diseases, such as pulmonary fibrosis, airway remodeling, liver cirrhosis, and atherosclerosis.

Acknowledgments. This study was supported in part by a Grant-in-Aid for Scientific Research from the Ministry of Education, Science, Culture, and Sports of Japan; Japan Society for the Promotion of Science; and the Mie University COE Project Fund.

References

1. Chapman HA (2004) Disorders of lung matrix remodeling. J Clin Invest 113:148–157
2. Wynn TA (2007) Common and unique mechanisms regulate fibrosis in various fibro-proliferative diseases. J Clin Invest 117:524–529
3. Strieter RM, Belperio JA, Keane MP (2003) CXC chemokines in vascular remodeling related to pulmonary fibrosis. Am J Respir Cell Mol Biol 29:S67–S69
4. Keane MP, Belperio JA, Strieter RM (2003) Cytokine biology and the pathogenesis of interstitial lung disease. In: Schwarz MI, King TE (eds) Interstitial Lung Disease. BC Decker, Hamilton, Ontario, Canada, pp 245–275
5. Sheppard D (2006) Transforming growth factor beta: a central modulator of pulmonary and airway inflammation and fibrosis. Proc Am Thorac Soc 3:413–417
6. Chambers RC (2003) Role of coagulation cascade proteases in lung repair and fibrosis. Eur Respir J Suppl 44:33s–35s
7. Mann KG (2003) Thrombin formation. Chest 124:4S–10S
8. Gabazza EC, Taguchi O, Kamada H, et al (2004) Progress in the understanding of protease-activated receptors. Int J Hematol 79:117–122
9. Traynelis SF, Trejo J (2007) Protease-activated receptor signaling: new roles and regulatory mechanisms. Curr Opin Hematol 14:230–235
10. Coughlin SR (2005) Protease-activated receptors in hemostasis, thrombosis and vascular biology. J Thromb Haemost 3:1800–1814
11. Nakamura M, Gabazza EC, Imoto I, et al (2005) Anti-inflammatory effect of activated protein C in gastric epithelial cells. J Thromb Haemost 3:2721–2729
12. Yasui H, Gabazza EC, Tamaki S, et al (2001) Intratracheal administration of activated protein C inhibits bleomycin-induced lung fibrosis in the mouse. Am J Respir Crit Care Med 163:1660–1668
13. Shimizu S, Gabazza EC, Taguchi O, et al (2003) Activated protein C inhibits the expression of platelet-derived growth factor in the lung. Am J Respir Crit Care Med 167:1416–1426
14. Fujimoto H, Gabazza EC, Hataji O, et al (2003) Thrombin-activatable fibrinolysis inhibitor and protein C inhibitor in interstitial lung disease. Am J Respir Crit Care Med 167:1687–1694

15. Bastarache JA, Ware LB, Bernard GR (2006) The role of the coagulation cascade in the continuum of sepsis and acute lung injury and acute respiratory distress syndrome. Semin Respir Crit Care Med 27:365–376

16. Chambers RC, Laurent GJ (2002) Coagulation cascade proteases and tissue fibrosis. Biochem Soc Trans 30:194–200

17. Kanfer A (1989) Role of coagulation in glomerular injury. Toxicol Lett 46:83–92

18. Steffel J, Luscher TF, Tanner FC (2006) Tissue factor in cardiovascular diseases: molecular mechanisms and clinical implications. Circulation 113:722–731

19. Liu X, Piela-Smith TH (2000) Fibrin(ogen)-induced expression of ICAM-1 and chemokines in human synovial fibroblasts. J Immunol 165:5255–5261

20. Hamaguchi M, Morishita Y, Takahashi I, et al (1991) FDP D-dimer induces the secretion of interleukin-1, urokinase-type plasminogen activator, and plasminogen activator inhibitor-2 in a human promonocytic leukemia cell line. Blood 77:94–100

21. Thompson WD, Stirk CM, Melvin WT, et al (1996) Plasmin, fibrin degradation and angiogenesis. Nat Med 2:493

22. Fujimoto H, Gabazza EC, Taguchi O, et al (2006) Thrombin-activatable fibrinolysis inhibitor deficiency attenuates bleomycin-induced lung fibrosis. Am J Pathol 168:1086–1096

23. Suzuki K, Gabazza EC, Hayashi T, et al (2004) Protective role of activated protein C in lung and airway remodeling. Crit Care Med 32:S262–S265

24. Ploplis VA, Wilberding J, McLennan L, et al (2000) A total fibrinogen deficiency is compatible with the development of pulmonary fibrosis in mice. Am J Pathol 157:703–708

25. Shimizu S, Gabazza EC, Hayashi T, et al (2000) Thrombin stimulates the expression of PDGF in lung epithelial cells. Am J Physiol Lung Cell Mol Physiol 279:L503–L510

26. Howell DC, Laurent GJ, Chambers RC (2002) Role of thrombin and its major cellular receptor, protease-activated receptor-1, in pulmonary fibrosis. Biochem Soc Trans 30:211–216

27. Okajima K (2001) Regulation of inflammatory responses by natural anticoagulants. Immunol Rev 184:258–274

28. Kijiyama N, Ueno H, Sugimoto I, et al (2006) Intratracheal gene transfer of tissue factor pathway inhibitor attenuates pulmonary fibrosis. Biochem Biophys Res Commun 339:1113–1119

29. Nishii Y, Gabazza EC, Fujimoto H, et al (2006) Protective role of protein C inhibitor in monocrotaline-induced pulmonary hypertension. J Thromb Haemost 4:2331–2339

30. Gabazza EC, Taguchi O, Suzuki K (2006) Coagulation cascade: protein C and protein S. In: Laurent G, Shapiro S (eds) Encyclopedia of Respiratory Medicine. Elsevier, Amsterdam, pp 519–525

31. Kobayashi H, Gabazza EC, Taguchi O, et al (1998) Protein C anticoagulant system in patients with interstitial lung disease. Am J Respir Crit Care Med 157:1850–1854

32. Hataji O, Taguchi O, Gabazza EC, et al (2002) Activation of protein C pathway in the airways. Lung 180:47–59

33. Yamaguchi M, Gabazza EC, Taguchi O, et al (2006) Decreased protein C activation in patients with fulminant hepatic failure. Scand J Gastroenterol 41:331–337

34. Wang L, Bastarache JA, Wickersham N, et al (2007) Novel role of the human alveolar epithelium in regulating intra-alveolar coagulation. Am J Respir Cell Mol Biol 36:497–503

35. Fujii K, Kishiwada M, Hayashi T, et al (2006) Differential regulation of protein S expression in hepatocytes and sinusoidal endothelial cells in rats with cirrhosis. J Thromb Haemost 4:2607–2615

36. Hayashi T, Kishiwada M, Fujii K, et al (2006) Lipopolysaccharide-induced decreased protein S expression in liver cells is mediated by MEK/ERK signaling and NFkappaB activation: involvement of membrane-bound CD14 and toll-like receptor-4. J Thromb Haemost. 4:1763–1773

37. Yoshikawa A, Kaido T, Seto S, et al (2000) Activated protein C prevents multiple organ injury following extensive hepatectomy in cirrhotic rats. J Hepatol 33:953–960
38. Yasui H, Gabazza EC, Taguchi O, et al (2000) Decreased protein C activation is associated with abnormal collagen turnover in the intraalveolar space of patients with interstitial lung disease. Clin Appl Thromb Hemost 6:202–205
39. Hayashi T, Nishioka J, Nakagawa N, et al (2007) Protein C inhibitor directly and potently inhibits activated hepatocyte growth factor activator. J Thromb Haemost 5:1477–1485
40. Hamada T, Kamada H, Hayashi T, et al (2008) Protein C inhibitor regulates hepatocyte growth factor activator-mediated liver regeneration in mice. Gut (in press)

Structure and Function of the Endothelial Cell Protein C Receptor

Kenji Fukudome

Summary. The endothelial cell protein C receptor (EPCR) is a cell surface transmembrane glycoprotein and a novel member of the CD1/major histocompatibility complex class 1 superfamily. It was originally found to express specifically in endothelial cells, but recent studies demonstrated wide distribution of the molecule. EPCR is capable of high-affinity binding specific for protein C (PC) and activated protein C (APC). PC binding to EPCR results in acceleration of conversion to APC mediated by the thrombin/thrombomodulin complex, and it is required for regulation of blood coagulation. APC also binds to EPCR, and the APC–EPCR complex was found to promote activation of protease-activated receptor-1 (PAR-1). Binding to EPCR modulates the enzymatic specificity of APC and enables proteolysis of PAR-1, leading to intracellular signal transduction. The EPCR–APC–PAR-1 signaling induces various genes and explains the antiinflammatory, antiapoptotic, and wound healing effects of APC.

Key words. Endothelial cell · Protein C · Receptor

Structure

The endothelial cell protein C receptor (EPCR) was identified as an endothelial cell surface protein serving high-affinity binding sites specific for protein C (PC) and activated protein C (APC) [1]. The structural features are well conserved among species [2]. Mature human EPCR is composed of 211 amino acids, and the extracellular region contains four *N*-glycosylation sites and three cysteine sites. The first exists as free, and the other two cysteines form a disulfate bond. EPCR itself is not a signaling receptor because it entirely lacks a cytoplasmic domain. However, recent studies demonstrated that EPCR contributed to intracellular signaling by cooperating with other cell surface signaling receptors (described later). The extracellular region of EPCR is homologous to that of molecules belonging to the CD1/major histocompatibility complex class 1(MHC-1) superfamily. The extracellular region of the CD1/MHC-1 molecules contains three domains known as α1, α2, and α3. On the other hand, EPCR lacks the α3 domain, which is known as an immunoglobulin domain (Fig. 1). Gene clusters of CD1 and MHC-1 localize on chromosomes 1 and 6, respectively; and EPCR localizes on chromosome 20 [3]. Although genes locate on distinct chromosomes, the exon/intron boundaries for the α1 and α2 domains of these molecules are

211

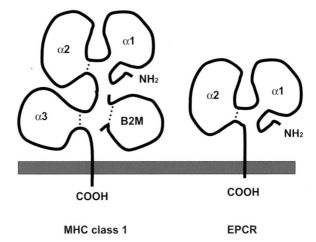

FIG. 1. Structure of the endothelial cell protein C receptor (*EPCR*). The extracellular region of major histocompatibility complex (*MHC*) class 1 molecule is composed of α1, α2, and α3 domains (*left*). The α1 and α2 domains form a glove structure that is capable of binding antigen peptide. It carries an association molecule, β_2-microglobulin (*B2M*). The extracellular region of EPCR is composed of two domains, which correspond to the α1 and α2 domains of MHC class 1 molecules (*right*)

identical. EPCR appears to originate from a common ancestor of the superfamily. The α1 and α2 domains of the CD1/MHC-1 molecules are known to form a groove structure, which serves as a binding site for antigen peptides. The similar structure of EPCR, composed of the α1 and α2 domains, is important for ligand binding because mutations in the α1 or α2 domain resulted in loss of the binding function for PC/APC [4].

Binding Function

CD1d is the most similar to EPCR in the Cd1/MHC-1 family [1], and phospholipids have been identified as the endogenous ligands for CD1d [5]. Crystal structure analysis demonstrated a striking similarity between the EPCR structure and that of CD1d, and it detected a phospholipid molecule in the groove structure [6]. Although biochemical experiments demonstrated a contribution of the phospholipid in EPCR to the binding function, direct interaction between the ligand and phospholipid was not shown by structural analysis [6]. The Gla domain of PC/APC was required for the binding to EPCR [1], and the Gla domain itself was shown to bind to the receptor [7]. The Gla domain requires calcium ions for the binding function, and PC/APC binding to EPCR was also calcium-dependent [1]. Crystal structure analysis showed that binding of calcium ions was required for the protein–protein interaction between the Gla domain and EPCR [6]. Structural modulation by calcium binding enabled close interaction of the residues in the Gla domain and those in EPCR, which were shown to be critical for the binding by mutation analysis [8], as shown in Fig. 2.

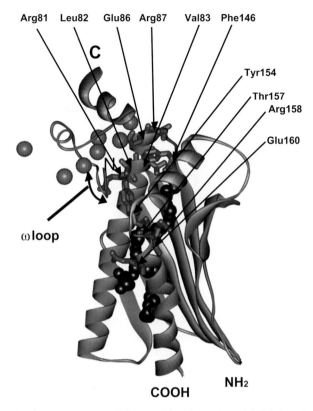

Fig. 2. Interaction between EPCR and the γ-carboxyglutamic acid (Gla) domain of protein C. EPCR is represented as a ribbon model in yellow, and the Gla domain of PC is green. Bound calcium ions and phospholipid are represented as magenta and blue balls, respectively. Amino acid residues of EPCR, which were shown to be important for the binding function by alanine scanning, are indicated

EPCR Function for PC Activation

Protein C circulates as a zymogen form of serine protease, and conversion to the active protease is required to demonstrate the physiological functions. PC conversion to APC is promoted by selective cleavage by thrombin [9], and the catalytic reaction was found to be greatly accelerated on endothelial cells [10]. TM was identified as the molecule responsible for the phenomenon [11, 12]; however, the extreme efficiency of the catalytic reaction could not be explained by the thrombin–TM complex. In fact, only moderate PC activation was detected on the cells transfected with TM alone. The Kd value for PC binding to EPCR was approximately 30 nM, which is lower than that of the blood concentration of PC (65 nM). This fact suggests that PC is concentrated on the endothelial cell surface by the EPCR function, enabling rapid conversion to APC by the thrombin–TM complex. In fact, efficient PC activation was demonstrated on transfected cells with EPCR in addition to TM, and it was comparable to that on

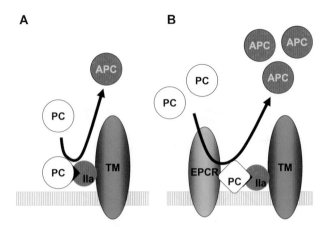

FIG. 3. Protein C (*PC*) activation. The catalytic reaction by the thrombin–thrombomodulin (TM) complex is not efficient (**A**). EPCR recruits PC from the bloodstream and presents it to the complex, enabling efficient activation (**B**). *APC*, activated protein C; *IIa*, thrombin

endothelial cells [13]. Furthermore, function-blocking anti-EPCR monoclonal antibodies specifically suppressed the catalytic reaction on primary cultured endothelial cells [14]. PC/APC binding to EPCR is relatively high affinity, but they are replaced within 5 min under physiological conditions. This reversible binding appears to be beneficial for generating a significant amount of APC. In large arterial vessels, circulating PC has less chance to contact the endothelial cell surface, suggesting inefficiency of PC activation. Interestingly, EPCR expression was found to be abundant in large arteries but not so abundant in small veins [14]. PC activation appears to be possible in various vessels because of the EPCR function (Fig. 3).

Signal Transduction Induced by the APC/EPCR Complex

Activated PC has been shown to have antiinflammatory activities [15], and administration of recombinant APC significantly reduced mortality in patients with severe sepsis [16]. The activities were demonstrated only by the active enzyme form of APC but not by the zymogen form of PC. Although the molecular mechanism for antiinflammatory activities of APC had long been unidentified, the capability of APC to induce signal transduction via PAR-1 was recently identified [17]. PAR-1 was originally identified as a cell surface signaling receptor specific for thrombin [18]. Thrombin cleaves off the amino terminal peptide from the receptor, resulting in signal transduction mediated by G-proteins. PAR-1 is not a substrate for stand-alone APC but is for APC coupled with EPCR. Binding to EPCR modulates the enzyme specificity of APC and enables cleavage, leading to PAR-1 activation. Monocyte chemoattractant protein-1 (MCP-1) was found to be induced in endothelial cells specifically by the signaling pathway [17]. MCP-1 is a CC chemokine with chemoattractant activity for

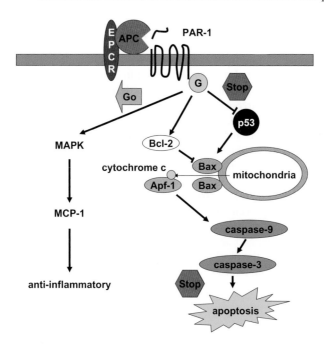

FIG. 4. EPCR–APC–PAR-1 signal pathway. APC-binding to EPCR modulates enzyme specificity of APC, enabling a catalytic reaction for activation of PAR-1. The signal is transduced to nucleoli mediated by the MAPK pathway and induces expression of MCP-1 (*left*). Hypoxia results in p53-mediated apoptosis in endothelial cells. The EPCR–APC–PAR-1 signal prevents induction of the p53 pathway and protects from apoptosis (*right*)

T cells, mast cells, and basophils [19, 20]. It is produced in various cell types by stimulation with proinflammatory cytokines and also with lipopolysaccharide (LPS) [20, 21]. Elevation of MCP-1 levels was detected in patients with sepsis [21], and it appears to be required for protection from its lethal effects. In murine models, neutralization of MCP-1 by antibodies enhanced the lethal effects of LPS, and administration of recombinant MCP-1 contrarily protected the animals [22–24]. The molecular mechanism of the protective effects of APC against sepsis can be at least partially explained by induction of MCP-1 via the signaling pathway mediated by EPCR and PAR-1 (Fig. 4).

The signaling from the APC–EPCR complex was also found to protect endothelial cells from apoptosis in ischemic injury [25]. Hypoxia increased transient transcription of p53 and its target Bax and decreased that of an inhibitor, Bcl-2. Induction of p53 and Bax resulted in activation of caspases leading to apoptosis. APC prevented these responses against hypoxia in an EPCR- and PAR-1-dependent fashion, and protected the cells form apoptosis (Fig. 4). The protective effects of APC administration in focal ischemic stroke were also demonstrated in animal experiments, suggesting potential therapeutic approaches for neurodegenerative disorders.

EPCR Functions in Other Cell Types

EPCR was originally found to be expressed specifically in endothelial cells; however, recent studies demonstrated that it is expressed in other cell types as well. Functionally active EPCR was detected in several cell lines derived from monoblastic leukemia, erythroleukemia, osteosarcoma, and glioblastoma [26] as well as in normal cells. Skin keratinocytes were found to express EPCR [27] and synthesize the ligand, PC [28]. PC was shown to function as an autocrine growth factor for keratinocytes, and it was specifically prevented by a functional blocking monoclonal antibody [14]. EPCR was also detected in vascular smooth muscle cells and functioned for proliferation [29]. Therefore, APC–EPCR–PAR-1 signaling appears to function in wound healing. Primary isolated monocytes [30] and neutrophils [31] were also positive for the receptor. PC/APC prevented neutrophil chemotaxis, and a function-blocking monoclonal antibody diminished the inhibitory effects of the ligands. Interestingly, PC and APC functioned equally on neutrophils; and anti-EPCR monoclonal antibodies prevented the functions independently of the blocking function for the PC/APC binding. This suggests an unidentified signaling pathway distinct from the APC–EPCR–PAR-1 signaling.

References

1. Fukudome K, Esmon CT (1994) Identification, cloning, and regulation of a novel endothelial cell protein C/activated protein C receptor. J Biol Chem 269:26486–26491
2. Fukudome K, Esmon CT (1995) Molecular cloning and expression of murine and bovine endothelial cell protein C/activated protein C receptor (EPCR): the structural and functional conservation in human, bovine, and murine EPCR. J Biol Chem 270:5571–5577
3. Simmonds RE, Lane DA (1999) Structural and functional implications of the intron/exon organization of the human endothelial cell protein C/activated protein C receptor (EPCR) gene: comparison with the structure of CD1/major histocompatibility complex alpha1 and alpha2 domains. Blood 94:632–641
4. Fukudome K, Kurosawa S, Stearns-Kurosawa DJ, et al (1996) The endothelial cell protein C receptor: cell surface expression and direct ligand binding by the soluble receptor. J Biol Chem 271:17491–17498
5. Brutkiewicz RR (2006) CD1d ligands: the good, the bad, and the ugly. J Immunol 177:769–775
6. Oganesyan V, Oganesyan N, Terzyan S, et al (2002) The crystal structure of the endothelial protein C receptor and a bound phospholipid. J Biol Chem 277:24851–24854
7. Regan LM, Mollica JS, Rezaie AR, et al (1997) The interaction between the endothelial cell protein C receptor and protein C is dictated by the gamma-carboxyglutamic acid domain of protein C. J Biol Chem 272:26279–26284
8. Liaw PC, Mather T, Oganesyan N, et al (2001) Identification of the protein C/activated protein C binding sites on the endothelial cell protein C receptor: implications for a novel mode of ligand recognition by a major histocompatibility complex class 1-type receptor. J Biol Chem 276:8364–8370
9. Kisiel W, Canfield WM, Ericsson LH, et al (1977) Anticoagulant properties of bovine plasma protein C following activation by thrombin. Biochemistry 16:5824–5831
10. Esmon CT, Owen WG (1981) Identification of an endothelial cell cofactor for thrombin-catalyzed activation of protein C. Proc Natl Acad Sci U S A 78:2249–2252

11. Jackman RW, Beeler DL, VanDeWater L, et al (1986) Characterization of a thrombo-modulin cDNA reveals structural similarity to the low density lipoprotein receptor. Proc Natl Acad Sci U S A 83:8834–8838

12. Suzuki K, Kusumoto H, Deyashiki Y, et al (1987) Structure and expression of human thrombomodulin, a thrombin receptor on endothelium acting as a cofactor for protein C activation. EMBO J 6:1891–1897

13. Fukudome K, Ye X, Tsuneyoshi N, et al (1998) Activation mechanism of anticoagulant protein C in large blood vessels involving the endothelial cell protein C receptor. J Exp Med 187:1029–1035

14. Ye X, Fukudome K, Tsuneyoshi N, et al (1999) The endothelial cell protein C receptor (EPCR) functions as a primary receptor for protein C activation on endothelial cells in arteries, veins, and capillaries. Biochem Biophys Res Commun 259:671–677

15. Taylor FB Jr, Chang A, Esmon CT, et al (1987) Protein C prevents the coagulopathic and lethal effects of *Escherichia coli* infusion in the baboon. J Clin Invest 79:918–925

16. Bernard GR, Vincent JL, Laterre PF, et al (2001) Efficacy and safety of recombinant human activated protein C for severe sepsis. N Engl J Med 344:699–709

17. Riewald M, Petrovan RJ, Donner A, et al (2002) Activation of endothelial cell protease activated receptor 1 by the protein C pathway. Science 296:1880–1882

18. Vu TK, Hung DT, Wheaton VI, et al (1991) Molecular cloning of a functional throm-bin receptor reveals a novel proteolytic mechanism of receptor activation. Cell 64:1057–1068

19. Leonard EJ, Yoshimura T (1990) Human monocyte chemoattractant protein-1 (MCP-1). Immunol Today 11:97–101

20. Chensue SW, Warmington KS, Ruth JH, et al (1996) Role of monocyte chemoattractant protein-1 (MCP-1) in Th1 (mycobacterial) and Th2 (schistosomal) antigen-induced granuloma formation: relationship to local inflammation, Th cell expression, and IL-12 production. J Immunol 157:4602–4608

21. Bossink AW, Paemen L, Jansen PM, et al (1995) Plasma levels of the chemokines mono-cyte chemotactic proteins-1 and -2 are elevated in human sepsis. Blood 86:3841–3847

22. Zisman DA, Kunkel SL, Strieter RM, et al (1997) MCP-1 protects mice in lethal endo-toxemia. J Clin Invest 99:2832–2836

23. Matsukawa A, Hogaboam CM, Lukacs NW, et al (1999) Endogenous monocyte che-moattractant protein-1 (MCP-1) protects mice in a model of acute septic peritonitis: cross-talk between MCP-1 and leukotriene B$_4$. J Immunol 163:6148–6154

24. Bone-Larson CL, Hogaboam CM, Steinhauser ML, et al (2000) Novel protective effects of stem cell factor in a murine model of acute septic peritonitis: dependence on MCP-1. Am J Pathol 157:1177–1186

25. Cheng T, Liu D, Griffin JH, et al (2003) Activated protein C blocks p53-mediated apop-tosis in ischemic human brain endothelium and is neuroprotective. Nat Med 9:338–342

26. Tsuneyoshi N, Fukudome K, Horiguchi S, et al (2001) Expression and anticoagulant function of the endothelial cell protein C receptor (EPCR) in cancer cell lines. Thromb Haemost 85:356–361

27. Xue M, Campbell D, Sambrook PN, et al (2005) Endothelial protein C receptor and protease-activated receptor-1 mediate induction of a wound-healing phenotype in human keratinocytes by activated protein C. J Invest Dermatol 125:1279–1285

28. Xue M, Campbell D, Jackson CJ (2007) Protein C is an autocrine growth factor for human skin keratinocytes. J Biol Chem 282:13610–13616

29. Bretschneider E, Uzonyi B, Weber AA, et al (2007) Human vascular smooth muscle cells express functionally active endothelial cell protein C receptor. Circ Res 100:255–262

30. Galligan L, Livingstone W, Volkov Y, et al (2001) Characterization of protein C receptor expression in monocytes. Br J Haematol 115:408–414

31. Sturn DH, Kaneider NC, Feistritzer C, et al (2003) Expression and function of the endothelial protein C receptor in human neutrophils. Blood 102:1499–1505

New Aspects of Antiinflammatory Activity of Antithrombin: Molecular Mechanism(s) and Therapeutic Implications

Naoaki Harada and Kenji Okajima

Summary. Antithrombin (AT), an important natural anticoagulant, has been shown to reduce various organ failures as well as coagulation abnormalities in animal sepsis models and in patients with severe sepsis. Proinflammatory cytokines, such as tumor necrosis factor (TNF), play critical roles in the development of the multiple organ failure including disseminated intravascular coagulation by inducing endothelial cell injury through neutrophil activation during sepsis. AT increases the endothelial production of prostacyclin, a potent inhibitor of TNF production, thereby attenuating inflammatory responses in experimental animals given endotoxin and in those subjected to organ ischemia/reperfusion. AT increases the endothelial production of prostacyclin via promotion of calcitonin gene-related peptide (CGRP) release from sensory neurons. CGRP has been shown to increase the production of insulin-like growth factor-I (IGF-I), a potent antiapoptotic factor, in various organs in mice. AT increases IGF-I production via enhancing sensory neuron activation, thereby preventing reperfusion-induced hepatic apoptosis in mice. Because IGF-I has various important biological activities, such as promoting differentiation of various cell types and an anabolic effect in addition to potent antiapoptotic activity, AT might exert novel biological activities other than anticoagulant activities by promoting IGF-I production. These functional properties of AT might explain at least in part its therapeutic efficacy in patients with severe sepsis.

Key words. Antithrombin · Insulin-like growth factor-I · Sensory neurons · Severe sepsis · Tumor necrosis factor

Introduction

Antithrombin (AT) is an important natural anticoagulant that inhibits serine proteases generated during activation of the coagulation cascade. The physiological significance of AT is clearly illustrated by the development of recurrent thrombosis in patients with congenital AT deficiency. In addition, AT improves microcirculatory disturbances by attenuating inflammatory responses in experimental animals given endotoxin or in those subjected to organ ischemia/reperfusion. Furthermore, AT improves the outcome of patients with severe sepsis when they are associated with disseminated intravascular coagulation. Because both inflammation and microthrombi formation are critically involved in the development of microcirculatory disturbances in the pathological condition of severe sepsis, AT might play critical roles in maintenance of proper microcirculation by regulating inflammatory responses and coagulation abnormalities.

This chapter describes novel molecular mechanism(s) by which AT regulates microcirculation and further mentions the possible therapeutic applications of AT to improve microcirculatory conditions in various disease states including severe sepsis.

Anticoagulant Activity of AT

A glycoprotein, AT has a molecular weight of 58 200 Da with 432 amino acids and is mainly synthesized in the liver. AT is a physiological serine protease inhibitor that inhibits activated coagulation factors such as thrombin and factor Xa. The reactive site loop of AT includes a P1-P′1 (Arg393-Ser394) bond (Fig. 1). When thrombin cleaves this bond that resembles the substrate of thrombin, the protease is covalently linked to P1 residue. Inhibition of these serine proteases by AT is accelerated approximately 1000-fold by binding of heparin to arginine residues located at the heparin-binding site of AT (Fig. 1), with the P1-P′1 reactive center. Amino acid residues other than arginine shown in Fig. 1 have also been found to be critical for interaction with heparin. AT is activated on the endothelial surface, where thrombin generation is increased through binding to heparan sulfate molecules of ryudocan or syndecan. The physiological importance of AT is well illustrated in patients with congenital AT deficiency who developed recurrent thrombosis during their youth; 70% of the patients developed thrombosis before 35 years of age. Congenital AT deficiency usually presents as a heterozygous state associated with venous thrombosis, with the homozygous state extremely rare, probably because of its presentation as lethal neonatal thrombosis.

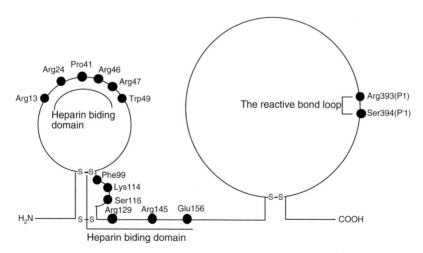

FIG. 1. Localization of heparin-binding domains and the reactive site in the primary structure of antithrombin

Similar observations have been reported for the pathologic sequelae in AT knockout mice [1]. The homozygous state of AT deficiency was reported only in patients who have variant AT molecules with heparin-binding defects, and patients with such variant molecules have arterial thrombosis as well as venous thrombosis [2]. Possible mechanisms by which arterial thrombosis occurs in these patients are discussed later in this review. These observations strongly suggested that AT plays a critical role in regulation of the coagulation system by inhibiting thrombin and other serine proteases, and the interaction of AT with the endothelial cell surface heparin-like substances is quite important for rapid, effective inhibition of such coagulation factors.

Antiinflammatory Activities of AT

Activation of the coagulation system leading to disseminated intravascular coagulation (DIC) is frequently seen as a part of inflammatory responses in pathological conditions such as sepsis and circulatory shock. Proinflammatory cytokines such as tumor necrosis factor (TNF) play an important role in these inflammatory responses [3]. Although TNF is implicated in the activation of the coagulation system during sepsis, it also activates neutrophils, thereby promoting the release of a wide variety of inflammatory mediators such as neutrophil elastase and oxygen free radicals that are capable of damaging endothelial cells. The resultant endothelial cell damage increases microvascular permeability, leading to microcirculatory disturbance due to hemoconcentration. Such microcirculatory disturbance precedes microthrombus formation [4]. Microthrombus formation further increases TNF production, thereby exacerbating the inflammatory response to form a vicious cycle of progression of microcirculatory disturbances [5] (Fig. 2).

We previously reported that AT reduced pulmonary endothelial injury by inhibiting neutrophil accumulation in the lung of rats given endotoxin. AT also reduced

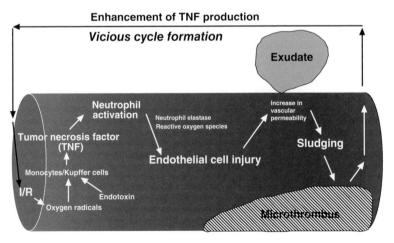

FIG. 2. Pathological mechanism(s) leading to the microcirculatory disturbance associated with sepsis or circulatory shock. *I/R*, ischemia/reperfusion

endotoxin-induced hypotension in rats by inhibiting TNF production [6]. These therapeutic effects of AT could not be explained by the anticoagulant activity but by its promotion of endothelial production of prostacyclin (PGI_2), which potently inhibits leukocyte activation. Interaction of AT with glycosaminoglycans might be critical for promotion of endothelial production of PGI_2. Ischemia/reperfusion (I/R) is an important pathological mechanism leading to the development of coagulation abnormalities and organ failure seen in patients with circulatory shock [7]. TNF is also critically involved in this pathological process. AT increased the hepatic tissue blood flow by inhibiting neutrophil activation in rats subjected to hepatic I/R. AT also increased renal tissue blood flow by inhibiting neutrophil activation through inhibition of TNF production, thereby reducing renal injury [8]. These effects of AT are independent of its anticoagulant activity but dependent on its capacity to promote endothelial production of PGI_2. Although AT itself has been shown to inhibit TNF production by monocytes stimulated with endotoxin in vitro, inhibition by AT of TNF production in vivo is not observed when animals are pretreated with indomethacin, which inhibits prostaglandin biosynthesis. Iloprost, a stable derivative of PGI_2, produces effects similar to those of AT in these animal models of sepsis and in those subjected to tissue I/R. These observations strongly suggest that antiinflammatory activity of AT might be mediated by PGI_2 released from endothelial cells.

Molecular Mechanism(s) of the AT-Induced Increase in Endothelial Production of PGI_2

To elucidate the precise molecular mechanism(s) by which AT exerts antiinflammatory activity, we examined the effect of AT on the endothelial production of PGI_2 using cultured endothelial cells. However, AT did not directly increase endothelial production of PGI_2 in cultured endothelial cells. Thus, the mechanism(s) by which AT promotes endothelial release of PGI_2 in vivo might involve unknown factors other than endothelial cells.

Capsaicin-sensitive sensory neurons are nociceptive neurons that are activated by a wide variety of noxious physical and chemical stimuli. Because ablation of sensory fibers can result in a marked increase in the severity of inflammation, the sensory neurons have been shown to play a role in the maintenance of tissue integrity by regulating local inflammatory responses. On activation, the sensory neurons release calcitonin gene-related peptide (CGRP), which can increase endothelial production of PGI_2 in vitro. Because various noxious stimuli that activate the sensory neurons to release CGRP are capable of inducing tissue damage, the CGRP-induced increase in endothelial production of PGI_2 might contribute to attenuation of local inflammatory responses, thereby reducing the tissue damage. Consistent with this hypothesis, we previously reported that capsaicin-sensitive sensory neurons were activated during hepatic I/R or water-immersion restraint stress in rats, leading to an increase in the endothelial production of PGI_2 via activation of both endothelial nitric oxide synthase (NOS) and cyclooxygenase (COX)-1 [9] (Fig. 3).

We previously reported that AT reduced I/R-induced liver injury in rats by increasing endothelial production of PGI_2. However, the mechanism(s) underlying this

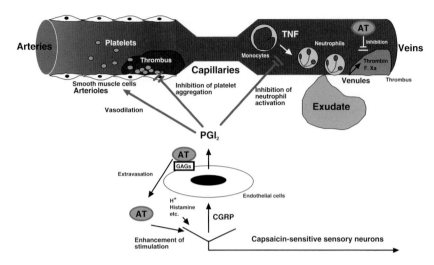

FIG. 3. Regulation of the microcirculation by antithrombin. *AT*, antithrombin; *GAGs*, glycosaminoglycans; *TNF*, tumor necrosis factor; *CGRP*, calcitonin gene-related peptide; *F. Xa*, activated form of coagulation factor X; *PGI₂*, prostacyclin

phenomenon remains to be fully elucidated. AT significantly enhanced the I/R-induced increase in hepatic tissue levels of CGRP in rats. The increase in hepatic tissue levels of 6-keto-PGF$_{1\alpha}$, a stable metabolite of PGI$_2$, increase in hepatic tissue blood flow, and attenuation of both hepatic local inflammatory responses and liver injury in rats administered AT are completely reversed by administration of capsazepine, an inhibitor of sensory neuron activation, and CGRP(8-37), a CGRP antagonist. Although AT itself does not increase CGRP release from cultured dorsal root ganglion (DRG) neurons, it significantly enhances CGRP release from DRG neurons in the presence of anandamide, an endogenous agonist of vanilloid receptor-1 activation produced by macrophages [10]. Therefore, AT might extravasate at the site of endothelial injury, thereby enhancing the capsaicin-sensitive sensory neurons, leading to an increase in PGI$_2$ levels in hepatic tissue [11] (Fig. 3). The AT-induced increase in CGRP release from DRG neurons in the presence of anandamide is inhibited by KT5720, an inhibitor of protein kinase A. In addition, AT increased cAMP levels in DRG neurons, and the AT-induced increase is more marked in the presence of anandamide. Because anandamide increases CGRP release from sensory neurons by activating protein kinase A through increases in intracellular cAMP levels [12], AT might enhance anandamide-induced increases in cAMP levels in DRG neurons, thereby enhancing protein kinase A activation. Although the mechanism(s) by which AT increases cAMP levels in DRG neurons is not known at present, AT has also been reported to increase cAMP levels in cultured endothelial cells [13]. These observations raise the possibility that AT sensitizes sensory neurons against activation by endogenous agonist that are capable of activation of vanilloid receptor-1. Mechanism(s) underlying this effect of AT should be elucidated by further investigations in the near future.

Roles of Antiinflammation and Anticoagulation by AT in Regulation of Microcirculation

Tumor necrosis factor has an important role in the development of the microcirculatory disturbance associated with sepsis or circulatory shock—primarily by activating neutrophils, thereby inducing endothelial cell injury leading to microcirculatory disturbance due to local hemoconcentration, and secondarily by inducing microthrombus formation, which precipitates the preexisting microcirculatory disturbance as described earlier [5] (Fig. 2). Thus, inhibition of both neutrophil activation and microthrombus formation might be important for maintenance of proper microcirculation. AT increases the endothelial production of PGI_2 via enhancement of capsaicin-sensitive sensory neuron activation. PGI_2 is a vasodilator and has a potent inhibitory effect on platelet activation [14]. Because platelet activation and subsequent vasospasm are critical factors in the development of arterial thrombosis [15], AT might prevent arterial thrombosis by increasing the endothelial production of PGI_2 through interaction with heparin-like substances on endothelial cells, contributing to maintenance of proper microcirculation in the arteries (Fig. 3). This hypothesis is consistent with the observations that both arterial thrombosis and venous thrombosis are frequently found in patients in a homozygous state with variant AT molecules that do not have heparin affinity. In addition, PGI_2 has inhibitory effects on monocytic TNF production and neutrophil activation [16, 17]. Activated neutrophils are critically involved in the development of microcirculatory disturbance in the postcapillary venules due to the increase in microvascular permeability, leading to reduced blood flow resulting from local hemoconcentration [18]. In addition to stasis, activated coagulation factors such as factor Xa are essential for development of venous thrombosis [19] (Fig. 3). Thus, AT might prevent thrombus formation by promoting endothelial production of PGI_2 in postcapillary venules where activated neutrophil-induced endothelial damage is frequently induced and also in veins by its inherent anticoagulant activity. Thus, both antiinflammatory and anticoagulant activities of AT might contribute to the maintenance of proper microcirculation.

Possible Therapeutic Applications of AT for Microcirculatory Disturbances

As described earlier, intravenous administration of AT reduces pulmonary vascular injury and hypotension because of its antiinflammatory activity in rats given endotoxin [6, 20], suggesting that AT supplementation might be useful for treating acute respiratory distress syndrome and shock associated with sepsis. Administration of AT is also effective in the treatment of I/R-induced liver and kidney injuries [8, 21] and stress-induced gastric mucosal injury in rats based on its capacity to promote endothelial production of PGI_2 [22], suggesting that AT might be useful for treating various organ failures associated with circulatory shock.

Reperfusion-induced spinal cord injury is an important pathological mechanism for the development of paraplegia after operations on the descending thoracic and thoracoabdominal aorta due to interruption of the intercostals and lumber arteries feeding the spinal cord [23]. Intravenous administration of AT significantly alleviates motor disturbances by inhibiting the reduction of the number of motor neurons in rats subjected to transient spinal cord ischemia [24]. Because both local inflammatory responses and spinal cord microinfarction are significantly reduced through increases in spinal cord tissue levels of PGI_2 in animals treated with AT, both antiinflammatory and anticoagulant activities of AT might be critical for the therapeutic effect of AT. These observations strongly suggest that AT might be a useful neuroprotective agent for prevention of spinal cord injury after surgery to repair aortic aneurysms. These possibilities should be examined in the clinical setting in the near future.

Antiapoptotic Activity of AT by Increasing Insulin-Like Growth Factor-I Production

Insulin-like growth factor-I is a basic peptide of 70 amino acids with a rather ubiquitous distribution among various tissues and cells; it mediates the growth-promoting actions of growth hormone (GH) and plays an important role in postnatal and adolescent growth [25]. Furthermore, IGF-I has been shown to have various important biological activities, such as promoting differentiation of various cell types, potent antiapoptotic activity, and an anabolic effect [26]. Among these activities, the antiapoptotic activity of IGF-I has been shown to be important for reducing reperfusion-induced tissue injury by attenuating inflammatory responses [27]. Transduction of signals through the IGF-I receptor leads to multiple series of intracellular phosphorylation events and the activation of several signaling pathways, thereby preventing cell death [26]. Because apoptosis is known to be an important etiological factor for the development of various disease states [27], its prevention by increasing production of IGF-I might have therapeutic potential. Although it is well known that GH increases IGF-I synthesis in the liver [28], little is known about the mechanism by which IGF-I production is increased in various tissues independent of GH action. Vignery and McCarthy [29] previously reported that CGRP increases IGF-I production by increasing cAMP levels in primary fetal rat osteoblasts, suggesting that sensory neuron activation might increase IGF-I production. As sensory neurons are widely distributed in most tissues, we hypothesized that activation of sensory neurons leading to release of CGRP might increase IGF-I production in various tissues, thereby reducing I/R-induced tissue injury.

We have demonstrated that AT reduced I/R-induced liver injury by attenuating proinflammatory responses in mice [30]. The mechanism underlying this effect of AT depends mainly on promotion of sensory neuron activation, leading to CGRP release from nerve endings [30]. Hepatic apoptosis has been shown to play a primary role in triggering proinflammatory responses observed during reperfusion-induced tissue injury [27]. We recently reported that CGRP increases hepatic production of IGF-I, a potent antiapoptotic factor [31], thereby reducing reperfusion-induced liver injury by preventing hepatic apoptosis [32]. We further reported that AT prevented hepatic apoptosis by enhancing IGF-I production through promotion of sensory neuron activation, leading to reduction of reperfusion-induced liver injury [33] (Fig. 4).

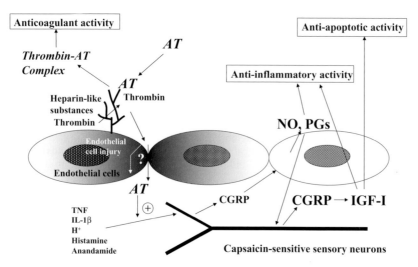

FIG. 4. Contribution of antithrombin-induced increase in insulin-like growth factor-I production in cytoprotection. *IGF-I*, insulin-like growth factor-I; *IL-1*, interleukin-1; *NO*, nitric oxide; *PGs*, prostaglandins

Conclusions and Perspectives

Although AT is an important natural anticoagulant, it might attenuate inflammatory responses by inhibiting the production of TNF. AT also might inhibit apoptosis through the promotion of IGF-I production. Because apoptosis is a crucial event that can initiate reperfusion-induced inflammation and subsequent tissue injury [27], it is possible that preventing hepatic apoptosis by AT via enhancement of increased hepatic IGF-I production contributes to the reduction of I/R-induced liver injury through attenuation of proinflammatory responses.

Apoptosis plays a critical role in the development of organ failure in sepsis [33] as well as in reperfusion-induced tissue injury [34]. Therefore, prevention of apoptosis by AT through enhancement of IGF-I production might contribute to therapeutic efficacy of AT in patients with severe sepsis.

It has been proposed that apoptotic lymphocytes that have accumulated in the spleen, thymus, and other organs might, at least in part, contribute critically to the pathogenesis of organ damage during severe sepsis [35]. Because IGF-I has been shown to prevent lymphocyte apoptosis by activating the PI3K-dependent pathway [36], it is possible that AT inhibits lymphocyte apoptosis by increasing IGF-I production, thereby reducing organ damage during severe sepsis. We are currently investigating this possibility using an experimental sepsis model in mice.

References

1. Ishiguro K, Kojima T, Kadomatsu K, et al (2000) Complete antithrombin deficiency in mice results in embryonic lethality. J Clin Invest 106:873–878

2. Hirsh J, Piovella F, Pini M (1989) Congenital antithrombin III deficiency: incidence and clinical features. Am J Med 87:34S–38S
3. Okajima K (2001) Regulation of inflammatory responses by natural anticoagulants. Immunol Rev 184:258–274
4. Okajima K, Harada N, Kushimoto S, et al (2002) Role of microthrombus formation in the development of ischemia/reperfusion-induced liver injury in rats. Thromb Haemost 88:473–480
5. Okajima K, Harada N, Uchiba M (2003) Microthrombus formation enhances tumor necrosis factor-alpha production in the development of ischemia/reperfusion-induced liver injury in rats. J Thromb Haemost 1:1316–1317
6. Isobe H, Okajima K, Uchiba M, et al (2002) Antithrombin prevents endotoxin-induced hypotension by inhibiting the induction of nitric oxide synthase in rats. Blood 99:1638–1645
7. Keller GA, West MA, Cerra FB, et al (1985) Macrophage-mediated modulation of hepatic function in multiple-system failure. J Surg Res 39:555–563
8. Mizutani A, Okajima K, Uchiba M, et al (2003) Antithrombin reduces ischemia/reperfusion-induced renal injury in rats by inhibiting leukocyte activation through promotion of prostacyclin production. Blood 101:3029–3036
9. Harada N, Okajima K, Uchiba M, et al (2003) Contribution of capsaicin-sensitive sensory neurons to stress-induced increases in gastric tissue levels of prostaglandins in rats. Am J Physiol Gastrointest Liver Physiol 285:G1214–G1224
10. Hogestatt ED, Zygmunt PM (2002) Cardiovascular pharmacology of anandamide. Prostaglandins Leukot Essent Fatty Acids 66:343–351
11. Harada N, Okajima K, Yuksel M, et al (2005) Contribution of capsaicin-sensitive sensory neurons to antithrombin-induced reduction of ischemia/reperfusion-induced liver injury in rats. Thromb Haemost 93:48–56
12. De Petrocellis L, Harrison S, Bisogno T, et al (2001) The vanilloid receptor (VR1)-mediated effects of anandamide are potently enhanced by the cAMP-dependent protein kinase. J Neurochem 77:1660–1663
13. Uchiba M, Okajima K, Kaun C, et al (2004) Inhibition of the endothelial cell activation by antithrombin in vitro. Thromb Haemost 92:1420–1427
14. Weksler BB, Ley CW, Jaffe EA (1978) Stimulation of endothelial cell prostacyclin production by thrombin, trypsin, and the ionophore A 23187. J Clin Invest 62:923–930
15. Drouet L, Bal Dit Sollier C, Ruton S, et al (1990) Role of serotonin in arteriolar thrombosis and secondary vasospasm. J Cardiovasc Pharmacol 16(Suppl 3):S49–S53
16. Eisenhut T, Sinha B, Grottrup-Wolfers E, et al (1993) Prostacyclin analogs suppress the synthesis of tumor necrosis factor-alpha in LPS-stimulated human peripheral blood mononuclear cells. Immunopharmacology 26:259–264
17. Kainoh M, Imai R, Umetsu T, et al (1990) Prostacyclin and beraprost sodium as suppressors of activated rat polymorphonuclear leukocytes. Biochem Pharmacol 39:477–484
18. Van Nieuw Amerongen GP, van Hinsbergh VW (2002) Targets for pharmacological intervention of endothelial hyperpermeability and barrier function. Vasc Pharmacol 39:257–272
19. Walenga JM, Petitou M, Lormeau JC, et al (1987) Antithrombotic activity of a synthetic heparin pentasaccharide in a rabbit stasis thrombosis model using different thrombogenic challenges. Thromb Res 46:187–198
20. Uchiba M, Okajima K, Murakami K, et al (1996) Attenuation of endotoxin-induced pulmonary vascular injury by antithrombin III. Am J Physiol 270:L921–L930
21. Harada N, Okajima K, Kushimoto S, et al (1999) Antithrombin reduces ischemia/reperfusion injury of rat liver by increasing the hepatic level of prostacyclin. Blood 93:157–164
22. Isobe H, Okajima K, Liu W, et al (1999) Antithrombin prevents stress-induced gastric mucosal injury by increasing the gastric prostacyclin level in rats. J Lab Clin Med 133:557–565

23. Svensson LG, Crawford ES, Hess KR, et al (1993) Experience with 1509 patients undergoing thoracoabdominal aortic operations. J Vasc Surg 17:357–368; discussion 368–370
24. Hirose K, Okajima K, Uchiba M, et al (2004) Antithrombin reduces the ischemia/reperfusion-induced spinal cord injury in rats by attenuating inflammatory responses. Thromb Haemost 91:162–170
25. Daughaday WH, Rotwein P (1989) Insulin-like growth factors I and II: peptide, messenger ribonucleic acid and gene structures, serum, and tissue concentrations. Endocr Rev 10:68–91
26. Carroll PV (2001) Treatment with growth hormone and insulin-like growth factor-I in critical illness. Best Pract Res Clin Endocrinol Metab 15:435–451
27. Daemen MA, van 't Veer C, Denecker G, et al (1999) Inhibition of apoptosis induced by ischemia-reperfusion prevents inflammation. J Clin Invest 104:541–549
28. Jones J, Clemmons D (1995) Insulin-like growth factors and their binding proteins: biological actions. Endocr Rev 16:3–34
29. Vignery A, McCarthy TL (1996) The neuropeptide calcitonin gene-related peptide stimulates insulin-like growth factor I production by primary fetal rat osteoblasts. Bone 18:331–335
30. Harada N, Okajima K, Uchiba M, et al (2006) Antithrombin reduces reperfusion-induced liver injury in mice by enhancing sensory neuron activation. Thromb Haemost 95:788–795
31. Vincent AM, Feldman EL (2002) Control of cell survival by IGF signaling pathways. Growth Horm IGF Res 12:193–197
32. Harada N, Okajima K, Kurihara H, et al (2007) Stimulation of sensory neurons by capsaicin increases tissue levels of IGF-I, thereby reducing reperfusion-induced apoptosis in mice. Neuropharmacology 52:1303–1311
33. Harada N, Okajima K, Kurihara H, et al (In press) Antithrombin prevents reperfusion-induced hepatic apoptosis by enhancing insulin-like growth factor-I production in mice. Crit Care Med
34. Haimovitz-Friedman A, Cordon-Cardo C, Bayoumy S, et al (1997) Lipopolysaccharide induces disseminated endothelial apoptosis requiring ceramide generation. J Exp Med 186:1831–1841
35. Endres M, Namura S, Shimizu-Sasamata M, et al (1998) Attenuation of delayed neuronal death after mild focal ischemia in mice by inhibition of the caspase family. J Cereb Blood Flow Metab 18:238–247
36. Hotchkiss RS, Chang KC, Swanson PE, et al (2000) Caspase inhibitors improve survival in sepsis: a critical role of the lymphocyte. Nat Immunol 1:496–501
37. Jimenez Del Rio M, Velez-Pardo C (2006) Insulin-like growth factor-1 prevents Abeta[25-35]/(H₂O₂)- induced apoptosis in lymphocytes by reciprocal NF-kappaB activation and p53 inhibition via PI3K-dependent pathway. Growth Factors 24:67–78

Part 3 Platelets

Platelet Collagen Receptors

STEPHANIE M. JUNG AND MASAAKI MOROI

Summary. The physiological collagen receptors integrin $\alpha_2\beta_1$ (previously known as GP Ia/IIa) ($\alpha_2\beta_1$) and glycoprotein VI (GPVI) were identified by analyzing patients' platelets deficient in one of these proteins. Stimulant-induced platelet activation converts $\alpha_2\beta_1$ to its active form, which can bind tightly to its ligand collagen. Collagen interacts with $\alpha_2\beta_1$ mainly through a Glu residue (part of the receptor-binding sequence GFOGER) that interacts with the MIDAS (metal ion-dependent motif) coordinated divalent cation (Mg^{2+}, physiologically), located in the I domain of the α-subunit of this integrin. GPVI binds more weakly to collagen than activated $\alpha_2\beta_1$, but this interaction is sufficient to initiate a strong signaling cascade to transduce signals, among which tyrosine phosphorylation is the most important, resulting in full platelet activation. GPVI recognizes repeats of the Gly-Pro-hydroxyPro (GPO) sequence in collagen and crosslinking of GPVI induces the initial signaling reaction. Analysis of platelet adhesion and aggregation to a collagen surface under blood flow, which approximates physiological conditions, indicates that $\alpha_2\beta_1$ mainly contributes to the initial adhesion to collagen, and GPVI contributes to large aggregate formation after the initial adhesion. Thus, $\alpha_2\beta_1$ and GPVI have different properties as collagen receptors, and both contribute to physiological platelet plug formation.

Key words. Collagen receptor · GPVI · Integrin $\alpha_2\beta_1$ · Collagen · vWf

Introduction

Under normal physiological conditions, platelets circulating in the bloodstream are in the resting, nonactivated state. They do not form aggregates by themselves, nor do they adhere to endothelial cells covering the inner surface of blood vessels. Damage to a blood vessel exposes its subendothelium to the bloodstream, so platelets can react with the subendothelial collagen, adhere to it, and form aggregates on the damaged surface. Platelets activated through these events can readily adhere to the subendothelial collagen, enabling them to form aggregates easily. This ability to become activated is the most important property of platelets because they must not easily adhere and aggregate to avoid thrombus formation but still must be able to be activated quickly to perform their physiological function in thrombus formation if blood vessels become damaged. Furthermore, platelet activation must be closely regulated because functional overactivity or deficiency would result in thrombosis or bleeding tendencies, respectively. Numerous substances are known to activate platelets, including thrombin, adenosine diphosphate (ADP), collagen, and epinephrine. Among them, collagen is a highly important physiological activator because of its strong activity to induce platelet aggregation and adhesion and its high content in the subendothelium. Upon activation, a reaction cascade is set off inside these cells, whose function is to activate the integrins, adhesive proteins, which in turn leads to platelet adhesion and aggregation.

The importance of the abundant subendothelial component collagen in hemostasis has driven the search to identify specific platelet receptors for collagen. Many proteins have been proposed to be the platelet collagen receptor, including collagen:glycosyl transferase, fibronectin, CD36, factor XIII, and pro-polypeptide of von Willebrand factor (vWF), only to be abandoned as further studies brought new data inconsistent with their possible role as a physiological receptor on the platelet surface. These unsuccessful attempts are not surprising considering that collagen is a multitype insoluble macromolecular protein that has affinity to many different proteins as well as numerous possible points of interaction. Also, platelet activation by collagen is a complicated reaction, making it difficult to show which proteins are really functioning as collagen receptors. Finally, two physiological collagen receptors were identified by analyzing patients' platelets specifically lacking reactivity toward collagen. In 1985, Nieuwenhuis et al. described patient's platelets lacking collagen reactivity that contained only 15%–20% the normal level of glycoprotein Ia (GPIa, integrin subunit α_2) [1], providing the first evidence that GPIa is a physiologically active collagen receptor. GPIa was later shown to form a complex with GPIIa; and the same complex was also identified as a collagen receptor in many other cells. Thus, the ubiquitous membrane protein GPIa/IIa, integrin $\alpha_2\beta_1$ ($\alpha_2\beta_1$), was generally recognized as a physiological collagen receptor in platelets and other cells. Soon after this report, several Japanese patients whose platelets also lacked collagen-induced activation were reported by Sugiyama et al. [2] and Moroi et al. [3]. These patients' platelets contained a normal level of $\alpha_2\beta_1$ but lacked a 62-kDa protein normally present in platelets. Although this protein band had been previously identified and designated glycoprotein VI (GPVI) without further functional characterization, its true function was finally revealed using GPVI-deficient platelets and autologous antibody from one patient; thus, GPVI was recognized as a new physiological collagen receptor unique to platelets.

This chapter focuses on integrin $\alpha_2\beta_1$ and GPVI, the two functionally established platelet collagen receptors. Other reported collagen receptor proteins, such as GPIV (CD36), platelet receptor for type I collagen, platelet receptor for type III collagen, and type III collagen binding protein are not discussed here because there are insufficient data to establish their physiological roles conclusively. In addition, plasma vWF binds to collagen and the platelet GPIb/V/IX complex binds to the immobilized vWF, so this complex also functions as a collagen receptor. The vWF–GPIb interaction is particularly important in platelet adhesion under the high-shear flow in small arteries and arterioles.

Integrin $\alpha_2\beta_1$

As described above, GPIa/IIa was identified as a collagen receptor from studying patient's platelets lacking the collagen-induced aggregation response [1]. Around the same time, GPIa was suggested to be a collagen receptor by Santoro, who detected a protein specifically interacting with collagen. Other groups identified VLA-2 and ECMR-II as the antigen of monoclonal antibodies that inhibit the interaction of collagen with other cells, establishing their function as a collagen receptor. Later, GPIa/IIa, VLA-2, and ECMR-II were found to be the same protein and were renamed integrin $\alpha_2\beta_1$ because it belongs to the integrin family of cell membrane proteins.

Structure of Integrin $\alpha_2\beta_1$ and Its Activation Mechanism

The main properties of $\alpha_2\beta_1$ are similar to those of the other integrin family members: structural homology, heterodimer of one α subunit and one β subunit, divalent metal ion-dependent activity, and similar activation mechanism. The α_2 subunit consists of a large extracellular domain, transmembrane domain, and short intracellular domain (residues 1160–1181). On its N-terminal side, the α_2 chain contains seven repeats (with 60 amino acids in each) of a similar structure, which can be modeled as a seven-bladed β-propeller. For a thorough, detailed description of integrin structure, readers can refer to the excellent article by Humphries [4].

In all the integrin α subunits constructed from one peptide chain (e.g., α_1, α_2, α_L, α_M, α_X), an I (inserted) domain (residues 140–359 in the case of the α_2 chain) is located between the second and third blade of the propeller. Because of its homology to the vWF A domains, the I domain is sometimes called the A domain. The recombinant α_2 subunit I domain bound specifically to collagen, and epitopes of all monoclonal antibodies that specifically inhibit $\alpha_2\beta_1$ binding to collagen are residues within the I domain, indicating that it contains all the components necessary for collagen binding. The crystal structure of the I domain and its complex with a collagen model peptide indicate that a collagen-binding site is created upon complex formation with the peptide when several changes are induced in the structure of the I domain [5], allowing the MIDAS (metal ion-dependent adhesion site)-coordinated divalent cation to move up and coordinate with the glutamate of collagen (Fig. 1).

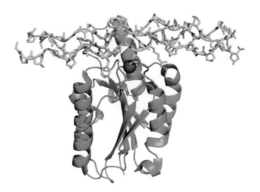

FIG. 1. Structure of the integrin $\alpha_2\beta_1$ I-domain–collagen complex. In this complex, the divalent metal ion (depicted as a red ball, Co^{2+} in this case) in MIDAS (metal ion-dependent adhesion site) forms a "positively charged hole," and the Glu side chain of the GFOGER collagen model peptide inserts into this hole. The complex is stabilized by hydrophobic interactions between the collagen peptide and the I domain. The model is drawn from the protein database (PDB: 1DZI) [5]

Detailed analyses of the divalent ion dependence of $\alpha_2\beta_1$ activity showed that Mg^{2+} is the physiological metal ion coordinated with MIDAS, and its replacement by Ca^{2+} or Mn^{2+}, respectively, inhibits or stimulates $\alpha_2\beta_1$ activity. Propeller blades 5–7 also contain other divalent cation binding motifs, so binding of a divalent cation, particularly Ca^{2+}, to these sites would affect $\alpha_2\beta_1$ binding activity. The β_1 subunit also contains a structure similar to the I domain, the I-like domain that contains a MIDAS-like structure. The function of the β_1 I-like domain is not established; but in other integrins lacking an I domain, the I-like domain may act as a ligand-binding site. Binding activity of the α subunit was reduced by mutation of the β_2 integrin I-like domain, so it may regulate binding activity by interacting with the α subunit I domain.

By analyzing collagen fragments to determine which ones have $\alpha_2\beta_1$-binding activity, Barnes' group at Cambridge identified the GFOGER sequence (O = hydroxyproline) as the $\alpha_2\beta_1$ binding site in collagen [6]. Farndale's group expanded these findings by synthesizing collagen-mimetic peptides inserted with sequences from type I and III collagens and identified the following binding motifs: GFOGER, GLOGER, and GASGER in type I collagen and GROGER, GMOGER, GLKGEN, and GLOGEN in type III collagen. Their results confirmed that the Glu residue is important for collagen binding to $\alpha_2\beta_1$ and indicate that collagen contains multiple binding sites.

Like other integrins, $\alpha_2\beta_1$ must be activated to express its full binding activity for collagen. However, the analysis of $\alpha_2\beta_1$ activation was hindered by the fact that collagen is an insoluble macromolecule. We overcame this problem by developing an assay that uses the binding of soluble collagen to platelets as a measure of $\alpha_2\beta_1$-binding affinity toward collagen. Resting platelets do not bind to soluble collagen, but platelets activated by stimulants such as ADP, thrombin, or collagen-related peptide (CRP) or by the $\alpha_2\beta_1$ activating antibody TS2/16 show strong binding to this ligand, indicating that the $\alpha_2\beta_1$ of resting platelets is in a nonactive conformation and the stimulant-induced activation of platelets converts $\alpha_2\beta_1$ to its active form, which readily binds collagen [7]. Notably, the collagen binding to platelets activated by ADP and that in thrombin-activated platelets are different: The binding constant (Kd) of thrombin-activated platelets is 9.96×10^{-9} M and that of ADP-activated platelets is 5.85×10^{-8} M, but the two types of platelets show a similar number of binding sites (Bmax). Thus, there are two active conformations of activated $\alpha_2\beta_1$—a low-affinity one and a high-affinity one—that, respectively, result from activation by a "weak" agonist (e.g., ADP) and a "strong agonist" (e.g., thrombin or CRP). The two active forms are produced through different mechanisms because signaling-protein–specific inhibitors had different effects on ADP-induced activation and thrombin- or CRP-induced activation [7]. Outside-in signaling through integrin $\alpha_{IIb}\beta_3$ was suggested to contribute to $\alpha_2\beta_1$ activation induced by convulxin (Cvx), a snake venom toxin that activates GPVI.

Physiological Function of Integrin $\alpha_2\beta_1$

Traditionally, $\alpha_2\beta_1$ has been thought to contribute to platelet adhesion to collagen. However, as we demonstrated, $\alpha_2\beta_1$ requires activation to express its strong affinity for collagen (refer to Fig. 4); thus, platelets must be activated by a stimulant for firm adhesion to occur. In the case of platelet adhesion to collagen, this requisite $\alpha_2\beta_1$ activation is provided by another collagen receptor, GPVI, which binds to collagen via a relatively low-affinity interaction but is nonetheless sufficient to initiate signaling pathways leading to the activation of $\alpha_2\beta_1$. Once activated, $\alpha_2\beta_1$ binds to collagen with high affinity, thereby "arresting" the platelets on the collagen surface, which facilitates the interaction between collagen and GPVI, causing further activation. Thus $\alpha_2\beta_1$ and GPVI together would activate platelets more efficiently that either receptor alone. For platelet adhesion to collagen under flow at shear rates, such as those encountered physiologically, the situation is more complex. Plasma vWF binds to collagen, and platelets adhere to the immobilized vWF under flow, but this situation is complicated by the fact that vWF binding is collagen type-specific, as we recently showed. vWF binds strongly to type III collagen but only weakly to type I. In the case of type I collagen, as a result of the low binding of vWF to this substrate $\alpha_2\beta_1$ directly contributes to the platelet adhesion to collagen. Thus anti-$\alpha_2\beta_1$ antibody inhibited platelet adhesion to type I collagen under flow but did not affect the adhesion to type III [8]. This result explains the discrepancies of the published data, with some articles reporting that $\alpha_2\beta_1$ contributes to platelet adhesion to collagen under flow and others reporting that it does not. The discrepancy would be ascribed to the different types of collagen they used.

The above studies suggest that variations in platelet surface density of $\alpha_2\beta_1$ might affect thrombogenicity, but do the clinical data support this hypothesis? Kunicki et al. found linked, allelic polymorphism in the α_2 gene that correlates with $\alpha_2\beta_1$ density [9]. Three α_2 alleles were associated with changes in the levels of $\alpha_2\beta_1$: Allele 1 (807T/1648G/2531C) is associated with increased $\alpha_2\beta_1$ levels, and allele 2 (807C/1648G/2531C) and allele 3 (807C/1648A/2531C) are associated with decreased levels. Although these are silent mutations and the expressed $\alpha_2\beta_1$ is a normal protein, the different $\alpha_2\beta_1$ densities affect the platelet adhesion to type I collagen under a high shear rate. In fact, a correlation between the genotype with a low level of $\alpha_2\beta_1$ and bleeding tendency in type I von Willebrand disease was found [10], and a relation between allele 1 and myocardial infarction in young individuals was indicated [11]. Thus, the platelet $\alpha_2\beta_1$ expression level may influence hemostasis and thrombus formation in some patients. However, because some investigators found an association between inheritance of $\alpha_2\beta_1$ allele 1 and risk for coronary artery disease and myocardial infarction but others did not, the association of $\alpha_2\beta_1$ expression level with risk for thrombotic or hemostatic episodes in these diseases remains controversial.

Glycoprotein VI

Glycoprotein VI (GPVI) was first identified as a protein deficient in a small number of Japanese patients whose platelets did not aggregate in response to collagen [2, 3] but aggregated normally in response to other agonists, including thrombin, ADP, and ristocetin. These observations suggested that GPVI acts in a critical step specifically related to the reaction with collagen. However, early studies on GPVI progressed slowly because GPVI could be identified only when using GPVI-deficient platelets or the limited-availability autoantibody to GPVI from one of the deficient patients. During 1995–1997, research on GPVI was greatly accelerated by the discovery of GPVI-specific agonists. While searching for the specific structure of collagen that induces platelet activation, Barnes' laboratory found that a triple helical collagen-mimetic peptide containing 10 repeats of the Gly-Pro-Hyp sequence (CRP, collagen-related peptide) induced platelet aggregation independently of $\alpha_2\beta_1$ [12]. CRP-induced activation was inhibited by the Fab fragment of an anti-GPVI antibody, verifying that CRP activates platelets by specifically reacting with GPVI. The tropical rattlesnake *Crotalus durissus terrificus* has a platelet-activating venom protein, con-vulxin (Cvx), that acts as a specific agonist of GPVI [13]. Similar to collagen, these two agonists (CRP and Cvx) induce many of the platelet activation reactions, but they react only with GPVI, so the function of GPVI could be easily differentiated from that of $\alpha_2\beta_1$.

The platelet-specific receptor GPVI has a unique physiological function in throm-bus formation. It specifically reacts with collagen fibrils and activates platelets through a tyrosine phosphorylation-dependent pathway. Clinical and animal studies indicate that GPVI deficiency does not cause any severe bleeding tendency, but GPVI-deficient mice have impaired thrombus formation. Thus, specific inhibitors against GPVI are potentially ideal antithrombotic drugs without significant bleeding as a side effect, and finding a specific GPVI inhibitor suitable for clinical use is of high priority.

Structure of GPVI

GPVI is a glycoprotein composed of 319 amino acid residues and a signal sequence of 20 amino acids [14]. Its extracellular region contains two immunoglobulin (Ig)-like domains; and a mucin-like Ser/Thr-rich domain is present between the Ig-like and transmembrane domains. The Ig-like domain is responsible for the binding of GPVI to collagen. Many O-linked carbohydrate chains are expected to conjugate with the Ser/Thr residues of the Ser/Thr-rich domain. GPVI contains a positively charged residue Arg252 in its transmembrane domain. The transmembrane-domain charged amino acid residue is characteristic of proteins that associate with the Fc receptor γ-chain (FcRγ), which contains a negatively charged Asp residue in its transmembrane domain. Thus, a salt bridge formed between GPVI and FcRγ stabilizes the receptor complex.

The two Ig-like domains of GPVI show homology to other proteins belonging to the paired Ig-like receptor family, including the Fcα receptor, mouse mast cell receptor, and both inhibitory and activatory members of the natural killer cell receptors. The recently determined crystal structure of the Ig-like domains of GPVI [15] clearly indicates that GPVI forms a back-to-back dimmer, and the docking algorithm identi-fies two parallel collagen-binding sites on the GPVI dimer surface. Compared to the structures of other homologous proteins, GPVI lacks 11–13 amino acids between Glu49 and Gln50, and this deletion induces a unique shallow hydrophobic groove on the surface of the Ig-like domain 1 (D1) of GPVI. Simulation of the binding of a collagen model peptide and GPVI indicated the collagen-binding site in this groove (Fig. 2). Mutational analysis indicated that amino acids Lys41, Lys59, Arg60, and Arg166 are involved in collagen binding, and these residues are all oriented around the contact sites on the collagen model peptides, thus confirming the collagen-binding site. Val34 and Leu36 have also been reported to bind to collagen or CRP, but these residues are present at the distal end of D1. The mutational effects of the single resi-dues are rather small, suggesting that these residues might form another binding site

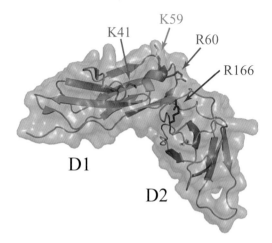

FIG. 2. Structure of glycoprotein VI (GPVI) immunoglobulin (Ig)-like domain. Ig-like domains (*D1* and *D2*) of GPVI were crystallized, and their structures were determined. The collagen model peptide (poly-GPO) to fit this structure and the collagen-binding sites were simulated. The amino acid residues that were simulated to interact with collagen and also indicated to contribute to collagen binding are indicated. The model is from database PDB: 2GI7 [15]

for collagen, rather than being the primary one. The model of the collagen-binding site depicts a back-to-back dimer of the GPVI Ig-like domain, so two nearly parallel putative collagen-binding sites are separated by a distance of 55 Å. This space is sufficient to fit three triple helical collagen molecules between these two collagen-binding sites. Thus, the model can explain why GPVI binds preferably to polymeric collagen fibrils rather than monomeric collagen.

The 51-amino-acid cytoplasmic "tail" of GPVI can be divided into four regions: juxtamembrane, basic, proline-rich, and C-terminal regions. The juxtamembrane region is necessary for interaction with the associated FcRγ subunit. The basic region next to the juxtamembrane region is rich in basic amino acids, making it favorable for calmodulin binding. This region is also necessary for the interaction with FcRγ. The adjacent Pro-rich region contains a consensus Pro-rich motif that binds to the Src homology 3 (SH3) domain. The Src family tyrosine kinases Fyn and Lyn were also shown to bind to this region. The C-terminal region is thought to have no function because mouse GPVI lacks this region but has activity similar to human GPVI.

GPVI-Induced Signal Transduction

Because all the known GPVI agonists have a polyvalent GPVI-binding capability, platelet activation induced by GPVI may occur through crosslinking of GPVI molecules on the platelet surface. An anti-GPVI antibody that induces platelet activation can be changed to a GPVI inhibitor when it is converted to monovalent Fab fragments. The platelet-aggregating activity of CRP is greatly enhanced when it is crosslinked. Collagen that is physiologically active exists as polymerized fibrous collagen under physiological conditions, whereas soluble monomeric collagen has little or no activity

[16]. GPVI is present as a complex with the FcRγ homodimer. Thus, crosslinking the GPVI–FcRγ complex to stabilize it, or closing the complex, would initiate the activating reaction by GPVI, but its exact mechanism must still be elucidated.

Figure 3 summarizes these signaling pathways induced by GPVI–FcRγ. S–S-bridged homodimeric FcRγ is essential for both the expression and function of GPVI. Each subunit of FcRγ contains one copy of an immunoreceptor tyrosine–based activation motif (ITAM) with two Tyr residues that undergo phosphorylation when activation is induced through GPVI. Upon receptor crosslinking, the Src family protein kinases Fyn and Lyn phosphorylate the ITAM Tyr residues, thereby initiating the GPVI-mediated signaling pathway [17]. The SH3 domains of these kinases have been shown to associate with the Pro-rich domain of the GPVI cytoplasmic tail upon stimulation by a GPVI agonist. Thus, the association of Fyn and Lyn with the cytoplasmic domain of GPVI can be the first step of the activation mechanism induced by a GPVI agonist. However, this step would be more complicated because a study using Fyn- or Lyn-deficient mouse platelets suggested that the contribution of another factor.

The tyrosine kinase Syk binds to phosphorylated ITAM, which induces phosphorylation and activation of Syk. The activated Syk initiates the downstream signaling cascade by phosphorylating other proteins, including LAT, SLP-76, and phospholipase Cγ2 (PLCγ2). LAT has many phosphorylation sites, and phosphorylated LAT binds numerous proteins containing the SH2 domain, including PLCγ2, PI 3-kinase, Gads, Btk, Vav 1, and Vav3 [18]. Thus, using LAT as a scaffold, these proteins form a large

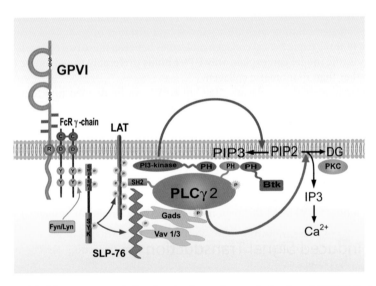

FIG. 3. Model of the GPVI/FcRγ-induced activation pathways. Activation of GPVI/FcRγ induces tyrosine phosphorylation, which facilitates interactions between many signaling molecules, resulting in the formation of a large signaling complex, the so-called LAT signalosome. Tyrosine phosphorylation is indicated by *P in asterisk*, and each *arrow* indicates an enzyme reaction. *PH*, pleckstrin-homology domain; *SH2*, Src-homology domain 2 domain; *LAT*, linker for activation of T cells; *SLP-76*, SH2 domain-containing leukocyte protein of 76 kDa; *C*, Cys; *R*, Arg; *D*, Asp; *Y*, Tyr. (From Moroi and Jung [27], with permission)

complex, the so-called LAT signalosome [18]. Among the components of the complex, PLCγ2 is thought to be the most important one because of its enzymatic activity. The membrane-localized and activated PLCγ2 degrades phosphatidyl inositol 4,5 diphosphate (PIP2) to 1,2 diacylglycerol (DG) and inositol 1,4,5-triphosphate (IP3). DG is recognized as an activator of protein kinase C (PKC), whereas IP3 binds to IP3 receptors and induces Ca^{2+} release. The activation of PKC and intracellular Ca^{2+} release cause the release reaction and integrin-mediated activation of platelets. Platelets deficient in FcRγ, Syk, LAT, PLCγ, or SLP-76 lack collagen-induced platelet aggregation without major effects on aggregation induced by thrombin, confirming that all these proteins contribute to GPVI-induced platelet activation.

Physiological function of GPVI

GPVI-deficient patients have mildly prolonged bleeding times [2, 3] and GPVI-depleted mice have moderately increased or normal bleeding times [19], suggesting that in vivo thrombus formation can be induced not only by the GPVI–collagen interaction but also by activation of other signaling pathways. The contribution of GPVI to platelet adhesion on an immobilized-collagen surface under blood flow was analyzed with GPVI-deficient human or mouse blood or human blood in the presence of anti-GPVI antibody. These data all indicated that formation of large platelet aggregates on the collagen surface is severely impaired if GPVI is defective or inhibited. However, conflicting results have been reported: Some laboratories indicated that GPVI-deficient platelets can adhere to collagen, whereas others reported little or no adhesion to a collagen surface. Because the residual (although very low level) platelet adhesion or aggregation of GPVI-deficient platelets is $\alpha_2\beta_1$-dependent, it is reasonable to suggest that through mediation by $\alpha_2\beta_1$ GPVI-deficient platelets can adhere to collagen but cannot become activated because of the lack of GPVI, thus preventing formation of large aggregates. Figure 4 summarizes the mechanism of platelet adhesion on collagen under high-shear blood flow. The first reaction is the binding of vWF in blood to the collagen surface (reaction 1). The flowing platelets bind to the immobilized vWF through GPIb/V/IX, but the interaction between vWF and GPIb is not a strong one. Thus, the bound platelets can slowly move on the collagen surface as the interaction between vWF and GPIb is broken and reformed again in response to the shear force; this reaction (Fig. 4, reaction 2) is called "tethering." While the platelets are tethered, integrin $\alpha_2\beta_1$ on the platelets can bind to collagen, and this further interaction stops the platelets from moving (firm adhesion). The flowing platelets can also bind directly to collagen through $\alpha_2\beta_1$ (reaction 3). GPVI on the adhered or tethered platelets react with collagen and activate the platelets (reaction 4). The activated platelets express activated integrin $\alpha_{IIb}\beta_3$ and P selectin on their surfaces and also secrete their granule contents, which include ADP, fibrinogen, and Ca^{2+} (reaction 5). This reaction induces platelet aggregation through activated integrin $\alpha_{IIb}\beta_3$ and also activates the flowing resting platelets by the secreted substances, inducing the formation of a platelet aggregate on the collagen surface (reaction 6).

In vivo analysis of thrombus formation caused by laser injury of blood vessels using control and GPVI-deficient mice showed that this process depends on P-selectin and thrombin but not GPVI. In contrast, when blood vessels are injured by $FeCl_3$, which

FIG. 4. Model of the platelet–collagen interaction under high-shear blood flow. Vessel injury exposes the subendothelial collagen, with which circulating platelets can interact through various reactions (for details, see the text). The receptors indicated on the platelets are as follows: collagen receptor GPVI (green); integrin $\alpha_2\beta_1$ (green/blue); activated integrin $\alpha_2\beta_1$ (green/blue with a yellow outline); GPIb/IX/V complex (black). von Willebrand factor is represented by the spiral-shaped gray shapes, and endothelial collagen is represented by the wavy lines

damages the endothelial cells and thereby exposes the subendothelium, thrombus formation is GPVI-dependent [20]. It is also indicated that thrombin makes a significant contribution to platelet adhesion to collagen under flow when blood is not anticoagulated [21]. These results indicate that GPVI contributes to in vivo thrombus formation when collagen fibrils are exposed to the flowing blood, which is the case in atherosclerotic plaque rupture. On the other hand, GPVI would not be a major factor in thrombus formation after mechanical injury of vessels, such as with the usual method employed to measure bleeding time.

Several GPVI-deficient patients have been reported [3, 22], but the details of their genetic defects have not yet been analyzed. Also, some patients with autoimmune disease were reported to have antibodies to GPVI, and their platelets are GPVI-deficient [2, 23]. Such acquired GPVI deficiency can be experimentally induced in mice by injecting anti-GPVI antibody [24]. Although the molecular mechanism for GPVI depletion is still unclear, the same mechanism would likely operate in patients with autoantibodies against GPVI. Furthermore, two patients with idiopathic thrombocytopenic purpura (ITP) were found to have GPVI-deficient platelets but no detectable anti-GPVI antibody in their plasma [25, 26]. The platelets of these patients would have been initially exposed to antibody against GPVI and thereby depleted of GPVI and remain depleted of this receptor even after the antibody was cleared from their blood. Thus, exposure to the antibody would induce a long-lasting depletion of GPVI from the platelets.

GPVI is the main collagen receptor activating platelets, and its inhibition seems not to induce severe bleeding. Thus an inhibitor of the GPVI-collagen interaction would have great clinical potential as an antithrombotic drug. To date, the reported GPVI inhibitors are all antibodies, with the exception of triplatin, a protein from the salivary glands of the assassin bug *Triatoma infestans*. Injection of the Fab fragment of an anti-GPVI antibody into *Cynomolgus* monkeys has been reported to inhibit collagen-induced platelet aggregation for a longer time than an anti-integrin $\alpha_{IIb}\beta_3$ antibody, Abciximab; and these animals showed no bleeding tendency. These findings suggest the clinical applicability of GPVI inhibitors, but more studies are needed to determine their function under different pathological conditions. In addition, small molecule inhibitors seem to be more suitable for antithrombotic drug design, and development of such compounds has been recently facilitated by the reported crystal structure of GPVI.

References

1. Nieuwenhuis HK, Akkerman JWN, Houdijk WPM, et al (1985) Human blood platelets showing no response to collagen fail to express surface glycoprotein Ia. Nature 318:470–472
2. Sugiyama T, Okuma M, Ushikubi F, et al (1987) A novel platelet aggregating factor found in a patient with defective collagen-induced platelet aggregation and autoimmune thrombocytopenia. Blood 69:1712–1720
3. Moroi M, Jung SM, Okuma M, et al (1989) A patient with platelets deficient in glycoprotein VI that lack both collagen-induced aggregation and adhesion. J Clin Invest 84:1440–1445
4. Humphries MJ (2000) Integrin structure. Biochem Soc Trans 28:311–339
5. Emsley J, Knight CG, Farndale RW, et al (2000) Structural basis of collagen recognition by integarin $\alpha_2\beta_1$. Cell 100:47–56
6. Knight CG, Morton LF, Onley DJ, et al (1998) Identification in collagen type I of an integrin $\alpha_2\beta_1$-binding site containing an essential GER sequence. J Biol Chem 273:33287–33294
7. Jung SM, Moroi M (2000) Activation of the platelet collagen receptor integrin $\alpha_2\beta_1$: its mechanism and participation in the physiological functions of platelets. Trends Cardiovasc Med 10:285–292
8. Moroi M, Jung SM (2007) A mechanism to safeguard platelet adhesion under high-shear flow: von Willebrand factor-glycoprotein Ib and integrin $\alpha_2\beta_1$-collagen interactions make complementary, collagen-type-specific contributions to adhesion. J Thromb Haemost 5:797–803
9. Kunicki TJ, Kritzik M, Annis DS, et al (1997) Hereditary variation in platelet integrin $\alpha_2\beta_1$ density is associated with two silent polymorphisms in the $\alpha2$ gene coding sequence. Blood 89:1939–1943
10. Di Paola J, Federici AB, Mannucci PM, et al (1999) Low platelet $\alpha_2\beta_1$ levels in type I von Willebrand disease correlate with impaired platelet function in a high shear stress system. Blood 93:3578–3582
11. Santoso S, Kunicki TJ, Kroll H, et al (1999) Association of the platelet glycoprotein Ia $C_{807}T$ gene polymorphism with myocardial infarction in younger patients. Blood 93:2449–2453
12. Morton LF, Hargreaves PG, Farndale RW, et al (1995) Integrin $\alpha_2\beta_1$-independent activation of platelets by simple collagen-like peptides: collagen tertiary (triple-helical) and quaternary (polymeric) structures are sufficient alone for $\alpha_2\beta_1$-independent platelet reactivity. Biochem J 306:337–344
13. Polgar J, Clemetson JM, Kehrel BE, et al (1997) Platelet activation and signal transduction by convulxin, a C-type lectin from *Crotalus durussus terrificus* (tropical rattlesnake) venom via the p62/GPVI collagen receptor. J Biol Chem 272:13576–13583

14. Clemetson JM, Polgar J, Magnenat E, et al (1999) The platelet collagen receptor glycoprotein VI is a member of the immunoglobulin superfamily closely related to FcαR and the natural killer receptors. J Biol Chem 274:29019–29024
15. Horii K, Kahn ML, Herr AB (2006) Structural basis for platelet collagen responses by immune-type receptor glycoprotein VI. Blood 108:936–942
16. Jung SM, Moroi M (1998) Platelets interact with soluble and insoluble collagens through characteristically different reactions. J Biol Chem 273:14827–14837
17. Quek LS, Pasquet J-M, Hers I, et al (2000) Fyn and Lyn phosphorylate the Fc receptor γ chain downstream of glycoprotein VI in murine platelets, and Lyn regulates a novel feedback pathway. Blood 96:4246–4253
18. Watson SP, Auger JM, McCarty OJT, et al (2005) GPVI and integrin αIIbβ3 signaling in platelets. J Thromb Haemost 3:1752–1762
19. Nieswandt B, Schulte V, Bergmeier W, et al (2001) Long-term antithrombotic protection by in vivo depletion of platelet glycoprotein VI in mice. J Exp Med 193:459–469
20. Furie B, Furie BC (2005) Thrombus formation in vivo. J Clin Invest 115:3355–3362
21. Mangin P, Yap CL, Nonne C, et al (2006) Thrombin overcomes the thrombosis defect associated with platelet GPVI/FcRγ deficiency. Blood 107:4346–4353
22. Arai M, Yamamoto N, Moroi M, et al (1995) Platelets with 10% of the normal amount of glycoprotein VI have an impaired response to collagen that results in a mild bleeding tendency. Br J Haematol 89:124–130
23. Takahashi H, Moroi M (2001) Antibody against platelet membrane glycoprotein VI in a patient with systemic lupus erythematosus. Am J Hematol 67:262–267
24. Schulte V, Rabie T, Prostredna M, et al (2003) Targeting of the collagen-binding site on glycoprotein VI is not essential for in vivo depletion of the receptor. Blood 101:3948–3952
25. Kojima H, Moroi M, Jung SM, et al (2006) Characterization of a patient with glycoprotein (GP) VI deficiency possessing neither anti-GPVI autoantibody nor genetic aberration. J Thromb Haemost 4:2433–2442
26. Boylan B, Chen H, Rathore V, et al (2004) Anti-GPVI-associated ITP: an acquired platelet disorder caused by autoantibody-mediated clearance of the GPVI/FcRγ-chain complex from the human platelet surface. Blood 104:1350–1355
27. Moroi M, Jung SM (2007) Platelet glycoprotein VI. In: Sigalov AB (ed) Multichain immune recognition receptor signaling: from spatiotemporal organization to human disease. Landes Bioscience, Georgetown, in press

Positive and Negative Regulation of Integrin Function

Yoshiaki Tomiyama, Masamichi Shiraga, and Hirokazu Kashiwagi

Summary. Platelet integrin $\alpha_{IIb}\beta_3$, a prototypic non-I domain integrin, plays an essential role in platelet aggregation. The structure and function of $\alpha_{IIb}\beta_3$ is dramatically changed during platelet plug formation and pathological thrombus formation. The function of this integrin is regulated by the balance of actions of positive and negative regulatory factors. Several novel regulators have emerged from recent studies. As a positive regulator, the $P2Y_{12}$ plays a critical role in thrombus stability; and continuous interaction between ADP and $P2Y_{12}$ is essential for sustained $\alpha_{IIb}\beta_3$ activation. Semaphorin 3A and SHPS-1 have been identified as negative regulators. These molecules are secreted from or expressed on endothelial cells and inhibit the function of platelets as well as $\alpha_{IIb}\beta_3$. Investigation on these positive and negative regulatory factors should provide a new insight into the treatment of pathological thrombosis.

Key words. Inside-out signaling · Outside-in signaling · $P2Y_{12}$ · Semaphorin 3A · SHPS-1

Introduction

Platelets play a crucial role not only in hemostatic plug formation but also in a pathological thrombus formation, particularly in atherosclerotic arteries subjected to high shear stress [1, 2]. Moreover, recent studies have revealed that the platelet is a major player in the initiation of vascular remodeling as well as atherosclerotic lesion formation [3, 4]. As an initial step in thrombogenesis, platelets adhere to altered vascular surfaces or exposed subendothelial matrices and then become activated and aggregate with each other. As summarized in Fig. 1, it has been well documented that these processes are primarily mediated by platelet surface glycoproteins: GPIb-IX-V, integrin $\alpha_2\beta_1$ (also known as GPIa-IIa), GPVI, and integrin $\alpha_{IIb}\beta_3$ (GPIIb-IIIa) [5, 6].

Integrins comprise a family of heterodimeric adhesion receptors that mediate cellular attachment to the extracellular matrix and cell cohesion [7–9]. Platelets express at least five integrins on their surface: $\alpha_2\beta_1$(GPIa-IIa); $\alpha_5\beta_1$(GPIc-IIa); $\alpha_6\beta_1$(GPIc′-IIa); $\alpha_{IIb}\beta_3$(GPIIb-IIIa); $\alpha_v\beta_3$. Platelet integrin $\alpha_{IIb}\beta_3$ is a prototypic non-I domain integrin and plays an essential role in platelet aggregation as a physiological receptor for fibrinogen and von Willebrand factor. The importance of this integrin has been well documented by the clinical features of a congenital bleeding disorder, Glanzmann

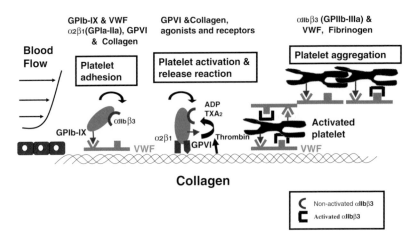

FIG. 1. Mechanisms of platelet plug formation and pathological thrombus formation. These processes depend primarily on platelet adhesive interactions with both platelet surface glycoproteins including integrins and extracellular matrix proteins. Platelet adhesion (or tethering) is mainly mediated by glycoprotein (*GP*)Ib-IX and von Willebrand factor (*VWF*) especially under high shear rates; and platelet aggregation is mediated by $\alpha_{IIb}\beta_3$ and VWF and fibrinogen. Platelet activation and released factors such as adenosine diphosphate (ADP) play a critical role in thrombus stability. *TXA_2*, thromboxane A_2

thrombasthenia (GT) [10, 11]. The crystal structure of $\alpha_{IIb}\beta_3$ revealed that the ligand-binding head is formed by a seven-bladed β-propeller domain from α_{IIb} and a β I-domain from β_3 [12, 13]. Despite the presence of integrin $\alpha_{IIb}\beta_3$ and its ligands, single platelets circulate freely within the vascular tree that is lined by an intact monolayer of endothelial cells. Thus, the function of integrin $\alpha_{IIb}\beta_3$ is regulated by the balance of actions of positive and negative regulatory factors. During thrombogenesis, the affinity of $\alpha_{IIb}\beta_3$ for macromolecular ligands is dynamically changed [8, 9]. In resting platelets, $\alpha_{IIb}\beta_3$ is in a low-affinity state and does not bind soluble macromolecular ligands. However, after exposure to subendothelial matrix and several mediators such as adenosine 5′-diphosphate (ADP), thromboxane A_2, and thrombin, platelets become activated, and activation signals (inside-out signaling) that induce a high-affinity state of $\alpha_{IIb}\beta_3$ for soluble ligands ($\alpha_{IIb}\beta_3$ activation) are generated. After ligand binding to $\alpha_{IIb}\beta_3$, postligand-binding signals (outside-in signaling) that induce tyrosine phosphorylation and cytoskeletal reorganization are further generated, leading to full expression of $\alpha_{IIb}\beta_3$ function. Molecular characterization of GT due to a dysfunctional $\alpha_{IIb}\beta_3$ (referred as variant GT) provides strong evidence that the cytoplasmic domain of β_3 is involved in inside-out signaling [14, 15]. Indeed, specific binding of the cytoskeletal protein talin to integrin β subunit cytoplasmic tails leads to $\alpha_{IIb}\beta_3$ activation as a final common step in integrin activation [16]. Major advances have been made regarding the structural basis of $\alpha_{IIb}\beta_3$ activation, resulting in the proposal of the "switchblade" model [17]. However, much remains to be elucidated about factors (or molecules) surrounding platelets that positively or negatively regulate $\alpha_{IIb}\beta_3$ function. In this review, we focus on recently identified factors and/or mechanisms that regulate $\alpha_{IIb}\beta_3$ function.

Positive Regulators for $\alpha_{IIb}\beta_3$ Function

In vivo fluorescence microscopy reveals that a few platelets are tethered to the intact vascular wall even under physiological conditions [18]. However, ~100% of these platelets were displaced from the vascular wall without firm arrest. Thus, a threshold for further platelet activation and the initiation of thrombus formation seems to exist, and $\alpha_{IIb}\beta_3$ function should be dynamically controlled by the balance of positive and negative regulators. A number of factors have been identified as a positive regulator for $\alpha_{IIb}\beta_3$ function (Table 1). These factors contribute to stabilize the platelet thrombus as well as initiate thrombus formation. ADP, collagen, and thrombin are classic, well-known factors that initiate thrombus formation by inducing $\alpha_{IIb}\beta_3$ activation. In contrast, serotonin acts as a potentiator, rather than an initiator, for $\alpha_{IIb}\beta_3$ activation. Recently, several factors that contribute to stabilize platelet thrombus have been identified: CD40L and $\alpha_{IIb}\beta_3$, Eph kinases and ephrins, Gas6 and its receptors, and ADP and P2Y$_{12}$ receptor (for review see ref. 19). CD40L, a member of the tumor necrosis factor (TNF) family, is expressed on the platelet surface after platelet activation, and a soluble form of CD40L (sCD40L) is generated by the activation as well. Although CD40 is known to be a receptor for CD40L, the effect of CD40L (and sCD40L) on platelets is mediated by $\alpha_{IIb}\beta_3$ but not by CD40. The interaction of CD40L and $\alpha_{IIb}\beta_3$ contributes to thrombus stability, probably via augmentation of $\alpha_{IIb}\beta_3$-mediated outside-in signaling [20, 21]. Eph kinases and ephrins also augment $\alpha_{IIb}\beta_3$ outside-in signaling [22, 23]. Platelets express the Eph receptor kinase (EphA4 and EphB1) and the Eph kinase ligand, ephrinB1; and blockade of the Eph/Ephrin interactions causes platelet disaggregation induced by low concentrations of ADP and decreased platelet thrombus volume on a collagen-coated surface at high shear rates. Gas6 is a secreted protein localized in α-granules; and its receptors Axl, Sky, and Mer are also expressed

TABLE 1. Regulators for $\alpha_{IIb}\beta_3$ function

Positive regulators
ADP
Collagen
Thrombin
Epinephrine
PAF
Serotonin
CD40L
Eph kinases/ephrins
Gas6
Leptin
Negative regulators
Prostacyclin
Nitric oxide
CD39 (NTPDase1)
PECAM-1
Semaphorin 3A
SHPS-1 (SIRPα1)

on platelets. It has been demonstrated that secreted Gas6 binds to its receptors, leading to the promotion and stabilization of platelet plug formation via $\alpha_{IIb}\beta_3$ outside-in signaling [24]. Thus, these newly identified factors may play a role in the stability of platelet aggregation in vivo. However, recent studies have revealed that the interaction between ADP and its receptor $P2Y_{12}$ play a critical role in the stability of platelet thrombus.

Role of the Interaction Between ADP and $P2Y_{12}$ in the Maintenance of $\alpha_{IIb}\beta_3$ Activation

ADP is stored within platelet dense granules and actively secreted upon platelet activation; approximately 2.5 μmol ADP exists in 10^{11} platelets [25]. Platelets have at least two major G protein-coupled ADP receptors: $P2Y_1$ is a G_q-coupled receptor responsible for mediating platelet shape change and reversible platelet aggregation through intracellular calcium mobilization, whereas $P2Y_{12}$ is a G_i-coupled receptor responsible for mediating the inhibition of adenylyl cyclase and sustained platelet aggregation [26]. $P2Y_{12}$ consists of 342 amino acid residues with seven transmembrane domains. The importance of $P2Y_{12}$ is well documented by the clinical feature of congenital bleeding disorder due to $P2Y_{12}$ deficiency [27–29]. We have identified a Japanese patient with $P2Y_{12}$ deficiency, OSP-1, caused by a point mutation in the translation initiation codon (ATG to AGG) [30]. $P2Y_{12}$-mediated signaling evoked by endogenous ADP plays a major role in platelet aggregation induced by low concentrations of collagen, U46619, and PAR1 TRAP in vitro. We and others have demonstrated impaired thrombus stability under flow conditions [29, 30]. Employing whole blood obtained from OSP-1, real-time analysis of thrombogenesis on a type I collagen-coated surface under a high shear rate ($2000 s^{-1}$) revealed that $P2Y_{12}$ deficiency led to loosely packed thrombus and impaired thrombus growth with enhancing adhesion to collagen. The increase in platelet adhesion to collagen was probably due to the impaired platelet consumption by the growing thrombi. Moreover, our real-time observation indicated that the loosely packed aggregates were unable to resist against high shear stress, and most of the aggregates at the apex of the thrombi came off the thrombi [30]. In a mesenteric artery injury model $P2Y_{12}$-knockout mice also demonstrated the instability of thrombus formation [31]. Thus, the ADP-$P2Y_{12}$ interaction plays a major role in the stability of thrombus.

We assessed the $\alpha_{IIb}\beta_3$ activation on OPS-1 platelets in vitro by the binding of ligand-mimetic monoclonal antibody, PAC-1. Interestingly, $\alpha_{IIb}\beta_3$ activation is markedly impaired by stimulation with PAR1-TRAP, PAR4-TRAP, or U46619 in the absence of $P2Y_{12}$ [30]. On the other hand, PAR1-TRAP and U46619 are able to induce transient aggregation of OSP-1 platelets, indicating that $\alpha_{IIb}\beta_3$ could be transiently activated with these agonists. Based on these findings, we assume that $\alpha_{IIb}\beta_3$ activation may be too short and unstable to be detected by the PAC1 binding assay on OSP-1 platelets and that released ADP and $P2Y_{12}$-mediated signaling may play a critical role in the maintenance of $\alpha_{IIb}\beta_3$ activation. Employing modified ligand-binding assays, we have analyzed the mechanism of sustained $\alpha_{IIb}\beta_3$ activation induced by thrombin. After completion of $\alpha_{IIb}\beta_3$ activation and induction of α-granule secretion, a $P2Y_{12}$ antagonist (AR-C69931MX) was added to the activated platelets [32]. Under these conditions, the stimulated platelets showed long-lasting $\alpha_{IIb}\beta_3$ activation. However, the addition of 1 μM AR-C69931MX at any time tested after thrombin stimulation disrupted the sustained $\alpha_{IIb}\beta_3$ activation without inhibiting CD62P expression (Fig. 2). Neither yohimbine (an adrenergic receptor antagonist), MIC-9042 (a 5-HT$_2$ receptor antagonist), nor SQ-29548 (a thromboxane A$_2$ receptor antagonist) inhibited sustained $\alpha_{IIb}\beta_3$ activation. Dilution of platelet concentrations from 50 000 platelets/μl to 500 platelets/ μl also abolished sustained $\alpha_{IIb}\beta_3$ activation, and disruption of $\alpha_{IIb}\beta_3$ activation by the dilution was abrogated by the addition of small amounts of "exogenous" ADP. Thus, the continuous interaction between secreted ADP with P2Y$_{12}$ is necessary for sustained $\alpha_{IIb}\beta_3$ activation induced by thrombin; and substantial amounts of ADP (= substantial platelets) are needed to maintain $\alpha_{IIb}\beta_3$ activation. The critical role of the interaction between ADP and P2Y$_{12}$ is also evident in the sustained $\alpha_{IIb}\beta_3$ activation induced by U46619 (TXA$_2$ analogue) [32]. Even in the absence of P2Y$_{12}$, platelets can transiently aggregate with each other. However, platelets lacking G$_q$ and G$_{13}$ are completely unresponsive to thrombin, and the activation of G$_i$-mediated signaling alone is not sufficient to induce platelet aggregation [33]. Thus, it is likely that once $\alpha_{IIb}\beta_3$ is activated by G$_q$ and/or G$_{13}$-mediated signaling the ADP-P2Y$_{12}$ may prevent the shift from the activated $\alpha_{IIb}\beta_3$ to the resting $\alpha_{IIb}\beta_3$ (Fig. 2).

Recent in vivo observations demonstrated that during platelet thrombus formation circulating platelets were tethered to the luminal surface of growing thrombi by VWF–GPIb interaction. However, more than 95% of tethered platelets were subsequently translocated and/or detached [18]. Activated $\alpha_{IIb}\beta_3$ on the detached platelets should become inactivated because the released ADP is immediately diluted by the

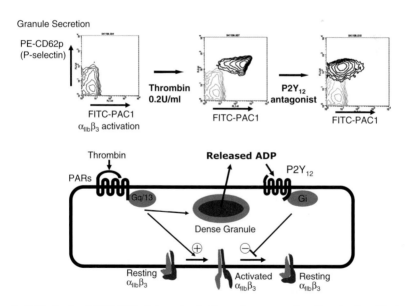

FIG. 2. Critical role of ADP–P2Y$_{12}$ interaction in the maintenance of $\alpha_{IIb}\beta_3$ activation. Blockade of ADP–P2Y$_{12}$ interaction at any time after thrombin stimulation disrupts $\alpha_{IIb}\beta_3$ activation. Once $\alpha_{IIb}\beta_3$ is activated by G$_q$- and/or G$_{13}$-mediated signaling, the ADP–P2Y$_{12}$ may prevent the shift from activated $\alpha_{IIb}\beta_3$ to resting $\alpha_{IIb}\beta_3$. *FITC-PAC1*, fluorescein isothiocyanate-conjugated PAC1

blood flow. At the luminal surface, activated $\alpha_{IIb}\beta_3$ on the tethered platelets would be maintained only when the platelets are continuously exposed to ADP released from adjacent activated platelets. At the inside of growing thrombi, it appears that platelets are constantly exposed to such high concentrations of released ADP that $\alpha_{IIb}\beta_3$ can be maintained in its high-affinity state in concert with the effects of thrombin and TXA_2. It is possible that ADP concentrations surrounding platelets may largely influence whether platelets participate in thrombus formation. Thus, $P2Y_{12}$ may serve as a sensor for thrombogenic status surrounding individual platelets, and the interaction between ADP and $P2Y_{12}$ likely determines thrombus size.

Negative Regulators for $\alpha_{IIb}\beta_3$ Function

Prostacyclin and nitric oxide produced by endothelial cells are well-known negative regulators for the platelet function [34]. In addition to these molecules several negative regulators have been emerged in recent studies (Table 1). We have identified that semaphorin 3A and SHPS-1 act as negative regulators for $\alpha_{IIb}\beta_3$ function [35, 36].

Semaphorin 3A as a Negative Regulator for Platelet Function

The semaphorin family comprises soluble and membrane-bound proteins that are defined by the presence of a conserved 500-amino-acid semaphorin domain at their amino termini. Class 3 semaphorins are secreted disulfide-bound homodimeric molecules; and Sema3A, a prototypic class 3 semaphorin, causes growth cone collapse and provides chemorepulsive guidance for migrating axons. Cell surface receptor for Sema3A consists of a complex of two distinct transmembrane receptors, neuropilin-1 and plexin A (A1-A3). It has been demonstrated that Sema3A is produced by endothelial cells and inhibits integrin function on endothelial cells in an autocrine manner [37]. Employing two distinct Sema3A chimera proteins, we have demonstrated that Sema3A has extensive inhibitory effects on platelet function [35]. Sema3A inhibited agonist-induced $\alpha_{IIb}\beta_3$ activation dose-dependently. Moreover, Sema3A inhibited granular secretion as well as platelet spreading on immobilized fibrinogen. However, Sema3A did not show any effects on the levels of cAMP or cGMP or thrombin-induced increase in intracellular Ca^{2+} concentrations. It is likely that Sema3A inhibits cytoskeletal reorganization in activated platelets as Sema3A inhibits platelet spreading and granule secretion.

Indeed, Sema3A inhibited agonist-induced elevation of filamentous actin (F-actin) contents and Rac1 activation. Rac1 activation is necessary for platelet actin assembly and lamellipodia formation after agonist stimulation. Therefore, marked impairment of Rac1 activation is likely to account for the Sema3A-induced impairment of actin rearrangement and spreading in platelets. There were two major downstream effectors of Rac1 identified: PAK and WAVEs [Wiskott-Aldrich syndrome protein (WASP) family verprolin-homologous proteins]. Several PAK substrates or binding partners have been implicated in the effects of PAK, including filamin, LIM kinase, myosin, and paxillin. Among them, LIM kinase phosphorylates and inactivates cofilin, a protein that promotes severing and depolymerization of F actin. Consistent with the inhibition of Rac1 activation, Sema3A inhibited phosphorylation of cofilin in both resting

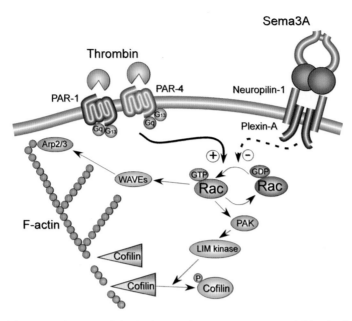

FIG. 3. Inhibitory mechanisms of platelet function by Sema 3A. Sema 3A inhibits platelet spreading and granular secretion as well as $\alpha_{IIb}\beta_3$ activation. The inhibitory effects are mediated in part by the inhibition of agonist-induced Rac1 activation and phosphorylation of cofilin. This inhibition leads to the inhibition of F-actin elevation and cytoskeleton rearrangement

and activated platelets, suggesting that Sema3A increases severing and depolymerization of F-actin by keeping cofilin in the activated state (Fig. 3). In addition to Rac1 inactivation, our recent data showed that Sema3A inhibited the PI3 kinase pathway, including Rap1B, which may account for the inhibition of $\alpha_{IIb}\beta_3$ activation (unpublished data).

SHPS-1 as a Negative Regulator for Platelet Function

SHPS-1 (Src homology 2 domain-containing protein tyrosine phosphatase substrate-1), also known as signal regulatory protein α1 (SIRP α1), is a membrane glycoprotein with three extracellular immunoglobulin (Ig)-like domains, a single transmembrane domain, and an intracellular domain containing two immunoreceptor tyrosine-based inhibitory motifs (ITIM) and expressed on endothelial cells and leukocytes. CD47 (integrin-associated protein, or IAP) is a ubiquitously expressed 50-kDa membrane glycoprotein with an extracellular Ig domain, five membrane-spanning domains, and a short cytoplasmic tail. CD47 physically associates with $\alpha_{IIb}\beta_3$, $\alpha_v\beta_3$, and $\alpha_2\beta_1$ and modulates a variety of cell functions [38]. Two ligands are known to bind to CD47: thrombospondin-1 (TSP-1) and SHPS-1. The TSP-1–CD47 interaction has been believed to augment integrin-mediated platelet function. On the other hand, SHPS-1–Ig, a fusion protein consisting of the extracellular domain of SHPS-1 and human Ig Fc domain, impaired secondary platelet aggregation induced by a low concentration of ADP (2.5 μM). Moreover, SHPS-1–Ig markedly impaired $\alpha_{IIb}\beta_3$-mediated platelet spreading onto immobilized fibrinogen. The inhibition of platelet spreading is CD47-specific because it was not observed in CD47-deficient (CD47$^{-/-}$) murine platelets. Of particular interest is that SHPS-1 inhibits $\alpha_{IIb}\beta_3$-mediated platelet spreading without disturbing Syk and FAK tyrosine phosphorylation. SHPS-1 did inhibit tyrosin phosphorylation of α-actinin, a downstream effector of FAK. Thus, SHPS-1 negatively regulates platelet function through CD47, especially $\alpha_{IIb}\beta_3$-mediated outside-in signaling, by interfering with the downstream pathway of FAK.

Conclusion

Thrombogenesis is a complex process regulated by the balance of positive and negative regulatory proteins (or molecules). Further investigations of these regulatory molecules would provide a new insight into the more effective prevention of pathological thrombosis.

Acknowledgments. This work was supported in part by a Grant-in-Aid for Scientific Research from the Ministry of Education, Culture, Sports, Science, and Technology in Japan; the Ministry of Health, Labor, and Welfare in Japan; and the "Academic Frontier" Project in Japan; and Mitsubishi Pharma Research Foundation.

References

1. Fuster V, Badimon L, Badimon JJ, et al (1992) The pathogenesis of coronary artery disease and the acute coronary syndromes. N Engl J Med 326:242–250
2. Antithrombotic Trialists' Collaboration (2002) Collaborative meta-analysis of randomised trials of antiplatelet therapy for prevention of death, myocardial infarction, and stroke in high risk patients. BMJ 324:71–86
3. Massberg S, Brand K, Grüner S, et al (2002) A critical role of platelet adhesion in the initiation of atherosclerotic lesion formation. J Exp Med 196:887–896
4. Massberg S, Konrad I, Schürzinger K, et al (2006) Platelets secrete stromal cell-derived factor 1α and recruit bone marrow-derived progenitor cells to arterial thrombi in vivo. J Exp Med 203:1221–1233
5. Tomiyama Y, Shiraga M, Shattil SJ (2002) Platelet membrane proteins as adhesion receptors. In: Gresele P, Page C, Fuster V, et al (eds) Platelets in thrombotic and nonthrombotic disorders: pathophysiology, pharmacology and therapeutics. Cambridge University Press, Cambridge, UK, pp 80–92
6. Savage B, Almus-Jacobs F, Ruggeri ZM (1998) Specific synergy of multiple substrate-receptor interactions in platelet thrombus formation under flow. Cell 94:657–666
7. Hynes RO (2002) Integrins: bidirectional, allosteric signaling machines. Cell 110:673–687
8. Phillips DR, Charo IF, Scarborough RM (1991) GPIIb-IIIa: the responsive integrin. Cell 65:359–362
9. Shattil SJ, Newman PJ (2004) Integrins: dynamic scaffolds for adhesion and signaling in platelets. Blood 104:1606–1615
10. George JN, Caen JP, Nurden AT (1990) Glanzmann's thrombasthenia: the spectrum of clinical disease. Blood 75:1383–1395
11. Tomiyama Y (2000) Glanzmann thrombasthenia: integrin $\alpha_{IIb}\beta_3$ deficiency. Int J Hematol 72:448–454

12. Xiong JP, Stehle T, Diefenbach B, et al (2001) Crystal structure of the extracellular segment of integrin $\alpha_v\beta_3$. Science 294:339–345

13. Xiao T, Takagi J, Coller BS, et al (2004) Structural basis for allostery in integrins and binding to fibrinogen-mimetic therapeutics. Nature 432:59–67

14. Chen Y-P, Djaffar I, Pidard D, et al (1992) Ser-752 → Pro mutation in the cytoplasmic domain of integrin β_3 subunit and defective activation of platelet integrin $\alpha_{IIb}\beta_3$ (glycoprotein IIb-IIIa) in a variant of Glanzmann thrombasthenia. Proc Natl Acad Sci U S A 89:10169–10173

15. Wang R, Shattil SJ, Ambruso DR, et al (1997) Truncation of the cytoplasmic domain of β_3 in a variant form of Glanzmann thrombasthenia abrogates signaling through the integrin $\alpha_{IIb}\beta_3$. J Clin Invest 100:2393–2403

16. Tadokoro S, Shattil SJ, Eto K, et al (2003) Talin binding to integrin beta tails: a final common step in integrin activation. Science 302:103–106

17. Takagi J, Petre BM, Walz T, et al (2002) Global conformational rearrangements in integrin extracellular domains in outside-in and inside-out signaling. Cell 110:599–611

18. Massberg S, Gawaz M, Gruner S, et al (2003) A crucial role of glycoprotein VI for platelet recruitment to the injured arterial wall in vivo. J Exp Med 197:41–49

19. Brass LF, Zhu L, Stalker TJ (2005) Minding the gaps to promote thrombus growth and stability. J Clin Invest 115:3385–3392

20. André P, Prasad KS, Denis CV, et al (2002) CD40L stabilizes arterial thrombi by a β_3 integrin-dependent mechanism. Nat Med 8:247–252

21. Prasad KS, André P, He M, et al (2003) Soluble CD40 ligand induces β_3 integrin tyrosine phosphorylation and triggers platelet activation by outside-in signaling. Proc Natl Acad Sci U S A 100:12367–12371

22. Prévost N, Woulfe D, Tanaka T, et al (2002) Interactions between Eph kinases and ephrins provide a mechanism to support platelet aggregation once cell-to-cell contact has occurred. Proc Natl Acad Sci U S A 99:9219–9224

23. Prévost N, Woulfe DS, Jiang H (2005) Eph kinases and ephrins support thrombus growth and stability by regulating integrin outside-in signaling in platelets. Proc Natl Acad Sci USA 102:9820–9825

24. Angelillo-Scherrer A, de Frutos P, Aparicio C, et al (2001) Deficiency or inhibition of Gas6 causes platelet dysfunction and protects mice against thrombosis. Nat Med 7:215–221

25. D'Souza L, Glueck HI (1997) Measurement of nucleotide pools in platelets using high pressure liquid chromatography. Thromb Haemost 38:990–1001

26. Gachet C (2001) ADP receptors of platelets and their inhibition. Thromb Haemost 86:222–232

27. Cattaneo M, Lecchi A, Randi AM, et al (1992) Identification of a new congenital defect of platelet function characterized by severe impairment of platelet responses to adenosine diphosphate. Blood 80:2787–2796

28. Nurden P, Savi P, Heilmann E, et al (1995) An inherited bleeding disorder linked to a defective interaction between ADP and its receptor on platelets: its influence on glycoprotein IIb-IIIa complex function. J Clin Invest 95:1612–1622

29. Remijn JA, Wu YP, Jeninga EH, et al (2002) Role of ADP receptor P2Y$_{12}$ in platelet adhesion and thrombus formation in flowing blood. Arterioscler Thromb Vasc Biol 22:686–691

30. Shiraga M, Miyata S, Kato H, et al (2005) Impaired platelet function in a patient with P2Y$_{12}$ deficiency caused by a mutation in the translation initiation codon. J Thromb Haemost 3:2315–2323

31. Andre P, Delaney SM, LaRocca T, et al (2003) P2Y$_{12}$ regulates platelet adhesion/activation, thrombus growth, and thrombus stability in injured arteries. J Clin Invest 112:398–406

32. Kamae T, Shiraga M, Kashiwagi H, et al (2006) Critical role of ADP interaction with P2Y$_{12}$ receptor in the maintenance of $\alpha_{IIb}\beta_3$ activation: association with Rap1B activation. J Thromb Haemost 4:1379–1387

33. Moers A, Wettschureck N, Grüner S, et al (2004) Unresponsiveness of platelets lacking both Gα_q and Gα_{13}: implications for collagen-induced platelet activation. J Biol Chem 279:45354–45359

34. Wu KK, Thiagarajan P (1996) Role of endothelium in thrombosis and hemostasis. Annu Rev Med 47:315–331

35. Kashiwagi H, Shiraga M, Kato H, et al (2005) Negative regulation of platelet function by a secreted cell repulsive protein, semaphorin 3A. Blood 106:913–921

36. Kato H, Honda S, Yoshida H, et al (2005) SHPS-1 negatively regulates integrin $\alpha_{IIb}\beta_3$ function through CD47 without disturbing FAK phosphorylation. J Thromb Haemost 3:763–774

37. Serini G, Valdembri D, Zanivan S, et al (2003) Class 3 semaphorins control vascular morphogenesis by inhibiting integrin function. Nature 424:391–397

38. Brown EJ, Frazier WA (2001) Integrin-associated protein (CD47) and its ligands. Trends in Cell Biol 11:130–135

GPIb-Related Signaling Pathways in Platelets

Yukio Ozaki

Summary. An accumulating body of evidence suggests that tyrosine kinases and related signaling molecules constitute an essential signal transduction pathway related to the interaction between platelet GPIb-IX-V complex (GPIb) and von Willebrand factor. Src family kinases, Lyn and Src, Syk, PLCγ2 activation, and Ca^{2+} mobilization appear to be the mainstream sequential signaling events, whereas PLA_2 with subsequent thromboxane A_2 production and FcRγ-chain/FcγRIIA enhance platelet activation related to GPIb. Although PI-3K serves to potentiate various signaling events culminating in $\alpha_{IIb}\beta_3$ activation, PI-3K activity may be dispensable for Src-PLCγ2 activation. Glycosphingolipid-enriched microdomains (GEMs) appear to provide platforms for the signal transduction pathway related to GIb-IX-V as the interaction between GPIb-IX-V and Src occurs exclusively in GEMs.

Key words. Platelets · GPIb-IX-V · Tyrosine kinases · PLCγ2 · GEM

Introduction

Platelets play an important role in the physiological process of hemostasis and are also closely involved in pathological thrombus formation. At sites of vessel injury, the initial event, particularly under the conditions of high shear stress, is the interaction between plasma von Willebrand factor (vWF) and collagen, one of the major components of the subendothelial matrices (SEMs), which are exposed upon endothelial cell detachment. The changes in vWF configuration induced by the interaction with collagen, or upon exposition to high shear rates, gives vWF the capacity to bind the glycoprotein GPIb-IX-V complex on platelet membranes [1]. Other components in SEM, such as laminin, appear to play a role similar to that of collagen in terms of mediating vWF–platelet interaction [2]. Because the interaction between vWF and platelet GPIb-IX-V complex is a process with rapid on/off rates, platelets rolls on vWF-coated surfaces for a short period of time under high shear conditions and are eventually released again into the circulation (platelet tethering). However short the time frame of platelet adhesion to vWF is, it allows platelets to interact with various components of the SEMs through adhesive receptors on platelet membranes. The collagen receptors on platelet membranes, integrin $\alpha_2\beta_1$ and glycoprotein VI (GPVI), interact with collagen exposed at sites of endothelial cell damage. It is also likely that other components in SEMs (e.g., laminin, vitronectin) contribute to platelet adhesion under

253

certain conditions [2]. Adherent platelets are activated by intracellular signaling pathways elicited by receptor–ligand interactions, and resultant activation of integrin $\alpha_{IIb}\beta_3$ on platelet membranes leads to platelet aggregation by its interaction with vWF [3, 4] or with fibrinogen [5]. Although it has been long recognized that the interaction between collagen receptors and collagen can elicit intracellular activation signals that finally culminate in integrin $\alpha_{IIb}\beta_3$ activation [6], the role of GPIb-IX-V for intracellular signaling and integrin $\alpha_{IIb}\beta_3$ activation has remained controversial for a relatively long time. This is due to the fact that platelets fixed with paraformaldehyde can form platelet aggregates when mixed with vWF and vWF-modulating agents such as ristocetin or botrocetin [7], and that the rapid on/off rate between GPIb-IX-V and VWF may not allow enough time for initiation of efficient activation signaling [5, 8]. Thus, it was assumed that the GPIb-IX-V–VWF interaction only provides physical force that fixes platelets to the SEMs, thereby allowing enough time for the interaction of collagen in SEMs and the collagen receptors to elicit intracellular activation signals. A series of recent reports have corrected this concept, and now it is evident that GPIb-IX-V mediates intracellular signaling that leads to full activation and aggregate formation of platelets with integrin $\alpha_{IIb}\beta_3$ activation.

Intracellular Signals Investigated by Early Research on GPIb-Related Signal Transduction

Investigation on the signal transduction pathways and signaling molecules related to the GPIb-IX-V–VWF interaction started during the late 1980s [9]. Because phospholipase C with its two products—diacylglycerol with protein kinase C (PKC) activation and inositol trisphosphate with Ca^{2+} release—constituted the mainstream of signaling pathways in various cells at that time, the involvement of this pathway was also intensively investigated in GPIb-related signal transduction. PKC activation and a rise in intracellular Ca^{2+} concentration ($[Ca^{2+}]i$) was observed with platelets treated with ristocetin plus vWF, which suggested that phospholipase C (PLC) activation indeed occurs during GPIb-mediated platelet activation [9]. The signaling molecules first explored, which lie upstream of PLC, was thromboxane A_2 (TXA_2), a metabolite of arachidonic acid with potent ability to activate platelets. Cyclooxygenase inhibitors, which block conversion of arachidonic acid to TXA_2, impair a wide variety of platelet reactions elicited by the GPIb-IX-V–vWF interaction; and production of TXB_2, a stable metabolite of TXA_2, upon GPIb-IX-V stimulation has been documented in a few reports [10]. Activation of p38MAPK and phospholipase A_2 (PLA_2), which lead to TXA_2 production, has been also documented upon GPIb-IX-V–vWF interaction [11]. Thus, it was suggested that PLA_2 lies mostly upstream, proximal to GPIb-IX-V, followed by TXA_2, with resultant activation of PLC, which then induces PKC activation and a rise in $[Ca^{2+}]i$.

PLA_2 Activation

One may then ask what leads to PLA_2 activation downstream of the GPIb-IX-V–vWF interaction. The cGMP–PKG pathway and MAP kinase were suggested to play an important stimulatory role in PLA_2 activation mediated by GPIb-IX-V [12].

Expression of recombinant cGMP-dependent protein kinase (PKG) in a cell model enhanced vWF-induced activation of $\alpha_{IIb}\beta_3$ with MAP kinase activation. MAP kinase inhibitors blocked PLA$_2$ and $\alpha_{IIb}\beta_3$ activation [13]. This issue is intriguing as it is generally accepted that an elevated level of cGMP and PKG inhibits platelet activation [14]. There is a report that contradicts these findings in that the vWF–GPIb-IX-V interaction neither increases cGMP levels in platelets nor induces MAP kinase activation under high shear conditions, which involve vWF–GPIb-IX-V interaction [15]. Although there have been several contradictory publications on this issue, recent reports support the idea that the MAP kinase pathway indeed lies upstream of PLA$_2$ activation but that cGMP-PKG is not involved in this process. Src family kinases and Syk, to be explained in detail later, appear to lie upstream of PLA$_2$ [16, 17].

Thus, there is a large body of evidence to suggest that PLA$_2$ activation with resultant production of TXA$_2$ plays an important role in platelet activation mediated by GPIb-IX-V. For instance, $\alpha_{IIb}\beta_3$ activation and platelet aggregation, following platelet agglutination induced by botrocetin plus vWF, is dependent on TXA$_2$ production [18]. On the other hand, a number of reports evaluating platelet activation and aggregation induced by shear stress observed that the shear-induced platelet response is insensitive to aspirin, suggesting that TXA$_2$ production is irrelevant to GPIb-mediated activation of $\alpha_{IIb}\beta_3$ and subsequent platelet aggregation under shear stress [19–21]. Thus, it is likely that the mode of GPIb-IX-V-related signaling appears to be distinct between shear-induced and agglutination (botrocetin–vWF)-induced platelet activation, and that GPIb-IX-V-mediated platelet activation involves TXA$_2$-dependent and TXA$_2$-independent signaling pathways.

Ca^{2+} Mobilization

Mobilization of Ca^{2+} has been one of the most controversial issues with GPIb-IX-V-related platelet activation. One of the earliest reports observed an increase in intracellular Ca^{2+} in platelets treated with vWF and ristocetin [9]. During subsequent years, a number of studies supported the idea that intracellular Ca^{2+} mobilization does occur upon GPIb-IX-V–vWF interaction. A considerable number of studies have suggested that intracellular Ca^{2+} mobilization mediated by GPIb-IX-V is attributed to Ca^{2+} influx rather than to Ca^{2+} release from intracellular Ca^{2+} stores [3, 19, 22]. Some reported Ca^{2+} release from intracellular Ca^{2+} stores, which suggests phospholipase C (PLC) activation [23]. On the other hand, a considerable number of articles failed to detect any Ca^{2+} mobilization [24]. Attempts to measure inositol trisphosphate, a PLC product, directly produced contradictory results [9, 25, 26].

Regardless of whether these discrepancies could be ascribed to the different techniques or agents used in these studies, it could be concluded that GPIb-IX-V–vWF interaction would induce only a weak level of Ca^{2+} mobilization; moreover, if it indeed did activate PLC with inositol trisphosphate production and Ca^{2+} release from intracellular Ca^{2+} stores, it should be at a level far less than those of G protein-coupled receptors or collagen. Ca^{2+} release from intracellular Ca^{2+} induced by the GPIb-IX-V–vWF interaction was finally established by microscopic analysis of single-platelet Ca^{2+} oscillation profiles, which clearly showed that at least a portion of Ca^{2+} mobilization mediated by GPIb-IX-V is attributed to intracellular Ca^{2+} release [26, 27]. This premise has been further substantiated by a recent article that demonstrated Ca^{2+} release induced by dimeric vWF A1 domain in platelets from human GPIb-transgenic mice [28].

Release of Ca^{2+} from intracellular Ca^{2+} storage sites is mediated by inositol triphosphate, a product of PLC. Although agonists such as ADP and thromboxane A_2, which bind to seven-transmembrane receptors, activate members of the PLCβ subfamily, there has been an accumulating body of evidence to suggest that PLCγ2 (instead of PLCβ) is activated in GPIb-IX-V-related signaling. Shape change on vWF-coated surfaces occurs normally with $G\alpha_q$-deficient mice, excluding a role of PLCβ in this process [26]. In a wide variety of cell types, PLCγ2 activity is regulated by its association with tyrosine kinases and subsequent tyrosine phosphorylation of PLCγ2 [29]. The GPIb-IX-V–vWF interaction in platelets indeed induces a considerable level of PLCγ2 tyrosine phosphorylation [21, 24–26]. Furthermore, Ca^{2+} mobilization mediated by GPIb-IX-V is significantly reduced in mice lacking PLCγ2 [26]. However, a residual level of Ca^{2+} mobilization in PLCγ2$^{-/-}$ knockout mice suggests that other isoforms of PLC may also be involved. In this context, PLCγ1 has been reported to play a role in GPVI-activated PLCγ2$^{-/-}$ knockout platelets [30]. It is also of interest that inositol trisphosphate production and Ca^{2+} mobilization in GPIb-mediated platelets activation is only minimal, in great contrast to considerable levels of PLCγ2 tyrosine phosphorylation, which presumably represents its activity. In this respect, it has been reported that tyrosine phosphorylation sites of PLCγ2 induced by GPVI is distinct from that of PLCγ2 induced by GPIb [31]; virtually equal levels of PLCγ2 tyrosine phosphorylation between GPIb-mediated or GPVI-mediated platelet activation reported in a few articles may be attributed to the use of anti-phosphotyrosine antibodies that cannot differentiate specific phosphotyrosine residues.

However limited it may be, PLCγ activation, subsequent Ca^{2+} release, and oscillation constitute an essential signal transduction pathway related to GPIb-IX-V as platelet responses to VWF including filopodia formation are almost completely abrogated in PLCγ2$^{-/-}$ platelets or chelation of intracellular Ca^{2+} by BAPTA [26].

Src Family Tyrosine Kinases and Signaling Molecules Related to Tyrosine Phosphorylation

The main stream of the signal transduction pathways related to the interaction between GPIb-IX-V and vWF is now considered to be the signaling events related to tyrosine phosphorylation, and one of the signaling molecules most proximal to GPIb-IX-V is the Src family kinases. Initial reports suggesting a role for signaling molecules related to tyrosine phosphorylation date back to 1994 and 1995 when the cytoskeletal association of Src and the appearance of multiple tyrosine-phosphorylated proteins was observed in GPIb-IX-V-mediated platelet activation [19, 32, 33]. Later, Syk, another tyrosine kinase, and shc, an adaptor protein, were reported to be tyrosine phosphorylated; and tyrosine kinase activity, though not identified, was associated with the GPIb-IX-V complex upon vWF stimulation [20]. In 1999, it was reported that a snake venom, alboaggregin A, which presumably interacted with the GPIb-IX-V complex, induced tyrosine phosphorylation of FcRγ-chain, Syk activation, PLCγ2 tyrosine phosphorylation, and complex formation between GPIb and two Src family tyrosine kinases, Lyn and Fyn [34]. This report was a great surprise to the investigators involved in the GPIb-IX-V-mediated signaling pathways as the proposed model of signal transduction was exactly the same as that of the collagen receptor, GPVI. GPVI-mediated platelet activation involves the sequential activation of signaling molecules, Src family kinases, Lyn and Fyn, FcRγ-chain, Syk, and PLCγ2 [35]. However, it was later found that alboaggregin A also interacts with GPVI, and the signal transduction pathway characteristic of the GPIb-IX-V remained to be determined [36].

In 2001, using the combination of vWF and a vWF modulator, botrocetin, which is known to react with GPIb-IX-V but not with GPVI, it was found that GPIb-IX-V-mediated platelet activation induces tyrosine phosphorylation of FcRγ-chain, Syk, LAT, and PLCγ2 [25]. Src kinase inhibition markedly suppressed these events; and the Src kinases Src and Lyn formed a complex with FcRγ-chain and Syk upon GPIb-IX-V–vWF interaction, suggesting an important role of Src kinases in these processes [37]. It was also reported that FcγRIIA, another ITAM-containing molecule, undergoes tyrosine phosphorylation upon platelet activation induced by the addition of vWF and ristocetin, followed by Syk and PLCγ2 activation [38]. A selective Src kinase inhibitor PP1 severely abrogated these events. One of the most proximal signaling molecules downstream of GPIb-IX-V is suggested to be Src [37]. The p85 subunit of PI-3K constitutively associates with GPIb-IX-V, and this binding is not affected by PI-3K inhibitors [39]. Upon platelet activation with vWF–GPIb-IX-V interaction, Src with its SH3 domain binds GPIb-associated p85, the regulatory subunit of PI-3K [37]. The role of Src kinases and their downstream signaling molecule PLCγ2 was also confirmed with a number of platelet responses, including spreading and Ca^{2+} mobilization on vWF-coated surfaces [15, 16, 24, 26]. Although there has been an accumulating body of evidence in addition to the studies described above to suggest that tyrosine kinases, Syk, Src family kinases, and PLCγ2 are involved in GPIb-IX-V-mediated platelet activation, there remained some room for criticism that vWF modulators such as botrocetin or ristocetin might interact with certain membrane molecules or that vWF through its C1 domain might interact with $\alpha_{IIb}\beta_3$, and the outside-in signals elicited by $\alpha_{IIb}\beta_3$ might confound the analysis of GPIb-IX-V-mediated signaling. Utilizing dimeric A1 domains of vWF and human GPIbα transgenic mice, it was confirmed that GPIb-IX-V itself can indeed signal to activate $\alpha_{IIb}\beta_3$ through sequential actions of Src kinases and Ca^{2+} oscillation, a marker of PLC activation [28]. Another study using several lines of knockout mice has also demonstrated that platelet activation induced by vWF–GPIb-IX-V interaction is dependent upon Lyn, enhanced by Src, and propagated through Syk, SLP-76, PI-3K, and PLCγ2 [40]. More recently, the involvement of Bruton tyrosine kinase and ADAP (SLP-130) has also been demonstrated in GPIb-IX-V-related signal transduction, especially for $\alpha_{IIb}\beta_3$ activation and TXA$_2$ production [41, 42].

On the whole, it can be concluded that there are striking similarities in signal transduction pathways between GPIb-IX-V and GPVI except for several points: Src and Lyn appear to be recruited to GPIb-IX-V upon platelet activation, whereas Lyn and Fyn constitutively associate with GPVI. GPVI activation induces a robust level of inositol phosphate production and PLCγ2 activity, whereas with GPIb-IX-V activation PLCγ2 activation is only modest and the tyrosine phosphorylation sites of PLCγ2 are distinct from those of GPVI stimulation [31].

FcRγ Chain and FcγRIIA

GPIb-IX-V-mediated activation of platelets leads to tyrosine phosphorylation of two ITAM-containing molecules, FcRγ-chain and FcγRIIA [21, 25]. FcRγ-chain forms a complex with Syk, and GPIb-IX-V and FcRγ-chain are co-precipitated with Brij 35 lysates of platelets, suggesting a functional link between GPIb-IX-V. Some of the activation signals are attenuated in FcRγ-chain knockout mice [25, 28]. A physical proximity of less than 10 nm between GPIb-IX-V and FcγRIIA was also shown by fluorescent energy transfer; and indeed they may associate on the platelet membrane, based on the results of a two-hybrid system [43, 44]. On the other hand, normal shape change and Ca^{2+} mobilization was observed in platelets treated with anti-FcγRIIA antibodies or in FcRγ-chain-deficient platelets [26]. Another study observed only slight reduction in Ca^{2+} oscillation and $\alpha_{IIb}\beta_3$ activation in FcRγ-chain-deficient mice, while confirming tyrosine phosphorylation of FcRγ-chain upon GPIb-IX-V stimulation [28]. However, a more recent study has demonstrated an important role of FcRγ-chain for granule release in platelet activation mediated by GPIb-IX-V [40] These studies taken together suggest that FcRγ-chain and/or FcγRIIA is dispensable for GPIb-IX-V-mediated signal transduction, although it may have a limited potentiating effect on downstream signals.

PI-3K

There is a large body of evidence to suggest that PI-3K is involved in GPIb-IX-V-mediated platelet activation. PI-3K is activated by vWF in the presence of ristocetin or in platelets adhering to vWF-coated surfaces [45]. PI-3K activity is also increased by high shear stress, as assessed by PIP3 production [46]. PI-3K inhibition leads to decreased platelet spreading and aggregate formation under flow conditions [45, 47]. On the other hand, there are several cellular events unaffected by the PI-3K inhibitors wortmannin or LY294002. Irrespective of shear stress, filopodia formation and Ca^{2+} spikes are insensitive to PI-3K inhibition [45, 48]. Under static conditions, wortmannin did not inhibit tyrosine phosphorylation of Src and PLCγ2 tyrosine phosphorylation [37] or Ca^{2+} oscillations [28]. Thus, at least under static conditions, the GPIb-IX-V-mediated signal transduction pathway sequentially involving Src, PLCγ2 activation, and Ca^{2+} oscillations is unaffected by the PI-3K activity, and platelet aggregation and spreading on vWF-coated surfaces supported by $\alpha_{IIb}\beta_3$ is dependent on PI-3K. Under high shear stress, the roles of PI-3K are somewhat at variance, probably due to the various degrees of involvement of $\alpha_{IIb}\beta_3$ outside-in signaling in experimental settings. By activating Src and Syk, $\alpha_{IIb}\beta_3$ leads to PLCγ2 activation, and the resulting Ca^{2+} mobilization along with secondary mediators such as ADP and thromboxane A_2 potentiate PI-3K activity, which can then regulate PLCγ activity [48, 49]. Recent studies using knockout mice demonstrate an essential role of Syk for PI-3K activation [40].

PI-3K constitutively associates with GPIb-IX-V, and this binding is not affected by PI-3K inhibitors [39]. It is suggested that Src binds to GPIb-IX-V upon vWF–GPIb-IX-V interaction [37, 50–52]. Because GPIb-IX-V has no direct binding sites for Src, it appears that the interaction between Src SH3 and the proline-rich domain of PI-3K mediates indirect binding of Src to GPIb-IX-V.

Glycosphingolipid-Enriched Microdomains

Lipids and proteins on cell membranes are unequally distributed and form distinct microdomains with specific lipid and protein components. Glycolipid-enriched microdomains (GEMs), also known as rafts, which are rich in glycosphingolipids, saturated phospholipids, and cholesterol, have been identified in many cell types. Molecules present in GEMs have diffusion velocities much lower than those present in non-GEM areas of cell membranes, and thus GEMs appear to provide organized milieu on cell membranes that otherwise are chaotic [49]. GEMs appear to act as platforms for signal transduction and ligand localization, selectively recruiting a certain set of signaling molecules while excluding others [50–52].

There is an increasing body of evidence to suggest that GEMs also have functional roles in platelets. Phosphatidylinositol 3,4,5-triphosphate is produced in platelet GEMs [53]. GEMs accumulate at the extended tips of the formed filopodia, and this concentration process of GEMs is accompanied by the simultaneous enrichment of Src and the tetraspanin CD63 [54]. GPVI co-localizes with FcRγ-chain in GEMs, and destruction GEMs in platelets leads to a lower response to GPVI agonists [55]. With regard to GPIb-IX-V, it appears that a minor portion of GPIb-IX-V molecules (8%) on platelet membranes reside in GEMs in the resting state, and this portion increases three- to six-fold with platelet activation by vWF–GPIb-IX-V interaction [56].

The vWF/GPIb-IX-V interaction involves a set of signaling molecules, Src, FcRγ-chain, Syk, PLCγ2, and PI-3K [25, 37]. It is of interest to investigate whether these signaling molecules play their roles in GEMs. The association between GPIb-IX-V and PI-3K occurs constitutively regardless of their localization [37]. However, Src association with GPIb-IX-V induced by VWF–GPIb-IX-V interaction is restricted to GEMs, and this recruitment of Src is confined to the activated form of Src, with its 416-tyrosine residue phosphorylated. Syk and PLCγ2 are present in both GEMs and non-GEMs, However, PLCγ2 in GEMs is more intensively tyrosine-phosphorylated by vWF–GPIb-IX-V interaction, and their tyrosine phosphorylation is marked suppressed by methyl-β-cyclodextrin treatment of platelets, which disrupts GEMs [50–52]. Thus, the association between GPIb-IX-V and active Src, tyrosine phosphorylation of Syk, and PLCγ2 tyrosine phosphorylation all occur in GEMs and are dependent on the intact structure of GEMs. These findings suggest that GEMs indeed provide platforms for the signal transduction pathway related to GIb-IX-V.

14-3-3ζ

14-3-3ζ belongs to a family of proteins involved in regulation of a diverse number of intracellular signaling proteins through their interaction with serine-phosphorylated signaling molecules [57]. A wide variety of proteins, including Raf-1 kinase, Bad, and PI-3K, associate with 14-3-3ζ [58]. GPIb-IX-V has several specific binding sites for 14-3-3ζ, and phosphorylation of these sites ensures constitutive association between 14-3-3ζ and GPIb-IX-V [59]. It has been reported that GPIb-IX-V-mediated activation of $\alpha_{IIb}\beta_3$ requires 14-3-3ζ binding to the cytoplasmic domain of GPIbα [60]. The heterotrimeric complex of GPIb-IX-V, 14-3-3ζ, and p85 subunit of PI-3K is present in resting platelets [39]. Because GPIb-IX-V has no apparent binding sites for PI-3K, it is most likely that p85 PI-3K binds to GPIb-IX-V via 14-3-3ζ. Based on the findings that Src associates with p85 PI-3K bound to GPIb-IX-V upon vWF–GPIb-IX-V interaction [39] and that the downstream signaling pathways of Src and PLCγ2 activate $\alpha_{IIb}\beta_3$, the requirement for 14-3-3ζ in GPIb-IX-V-mediated $\alpha_{IIb}\beta_3$ activation, as shown in a previous paper, may be explained as its adaptor role for sequentially binding p85, PI-3K, and then Src to GPIb-IX-V.

In addition to its role as an adaptor protein to recruit PI-3K and Src to GPIb-IX-V, another functional role has been recently suggested for 14-3-3ζ. Shear stress induces dissociation of 14-3-3ζ from GPIb-IX-V, concomitant with dephosphorylation of 14-3-3ζ-binding sites of GPIb-IX-V [59]. Released 14-3-3ζ somehow activates Rac and Cdc42, which regulate cytoskeletal reorganization during integrin-dependent cell activation [61]. It is conceivable that the dephosphorylation process of GPIb-IX-V and release of 14-3-3ζ appears not to be required during the initial stage of platelet adhesion to a vWF matrix, but may be required for the phase of platelet spreading supported by $\alpha_{IIb}\beta_3$ activation. Thus, the role of 14-3-3ζ in GPIb-IX-V-mediated platelet activation needs to be considered in a dynamic mode, with its role changing at different stages of platelet activation.

Conclusion

Although the signaling pathways related to GPIb-IX-V have not been fully elucidated, an accumulating body of evidence suggests that Src family kinase- and PLCγ2-related signaling plays an important role in GPIb-IX-V-mediated platelet activation. Figure 1 illustrates the hypothetical signal transduction pathway mediated by GPIb-IX-V.

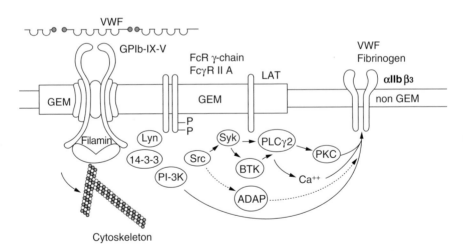

FIG. 1. Hypothetical signal transduction pathway mediated by glycoprotein (*GP*)*Ib-IX-V*, leading to $\alpha_{IIb}\beta_3$ activation. GPIb-IX-V constitutively associates with p85 subunit of PI-3K via 14-3-3ζ. The interaction between GPIb-IX-V and von Willebrand factor (*VWF*) induces the binding between P85 PI-3K and Src, which then elicits downstream signals leading to PLCγ2 activation. These processes appear to take place predominantly in glycosphingolipid-enriched microdomains (*GEM*). *LAT*, linker for activation of T cell; *ADAP*, adhesion and degranulation adaptor protein; *PKC*, protein kenase C

References

1. Wu YP, Vink T, Schiphorst M, et al (2000) Platelet thrombus formation on collagen at high shear rates is mediated by von Willebrand factor-glycoprotein Ib interaction and inhibited by von Willebrand factor-glycoprotein IIb/IIIa interaction. Arterioscler Thrombo Vasc Biol 20:1661–1667
2. Inoue O, Suzuki-Inoue K, McCarty OJ, et al (2006) Laminin stimulates spreading of platelets through integrin alpha6beta1-dependent activation of GPVI. Blood 107:1405–1412
3. Ikeda Y, Handa M, Kamata T, et al (1993) Transmembrane calcium influx associated with von Willebrand factor binding to GP Ib in the initiation of shear-induced platelet aggregation. Thromb Haemost 69:496–502
4. Konstantopoulos K, Chow TW, Turner NA, et al (1997) Shear stress-induced binding of von Willebrand factor to platelets. Biorheology 34:57–71
5. Savage B, Saldívar E, Ruggeri ZM (1996) Initiation of platelet adhesion by arrest onto fibrinogen or translocation on von Willebrand factor. Cell 84:289–297
6. Watson D, Berlanga O, Best D, et al (2000) Update on collagen receptor interactions in platelets: is the two-model still valid? Platelets 11:252–258
7. Allain JP, Cooper HA, Wagner RH, et al (1975) Platelets fixed with paraformaldehyde: a new reagent for assay of von Willebrand factor and platelet aggregating factor. J Lab Clin Med 85:318–325
8. Goto S, Salomon DR, Ikeda Y, et al (1995) Characterization of the unique mechanism mediating the shear-dependent binding of soluble von Willebrand factor to platelets. J Biol Chem 270:23352–23361
9. Kroll MH, Harris TS, Moake JL, et al (1991) von Willebrand factor binding to platelet GPIb initiates signals for platelet activation. J Clin Invest 88:1568–1573
10. Francesconi M, Casonato A, Pontara E, et al (1995) Type B von Willebrand factor induces phospholipase A_2 activation and cytosolic Ca^{2+} increase in platelets. Biochem Biophys Res Commun 214:102–109
11. Li Z, Xi X, Du X (2001) A mitogen-activated protein kinase-dependent signaling pathway in the activation of platelet integrin alphaIIb beta3. J Biol Chem 276:42226–42232
12. Li Z, Xi X, Gu M et al (2003) A stimulatory role for cGMP-dependent protein kinase in platelet activation. Cell 112:77–86
13. Li Z, Zhang G, Marjanovic JA, et al (2004) A platelet secretion pathway mediated by cGMP-dependent protein kinase. J Biol Chem 279:42469–42475
14. Haslam RJ, Dickinson NT, Jang EK (1999) Cyclic nucleotides and phosphodiesterases in platelets. Thromb Haemost 82:412–423
15. Marshall SJ, Senis YA, Auger JM, et al (2004) GPIb-dependent platelet activation is dependent on Src kinase but not MAP kinase or cGMP-dependent kinase. Blood 103:2601–2609
16. Canobbio I, Reineri S, Sinigaglia F, et al (2004) A role for p38 MAP kinase in platelet activation by von Willebrand factor. Thromb Haemost 91:102–110
17. Garcia A, Quinton TM, Dorsam RT, et al (2005) Src family kinase-mediated and Erk-mediated thromboxane A_2 generation are essential for VWF/GPIb-induced fibrinogen receptor activation in human platelets. Blood 106:3410–3414
18. Liu J, Pestina TI, Berndt MC, et al (2004) The role of ADP and TXA2 in botrocetin/VWF-induced aggregation of washed platelets. J Thromb Haemost 2:2213–2222
19. Ozaki Y, Satoh K, Yatomi Y, et al (1995) Protein tyrosine phosphorylation in human platelets induced by interaction between glycoprotein Ib and von Willebrand factor. Biochim Biophys Acta 1243:482–488
20. Asazuma N, Ozaki Y, Satoh K, et al (1997) Glycoprotein Ib-von Willebrand factor interactions activate tyrosine kinases in human platelets. Blood 90:4789–4798

21. Canobbio I, Bertoni A, Lova P, et al (2001) Platelet activation by von Willebrand factor requires coordinated signaling through thromboxane A_2 and Fc gamma IIA receptor. J Biol Chem 276:26022–26029

22. Mazzucato M, De Marco L, Pradella P, et al (1996) Porcine von Willebrand factor binding to human platelet GPIb induces transmembrane calcium influx. Thromb Haemost 75:655–660

23. Milner EP, Zheng Q, Kermode JC, et al (1998) Ristocetin-mediated interaction of human von Willebrand factor with platelet glycoprotein Ib evokes a transient calcium signal: observations with Fura-PE3. J Lab Clin Med 131:49–62

24. Marshall SJ, Asazuma N, Best D, et al (2002) Glycoprotein IIb-IIIa-dependent aggregation by glycoprotein Ibα is reinforced by a Src family kinase inhibitor (PP1)-sensitive signaling pathway. Biochem J 361(Pt 2):297–305

25. Wu Y, Suzuki-Inoue K, Satoh K (2001) Role of Fc receptor gamma-chain in platelet glycoprotein Ib-mediated signaling. Blood 97:3836–3845

26. Mangin P, Yuan Y, Goncalves I, et al (2003) Signaling role for phospholipase Cγ2 in platelet glycoprotein Ibα calcium flux and cytoskeletal reorganization. J Biol Chem 278:32880–32891

27. Mazzucato M, Pradella P, Cozzi MR, et al (2002) Sequential cytoplasmic calcium signals in a 2-stage platelet activation process induced by the glycoprotein Ibα mechanoreceptor. Blood 100:2793–2800

28. Kasirer-Friede A, Rita Cozzi M, Mazzucato M, et al (2004) Signaling through GPIb-IX-V activates αIIbβ3 independently of other receptors. Blood 103:3403–3411

29. Watanabe D, Hashimoto S, Ishii M, et al (2001) Four tyrosine residues in phospholipase C-γ2 identified as Btk-dependent phosphorylation sites, are required for B cell antigen receptor-coupled calcium signaling. J Biol Chem 276:38595–38601

30. Suzuki-Inoue K, Inoue O, Frampton J, et al (2003) Murine GPVI stimulates weak integrin activation in PLCγ2–/– platelets: involvement of PLCγ1 and PI3-kinase. Blood 102:1367–1373

31. Suzuki-Inoue K, Wilde JI, Andrews RK, et al (2004) Glycoprotein VI and Ib-IX-V stimulate tyrosine phosphorylation of tyrosine kinase Syk and phospholipase Cγ2 at distinct sites. Biochem J 378:1023–1029

32. Jackson SP, Schoenwaelder SM, Yuan Y, et al (1994) Adhesion receptor activation of phosphatidylinositol 3-kinase: von Willebrand factor stimulates the cytoskeletal association and activation of phosphatidylinositol 3-kinase and pp60c-src in human platelets. J Biol Chem 269:27093–27099

33. Oda A, Yokoyama K, Murata M, et al (1995) Protein tyrosine phosphorylation in human platelets during shear stress-induced platelet aggregation (SIPA) is regulated by glycoprotein (GP)Ib/IX as well as GPIIb/IIIa and requires intact cytoskeleton and endogenous ADP. Thromb Haemost 74:736–742

34. Falati S, Edmead CE, Poole AW (1999) Glycoprotein Ib-IX-V, a receptor for von Willebrand factor, couples physically and functionally with the Fc receptor γ chain, Fyn and Lyn to activate human platelets. Blood 94:1648–1656

35. Briddon SJ, Watson SP (1999) Evidence for the involvement of p59fyn and p53/56lyn in collagen receptor signalling in human platelets. Biochem J 338:203–209

36. Dormann D, Clemetson JM, Navdaev A, et al (2001) Alboaggregin A activates platelets by a mechanism involving glycoprotein VI as well as glycoprotein Ib. Blood 97:929–936

37. Wu Y, Asazuma N, Satoh K, et al (2003) Interaction between von Willebrand factor and glycoprotein Ib activates Src kinase in human platelets: role of phosphoinositide 3-kinase. Blood 101:3469–3476

38. Torti M, Bertoni A, Canobbio I, et al (1999) Rap1B and Rap2B translocation to the cytoskeleton by von Willebrand factor involves FcγII receptor-mediated protein tyrosine phosphorylation. J Biol Chem 274:13690–13697

39. Munday AD, Berndt MC, Mitchell CA (2000) Phosphoinositide 3-kinase forms a complex with platelet membrane glycoprotein Ib-IX-V complex and 14-3-3zeta. Blood 96:577–584

40. Liu J, Pestina TI, Berndt MC, et al (2005) Botrocetin/VWF-induced signaling through GPIb-IX-V produces TXA2 in a αIIbβ3- and aggregation-independent manner. Blood 106:2750–2756

41. Liu J, Fitzgerald ME, Berndt MC, et al (2006) Bruton tyrosine kinase is essential for botrocetin/VWF-induced signaling and GPIb-dependent thrombus formation in vivo. Blood 108:2596–2603

42. Kasirer-Friede A, Moran B, Nagrampa-Orje J, et al (2007) ADAP is required for normal αIIbβ3 activation by VWF/GPIb-IX-V and other agonists. Blood 109:1018–1025

43. Sullam PH, Hyun WC, Szollosi J, et al (1998) Physical proximity and functional interplay of the glycoprotein Ib-IX-V complex and the Fc receptor FcγRIIA on the platelet plasma membrane. J Biol Chem 273:5331–5336

44. Sun B, Li J, Kambayashi J (1999) Interaction between GPIbα and FcγIIA receptor in human platelets. Biochem Biophys Res Commun 266:24–27

45. Yap CL, Anderson KE, Hughan SC (2002) Essential role for phosphoinositide 3-kinase in shear-dependent signaling between platelet glycoprotein Ib/V/IX and integrinα$_{IIb}$β$_3$. Blood 99:151–158

46. Resendiz JC, Feng S, Ji G (2004) von Willebrand factor binding to platelet glycoprotein Ib-IX-V stimulates the assembly of an α-actinin-based signaling complex. J Thromb Haemost 2:161–169

47. Kuwahara M, Sugimoto M, Tsuji S, et al (2002) Platelet shape changes and adhesion under high shear flow. Arterioscler Thromb Vasc Biol 22:329–334

48. Nesbitt WS, Kulkarni S, Giuliano S, et al (2002) Distinct glycoprotein Ib/V/IX and integrinα$_{IIb}$β$_3$-dependent calcium signals cooperatively regulate platelet adhesion under flow. J Biol Chem 277:2965–2972

49. Obergfell A, Eto K, Mocsai A, et al (2002) Coordinate interactions of Csk, Src, and Syk kinases with α$_{IIb}$β$_3$ initiate integrin signaling to the cytoskeleton. J Cell Biol 157:265–275

50. Jin W, Inoue O, Tamura N, et al (2007) A role for glycosphingolipid-enriched microdomains in platelet glycoprotein Ib-mediated platelet activation. J Thromb Haemost 5:1034–1040

51. Simons K, Ikonen E (1997) Functional rafts in cell membranes. Nature 387:569–572

52. Melkonian KA, Ostermeyer AG, Chen JZ, et al (1999) Role of lipid modifications in targeting proteins to detergent-resistant membrane rafts: many raft proteins are acylated, while few are prenylated. J Biol Chem 274:3910–3917

53. Bodin S, Giuriato S, Ragab J, et al (2001) Production of phosphatidylinositol 3,4,5-trisphosphate and phosphatidic acid in platelets rafts: evidence for a critical role of cholesterol-enriched domains in human platelet activation. Biochemistry 40:15290–15299

54. Heijnen HFG, van Lier M, Waaijenborg S, et al (2003) Concentration of rafts in platelet filopodia correlates with recruitment of c-Src and CD63 to these domains. J Thromb Haemost 1:1161–1173

55. Ezumi Y, Kodama K, Uchiyama T, et al (2002) Constitutive and functional association of the platelet collagen receptor glycoprotein VI-Fc receptor γ-chain complex with membrane rafts. Blood 99:3250–3255

56. Shrimpton CN, Borthakur G, Larucea S, et al (2002) Localization of the adhesion receptor glycoprotein Ib-IX-V complex to lipid rafts is required for platelet adhesion and activation. J Exp Med 196:1057–1106

57. Morrison D (1994) 14-3-3: Modulators of signaling proteins? Science 266:56–57

58. Bonneboy-Berard N, Liu YC, von Willebrand M, et al (1995) Inhibition of phosphatidylinositol 3-kinase activity by association with 14-3-3 proteins in T cells. Proc Natl Acad Sci U S A 92:10142–10146
59. Feng S, Christodoulides N, Resendiz JC, et al (2000) Cytoplasmic domains of GPIbα and GPIbβ regulate 14-3-3ζ binding to GPIb/IX/V. Blood 95:551–557
60. Gu M, Xi X, Englund GD, Berndt MC, et al (1999) Analysis of the roles of 14-3-3 in the platelet glycoprotein Ib-IX-mediated activation of integrin $\alpha_{IIb}\beta_3$ using a reconstituted mammalian cell expression model. J Cell Biol 147:1085–1096
61. Bialkowska K, Zaffran Y, Meyer SC, et al (2003) 14-3-3 Mediates integrin-induced activation of Cdc42 and Rac. J Biol Chem 278:33342–33350

Sphingosine 1-Phosphate as a Platelet-Derived Bioactive Lipid

Yutaka Yatomi

Summary. Sphingosine 1-phosphate (S1P) is now established as a multifunctional bioactive lipid that plays important (patho)physiological roles; this is especially true in the fields of vascular biology and immunology. Blood platelets are unique in that they store S1P abundantly (possibly due to the existence of highly active sphingosine kinase and a lack of S1P lyase) and release this lysophospholipid mediator extracellularly upon stimulation. Vascular cells, including endothelial cells and smooth muscle cells, respond dramatically to S1P mainly through a family of G protein-coupled receptors. In fact, the importance of S1P in platelet–vascular cell interactions has been shown in a number of studies, although red blood cells have been recently attracting attention as another source for blood S1P. It is likely that control of S1P biological activities is important for the therapeutic purpose of regulating vascular disorders.

Key words. Endothelial cell · Platelet · Smooth muscle cell · Sphingosine 1-phosphate · Vascular biology

Introduction

Sphingosine 1-phosphate (S1P) is a bioactive lysophospholipid capable of inducing a wide spectrum of biological responses, including but not limited to cell survival, growth, migration, and contraction [1–4]. Although S1P has been proposed to act as both an intracellular second messenger and an extracellular mediator [1–4], this bioactive lipid seems to play an important role as the latter in the area of vascular biology (and immunology). Blood platelets are an established source of plasma or serum S1P [5, 6], although red blood cells (RBCs) have been attracting attention recently as a new source for blood S1P [7, 8]. On the other hand, vascular cells such as endothelial cells (ECs) and smooth muscle cells (SMCs) express some of a family of G protein-coupled S1P receptors named $S1P_1$ through $S1P_5$, originally referred to as EDG-1, 5, 3, 6, and 8, respectively [3, 4, 9]. As expected, the importance of platelet–vascular cell interactions through S1P and the vascular responses elicited by S1P have been extensively studied, and it is likely that regulation of S1P biological activities is important for the therapeutic purpose of controlling vascular disorders.

FTY720 (fingolimod), a derivative of the natural product myriocin that bears structural similarity to sphingosine (Sph), has been effectively phosphorylated by Sph kinase-2 in vivo [10]. The phosphorylated compound targets S1Ps, especially $S1P_1$, on lymphocytes and ECs, thereby inhibiting the egress of T and B cells from secondary lymphoid organs into the blood and their recirculation to the sites of inflamed tissues or where tissues are transplanted [11, 12]. It is now believed that FTY720 is promising as a new type of immunomodulator based on fine preclinical test results in transplantation and autoimmune disease models. This unique compound has been shown to be promising also in the area of vascular biology (i.e., to control vascular disorders) [13, 14]. Because the themes of this book are thrombosis and hemostasis, I focus on S1P as a platelet-derived bioactive lipid and on its importance in vascular biology.

S1P Release from Activated Platelets

S1P is formed intracellularly through phosphorylation of Sph catalyzed by Sph kinase [1, 2]. On the other hand, S1P lyase, which degrades S1P into phosphoethanolamine and fatty aldehyde, seems the most important enzyme for degradation of S1P [1, 2]. Platelets possess a highly active Sph kinase (Fig. 1A) but are devoid of S1P lyase [15, 16]. As a result, platelets store S1P abundantly [5, 6]. Furthermore, S1P, abundantly stored in platelets, is released extracellularly following stimulation with physiological agonists such as thrombin [15, 16] (Fig. 1B). Recently, platelets were shown to release S1P also by shear stress, which may be important in

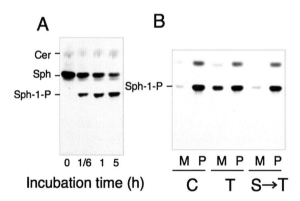

FIG. 1. Sphingosine 1-phosphate (*Sph-1-P*) formation in and release from platelets. **A** Platelet suspensions were incubated with radiolabeled sphingosine (*Sph*) for the indicated durations, followed by lipid extraction. Metabolic changes of Sph in human platelets were then analyzed by thin-layer chromatography (TLC) autoradiography. **B** Platelet release of newly formed radiolabeled S1P. Platelets labeled with Sph were preincubated without or with (*S*) staurosporine and further stimulated without (*C*) or with (*T*) thrombin. Lipids were then extracted from the platelet pellet (*P*) and medium (*M*) after separation by centrifugation and analyzed by TLC autoradiography

view of established roles of shear stress activation of platelets in thrombosis and hemostasis [17].

Although the precise mechanism for extracellular release of S1P following platelet activation remains to be fully clarified, it must require the function of specific transporter(s) because S1P possesses the polar nature of the head group. Recently, important progress has been reported on S1P release from platelets using streptolysin O, which permeabilizes the platelet plasma membrane, and α-toxin, which makes smaller pores in the plasma membrane than streptolysin O and depletes cytosolic ATP [18]. With the use of these agents, it was indicated that two independent S1P release systems might exist in the platelet plasma membrane: an ATP-dependent system stimulated by thrombin and an ATP-independent system stimulated by Ca^{2+} [18].

Sph supply (for S1P formation) is also a big problem because platelets lack the de novo sphingolipid biosynthesis necessary to provide the substrate, although they have high Sph kinase activity. Recently, a generation pathway for Sph, the precursor of S1P, in human platelets has been revealed [19]. It has been reported that Sph formed from plasma sphingomyelin by bacterial sphingomyelinase and neutral ceramidase is rapidly incorporated into platelets and converted to S1P, suggesting that platelets use extracellular Sph as a source of S1P. Platelets abundantly express sphingomyelin, possibly supplied from plasma lipoproteins, at the cell surface; and treating platelets with bacterial sphingomyelinase resulted in Sph generation at the cell surface, conceivably by the action of membrane-bound neutral ceramidase. Simultaneously, a time-dependent increase in S1P levels was observed. Furthermore, it was demonstrated that secretory acid sphingomyelinase also induces S1P increases in platelets. Accordingly, it was suggested that platelet Sph may be supplied from at least two sources: generation in the plasma followed by incorporation, and generation at the outer leaflet of the plasma membrane, initiated by cell surface sphingomyelin degradation [19, 20], although further study is needed to prove it.

RBCs as Another Source of Plasma or Serum S1P

As described above, platelets store a large amount of S1P and release it into the plasma in a stimuli-dependent manner. It was also known that RBCs are capable of incorporating Sph, converting it to S1P, and releasing it; S1P release from RBCs is stimuli-independent [21]. Recently, measurement of the S1P and Sph levels in RBCs by high-performance liquid chromatography (HPLC) has been reported, and it was found that the contribution of RBC S1P to the whole blood S1P level is greater than that of platelets [22]. In vitro assays demonstrated that RBCs have much weaker Sph kinase activity compared to platelets but lack the S1P-degrading activities such as the lyase and phosphatase. It was speculated that this combination might enable RBCs to maintain a high S1P content relative to Sph; the absence of both S1P-metabolizing enzymes has not been reported for other cell types [22]. Furthermore, the volume of the RBC is much larger than that of the platelet. Thus, RBCs may be specialized cells for storing and supplying plasma S1P. Although platelets are no more an only source for plasma or serum S1P [7, 8, 22], it seems likely that S1P locally and abundantly released from stimulated or aggregating platelets may play important roles in vivo. Presumably, RBCs may be responsible for the basal level of plasma S1P.

S1P: A Normal Constituent of Human Plasma and Serum

As expected from the fact that S1P is released from activated platelets (and RBCs), S1P was identified as a normal constituent of human plasma and serum [5], which we reported for the first time. S1P was extracted into the aqueous phase to separate it from other lipids under alkaline conditions and then reextracted it into the organic phase under acidic conditions [23]. S1P, thus partially purified, was acetylated with radioactive acetic anhydride; the product, radioactive C_2-ceramide 1-phosphate, was subjected to thin-layer chromatography autoradiography and quantified [23]. By this method, the S1P levels in plasma and serum were first reported to be about 200 and 500 nM, respectively [5]. The serum S1P levels reported have always been higher than those in plasma [5], and it is most likely that the source of discharged S1P during blood clotting is platelets, as they abundantly store S1P compared with other blood cells, and release the stored S1P extracellularly upon stimulation with thrombin, a product of the coagulation cascade. However, the methods by which S1P is measured are far from being standardized, and the procedures by which suitable plasma samples are prepared remain to be determined; platelets are easily activated by in vitro manipulation (i.e., after venipuncture). In fact, when we measured plasma S1P by its derivatization with o-phthaldialdehyde, followed by HPLC separation and the resultant fluorescence monitoring, it was found that the plasma S1P concentrations were significantly higher in men (413 ± 52 nM; mean $\pm$ SD) than in women (352 ± 40 nM), whereas the plasma S1P concentrations varied substantially depending on the reports from various groups. Because of the established importance of S1P in the area of vascular biology and immunology, information on the plasma levels of S1P and the factor(s) regulating them may give insight into the (patho)physiology of various disorders. Accordingly, introduction of a plasma S1P assay for clinical laboratory testing is important and is in progress in our laboratory.

S1P Effects on Platelets

When added exogenously, S1P with its micromolar concentrations induces platelet shape change and reversible aggregation, possibly via intracellular Ca^{2+} mobilization [15]. Irreversible aggregation with granule secretion can be induced when platelets are strongly activated, but S1P fails to induce this [15]. That is, S1P can be classified as a weak platelet agonist, although $S1P_4$ (EDG-6) was found to be expressed in platelets [24]. The physiological role of S1P as an autocrine stimulator of platelets was first proposed [6, 15], but this possibility seems remote; the stimulatory effect of S1P on platelet activation is much weaker than that of thromboxane A_2 or lysophosphatidic acid. It is likely that platelets are important as the generator of blood S1P (Fig. 2) not the target of S1P.

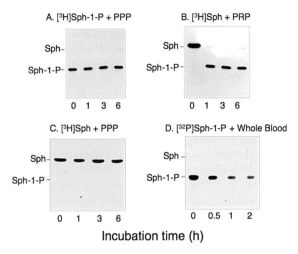

FIG. 2. Metabolism of radiolabeled Sph and S1P in platelet-poor plasma (PPP), platelet-rich plasma (PRP), and whole blood. **A** PPP was incubated with [^{3}H]. **B** PRP challenged with collagen was incubated with [^{3}H]Sph, and PPP was obtained by centrifugation. **C** PPP was incubated with [^{3}H]S1P. The lipids in these PPP samples (**A–C**) were extracted and analyzed by TLC autoradiography. **D** Whole blood samples were incubated with [^{32}P]S1P, and the lipids were then extracted and analyzed for the metabolism of [^{32}P]S1P by TLC autoradiography. Both Sph and S1P are metabolically stable in the plasma, whereas activated platelets convert Sph to S1P. S1P is degraded possibly by the ectoenzyme lipid phosphate phosphatase

S1P$_x$ Expression on ECs

S1P$_1$ (EDG-1), S1P$_2$ (EDG-5), S1P$_3$ (EDG-3), S1P$_4$ (EDG-6), and S1P$_5$ (EDG-8) exhibit overlapping, as well as distinct patterns of expression in various tissues as S1P receptors; the eventual cellular responses to extracellular S1P depend on the types of S1P receptor expressed [1–3, 9]. It was reported that S1P$_1$ is abundantly expressed in human umbilical vein ECs, whereas the S1P$_3$ transcript is expressed at a lower level [3, 9, 25]. Consistent with expression of the high-affinity S1P receptors on EC cells, nanomolar S1P reportedly induces a variety of EC responses [3, 4, 9, 25, 26].

To gain insight into the pathophysiological role(s) of S1P in the vascular system, the mechanism(s) by which EC S1P expression is regulated seems important. In fact, expression of S1P$_1$, the most important S1P receptor on vascular ECs, is dynamically regulated. S1P$_1$ was originally isolated as a phorbol ester-inducible immediate early transcript from vascular ECs; the S1P$_1$ protein was assumed to play a role in EC differentiation and angiogenesis as phorbol esters induce morphogenetic differentiation of ECs into capillary-like tubules [27]. The mechanical force associated with blood flow (i.e., fluid shear stress) modulates vascular structure and function and plays an important role in the pathogenesis of vascular diseases, including atherosclerosis, hypertension, and restenosis [28]. In this context, the fluid flow has been reported to enhance EC gene and protein expression [17, 29], suggesting involvement of S1P/S1P$_1$ signaling in the flow-mediated responses. Furthermore, it was shown that S1P/S1P signaling could be modulated by the inflammatory cytokine tumor necrosis factor-α

(TNFα), possibly through regulation of $S1P_3$ expression [30]. Recently, $S1P_1$ up-regulation by vascular endothelial growth factor (VEGF) (within its physiological concentration range) and the resultant VEGF sensitization of ECs to the S1P effects were reported [31]. The cross-talk between these two stimuli of completely different characteristics may be potentially important in the vasculature. Finally, it was also found that thrombin and S1P itself rapidly up-regulate the expression of $S1P_1$ and $S1P_3$ [32].

The critical requirement of S1P in vascular system development was shown by $S1P_1$-deficient mice [33]. Homozygous $S1P_1^{-/-}$ mice reportedly die during embryonic development at 12.5–14.5 days of gestation owing to hemorrhage [33]. Unexpectedly, both vasculogenesis and angiogenesis were not affected in the mutant embryos, but vascular maturation was incomplete; the vessels were unstable and resulted in disruption, edema, and bleeding. Histological analysis suggested that blood vessels are incompletely covered by smooth muscle cells (SMCs). When $S1P_1$ was disrupted solely in ECs by using the Cre/loxP system, the phenotype of the conditional mutant embryos was found to mimic that obtained in the embryos globally deficient in $S1P_1$ [34]. Accordingly, vessel coverage by SMCs is not directly induced by S1P but is directed by S1P interaction with $S1P_1$ receptors on ECs.

S1P Effects on ECs

It is well established that S1P stimulates EC survival or proliferation through a Gi-coupled receptor, mainly via $S1P_1$ [3, 4, 9, 25]. Furthermore, S1P induces migration, adherens junction assembly, capillary tube formation, and the resultant angiogenesis via $S1P_1$ (coupled with Gi) and $S1P_3$ (coupled with G_i, G_{13}, and G_q) [3, 9, 25, 35]. Consistent with the fact that S1P, as well as serum, is an angiogenic factor and is released from platelets, platelet-released S1P was reported to be responsible for the serum angiogenic activity [36].

Through $S1P_1$ and $S1P_3$-dependent cytoskeletal rearrangement, S1P also promotes EC barrier integrity, which might be important for maintaining vascular homeostasis [36–38]. S1P reportedly elicits rapid, sustained, dose-dependent increases in trans-monolayer electrical resistance across both human and bovine pulmonary artery and lung microvascular ECs [37]. S1P also reverses barrier dysfunction elicited by thrombin [37]. The previous in vitro and in vivo evidence indicates that circulating platelets affect both vascular integrity, and now it seems highly possible that a major platelet-derived vascular barrier-protective factor is S1P [4, 38]. This beneficial effect of S1P may lead to therapeutic regulation of vascular permeability in pathological conditions such as sepsis, hypoxia, and solid tumor growth [13, 14].

It is well established that NO production plays an important role in endothelial survival, maintenance of the endothelial function, and regulation of vascular tone. S1P, through binding to $S1P_1$ and the resultant activation of phosphoinositide 3-kinase and then Akt, phosphorylates (at Ser1179) and activates the endothelial isoform of NO synthase (eNOS) in ECs [39]. These signaling pathways elicited by S1P/S1P and endothelium-dependent vasodilation by S1P were reproduced in intact vessels [40]. S1P is expected to play an important protective role in the vasculature through NO production and limit vascular occlusion in vivo at least in an intact, undamaged endothelium.

Atherogenic Effects of S1P

As described above, many S1P-induced EC responses are considered to be antiatherogenic. However, there are some considered to be proatherogenic. Atherosclerosis is closely related to a hypercoagulable state in the vasculature, and thrombus formation at critical sites on vulnerable plaques are directly involved in the pathogenesis of acute coronary syndrome. In this context, it should be noted that S1P strongly potentiates thrombin-induced tissue factor expression in ECs, although S1P by itself fails to induce this response [32]. S1P enhances tissue factor expression at the transcriptional level, possibly via promoting the ERK1/2 MAP kinase-dependent activation of transcription factors NFkB and Egr-1 [32]. This synergistic effect of S1P on thrombin-induced tissue factor expression in ECs may propagate a positive feedback reaction involving thrombus formation.

Adhesion of circulating leukocytes, especially monocytes, to activated ECs, leading to the resultant transendothelial migration into the subendothelium, is an initial and important step in atherosclerosis. Because adhesion molecules such as E-selectin, VCAM-1, and ICAM-1 mediate this process, the up-regulation of these molecules is related to the development of atherosclerosis. It is well known that the inflammatory responses are strongly induced by TNFα, and it was postulated that intracellular S1P serves as a second messenger in mediating TNFα-induced adhesion molecule expression [41, 42]. It was also reported that high density lipoprotein (HDL) inhibits the expression of these adhesion molecules by inhibiting Sph kinase [43]. Although it is possible to speculate that intracellular S1P (as an atherogenic mediator) and extracellular S1P (as an antiatherogenic mediator) may function in the opposite way in terms of atherosclerosis [44], intracellular S1P targets remain to be elucidated. It was reported that S1P even inhibits TNFα-induced inflammatory responses [45, 46].

S1P Carrier in the Plasma

Plasma S1P is metabolically stable (Fig. 2A) and was once thought to be bound to albumin [15]. It was later reported, however, that when expressed as the per-unit amount of protein, S1P is concentrated in lipoprotein fractions with a rank order of HDL > low-density lipoprotein (LDL) > very low-density lipoprotein (VLDL) > lipoprotein-deficient plasma (mainly albumin), among plasma and serum components [44]. Recently, the importance of lipoproteins, especially HDL, as a S1P carrier in the plasma has been established. S1P, as well as HDL, protected ECs from serum depletion- or oxidized LDL-induced cytotoxicity, which were associated with ERK activation [44]. Several lines of evidence revealed that the HDL-induced cytoprotective actions were mediated by S1P activation of S1P$_1$ and S1P$_3$ receptors [44, 47]. As is well known, LDL (especially oxidized LDL) is closely correlated and HDL inversely correlated with the risk of atherosclerosis or cardiovascular disease. Among the lipid components associated with (oxidized) LDL, lysophosphatidylcholine (LPC) has been shown to mimic a variety of oxidized LDL-induced actions including monocyte migration, cytotoxic action toward ECs, expression of adhesion molecules and production of cytokines in ECs, and promotion of migration and proliferation of SMCs [44]. S1P content in LDL during its oxidation was measured; CuSO$_4$ treatment of LDL resulted in 70%–80% reduction of S1P content, which was associated with a several-fold increase in LPC content [47]. This balance was suggested to be an important determinant for lipoproteins to be atherogenic or antiatherogenic [44, 47]. It seems likely that antiatherogenic characteristics of HDL are at least partly ascribed to S1P associated with it [44, 46, 47].

S1P Effects on SMCs

Like ECs, SMCs express S1Ps and respond dramatically to S1P. However, the pattern of S1P expression is different from that in ECs. $S1P_1$, abundantly expressed in ECs, is expressed only in the fetal/intimal phenotype of SMCs, whereas both $S1P_2$ and $S1P_3$ receptors are expressed in adult as well as fetal/intimal phenotypes [4]. It is now evident that Rac activation is an important mechanism underlying the cell migration by S1P and that expression of S1P isoform-specific negative or positive regulation of cellular Rac activity is critically involved in S1P-mediated bimodal regulation of cell motility [9, 48]. That is, whereas $S1P_1$ (and $S1P_3$) act as typical chemotactic receptors, $S1P_2$ uniquely acts as a chemorepellant receptor; $S1P_1/S1P_3$ and $S1P_2$ regulate the activity of the Rho family GTPase Rac positively and negatively, respectively [9, 48]. Accordingly, S1P inhibits migration of SMCs, whereas it stimulates that of ECs [3, 9, 49].

The importance of differential expression of S1P receptor isoforms for cellular migratory responses is true within SMC responses under different situations. The presence of $S1P_1$ receptors has been shown to be critical for the stimulatory effect of S1P on SMC chemotaxis, and, in fact, S1P induces migration of intimal SMCs, which express $S1P_1$ [3, 9]. In addition, $S1P_1$ promotes SMC proliferation via the Gi/phosphoinositide 3-kinase pathway [3, 50]. Accordingly, the expression of $S1P_1$ in SMCs might be critical in causing neointima formation. Also by studying pup SMCs, which have properties similar to those of rat neointimal cells, it was observed that $S1P_1$ expression mediates S1P-induced migration and proliferation [3]. In contrast, the low $S1P_1/S1P_2$ ratio results in inhibition of migration in normal quiescent SMCs [9, 49–51]. Although the (patho)physiological significance of the predominant expression of $S1P_2$ in medial SMCs and the resultant inhibition of migration in response to S1P remains to be clarified, this may be one of the antiatherogenic effects S1P exerts.

Recently, the roles of S1P in vasoconstriction have been established [52]. Using a well-established canine isolated heart model, S1P was shown to decrease coronary blood flow and ventricular contractions [53]. The coronary vasoconstriction effect of S1P may be related to its contractile effect on coronary SMCs, and the observed negative inotropic effect may be at least partly induced by transient myocardial ischemia [53]. Consistent with this S1P-elicited coronary vasoconstriction in vivo, S1P strongly induces coronary artery SMC contraction, which was inhibited by a specific antagonist of $S1P_2$ and C3 exoenzyme or the Rho kinase inhibitor Y-27632 [54], indicating involvement of $S1P_2$/Rho/Rho kinase (Fig. 3). Rho kinase is now considered to be a key molecule in (coronary) artery hypercontraction induced by vasospastic agents (including S1P).

Platelet-derived mediators, especially bioactive lipids, play an important role in coronary artery constriction and hence the pathogenesis of ischemic heart diseases; inhibition of the spastic effects of these mediators would be useful therapeutically.

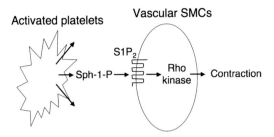

FIG. 3. $S1P_2$-mediated vascular smooth muscle cell (SMC) contraction induced by S1P released from activated platelets. See text for details

Thromboxane A_2 is an established example. As described above, it was shown that S1P induces coronary artery SMC contraction in vivo and in vitro, and at least the latter can be effectively inhibited by a specific antagonist of $S1P_2$ [28]. Importantly, the supernatant from activated platelet suspensions also induces SMC contraction, and a specific antagonist of $S1P_2$ strongly inhibits this response [55]. These findings suggest that platelet-derived S1P may be involved in the pathophysiology of coronary artery spasm through $S1P_2$ and that antagonists of this receptor might be useful therapeutically (Fig. 3). S1P is not the only vasoconstrictor released from activated platelets, and thromboxane A_2 is a much more established one. Although this eicosanoid is highly unstable, S1P is stable at least in the plasma [5] (Fig. 2A). It is possible to speculate that platelets release the lipid vasoconstrictors with different properties. Development of specific antagonists against $S1P_2$ receptor may lead to a new therapeutic approach to block coronary artery diseases and may supplement the therapeutic effect of thromboxane A_2 blockers, which are now widely used for this purpose.

Future directions

It has been established that S1P plays critical roles in the field of vascular biology (and immunology), and manipulation of S1P activities may greatly influence vascular events, including thrombosis, hemostasis, atherosclerosis, angiogenesis, and wound healing. Selective S1P agonists or antagonists may have great potential as drugs to control vascular disorders [4, 13, 14, 26]. Platelets, which play central roles in thrombosis and hemostasis, are an important source for plasma S1P, although RBCs are recently attracting attention as another source. Furthermore, it was reported recently that platelets efficiently convert FTY720 to FTY720 phosphate and release the latter extracellularly, suggesting that platelets are a major source for plasma FTY720 phosphate [56]. This may be worthy of notice considering the growing importance of the novel drug FTY720. Platelets will remain the focus of S1P research.

References

1. Spiegel S, Milstien S (2003) Sphingosine-1-phosphate: an enigmatic signalling lipid. Nat Rev Mol Cell Biol 4:397–407

2. Spiegel S, Milstien S (2007) Functions of the multifaceted family of sphingosine kinases and some close relatives. J Biol Chem 282:2125–2129
3. Kluk Mj, Hla T (2002) Signaling of sphingosine-1-phosphate via the S1P/EDG-family of G-protein-coupled receptors. Biochim Biophys Acta 1582:72–80
4. Schubert R (2006) Sphingosine-1-phosphate in the circulatory system: cause and therapeutic target for vascular dysfunction? Cardiovasc Res 70:9–11
5. Yatomi Y, Igarashi Y, Yang L, et al (1997) Sphingosine 1-phosphate, a bioactive sphingolipid abundantly stored in platelets, is a normal constituent of human plasma and serum. J Biochem (Tokyo) 121:969–973
6. Yatomi Y, Ozaki Y, Ohmori T, et al (2001) Sphingosine 1-phosphate: synthesis and release. Prostaglandins Other Lipid Mediat 64:107–122
7. Pappu R, Schwab SR, Cornelissen I, et al (2007) Promotion of lymphocyte egress into blood and lymph by distinct sources of sphingosine-1-phosphate. Science 316:295–298
8. Hanel P, Andreani P, Graler MH (2007) Erythrocytes store and release sphingosine 1-phosphate in blood. FASEB J 21:1202–1209
9. Takuwa Y (2002) Subtype-specific differential regulation of Rho family G proteins and cell migration by the Edg family sphingosine-1-phosphate receptors. Biochim Biophys Acta 1582:112–120
10. Billich A, Bornancin F, Devay P, et al (2003) Phosphorylation of the immunomodulatory drug FTY720 by sphingosine kinases. J Biol Chem 278:47408–47415
11. Martini S, Peters H, Bohler T, et al (2007) Current perspectives on FTY720. Expert Opin Invest Drugs 16:505–518
12. Chiba K, Matsuyuki H, Maeda Y, et al (2006) Role of sphingosine 1-phosphate receptor type 1 in lymphocyte egress from secondary lymphoid tissues and thymus. Cell Mol Immunol 3:11–19
13. Tolle M, Levkau B, Kleuser B, et al (2007) Sphingosine-1-phosphate and FTY720 as anti-atherosclerotic lipid compounds. Eur J Clin Invest 37:171–179
14. Brinkmann V, Baumruker T (2006) Pulmonary and vascular pharmacology of sphingosine 1-phosphate. Curr Opin Pharmacol 6:244–250
15. Yatomi Y, Ruan F, Hakomori S, et al (1995) Sphingosine-1-phosphate: a platelet-activating sphingolipid released from agonist-stimulated human platelets. Blood 86:193–202
16. Yatomi Y, Yamamura S, Ruan F, et al (1997) Sphingosine 1-phosphate induces platelet activation through an extracellular action and shares a platelet surface receptor with lysophosphatidic acid. J Biol Chem 272:5291–5297
17. Aoki S, Osada M, Kaneko M, et al (2007) Fluid shear stress enhances the sphingosine 1-phosphate responses in cell-cell interactions between platelets and endothelial cells. Biochem Biophys Res Commun 358:1054–1057
18. Kobayashi N, Nishi T, Hirata T, et al (2006) Sphingosine 1-phosphate is released from the cytosol of rat platelets in a carrier-mediated manner. J Lipid Res 47:614–621
19. Tani M, Sano T, Ito M, et al (2005) Mechanisms of sphingosine and sphingosine 1-phosphate generation in human platelets. J Lipid Res 46:2458–2467
20. Tani M, Ito M, Igarashi Y (2007) Ceramide/sphingosine/sphingosine 1-phosphate metabolism on the cell surface and in the extracellular space. Cell Signal 19:229–237
21. Yang L, Yatomi Y, Miura Y, et al (1999) Metabolism and functional effects of sphingolipids in blood cells. Br J Haematol 107:282–293
22. Ito K, Anada Y, Tani M, et al (2007) Lack of sphingosine 1-phosphate-degrading enzymes in erythrocytes. Biochem Biophys Res Commun 357:212–217
23. Yatomi Y, Ruan F, Ohta J, et al (1995) Quantitative measurement of sphingosine 1-phosphate in biological samples by acylation with radioactive acetic anhydride. Anal Biochem 230:315–320

24. Motohashi K, Shibata S, Ozaki Y, et al (2000) Identification of lysophospholipid receptors in human platelets: the relation of two agonists, lysophosphatidic acid and sphingosine 1-phosphate. FEBS Lett 468:189–193
25. Lee Mj, Thangada S, Claffey KP, et al (1999) Vascular endothelial cell adherens junction assembly and morphogenesis induced by sphingosine-1-phosphate. Cell 99:301–312
26. Yatomi Y (2006) Sphingosine 1-phosphate in vascular biology: possible therapeutic strategies to control vascular diseases. Curr Pharm Des 12:575–587
27. Hla T, Maciag T (1990) An abundant transcript induced in differentiating human endothelial cells encodes a polypeptide with structural similarities to G-protein-coupled receptors. J Biol Chem 265:9308–9313
28. Gimbrone Ma Jr, Nagel T, Topper JN (1997) Biomechanical activation: an emerging paradigm in endothelial adhesion biology. J Clin Invest 99:1809–1813
29. Takada Y, Kato C, Kondo S, et al (1997) Cloning of cDNAs encoding G protein-coupled receptor expressed in human endothelial cells exposed to fluid shear stress. Biochem Biophys Res Commun 240:737–741
30. Osada M, Yatomi Y, Ohmori T, et al (2002) Modulation of sphingosine 1-phosphate/EDG signaling by tumor necrosis factor-alpha in vascular endothelial cells. Thromb Res 108:169–174
31. Igarashi J, Erwin PA, Dantas AP, et al (2003) VEGF induces S1P1 receptors in endothelial cells: implications for cross-talk between sphingolipid and growth factor receptors. Proc Natl Acad Sci U S A 100:10664–10669
32. Takeya H, Gabazza EC, Aoki S, et al (2003) Synergistic effect of sphingosine 1-phosphate on thrombin-induced tissue factor expression in endothelial cells. Blood 102:1693–1700
33. Liu Y, Wada R, Yamashita T, et al (2000) Edg-1, the G protein-coupled receptor for sphingosine-1-phosphate, is essential for vascular maturation. J Clin Invest 106:951–961
34. Allende Ml, Yamashita T, Proia RL (2003) G-protein-coupled receptor S1P1 acts within endothelial cells to regulate vascular maturation. Blood 102:3665–3667
35. Ohmori T, Yatomi Y, Okamoto H, et al (2001) G(i)-mediated Cas tyrosine phosphorylation in vascular endothelial cells stimulated with sphingosine 1-phosphate: possible involvement in cell motility enhancement in cooperation with Rho-mediated pathways. J Biol Chem 276:5274–5280
36. English D, Brindley DN, Spiegel S, et al (2002) Lipid mediators of angiogenesis and the signalling pathways they initiate. Biochim Biophys Acta 1582:228–239
37. Garcia JG, Liu F, Verin AD, et al (2001) Sphingosine 1-phosphate promotes endothelial cell barrier integrity by Edg-dependent cytoskeletal rearrangement. J Clin Invest 108:689–701
38. Schaphorst KL, Chiang E, Jacobs KN, et al (2003) Role of sphingosine-1 phosphate in the enhancement of endothelial barrier integrity by platelet-released products. Am J Physiol Lung Cell Mol Physiol 285:L258–L267
39. Morales-Ruiz M, Lee MJ, Zollner S, et al (2001) Sphingosine 1-phosphate activates Akt, nitric oxide production, and chemotaxis through a Gi protein/phosphoinositide 3-kinase pathway in endothelial cells. J Biol Chem 276:19672–19677
40. Dantas AP, Igarashi J, Michel T (2003) Sphingosine 1-phosphate and control of vascular tone. Am J Physiol Heart Circ Physiol 284:H2045–H2052
41. Xia P, Gamble JR, Rye KA, et al (1998) Tumor necrosis factor-alpha induces adhesion molecule expression through the sphingosine kinase pathway. Proc Natl Acad Sci U S A 95:14196–14201
42. Xia P, Wang L, Gamble JR, et al (1999) Activation of sphingosine kinase by tumor necrosis factor-alpha inhibits apoptosis in human endothelial cells. J Biol Chem 274:34499–34505

43. Xia P, Vadas MA, Rye KA, et al (1999) High density lipoproteins (HDL) interrupt the sphingosine kinase signaling pathway: a possible mechanism for protection against atherosclerosis by HDL. J Biol Chem 274:33143–33147

44. Okajima F (2002) Plasma lipoproteins behave as carriers of extracellular sphingosine 1-phosphate: is this an atherogenic mediator or an anti-atherogenic mediator? Biochim Biophys Acta 1582:132–137

45. Aoki S, Yatomi Y, Shimosawa T, et al (2007) The suppressive effect of sphingosine 1-phosphate on monocyte-endothelium adhesion may be mediated by the rearrangement of the endothelial integrins alpha(5)beta(1) and alpha(v)beta(3). J Thromb Haemost 5:1292–1301

46. Kimura T, Tomura H, Mogi C, et al (2006) Role of scavenger receptor class B type I and sphingosine 1-phosphate receptors in high density lipoprotein-induced inhibition of adhesion molecule expression in endothelial cells. J Biol Chem 281:37457–37467

47. Kimura T, Sato K, Kuwabara A, et al (2001) Sphingosine 1-phosphate may be a major component of plasma lipoproteins responsible for the cytoprotective actions in human umbilical vein endothelial cells. J Biol Chem 276:31780–31785

48. Sugimoto N, Takuwa N, Okamoto H, et al (2003) Inhibitory and stimulatory regulation of Rac and cell motility by the G12/13-Rho and Gi pathways integrated downstream of a single G protein-coupled sphingosine-1-phosphate receptor isoform. Mol Cell Biol 23:1534–1545

49. Ryu Y, Takuwa N, Sugimoto N, et al (2002) Sphingosine-1-phosphate, a platelet-derived lysophospholipid mediator, negatively regulates cellular Rac activity and cell migration in vascular smooth muscle cells. Circ Res 90:325–332

50. Tamama K, Kon J, Sato K, et al (2001) Extracellular mechanism through the Edg family of receptors might be responsible for sphingosine-1-phosphate-induced regulation of DNA synthesis and migration of rat aortic smooth-muscle cells. Biochem J 353:139–146

51. Bornfeldt KE, Graves LM, Raines EW, et al (1995) Sphingosine-1-phosphate inhibits PDGF-induced chemotaxis of human arterial smooth muscle cells: spatial and temporal modulation of PDGF chemotactic signal transduction. J Cell Biol 130:193–206

52. Watterson KR, Ratz PH, Spiegel S (2005) The role of sphingosine-1-phosphate in smooth muscle contraction. Cell Signal 17:289–298

53. Sugiyama A, Yatomi Y, Ozaki Y, et al (2000) Sphingosine 1-phosphate induces sinus tachycardia and coronary vasoconstriction in the canine heart. Cardiovasc Res 46:119–125

54. Ohmori T, Yatomi Y, Osada M, et al (2003) Sphingosine 1-phosphate induces contraction of coronary artery smooth muscle cells via S1P2. Cardiovasc Res 58:170–177

55. Ohmori T, Yatomi Y, Osada M, et al (2004) Platelet-derived sphingosine 1-phosphate induces contraction of coronary artery smooth muscle cells via S1P2. J Thromb Haemost 2:203–205

56. Anada Y, Igarashi Y, Kihara A (2007) The immunomodulator FTY720 is phosphorylated and released from platelets. Eur J Pharmacol 568:106–111

Polymorphisms of Platelet Membrane Glycoproteins

Yumiko Matsubara, Mitsuru Murata, and Yasuo Ikeda

Summary. Platelet glycoproteins (GPs) act as membrane receptors. Functions of GPs were well studied in GPIa/IIa (integrin $\alpha_2\beta_1$), GPIb/IX/V, GPIIb/IIIa (integrin $\alpha_{IIb}\beta_3$), and GPVI. Each GP has a different role in the processes of physiological hemostasis and thrombus formation, such as platelet adhesion to the subendothelial matrix, firm adhesion with subsequent platelet activation, and platelet aggregation. In 1996, Weiss et al. first reported an association between the GPIIIa PlA1/A2 polymorphism and coronary thrombosis; since then a large number of studies have examined the effects of GP polymorphisms on susceptibility to atherothrombosis and receptor function. Recently, studies on GP polymorphisms have been highlighted as being of pharmacogenetic interest. Accumulating data indicate that platelet GP polymorphisms are likely targets for genetic-based individualization for the prevention and treatment of coronary artery disease, ischemic stroke, and complications following percutaneous coronary intervention, although genotyping of platelet GP polymorphisms is currently considered to be not useful for routine clinical evaluation for personalized management of atherothrombotic disorders. In this chapter, we review clinical and functional studies pertaining to each polymorphism of the platelet GP: GPIa/IIa, GPIb/IX/V, GPIIb/IIIa, and GPVI. Also we discuss issues and future directions in the research field of polymorphisms associated with susceptibility to athrothrombosis, variability in platelet function, and interindividual variations in the response to antiplatelet drugs.

Key words. Polymorphism · Platelet · Glycoprotein · Atherothrombosis · Pharmacogenetics

Introduction

Activated platelets play a crucial role in physiological hemostasis and thrombus formation at the site of vascular injury, occluding the vasculature and leading to atherothrombosis, such as coronary artery disease or ischemic stroke [1, 2]. The pathways of platelet activation are mainly mediated by platelet membrane receptors, especially glycoprotein (GP) Ia/IIa (integrin $\alpha_2\beta_1$), GPIb/IX/V, GPIIb/IIIa (integrin $\alpha_{IIb}\beta_3$), and GPVI. These GPs contribute to a variety of platelet functions in the physiological hemostasis and atherothrombotic processes, such as platelet adhesion to the subendothelial matrix, firm adhesion with subsequent platelet activation, and platelet aggregation. Accumulating evidence indicates that these platelet GPs have polymorphisms associated with the risk of atherothrombosis, variability in platelet function, and interindividual variations in platelet sensitivity to antiplatelet drugs [3–5]. The clinical significance and functional effects of platelet GP polymorphisms are currently being evaluated with the aim to prevent atherothrombosis or to produce a therapeutic benefit.

Since the first report of an association between the GPIIIa PlA1A2 polymorphism and coronary thrombosis by Weiss et al. in 1996 [6], abundant studies have examined the relation between platelet GP polymorphisms and susceptibility to atherothrombosis or receptor function. Regarding platelet sensitivity to antiplatelet drugs, some individuals have unexpected platelet reactivity to antiplatelet drugs, such as aspirin and clopidogrel. This subpopulation, called antiplatelet drug-resistant, has poor clinical outcomes. The impact of GP polymorphisms on the variability in platelet response to antiplatelet drugs has recently become evident in ex vivo and in vitro studies [3–5].

The association between GP polymorphisms and susceptibility to atherothrombotic disorders has been analyzed in case–control or cross-sectional studies in which genotype distributions among groups were compared. These clinical studies were mostly historical (retrospective) studies, which might bias the case ascertainment and heterogeneity of the study populations. Observations from different historical (retrospective) studies are therefore likely to generate conflicting views of the clinical significance of GP polymorphisms. To assess the relation between the GP polymorphisms and susceptibility to athrothrombosis reliably, a meta-analysis of the combined findings of several clinical studies or prospective studies are performed.

Glycoprotein polymorphisms might affect the quantity and/or quality of the protein product, which seems to alter platelet function. These functional effects of GP polymorphisms are examined using intact platelets and/or recombinant proteins. In the functional assays, various assay systems, such as agonist-induced platelet aggregation using aggregometer, interaction with ligands under flow conditions, or a platelet function analyzer-100 (PFA-100), are used. The PFA-100 system is a relatively new assay system using the collagen/epinephrine (CEPI) or collagen/ adenosine diphosphate (CADP) cartridge for assessing platelet function, by closure time (CT) under high shear flow conditions (5000–6000/s) [7]. The maximum CT measurement is 300 s, and longer CTs indicate lower platelet function. The CEPI-CT is sensitive to the effects of aspirin.

There have been inconsistent results among studies on GP polymorphisms, which might be caused by differences in the assays, study design, and/or donor characteristics. Thus, data on the effects of GP polymorphisms on function as well as clinical significance should be interpreted with caution. In this chapter, we review clinical and functional studies pertaining to each polymorphism of the platelet GP: GPIa/IIa, GPIb/IX/V, GPIIb/IIIa, and GPVI.

GPIa/IIa ($\alpha_2\beta_1$ Integrin)

The $\alpha_2\beta_1$ integrin acts as a receptor for collagen and laminin [8]. This integrin is present on platelets as GPIa/IIa. The α_2 integrin represents GPIa, and the β_1 integrin represents GPIIa. Integrins consist of noncovalently linked heterodimers of α and β

subunits [9]. They are usually present on the cell surface in a low- or high-affinity state. The GPIa/IIa and another collagen receptor, GPVI, are required for a full platelet response to collagen. Approximately 1000–2000 copies of GPIa/IIa are present on the platelet surface, and the interindividual variations in platelet expression levels are controlled by polymorphisms of the α_2 integrin [10]. The gene encoding α_2 integrin has several single nucleotide polymorphisms, one of which, the ^{807}T/C polymorphism at residue ^{224}Phe, is discussed below.

Clinical Studies

Since 1999, there have been at least 30 original reports of the analysis of data on the association between the ^{807}T/C polymorphism and atherothrombosis. Although some studies reported that the ^{807}T-allele is a risk factor for myocardial infarction (MI) or ischemic stroke, there are conflicting reports concerning the clinical significance of the ^{807}T/C polymorphism. Recently, a meta-analysis of 19 studies (total sample of 13835 subjects) demonstrated that the ^{807}T/C polymorphism is not a significant risk factor for coronary artery disease: ^{807}C allele versus ^{807}T allele, with an odds ratio (OR) of 0.998 [95% confidence interval (CI) 0.937–1.064] [11]. Another meta-analysis of 15 studies (total sample 14146 subjects) showed no effect of the ^{807}T/C polymorphism on coronary disease [12]. Nikolopoulos et al. reported that a meta-analysis of seven studies (total sample of 1848 subjects) showed no relation between the ^{807}T allele and risk of ischemic stroke: ^{807}C allele versus ^{807}T allele, OR 1.11 (95% CI 0.827–1.499) [13]. Therefore, it is currently thought that this polymorphism is not associated with atherothrombosis in some populations. Also, the ^{807}T-allele was reported to be a risk factor for diabetic retinopathy [14]. di Paola et al. demonstrated an association between the ^{807}T-allele and a lower risk of severe bleeding event in patients with von Willebrand disease [15].

Effect of the ^{807}T/C Polymorphism on Receptor Function

There are wide interindividual variations in platelet surface GPIa/IIa expression levels. In 1997, Kunicki et al. reported that the ^{807}T/C and ^{873}A/G polymorphisms, which are in linkage disequilibrium, are associated with GPIa/IIa levels: The ^{807}T(^{873}A)-allele has higher GPIa/IIa levels [10]. Later, linkage disequilibrium between the ^{807}T/C and $^{-52}$C/T within the promoter region was shown, and the $^{-52}$C/T affects the transcriptional activity of integrin α_2 [16]. Some observations suggest that the $^{-52}$C/T is responsible for the variability in GPIa/IIa levels. Further functional characterization demonstrated the effects of the ^{807}T/C polymorphism and platelet adhesion to immobilized collagen under flow conditions. The relation between the ^{807}T/C polymorphism and collagen-mediated platelet aggregation was shown to have inconsistent results. Accordingly, the ^{807}T/C polymorphism affects the GPIa/IIa levels, and the relation between the ^{807}T/C polymorphism and platelet function remains controversial.

Effect of the ^{807}T/C Polymorphisms on Platelet Sensitivity to Antiplatelet Drugs

Table 1 shows the reports on the association of GP polymorphisms with platelet sensitivity to antiplatelet drugs. Angiolillo et al. reported that the subjects with the

Table 1. Association between platelet glycoprotein polymorphisms and platelet sensitivity to antiplatelet drugs

Polymorphism	Effect on platelet sensitivity Yes/No	Resistant allele or genotype	Study subjects	Antiplatelet drugs	Assay	Method of comparison between genotypes	Reference
GPIa 807TC	No		601 Patients (ACS)	Clopidogrel 600 mg and aspirin 250 mg (at least 12 h before coronary angiography)	Aggregometer (ADP 10 uM) VASP phosphorylation, P-selectin expression	Maximam aggregation >70% Mean value	20
GPIa 807TC	Yes	807T	82 Patients (coronary heart disease)	Aspirin 100 mg/day and clopidogrel 75mg/ day (≥1 month)	Aggregometer (ADP 20 uM, collagen 6 ug/ ml, epinephrine 20 uM)	Mean value	17
GPIa 807TC GPIIIa PlA1/A2	No No		24 Healthy subjects	Aspirin 100 mg/day (2 days)	Aggregometer (arachidonic acid 1 mM, collagen 10 ug/ml) PFA-100 (CEPI-CT)	Mean value	19
GPIbα -5TC GPIa 807TC GPIIIa PlA1/A2	No No Yes	A1A1	98 Patients (stable angina)	Aspirin 160 mg/day (at least 1 month)	PFA-100 (CEPI-CT)	CEPI-CT <186 s	18
GPIbα -5TC, VNTR	Yes	VNTR CC+5TT	84 Patients (MI within the previous 6 months)	Aspirin 75 mg/day	PFA-100 (CADP-CT)	Mean value	31
GPIbα -5TC, 145Thr/Met	No No		22 Patients (8 patients with CAD, 14 patients without CAD) 77 Patients (45 patients with CAD, 32 patients without CAD)	Aspirin 81–325 mg within 24h prior	Aggregometer (ADP 0.5 uM, 2 uM) (epinephrine 0.4 uM, 4 uM) PFA-100 (CEPI-CT) PFA-100 (CADP-CT)	Mean value CEPI-CT >167 s Mean value	32

GPIIIa PlA1/A2	Yes	PlA2	40 Symptomless men	Aspirin 75 mg/day (1 week)	Thrombin generation	Mean value	41
GPIIIa PlA1/A2	Yes	PlA2	80 Healthy subjects	Aspirin 300 mg	Bleeding time	Mean value	49
GPIIIa PlA1/A2	Yes	PlA2	24 Healthy subjects	Aspirin 75 mg/day (1 week)	Prothrombin comsumption Fibrinogen consumption Factor Va generation Factor XIII activation Fibrinopeptides A and B	Mean value	50
GPIIIa PlA1/A2	Yes	PlA2	1014 Patients (unstable coronary syndromes, MI)	• Aspirin 150–162 mg + orbofiban 50 mg/30 mg or • Aspirin 150–162 mg + orbofiban 50 mg/50 mg or • Placebo (1–15 months)	Event Death MI Recurrent ischemia at rest leading to rehospitalization urgent revascularization Stroke		51
GPIIIa PlA1/A2	Yes	PlA2	87 Patients (undergoing percutaneous coronary interventions)	Abcixmab 0.25 mg/kg bolus then 10 ug/min (12 h)	Aggregometry (ADP 20 uM) Abciximab binding assays Ultegra rapid platelet function assay	Mean value	52
GPIIIa PlA1/A2	Yes	PlA2	38 Patients (undergoing coronary stent implantation)	Choronic treatment aspirin 100–250 mg Clopidogrel 300 mg at intervention time → Clopidogrel 75 mg/day (1 month)	Fibrinogen binding (ADP stimulation) P-selectin expression (ADP stimulation)	Mean value	53

TABLE 1. *Continued*

Polymorphism	Effect on platelet sensitivity		Study subjects	Antiplatelet drugs	Assay	Method of comparison between genotypes	Reference
	Yes/No	Resistant allele or genotype					
GPIIIa PIA1/A2	No		204 patients treated with intercoronary stent implantation (102 patients in-stent restenosis)	Aspirin 100–300 mg/day since the implantation date of intracoronary stent	PFA100 (CEPI-CT)	CEPI-CT <186 s	54
GPIIIa PIA1/A2	Yes	PIA1	28 Healthy subjects	(In vitro addition) Tirofiban 0–70 ng/ml Eptifibatide 0–1 ug/ml Abciximab 0–3 ug/ml	PFA100 (CEPI-CT)	Mean value	55
GPIIIa PIA1/A2	Yes	PIA1	48 Patients (CAD)	Clopidogrel 75 mg (2 weeks)	Bleeding time Thrombin generation PFA100 (CADP-CT) P-selectin PAC-1 binding	Mean value	56
GPIIIa PIA1/A2	No		120 Patients (undergoing percutaneous coronary intervention who received aspirin 81–325 mg for 1≥ week)	Clopidogrel 300 mg before/after 20–24 h	ADP 5 uM Arachidonic acid 1.6 uM ADP 5 uM ADP 20 uM	Maximam aggregation >70% Baseline minus posttreatment aggregation ≤10%	57
GPIIIa PIA1/A2	Yes	PIA2	28 Patients (CAD)	Aspirin 300 mg (2 weeks)	Thrombin generation Bleeding time	Mean value	58

GP, glycoprotein; ACS, acute coronary syndrome; CAD, coronary artery disease; MI, myocardial infarction; PFA-100 + A79, platelet function analyzer-100; CEPI-CT, collagen/epinephrine cartridge-closure time; CADP-CT, collagen/adenosine diphosphate cartridge-closure time

[807]T-allele had a significantly higher degree of ADP-, collagen-, and epinephrine-induced platelet aggregation among aspirin/clopidogrel-treated patients [17]. Macchi et al. [18], Gonzalez-Conejero et al. [19], and Cuisset et al. [20], however, concluded that there was no association between the [807]T/C polymorphism and platelet reactivity to aspirin and/or clopidogrel.

GPIb/IX/V

GPIb/IX/V is a receptor for von Willebrand factor (VWF); and GPIbα, the largest subunit of this complex, contains the VWF-binding site. Interaction between VWF and the GPIb/IX/V complex is critical in the adhesive process at the site of vascular injury, especially under high shear conditions [21, 22]. Approximately 25 000 copies of the GPIbα, GPIbβ, or GPIX and 12 000 copies of GPV are present on platelets. Each subunit of the complex, GPIbα, GPIbβ, GPIX, and GPV is encoded by a different gene.

Several genetic polymorphisms of GPIbα in relation to atherothrombosis have been reported: the [-5]T/C, [145]Thr/Met, and variable number tandem repeat (VNTR) polymorphisms. [-5]T/C polymorphism is located in the Kozak sequence. [145]Thr/Met polymorphism influences the antigenicity of the platelets, known as HPA-2. [145]Thr/Met and VNTR [1–4 repeats (1R to 4R, or D to A)] polymorphisms of the 13-amino-acid sequence are in linkage disequilibrium. The [145]Met-allele is tightly linked to the 3R (B)- or 4R (A)-allele, and there is a strong linkage between the [145]Thr-allele and the 1R (D)- or 2R (C)-allele. There is a race difference in the genotype distribution of the VNTR polymorphism.

Clinical Studies

In 1997, we reported that the [145]Met and 4R are associated with susceptibility to coronary artery disease in age <60 years population [23]. Since then, more than 30 studies have examined whether the GPIbα polymorphisms affect the susceptibility to coronary artery disease, acute coronary syndromes, and ischemic stroke. One large prospective study demonstrated an association between the [-5]T/C polymorphism and MI [24]. A meta-analysis of 14 studies (total sample 11 840 subjects) demonstrated no effect of the [-5]T/C polymorphism on the susceptibility to coronary disease [12]. In summary, several studies have reported that the [-5]C, [145]Met, and/or VNTR 3R 4R sequences are risk factor(s) for arterial vascular events, although these findings are somewhat controversial.

Effect of the GPIbα Polymorphisms on Receptor Function

Afshar-Kharghan et al. reported substantially increased GPIb/IX/V density on platelets with the [-5]C allele [25]. Cadroy et al. showed that the platelet deposition rate in subjects with the [-5]CT genotype was higher than that in subjects with the [-5]TT genotype [26], although there are reports of conflicting results. Jilma-Stohlawetz et al. reported a significant association of [-5]T/C or VNTR polymorphisms with CEPI assessed by PFA-100 in case of an analysis [-5]TT versus [-5]TC or VNTR C/D versus CC among 233 healthy individuals [27]. Ulrichts et al. demonstrated that the recombinant [145]Thr fragment or platelets from subjects with the [145]Thr/Thr genotype have a higher affinity for VWF binding under static conditions in the presence of ristocetin or botrocetin [28]. With regard to the discrepancy between the clinical association studies and these findings, Ulrichts et al. speculated that subjects with the [145]Thr-sequence might have a mild platelet-type von Willebrand disease with gain-of-function platelets. Patients with von Willebrand disease show spontaneous binding of high multimers of VWF to GPIbα, and thus plasma VWF is cleared, eventually resulting in bleeding problems [29]. Recently, we investigated the functional effects of the [145]Thr/Met and VNTR polymorphisms using Chinese hamster ovary cells expressing GPIbαβ/IX with different sequences of GPIbα [145]Thr/Met and VNTR polymorphisms. Our findings indicated that GPIbα proteins with both [145]Met and 4R sequences had a greater ability to interact with immobilized VWF under flow conditions [30]. The data are compatible with previous speculations that GPIbα with 4R is longer and thus places the VWF-binding global domain further away from the platelet plasma membrane. Thus, VWF would be more easily accessible to the binding site on the receptor under high shear conditions.

Effect of the GPIbα Polymorphisms on Platelet Sensitivity to Antiplatelet Drugs

Only a few published studies have addressed the effects of GPIbα polymorphisms on variability in the platelet response to antiplatelet drugs (Table 1). Macchi et al. reported no relation between the $^{-5}$T/C polymorphism and aspirin resistance status [18]. Douglas et al. reported a significant difference in CADP-CT between the VNTR CC + $^{-5}$TT and VNTR CD + $^{-5}$TC genotypes [31]. Recently, Williams et al. reported that the $^{-5}$TC and [145]Thr/Met polymorphisms are not associated with aspirin resistance status [32].

GPIIb/IIIa ($\alpha_{IIb}\beta_3$ Integrin, CD41/CD61)

The $\alpha_{IIb}\beta_3$ integrin is present on platelets as GPIIb/IIIa. The α_{IIb} integrin represents GPIIb, and the β_3 integrin represents GPIIIa. GPIIb/IIIa is the most abundant GP in platelets (80 000 copies/platelet) and acts as a receptor for fibrinogen and VWF [33]. Interactions between these molecules are essential for firm platelet aggregation. GPIIIa has a PlA1/A2 polymorphism (also known as HPA-1a/b) with amino acid substitution Leu/Pro at residue 33. The PlA1/A2 polymorphism is the most studied polymorphism of platelet GPs. The frequency of PlA1/A2 in Caucasians is approximately 72%, 26%, and 2% for the PlA1/A1, PlA1/A2, and PlA2/A2 polymorphisms, respectively. The PlA2-allele, however, is reported to be extremely rare in the Japanese population.

Clinical Studies

In 1996, the first case–control study (71 patients with acute coronary syndromes versus 68 age- and sex-matched controls) demonstrated that the PlA2-allele is a risk factor for acute coronary syndromes [6]. Since then, a large number of studies have examined the relation between the PlA1/A2 polymorphism and various forms of vascular disease. There were at least six studies in a meta-analysis regarding the association between the PlA1/A2 polymorphism and atherothrombotic disorders [12, 34–38]. The PlA2-allele is currently considered a modest risk factor for coronary thrombosis, but not stroke.

Effect of the PlA1/A2 Polymorphism on Receptor Function

Experimental studies on the immune response suggest that the PlA1/A2 polymorphism might induce structural changes in the platelets. The largest study ($n = 1422$) of ex vivo platelet function from the Framingham Offspring Study demonstrated that the increase in the ability of platelet aggregation induced by ADP and epinephrine was PlA2 dose-dependent [39]. Experimental studies of platelets and/or cultured cells expressing the PlA1 or PlA2 sequence demonstrated that the PlA1/A2 polymorphism is associated with alterations in platelet adhesion, aggregation, and secretion, or in GPIIb/IIIa-mediated adhesion, spreading, migration, and clot retraction, although there have been inconsistent results. There is an excellent review describing the cellular and molecular mechanisms underlying the association between the PlA1/A2 polymorphism and platelet functions or GPIIb/IIIa outside-in signaling function [40].

Effect of the PlA1/A2 Polymorphism on Platelet Sensitivity to Antiplatelet Drugs

In 1999, Undas et al. first demonstrated that aspirin treatment was less effective in patients with the PlA2 allele (75 mg/day for 7 days): thrombin generation was reduced in 23 of 25 PlA1/A1 carriers, but in only 9 of 15 PlA2 carriers [41]. The impact of the PlA1/A2 polymorphism on variability in the platelet response to antiplatelet drugs has been widely studied in vitro, ex vivo, and in vivo (Table 1). These studies had various study designs, and the results were inconsistent. Thus, it is difficult to interpret the impact of the PlA1/A2 polymorphism on platelet sensitivity to antiplatelet drugs.

GPVI

GPVI belongs to a member of the immunoglobulin superfamily and acts as a collagen receptor in addition to GPIa/IIa [42]. The expression of GPVI requires FcRg, whereas GPVI is not needed for FcRg expression. Both GPVI and GPIa/IIa are required for a full platelet response to collagen. There is a "two-step, two site" model of collagen activation in which binding to GPVI activates platelets and up-regulates GPIa/IIa. In 2001, Croft et al. screened GPVI single nucleotide polymorphisms in 21 healthy subjects and demonstrated five polymorphisms with amino acid substitutions: S219P, K237E, T249A, Q317L, and H322N [43]. There are relatively few reports of studies pertaining to GPVI polymorphisms.

Clinical Studies

The relation between these five polymorphisms and MI was examined in a case–control study comparing 525 MI patients and 474 controls. The results suggested that the S219P polymorphism is associated with susceptibility to MI [43]. Takagi et al.

reported that the T249A polymorphism affects MI among the Japanese population [44]. Ollikainen et al. showed that the S219P polymorphism is associated with coronary thrombosis and the area of complicated coronary lesions, as assessed by autopsy [45]. Taken together, clinical studies suggest that the [219]Pro-allele and [249]Thr-allele are risk factors for coronary thrombosis.

Effect of the GPVI Polymorphism on Platelet Function

Furihata et al. reported interindividual variations in GPVI content, as assessed by a ligand (convulxin) blotting assay; and more thrombin generation in subjects with a higher GPVI content was shown in assays of convulxin-induced prothrombinase activity [46]. Platelets from healthy subjects with the [219]Pro-allele express lower levels of GPVI and a reduced response to crosslinked collagen-related peptide in platelet aggregation, fibrinogen binding, platelet activation, and signaling [47, 48]. Although the [13254]C-allele ([219]Pro-allele) of the S219P polymorphism is suggested to be a risk factor for coronary thrombosis, this allele is associated with lower GPVI expression and reduced platelet function. Further studies of larger samples are needed to resolve these conflicting views.

Conclusion

Increasing evidence indicates that platelet GP polymorphisms are likely targets for genetics-based individualization for the prevention and treatment of coronary artery disease, ischemic stroke, and complications following percutaneous coronary intervention. Most experts in the field of polymorphisms associated with atherothrombotic disorders, however, agree that genotyping assays of platelet GPs polymorphisms are currently not useful for routine clinical evaluation for personalized management of atherothrombotic disorders. Several issues related to the study of polymorphisms remain to be addressed. Heterogeneity of the study population, inadequate sample size, and heterogeneity of assay systems might lead to inconsistent results. For further advances, it is necessary to perform well-designed, large-scale, prospective studies of the association between platelet GP polymorphisms and susceptibility to atherothrombosis. In addition, a standardized assay system is required to assess reliably the relation between GP polymorphisms and platelet function. The data obtained from carefully designed studies will contribute to optimize individualized patient care.

References

1. Ross R (1999) Atherosclerosis—an inflammatory disease. N Engl J Med 340:115–126
2. Jurk K, Kehrel BE (2005) Platelets: physiology and biochemistry. Semin Thromb Hemost 31:381–392
3. Yee DL, Bray PF (2004) Clinical and functional consequences of platelet membrane glycoprotein polymorphisms. Semin Thromb Hemost 30:591–600
4. Meisel C, Lopez JA, Stangl K (2004) Role of platelet glycoprotein polymorphisms in cardiovascular diseases. Naunyn Schmiedebergs Arch Pharmacol 369:38–54
5. Rozalski M, Boncler M, Luzak B, et al (2005) Genetic factors underlying differential blood platelet sensitivity to inhibitors. Pharmacol Rep 57:1–13

6. Weiss EJ, Bray PF, Tayback M, et al (1996) A polymorphism of a platelet glycoprotein receptor as an inherited risk factor for coronary thrombosis. N Engl J Med 334:1090–1094

7. Favaloro EJ (2002) Clinical application of the PFA-100. Curr Opin Hematol 9:407–415

8. Santoro SA, Zutter MM (1995) The alpha 2 beta 1 integrin: a collagen receptor on platelets and other cells. Thromb Haemost 74:813–821

9. Hynes RO (2002) Integrins: bidirectional, allosteric signaling machines. Cell 110:673–687

10. Kunicki TJ, Kritzik M, Annis DS, et al (1997) Hereditary variation in platelet integrin alpha 2 beta 1 density is associated with two silent polymorphisms in the alpha 2 gene coding sequence. Blood 15:1939–1943

11. Tsantes AE, Nikolopoulos GK, Bagos PG, et al (2007) Lack of association between the platelet glycoprotein Ia C807T gene polymorphism and coronary artery disease: a meta-analysis. Int J Cardiol 118:189–196

12. Ye Z, Liu EHC, Higgins JPT, et al (2006) Seven haemostatic gene polymorphisms in coronary disease: meta-analysis of 66,155 cases and 91,307 controls. Lancet 367:651–658

13. Nikolopoulos GK, Tsantes AE, Bagos PG, et al (2007) Integrin, alpha 2 gene C807T polymorphism and risk of ischemic stroke: a meta-analysis. Thromb Res 119:501–510

14. Matsubara Y, Murata M, Maruyama T, et al (2000) Association between diabetic retinopathy and genetic variations in alpha2beta1 integrin, a platelet receptor for collagen. Blood 95:1560–1564

15. Di Paola J, Federici AB, Mannucci PM, et al (1999) Low platelet alpha2beta1 levels in type I von Willebrand disease correlate with impaired platelet function in a high shear stress system. Blood 93:3578–3582

16. Jacquelin B, Tarantino MD, Krizik M, et al (2001) Allele-dependent transcriptional regulation of the human integrin alpha 2 gene. Blood 97:1721–1726

17. Angiolillo DJ, Fernandez-Oritiz A, Bernardo E, et al (2005) Variability in platelet aggregation following sustained aspirin and clopidogrel treatment in patients with coronary heart disease and influence of the 807 C/T polymorphism of the glycoprotein Ia gene. Am J Cardiol 96:1095–1099

18. Macchi L, Christiaens L, Brabant S, et al (2003) Resistance in vitro to low-dose aspirin is associated with platelet PlA1 (GP IIIa) polymorphism but not with C807T(GP Ia/IIa) and C-5T Kozak (GP Ibalpha) polymorphisms. J Am Coll Cardiol 42:1115–1119

19. Gonzalez-Conejero R, Rivera J, Corral J, et al (2005) Biological assessment of aspirin efficacy on healthy individuals: heterogeneous response or aspirin failure? Stroke 36:276–280

20. Cuisset T, Frere C, Quilici J, et al (2007) Lack of association between the 807 C/T polymorphism of glycoprotein Ia gene and post-treatment platelet reactivity after aspirin and clopidogrel in patients with acute coronary syndrome. Thromb Haemost 97:212–217

21. Clemetson KJ (1997) Platelet GPIb-V-IX complex. Thromb Haemost 78:266–270

22. Berndt MC, Shen Y, Dpheide SM, et al (2001) The vascular biology of the glycoprotein Ib-IX-V complex. Thromb Haemost 86:178–188

23. Murata M, Matsubara Y, Kawano K, et al (1997) Coronary artery disease and polymorphisms in a receptor mediating shear stress-dependent platelet activation. Circulation 96:3281–3286

24. Kenny D, Muckian C, Fitzgerald DJ, et al (2002) Platelet glycoprotein Ib alpha receptor polymorphisms and recurrent ischemic events in acute coronary syndrome patients. J Thromb Thrombolysis 13:13–19

25. Afshar-Kharghan V, Li CQ, Khoshnevis-Asl M, et al (1999) Kozak sequence polymorphism of the glycoprotein (GP) Ibalpha gene is a major determinant of the plasma membrane levels of the platelet GP Ib-IX-V complex. Blood 94:186–191
26. Cadroy Y, Sakariassen KS, Charlet JP, et al (2001) Role of 4 platelet membrane glycoprotein polymorphisms on experimental arterial thrombus formation in men. Blood 98:3159–3161
27. Jilma-Stohlawetz P, Homoncik M, Jilma B, et al (2003) Glycoprotein Ib polymorphisms influence platelet plug formation under high shear rates. Br J Haematol 120:652–655
28. Ulrichts H, Vanhoorelbeke K, Cauwenberghs H, et al (2003) von Willebrand factor but not alpha-thrombin binding to platelet glycoprotein Ibalpha is influenced by the HPA-2 polymorphism. Arterioscler Thromb Vasc Biol 23:1302–1307
29. Miller JL (1996) Platelet-type von Willebrand disease. Thromb Haemost 75:865–869
30. Matsubara Y, Murata M, Hayashi T, et al (2005) Platelet glycoprotein Ib alpha polymorphisms affect the interaction with von Willebrand factor under flow conditions. Br J Haematol 128:533–539
31. Douglas H, Davies GJ, Michaelides K, et al (2006) Detection of functional differences between different platelet membrane glycoprotein Ibalpha variable number tandem repeat and Kozak genotypes as shown by the PFA-100 system. Heart 92:676–678
32. Williams MS, Ng'alla LS, Vaidya D (2007) Platelet functional implications of glycoprotein Ibalpha polymorphisms in African Americans. Am J Hematol 82:15–22
33. Phillips DR, Charo IF, Parise LV, et al (1988) The platelet membrane glycoprotein IIb-IIIa complex. Blood 71:831–843
34. Zhu MM, Weedon J, Clark LT (2000) Meta-analysis of the association of platelet glycoprotein IIIa PlA1/A2 polymorphism with myocardial infarction. Am J Cardiol 86:1000–1005
35. Di Castelnuovo A, de Gaetano G, Donati MB, et al (2001) Platelet glycoprotein receptor IIIa polymorphism PLA1/PLA2 and coronary risk: a meta-analysis. Thromb Haemost 85:626–633
36. Wu AB, Tsongalis GJ (2001) Correlation of polymorphisms to coagulation and biochemical risk factors for cardiovascular diseases. Am J Cardiol 87:1361–1366
37. Burr D, Doss H, Cooke GE, et al (2003) A meta-analysis of studies on the association of the platelet PlA polymorphism of glycoprotein IIIa and risk of coronary heart disease. Stat Med 22:1741–1760
38. Wiwanitkit V (2006) PIA1/A2 polymorphism of the platelet glycoprotein receptor IIb/IIIIa and its correlation with myocardial infarction: an appraisal. Clin Appl Thromb Hemost 12:93–95
39. Feng D, Lindpaintner K, Larson MG, et al (1999) Increased platelet aggregability associated with platelet GPIIIa PlA2 polymorphism: the Framingham Offspring Study. Arterioscler Thromb Vasc Biol 19:1142–1147
40. Vijayan KV, Bray PF (2006) Molecular mechanisms of prothrombotic risk due to genetic variations in platelet genes: enhanced outside-in signaling through the Pro33 variant of integrin beta3. Exp Biol Med (Maywood) 231:505–513
41. Undas A, Sanak M, Musial J, et al (1999) Platelet glycoprotein IIIa polymorphism, aspirin, and thrombin generation. Lancet 353:982–983
42. Clemetson JM, Polgar J, Mognenat E, et al (1999) The platelet collagen receptor glycoprotein VI is a member of the immunoglobulin superfamily closely related to FcαR and the natural killer receptors. J Biol Chem 274:29019–29024
43. Croft SA, Samani NJ, Teare MD, et al (2001) Novel platelet membrane glycoprotein VI dimorphism is a risk factor for myocardial infarction. Circulation 104:1459–1463
44. Takagi S, Iwai N, Baba S, et al (2002) A GPVI polymorphism is a risk factor for myocardial infarction in Japanese. Atherosclerosis 165:397–398

45. Ollikainen E, Mikkelsson J, Perola M (2004) Platelet membrane collagen receptor glycoprotein VI polymorphism is associated with coronary thrombosis and fatal myocardial infarction in middle-aged men. Atherosclerosis 176:95–99
46. Furihata K, Clemetson KJ, Deguchi H, et al (2001) Variation in human platelet glycoprotein VI content modulates glycoprotein VI-specific prothrombinase activity. Arterioscler Thromb Vasc Biol 21:1857–1863
47. Best D, Senis YA, Jarvis GE, et al (2003) GPVI levels in platelets: relationship to platelet function at high shear. Blood 102:2811–2818
48. Joutsi-Korhonen L, Smethurst PA, Rankin A, et al (2003) The low-frequency allele of the platelet collagen signaling receptor glycoprotein VI is associated with reduced functional responses and expression. Blood 101:4372–4379
49. Szczeklik A, Undas A, Sanak M, et al (2000) Relationship between bleeding time, aspirin and the PlA1/A2 polymorphism of platelet glycoprotein IIIa. Br J Haematol 110:965–967
50. Undas A, Brummel K, Musial J, et al (2001) Pl(A2) polymorphism of beta(3) integrins is associated with enhanced thrombin generation and impaired antithrombotic action of aspirin at the site of microvascular injury. Circulation 104:2666–2672
51. O'Connor FF, Shields DC, Fitzgerald A, et al (2001) Genetic variation in glycoprotein IIb/IIIa (GPIIb/IIIa) as a determinant of the responses to an oral GPIIb/IIIa antagonist in patients with unstable coronary syndromes. 98:3256–3260
52. Wheeler GL, Braden GA, Bray PF, et al (2002) Reduced inhibition by abciximab in platelets with the PlA2 polymorphism. Am Heart J 143:76–82
53. Angiolillo DJ, Fernandez-Ortiz A, Bernardo E, et al (2004) PlA polymorphism and platelet reactivity following clopidogrel loading dose in patients undergoing coronary stent implantation. Blood Coagul Fibrinolysis 15:89–93
54. Pamukcu B, Oflaz H, Nisanci Y (2005) The role of platelet glycoprotein IIIa polymorphism in the high prevalence of in vitro aspirin resistance in patients with intracoronary stent restenosis. Am Heart J 149:675–680
55. Aalto-Satala K, Karhunen PJ, Mikkelsson J, et al (2005) The effect of glycoprotein IIIa PIA 1/A2 polymorphism on the PFA-100 response to GP IIb IIa receptor inhibitors—the importance of anticoagulants used. J Thromb Thrombolysis 20:57–63
56. Dropinski J, Musial J, Jakiela B, et al (2005) Anti-thrombotic action of clopidogrel and P1(A1/A2) polymorphism of beta3 integrin in patients with coronary artery disease not being treated with aspirin. Thromb Haemost 94:1300–305
57. Lev EI, Patel RT, Guthikonda S, et al (2007) Genetic polymorphisms of the platelet receptors P2Y(12), P2Y(1) and GP IIIa and response to aspirin and clopidogrel. Thromb Res 119:355–360
58. Dropinski J, Musial J, Sanak M, et al (2007) Antithrombotic effects of aspirin based on PLA1/A2 glycoprotein IIIa polymorphism in patients with coronary artery disease. Thromb Res 119:301–303

Platelet Procoagulant Activity Appeared by Exposure of Platelets to Blood Flow Conditions

Shinya Goto

Summary. Platelets can be activated by interaction of their receptor proteins GPIbα and GPIIb/IIIa with von Willebrand factor under conditions of high shear stress. Platelets activated thus exhibit procoagulant activity through surface expression of negatively charged phospholipid and the release of procoagulant microparticles. Accumulation of tissue factor-bearing microparticles released from leukocytes around activated platelets may also contribute in some part to platelet-derived procoagulant activity. We developed an assay system to demonstrate the appearance of activated platelet-derived procoagulant activity on platelet thrombi formed on collagen fibrils under blood flow conditions. In this system, platelet thrombus formation as well as the appearance of fibrin monomers around the platelet thrombi can be visualized simultaneously with a multicolor fluorescence imaging system. We were able to provide evidence to suggest that some antiplatelet agents may have the potential to block activation of the coagulant cascade through inhibiting the procoagulant activity derived from activated platelets. In summary, it is important to interrupt the positive feedback loop between activation of platelets and the coagulant cascade for preventing the onset of arterial thrombotic diseases, such as acute myocardial infarction.

Key words. Platelet · Procoagulant activity · Thrombin · von Willebrand factor

Relation Between Platelets and Fibrin Formation

Platelet function has been assessed independently from that of activation of the coagulation cascade. Many expert investigators focusing on platelet function have contributed to detailed clarification of the mechanism of platelet activation. The same is true for clarification of the functioning of the coagulant cascade. Indeed, traditional platelet function tests, such as the platelet aggregation test, involve exogenous addition of soluble agonists, including ADP and collagen, to platelet-rich plasma, where the functioning of the coagulation cascade is blocked by the anticoagulant citrate. Similarly, coagulant function testing, such as measurement of the prothrombin time (PT), has been conducted in the absence of blood cells, including platelets. Until now, only a few investigators have exhibited interest in the interplay between platelets and the coagulant cascade.

Accordingly, most antithrombotic agents have been developed either as antiplatelet agents or anticoagulants. Screening for these drugs was conducted with traditional function tests, including tests of platelet aggregation or blood clotting. Our recent experience has indicated that the effects observed in these function tests do not directly translate into effective prevention of thrombotic events. The most striking experience was obtained with the clinical development of anti-GPIIb/IIIa agents. Because platelet aggregation is mediated by binding of the plasma ligands fibrinogen or von Willebrand factor with activated GPIIb/IIIa, anti-GPIIb/IIIa agents were developed as strong inhibitors of platelet aggregation [1]. When anti-GPIIb/IIIa agents were developed, many investigators believed that arterial thrombotic diseases, such as acute myocardial infarction, could be efficiently prevented by these agents [2, 3]. Indeed, during the initial experience, these agents were found to prevent thrombotic complications after coronary intervention [4]. However, orally available anti-GPIIb/IIIa agents, which could efficiently block platelet aggregation, were not effective at preventing the onset of arterial thrombotic diseases, including acute myocardial infarction [5]. These experiences raise questions regarding the importance of platelet aggregation and platelet thrombus formation in the onset of arterial thrombotic diseases [6]. It was therefore speculated that other important functions of platelets, such as platelet-derived procoagulant activity, might be involved in the onset of arterial thrombotic diseases.

Recently, we demonstrated the components of coronary thrombi that were causing acute myocardial infarction using freshly obtained thrombus specimens from acute myocardial infarction patients with percutaneous aspiration devices [7, 8]. As reported previously [7] and shown in Fig. 1, we demonstrated the consistent presence of platelets, fibrin, and leukocytes in the thrombus specimens [7]. Moreover, we demonstrated the co-localization of fibrin with activated platelets in most of the thrombus specimens (Fig. 1) [7]. Immunohistochemical staining of fibrin (left panel, Fig. 1) and platelets (right panel, Fig. 1) revealed the localization of fibrin fibrils around the platelets. These results strongly indicate the importance of platelets in the formation

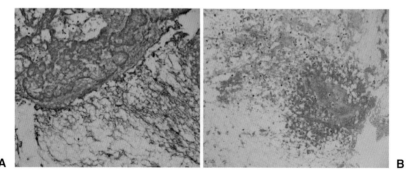

A   **B**

Fig. 1. Histochemical staining of freshly obtained coronary thrombi from patients with acute myocardial infarction using a percutaneous thrombus aspiration device. Histochemical staining of fibrin (**A**) and platelets (**B**) are shown. Co-localization of platelets and fibrin can be recognized in both panels of the figure (Prepared in collaboration with Prof. Asada, Miyazaki University)

of fibrin thrombi, especially in the formation of pathological thrombi causing acute myocardial infarction. We think that activation of the coagulant cascade around activated platelets, which was previously named "platelet-derived procoagulant activity" [9], may play a more important role in the onset of arterial thrombosis than speculated previously.

The concept of platelet-derived procoagulant activity was established a long time ago [10]. However, the exact mechanism of platelet-derived procoagulant activity is still under investigation [9]. In experiments conducted with citrated platelet-rich plasma, it was found that the recalcification clotting time was significantly shortened in the presence of activated platelets [11]. Although the exact mechanisms have not yet been clarified, there are obviously several important players in this observed shortening of the clotting time, such as surface expression of negatively charged phospholipid (which can be detected by binding of annexin V) [11, 12], microparticle release [11], and formation of the prothrombinase complex on the surface of activated platelets [9], among others. It is important to note that some investigators also believe that there is a contributory role of tissue factor-carrying microparticles (presumably released from leukocytes and endothelial cells) in the expression of platelet-derived procoagulant activity [13, 14].

Initially, platelet-derived procoagulant activity was tested by demonstrating the shortening of the recalcification clotting time in the presence of activated platelets and a contact activation factor such as kaolin. Thus, initially recognized platelet-derived procoagulant activity represented augmentation of the intrinsic pathway of blood coagulation [11]. Recently, the important role of activated platelets in the formation of fibrin fibrils was also demonstrated with in vivo animal models [15]. Real-time in vivo imaging with the use of ultra-fast laser confocal microscopy was effective for demonstrating the simultaneous accumulation of platelets, tissue factor, and fibrin at the site of laser damage in the cremasteric arteries of mice [15]. These animal models, along with our clinical observations, strongly support the notion of the importance of a relation between platelet activation and fibrin formation in the onset of arterial thrombotic diseases.

Shear-Induced Platelet Activation and Aggregation

Unlike classic agonist-induced platelet activation and aggregation established during the 1960s [16], it was found that platelet activation and aggregation can be induced without exogenous addition of any agonists when platelet-rich plasma is exposed to high shear rate conditions [17–19]. We demonstrated that shear-induced platelet activation is mediated through interaction of both the platelet GPIbα and GPIIb/IIIa complexes with von Willebrand factor (VWF) (Fig. 2) [19]. The former appears to be especially important to capture platelets under high shear stress conditions [20]. Indeed, platelet aggregation induced by the addition of agonists such as ADP, collagen, and thrombin can be easily reversed if the interaction of GPIbα with VWF is blocked by specific inhibitors [20]. It is worthy of note that ADP released with continuous stimulation of the ADP receptor $P2Y_{12}$ and the subsequent increase in the intracellular calcium ion concentration is also crucial to keep the platelet aggregates and thrombi stable under high shear stress conditions [21].

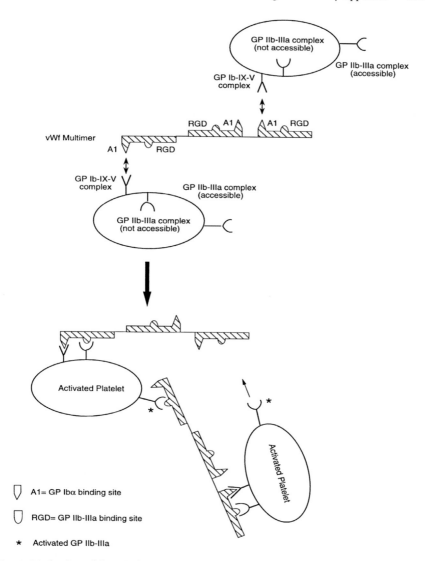

Fig. 2. Mechanism of shear-induced platelet aggregation. Platelets transiently interact with von Willebrand factor (*VWF*) through GPIbα under high shear stress conditions. This causes stretching of the VWF molecules by several platelets attached to the same VWF molecules. GPIIb/IIIa appeared to induce aggregation by some unknown mechanism (From Goto et al. [19], with permission)

When platelets are activated by high shear stress, platelet-derived procoagulant activity is also presumed to increase because platelets expressing negatively charged phospholipid (detected by annexin V) release procoagulant microparticles [11]. Other vasoactive substances, such as CD40 ligand [22], serotonin, certain chemokines such as RANTES, and some as yet unclarified vasoactive proteins [23] are also released upon exposure of platelets to high shear stress. Because platelets play a crucial role in the formation of thrombi under blood flow conditions, this shear-induced activation of platelets may play an important role in the onset of arterial thrombotic diseases [6].

Regulation of Shear-Induced Expression of Platelet-Derived Procoagulant Activity

We have previously shown that the clotting time of citrated plasma is shortened after the addition of calcium and the contact activation factor kaolin in the presence of platelets exposed to high shear stress [11]. In the same study, we also showed increased microparticle release upon exposure of platelet-rich plasma to high shear stress [11]. Most of the platelet-derived microparticles we measured in this study exhibited procoagulant potential, as they showed positive annexin V binding [21]. Accordingly, we suppose that platelets begin to exhibit procoagulant activity after exposure to the high shear stress of blood flow conditions such as occurs at the site of atherosclerotic stenosis or the site of stent implantation [24].

This shear-induced increase in platelet procoagulant activity appears to be mediated by the binding of GPIbα to VWF because specific antagonists blocking this interaction completely inhibit the shear-induced platelet-derived procoagulant activity [21]. It is noteworthy that some anti-GPIIb/IIIa agents, such as abciximab, also have the potential to inhibit the shear-induced increase of the procoagulant activity of platelets because they can block not only GPIIb/IIIa but also another integrin of $\alpha_v\beta_3$ [11].

Another important finding was that the shear-induced procoagulant microparticle release was prevented when the ADP receptor, known as a target of the active metabolite(s) of the commonly used antiplatelet agents ticlopidine and clopidogrel (P2Y$_{12}$), was blocked by specific inhibitors [21]. It is reasonable to suppose that the clinically proven antithrombotic effects of ticlopidine and clopidogrel [25, 26] may be dependent, in part, on the effects of these drugs in preventing shear-induced induction of platelet-derived procoagulant activity.

New Methodology to Detect Procoagulant Activity on Platelet Thrombi Formed on Collagen Fibrils Under High Shear Stress Conditions

Recently, we developed a method to detect the three-dimensional growth of platelet thrombi on collagen fibrils under blood flow conditions [27, 28]. The advantage of this method over the use of animal models is that the experiments can be conducted with human blood specimens. With the use of relatively weak and specific thrombin inhibitors, such as argatroban, activation of the coagulant cascade on the surface of the platelet thrombi can be detected with the use of a fluorescinated antibody against fibrin monomers (Fig. 3). This method may provide clinically relevant data because the mechanism of platelet thrombus formation and fibrin formation in this method

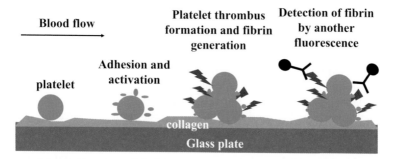

Fɪɢ. 3. Methods to detect the activated platelet-derived procoagulant activity appearing on the surface of platelets forming thrombi on collagen fibrils under blood flow conditions. Flowing platelets rendered fluorescent by the addition of mepacrine adhere to the collagen fibrils through the vWF molecules bound on them. This induces platelet activation and the formation of thrombi. A small amount of fibrin monomer formation around the platelet thrombi was detected using a monoclonal antibody against fibrin monomers that had been rendered fluorescent with a wavelength other than that of mepacrine

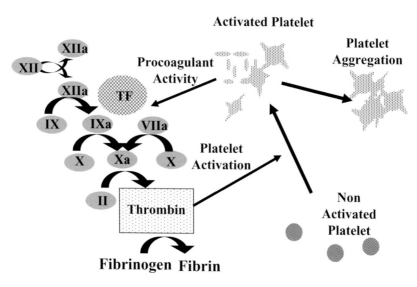

Fɪɢ. 4. Relation between platelets and the coagulation cascade. Once platelets are activated, they provide a procoagulant surface for the formation of thrombin and fibrin. Because platelets have thrombin receptors, such as protease activated receptor 1 (PARR1), platelets forming thrombi are also stimulated by thrombin generated on the surface of the activated platelets. This interaction between platelets and the coagulant cascade may play an important role in the formation of pathological thrombi, causing arterial thrombotic diseases, including myocardial infarction and ischemic stroke

is similar to that believed to occur at sites where the subendothelial matrix is exposed to the blood stream at the site of atheroma rupture [29].

With the use of this newly developed assay system, we demonstrated that not only fibrin monomer formation but also the three-dimensional growth of platelet thrombi can be inhibited by blocking the functions of thrombin generated on the surface of platelet thrombi (N. Tamura et al., unpublished findings). These results suggest that anticoagulant agents may inhibit not only the coagulant cascade but also the growth of platelet thrombi by blocking the effect of thrombin on the activation of platelet thrombi; thus, they can also be considered antiplatelet agents. Similarly, some antiplatelet agents may also act as anticoagulants because they prevent procoagulant activity by blocking platelet activation. For the prevention of arterial thrombotic diseases (including acute myocardial infarction and ischemic stroke [30, 31]), it is important to interrupt the positive feedback loop existing between platelets and the coagulation cascade (Fig. 4). Inhibition of platelet-derived procoagulant activity might thus represent a valid target in the future development of antithrombotic agents.

References

1. Yeghiazarians Y, Braunstein JB, Askari A, et al (2000) Unstable angina pectoris. N Engl J Med 342:101–114
2. Coller BS, Peerschke EI, Scudder LE, et al (1983) A murine monoclonal antibody that completely blocks the binding of fibrinogen to platelets produces a thrombasthenic-like state in normal platelets and binds to glycoproteins IIb and/or IIIa. J Clin Invest 72:325–338
3. Boersma E, Harrington RA, Moliterno DJ, et al (2002) Platelet glycoprotein IIb/IIIa inhibitors in acute coronary syndromes: a meta-analysis of all major randomised clinical trials. Lancet 359:189–198
4. (1997) Platelet glycoprotein IIb/IIIa receptor blockade and low-dose heparin during percutaneous coronary revascularization. The EPILOG Investigators. N Engl J Med 336:1689–1696
5. Gurbel PA, Serebruany VL (2000) Oral platelet IIb/IIIa inhibitors: from attractive theory to clinical failures. J Thromb Thrombolysis 10:217–220
6. Goto S (2004) Understanding the mechanism of platelet thrombus formation under blood flow conditions and the effect of new antiplatelet agents. Curr Vasc Pharmacol 2:23–32
7. Hoshiba Y, Hatakeyama K, Tanabe T, et al (2006) Co-localization of von Willebrand factor with platelet thrombi, tissue factor and platelets with fibrin, and consistent presence of inflammatory cells in coronary thrombi obtained by an aspiration device from patients with acute myocardial infarction. J Thromb Haemost 4:114–120
8. Yamashita A, Asada Y, Sugimura H, et al (2003) Contribution of von Willebrand factor to thrombus formation on neointima of rabbit stenotic iliac artery under high blood-flow velocity. Arterioscler Thromb Vasc Biol 23:1105–1110
9. Monroe DM, Hoffman M, Roberts HR (2002) Platelets and thrombin generation. Arterioscler Thromb Vasc Biol 22:1381–1389
10. Niemetz J, Marcus AJ (1974) The stimulatory effect of platelets and platelet membranes on the procoagulant activity of leukocytes. J Clin Invest 54:1437–1443
11. Goto S, Tamura N, Li M, et al (2003) Different effects of various anti-GPIIb-IIIa agents on shear-induced platelet activation and expression of procoagulant activity. J Thromb Haemost 1:2022–2030

12. Miyazaki Y, Nomura S, Miyake T, et al (1996) High shear stress can initiate both platelet aggregation and shedding of procoagulant containing microparticles. Blood 88: 3456–3464

13. Giesen PL, Rauch U, Bohrmann B, et al (1999) Blood-borne tissue factor: another view of thrombosis. Proc Natl Acad Sci U S A 96:2311–2315

14. Falati S, Liu Q, Gross P, et al (2003) Accumulation of tissue factor into developing thrombi in vivo is dependent upon microparticle P-selectin glycoprotein ligand 1 and platelet P-selectin. J Exp Med 197:1585–1598

15. Falati S, Gross P, Merrill-Skoloff G, et al (2002) Real-time in vivo imaging of platelets, tissue factor and fibrin during arterial thrombus formation in the mouse. Nat Med 8:1175–1181

16. Born GV (1962) Aggregation of blood platelets by adenosine diphosphate and its reversal. Nature 194:927–929

17. Ikeda Y, Handa M, Kawano K, et al (1991) The role of von Willebrand factor and fibrinogen in platelet aggregation under varying shear stress. J Clin Invest 87: 1234–1240

18. Goto S, Ikeda Y, Murata M, et al (1992) Epinephrine augments von Willebrand factor-dependent shear-induced platelet aggregation. Circulation 86:1859–1863

19. Goto S, Salomon DR, Ikeda Y, et al (1995) Characterization of the unique mechanism mediating the shear-dependent binding of soluble von Willebrand factor to platelets. J Biol Chem 270:23352–23361

20. Goto S, Ikeda Y, Saldivar E, et al (1998) Distinct mechanisms of platelet aggregation as a consequence of different shearing flow conditions. J Clin Invest 101:479–486

21. Goto S, Tamura N, Eto K, et al (2002) Functional significance of adenosine 5'-diphosphate receptor (P2Y(12)) in platelet activation initiated by binding of von Willebrand factor to platelet GP Ibalpha induced by conditions of high shear rate. Circulation 105:2531–2536

22. Tamura N, Yoshida M, Ichikawa N, et al (2002) Shear-induced von Willebrand factor-mediated platelet surface translocation of the CD40 ligand. Thromb Res 108:311–315

23. Hagihara M, Higuchi A, Tamura N, et al (2004) Platelets, after exposure to a high shear stress, induce IL-10-producing, mature dendritic cells in vitro. J Immunol 172: 5297–5303

24. Sakakibara M, Goto S, Eto K, et al (2002) Application of ex vivo flow chamber system for assessment of stent thrombosis. Arterioscler Thromb Vasc Biol 22:1360–1364

25. Schomig A, Neumann FJ, Kastrati A, et al (1996) A randomized comparison of anti-platelet and anticoagulant therapy after the placement of coronary-artery stents. N Engl J Med 334:1084–1089

26. Bhatt DL, Fox KA, Hacke W, et al (2006) Clopidogrel and aspirin versus aspirin alone for the prevention of atherothrombotic events. N Engl J Med 354:1706–1717

27. Goto S, Tamura N, Ishida H, et al (2006) Dependence of platelet thrombus stability on sustained glycoprotein IIb/IIIa activation through adenosine 5'-diphosphate receptor stimulation and cyclic calcium signaling. J Am Coll Cardiol 47:155–162

28. Goto S, Tamura N, Ishida H (2004) Ability of anti-glycoprotein IIb/IIIa agents to dissolve platelet thrombi formed on a collagen surface under blood flow conditions. J Am Coll Cardiol 44:316–323

29. Goto S, Handa S (1998) Coronary thrombosis. Effects of blood flow on the mechanism of thrombus formation. Jpn Heart J 39:579–596

30. Steg PG, Bhatt DL, Wilson PW, et al (2007) One-year cardiovascular event rates in outpatients with atherothrombosis. Jama 297:1197–1206

31. Bhatt DL, Steg PG, Ohman EM, et al (2006) International prevalence, recognition, and treatment of cardiovascular risk factors in outpatients with atherothrombosis. Jama 295:180–189

Part 4 Tissue Proteolysis

Regulation of Cellular uPA Activity and Its Implication in Pathogenesis of Diseases

Soichi Kojima

Summary. Tissue fibrinolysis or tissue-associated fibrinolysis is important not only for local fibrinolysis but also for maintenance of tissue homeostasis and remodeling by means of the activation of other proteases such as procollagenases and growth factors/cytokines, including insulin-like growth factor I (IGF-I), hepatocyte growth factor (HGF), platelet-derived growth factor (PDGF), and transforming growth factor-β (TGFβ). The major plasminogen activator (PA) involved in this process is urokinase-type PA (uPA). Therefore, the expression and activity of uPA and its cell surface binding protein, uPA receptor (uPAR) are rigidly regulated. Impairment of this regulation results in the pathogenesis of numerous diseases including hepatic diseases, atherosclerosis, and cancer metastasis.

Key words. Tissue fibrinolysis · Urokinase-type plasminogen activator · Cell surface reaction · Matrix metalloproteinases · Growth factors

Introduction

Fibrinolysis was originally found as a biophylactic reaction in which coagulated blood clots are degraded into soluble degradation products by a serine protease, plasmin. Plasmin, a two-chain glycoprotein, emerges from its single-chain zymogen, plasminogen (plasma concentrations, 130–180 μg/ml in humans) by the action of plasminogen activators (PAs) (Fig. 1). There are two physiological PAs: tissue-type PA (tPA) and urokinase-type PA (uPA). They cleave the Arg561-Val562 peptide bond in plasminogen and convert it to plasmin containing the active site triad composed of His603, Asp646, and Ser741. In addition to tPA and uPA, a couple of bacterial proteins, staphylokinase and streptokinase, are known to activate plasminogen to plasmin by forming complexes with plasminogen; they thus have been used for clinical treatment.

From early in the history of studying fibrinolysis, vascular endothelial cells have been noted as a major source of PAs, and fibrin has been studied extensively as a primary substrate. However, along with progress in understanding the molecular and cellular biology of fibrinolysis, it has become evident that most cell and tissue types produce tPA and/or uPA, and that fibrinolytic reactions, namely the PA–plasmin system, is involved not only in original thrombolysis but also in regulation of cell growth, migration, invasion, and differentiation, thereby participating in tissue

Fibronolytic (PA/plasmin) System

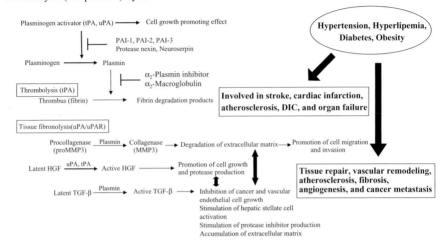

FIG. 1. Fibrinolytic [plasminogen activator (*PA*)/plasmin] system. *tPA*, *uPA*, tissue- and uroki-nase-type PA; *uPAR*, uPA receptor; *MMP*, matrix metalloproteinase; *PAI*, PA inhibitor; *HGF*, hepatocyte growth factor; *TGF*, transforming growth factor; *DIC*, disseminated intravascular coagulation

formation and remodeling as well as related diseases such as cancer metastasis and, more recently discovered, angiogenesis and liver fibrosis. We collectively call these fibrinolytic reactions "tissue fibrinolysis" (Fig. 1).

Players of Tissue Fibrinolysis

tPA and uPA

It has long been known that various mammalian tissues have PAs, and early on (before the late 1970s) they were classified into tissue, vascular, blood, and urine PAs based on their source. However, upon cloning these enzymes, it was disclosed that these PAs are either tPA, uPA, or mixture of tPA and uPA (Table 1).

Tissue PA is the major PA responsible for thrombolysis, which is accomplished by means of its fibrin-binding activity and conforming the fibrin-specific PAs together with single-chain uPA (scuPA). The tPA antigen concentration in human plasma is 4–10 ng/ml [1]. The human tPA gene (*PLAT*) consists of 14 exons separated by 13 introns, which code for a 1686-nucleotide transcript and thus a 562-aminoacid single-chain glycoprotein with molecular weight of 70 000. It has heterogeneity in the N-terminal region (Gly1-Ala2-Arg3 and Ser4-Tyr5-Arg6, corresponding to deletion of the three terminal amino acids from the sequence originally reported) [2]. Human tPA can be found in a two-chain form (tstPA) that has a proteolytic processing site (Arg275-Ile276) formed by plasmin, thrombin, tissue kallikrein, and factor Xa and consists of N-terminal A chain and C-terminal B chain. The A chain contains one finger domain homologous to fibronectin type I, one epidermal growth factor (EGF) domain, and two kringle domains, whereas the B chain has the active site triad

TABLE 1. Summary of human fibrinolytic components

Protein	Gene symbol	Chromosome	AA	Mr	Plasma levels	Homozygous deficiency in mice	
						Embryonic lethal	Spontaneous thrombosis
Plasminogen	PLG	6	810	~88 000	130–180 µg/ml	No	Yes
tPA	PLAT	8	562	~70 000	4–10 ng/ml	No	No
uPA	PLAU	10	431		3–5 ng/ml	No	No
High MW				~55 000			
Low MW				~31 500			
PAI-1	SERPINE1	7	402	~50 000	4–10 ng/ml	No	No
PAI-2	SERPINE2	18	415	~60 000[a]	Not reported	No	No
				~43 000[b]			
uPAR	PLAUR	19	335	~55 000	Not reported	No	No
Annexin II	ANXA2	15	339	~39 000	Not reported	No	No

[a]Extracellular glycosylated from
[b]Intercellular form

composed of His322, Asp371, and Ser478. Plasminogen activation by tPA is stimulated 160-folds with 0.32 μM cyanogen bromide-digested fibrinogen.

uPA is found in urine and is secreted by various cell types as a 431-amino-acid single-chain glycoprotein, scuPA or pro-uPA, having a molecular weight of 55 000. Normal human urinary uPA antigen concentration is 3–5 ng/ml [1]. The human uPA gene (*PLAU*; accession no. AF377330) is about 10 kb consisting of 10 exons and 9 introns (Fig. 2). Proteolytic cleavage of the Lys158-Ile159 peptide bond by plasmin or plasma kallikrein converts scuPA to two-chain uPA (tcuPA or called just uPA or urokinase) composed of an N-terminal A-chain with a molecular weight of 22 000 and a C-terminal B-chain with a molecular weight of 33 000. Further processing by plasmin of Lys135-Lys136 and Arg156-Phe157 peptide bonds results in formation of the low-molecular-weight uPA (31 500). The A-chain contains one EGF domain (Cys11-Cys50) and one kringle domain (Cys50-Cys131), whereas the B chain contains the active site triad composed of His204, Asp255, and Ser356. tcuPA is the non-fibrin-specific PA. Instead, in the most cases, it functions via binding to a specific glycolipid-anchored cell surface receptor, uPA receptor (uPAR), making uPA the membrane-bound major fibrinolytic enzyme responsible for tissue fibrinolysis [3].

Inhibitors of Fibrinolysis

Fibrinolysis is regulated at two steps: inhibition of plasmin activity by α_2-antiplasmin and inhibition of PA activity by PA inhibitors (PAIs) (Fig. 1, Table 1). All of these fibrinolysis inhibitors belong to the serpin family of serine protease inhibitors, sharing 30%–45% homology.

α_2-Antiplasmin is the fast-acting specific plasmin inhibitor composed of 452 amino acids and has a molecular weight of 67 000. α_2-Antiplasmin is produced in and secreted from the liver, and its plasma concentration is 70 mg/l (~1 μM)—about half of the plasma concentration of plasminogen. N-terminal Gln2 is crosslinked to fibrin via factor XIIIa (plasma transglutaminase; see the chapter by Ichinose in this book).

PAI-1 is a single-chain glycoprotein consisting of 397 amino acids and has a molecular weight of 52 000. PAI-1 is the most important PAI. It readily forms a 1:1 complex both with tPA and uPA and inhibits their activity with a half-time ($t_{1/2}$) of about 100 s. PAI-1 is known to become the latent form with a $t_{1/2}$ of about 90 min. Binding to vitronectin extends this half-time two folds. The physiological mechanism

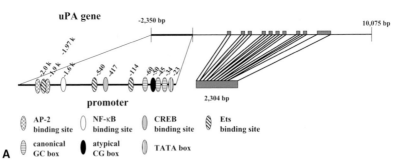

FIG. 2. Structure of uPA. A Gene structure. B cDNA sequence. C Domain structure

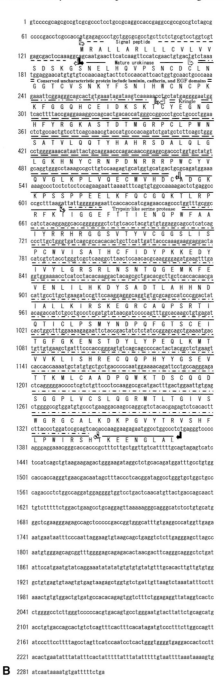

B

FIG. 2. *Continued*

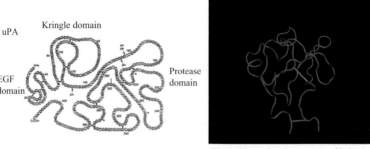

C

Three-dimensional structure of kringle

FIG. 2. *Continued*

for conversion of its latent form to its active form is unknown. However, in vitro detergents are known to activate latent PAI-1. Although various cell types, including vascular endothelial cells, smooth muscle cells, and adipocytes, are known to produce and secrete PAI-1 in response to numerous stimuli, PAI-1 in the blood circulation is mainly derived from endothelial cells and/or α-granules of platelets.

PAI-2 is produced from various tissues, including placenta and monocytes. It is synthesized as a single-chain polypeptide consisting of 415 amino acids with a molecular weight of 43 000 (intracellular storage form). It then undergoes glycosylation to become the secreted form with a molecular weight of 60 000. Its concentration in normal plasma is negligible but increases in women in late pregnancy. PAI-2 also forms a complex with both tPA and uPA 20-fold more slowly than PAI-1 and preferentially blocks uPA.

Furthermore, α_2-macroglobulin and histidine-rich glycoprotein are known to inhibit plasmin activity, whereas protein C inhibitor (also called PAI-3) (found as an inhibitor to activated protein C), protease nexin (derived from fibroblast-conditioned medium), and neuroserpin (found in urine) are known to inhibit PA activity.

Cell-associated Fibrinolysis

During the late 1980s to the early 1990s, a series of articles reported that fibrinolytic reactions proceed 100- to 1000-fold more efficiently on the cell surface or extracellular matrix (ECM) surface comparing to reactions in blood or tissue fluids. Currently, cell or ECM surface reactions are thought to comprise physiological fibrinolytic reactions and are called cell-associated fibrinolysis, which is the basis of tissue fibrinolysis. For example, upon binding to the endothelial cell or extracellular matrix surface, tPA and uPA activate plasminogen 12.7-fold and 100-fold faster than liquid phase reactions, and plasmin generated by this system is protected from inhibition by α_2-antiplasmin (Fig. 3).

A key molecule in the cell-associated fibrinolysis is uPAR [3]. uPAR is a high-affinity (Kd 10^{-9}–10^{-10}) glycosyl phosphatidylinositol (GPI)-anchored protein. The human and mouse proteins share 62% amino acid identity [4], glycosylation pattern, GPI anchor, and ligand-binding properties and both bind active uPA and pro-uPA. Despite these similarities, binding of uPA to uPAR (and thus the related tissue fibrinolysis reaction

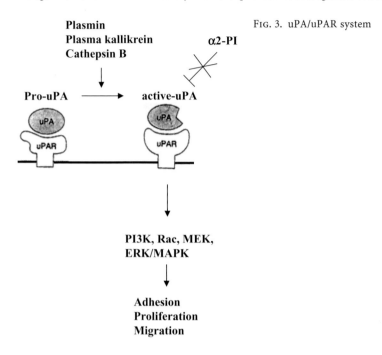

FIG. 3. uPA/uPAR system

via uPA/uPAR) is species-specific [5]. A soluble form of human uPAR, which lacks the amino-terminal domain, has been identified in cell lines, tumors, and some disease states. In addition to uPA/pro-uPA, plasma kallikrein has been known to bind to uPAR [6]. Furthermore, uPAR functions not only as a protease receptor but also as a cell-surface signaling receptor, and it transmits intracellular signals to regulate cell growth, differentiation, adhesion, and migration [3].

On the other hand, tPAR as well as several cell-surface plasminogen-binding proteins such as α-enolase and annexin II are known to serve as cell-surface binding proteins for tPA [7, 8]. Furthermore, most recently, histone H2B has been shown to function as a cell-surface plasminogen receptor on macrophages [9].

Regulation of Tissue Fibrinolysis

Because plasma levels of plasmin and α_2-antiplasmin are not changed much except for the case of certain genetic disorders and liver dysfunction, tissue fibrinolysis is mainly regulated by local levels of tPA and uPA as well as levels of their inhibitors and cell surface binding proteins such as uPAR, whose levels are changed in accordance with various physiological and pathological settings. Table 2 summarizes factors reported to up- or down-regulate the expression of tPA, uPA, PAI-1, and uPAR. Basic fibroblast growth factor (bFGF), vascular endothelial growth factor (VEGF), and retinoids up-regulate uPA, uPAR, and PAI-1 in the order uPA > uPAR > PAI-1, thereby increasing total fibrinolytic activity in endothelial cells. Interleukin-1 (IL-1) and tumor necrosis

TABLE 2.

	Factor	Substance
tPA		
Promotion	PKC, PKA-activators	TPA (e.g., HUVEC, melanoma), cAMP (e.g., HUVEC, F9 cell), butyrate (e.g., HUVEC)
	Hormones, growth factors	Thrombin (HUVEC); histamine (HUVEC); VEGF (BCEC); pituitary hormone and gonadotropic hormone (e.g., LH, FSH, prolactin; pituitary gland-derived cell, granulosa cell); bFGF, EGF (bone marrow stromal fibroblast)
	Nuclear receptor ligands	Retinoid (t.g., HUVEC, F9 cell); glucocorticoid (HT-1080); estrogen (breast cancer cell)
	Releasing enhancers	A23187, (nor)epinephrine, endothelin (e.g., HUVEC); substances generated as a result of coagulation cascade and platelet activation (e.g., factor Xa, active protein C, bradykinin, prostaglandin, leukotriene, ADP, serotonin, PAF; HUVEC)
	Others	Triazolobenzodiazepine (HUVEC); stress (shear stress, oxygen radicals, hyperosmotic pressure; HUVEC)
Inhibition		LPS (BAEC), IL-1 (HUVEC), TNFα (HUVEC), oxidized LDL (HUVEC)
uPA		
Promotion	PKC, PKA activators	TPA (e.g., keratinocyte, LLC-PK1 cell), dbcAMP (e.g., L cell)
	Growth factors, Cytokines	bFGF (e.g., BAEC, BCEC), VEGF (BCEC), HGF (MDCK cell), IL-1 (A549, bone marrow stromal fibroblast), EGF (epidermal keratinocyte); TNFα (HUVEC, A549, keratinocyte); M-CSF (macrophage); G-CSF (BAEC); oncostatin M (BAEC, synovial fibroblast); TGFβ (bone marrow stromal fibroblast); calcitonin (LLC-PK1 cell)
	Nuclear receptor ligands	Retinoid (e.g., BAEC, keratinocyte); sitosterol (BAEC); fucosterol (BAEC); vitamin D_3 (e.g., BAEC)
	Cytoskeletal organization inhibitors	Colchicine, cytochalasin (LLC-PK1 cell)
	Releasing enhancers	Saponin (BAEC)
	Others	Ethanol, active protein C, vitamin C, sulfonylurea (BAEC); okadaic acid (keratinocyte)
Inhibition		LPS, TGFβ, thrombin (BAEC); IFNγ (HUVEC, HCEC); glucocorticoid (e.g., A549, LLC-PK1 cell)
PAI-1		
Promotion		LPS (e.g., HUVEC, BAEC, monocyte, macrophage); TGFβ (BAEC, CCL-64, WI-38, AKR-2B, fibroblast); IL-1 (HUVEC, BAEC); TNFα (HUVEC, BAEC); thrombin (HUVEC); glucocorticoid (e.g., BAEC, HT-1080, fibroblast); retinoid (BAEC); bFGF (BCEC); VEGF (BCEC); oxidized LDL (HUVEC)
Inhibition		Active protein C (BAEC); midkine/pleiotrophin (BAEC); parathyroid hormone (osteoblast)
uPAR		
Promotion		bFGF, VEGF, retinoid (BAEC); IFNγ (mononuclear phagocytic cell); HGF (MDCK cell); heat stress (HUVEC)

HUVEC, human umbilical vein endothelial cell; BAEC, bovine aortic endothelial cell; BCEC, bovine capillary endothelial cell; LLC-PK1, pig kidney; MDCK, Madin-Darby canine kidney; A549, human lung carcinoma; TPA, 12-O-tetradecanoylphorbol-13-acetate; FSH, follicle-stimulating hormone; LH, lutenizing hormone; PAF, platelet-activating factor

factor-α (TNFα) moderately up-regulate uPA but down-regulate tPA, and strongly upregulate PAI-1, thereby decreasing total fibrinolytic activity. Conversion of pro-uPA to tsuPA by plasmin, plasma kallikrein, and cathepsin is also an important step for regulation of tissue fibrinolysis [6]. Moreover, formation of a 1:1 complex between tPA/uPA and PAIs is important not only for direct regulation of tPA/uPA activity but also for clearance of tPA/uPA from the cell surface through low density lipoprotein (LDL) receptor-related protein (LRP)-mediated internalization [10].

Because the uPA/uPAR system plays a major role in tissue fibrinolysis, and the expression and activity of uPA and uPAR are rigidly regulated, here we mainly introduce and discuss the regulation of cellular uPA activity. Impairment of this regulation results in the pathogenesis of numerous diseases, including cancer angiogenesis, invasion, and metastasis, as well as atherosclerosis and hepatic diseases.

As shown in Fig. 2, various transcription factor binding sites, including AP-2-binding sites, Ets-binding sites, an NF-κB-binding site, a CREB-binding site, several GC boxes, and a TATA box exist within the promoter of the uPA gene and mediate its transcriptional regulation under various stimuli. It was found more than three decades ago that in certain cell types cell growth and/or differentiation parallels an increase in cellular fibrinolytic activity; therefore, tPA/uPA has been measured as a molecular marker for their differentiation [3]. This is because many growth factors, such as bFGF and VEGF, produced from monocytic cells as well as tumor cells up-regulate the expression of uPA and uPAR simultaneously when they stimulate cell growth and differentiation and induce angiogenesis or tissue remodeling. In turn, plasmin directly cleaves various matrix proteins, including fibronectin and laminin, or indirectly degrades collagen via activation of prometalloproteinases (proMMPs) and releases bound cytokines such as insulin-like growth factor I (IGF-I), bFGF, and VEGF [11]. Furthermore, the plasma kallikrein-dependent plasminogen cascade is shown to be required for adipocyte differentiation by degradation of the fibronectin-rich preadipocyte stromal matrix [12].

These growth factors stimulate their cognate receptor tyrosine kinases and downstream MAP kinase pathway, resulting in up-regulation and stabilization of uPA and uPAR mRNAs via phospho-activated Ets [13]. On the other hand, constitutive expression of uPA and uPAR is governed by Sp1, a general transcription factor that contains a DNA-binding site composed of three Cys_2-His_2 type zinc-fingers. Interaction among transcription factors, target DNA promoter, co-activators/co-repressors, and fundamental transcription machinery determines the transcriptional status of the uPA gene. Accordingly, every factor that affects one of these interactions regulates uPA expression and tissue fibrinolysis. For example, methylation and demethylation of the uPA promoter effectively control uPA gene expression in certain cancer cells [14].

It is important to determine if the status of uPA and uPAR expression is directly regulated by a certain bioactive substance or under a certain stimulus or is indirectly regulated via induction of mediating factors. Upon receiving an initial stimulus, the first, direct regulation and the secondary, indirect regulation together determine the transcriptional status of the uPA gene at each time point. Furthermore, many initial stimuli occur simultaneously in our body; and the numerous balances between the direct and indirect regulation due to different stimuli converge and appear as the resultant tissue fibrinolysis. Below are examples for regulation of uPA expression and activity in the context of its physiological and pathological roles. In the most cases, its levels are changed during the process of tissue remodeling.

Angiogenesis

Angiogenesis is the growth and sprouting of new blood vessels from preexisting blood vessels [15]. Both endothelial cells derived from preexisting endothelial cells and endothelial cells newly born through differentiation of the vascular progenitor cells invade avascular space and form new blood vessels by means of their peripheral fibrinolytic potential. They express uPA and uPAR at the edge of new blood vessel formation [16] and c-Ets-1 transcription factor plays a role in their transcriptional activation. In contrast, kringle-containing fragments of several molecules, including plasmin(ogen), uPA, and hepatocyte growth factor (HGF), are known to be potent inhibitors of angiogenesis and feedback-control the tissue fibrinolysis [17, 18].

Cancer Invasion and Metastasis

Similar to migrating endothelial cells during the process of angiogenesis, in response to stimuli of growth factors many cancer cells express uPA/uPAR on their surface and gain higher activity to dissolve extracellular matrices. Therefore, levels of their fibrinolytic activity correlate with their activity of invasion and metastasis. Trials have been conducted to prevent cancer invasion and metastasis by suppressing the fibrinolytic activity of cancer cells [14].

Inflammation and Immune Reaction

Inflammation is a key event observed during the first phase of cirrhosis. AP-1 and NFκB—two major transcription factors that participate in the process of inflammation and immune reactions directly or indirectly via production of IL-1 and TNFα—up-regulate the expression of uPA and uPAR [19]. AP-1 is heterodimer of c-Fos and c-Jun family proteins that contain a leucine zipper structure, and NFκB is composed of p65 and p50 subunits. Dissociation of I-κB or p105 due to phosphorylation releases active NFκB from the latent complex and allows it to access its target promoter, which stimulates its transactivation. Recently, uPA was shown to be involved in the regulation of natural killer T-cell function by virtue of its role in the activation of latent transforming growth factor-β (TGFβ) [20].

Wound Healing

Tissue fibrinolysis participates in the wound-healing process. The expression of uPA/uPAR is up-regulated in injured tissues. For example, balloon injury of the rat carotid artery induces transient expression of Egr-1 transcription factor at the site of the injury within half an hour after making scars [21]. Egr-1 is a transcription factor containing the zinc-finger domainlike Sp1 but is not constitutively expressed. It is transiently expressed soon after receiving stimuli with serum and/or growth factors, thereby functioning as an early-response gene. It substitutes with Sp1 and strongly stimulates the expression of uPA. Thereafter, another transcription factor, KLF6, a member of the Sp1 transcription factor family sharing three Cys_2-His_2 type zinc-fingers, is induced at and around the site of injury and synergizes with Sp1 in enhancing the expression of uPA, which results in induction of proteolytic activation of TGFβ [22]. In addition to Egr-1 and KLF6, other transcription factors (e.g., c-Ets-1, AP-1, NFκB, Sp1) are known to be under redox control and play roles in the up-regulation of uPA/uPAR by oxidative stress and inflammation, as described above.

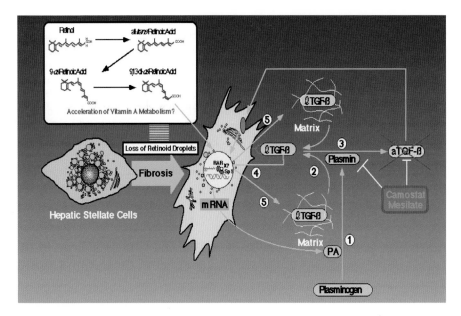

FIG. 4. Proteolytic activation of latent TGFβ during formation of liver fibrosis. *1*, Conversion of plasminogen to plasmin; *2*, release of latent TGFβ from the extracellular matrix (*Matrix*); *3*, proteolytic activation of TGFβ; *4*, autoinduction of TGFβ; *5*, production of an excess of matrix proteins

An excess of this injury response causes cirrhotic disease, including hepatic fibrosis and atherosclerosis. In the damaged liver, stored vitamin A is converted to its active metabolite, retinoic acid. Retinoic acid stimulates the expression of tPA/uPA/uPAR as well as TGFβ and its receptors in hepatic stellate cells, culminating in plasmin-dependent activation of TGFβ and TGFβ-dependent deposition of ECM proteins (Fig. 4) [23]. In addition, plasma kallikrein has been shown to activate TGFβ and play a role in impaired liver regeneration [24]. TGFβ activation is also implicated in the early stage of atherosclerosis formation [25].

On the other hand, uPA is known to convert pro-HGF to active HGF (see the chapter by Okada and Matsuo in this book) and latent PDGF-DD to its active form [26]. Hence, in uPA-knockout mice, regeneration of the partially hepatectomized liver is retarded due to reduced production of active HGF. Furthermore, plasmin is an activator of procollagenases (MMP1 and MMP3), which may function as a feedback mechanism to dissolve excessively produced ECM (Fig. 4).

Conclusion

Tissue fibrinolysis plays opposing roles. It parallels cell growth under common growth factor stimuli and, in turn, releases and activates several growth factors such as VEGF and bFGF and promotes cell migration (angiogenesis and cancer invasion/metastasis) via degradation of the ECM. On the other hand, it activates TGFß and promotes deposition of matrix proteins. These opposite activities are well orchestrated and maitain homeostasis. It is important to elucidate its underlying molecular mechanism. It appears that tissue fibrinolytic activity by migrating cells such as blood cells and epithelial cells (e.g., endothelial cells in angiogenesis, neuronal cells in neurite outgrowth—see the chapter by Urano in this book) or invasive cancer cells functions to degrade the tissue structure, whereas tissue fibrinolytic activity by nonmoving mesenchymal cells, such as activated vascular smooth muscle cells and hepatic stellate cells, functions to restore it by sustaning matrix deposition via generation of active TGFß (Fig. 1).

Another important subject to be resolved is posttranslational modification of tPA/uPA/uPAR, such as phosphorylation, which appears to be an additional mechanism for the regulation of tissue fibrinolysis [27]. Futheremore, more analyses on the cell surface complexes, including uPA/uPAR, integrin, vitronectin, and a uPAR-associated protein, are needed to deepen our understanding of fibrinolytic activity-independent functions of tissue fibrinolysis in the regulation of cell growth and apoptosis [27–32]. For example, inhibition of uPAR results in apoptosis via activation of p53 [31]; and, in turn, p53 stimulates the expression of uPA by stabilizing its mRNA [32].

Acknowledgments. This work was supported partly by the Program for Promotion of Fundamental Studies in Health Sciences of the National Institute of Biomedical Innovation (NIBI) to S.K. and a grant for the Chemical Biology Research Program from RIKEN (S.K.).

References

1. Rijken DC (1995) Plasminogen activators and plasminogen activator inhibitors: biochemical aspects. Baillieres Clin Haematol 8:291–312
2. Lijnen HR, van Hoef B, Beelen V, et al (1994) Characterization of the murine plasma fibrinolytic system. Eur J Biochem 224:863–871
3. Blasi F, Carmeliet P (2002) uPAR: a versatile signalling orchestrator. Nat Rev Mol Cell Biol 3:932–943
4. Kristensen P, Eriksen J, Blasi F, et al (1991) Two alternatively spliced mouse urokinase receptor mRNAs with different histological localization in the gastrointestinal tract. J Cell Biol 115:1763–1771
5. Quax PH, Grimbergen JM, Lansink M, et al (1998) Binding of human urokinase-type plasminogen activator to its receptor: residues involved in species specificity and binding. Arterioscler Thromb Vasc Biol 18:693–701
6. Antikainen J, Kuparinen V, Lahteenmaki K, et al (2007) Enolases from Gram-positive bacterial pathogens and commensal lactobacilli share functional similarity in virulence associated traits. FEMS Immunol Med Microbiol 51:526–534
7. Ling Q, Jacovina AT, Deora A, et al (2004) Annexin II regulates fibrin homeostasis and neoangiogenesis in vivo. J Clin Invest 113:38–48
8. Das R, Burke T, Plow EF (2007) Histone H2B as a functionally important plasminogen receptor on macrophages. Blood 110:3763–3772
9. Lin Y, Harris RB, Yan W, et al (1997) High molecular weight kininogen peptides inhibit the formation of kallikrein on endothelial cell surfaces and subsequent urokinase-dependent plasmin formation. Blood 90:690–697
10. Czekay RP, Kuemmel TA, Orlando RA, et al (2001) Direct binding of occupied urokinase receptor (uPAR) to LDL receptor-related protein is required for endocytosis of uPAR and regulation of cell surface urokinase activity. Mol Biol Cell 12:1467–1479
11. Booth BA, Boes M, Bar RS (1996) IGFBP-3 proteolysis by plasmin, thrombin, serum: heparin binding, IGF binding, and structure of fragments. Am J Physiol 271:E465–E470

12. Selvarajan S, Lund LR, Takeuchi T, et al (2001) A plasma kallikrein-dependent plasminogen cascade required for adipocyte differentiation. Nat Cell Biol 3:267–275

13. Dunn SE, Torres JV, Oh JS, et al (2001) Up-regulation of urokinase-type plasminogen activator by insulin-like growth factor-I depends upon phosphatidylinositol-3 kinase and mitogen-activated protein kinase kinase. Cancer Res 61:1367–1374

14. Pulukuri SM, Rao JS (2007) Small interfering RNA directed reversal of urokinase plasminogen activator demethylation inhibits prostate tumor growth and metastasis. Cancer Res 67:6637–6646

15. Carmeliet P (2005) Angiogenesis in life, disease and medicine. Nature 438:932–936

16. Prager GW, Breuss JM, Steurer S, et al (2004) Vascular endothelial growth factor receptor-2-induced initial endothelial cell migration depends on the presence of the urokinase receptor. Circ Res 94:1562–1570

17. Wahl ML, Kenan DJ, Gonzalez-Gronow M, et al (2005) Angiostatin's molecular mechanism: aspects of specificity and regulation elucidated. J Cell Biochem 96:242–261

18. Du W, Hattori Y, Yamada T, et al (2007) NK4, an antagonist of hepatocyte growth factor (HGF), inhibits growth of multiple myeloma cells: molecular targeting of angiogenic growth factor. Blood 109:3042–3049

19. Wang XQ, Bdeir K, Yarovoi S, et al (2006) Involvement of the urokinase kringle domain in lipopolysaccharide-induced acute lung injury. J Immunol 177:5550–5557

20. Sonoda KH, Nakamura T, Young HA, et al (2007) NKT cell-derived urokinase-type plasminogen activator promotes peripheral tolerance associated with eye. J Immunol 179:2215–2222

21. Khachigian LM (2006) Early growth response-1 in cardiovascular pathobiology. Circ Res 98:186–191

22. Botella LM, Sanchez-Elsner T, Sanz-Rodriguez F, et al (2002) Transcriptional activation of endoglin and transforming growth factor-β signaling components by cooperative interaction between Sp1 and KLF6: their potential role in the response to vascular injury. Blood 100:4001–4010

23. Okuno M, Akita K, Moriwaki H, et al (2001) Prevention of rat hepatic fibrosis by the protease inhibitor, camostat mesilate, via reduced generation of active TGF-β. Gastroenterology 120:1784–1800

24. Akita K, Okuno M, Enya M, et al (2002) Impaired liver regeneration in mice by lipopolysaccharide via TNF-α/kallikrein-mediated activation of latent TGF-β. Gastroenterology 123:352–364

25. Sakamoto Y, Miyazaki A, Tamagawa H, et al (2005) Specific interaction of oxidized low-density lipoprotein with thrombospondin-1 inhibits transforming growth factor-beta from its activation. Atherosclerosis 183:85–93

26. Ustach CV, Kim HR (2005) Platelet-derived growth factor D is activated by urokinase plasminogen activator in prostate carcinoma cells. Mol Cell Biol 25:6279–6288

27. Silvestri I, Longanesi Cattani I, et al (2002) Engaged urokinase receptors enhance tumor breast cell migration and invasion by upregulating avß5 vitronectin receptor cell surface expression. Int J Cancer 102:562–571

28. Madsen CD, Ferraris GM, Andolfo A, et al (2007) uPAR-induced cell adhesion and migration: vitronectin provides the key. J Cell Biol 177:927–939

29. Curino AC, Engelholm LH, Yamada SS, et al (2005) Intracellular collagen degradation mediated by uPARAP/Endo180 is a major pathway of extracellular matrix turnover during malignancy. J Cell Biol 169:977–985

30. Degryse B, Resnati M, Czekay RP (2005) Domain 2 of the urokinase receptor contains an integrin-interacting epitope with intrinsic signaling activity: generation of a new integrin inhibitor. J Biol Chem 280:24792–24803

31. Besch R, Berking C, Kammerbauer C, et al (2007) Inhibition of urokinase-type plasminogen activator receptor induces apoptosis in melanoma cells by activation of p53. Cell Death Differ 14:818–829

32. Shetty S, Velusamy T, Idell S, et al (2007) Regulation of urokinase receptor expression by p53: novel role in stabilization of uPAR mRNA. Mol Cell Biol 27:5607–5618

Role of tPA in the Neural System

Nobuo Nagai and Tetsumei Urano

Summary. In addition to triggering fibrinolysis cascades via plasminogen activation, recent studies have revealed new functions for the tissue plasminogen activator (tPA) in the neural system. In the central nervous system, tPA is expressed in both neurons and microglia and is involved in a variety of physiological and pathological processes, including synaptic plasticity, cell migration, neurite outgrowth, and neuronal cell death. Synaptic plasticity is related to learning, memory, stress, and rewarding, whereas cell migration and neurite outgrowth are important during development. In the peripheral nervous system, it has been shown that induction of tPA by both damaged neurons and Schwann cells results in enhanced axonal regeneration. The autonomic nervous system is reportedly modified by release of tPA from the sympathetic nerve terminal on the vessel wall. Some of these tPA functions depend on its plasminogen activation property and thus on subsequent degradation of extracellular proteins including laminin, neuronal cell adhesion molecules, and deposited fibrin. Other tPA functions, however, are independent of its plasminogen activation property and are associated with modification of receptor function including the N-methyl-D-aspartate receptor and lipoprotein receptor associated protein. Here, a variety of tPA functions in the nervous system and the underlying mechanisms are discussed.

Key words. Tissue plasminogen activator · Synaptic plasticity · Neuronal cell death · LRP · Brain-derived neurotropic factor (BDNF)

Introduction

Tissue plasminogen activator (tPA) and urokinase plasminogen activator (uPA), members of the serine proteases family, are physiological plasminogen activators (PAs) in mammals. Limited proteolysis of plasminogen by PAs results in the generation of the broad-spectrum serine protease plasmin, which then proteolytically cleaves a wide range of substrates, including fibrin, a major component of thrombus [1], matrix proteins, and cell adhesion molecules. The activities of these PAs are regulated by endogenous inhibitors including plasminogen activator inhibitor-1 (PAI-1), PAI-2, and neuroserpin, whereas plasmin activity is regulated by α_2-antiplasmin (α_2-plasmin inhibitor; α_2-AP) [2]. PAIs inhibit both tPA and uPA, whereas neuroserpin, isolated from brain cDNA, specifically inhibits tPA [3]. tPA and PAI-1 have key roles in intravascular fibrinolysis and maintain vascular patency [4]. Recent studies have shown that tPA also contributes to various physiological and pathological processes in the neural system, including the central nervous system (CNS), the peripheral nervous system (PNS), and the autonomic nervous system (ANS). This review describes the physiological and pathological roles of tPA in the neural system and the underlying mechanisms.

Expression and Secretion of tPA

In the CNS, tPA is highly expressed in both neurons and microglia in the hippocampus [5–7], cerebellum [8, 9], amygdala [10], and hypothalamus [7, 11, 12]. In these regions, expression of the *tPA* gene is enhanced by several physiological and pathological stimuli as follows. tPA, localized to axon terminals [13] and dendrites [14, 15], is released as a result of neuronal depolarization [7, 13]. Released tPA modifies neural functions such as synaptic plasticity [13, 16–28], neuronal migration [29, 30], and neurite outgrowth during the development of the neuronal system [31] and pathological synaptogenesis [32]. Similarly, expression of the *tPA* gene is up-regulated in microglia [33] in parallel with its activation through brain injury [34]. *tPA* gene expression is also increased in the PNS after nerve injury. Sciatic nerve damage has been shown to induce *tPA* gene expression in both dorsal root ganglion (DRG) neurons [35, 36] and Schwann cells [37] at the site of damage. tPA induced in DRG neurons and Schwann cells has a role in axonal regeneration [35], protection of axonal degeneration [37], and relating neuropathic pain [36]. tPA that localizes to the axonal terminals of vascular sympathetic nerves is released through nerve depolarization [38–40] and appears to modulate sympathetic nerve function [41] as well as fibrinolytic activity in the systemic circulation [42].

Plasminogen is also present in the neural system [43–45], and thus tPA can act through both plasminogen-dependent [10, 29, 32, 44, 46, 47] and plasminogen-independent mechanisms [13, 27, 48–50] to modify a variety of neural functions as described below. In addition, PAI-1 [10, 51–53] and neuroserpin [3, 54] are also present in the neural system, and thus the actions of tPA may be regulated by these inhibitors.

Role in Synaptic Plasticity

Long-Term Potentiation

tPA was identified as an immediate to early gene whose mRNA transcription was quickly induced in the hippocampus after neuronal stimulation associated with seizure, kindling and long-term potentiation (LTP) [6]. LTP is a phenomenon that describes the long-lasting potentiation of synaptic transmission after frequent presynaptic stimuli, and is closely related to memory. The role of tPA in LTP was demonstrated in studies showing that an increase in *tPA* gene expression in neurons was directly related to the acceleration of LTP and synaptic growth [16] (Table 1). This was further supported by reports showing that late-phase LTP (L-LTP) in the hippocampus [13, 17–19], two-way active avoidance memory [18] and passive avoidance memory [20], was suppressed in *tPA* gene-deficient (tPA$^{-/-}$) mice.

TABLE 1. Process roles of t-PA in the neural system

Site		Function	Involvement of plasmin	References
CNS				
Hippocampus	•	L-LTP	Yes	6, 13, 16–19
	•	Avoidance memory	Yes/no	18, 20
	•	Cognitive impairment	Yes	55
	•	Excitotoxicity	Yes/no	27, 28, 33, 44, 47, 63
Cortex	•	Ocular dominance plasticity	?	21, 22
	•	Ischemic neuronal death	No	73–77
	•	Ischemic hemorrhage	No	48, 80–85, 87, 90
	•	Alzheimer's disease	Yes	92–94
Amygdala	•	Fear response	?	10, 56
Cerebellum	•	Motor learning	?	9, 57
	•	Neuronal migration and neurite outgrowth	?	29–32, 70
NAc	•	Rewarding	Yes	46, 58
Spinal cord	•	Neuronal death	?	78, 79
PNS				
DRG	•	Axonal regeneration	Yes	35, 37, 60
	•	Pain response	?	36, 53
ANS				
Sympathetic n.	•	Release to blood	—	38, 39, 61
	•	Increase in tonus	?	41

t-PA, tissue plasminogen activator; L-LTP, late-phase long-term potentiation; NAc, nucleus accumbens; PNS, peripheral nervous system; ANS, autonomic nervous system; DRG, dorsal root ganglion

Role in Synaptic Remodeling

Mataga et al. demonstrated the involvement of tPA in ocular dominance plasticity (ODP) [21]. ODP is the visual cortical plasticity induced by rapid functional disconnection from an eye with monocular deprivation (MD), followed by anatomical rearrangement of responsiveness from the nondeprived eye in the binocular zone (BZ) in the visual cortex. After MD, increased tPA activity in the BZ paralleled ODP. Furthermore, ODP by MD was impaired in tPA$^{-/-}$ mice [21]. Further studies have shown that in tPA$^{-/-}$ mice, the loss of protrusion in the apical dendrites of BZ neurons that paralleled ODP by MD, was abolished [22]. These findings suggested that the role of tPA in ODP by MD may involve clearance of established synapses, leading to new synapse generation. Interestingly, the cognitive impairment induced by posttraumatic stress disease (PTSD) during major depression and chronic stress was associated with dendritic spine loss in pyramidal neurons of the hippocampal CA1 region. However, in tPA$^{-/-}$ and plasminogen gene-deficient (Plg$^{-/-}$) mice, this cognitive impairment was suppressed [55], also suggesting that the tPA–plasmin system is involved in the clearance of established synapses. Excess activation of the tPA/plasmin system appears to result in cognitive impairment.

Role in Amygdala-Dependent Memory in the Fear Response

tPA also participates in amygdala-dependent memory during the fear response. Pawlak et al. showed that the increase in anxiety-like behavior after acute restraint stress was associated with an increase in tPA activity and neuronal plasticity in the amygdala and that this anxiety-like behavior and neural plasticity were impaired in tPA$^{-/-}$ and Plg$^{-/-}$ mice [10]. Intracerebroventricular injection of corticotrophin releasing factor (CRF), a critical hormone released during the stress response, also increased both anxiety-like behavior and tPA activity in the amygdala [56], suggesting that during the stress response CRF regulates the increase in tPA together with neural plasticity in the amygdala.

Role in Motor Learning in the Cerebellum

tPA is involved in regulation of motor learning in the cerebellum. Seeds et al. showed that tPA mRNA and protein were induced in granular/Purkinje neurons in the cerebellum during the learning of a complex motor skill [9]. Learning was abolished by genetic or pharmacological inhibition of tPA [57], showing the importance of the proteolytic activity of tPA during this learning process.

Role in the Rewarding Pathway in the Mesolimbic System

In the nucleus accumbens (NAc), the tPA–plasmin system plays a role on the rewarding effects of abuse drugs such as morphine and nicotine in the mesolimbic system. Morphine treatment induced tPA mRNA and protein in the NAc, where specific dopamine release was observed [46]. Morphine-induced conditioned place preference (a rewarding behavior), hyperlocomotion, and dopamine release in the NAc were diminished in tPA$^{-/-}$ and Plg$^{-/-}$ mice. However, these responses were restored by microinjection of exogenous tPA or plasmin into the NAc. Nicotine also evoked dopamine release in the NAc, which was associated with tPA release into the extracellular space [58]. This dopamine release was suppressed in tPA$^{-/-}$ mice and restored by microinjection of either tPA or plasmin into the NAc [58]. It has been proposed that this mechanism is regulated by plasmin-mediated activation of PAR-1, a member of a family of receptors that are activated by limited cleavage of N-terminal by extracellular proteases [58]. Thus, the tPA–plasmin system seems to be strongly involved in synaptic plasticity and remodeling in the mesolimbic system, which is essential for the rewarding effects associated with drug abuse [59].

Role in Axonal Degeneration in the Peripheral Nervous System

tPA participates in axonal regeneration after sciatic nerve injury in the PNS. Induction of tPA in DRG neurons was observed after sciatic nerve crush, with a peak at 3–7 days [35, 36]. tPA was also induced in Schwann cells at the site of damage after sciatic nerve crush [37]. Axonal damage after sciatic nerve crush was accelerated in both tPA$^{-/-}$ and Plg$^{-/-}$ mice. However, the damage was recovered by genetic or pharmacological depletion of deposited fibrin at the damaged site [37]. These findings suggest that the protective effect of the tPA–plasmin system on peripheral nerve crush is mediated by the removal of deposited fibrin. Furthermore, exogenous tPA alone or together with plasminogen enhanced nerve regeneration and functional recovery after sciatic nerve crush [60], suggesting that tPA–plasmin also accelerates axonal regeneration. tPA expression was also observed in astrocytes during radiculopathy pain behavior induced by either dorsal root injury or neural stimulation in the dorsal horn of the spinal cord where DRG neurons terminate [36]. Intrathecal administration of a tPA inhibitor (tPA STOP) diminished the pain behavior, suggesting that tPA may be involved in regulation of neural hypersensitivity. Interestingly, both PAI-1 and PAI-2 were induced in DRG neurons after nerve injury [53] and may therefore participate in neural hypersensitivity by controlling tPA activity.

Role in the Modification of the Autonomic Nervous Function

Tissue PA, synthesized in vascular sympathetic neurons that terminate at the resistance vessel wall [38, 39] may be involved in vessel remodeling via plasmin activation, as well as in fibrinolysis in the circulation. In fact, coronal tPA was increased after cardiac sympathetic nerve stimulation [61], and 60% of tPA activity in the blood was reduced by chemical sympathectomy [42]. A role for tPA during the modulation of sympathetic nerve functions was demonstrated by Schaefer et al., who reported that both the contractile response of vasa deferens and norepinephrine release from isolated heart synaptosome by electrical stimulation were suppressed by pharmacological or genetic inhibition of tPA. In contrast, these responses were accelerated by *PAI-1* gene deficiency [41]. Interestingly, these responses were not observed in Plg$^{-/-}$ mice [41]. This finding suggests that the mechanism does not involve plasmin generation, which is outlined below.

Underlying mechanisms in synaptic plasticity

Several mechanisms for tPA have been proposed in the neural system. These may be either catalytic activity dependent or independent and have been demonstrated mainly in hippocampal LTP and neurotoxicity. They are outlined below and summarized in Fig. 1.

Cleavage and Maturation of BDNF

The first proposed mechanism involves limited cleavage and maturation of brain-derived neurotrophic factor (BDNF) from its immature precursor by the tPA–plasmin system (Table 2). The essential functions of BDNF, tPA, and plasminogen in L-LTP have been confirmed by impaired L-LTP in gene-deficient animals for any of these proteins [24]. A close relation between these three molecules has been demonstrated by the fact that plasmin is one of the few extracellular proteases that converts precursor BDNF to its mature form [62]. The physiological relevance of this catalytic reaction was confirmed by the fact that effective cleavage and maturation of precursor BDNF was observed in hippocampal lysates from wild-type mice but not in those from tPA$^{-/-}$ or Plg$^{-/-}$ mice [24]. It has also been shown that perfusion of mature BDNF rescued impaired L-LTP not only in BDNF-deficient mice but also in tPA$^{-/-}$ and Plg$^{-/-}$ mice, suggesting the physiological relevance of a BDNF-activating mechanism by the tPA–plasmin system in L-LTP [24].

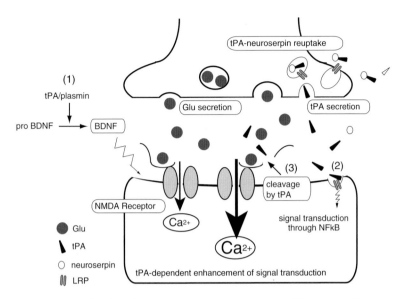

FIG. 1. Suggested mechanisms for tissue plasminogen activator (tPA) during enhancement of long-term potentiation (LTP): (1) activation of brain-derived neurotrophic factor (BDNF); (2) modification of LDL receptor related protein (LRP) function; (3) modification of N-methyl-D-aspartate (NMDA) receptor-dependent calcium influx. Furthermore, cleavage of matrix proteins and intercellular adhesion molecules (not included in the figure) are thought to be involved in tPA-dependent enhancement of LTP

TABLE 2. Biochemical roles of t-PA in the neural system

Function	Involvement of plasmin	References
Laminin degradation	Yes	47, 63
NCAM degradation	Yes	26
Aβ degradation	Yes	92, 93
BDNF maturation	Yes	24, 62
NMDA receptor pathway	No	27, 28, 69
Annexin II pathway	No	50
LRP pathway	No	13, 48, 49

NCAM, neural cell adhesion molecule; Aβ, β-amyloid; BDNF, brain-derived neurotrophic factor; NMDA, N-methyl-D-aspartate; LRP, receptor-related protein

Cleavage of Extracellular Matrix Proteins and Membrane-Bound Proteins

Proteolytic cleavage and degradation of matrix proteins and/or intercellular adhesion molecules by the tPA–plasmin system seem to be essential in synaptic plasticity as well as memory. Laminin is an extracellular matrix (ECM) component in the hippocampus and is degraded by the tPA–plasmin system [47, 63]. Interestingly, LTP was affected by the degradation of laminin after plasmin pretreatment of slice-cultured hippocampus [25]. These findings suggest that strictly controlled cleavage of laminin in a space- and time-dependent manner by the tPA–plasmin system is essential for LTP. The tPA–plasmin system has also been reported to cleave neuronal cell adhesion molecule (NCAM) [26], which is involved in synaptic plasticity and memory [64].

Modification of Receptor Function

LDL Receptor-Related Protein

During LTP, modification of relevant receptor functions by the tPA–plasmin system has also been reported. Low density lipoprotein (LDL) receptor-related protein (LRP) is a member of a family of scavenger receptors of lipoproteins and other proteins including tPA or the tPA/PAI-1 complex [65]. Because tPA is secreted into the synaptic space and is reuptaken through LRP, its functional modification is possibly related to changes in synaptic function. This was confirmed in a study showing that treatment with an LRP antagonist, receptor-associated protein (RAP), suppressed L-LTP in both wild-type mice and tPA$^{-/-}$ mice supplemented with exogenous tPA [13]. Direct interaction of tPA with LRP and the associated signal transduction has also been reported [48].

N-Methyl-D-Aspartate Receptor

Modification of the N-methyl-D-aspartate (NMDA) receptor, which plays an essential role in synaptic plasticity, by tPA has also been reported. The NMDA receptor is a Ca^{2+} channel-coupled glutamate receptor that is involved in synaptic plasticity, learning, and excitotoxicity. Agonists of the NMDA receptor including glutamate or its analogues have been shown to induce NMDA receptor-dependent excitotoxicity [66]. tPA was reported to increase the NMDA-mediated Ca^{2+} influx via direct cleavage of the NR1 subunit of the NMDA receptor at Arg260 [27, 28], although this mechanism is still controversial [67, 68]. Furthermore, tPA also increased gene expression of the NMDA receptor and its phosphorylation at Tyr-1472 of the NR2B subunit in the presence of chronic ethanol intoxication as well as during seizures associated with ethanol withdrawal [69]. NMDA-mediated Ca^{2+} influx also plays a role in neurotoxicity, so the enhanced channel function by tPA is naturally related to neuronal death, as discussed later.

Neuronal Migration and Neurite Outgrowth

tPA is strongly expressed during cell migration in the brain [29]. In the developing cerebellum, *tPA* gene expression was reported in granule neurons as they started to migrate through the cerebellar molecular layer to the internal granule cell layer [70]. Neuronal migration speed was decreased in tPA$^{-/-}$ mice, resulting in an increase in the number of migrating neurons in the cerebellar molecular layer [30]. Localization of tPA activity at the growth cone of cultured neurons from neonatal animals suggests that tPA is involved in the regulation of neurite outgrowth during the developmental stage [31]. In addition, hippocampal neurite outgrowth associated with seizure formation was suppressed in tPA$^{-/-}$ mice, further demonstrating the involvement of tPA in neurite outgrowth in the adult brain [32].

Excitotoxicity and Ischemic Neuronal Cell Death

Tsirka et al. showed that *tPA* and *Plg* gene deficiencies resulted in the resistance of hippocampal CA1 region neurons against excitotoxicity [33, 44]. Increased tPA levels, associated with preceding laminin loss, were observed in the hippocampus during the excitotoxic neuronal degeneration. Reduction of both neuronal degeneration and laminin loss was also demonstrated by genetic or pharmacological inactivation of tPA or plasmin [47, 63]. Because laminin is important for neuronal survival [71], tPA is thought to induce neuronal cell death by laminin degradation via activation of plasmin. Similarly, loss of ganglion cells during retinal damage has been associated with a catalytic reaction by the PA–plasmin system [72]. As described earlier, enhancement of NMDA-mediated Ca^{2+} influx by tPA also plays a role in tPA-dependent neuronal death [27].

Because recombinant tPA has been used as a thrombolytic agent for ischemic stroke, the toxic effect of tPA has been studied extensively. Reduction of the ischemic stroke size in tPA$^{-/-}$ mice demonstrated the deteriorative effect of tPA on ischemic neuronal death [73, 74]. The deteriorative effect of tPA has also been shown in cortical neurons cultured from tPA$^{-/-}$ mice: These neurons were less vulnerable against oxygen-glucose deprivation (OGD) [75], an ex vivo model mimicking ischemia. Interestingly, ischemic stroke size was conversely expanded in Plg$^{-/-}$ mice [73], suggesting that this neurotoxic effect of tPA was independent of plasmin generation. Both PAI-1 [73] and neuroserpin [76] counteract tPA-dependent neurotoxicity via inhibition of tPA. In support of these findings, tPA is also involved in delayed hippocampal neuronal cell death induced by global ischemia [77] and neuronal damage after spinal cord contusion [78]. However, motor neuron degeneration by excitotoxicity was not affected by tPA gene deficiency [79], suggesting that the involvement of tPA in motor neuronal death is not associated with enhanced excitotoxicity.

Although the beneficial effect of tPA treatment in patients with ischemic stroke was demonstrated in a large-scale clinical trial, an increased risk of intracranial bleeding (ICB) [80], especially when the treatment was delayed, was also reported [81, 82]. Recent studies have demonstrated that matrix metalloproteinases (MMPs), which are induced by tPA, play key roles in ICB during tPA treatment for ischemic stroke patients [83–85]. MMPs are extracellular metalloproteinases that degrade extracellular proteins, and some of them are activated by plasmin [86]. After ischemic stroke, MMP-9 has been reported to be induced at the damaged site through the LRP-NFκB pathway [48], whereas it was suppressed in tPA$^{-/-}$ mice [48]. Suzuki et al., however, have reported the importance of MMP-3, rather than MMP-9, in tPA-induced ICB [87]. The involvement of LRP activation by tPA has also been suggested by the increased permeability of the blood-brain barrier (BBB) in the ischemia-damaged area [49], which is closely related to vasogenic edema, intracranial hypertension [88], and probably ICB [89]. Interestingly, activated protein C (aPC), an anticoagulatory serine protease that cleaves and inactivates factors Va and VIIIa, was shown to inhibit NFκB activation as well as MMP-9 induction [90], probably through the protease activatable receptor (PAR) pathway.

Alzheimer's Disease

Alzheimer's disease (AD), the most common cause of dementia, is characterized by the deposition of β-amyloid (Aβ) in the CNS. tPA preferentially binds to and proteolytically cleaves the crossed β-sheet structure, which is common to denatured proteins including fibrin and Aβ [91]. Degradation of Aβ by the tPA–plasmin system results in clearance of Aβ [92, 93]. In addition, increased PAI-1 expression during an inflammatory response in the CNS caused by Aβ deposition [93] resulted in insufficient degradation of Aβ. In contrast, tPA-dependent enhancement of Aβ toxicity has also been implicated through Erk1/2 activation in a protease activity-independent mechanism [94]. Further studies are needed to clarify the implication of correlative distribution of tPA immunoreactivity with Aß in AD patients [94].

Conclusions

Recent studies highlighting the many roles of tPA in the neural system have extended the understanding of fundamental mechanisms associated with neural function, including synaptic plasticity and neuronal death. These findings have contributed to the development of new therapeutic strategies on stress-associated cognitive disorder, ischemic stroke, and AD. The understanding of finely tuned time- and space-dependent control mechanisms of the tPA-mediated neural modification system, including involvement of the tPA inhibitors, is prerequisite to establishing new therapeutic tactics.

References

1. Collen D (1999) The plasminogen (fibrinolytic) system. Thromb Haemost 82:259–270
2. Lijnen HR, Collen D (1995) Mechanisms of physiological fibrinolysis. Baillieres Clin Haematol 8:277–290
3. Hastings GA, Coleman TA, Haudenschild CC, et al (1997) Neuroserpin, a brain-associated inhibitor of tissue plasminogen activator is localized primarily in neurons: implications for the regulation of motor learning and neuronal survival. J Biol Chem 272:33062–33067
4. Urano T, Sumiyoshi K, Pietraszek MH, et al (1991) PAI-1 plays an important role in the expression of t-PA activity in the euglobulin clot lysis by controlling the concentration of free t-PA. Thromb Haemost 66:474–478
5. Salles FJ, Strickland S (2002) Localization and regulation of the tissue plasminogen activator-plasmin system in the hippocampus. J Neurosci 22:2125–2134
6. Qian Z, Gilbert ME, Colicos MA, et al (1993) Tissue-plasminogen activator is induced as an immediate-early gene during seizure, kindling and long-term potentiation. Nature 361:453–457
7. Sappino AP, Madani R, Huarte J, et al (1993) Extracellular proteolysis in the adult murine brain. J Clin Invest 92:679–685
8. Ware JH, Dibenedetto AJ, Pittman RN (1995) Localization of tissue plasminogen activator mRNA in adult rat brain. Brain Res Bull 37:275–281
9. Seeds NW, Williams BL, Bickford PC (1995) Tissue plasminogen activator induction in Purkinje neurons after cerebellar motor learning. Science 270:1992–1994

10. Pawlak R, Magarinos AM, Melchor J, et al (2003) Tissue plasminogen activator in the amygdala is critical for stress-induced anxiety-like behavior. Nat Neurosci 6:168–174

11. Teesalu T, Kulla A, Simisker A, et al (2004) Tissue plasminogen activator and neuroserpin are widely expressed in the human central nervous system. Thromb Haemost 92:358–368

12. Miyata S, Nakatani Y, Hayashi N, et al (2005) Matrix-degrading enzymes tissue plasminogen activator and matrix metalloprotease-3 in the hypothalamo-neurohypophysial system. Brain Res 1058:1–9

13. Zhuo M, Holtzman DM, Li Y, et al (2000) Role of tissue plasminogen activator receptor LRP in hippocampal long-term potentiation. J Neurosci 20:542–549

14. Shin CY, Kundel M, Wells DG (2004) Rapid, activity-induced increase in tissue plasminogen activator is mediated by metabotropic glutamate receptor-dependent mRNA translation. J Neurosci 24:9425–9433

15. Lochner JE, Honigman LS, Grant WF, et al (2006) Activity-dependent release of tissue plasminogen activator from the dendritic spines of hippocampal neurons revealed by live-cell imaging. J Neurobiol 66:564–577

16. Madani R, Hulo S, Toni N, et al (1999) Enhanced hippocampal long-term potentiation and learning by increased neuronal expression of tissue-type plasminogen activator in transgenic mice. EMBO J 18:3007–3012

17. Frey U, Muller M, Kuhl D (1996) A different form of long-lasting potentiation revealed in tissue plasminogen activator mutant mice. J Neurosci 16:2057–2063

18. Huang YY, Bach ME, Lipp HP, et al (1996) Mice lacking the gene encoding tissue-type plasminogen activator show a selective interference with late-phase long-term potentiation in both Schaffer collateral and mossy fiber pathways. Proc Natl Acad Sci U S A 93:8699–8704

19. Baranes D, Lederfein D, Huang YY, et al (1998) Tissue plasminogen activator contributes to the late phase of LTP and to synaptic growth in the hippocampal mossy fiber pathway. Neuron 21:813–825

20. Pawlak R, Nagai N, Urano T, et al (2002) Rapid, specific and active site-catalyzed effect of tissue-plasminogen activator on hippocampus-dependent learning in mice. Neuroscience 113:995–1001

21. Mataga N, Nagai N, Hensch TK (2002) Permissive proteolytic activity for visual cortical plasticity. Proc Natl Acad Sci U S A 99:7717–7721

22. Mataga N, Mizuguchi Y, Hensch TK (2004) Experience-dependent pruning of dendritic spines in visual cortex by tissue plasminogen activator. Neuron 44:1031–1041

23. Matys T, Pawlak R, Strickland S (2005) Tissue plasminogen activator in the bed nucleus of stria terminalis regulates acoustic startle. Neuroscience 135:715–722

24. Pang PT, Teng HK, Zaitsev E, et al (2004) Cleavage of proBDNF by tPA/plasmin is essential for long-term hippocampal plasticity. Science 306:487–491

25. Nakagami Y, Abe K, Nishiyama N, et al (2000) Laminin degradation by plasmin regulates long-term potentiation. J Neurosci 20:2003–2010

26. Endo A, Nagai N, Urano T, et al (1999) Proteolysis of neuronal cell adhesion molecule by the tissue plasminogen activator-plasmin system after kainate injection in the mouse hippocampus. Neurosci Res 33:1–8

27. Nicole O, Docagne F, Ali C, et al (2001) The proteolytic activity of tissue-plasminogen activator enhances NMDA receptor-mediated signaling. Nat Med 7:59–64

28. Fernandez-Monreal M, Lopez-Atalaya JP, Benchenane K, et al (2004) Arginine 260 of the amino-terminal domain of NR1 subunit is critical for tissue-type plasminogen activator-mediated enhancement of N-methyl-D-aspartate receptor signaling. J Biol Chem 279:50850–50856

29. Moonen G, Grau-Wagemans MP, Selak I (1982) Plasminogen activator-plasmin system and neuronal migration. Nature 298:753–755

30. Seeds NW, Basham ME, Haffke SP (1999) Neuronal migration is retarded in mice lacking the tissue plasminogen activator gene. Proc Natl Acad Sci U S A 96:14118–14123

31. Pittman RN, Ivins JK, Buettner HM (1989) Neuronal plasminogen activators: cell surface binding sites and involvement in neurite outgrowth. J Neurosci 9:4269–4286

32. Wu YP, Siao CJ, Lu W, et al (2000) The tissue plasminogen activator (tPA)/plasmin extracellular proteolytic system regulates seizure-induced hippocampal mossy fiber outgrowth through a proteoglycan substrate. J Cell Biol 148:1295–1304

33. Tsirka SE, Gualandris A, Amaral DG, et al (1995) Excitotoxin-induced neuronal degeneration and seizure are mediated by tissue plasminogen activator. Nature 377:340–344

34. Rogove AD, Tsirka SE (1998) Neurotoxic responses by microglia elicited by excitotoxic injury in the mouse hippocampus. Curr Biol 8:19–25

35. Siconolfi LB, Seeds NW (2001) Induction of the plasminogen activator system accompanies peripheral nerve regeneration after sciatic nerve crush. J Neurosci 21:4336–4347

36. Yamanaka H, Obata K, Fukuoka T, et al (2004) Tissue plasminogen activator in primary afferents induces dorsal horn excitability and pain response after peripheral nerve injury. Eur J Neurosci 19:93–102

37. Akassoglou K, Kombrinck KW, Degen JL, et al (2000) Tissue plasminogen activator-mediated fibrinolysis protects against axonal degeneration and demyelination after sciatic nerve injury. J Cell Biol 149:1157–1166

38. Wang Y, Jiang X, Hand AR, et al (2002) Additional evidence that the sympathetic nervous system regulates the vessel wall release of tissue plasminogen activator. Blood Coagul Fibrinolysis 13:471–481

39. Jiang X, Wang Y, Hand AR, et al (2002) Storage and release of tissue plasminogen activator by sympathetic axons in resistance vessel walls. Microvasc Res 64:438–447

40. Hao Z, Jiang X, Sharafeih R, et al (2005) Stimulated release of tissue plasminogen activator from artery wall sympathetic nerves: implications for stress-associated wall damage. Stress 8:141–149

41. Schaefer U, Machida T, Vorlova S, et al (2006) The plasminogen activator system modulates sympathetic nerve function. J Exp Med 203:2191–2200

42. Peng T, Jiang X, Wang Y, et al (1999) Sympathectomy decreases and adrenergic stimulation increases the release of tissue plasminogen activator (t-PA) from blood vessels: functional evidence for a neurologic regulation of plasmin production within vessel walls and other tissue matrices. J Neurosci Res 57:680–692

43. Nakajima K, Tsuzaki N, Nagata K, et al (1992) Production and secretion of plasminogen in cultured rat brain microglia. FEBS Lett 308:179–182

44. Tsirka SE, Rogove AD, Bugge TH, et al (1997) An extracellular proteolytic cascade promotes neuronal degeneration in the mouse hippocampus. J Neurosci 17:543–552

45. Matsuoka Y, Kitamura Y, Taniguchi T (1998) Induction of plasminogen in rat hippocampal pyramidal neurons by kainic acid. Neurosci Lett 252:119–122

46. Nagai T, Yamada K, Yoshimura M, et al (2004) The tissue plasminogen activator-plasmin system participates in the rewarding effect of morphine by regulating dopamine release. Proc Natl Acad Sci U S A 101:3650–3655

47. Chen ZL, Strickland S (1997) Neuronal death in the hippocampus is promoted by plasmin-catalyzed degradation of laminin. Cell 91:917–925

48. Wang X, Lee SR, Arai K, et al (2003) Lipoprotein receptor-mediated induction of matrix metalloproteinase by tissue plasminogen activator. Nat Med 9:1313–1317

49. Yepes M, Sandkvist M, Moore EG, et al (2003) Tissue-type plasminogen activator induces opening of the blood-brain barrier via the LDL receptor-related protein. J Clin Invest 112:1533–1540

50. Siao CJ, Tsirka SE (2002) Tissue plasminogen activator mediates microglial activation via its finger domain through annexin II. J Neurosci 22:3352–3358
51. Ahn MY, Zhang ZG, Tsang W, et al (1999) Endogenous plasminogen activator expression after embolic focal cerebral ischemia in mice. Brain Res 837:169–176
52. Hosomi N, Lucero J, Heo JH, et al (2001) Rapid differential endogenous plasminogen activator expression after acute middle cerebral artery occlusion. Stroke 32:1341–1348
53. Yamanaka H, Obata K, Fukuoka T, et al (2005) Induction of plasminogen activator inhibitor-1 and -2 in dorsal root ganglion neurons after peripheral nerve injury. Neuroscience 132:183–191
54. Osterwalder T, Contartese J, Stoeckli ET, et al (1996) Neuroserpin, an axonally secreted serine protease inhibitor. EMBO J 15:2944–2953
55. Pawlak R, Rao BS, Melchor JP, et al (2005) Tissue plasminogen activator and plasminogen mediate stress-induced decline of neuronal and cognitive functions in the mouse hippocampus. Proc Natl Acad Sci U S A 102:18201–18206
56. Matys T, Pawlak R, Matys E, et al (2004) Tissue plasminogen activator promotes the effects of corticotropin-releasing factor on the amygdala and anxiety-like behavior. Proc Natl Acad Sci U S A 101:16345–16350
57. Seeds NW, Basham ME, Ferguson JE (2003) Absence of tissue plasminogen activator gene or activity impairs mouse cerebellar motor learning. J Neurosci 23:7368–7375
58. Nagai T, Ito M, Nakamichi N, et al (2006) The rewards of nicotine: regulation by tissue plasminogen activator-plasmin system through protease activated receptor-1. J Neurosci 26:12374–12383
59. Berton O, Nestler EJ (2006) New approaches to antidepressant drug discovery: beyond monoamines. Nat Rev 7:137–151
60. Zou T, Ling C, Xiao Y, et al (2006) Exogenous tissue plasminogen activator enhances peripheral nerve regeneration and functional recovery after injury in mice. J Neuropathol Exp Neurol 65:78–86
61. Bjorkman JA, Jern S, Jern C (2003) Cardiac sympathetic nerve stimulation triggers coronary t-PA release. Arterioscler Thromb Vasc 23:1091–1097
62. Lee R, Kermani P, Teng KK, et al (2001) Regulation of cell survival by secreted proneurotrophins. Science 294:1945–1948
63. Nagai N, Urano T, Endo A, et al (1999) Neuronal degeneration and a decrease in laminin-like immunoreactivity is associated with elevated tissue-type plasminogen activator in the rat hippocampus after kainic acid injection. Neurosci Res 33:147–154
64. Hoffman KB (1998) The relationship between adhesion molecules and neuronal plasticity. Cell Mol Neurobiol 18:461–475
65. Herz J, Strickland DK (2001) LRP: a multifunctional scavenger and signaling receptor. J Clin Invest 108:779–784
66. Rothman SM, Olney JW (1995) Excitotoxicity and the NMDA receptor—still lethal after eight years. Trends Neurosci 18:57–58
67. Liu D, Cheng T, Guo H, et al (2004) Tissue plasminogen activator neurovascular toxicity is controlled by activated protein C. Nat Med 10:1379–1383
68. Matys T, Strickland S (2003) Tissue plasminogen activator and NMDA receptor cleavage. Nat Med 9:371–372; author reply 372–373
69. Pawlak R, Melchor JP, Matys T, et al (2005) Ethanol-withdrawal seizures are controlled by tissue plasminogen activator via modulation of NR2B-containing NMDA receptors. Proc Natl Acad Sci U S A 102:443–448
70. Ware JH, DiBenedetto AJ, Pittman RN (1995) Localization of tissue plasminogen activator mRNA in the developing rat cerebellum and effects of inhibiting tissue plasminogen activator on granule cell migration. J Neurobiol 28:9–22

71. Liesi P, Wright JM (1996) Weaver granule neurons are rescued by calcium channel antagonists and antibodies against a neurite outgrowth domain of the B2 chain of laminin. J Cell Biol 134:477–486
72. Zhang X, Chaudhry A, Chintala SK (2003) Inhibition of plasminogen activation protects against ganglion cell loss in a mouse model of retinal damage. Mol Vis 9:238–248
73. Nagai N, De Mol M, Lijnen HR, et al (1999) Role of plasminogen system components in focal cerebral ischemic infarction: a gene targeting and gene transfer study in mice. Circulation 99:2440–2444
74. Wang YF, Tsirka SE, Strickland S, et al (1998) Tissue plasminogen activator (tPA) increases neuronal damage after focal cerebral ischemia in wild-type and tPA-deficient mice. Nat Med 4:228–231
75. Nagai N, Yamamoto S, Tsuboi T, et al (2001) Tissue-type plasminogen activator is involved in the process of neuronal death induced by oxygen-glucose deprivation in culture. J Cereb Blood Flow Metab 21:631–634
76. Yepes M, Sandkvist M, Wong MK, et al (2000) Neuroserpin reduces cerebral infarct volume and protects neurons from ischemia-induced apoptosis. Blood 96:569–576
77. Takahashi H, Nagai N, Urano T (2005) Role of tissue plasminogen activator/plasmin cascade in delayed neuronal death after transient forebrain ischemia. Neurosci Lett 381:189–193
78. Abe Y, Nakamura H, Yoshino O, et al (2003) Decreased neural damage after spinal cord injury in tPA-deficient mice. J Neurotrauma 20:43–57
79. Vandenberghe W, Van Den Bosch L, Robberecht W (1998) Tissue-type plasminogen activator is not required for kainate-induced motoneuron death in vitro. Neuroreport 9:2791–2796
80. The National Institute of Neurological Disorders and Stroke rt-PA Stroke Study Group (1995) Tissue plasminogen activator for acute ischemic stroke. N Engl J Med 333:1581–1587
81. Clark WM, Wissman S, Albers GW, et al (1999) Recombinant tissue-type plasminogen activator (Alteplase) for ischemic stroke 3 to 5 hours after symptom onset: the ATLANTIS study: a randomized controlled trial—Alteplase Thrombolysis for Acute Noninterventional Therapy in Ischemic Stroke. JAMA 282:2019–2026
82. Suzuki Y, Nagai N, Collen D (2004) Comparative effects of microplasmin and tissue-type plasminogen activator (tPA) on cerebral hemorrhage in a middle cerebral artery occlusion model in mice. J Thromb Haemost 2:1617–1621
83. Sumii T, Lo EH (2002) Involvement of matrix metalloproteinase in thrombolysis-associated hemorrhagic transformation after embolic focal ischemia in rats. Stroke 33:831–836
84. Romanic AM, White RF, Arleth AJ, et al (1998) Matrix metalloproteinase expression increases after cerebral focal ischemia in rats: inhibition of matrix metalloproteinase-9 reduces infarct size. Stroke 29:1020–1030
85. Zhao BQ, Ikeda Y, Ihara H, et al (2004) Essential role of endogenous tissue plasminogen activator through matrix metalloproteinase 9 induction and expression on heparin-produced cerebral hemorrhage after cerebral ischemia in mice. Blood 103:2610–2616
86. Lijnen HR (2001) Plasmin and matrix metalloproteinases in vascular remodeling. Thromb Haemost 86:324–333
87. Suzuki Y, Nagai N, Umemura K, et al (2007) Stromelysin-1 (MMP-3) is critical for intracranial bleeding after t-PA treatment of stroke in mice. J Thromb Haemost 5:1732–1739
88. Klatzo I (1987) Blood-brain barrier and ischaemic brain oedema. Z Kardiol 76(suppl 4):67–69
89. Kawanishi M (2003) Effect of hypothermia on brain edema formation following intracerebral hemorrhage in rats. Acta Neurochir (Wien) 86:453–456

90. Cheng T, Petraglia AL, Li Z, et al (2006) Activated protein C inhibits tissue plasminogen activator-induced brain hemorrhage. Nat Med 12:1278–1285
91. Solomon B (2002) Anti-aggregating antibodies, a new approach towards treatment of conformational diseases. Curr Med Chem 9:1737–1749
92. Ledesma MD, Da Silva JS, Crassaerts K, et al (2000) Brain plasmin enhances APP alpha-cleavage and Abeta degradation and is reduced in Alzheimer's disease brains. EMBO Rep 1:530–535
93. Melchor JP, Pawlak R, Strickland S (2003) The tissue plasminogen activator-plasminogen proteolytic cascade accelerates amyloid-beta (Abeta) degradation and inhibits Abeta-induced neurodegeneration. J Neurosci 23:8867–8871
94. Medina MG, Ledesma MD, Dominguez JE, et al (2005) Tissue plasminogen activator mediates amyloid-induced neurotoxicity via Erk1/2 activation. EMBO J 24:1706–1716

Role of Fibrinolysis in the Nasal System

TAKAYUKI SEJIMA AND YOICHI SAKATA[†]

Summary. In this chapter, we show the presence of tissue-type plasminogen activator (t-PA), urokinase-type plasminogen activator (u-PA), and plasminogen activator inhibitor-1 (PAI-1) in nasal mucosa. It is suggested that t-PA synthesized in mucous cells is promptly secreted and modifies the watery nasal discharge in allergic rhinitis and that u-PA activity may help with the passage of large amounts of rhinorrhea by reducing its viscosity. Furthermore, we clarify the relation between fibrinolytic components and the pathology of allergy, particularly during the development of nasal allergy and nasal tissue changes. Wild-type (WT) mice can develop nasal allergy for ovalbumin (OVA) sensitization, but PAI-1-deficient mice (PAI-1$^{-/-}$) cannot. The production of specific immunoglobulins IgG1 and IgE in the serum and production of interleukins IL-4 and IL-5 in splenocyte culture supernatant increased significantly in WT-OVA mice. In PAI-1$^{-/-}$ mice, these reactions were absent, and specific IgG2a in serum and interferon-γ in splenocyte culture medium increased significantly. Histopathologically, there was marked goblet cell hyperplasia and eosinophil infiltration into the nasal mucosa in WT-OVA mice, but these were absent in PAI-1$^{-/-}$ mice. These results indicate that the immune response in WT-OVA mice can be classified as a dominant Th2 response, which would promote collagen deposition. In contrast, the Th2 response in PAI-1$^{-/-}$ mice was down-regulated and the immune response shifted from Th2-dominant reaction to a Th1-dominant one. Taken together, these findings suggest that PAI-1 plays an important role not only in thrombolysis but also in the immune response.

Key words. PAI-1 · t-PA · u-PA · Nasal allergy · Transgenic/knockout mice

Introduction

The fibrinolytic system is associated with not only intravascular fibrinolysis but also various reactions in tissues, including ovulation, arterial sclerosis, cell migration of keratinocytes and smooth muscle cells, and neovascularization [1–4]. Considering these reports about the actions of fibrinolytic components in various tissues, it is thought that fibrinolytic components act on various kinds of physiological functions related to inflammatory reaction such as cellular infiltration and tissue remodeling.

Although the effect of fibrinolytic components in various tissues has been elucidated, the action and the importance of fibrinolytic components in nasal mucosa have been barely reviewed to date. Oh et al. reported that up-regulation of plasminogen activator inhibitor-1 (PAI-1) synthesis occurs in lung and bronchoalveolar lavage fluids in the ovalbumin (OVA)-challenged murine asthma model and that PAI-1 promotes extracellular matrix (ECM) deposition in the airways and inhibits the activity of matrix metalloproteinase (MMP) and plasmin generation [5]. In addition, Gyetko et al. reported that fibrinolytic components have an influence on cytokine production and play a constant role in the immune response [6, 7]. Based on these reports, it is suggested that fibrinolytic components act on disease formation, and the tissue changes greatly during inflammation of the nasal mucosa, particularly in allergic rhinitis. In late years, an increase of morbidity in individuals with allergic rhinitis, including pollinosis, has become a serious social issue all over the world. Furthermore, it is an important problem clinically in that allergic rhinitis is difficult to cure. The mechanism of allergic rhinitis cannot be explained as a simple type I allergy. It is suggested that allergic rhinitis is a complicated process in which inflammation begins with sensitization, becomes chronic, and causes tissue change (Fig. 1).

In this chapter, we describe the involvement of the fibrinolytic components in serial allergic reactions, especially allergic rhinitis, and present data from the knockout mouse concerning the fibrinolytic component.

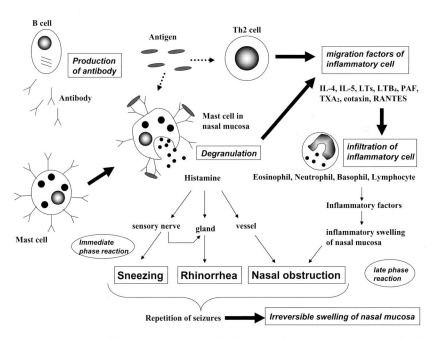

FIG. 1. Mechanism of allergic rhinitis. *IL*, interleukin, *LT*, leukotriene, *PAF*, platelet-activating factor, *TX*, thromboxane, *RANTES*, regulated upon activation of normal T cells expressed and secreted

Localization and Dynamics of Fibrinolytic Components in Nasal Mucosa

The presence of fibrinolytic components in nasal mucosa was identified by reverse transcription polymerase chain reaction (RT-PCR), in situ hybridization, and immunohistochemical staining; and it was shown that its expression changed in the presence of allergic inflammation (Table 1). t-PA was constitutively detected in the epithelium of nasal mucosa, and its expression was almost constant or had an attenuation tendency in the allergic state. Particularly, expression of u-PA and PAI-1 mRNA was significantly high in allergic nasal mucosa in comparison with normal mucosa [8]. Larsen et al. determined the t-PA and u-PA activities in nasal polyps and control nasal mucosa and indicated that the shift toward a higher u-PA/t-PA activity ratio and the higher levels of u-PA in nasal polyps suggested an inflammatory process [9]. Therefore, it is suggested that u-PA activity is similar in allergic rhinitis. u-PA is involved in such biological processes as tissue remodeling and cell migration; and tissues with allergic rhinitis or nasal polyps include tissue changes such as remodeling and infiltration of eosinophils and lymphocytes. On the other hand, considering that u-PA is inactivated by forming a complex with PAI-1, the expression of PAI-1 in serous cells may increase to regulate u-PA activity.

In our previous study, immunohistochemical staining for t-PA was negative in submucosal glands [8]. The presence of t-PA mRNA was noted in mucinous cells in the presence of allergic nasal mucosa, whereas it was not detected in mucinous cells of normal nasal mucosa. These results indicated that t-PA produced by mucinous cells is secreted promptly into the nasal cavity during allergic rhinitis. Furthermore, we reported that u-PA expression was noted in normal nasal mucosa, but compared with that in allergic nasal mucosa there was very little of it. In fibrinautography of nasal discharge, u-PA was markedly detected in allergic patients [8]. Åkerlund et al. measured fibrinogen in nasal discharge and reported that it was increased in the presence of viral upper respiratory inflammation; they supposed that fibrinolytic peptides were generated to participate in inflammatory and defense processes [10].

TABLE 1. Expression of fibrinolytic components in nasal mucosa (human)

Component	t-PA	u-PA	PAI-1
Control			
Epithelium	+	±	−
Gland (submucosal gland)			
Mucinous cells	−	±	−
Serous cells	−	−	−
Endothelium	+	−	−
Allergic rhinitis			
Epithelium	+	+	−
Gland (submucosal gland)			
Mucinous cells	+	++	−
Serous cells	−	−	++
Endothelium	+	−	−

++, strong, expression; +, moderate expression; ±, weak expression; −, no expression

Nasal allergy is characterized by large amounts of serous rhinorrhea. That production of u-PA increases in allergic rhinitis suggests that u-PA activity may help with the passage of large amounts of rhinorrhea by reducing its viscosity. Additionally, it is thought that t-PA acts as an adjunct to treat a large quantity of nasal discharge effectively because of the finding that t-PA has a rapid production turnover and secretion in allergic nasal mucosa. In our study of normal nasal mucosa, t-PA mRNA appeared in mucosal epithelia, and t-PA activity was noted in nasal discharge according to fibrinautography [8]. Thus, t-PA may constitutively adjust the viscosity of the discharge, even in nonallergic tissue. In fact, plasminogen, which is substrate of plasminogen activators (PAs), was present in nasal discharge with or without allergy [8].

In allergic rhinitis, it is interesting that only PAI-1 in fibrinolytic components was produced in serous cells and was not produced in epithelium [8]. PAI-1 has secretory modality and functions distinct from those of the other fibrinolytic components. It is conceivable that PAI-1 functions in circumferential tissue and promotes fibrosis in allergic nasal tissue. Thus, PAI-1 also is secreted into nasal cavity in allergic rhinitis and adjusts the fluidity of nasal discharge. Only u-PA increased in parallel with an increase of PAI -1 in allergic rhinitis, whereas the t-PA concentration did not change. It shows that not t-PA but u-PA acts while competing with PAI-1 in allergic rhinitis. These fibrinolytic components may play an important role in tissue fibrosis, maintaining the fluidity of the nasal discharge. The production of these components is regulated by several cytokines [11, 12], and allergic disease also involves many cytokines. It is possible that the metabolism of fibrinolytic components in allergic tissues is altered under the influence of cytokines.

Tissue Remodeling and Fibrinolytic Components in Allergic Rhinitis

In late years, physiological functions of tissue fibrinolysis, such as cell migration and remodeling, have been investigated regarding the fibrinolytic components. Fibrinolytic components act in the nasal mucosa tissue itself as well as nasal discharge. For allergic disease, it is said that morbid tissue change, or "remodeling," becomes an important factor that makes cure difficult. The phenomenon of remodeling has been studied mainly in bronchial asthma until now, It is characterized by submucosal fibrosis, deposition of ECM, and goblet cell hyperplasia in epithelium [13, 14], and it is thought that the fibrinolytic system is involved in ECM deposition and fibrosis in inflammatory tissues. PAs convert the inactive proenzyme plasminogen to the active form plasmin, a protease of fairly broad substrate specificity. Plasmin degrades fibrin and converts inactive pro-MMP into active MMP. Activated MMP degrades the ECM proteins, including collagen, which is the main protein component of fibrotic tissue in the airway [15]. Previous studies have demonstrated subepithelial depositions of collagen types I and III in bronchial biopsy specimens of asthma patients and allergic nasal mucosa, which correlates with airway hyperresponsiveness [16, 17].

To clarify the relation of fibrinolytic components to the pathology of allergy, particularly the development of nasal allergy and nasal tissue changes, we made a nasal allergy model with PAI-1$^{-/-}$ mice [18]. In our study, we employed a murine model of allergic rhinitis induced by ovalbumin (OVA). Excess amounts of type I and type III collagen are found in the nasal mucosa obtained from OVA-challenged wild-type (WT) mice in our system. In contrast, collagen deposition in the nasal mucosa from OVA-challenged PAI-1$^{-/-}$ mice appeared less significant than that in the OVA-challenged WT mice. Employing WT and PAI-1$^{-/-}$ mice, Hattori et al. in the bleomycin-induced pulmonary fibrosis model [19] and Oh et al. in the OVA-induced asthma model observed a similar effect of PAI-1 on excess fibrous material accumulation in mouse lung tissues [5].

Protection of PAI-1$^{-/-}$ Mice from Nasal Allergy

In addition, we demonstrated that WT mice can develop nasal allergy for OVA, but the PAI-1$^-$ mice cannot [18]. In type I allergy, including allergic rhinitis, increased serum immunoglobulin E (IgE) is characteristic of the immune reaction after exposure to an antigen. Based on the type of cytokine helper T-cells secrete, the helper T-cell is classified as Th1 type or Th2 type. Immune response characteristics are prescribed according to the predominant T-cell type. In mice, the Th2 response results in IgG1 and IgE production, whereas the Th1 response leads to IgG2a synthesis [20]. Therefore, allergic rhinitis is thought to be a result of Th2 cell activation. The significantly increased production of IgE and IgG1 and the low levels of IgG2a in WT-OVA mice immunized and challenged with OVA implicated the Th2 response against this allergen. In contrast, high levels of specific IgG1 and IgE were nearly absent in PAI-1$^{-/-}$-OVA mice. Furthermore, only the PAI-1$^{-/-}$-OVA group showed a significant increase in the level of specific IgG2a [18]. Thus, these results indicate that down-regulation of the Th2 immune response in PAI-1$^{-/-}$ mice brings about inappropriate overactivation of the Th1 immune response to the antigen, which would otherwise induce the Th2 response. Considering the importance of the Th2 phenotype in the development of fibrotic pulmonary and extrapulmonary complications [21], it is possible that the Th2 phenotype itself would promote collagen deposition in the WT compared to that in PAI-1$^{-/-}$ mice. The change in immune responsiveness in PAI-1$^{-/-}$ mice was also confirmed in the cytokine profiles of the nasal lavage fluid (local) and the supernatant of the cultured lymphocytes of the spleen (systemic) from mice challenged by OVA. The levels of IL-4 and IL-5 in the supernatant of OVA-stimulated lymphocyte cultures from WT-OVA mice were 10–20 times higher than those from PAI-1$^{-/-}$-OVA mice. OVA-stimulated cells from PAI-1$^{-/-}$-OVA mice had a 100-fold higher level of interferon-γ (IFNγ) than those from WT-OVA mice [18]. The tendencies of these cytokine profiles were also reflected in the splenocyte proliferation assay from the conditioned mice [18]. These results show that the sensitized group exhibited each immune reaction not only locally but also systemically, explaining the high level of IgE in the WT-OVA mice. The high level of IL-4 would induce high-level antibody production, including IgE as described [22, 23], in the nasal cavity, thereby contributing to the immediate-type allergic reaction after antigen inhalation. On the other hand, IL-5 has highly specific effects on eosiophilic proliferation, migration, activation, and survival [24, 25]. The high level of IL-5 would be responsible for the infiltration of eosinophils into the nasal mucosa. As for the hyperplasia of goblet cells in sensitized mice, a similar tissue change was observed in human allergic rhinitis [8]. Surprisingly, in PAI-1$^{-/-}$-OVA mice, however, these mucosal changes were not observed.

Mechanism by which Fibrinolytic Components Control Allergic Inflammation

The histological data and cytokine profiles indicated that mice deficient in PAI-1 fail to generate the Th2 immune response to the OVA challenge. PAI-1 works not only as a serine protease inhibitor to prevent ECM degradation but acts as a de-adhesion molecule to detach cells attached to vitronectin via integrins [26]. In addition, the de-adhesive activity of PAI-1 does not require its interaction with vitronectin. However, there is an absolute requirement for its binding to u-PA. Free u-PA or PAI-1 with vitronectin has only weak detachment activity [27]. The loss of detachment activity in PAI-1$^{-/-}$-mice might explain the inhibition of cell movement, including eosinophil infiltration. This detachment profile may change the signal transduction of leukocytes through cell-to-cell and cell-to-ligand interactions. Many reports suggest that malignant cells expressing more PAI-1 can metastasize more efficiently than tumors with less PAI-1 production [28]. In addition, there is clinical report that a higher level of PAI-1 in plasma due to the 4G/5G polymorphism of the PAI-1 gene is closely related to allergic disease [29, 30]. These reports, in conjunction with our study, support the idea that PAI-1 is a rather critical regulator of some immune response for antigen stimulation.

Conclusion

We have shown that allergic rhinitis is restrained in OVA sensitization of PAI-1–deficient mice, and the immune response characteristics tend to shift from a Th2-dominant reaction to a Th1-dominant reaction. These findings, with other previous works, demonstrate that fibrinolytic components, including PAI-1, play an important role not only in thrombolysis and proteolysis but also in the immune response by changing the balance between the Th2 reaction and the Th1 reaction.

It was thought that there seemed to be no relation between allergic disease and fibrinolytic components. Therefore, until now, a relation of fibrinolytic components in allergic rhinitis has been little studied. From the standpoints of fibrinolysis, immunology, allergology, and rhinology, a history of the study in this area is short. Thus, there are many points we must elucidate in the future. Figure 2 provides information derived from conventional and current knowledge regarding the mechanism of fibrinolytic factors in the nasal mucosa.

Moreover, from now on we may use fibrinolytic components to intervene in the mechanisms involved in disorders of the nasal mucosa. For example, we may control a local allergic reaction in nasal mucosa by using gene transfection or RNA interference (RNAi). Although the relation between the fibrinolytic system and the immune system is still unclear, it will be useful to study it for new pathological elucidation.

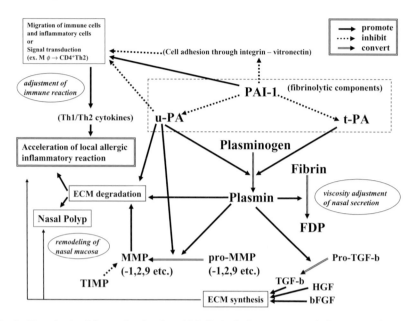

Fig. 2. Hypothesis of the mechanism by which fibrinolytic components influence nasal mucosa. *PAI*, plasminogen activator inhibitor; *u-PA*, *t-PA*, urokinase-type and tissue plasminogen activators; *ECM*, extracellular matrix; *FDP*, fibrin degradation product; *MMP*, metalloproteinase; *TGF-b*, transforming growth factor-β; *HGF*, hepatic growth factor; *bFGF*, basic fibroblast growth factor; *TIMP*, tissue inhibitor of MMPs

References

1. Liu YX, Peng XR, Ny T (1991) Tissue-specific and time-coordinated hormone regulation of plasminogen-activator-inhibitor type 1 and tissue-type plasminogen activator in the rat ovary during gonadotropin-induced ovulation. Eur J Biochem 195:549–555
2. Schneiderman J, Bordin GM, Engelberg I, et al (1995) Expression of fibrinolytic genes in atherosclerotic abdominal aortic aneurysm wall. J Clin Invest 96:639–645
3. Bugge TH, Kombrinck KW, Flick MJ, et al (1996) Loss of fibrinogen rescues mice from the pleiotropic effects of plasminogen deficiency. Cell 87:709–719
4. Carmeliet P, Moons L, Pioplis V, et al (1997) Impaired arterial neointima formation in mice with disruption of the plasminogen gene. J Clin Invest 99:200–208
5. Oh CK, Ariue B, Alban RF, et al (2002) PAI-1 promotes extracellular matrix deposition in the airways of a murine asthma model. Biochem Biophys Res Commun 294:1155–1160
6. Gyetko MR, Sud S, Chen GH, et al (2002) Urokinase-type plasminogen activator is required for the generation of a type 1 immune response to pulmonary *Cryptococcus neoformans* infection. J Immunol 168:801–809
7. Gyetko MR, Sud S, Chensue SW (2004) Urokinase-deficient mice fail to generate a type 2 immune response following schistosomal antigen challenge. Infect Immun 72:461–467
8. Sejima T, Madoiwa S, Mimuro J, et al (2004) Expression profiles of fibrinolytic components in nasal mucosa. Histochem Cell Biol 122:61–73

9. Larsen K, Maat MPM, Jespersen J (1997) Plasminogen activators in human nasal polyps and mucosa. Rhinology 35:175–177
10. Åkerlund A, Greiff L, Andersson M, et al (1993) Mucosal exudation of fibrinogen in coronavirus-induced common colds. Acta Otolaryngol 113:642–648
11. Sadwey M, Loskutoff DJ (1991) Regulation murine plasminogen activator inhibitor 1 gene expression in vivo: tissue specificity and induction by lipopolysaccharide, tumor necrosis factor-α, and transforming growth factor-β. J Clin Invest 88:1346–1353
12. Jensen PJ, Rodeck U (1993) Autocrine/paracrine regulation of keratinocyte urokinase plasminogen activator through the TGF-α/EGF receptor. J Cell Physiol 155:133–139
13. Holgate ST (1998) Airway remodeling. Eur Respir Rev 8:1007–1011
14. Bousquet J, Jeffery PK, Busse WB, et al (2000) Asthma: from bronchoconstriction to airway remodeling. Am J Respir Crit Care Med 61:1720–1745
15. Collen D (1999) The plasminogen (fibrinolytic) system. Thromb Haemost 82:259–270
16. Sanai A, Nagata H, Konno A (1999) Extensive interstitial collagen deposition on the basement membrane zone in allergic nasal mucosa. Acta Otolaryngol 119:473–478
17. Boulet L, Laviolette M, Turcotte H, et al (1997) Bronchial subepithelial fibrosis correlates with airway responsiveness to methacholine. Chest 112:45–52
18. Sejima T, Madoiwa S, Mimuro J, et al (2005) Protection of plasminogen activator inhibitor-1 deficient mice from nasal allergy. J Immunol 174:8135–8143
19. Hattori N, Degen JL, Sisson TH, et al (2000) Bleomycin-induced pulmonary fibrosis in fibrinogen-null mice. J Clin Invest 106:1341–1350
20. Mosmann TR, Cherwinski H, Bond MW, et al (1986) Two types of murine helper T cell clone. I. Definition according to profiles of lymphokine activities and secreted proteins. J Immunol 136:2348–2357
21. Sandler NG, Mentink-Kane MM, Cheever AW, et al (2003) Global gene expression profiles during acute pathogen-induced pulmonary inflammation reveal divergent roles for Th1 and Th2 responses in tissue repair. J Immunol 171:3655–3667
22. Del Prete G, Maggi E, Parronch P, et al (1988) IL-4 is an essential factor for the IgE synthesis induced in vitro by human T cell clones and their supernatants. J Immunol 140:4193–4198
23. Ueda A, Chandswang N, Ovary Z (1990) The action of interleukin-4 on antigen-specific IgG1 and IgE production by interaction in vivo primed B cells and carrier-specific cloned Th2 cells. Cell Immunol 128:31–34
24. Lopez AF, Sanderson CJ, Gamble JR, et al (1988) Recombinant human interleukin 5 is a selective activator of human eosinophil function. J Exp Med 167:219–224
25. Yamaguchi Y, Hayashi Y, Sugama Y, et al (1988) Highly purified murine interleukin 5 (IL-5) stimulates eosinophil function and prolongs in vitro survival: IL-5 as an eosinophil chemotactic factor. J Exp Med 167:1737–1742
26. Deng G, Curriden SA, Hu G, et al (2001) Plasminogen activator inhibitor-1 regulates cell adhesion by binding to the somatomedin B domain of vitronectin. J Cell Physiol 189:23–33
27. Tarui T, Andronicos N, Czekay RP, et al (2003) Critical role of integrin $\alpha_5\beta_1$ in urokinase (uPA)/urokinase receptor (uPAR, CD87) signaling. J Biol Chem 278:29863–29872
28. Dano K, Andreasen PA, Grondahl-Hansen J, et al (1985) Plasminogen activators, tissue degradation and cancer. Adv Cancer Res 44:139–266
29. Cho SH, Hall IP, Wheatley M, et al (2001) Possible role of the 4G/5G polymorphism of the plasminogen activator inhibitor 1 gene in the development of asthma. J Allergy Clin Immunol 108:212–214
30. Bucková D, Izakovicova Hollá L, Vácha J (2002) Polymorphism 4G/5G in the plasminogen activator inhibitor-1 (PAI-1) gene is associated with IgE-mediated allergic diseases and asthma in the Czech population. Allergy 57:446–448

Role of Fibrinolysis in Hepatic Regeneration

KIYOTAKA OKADA, SHIGERU UESHIMA, AND OSAMU MATSUO

Summary. Liver regeneration has been studied intensively, and it has been revealed that the major enzymes that degrade the extracellular matrix (ECM) are matrix metalloproteinases and fibrinolytic enzymes. This proteolytic activity regulates ECM assembly, editing of excess ECM components, remodeling of the ECM structure, and release of growth factors and cytokines during the liver regeneration process. The fibrinolytic factors act in the liver regeneration process in the following two ways: activation of growth factors and degradation of ECM components. Studies on these points have been drastically advanced by the use of genetic all-deficient mice. Plasminogen (Plg) and urokinase-type plasminogen activator (u-PA) are bound to hepatocytes from mouse liver. Binding of Plg to hepatocytes impairs the inhibitory effect of α_2-antiplasmin (α_2-AP); that is, activation of Plg by Plg activator (PA) in the presence of hepatocytes is more enhanced than in the absence of hepatocytes. u-PA activates hepatocyte growth factor (HGF) and induces the proliferation of hepatocytes. Because Plg cleared ECM in mouse models, the loss of Plg showed a delay in liver regeneration. Thus, fibrinolytic factors are thought to act in two ways in the process of healing and regeneration after liver injury. The u-PA/PA inhibitor type-1 (PAI-1) system regulates the proliferation of hepatocytes by controlling HGF activation. The plasmin/α_2-AP system regulates the microenvironment of liver regeneration by controlling ECM degradation. It is anticipated that in the future the regulation of proteases, which involve mainly fibrinolytic factors, will be a form of regenerative medicine for the liver.

Key words. Cellular fibrinolysis · Liver regeneration · ECM degradation · Growth factor activation · Knockout mice

Introduction

Liver tissue has multiple functions in metabolism and protein synthesis. Fibrinolytic factors, such as plasminogen (Plg) and α_2-antiplasmin (α_2-AP), are synthesized by normal hepatocytes [1]. The fibrinolytic system consists of a potential enzyme system, in which Plg is activated by Plg activators (PAs), such as tissue-type PA (t-PA) and urokinase-type PA (u-PA) [2]. The inhibitors for PAs and plasmin are PA inhibitors (PAIs) and α_2-AP, respectively; these inhibitors regulate enzymatic activity. Recently, it was demonstrated that the receptors and binding sites for these fibrinolytic factors were present on the surface of cells [3], and the cell-bound fibrinolytic factors express proteolytic activity on the cell circumference and cause intracellular signal transudation [4, 5]. A function of the fibrinolytic factor in these cell circumferences was said to be the "cellular fibrinolytic system." We distinguished it from the functional blood "fibrinolytic system" in conventional blood (Fig. 1) [6]. The cellular fibrinolytic system contributes to various physiological and pathophysiological phenomena such as neovascularization, tissue repair, or invasion and metastasis of tumors. Knowledge of this cellular fibrinolytic system was greatly advanced using gene-deficient (knockout, or KO) mice [7, 8].

Liver regeneration after physical or toxic injury requires both the well-orchestrated proliferation of liver cells and tissue remodeling events to restore hepatic architecture. Liver regeneration has been studied intensively, revealing that it is regulated by a variety of growth factors and cytokines, including epidermal growth factor (EGF), transforming growth factor-β (TGFβ), hepatocyte growth factor (HGF), tumor necrosis factor-α (TNFα), and interleukin-6 (IL-6) [9]. Fibrinolytic enzymes have the ability to activate HGF during liver regeneration after liver injury [10–12]. Also, TGFβ is activated by plasmin [13].

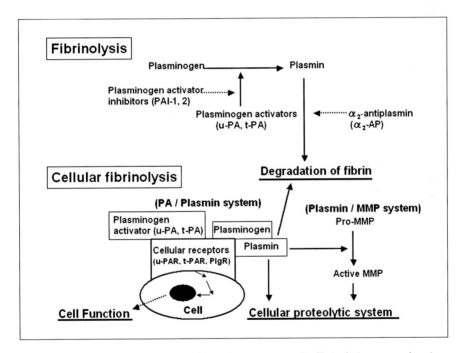

Fig. 1. Fibrinolytic system and cellular fibrinolytic system. In the fibrinolytic system, plasminogen (*Plg*) is activated by plasminogen activators (*PA*), such as tissue-type PA (*t-PA*) and urokinase-type PA (*u-PA*). The inhibitors of PAs and plasmin—PA inhibitor (*PAI*) and α_2-AP, respectively—regulate enzymatic activity. Plasmin degrades fibrin and induces thrombolysis in the blood vessels. The cellular fibrinolytic system is mediated by fibrinolytic factors on the cell surface. The receptors and binding sites for these fibrinolytic factors are present on the cell surface, and cell-bound fibrinolytic factors express proteolytic activity in the pericellular circumference and cause the control or induction of intracellular signaling. *MMP*, metalloproteinase

Proteolysis regulates extracellular matrix (ECM) assembly, the editing of excess ECM components, the remodeling of the ECM structure, and the release of growth factors and cytokines. The major enzymes, those that degrade ECM, are the metalloproteinases (MMPs) and tissue serine proteinases, such as t-PA, u-PA, and plasmin. Also, proteolysis of the ECM may play an important role during the early stages of liver regeneration [14, 15].

In this review, we represent the role of fibrinolytic factors in the liver regeneration process through the following three points: (1) activation of growth factors by fibrinolytic factors; (2) degradation of ECM components by fibrinolytic factors; (3) results obtained from analyses using KO mice of fibrinolytic factors.

Activation of Growth Factors by Fibrinolytic Factors

Activation of HGF by Fibrinolytic Factors

Hepatocyte growth factor, also known as scatter factor (SF), was first identified as a blood-derived mitogen for hepatocytes in culture [16–19]. HGF and its receptor c-Met are key components for liver growth and function. Studies using HGFKO mice or c-MetKO mice showed that HGF and its receptor are both essential for liver development [20, 21]. HGF is normally produced as an immature single-chain polypeptide. Then HGF is activated by proteolytic cleavage into its mature two-chain form. This activation of HGF is performed by proteases such as the HGF activator [22–24], a serine protease with homologous structural domains to those observed in blood coagulation factor XII, and fibrinolytic enzymes [10–12]. Also, plasmin can directly activate HGF on lung fibroblasts [17]. The activated two-chain form of HGF binds to c-Met with high affinity and triggers signal transudation in cells. Tyrosine phosphorylation of c-Met is seen several hours before the initiation of DNA synthesis [16], suggesting involvement of HGF in the generation of the mitogenic signal.

The HGF is localized on heparan sulfate proteoglycans constructing ECM of the liver, spleen, and kidney [25]. After 70% partial hepatectomy (PH), the HGF level in plasma increases within 2 h. This increase is due to ECM degradation [26].

Furthermore, the activity of u-PA rapidly increases within 1 min after 70% PH because of the augmentation of u-PA receptor (u-PAR) expression [26]. In addition to its role in activating plasminogen, u-PA is considered to be one of the major initiators of the MMP activation associated with ECM degradation [27]. u-PA is also a key molecule associated with the activation of single-chain HGF to two-chain HGF during liver regeneration [28].

Activation of TGFβ by Fibrinolytic Factors

Transforming growth factor-β (TGFβ) mediates key components in the injury response, down-regulating inflammation, modulating growth inhibition and differentiation, and enhancing ECM production in a variety of cell types [29, 30]. TGFβ is synthesized and secreted as a biologically latent form. Latent TGFβ consists of three components: the active TGFβ homodimer, the paired latency-associated peptide (LAP), and the latent TGFβ-binding protein. For its activation, the homodimeric active TGFβ molecule is released from the large complex [31]. Although the physiological mechanisms in activating latent TGFβ are not fully understood, three potential factors have been addressed recently. Three potential factors against TGFβ activation are thrombospondin 1 [32], integrin αvβ6 [33], and proteases such as plasmin [13] and plasma kallikrein [34]. In the normal liver, TGFβ is produced and secreted from sinusoidal endothelial cells and Kupffer cells at low levels. After PH, elevated production of TGFβ was seen first in all cell types of the liver and then mainly in hepatic parenchymal cells and hepatic stellate cells (HSCs), whereas elevated production of TGFβ was seen solely in HSCs after inflammation and fibrosis. Plasmin induces the release of latent TGFβ from the ECM and activates it by cleaving LAP from latent TGFβ molecules on the HSC surface [13]. In addition, Kojima et al. [35] reported that a Krüppel-like factor, Zf9, is up-regulated in acute liver injury in activated HSCs, and it up-regulates endogenous u-PA mRNA and activity, resulting in increased active TGFβ via enhancement of proteolytic activation of the latent molecule.

ECM Degradation by Fibrinolytic Factors

The ECM is composed of various macromolecules, including fibronectin, collagen, laminin, vitronectin, and proteoglycans. The ECM functions as a physical support for epithelial and endothelial cells, and it modulates cell differentiation, migration, growth, and apoptosis [36]. The ECM structure of the liver is distinct from that of all other organs [37]. In other organs, the interstitium acts as a diffusion barrier between plasma and the epithelial cells and contains a basement membrane; however, in the liver the ECM exists in Disse's (perisinusoidal) space, which is a small space between hepatocytes and sinusoidal endothelial cells that lacks a true basement membrane. This arrangement may facilitate maximal exchange of nutrients between circulating blood and hepatocytes. Due to this specific structure of the liver, liver regeneration is deeply related to ECM; therefore, smooth degradation of ECM is an important phenomenon in regenerating the liver [27].

After PH, plasmin degrades matrix components including fibronectin, laminin, and entactin [38] and it also indirectly participates in matrix degradation through the activation of MMP and interstitial collagenase [27]; therefore, plasmin(ogen) is assumed to relate to ECM degradation and remodeling. Hepatocytes have u-PAR and Plg binding sites on their membrane. Recently, it was demonstrated that Plg and u-PA bind to isolated hepatocytes from mouse liver [39]. Furthermore, the mouse hepatocytes enhanced Plg activation and impaired the inhibitory effect of α_2-AP. These results indicate that the PA/Plg system on the surface of mouse hepatocytes plays an important role in ECM degradation of the liver.

Analysis of Fibrinolytic Factors Using KO Mice

It is thought that a fibrinolytic factor contributes to activation of a growth factor and the degradation of ECM. Fibrinolytic factors, such as Plg and α_2-AP, are synthesized by normal hepatocytes [1]. The expression of both proteins is induced by cytokines [40] and hormones [41]; however, PAs and PAI-1a are not expressed in normal mammalian hepatocytes in vivo [42], and both primary cultures of animal hepatocytes [43] and established hepatoma cell lines [44, 45] synthesize PAs and PAI-1 in vitro.

These observations suggested a possible relation between expression of these protein and the switching of hepatocytes from the quiescent state to one in which a growth program has been initiated. PAI-1 and PAs belong to the immediate early growth response gene and delayed early response gene in regenerating hepatocytes, respectively [46]. On the other hand, these proteins are induced for production in the liver regeneration process after liver injury [26, 47]. Interaction between the expression of fibrinolytic factors and the liver regeneration process was analyzed using experiments of animal models [48] and cell cultures [49, 50].

Carbon tetrachloride (CCl_4) injection in mice induces the activation of HSCs, and activated HSCs produce ECM proteins, such as fibronectin and collagen (Table 1) [51]. Injection of a small dose of CCl_4 results in the death of hepatocytes adjacent to the central vein due to P450-dependent formation of toxic metabolites. In normal mice, the necrotic zones are cleared, and normal liver architecture is restored within 7–14 days. In this model that induced hepatic fibrosis, we have demonstrated that PlgKO and Plg/α_2-APKO mice showed delayed liver regeneration, dependent on the quantity of the remaining fibronectin in the ECM [52]. Furthermore, α_2-APKO mice rapidly restored fibronectin, which was increased by CCl_4 administration in comparison with WT mice. Thus, liver regeneration in α_2-APKO mice was more accelerated than in WT mice. These results suggest that the plasmin/α_2-AP system plays an important role in degradation and remodeling of ECM in liver regeneration. In addition, as Bezerra et al. reported, liver regeneration in PlgKO mice [53] and t-PA/u-PAKO mice [54] was delayed. In most wound areas, fibrin was a prominent feature in the damaged centrilobular tissues of PlgKO mice; therefore, one potential explanation for the failure

TABLE 1. Liver regeneration in fibrinolytic factor knockout mice in several liver injury models

Mice model	CCl_4	Fas-IgG	70% PH	Alcohol	BDL
u-PAKO	↓	↓	↓		
t-PAKO	→		→		↓
PAI-1KO		↑		↑	↑
PlgKO	↓		↓		
a$_2$-APKO	↑				
u-PARKO	→		→		
u-PA/t-PAKO	↓				
t-PA/u-PARKO	→				
Plg/a$_2$-APKO	↓				
FbgKO	→				
Plg/FbgKO	↓				
u-PATg	→				
u-PATg/PlgKO	↓				
Model of disease	Fibrosis	Apoptosis	Partial hepatectomy	Alcoholic liver disease	Cholestatic liver disease
References	52, 53, 54, 56, 57, 74, 75	28	62, 63, 64	65	67, 68

IgG, immunoglobulin G; PH, partial hepatectomy; BDL, bile duct ligation
Liver regeneration: ↑, increase; ↓, decrease; →, no change

to repair necrotic liver tissue in PlgKO mice was that fibrin in the damaged tissue formed a physical barrier for removal of necrotic debris. However, genetically super-imposing a systemic deficit in fibrin(ogen) did not correct the abnormal pattern of hepatic repair observed in PlgKO mice [53, 55]. Thus, the data indicate that plasmin-mediated proteolysis of one or more nonfibrin targets is critical in the repair of necrotic liver tissue. In addition, Ng et al. [56] suggested that the long-term presence of activated HSCs producing ECM is associated with delayed liver regeneration in injured PlgKO mice by CCl_4. Furthermore, the liver regeneration ability in u-PA/t-PAKO mice was same as that in PlgKO mice [54]. u-PARKO or t-PAR/u-PARKO mice had no impact on the clearance of centrilobular injury after CCl_4 injection [57]. u-PA activates Plg in the liver regeneration process and is more important than t-PA at the point of promoting proteolysis of ECM. u-PAR shows little importance in a u-PA-dependent Plg activation in this process. These results suggested that liver regeneration is developed by ECM degradation and that it is directly or indirectly induced by u-PA-mediated plasmin. We have investigated whether plasmin regulates the proapoptotic protein of Bim_{EK} in the primary culture of hepatocytes and in vivo [58]. The increased expression of Bim_{EK} in the liver after CCl_4 injection persisted in PlgKO mice in comparison with PlgWT. In addition, hepatocytes of PlgKO showed delayed phosphorylation of ERK1/2 after CCl_4 injection. These results were reconfirmed by in vitro culture of hepatocytes. Our data suggest that the Plg/plasmin system could decrease Bim_{EK} expression via the ERK1/2 signaling pathway during liver regeneration.

An intraperitoneal administration of anti-Fas antibody induced massive apoptosis of hepatocytes, which may reflect the pathogenesis of human fulminant hepatic failure [59]. Fas (APO-1/CD95), a type I membrane protein belonging to the TNF/nerve growth factor receptor family, is the major cell surface molecule that mediates external ligand-stimulated apoptosis [60], and is ubiquitously expressed in a variety of tissues including the liver, thymus, heart, lung, and ovary [61]. Thus, we defined the role of u-PA and PAI-1 in hepatic regeneration after massive hepatocyte death induced by anti-Fas antibody injection in u-PAKO and PAI-1KO mice [28]. u-PAKO mice showed delayed liver regeneration with delayed HGF activation. On the other hand, PAI-1KO mice showed the acceleration of HGF activation and of liver regeneration in this model. Based on these results, we suggest that the u-PA/PAI-1 system may play an important role in the early steps of liver regeneration via activation of HGF after Fas-mediated massive hepatocyte death.

The 70% PH model is produced by a simple operation in which two-thirds of the mouse liver is removed. Specific liver lobes are removed without damage to the left lobes. The residual lobes enlarge to make up for the mass of the removed lobes, although the respective lobes never grow back. It takes about 5–7 days for regeneration. The 70% PH model is most often used to study liver regeneration because this model is not associated with tissue injury or inflammation compared with other methods that use hepatic toxins (e.g., CCl_4 or Fas antibody); moreover, initiation of the regenerative stimulus is precisely defined in this model. We therefore explored the role of Plg in hepatic regeneration after PH using PlgKO mice [62]. Sequential recovery of the liver weight after 70% PH increased over 1–7 days and thereafter increased no more. The recovery of liver weight in PlgKO mice was significantly impaired at 10 and 14 days compared with that in PlgWT mice. The focal area of cellular loss with fibrin deposition was detected in PlgKO mice after 70% PH; therefore, the data presented in this study clearly show that liver regeneration after 70% PH begins normally but then deteriorates, demonstrating that Plg plays a critical role in the late stages of liver regeneration. This phenomenon may be caused by an obstruction of blood flow, which is induced by fibrin deposition.

In addition, in another report [63], DNA synthesis by hepatocytes during the regeneration process in the 70%PH model was lower in u-PAKO mice than in u-PAWT mice; however, the recovery rate of liver weight was not significantly different between u-PAKO and u-PAWT mice. The presence of focal areas of fibrin deposition and cellular loss in liver tissues were more severe in u-PAKO mice than in u-PAWT mice. In contrast, regeneration was not impaired in u-PARKO mice. These data indicate that u-PA plays an important role in liver regeneration in the 70% PH mouse model independent of its interaction with u-PAR.

Drixler et al. [64] investigated the role of Plg and u-PA in angiogenesis in liver regeneration using 70% PH mouse models. The recovery of liver weight was significantly reduced in PlgKO and u-PAKO mice compared with that in WT mice. The increase of microvessel density in the liver after 70% PH was impaired in PlgKO mice. These results suggested that Plg appears to be a major determinant in regeneration-associated hepatic angiogenesis.

Bergheim et al. [65] investigated the potential role of PAI-1 in liver damage using mouse models of acute and chronic alcohol exposure. Animal models of ethanol exposure have made it possible to produce pathological changes in rodent liver that resemble early alterations in human alcoholic-induced liver disease (ALD) [66]. Alcohol exposure induced TNFα-mediated PAI-1 expression in the liver of PAI-1WT mice [65]. It was also associated with lipid accumulation and induced steatosis and necrosis, whereas PAI-1KO mice were protected from alcohol-induced liver damage. These results suggest a novel pathway in alcohol-induced steatosis and liver damage that involves the induction of PAI-1 expression in the liver through a TNFα-dependent pathway.

Wang et al. [67, 68] investigated the potential role of the t-PA/PAI-1 system in liver injury after bile duct ligation (BDL) using t-PAKO mice or WT mice. Extrahepatic cholestasis can be modeled by surgical BDL in rodents. This method reliably induces stereotypical histopathological changes, including hepatocellular necrosis, neutrophil infiltration, cholangiocyte and hepatocyte proliferation, HSC activation, and progressive fibrosis [69]. BDL also causes early mortality in mice and mimics human cholestatic liver disease. After BDL, t-PAKO mice showed increased bile infarcts, apoptotic cell and neutrophil infiltration, reduced hepatocyte proliferation compared to these findings in WT mice [68]. In addition, the protective and proliferative effects in PAI-1KO mice after BDL were dramatically blocked by the synthetic t-PA inhibitor [67, 68]. The genetic loss of t-PA reduced HGF activation and c-Met phosphorylation in the liver after BDL.

Conclusion

The function of fibrinolytic factors in the regeneration process after liver injury is shown in Fig. 2. u-PA induces proliferation of hepatocytes by activating the HGF/c-Met system. Plasmin prepares the microenvironment for liver regeneration by the induction of ECM degradation.

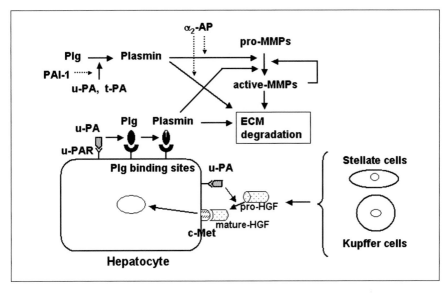

Fig. 2. Function of fibrinolytic factors in liver regeneration. u-PA bound to urokinase-type PA receptor (*u-PAR*) on hepatocytes activates pro-hepatic growth factor (*HGF*), and then activated HGF binds to c-Met (HGF receptor) and induces the trigger signal transduction associated with DNA synthesis in hepatocytes during liver regeneration. Furthermore, u-PA bound to u-PAR activates plasminogen on hepatocytes, and then activated plasmin induces the degradation of extracellular matrix (*ECM*) proteins and regulates the microenvironment during liver regeneration

Recently, u-PA gene therapy for liver cirrhosis was attempted in a rat model [70, 71]. In addition, humanized liver was prepared in u-PA-transgenic SCID mice for a trial of human hepatocytes [72, 73].

In the future, regulation of protease mainly on a fibrinolytic system factor is expected from the viewpoint of regenerative medicine for the liver.

References

1. Okada K, Yuasa H, Hagiya Y, et al (1995) Fibrinolytic activity in liver tissues of stroke-prone spontaneously hypertensive rats. Clin Exp Pharmacol Physiol 22:S275–S276
2. Collen D, Lijnen HR (1991) Basic and clinical aspects of fibrinolysis and thrombolysis. Blood 78:3114–3124
3. Haijar KA (1995) Cellular receptors in the regulation of plasmin generation. Thromb Haemost 74:294–301
4. Pepper MS (2001) Extracellular proteolysis and angiogenesis. Thromb Haemost 86:346–355
5. Bhat GJ, Gunaje JJ, Idell S (1999) Urokinase-type plasminogen actvator induces tyrosine phosphorylation of a 78-kDa protein in H-157 cells. Am J Physiol 277:L301–L309
6. Fukao H, Ueshima S, Okada K, et al (1997) The role of the pericellular fibrinolytic system in angiogenesis. Jpn J Physiol 47:161–171
7. Carmelie P (1997) Insights from gene-inactivation studies of the coagulation and plasminogen. Fibrinol Proteol 11:181–191

8. Degen JL (2001) Genetic interactions between the coagulation and fibrinolytic system. Thromb Haemost 86:130–137

9. Michalopoulos GK, DeFrances MC (1997) Liver regeneration. Science 276:60–66

10. Mars WM, Zarnegar R, Michalopoulos GK (1993) Activation of hepatocyte growth factor by the plasminogen activators uPA and tPA. Am J Pathol 143:949–958

11. Naldini L, Vignal E, Bardelli A, et al (1995) Biological activation of pro-HGF (hepatocyte growth factor) by urokinase is controlled by a stoichiometric reaction. J Biol Chem 270:603–611

12. Matsuoka H, Sisson TH, Nishiuma T, et al (2006) Plaminogen-mediated activation and release of hepatocyte growth factor from extracellular matrix. Am J Respir Cell Biol 35:705–713

13. Okuno M, Akita K, Moriwaki H, et al (2001) Prevention of rat hepatic fibrosis by the protease inhibitor, camostat mesilate, via reduced generation of active TGF-β. Gastroentrology 120:1784–1800

14. Werb Z (1997) ECM and cell surface proteolysis: regulating cellular ecology. Cell 91:439–442

15. Nagase H, Woessner JF Jr (1999) Matrix metalloproteinases. J Biol Chem 274:21491–21494

16. Nakamura T, Nawa K, Ichihara A (1984) Partial purification and characterization of hepatocyte growth factor from serum of hepatectomized rat. Biochem Biophys Res Commun 122:1450–1459

17. Michalopoulos G, Houck KA, Dolan ML, et al (1984) Control of hepatocyte replication by two serum factors. Cancer Res 44:4414–4419

18. Gherardi E, Stoker M (1990) Hepatocytes and scatter factor. Nature 346:228.

19. Naldini L, Vigna E, Narsimhan RP, et al (1991) Hepatocyte growth factor (HGF) stimulates the tyrosine kinase activity of the receptor encoded by the proto-oncogene c-met. Oncogene 6:501–504

20. Schmidt C, Bladt F, Goedecke S, et al (1995) Scatter factor/hepatocyte growth factor is essential for liver development. Nature 373:699–702

21. Uehara Y, Minowa O, Mori C, et al (1995) Placental defect and embryonic lethality in mice lacking hepatocyte growth factor/scatter factor. Nature 373:702–705

22. Shimomura T, Miyazawa K, Komiyama Y, et al (1995) Activation of hepatocyte growth factor by two homologous proteases, blood-coagulation factor XIIa and hepatocyte growth factor activator. Eur J Biochem 229:257–261

23. Miyazawa K, Shimomura T, Kitamura A, et al (1993) Molecular cloning and sequence analysis of the cDNA for a human serine protease responsible for activation of hepatocyte growth factor: structural similarity of the protease precursor to blood coagulation factor XII. J Biol Chem 268:10024–10028

24. Miyazawa K, Shimomura T, Kitamura A, et al (1996) Activation of hepatocyte growth factor in the injured tissues is mediated by hepatocyte growth factor activator. J Biol Chem 271:3615–3618

25. Yanagita K, Nagaike M, Ishibashi H, et al (1992) Lung may have an endocrine function producing hepatocyte growth factor in response to injury of distal organs. Biochem Biophys Res Commun 182:802–809

26. Mars WM, Liu M-L, Kitson RP, et al (1995) Immediate early detection of urokinase receptor after partial hepatectomy and its implications for initiation of liver regeneration. Hepatology 21:1695–1701

27. Olle EW, Ren X, McClintock SD, et al (2006) Matrix metalloproteinase-9 is an important factor in hepatic regeneration after partial hepatectomy in mice. Hepatology 44:540–549

28. Shimizu M, Hara A, Okuno M, et al (2001) Mechanism of retarded liver regeneration in plasminogen activator-deficient mice: impaired activation of hepatocyte growth factor after Fas-mediated massive hepatic apoptosis. Hepatology 33:569–576

29. Olaso E, Friedman SL (1998) Molecular regulation of hepatic fibrogenesis. J Hepatol 29:836–847

30. Border WA, Noble NA (1994) Transforming growth factor β in tissue fibrosis. N Engl J Med 331:1286–1292

31. Khalil N (1999) TGF-β latent to active. Microbes Infect 1:1255–1263

32. Crawford SE, Stellmach V, Murphy-Ullrich JE, et al (1998) Thrombospondin-1 is a major activator of TGF-β in vivo. Cell 93:1159–1170

33. Munger JS, Huang X, Kawakatsu H, et al (1999) The integrin $\alpha v \beta 6$ bind and activates latent TGF-β: a mechanism for regulating pulmonary inflammation and fibrosis. Cell 96:319–328

34. Akita K, Okuno M, Enya M, et al (2002) Impaired liver regeneration in mice by lipopolysaccharide via TNF-α/kallikrein-mediated activation of latent TGF-β. Gastroenterology 123:352–364

35. Kojima S, Hayashi S, Shimokado K, et al (2000) Transcriptional activation of urokinase by the Kruppel-like factor Zf9/COPEB activates latent TGF-β1 in vascular endothelial cells. Blood 95:1309–1316

36. Ashkenas J, Muschler J, Bisserll MJ, et al (1996) The extracellular matrix in epithelial biology: shared molecules and common themes in distant phyla. Dev Biol 180:433–444

37. Martinez-Hernandez A, Amenta PS (1995) The extracellular matrix in hepatic regeneration. FASEB J 9:1401–1410

38. Kim T-H, Mars WM, Stolz DB, et al (1997) Extracellular matrix remodeling at the early stages of liver regeneration in the rat. Hepatology 26:896–904

39. Okumura N, Seki T, Ariga T (2007) Cell surface-bound plasminogen regulates hepatocyte proliferation through a u-PA-dependent mechanism. Biosci Biotechnol Biochem 71:1542–1549

40. Jenkins GR, Seiffert D, Parmer RJ, et al (1997) Regulation of plasminogen gene expression by interleukin-6. Blood 89:2394–2403

41. Menoud P-A, Sappino N, Boudal-Khoshbeen M, et al (1996) The kidney is a major site of α_2-antiplasmin production. J Clin Invest 97:2478–2484

42. Ouax PHA, van den Hoogen CM, Verheijen JH, et al (1990) Endotoxin induction of plasminogen activator and plasminogen activator inhibitor type 1 mRNA in rat tissues in vivo. J Biol Chem 265:15560–15563

43. Uno S, Nakamura M, Ohomagari Y, et al (1998) Regulation of tissue-type plasminogen activator (t-PA) and type-1 plasminogen activator inhibitor (PAI-1) gene expression in rat hepatocytes in primary culture. J Biochem 123:806–812

44. Heaton JH, Gelehrter TD (1990) Cyclic nucleotide regulation of plasminogen activator and plasminogen activator inhibitor messenger RNAs in rat hepatoma cells. Mol Endocrinol 4:171–178

45. Imagawa S, Fujii S, Dong J, et al (2006) Hepatocyte growth factor regulates E box-dependent plasminogen activator inhibitor type 1 gene expression in HepG2 liver cell. Arterioscler Thromb Vasc Biol 26:2407–2413

46. Uno S, Nakamura M, Seki T, et al (1997) Induction of tissue-type plasminogen activator (t-PA) and type-1 plasminogen activator inhibitor (PAI-1) a early growth responses in rat hepatocytes in primary culture. Biosci Biotechol Biochem 9:123–128

47. Mueller L, Broering DC, Meyer J, et al (2002) The induction of the immediate-early-gene Egr-1, PAI-1 and PRL-1 during liver regeneration in surgical models is related to increased portal flow. J Hepatol 37:606–612

48. Nomura K, Miyagawa S, Ayukawa K, et al (2002) Inhibition of urokinase-type plasminogen activator delays expression of c-jun, activated transforming growth factor b1, and matrix metalloproteinase 2 during post-hepatectomy liver regeneration in mice. J Hepatol 36:637–644

49. Seki T, Healy AM, Fletcher DS, et al (1999) IL-1b mediates induction of hepatic type 1 plasminogen activator inhibitor in response to local tissue injury. Am J Physiol 277: G801–G809

50. Thornton AJ, Buruzdzinski CJ, Raper SE, et al (1994) Plasminogen activator inhibitor-1 is an immediate early response gene in regenerating rat liver. Cancer Res 54:1337–1343

51. Yazigi NA, Carrick TL, Bucuvalas JC, et al (1997) Expansion of transplanted hepatocytes during liver regeneration. Transplantation 64:816–820

52. Okada K, Ueshima S, Imano M, et al (2004) The regulation of liver regeneration by the plasmin/α_2-antiplasmin system. J Hepatol 40: 110–116

53. Bezerra JA, Bugge TH, Melin-Aldana H, et al (1999) Plasminogen deficiency leads to impaired remodeling after a toxic injury to the liver. Proc Natl Acad Sci USA 96:15143–15148

54. Bezerra JA, Currier AR, Melin-Aldana H, et al (2001) Plasminogen activators direct reorganization of the liver lobule after acute injury. Am J Pathol 158:921–929

55. Pohl JF, Melin-Aldana H, Sabla G (2001) Plasminogen deficiency leads to impaired lobular reorganization and matrix accumulation after chronic liver injury. Am J Pathol 159:2179–2186

56. Ng VL, Sabla GE, Melin-Aldana H (2001) Plasminogen deficiency results in poor clearance of non-fibrin matrix and persistent activation of hepatic stellate cells after an acute injury. J Hepatol 35:781–789

57. Shanmukhappa K, Sabla GE, Degen JL, et al (2006) Urokinase-type plasminogen activator supports liver repair independent of its cellular receptor. BMC Gastroenterol 6:1–9

58. Kawao N, Okada K, Kawata S, et al (2007) Plasmin decreases the BM3-only protein Bim_{EL} via the ERK1/2 signaling pathway in hepatocytes. Biochim Biophys Acta 1773:718–727

59. Ogasawara J, Watanabe-Fukunaga R, Adachi M, et al (1993) Lethal effect of the anti-Fas antibody in mice. Nature 364:806–809

60. Ito N, Yonehara S, Ishii A, at al (1991) The polypeptide encoded by the cDNA for human cell surface antigen Fas can mediate apoptosis. Cell 66:233–243

61. Watanabe-Fukunaga R, Brannan CI, Ito N, et al (1992) The cDNA structure, expression, and chromosomal assignment of the mouse Fas antigen. J Immunol 148:1274–1279

62. Tanaka M, Okada K, Ueshima S, et al (2001) Impaired liver regeneration after partial hepatectomy in plasminogen deficient mice. Fibrinolysis Proteolysis 15:2–8

63. Roselli HT, Su M, Washington K, et al (1998) Liver regeneration is transiently impaired in urokinase-deficient mice. Am J Physiol 275:G1472–G1479

64. Drixler TA, Vogten JM, Gebbink MFBG, et al (2003) Plasminogen mediates liver regeneration and angiogenesis after experimental partial hepatectomy. Br J Surg 90:1384–1390

65. Bergheim I, Guo L, Davis MA, et al (2006) Metformin prevents alcohol-induced liver injury in the mouse: critical role of plasminogen activator inhibitor-1. Gastoroentrology 130:2099–2112

66. Harrison SA, Diehl AM (2002) Fat and the liver—a molecular overview. Semin Gastrointest Dis 13:3–16

67. Wang H, Vohra BPS, Zhang Y, et al (2005) Transcriptional profiling after bile duct ligation indentifies PAI-1 as a contributor to cholestatic injury in mice. Hepatology 42:1099–1108

68. Wang H, Zhang Y, Heuckeroth RO (2007) Tissue-type plasminogen activator deficiency exacerbates cholestatic liver injury in mice. Hepatology 45:1527–1537

69. Miyoshi H, Rust C, Roberts PJ, et al (1999) Hepatocyte apoptosis after bile duct ligation in the mouse involves Fas. Gastroenterology 117:669–677

70. Salgado S, Garcia J, Vera J, et al (2000) Liver cirrhosis is reverted by urokinase-type plasminogen activator gene therapy. Mol Ther 2:545–551
71. Bueno M, Salgado S, Beas-Zarate C, et al (2006) Urokinase-type plasminogen activator gene therapy in liver cirrhosis is mediated by collagens gene expression down-regulation and up-regulation of MMPs, HGF and VEGF. J Gene Med 8:1291–1299
72. Tateno C, Yoshizane Y, Saito N, et al (2004) Near completely humanized liver in mice shows human-type metabolic responses to drugs. Am J Pathol 165:901–912
73. Katoh M, Sawada T, Soeno Y, et al (2007) In vivo drug metabolism model for human cytochrome P450 enzyme using chimeric mice with humanized liver. Assoc J Pharm Sci 96:428–437
74. Currier AR, Sabla G, Locaputo S, et al (2003) Plasminogen directs the pleiotropic effects of uPA in liver injury and repair. Am J Physiol Gastrointest Liver Physiol 284: G508–G515
75. Shanmukhappa K, Mourya R, Sabla GE, et al (2005) Hepatic to pancreatic switch defines a role for hemostatic factors in cellular plasticity in mice. Proc Natl Acad Sci U S A 102:10182–10187

Pathophysiology of Transglutaminases Including Factor FXIII

Akitada Ichinose

Summary. Transglutaminases are at least nine enzymes that crosslink a number of proteins. This type of reaction not only enhances or abolishes the original functions of substrate proteins but adds new functions to them. Factor XIII, a plasma transglutaminase circulating in blood as a heterotetramer, consists of two catalytic A subunits and two noncatalytic B subunits. It is a proenzyme to be activated by thrombin in the blood coagulation cascade. It plays an important role(s) in hemostasis, wound healing, and maintenance of pregnancy. Accordingly, a lifelong bleeding tendency, abnormal wound healing, and recurrent spontaneous miscarriage are common symptoms of factor XIII deficiency. Genetic and molecular mechanisms of congenital deficiencies have been analyzed in vitro. The mechanisms of these defects have also been studied in detail using factor XIII gene knockout (KO) mice in vivo. The factor XIII KO mice of either the A or B subunit demonstrated unforeseen cardiac pathologies only in males. Acquired factor XIII deficiency is not uncommon, and one must pay attention to it. Recently, meta-analyses have indicated that the Val34Leu polymorphism of FXIII-A is associated with thrombosis. Emerging data suggest that in addition to its extracellular role in hemostasis FXIII-A may function as a cellular transglutaminase and be involved in certain intracellular processes that include cytoskeletal remodeling and signal transduction.

Key words. Protein crosslinking · Bleeding disorder · Abnormal wound healing · Spontaneous miscarriage · Thrombosis

Introduction

Transglutaminases (TGases) are enzymes that catalyze the formation of ϵ-(γ-glutamyl)lysine bonds, in so-called protein crosslinking reactions, between a number of proteins [1, 2]. The crosslinking reactions not only augment or abolish the original functions of the substrate proteins, they add new properties to the substrates, which subsequently changes their functions. Protein crosslinking is involved in many physiological and pathological reactions, such as hemostasis, wound healing, tumor growth, skin formation, and apoptosis. Accordingly, the crosslinking reaction is one of the most important of the posttranslational modification reactions, such as proteolysis, phosphorylation, and glycosylation.

TGase: An Overview

TGase Family and TGase Superfamily

A total of 10 members of the TGase family were identified in the human genome project. Previously, nine TGases were known [2]: plasma TGase (the A subunit of factor FXIII, or FXIII-A), tissue (liver, red blood cells, endothelial cells) TGase, keratinocyte TGase, epidermal TGase, prostate TGase, TGases X, Y, and Z, and erythrocyte band 4.2 protein (see Fig. 1 in ref. 2). An additional homologous gene, TGase 3L (TGM3L, TGase3-like gene), has been identified and needs to be characterized in detail. It is of note that although all TGases exist in the cytoplasm inside cells (intracellular TGase), FXIII-A also circulates in blood (extracellular TGase) and works outside cells in the blood coagulation system. The B subunit of FXIII (FXIII-B) likely stabilizes the unstable FXIII-A in plasma by forming a complex.

After FXIII-A was cloned in 1986, a number of the genes in the animal TGase family have been identified in a wide range of vertebrates and invertebrates, including the horseshoe crab, grasshopper, sea squirt, and even slime mold. The last one (*Physarum polycephalum*) is the lowest organism in which an animal-type TGase was found.

Moreover, a superfamily of proteins homologous to the animal TGases (TGase homologue) has been discovered in archaea, bacteria, and even eukaryotes. Although three motifs—Cys, His, Asp residues—which form the catalytic triad in the TGases, are conserved in the members of this superfamily, their overall sequences differ from any known animal TGases or each other (Fig. 1).

Microbial TGase Homologues

During the last decade, various bacterial TGase homologues were identified and were found to have few or no structural similarities to their animal counterparts [3]. For

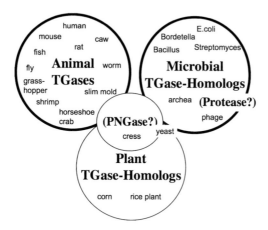

FIG. 1. Transglutaminase (TGase) family versus TGase superfamily including homologues. PNGase, peptide: N-Glycanase

example, TGase homologues cloned from two sporulating bacteria, *Streptomyces* sp. (formerly *Streptoverticillium*) and *Bacillus subtilis*, exhibit no homology with any known TGases or each other. These TGase homologues are present at various levels in the spore coat lattice and crosslink various coat proteins including GerQ.

It was shown by Horiguchi et al. in 1997 and Schmidt et al. in 1998 that the cytotoxic necrotizing factor 1 (CNF-1) from *Escherichia coli* and the homologous dermone-crotic toxin (DNT) from *Bordetella pertussis* also have TGase activity. CNF-1 deami-dates Gln-63 of RhoA, which is essential for GTP hydrolysis; and such modification prevents the targeted members of the Rho family of small GTPases from cycling back to their inactive GDP-bound states. DNT preferentially acts as a TGase, whereby it modifies Gln-63 of RhoA by crosslinking to primary amines, such as ethylenediamine, spermidine, and putresceine [4]. The persistent activation of RhoA by these toxins induces morphological and cytopathic effects on eukaryotic cells.

It is of interest that a protein identified in an archaeal phage of *Methanobacterium* was shown by Pfister et al. in 1998 to have protease activity rather than TGase activity, indicating that at least some of the prokaryotic TGase homologues may code for pro-teases. This pseudomurein endoisopeptidase of *Methonobacterium* phage psiM2 cata-lyzes lysis of the host cell wall. Because papain-like thiol proteases have the same catalytic triad (the nucleophilic Cys and His and Asp residues) as TGases and TGase-homologues, the proteolytic reaction of thiol enzymes and transglutamination may be the reverse of each other.

Plant TGase Homologues

To date, two plant TGase homologues have been relatively well characterized at the molecular level: one from maize (corn) chloroplast and another from *Arabidopsis thaliana* (thale cress). A recombinant protein of the former TGase homologue, which was expressed in *Escherichia coli* and demonstrated in the work of Villalobos et al. in 2004, the TGase catalytic activity. The TGase activity of a recombinant wild-type molecule of the latter TGase homologue (AtPNG1), which was expressed in E. coli, was also demonstrated by Della Mea et al. in 2004. However inactivation of this gene did not result in decreased TGase activity in the mutant plant, and recombinant AtPNG1 produced in yeast cells exhibited only a trace of TGase activity [5].

It is of note that AtPNG1 had been purified as a peptide: N-glycanase, which may contribute to the endoplasmic reticulum-associated degradation pathway (ERAD) by deglycosidation of misfolded proteins. A catalytic triad containing Cys, His, and Asp residues is employed in the catalytic mechanism of PNGases. *Olyza sativa* (rice plant) also has a TGase homologue sequence, which is also included as an N-glycanase. This class of PNGase is a highly conserved enzyme in eukaryotic cells in humans, mice, plants, fruit flies, nematodes, and fungi.

Biochemistry of Factor FXIII

Factor XIII (fibrin-stabilizing factor) circulates in blood as a heterotetramer consist-ing of two catalytic (FXIII-A) subunits and two noncatalytic B subunits (FXIII-B), A_2B_2.

Genes

The gene organization of TGase was first established for FXIII-A in 1988 by Ichinose and Davie, and the genes for other TGases followed. These studies revealed that members of the TGase family also share gene organization. With some exceptions, their genes are coded by 15 exons, interrupted by 14 introns, and the exon/intron boundaries of these TGases occur in the same sites among the primary structures. A major difference between these genes is their size. The gene for FXIII-A spans about 180 kb, whereas the keratinocyte TGase spans only 15 kb. The sizes of the introns in each TGase gene are responsible for this difference.

The gene for FXIII-A is located on chromosome 6 at bands p24-25. The gene encoding keratinocyte TGase is localized on chromosome 14q11.2-13 and that for prostate TGase (type IV, TGM4) on chromosome 3p21.33-p22. Thus, gene loci for these TGases are dispersed throughout the chromosomes; however, the remaining four and three TGases are clustered on chromosome 20q1 [tissue and epidermal TGases, TGM3L, and TGase Y] and 15q15 (TGases X and Z, band 4.2), respectively.

Primary Structures of FXIII

The primary structure of TGase was first established in 1986 by a combination of amino acid sequencing analysis and cDNA cloning for FXIII-A. FXIII-A consists of 731 amino acid residues. The amino acid sequences of other TGases have been determined, and their homology and evolutional relations established [1] (see Fig. 1 in ref. 2).

FXIII-B is composed of 641 amino acid residues and contains 8.5% carbohydrate. FXIII-B consists of 10 tandem repeats that have been designated as Sushi domains, GP-I structures, or short consensus repeats (SCRs) [1].

At least 3 of the 10 Sushi domains of FXIII-B have a distinct function—to form either a homodimer or a heterotetramer with FXIII-A—which can be ascribed to the difference in their amino acid sequences (Souri and Ichinose, manuscript in preparation).

The Human Genome Project has identified homologous Sushi domains in a total of 53 genes. Sushi domains have also been identified in the fruitfly and worms. Therefore, it may be a rather old protein module that plays important roles in biological protein–protein interactions.

Three-Dimensional Structure of FXIII

The three-dimensional structure of FXIII was demonstrated previously by electron microscopic studies by McDonough's group in 1989. FXIII-A and FXIII-B appear to be a globular particle and a filamentous strand, respectively.

FXIII-A of placenta and recombinant proteins have been crystallized. X-ray crystallography demonstrated that FXIII-A is composed of five distinct domains: an activation peptide, β-sandwich, central core, barrel 1, and barrel 2 regions (see Fig. 2 in ref. 2). It is of interest that the activation peptide, upon cleavage by thrombin, occupies the same position with respect to the rest of the molecule as it does in the zymogen structure; moreover, the activation peptide blocks the entrance to the catalytic cavity in the core domain. Examination by Sadasivan and Yee in 2000 of the X-ray crystal structure of human α-thrombin bound to a decapeptide (amino acid residues 28–37) of FXIII-A, revealed detailed atomic-level interactions between the activation peptide and thrombin. Calcium-binding sites were also identified in FXIII-A by X-ray crystallography.

To gain further insight into the structural changes that occur during the activation of FXIII, chemical modification involving the labeling of Cys and Lys side chains of the enzyme was performed by Maurer's group in 2004. They also carried out hydrogen deuterium exchange experiments after proteolytically activating FXIII-A by thrombin cleavage or nonproteolytically by exposure to a high calcium concentration. These approaches revealed that activation leads to local and long-range effects to FXIII-A and that many residues are influenced by calcium binding.

The X-ray crystal structure of FXIII-B has not been established. Nuclear magnetic resonance (NMR) studies revealed that Sushi domains in complement factor H consist of double- or triple-stranded β-sheets and several β-turns and cuboid forms.

Substrate Structures

Solution NMR methods by Marinescu et al. in 2002 revealed that the reactive Glns and residues located C-terminally of FXIII-A substrates come in direct contact with the enzyme's active-site region and adopt an extended conformation; moreover, the first possible substrate-binding site on FXIII-A involves its catalytic triad region where the reactive Gln and its nearest neighbors are anchored. When they were employed, N-terminal peptides of α_2-plasmin inhibitor, one of the best FXIII-A substrates, preferentially selected the Gln-2 residue for carrying out a crosslinking reaction. It was also suggested by Cleary et al. in 2006 that the Glu-3 and Gln-4 residues provide supporting roles in binding and that the Lys-12 provides an additional favorable site of interaction with the surface of FXIII-A. More recently, Hitomi et al. have identified several amino acid sequences that were preferred as glutamine donor substrates, most of which have a marked tendency for individual TGases: FXIII-A, GlnXXYXTrpPro; TGase 2, GlnXProYAsp(Pro), GlnXProY, and GlnXXYAspPro (where X and Y represent a non-conserved and a hydrophobic amino acid, respectively), using an unbiased M13 phage display random peptide library [6]. The substrate specificity of FXIII-A will provide a basis for the design of its inhibitors.

Physiology of FXIII

Biosynthesis of FXIII

Although tissue TGase exists in most tissues and organs, FXIII-A is mainly limited to plasma and megakaryocytes/platelets and monocytes/macrophages, as well as the placenta in pregnant females: keratinocyte TGase to skin keratinocytes, epidermal TGase to skin and hair follicular cells, and prostate TGase to prostate.

FXIII-A and other TGases share significant similarity in their gene organization; however, their nucleotide sequences in the 5′-flanking region differ from each other, and mechanisms for their gene regulation seem to be diverse as well. Naturally, nucleotide elements for transcription factors in one TGase gene are quite different from those in others. These transcription factors play a major role in the cell type-specific expression of each TGase (see Fig. 3 in ref. 2).

When we cloned the cDNA for FXIII-A, we first noted that there was no signal peptide for secretion; nor could we find any internal signal. Results of the expression of FXIII-A and/or FXIII-B in a mammalian cell system suggest that FXIII-A is not secreted through the conventional secretory pathway but is released from the cells by cell damage or apoptosis.

In contrast, because it has a typical signal peptide, FXIII-B is readily released from the synthesizing cells by the secretion pathway. The site of synthesis for the FXIII-B gene is the liver. The regulatory mechanism of FXIII-B expression with respect to its nucleotide sequence in the 5′-flanking region has not been characterized as yet.

Functions of FXIII

Factor XIII is a proenzyme activated to active enzyme called FXIIIa by thrombin that is generated in the final stage of the blood coagulation cascade. FXIII plays an important role(s) in hemostasis, wound healing, and maintenance of pregnancy. The enzyme promotes clot stability by forming covalent bonds between fibrin molecules and also by crosslinking fibrin with several proteins including α_2-plasmin inhibitor and fibronectin. These reactions lead to an increase in the mechanical strength, elasticity, and resistance to degradation by plasmin of fibrin clots, and promotion of wound healing by providing a scaffold for fibroblasts to proliferate and spread. It is suggested that interactions of FXIII with other cells, such as macrophages and platelets, play important roles in the physiological reactions. A number of other proteins are known to be substrates of FXIII—e.g., plasminogen activator inhibitor type-2 (PAI-2), osteopontin, lipoprotein(a), platelet vinculin, factor V, and thrombospondin; however, the physiological implications of these reactions are unclear.

Roles of FXIII in Hemostasis

It was reported by Dale et al. in 2002 that platelets stimulated simultaneously with collagen and thrombin express factor V derived from α-granules only on a subpopulation of platelets. These platelets were named COAT platelets because they bind additional α-granule proteins, such as fibrinogen, von Willebrand factor, thrombospondin, fibronectin, and α_2-plasmin inhibitor, all of which are also reported to be TGase substrates. Inhibitors of TGase prevent the binding of α-granule proteins and so reduce the production of COAT platelets. Analysis by high-performance liquid chromatography/mass spectrometry (HPLC/MS) showed that fibrinogen isolated from COAT platelets contained covalently crosslinked serotonin. Therefore, COAT platelets may use serotonin conjugation to enhance the retention of procoagulant proteins on their surface, which in turn accelerate thrombus formation.

The major platelet surface glycoprotein (GP) IIb/IIIa (integrin $\alpha_{II}\beta_3$) plays an essential role in this process by promoting platelet aggregation and by supporting thrombus growth. Kulkarni et al. in 2004 demonstrated that there is a novel mechanism down-regulating GPIIb/IIIa adhesive function, involving platelet FXIII-A and calpain, which serves to limit platelet aggregate formation and thrombus growth. Prolonged elevations in cytosolic calcium in collagen-adherent platelets induces this down-regulation mechanism, leading to gross changes in platelet morphology, such as membrane contraction, fragmentation, and microvesicle formation, as well as a significant reduction in GPIIb/IIIa adhesion.

Pregnancy

It is well known that FXIII-A is not required for fertilization or implantation of the human egg/embryo. However, FXIII is crucial for the healthy maintenance of the pregnancy after 5 weeks of gestation (wG). In cases of FXIII deficiency, spontaneous decidual bleeding occurs beginning at 5–6 wG, and miscarriage can follow if the woman is not treated properly with FXIII-A replacement therapy. In at least one patient with FXIII deficiency, the cytotrophoblastic shell and Nitabach's layers were poorly formed in the early implantation tissues [7], which must attach adequately to the uterine wall or the result is bleeding and miscarriage.

Wound Healing and Angiogenesis

Thymosin $\beta 4$ (TB4) is expressed ubiquitously in most cell types and is considered to be the major intracellular G-actin sequestering peptide. In contrast, extracellular TB4 enhances attachment and spreading of endothelial cells (ECs) on matrix components and stimulates EC migration. Huff et al. reported in 2002 that tissue TGase, and presumably FXIII, can covalently incorporate into fibrin(ogen) TB4, which is released after activation of human platelets by thrombin. Accordingly, the increased local concentration of TB4 surrounding sites of platelets/fibrin clots and tissue damage may contribute to wound healing, angiogenesis, and inflammatory processes.

It has also been shown that FXIII itself induces angiogenesis. Dardik et al. reported in 2005 that the proangiogenic effect of FXIII on EC is mediated by enhancement of crosslinked and noncovalent $\alpha_v\beta_3$/vascular endothelial growth factor receptor-2 (VEGFR-2) complex formation, tyrosine phosphorylation and activation of VEGFR-2, up-regulation of c-Jun and Egr-1 (early growth response-1), and down-regulation of thrombospondin-1, an antiangiogenic protein, induced indirectly by c-Jun through WT-1 (Wilm's tumor-1). These responses may indicate FXIII's roles in vascular remodeling and tissue repair.

Inflammation and Phagocytosis

In addition to its extracellular role in hemostasis, FXIII-A may function as a cellular TGase and be involved in certain intracellular processes that include cytoskeletal remodeling.

Crosslinking of cellular proteins such as lysosomal enzymes and histones by TGases may prevent leakage of these proteins from apoptotic cells and protect surrounding tissues from possible inflammation. Yamamoto's group found in 1996 that S19 ribosomal protein is crosslinked by FXIII-A and tissue TGase in apoptotic cells in rheumatoid tissues, and that the dimerized S19 ribosomal protein acquires chemotactic activity for monocytes. The apoptotic cells are removed by the infiltrated monocytes/macrophages. This reaction has been shown in both in vivo and in vitro experiments. Accordingly, TGases play important roles in the cleaning-up of apoptotic cells.

However, recruitment of monocytes from the circulation into the lesion may also enhance inflammatory reactions, which result in chronic, severe disease states.

Adany et al. suggested in 2004 that FXIII-A plays an important role in the Fcγ and complement receptor-mediated phagocytic activities of monocytes/macrophages. They also suggested in 2005 that FXIII-A expression is an intracellular marker for alternatively activated macrophages by Th2 cytokines, such as interleukin-4 (IL-4), which are involved in the disposal of soluble and particulate cell/tissue debris. Inversely, the absence of FXIII-A expression in macrophages may indicate their classically activated state by Th1 cytokines, such as interferon-γ (IFNγ), and microbes or microbial products. It is of note that tissue TGase contributes to neutrophil granulocyte differentiation, and tissue TGase knockout (KO) mice exhibited increased neutrophil phagocytic activity. Accordingly, FXIII-A may enhance phagocytosis of macrophages, whereas tissue TGase may suppress phagocytosis of neutrophils. However, no increased susceptibility to microbial infections was observed in patients with FXIII deficiency or in FXIII-A KO mice.

Vascular Remodeling

Quitterer et al. in 2005 reported that intracellular FXIII-A crosslinked agonist-induced AT1 receptor homodimers and that the appearance of the crosslinked dimers were correlated with enhanced signaling and desensitization in vitro and in vivo. Inhibition of angiotensin II release or of FXIII-A activity prevented the formation of crosslinked AT1 receptor dimers on monocytes of a control individual, and FXIII-A-deficient individuals lacked crosslinked AT1 dimers. Increased levels of crosslinked AT1 dimers were detected on monocytes of patients with hypertension in association with an enhanced angiotensin II-dependent monocyte adhesion to ECs. Because inhibition of angiotensin II generation or of intracellular FXIII-A activity significantly suppressed the rate of crosslinked AT1 receptors and atherosclerotic lesions in apolipoprotein E (ApoE)-deficient mice, increased levels of crosslinked AT1 receptor dimers on monocytes may contribute to atherogenesis.

It was also reported that tissue TGase plays an important role in the inward remodeling of small arteries associated with decreased blood flow. Moreover, adventitial monocytes/macrophages are a source of FXIII in tissue TGase-null mice and contribute to an alternative delayed mechanism of inward remodeling when tissue TGase is absent [8].

Most recently, Guiluy et al. in 2007 showed that serotonin-induced RhoA depletion was due to RhoA activation by the transamidation of intracellular serotonin by tissue TGase in vascular smooth muscle cells. The persistent activation of RhoA leads to its enhanced proteasomal degradation and Akt activation, which in turn is responsible for contraction inhibition. Transamidation of serotonin to RhoA in the pulmonary arteries of hypoxic rats suggests that this process could participate in pulmonary artery remodeling and hypertension.

Bone Maturation

Nurminskaya et al. in 2006 reported that both tissue TGase and FXIII are expressed in mineralized tissues of growth plate cartilage and bone. Immunohistochemistry and in situ hybridization methods confirmed that FXIII-A is expressed by osteoblasts and osteocytes in long bones and flat bones. It was demonstrated by Nakano et al. in 2007 that in bone and cultured osteoblasts FXIII-A is produced mainly as a small 37-kDa form, which is a product of posttranslational proteolytic processing. It is of note that forskolin inhibition of osteoblast differentiation revealed that this FXIII-A processing is regulated by the protein kinase A pathway. Although the 37-kDa form of FXIII-A appears to be associated with the osteoblast plasma membrane as part of the osteoblast differentiation process, no abnormality in bone was observed in patients with FXIII deficiency or in FXIII-A KO mice.

Pathology of FXIII

FXIII is known to be related to a number of disease states, such as thrombosis, dementia, and cancer. For example, it was reported by Naidu et al. in 1994 that a conditioned medium from type II human T-cell leukemia virus-infected T cells induces the conversion of endothelial cells to a Kaposi's sarcoma cell-like phenotype, and that the endothelial cells grown by hepatocyte growth factor in the culture medium acquired the ability to express FXIII-A.

Congenital FXIII Deficiency

More than 500 cases of FXIII deficiency have been identified. We classified FXIII deficiency at the DNA level: FXIII-A deficiency (former type II deficiency) and FXIII-B deficiency (former type I deficiency), and a possible combined deficiency of FXIII-A and FXIII-B (http://www.med.unc.edu/isth/99FXIII.html). Because the genes for FXIII-A and FXIII-B are localized at 6p and 1q, respectively, FXIII deficiency is inherited as an autosomal recessive trait and is caused by the absence of either subunit. Mutations causing FXIII deficiency are highly heterogeneous.

Mutations in the gene for FXIII-A include a variety of missense and nonsense mutations, small deletions and insertions with or without out-of-frame shift/premature termination, and splicing abnormalities (see Fig. 4 in ref. 2). A large deletion, exons IV to XI, has been reported in one case with FXIII-A deficiency. The effects of these mutations on FXIII-A biosynthesis have been confirmed in many cases in vitro.

More than one-third of examined cases with FXIII-A deficiency have point mutations that cause amino acid substitution. For example, rapid degradation of a novel Tyr283-Cys mutant has been ascribed to its instability characterized by our group in 2001 in a megakaryocytoid cell expression system. Gel filtration analysis revealed that the mutant was a monomer, whereas the wild-type formed a dimer.

A deletion of 4 bp in exon XI in one case leads to premature termination at codon 464 (see Fig. 5 in ref. 2). Reverse transcription polymerase chain reaction (RT-PCR) analysis demonstrated that the mRNA level was greatly reduced. Recombinant FXIII-A bearing the mutation showed that the mutant disappeared rapidly inside cells.

Also found were a 20-bp deletion at the boundary of exon I/intron A and an insertion of T in the invariant GT dinucleotide at the splicing donor site of exon IV/intron D in one patient. RT-PCR analysis demonstrated that only one kind of mRNA without exon IV was detected, although its level was greatly reduced to less than 5% of normal.

Thus, both mutations impaired the normal processing of mRNA for FXIII-A. Exon skipping has also been reported in other cases.

Two Japanese patients with FXIII-B deficiency have a previously described one-base deletion at the boundary between intron A/exon II heterozygously or homozygously. We proposed a founder effect for this mutation (see Fig. 6 in ref. 2). In one patient heterozygous for the above mutation, a novel mutation was also identified: deletion of guanosine in exon IX (delG) of the FXIII-B gene. A resultant truncated FXIII-B remained inside the cells and could not be secreted into the culture medium. Intracellular transportation of the truncated FXIII-B from the endoplasmic reticulum to the Golgi apparatus was impaired. Therefore, secretion of the truncated FXIII-B must also be impaired in vivo, leading to secondary FXIII-A deficiency in the patient.

It is important to examine closely the amino acid substitutions and the deletion/insertion of peptide regions to understand the structure/function relations of the FXIII molecule as well as its clinical implications in FXIII deficiency.

FXIII Knockout Mice

In vivo experiments are called for to understand the clinicopathological mechanisms of this disease. Accordingly, FXIII-A KO mice were independently produced by Dickneite's and our groups, and the phenotypes of these mice have been under investigation. In the German group's mouse model, plasma TGase activity was undetectable in homozygous FXIII-A KO mice, and no γ-dimerization of fibrin in the plasma of the FXIII-A KO mice could be detected. The mortality rate was higher among the FXIII-A KO mice than among normal mice because of serious bleeding episodes.

We have examined the two strains of FXIII-A KO mice for fertility. Although all FXIII-A KO female mice were capable of becoming pregnant, they frequently died owing to excessive vaginal bleeding during gestation. Abdominal incisions revealed that the uteri of the dead mice were filled with blood and that some embryos were much smaller than others within a single animal. Histopathological analyses of the small embryos revealed intrauterine fetal death (IUFD) among the dead FXIII-A KO female mice (Fig. 2). This was true regardless of the mouse strain or the genotypes of male mating partners or those of fetuses. These results indicate that the maternal FXIII plays a critical role in uterine hemostasis and maintaining the placenta during gestation.

We found that the FXIII-A KO mice demonstrated abnormal wound healing similar to that seen in human patients with FXIII deficiency (Koseki-Kuno and Ichinose, unpublished data), as reported by Inbal's group. Artificial wounds became deep ulcers in FXIII-A KO mice, and the healing process was retarded when compared to that of the wild-type mice.

Additionally, we established FXIII-B KO mice by homologous recombination in embryonic stem cells. Both homozygous and heterozygous KO mice showed no marked difference from the wild-type mice in general appearance (Souri and Ichinose, Int J Haematol., 2008). Recently, it was found that several FXIII-A-null male mice died of severe intrathoracic hemorrhage, and a large hematoma was observed in their hearts. Hemorrhage, hemosiderin deposition, and/or fibrosis were observed in the hearts of other dead FXIII-A-null males. Unexpectedly, fibrosis together with hemosiderin deposition was also found in the hearts of both FXIII-A-null and FXIII-B-null

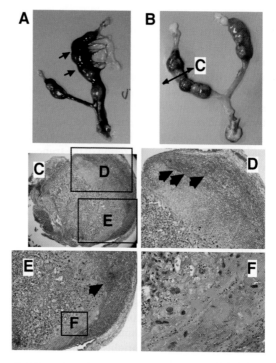

Fɪɢ. 2. Uteri of female FXIII-A knockout (KO) mice that spontaneously died of excessive vaginal bleeding. **A, B** Overview. **C–F** Histopathology of a fetal placenta (intrauterine fetal death). *Black* and *green arrows* indicate areas of bleeding and hemorrhagic necrosis, respectively

male mice, although the fibrotic lesion was not observed in the hearts of age-matched FXIII-deficient female mice or wild-type mice. Carditis in factor FXIII-deficient mice might be caused by an impaired or delayed repair process after hemorrhagic events in the heart tissue. It is important, therefore, to examine the possible cardiac involvement in human male patients with congenital factor FXIII deficiency.

In summary, the major symptoms of congenital FXIII deficiency can be monistically accounted for by the abnormal bleeding caused by defective hemostasis due to the lack of FXIII. Further analysis of these FXIII KO mice will lead to an improved understanding of the physiological and pathological functions of FXIII in vivo.

Acquired Deficiency

Acquired FXIII deficiency is frequently found secondary to disease states, such as disseminated intravascular coagulation (DIC) due to consumption of FXIII and hepatic disorders because of impaired synthesis of FXIII-B. One must also pay special attention to the occurrence of severe bleeding tendencies caused by autoantibodies against FXIII-A. Although relatively rare, it occurs in patients with autoimmune diseases, malignancy, and drug-induced disorders, among other conditions. According to the author's data, about one-third of cryptogenic FXIII deficiency cases are due to the presence of an anti-FXIII-A antibody.

Factor XIII has been reported to decrease in the plasma of patients with chronic inflammatory bowel diseases, and FXIII replacement therapy is useful for treating these patients. It is likely that FXIII is consumed in the inflammatory intestine, which results in impaired wound healing in the damaged tissue.

Factor XIII is also known to decrease in patients with Henoch-Schönlein purpura, especially with abdominal complications. In such cases, administration of FXIII concentrates often leads to remission of the disease. The mechanism(s) of this FXIII reduction is not known at present.

Thrombosis and Single Nucleotide Polymorphisms

Several allelic forms of FXIII-A have been identified in the normal population. The heterogeneity of these FXIII-A cases was confirmed by amino acid and DNA sequencing. All amino acid substitutions can be explained by single nucleotide polymorphisms in coding regions (cSNPs).

It was reported by Ariens et al. in 2002 and most recently by meta-analyses that the Val34Leu polymorphism is associated with myocardial infarction and venous thrombosis. A mechanism whereby Leu34 is protective against thrombosis has been investigated. The activation rate of the Leu34 type by thrombin seems to be faster than that of the Val34 type, although they have the same specific activity when fully activated. It is of interest that early covalent crosslinking of the fibrin clot by FXIII Leu34 reduced lateral aggregation of the fibrin fibers, leading to a reduction in fiber thickness and an alteration in rates of fibrinolysis of the fibrin clot. The effect of Val34Leu on the fibrin structure and function is dependent on fibrinogen levels and may alter the interaction with platelets.

In contrast to previous studies, we found that no healthy Japanese individuals have the polymorphism. This was also the case for Japanese patients with ischemic heart disease or cerebrovascular dementia; therefore, this polymorphism cannot be a discriminative factor for thrombosis among Japanese. The polymorphism was not detected in Korean or Chinese populations, whereas its gene frequency was 0.24 in Italians. Accordingly, theVal34Leu polymorphism is another example of ethnic specificity of gene polymorphisms analogous to factor V Leiden and prothrombin G20210A.

The heterogeneity of FXIII-B was also classified into several alleles. Recently, we have identified all the single nucleotide polymorphisms that are responsible for the differences in the amino acid sequence of FXIII-B and its genotypes as well as several nucleotide substitutions in noncoding parts of the gene (Kitano et al., manuscript in preparation). Reiner et al. suggested in 2003 that one of the cSNPs, His95Arg in the FXIII-B gene, may be related to a decreased risk of myocardial infarction, especially in the presence of the Val34Leu polymorphism of FXIII-A.

FXIII in Other Disease States

We found in 1998 that many FXIII-A-positive microglia were associated with primitive senile plaques in the brains of patients with Alzheimer's disease, whereas few or no FXIII-A-containing microglia were found associated with classic plaques. FXIII may play a role(s) in the early phase of Alzheimer's disease as the tau protein is reported to be crosslinked by TGases in affected brain tissue.

Wozniak et al. reported in 2001 that FXIII could protect endothelial barrier function and prevent the development of myocardial edema in children undergoing surgery for congenital heart disease. They identified a thrombin increase combined with a simultaneous decrease of FXIII-A as a probable cause. Because preoperative FXIII substitution showed a distinct effect in minimizing the incidence of myocardial swelling, they suggested that the clinical application of FXIII may have a valuable therapeutic benefit in cases of capillary leakage syndrome during extracorporeal circulation in congenital heart surgery.

Acknowledgments. This study was supported in part by research grants from The Ministry of Education, Science, and Culture, Japan; The NOVARTIS Foundation for the Promotion of Science (Japan); The Daiwa Health Foundation (Japan); The Nakatomi Foundation (Japan); and The Yamanouchi Foundation for Research on Metabolic Disorders (Japan). The author is grateful to Drs. S. Koseki-Kuno and M. Souri of Yamagata University for helpful discussions and to Ms. L. Boba for her help in the preparation of the manuscript.

References

1. Ichinose A, Bottenus RE, Davie EW (1990) Structure of transglutaminases. J Biol Chem 265:13411–13414
2. Ichinose A (2001) Physiopathology and regulation of factor XIII. Thromb Haemost 86:57–65
3. Makarova KS, Aravind L, Koonin EV (1999) A superfamily of archaeal, bacterial, and eukaryotic proteins homologous to animal transglutaminases. Protein Sci 8:1714–171
4. Horiguchi Y (2001) *Escherichia coli* cytotoxic necrotizing factors and *Bordetella* dermonecrotic toxin: the dermonecrosis-inducing toxins activating Rho small GTPases. Toxicon 39:1619–1627
5. Diepold A, Li G, Lennarz WJ, et al (2007) The *Arabidopsis* AtPNG1 gene encodes a peptide: N-glycanase. Plant J 52:94–104
6. Sugimura Y, Hosono M, Wada F, et al (2006) Screening for the preferred substrate sequence of transglutaminase using a phage-displayed peptide library: identification of peptide substrates for TGASE 2 and factor XIIIA. J Biol Chem 281:17699–17706
7. Asahina T, Kobayashi T, Okada Y, et al (2000) Maternal blood coagulation factor XIII is associated with the development of cytotrophoblastic shell. Placenta 21:388–393
8. Bakker EN, Pistea A, Spaan JA, et al (2006) Flow-dependent remodeling of small arteries in mice deficient for tissue-type transglutaminase: possible compensation by macrophage-derived factor XIII. Circ Res 99:86–92

Part 5 Vascular Development and Regeneration

A Linkage in the Developmental Pathway of Vascular and Hematopoietic Cells

JUN K. YAMASHITA

Summary. Blood vessels consist of at least three kinds of cell: endothelial cells lining the inside of the lumen to form tubes, mural cells (vascular smooth muscle cells and pericytes) supporting the endothelial tubes, and blood cells flowing inside. Blood and vascular cells are closely related to each other in their anatomical locations, origins, and differentiation processes. In addition to a long history of histological analyses, recent progress in stem cell biology using various genetic animal models, especially in vivo cell tracing technologies, and in vitro stem cell differentiation systems are now succeeding in providing molecular and cellular bases of the relation between these two cell populations. Accumulating data suggest that their differentiation processes are more complicated than was previously expected. That is, multiple origins of progenitor cells, multiple pathways of differentiation, and multiple molecular functions regulating cell fates exist and complicatedly interact with each other to complete the functional circulation system with blood and vessels. This chapter summarizes recent advances in the developmental processes and the relation of blood and vascular cells, especially between blood and endothelial cells.

Key words. Hematopoiesis · Endothelial cells · Hemogenic endothelium · Progenitor cells · Stem cells

Introduction

Blood cells and endothelial cells (ECs) directly contact each other and form an essential functional unit to maintain the blood supply for the whole body. These two anatomically and functionally related cells are also closely related in their origin and differentiation process. In the yolk sac of mouse embryos, mesoderm-derived cells form cell aggregates called blood islands [1]. The central cells differentiate into blood cells, and peripheral cells of blood islands develop into ECs and fuse with each other to form the initial vascular network [2]. This phenomenon symbolizes the close association and relation between the blood and EC differentiation process. Nevertheless, the actual differentiation pathway and mechanisms of blood and ECs seem not to be so simple and straightforward. Recent progress of developmental, molecular, and stem cell biology is revealing various novel aspects of their differentiation processes. This chapter aims to review the relation between blood and ECs in their differentiation and diversification process through recent advances.

Differentiation of Endothelial Cells

The origin of ECs is postulated to be mesoderm cells expressing Flk1 (also designated vascular endothelial growth factor receptor-2, VEGFR2) [2]. Flk1$^+$ cells in the periphery of blood islands and in the aorta-gonad-mesonephros (AGM) region in the embryo proper differentiate into ECs mainly by vascular endothelial growth factor (VEGF) signaling and simultaneously form an endothelial tube network called the primary plexus. Such direct formation of vascular structures from progenitor cells is called vasculogenesis. Vasculogenesis is followed by angiogenesis, in which neovessels are formed from preexisting vessels, followed by vascular remodeling with migration and attachment of mural cells to the vascular wall; mature blood vessels are then gradually formed [2, 3]. Diversification into arterial and venous ECs and formation of arteries and veins occur almost simultaneously with the initiation of vascular development [4]. The existence of arterial- or venous-specific progenitors is suggested but still has not been fully demonstrated [5]. Lymphatic vessels sprout and develop from a specific subset of venous ECs [6] (see the Chapter by P. Carmeliet, this volume). The sequential processes of EC differentiation and diversification (i.e., induction of Flk1$^+$ vascular progenitor cells, differentiation of ECs, and diversification into arterial, venous, and lymphatic ECs) are successfully reproduced in an embryonic stem (ES) cell differentiation system [7]. ES cell-derived Flk1$^+$ cells give rise to ECs by VEGF stimulation and mural cells by platelet-derived growth factor (PDGF) stimulation [8]. Whereas venous ECs are induced from Flk1$^+$ progenitors by VEGF treatment alone, arterial ECs are induced by the combinatory stimulation with VEGF and cyclic adenosine monophosphate (cAMP) [9]. Prox1$^+$ lymphatic ECs are induced from Flk1$^+$ progenitors by co-culture with OP9 stroma cells [10] or long-term culture of embryoid bodies [11, 12].

Differentiation of Hematopoietic Cells

Research into the differentiation pathway of blood cells (=hematopoietic cells, or HPCs) has a much longer and complicated history than that of ECs. Until recently, a model of the successive hematopoietic process (primitive and definitive hematopoiesis) has been largely accepted as a major pathway of hematopoietic development [13, 14]. That is, the first transient wave of hematopoiesis (primitive hematopoiesis), which generates primitive erythrocytes, occurs in the yolk sac at E7.0–7.5, but the primitive HPCs carrying the embryonic and fetal-type hemoglobins contribute only in the embryo, not in the adult. A second wave of hematopoiesis that generates the hematopoietic stem cells (HSCs) necessary throughout life (definitive hematopoiesis) originates in the embryo proper. Definitive HSCs arise from the AGM region, including the endothelium and surrounding mesenchyme of the dorsal aorta at E10.5; then the site of blood formation migrates to the fetal liver at E12.5 and finally to the bone marrow (Fig. 1a) [13, 15–17].

This model is currently being challenged and remodeled by various studies. In addition to the AGM region, HSCs appear in umbilical and vitelline arteries [18]. The placenta forms a large HSC pool around E11.5–12.5 through the rapid expansion of

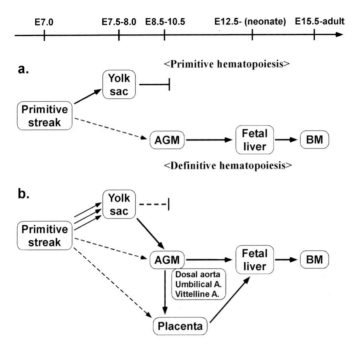

FIG. 1. Models of embryonic and adult hematopoiesis. **a** Conventional model. The first wave of hematopoiesis, called primitive hematopoiesis, occurs in the yolk sac at E7.0–7.5. Primitive hematopoietic cells (HPCs) carrying the embryonic and fetal-type hemoglobins are found only in the embryo, not in the adult. A second wave of hematopoiesis, called definitive hematopoiesis, occurs in the aorta-gonad-mesonephros (AGM) region and generates definitive HSCs that can contribute throughout life. Definitive HSCs colonize the fetal liver and finally the bone marrow (*BM*). **b** A multiorigin, multipathway model. The origin of yolk sac blood islands is polyclonal. Definitive HSCs in the AGM regions at least partly originate from yolk sac precursor cells. Major vessels in the embryo, dorsal aorta, umbilical artery, and vittelline artery as well as the placenta are another source of definitive HSCs

HSCs [19–21]. In the dorsal aorta, a subset of endothelium [15, 16] (hemogenic ECs; discussed later) as well as surrounding mesenchymal cells [22, 23] are postulated to be the origins of definitive HSCs. Recently, Ueno et al. revealed that the origin of yolk sac blood islands in vivo is polyclonal, using a novel direct clonal progenitor analysis in the mouse embryo [24]. Samokhvalov et al. demonstrated that specifically labeled Runx1[+] yolk sac precursor cells can develop into fetal lymphatic progenitors and even into adult HSCs [25], clearly indicating that yolk sac hematopoiesis can contribute to the adult. These reports indicate that HSCs develop through multiple origins, locations, and trafficking processes (Fig. 1). In addition, these various hematopoietic organs, yolk sac, AGM, fetal liver, and bone marrow serve as different microenvironments for HSC maturation [17]. Yolk sac-derived HPCs could not reconstitute hematopoiesis in the adult when transplanted into irradiated mouse models [26], even though they have an intrinsic potential of becoming adult HSCs [25]. On the other hand, yolk sac-derived HPCs could give rise to definitive HSCs after being injected into fetal liver [26], indicating that HSC niches in the embryo proper should serve as an appropriate microenvironment to achieve complete maturation of HSCs for engraftment and self-renewal activity.

Diversification of ECs and HPCs

From many decades ago, the close association of HPCs and ECs in their developmental process in the yolk sac led to the hypothesis that these cells originate from a common precursor, the hemangioblast [27]. Blast colony-forming cells (BL-CFCs), a putative hemangioblast that give rise to both HPCs and ECs, were first reported in differentiating embryonic stem cell culture in vitro [28]. BL-CFCs were enriched in the $Flk1^+$/ brachyury$^+$ mesoderm population. Recently, the existence of hemangioblasts in the gastrulating embryo has been demonstrated. Huber et al. revealed that hemangioblast comprises a subpopulation of mesoderm co-expressing Flk1 and brachyury. Highest dual differentiation potential to HPCs and ECs were observed in the posterior primitive streak but not in yolk sac, indicating that the first commitment of mesodermal progenitors to the HPC and EC lineages begins in the posterior primitive streak before the cells migrate into the yolk sac, prior to the formation of the blood islands [29]. The early segregation of HPC and EC lineages is also indicated by the fact that individual $Flk1^+$ precursors rarely contributed to both ECs and HPCs in blood islands by a direct clonal progenitor analysis in the mouse embryo [24] and that Runx1-marked yolk sac precursor cells can contribute to adult HSCs but not to ECs [25].

The next putative diverging point between ECs and HPCs is in the early ECs. During the early developmental stage of the dorsal aorta, attachment of HPC clusters to the EC layer is observed [30], inferring that HPCs including HSCs originate from a subset of ECs, termed hemogenic ECs, that possess the potential to give rise to HPCs. The hemogenic EC hypothesis is supported by several studies in addition to histological evidence. Embryonic as well as ES cell-derived vascular endothelial (VE) cadherin$^+$ ECs give rise to HPCs including T and B cells in vitro [31, 32]. Acetylated low density lipoprotein (Ac-LDL)-labeled ECs in the embryo give rise to $CD45^+$/Ac-LDL$^+$ HPCs in situ in chick and mouse embryos [33, 34]. A diverging point of hemogenic and non-hemogenic ECs was demonstrated using a transgenic embryo and ES cell line carrying the Flk1 promoter/enhancer [Flk(p/e)]-driven GFP gene. VE-cadherin$^+$ ECs first appeared as a GFP$^-$ population, but subsequently all ECs became VE-cadherin$^+$/GFP$^+$. Only GFP$^-$ ECs could give rise to definitive HPCs and were observed in hematopoietic cell clusters in the dorsal aorta, indicating that Flk(p/e)-GFP$^-$ ECs represent hemogenic ECs [35]. α_4 Integrin$^+$/VE-cadherin$^+$ ECs, but not α_4 integrin$^-$ ECs derived from mouse and primate ES cells, were reported to show hemogenic potential [36, 37]. VE-cadherin$^+$/CD45$^+$ cells at E9.5 of the mouse embryo, which give rise to definitive erythroid, myeloid, but not B lymphoid cells, may represent another intermediate cell type of hemogenic ECs [38]. These studies indicate that during the early developmental stage a subset of ECs show a differentiation window to give rise to HPCs. The hemogenic ECs should be at least one of the precursors for the definitive HSCs.

Molecular Machinery of the Fate Determination

Gene knockout technology has largely contributed to uncover the molecular mechanisms in differentiation of ECs and HPCs. In vitro cell differentiation systems using

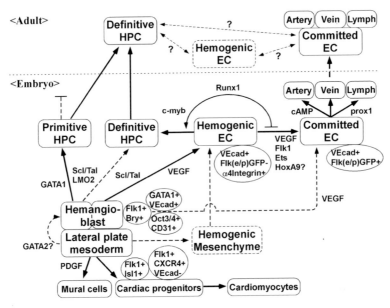

FIG. 2. Possible relation between ECs and HPCs during their differentiation. Cells constituting the circulation—ECs, HPCs, mural cells in the vascular wall, cardiomyocytes—mainly originate from lateral plate mesoderm. The putative relations, differentiation pathways, and regulatory mechanisms of these cells and their progenitors (hemangioblast and hemogenic ECs) are summarized. Cell populations are indicated in *rectangles*. Molecular markers are indicated in *ovals* attached to the corresponding cell populations. Issues not fully demonstrated are indicated by *dotted lines*

ES cells also play an important role in dissecting the molecular machinery for the cell differentiation and diversification process, especially at a single-cell level [39]. Various approaches using both in vitro and in vivo analyses have now demonstrated that many molecules are involved in the differentiation and diversification process of ECs and HPCs (Fig. 2).

Flk1/VEGFR2

Flk1 is one of the VEGF receptors (VEGFR2). As Flk1-deficient mice showed a defect in the development of ECs and HPCs [40], Flk1 is thought to be one of the earliest functional molecules for EC and HPC development and should mark their common progenitor, hemangioblasts. Flk1 is broadly expressed in lateral plate mesoderm and extraembryonic mesoderm [41, 42]. Flk1+ mesoderm cells are demonstrated to give rise to both ECs and HPCs [31]. Flk1+ cells also differentiate into mural cells in the vascular wall and cardiomyocytes [8, 43, 44]. During the differentiation, Flk1 expression is maintained only in ECs and disappears in mature HPCs, mural cells, and cardiomyocytes. Activation of Flk1 by VEGF is required to maintain Flk1 expression in Flk1+ mesoderm cells and drives their differentiation to EC lineage [8, 45]. Flk1 signaling through VEGF is the main inducer of ECs from their progenitors. Recently, phospholipase C-γ activation by VEGF through the autophosphorylation of tyrosine residue 1173 in Flk1 was revealed to be essential for vasculogenesis in the embryo [46]. Flk1 expression further serves various progenitor populations being combined with other molecular markers. Flk1[+]/Tal1[+] cells [47] and Flk1[+]/brachyury[+] cells [29] are reported as hemangioblasts. Flk1[+]/CXCR4[+]/VE-cadherin[−] cells [43] and Flk1[+]/isl1[+] cells [48] are shown to be progenitors for ECs and cardiomyocytes.

Scl/Tal1

Scl/Tal1 is a basic helix-loop-helix (bHLH) transcription factor expressed in yolk sac blood progenitors and ECs [49]. The most evident role of Scl/Tal1 is driving the commitment of progenitor cells to hematopoietic lineage. Scl/Tal1[−/−] mice showed a defect of embryonic hematopoiesis even though the primitive vascular network was normal [50, 51]. Scl/Tal1[−/−] ES cells did not contribute to hematopoietic tissues in chimeric animal models [52]. On the other hand, induced expression of Scl/Tal1 in the early differentiation stage of Scl/Tal1[−/−] ES cells restored HPC differentiation [53]. Flk1[+] mesoderm cells from Scl/Tal1[−/−] ES cells gave rise only to mural cell lineage, whereas cell lines expressing Scl/Tal1 in the Flk1 locus could make ECs and HPCs [47]. Phenotypes of Scl/Tal1-deficient mice resemble those of LIM-only protein, LMO2 [54] and GATA-binding protein, GATA-1 [55]. LMO2 is posited to bridge Scl/Tal1 and GATA proteins [56]. The Scl/Tal1-LMO2-GATA protein complex should contribute to specify mesoderm to HPC lineage [57].

GATA

Among GATA-binding transcription factors, GATA1, 2, 3 (considered hematopoietic GATAs) are posited to be involved in early hematopoiesis [58]. Disruption of GATA2, which is expressed in HSCs and progenitors, induces impaired primitive and definitive hematopoiesis in mice, indicating that GATA2 has a role in the proliferation and expansion of early HPCs [59, 60]. Recently, GATA2 induction during an early differentiation stage of ES cell differentiation resulted in an increase in hemangioblasts, erythroid cells, and ECs, suggesting a GATA2 role in the early stage of EC and HPC differentiation [61]. GATA1 is postulated to be involved mainly in erythrocyte differentiation. GATA1[−/−] mice show loss of erythrocytes [55]. ES cell-derived GATA1[+]/Flk1[−] cells give rise only to primitive erythrocytes [62]. Recently, GATA1[+]/VE-cadherin[+] cells in the early embryo (E7.0–7.5) were demonstrated to be a common precursor for both HPCs and ECs [63]. GATA3, which is expressed in hematopoietic progenitors as well as T cells, is necessary for the production of T-helper 2 cells [64]. GATA3 is also expressed in mesenchymal cells with stromal activity to generate definitive HPCs [65], suggesting that GATA3 may be involved in the formation of the microenvironment to support definitive HPC development.

Runx1

Runx1, also designated AML1 and CBFa2, is a transcription factor showing homology to the *Drosophila* pair-rule gene, *runt*. Runx1 plays pivotal roles in the development of definitive hematopoiesis. Runx1[−/−] mice showed normal primitive hematopoiesis but completely impaired definitive hematopoiesis with disappearance of budding hematopoietic clusters from ECs in dorsal aorta [66, 67]. In a chimeric animal model,

Runx1$^{-/-}$ ES cells failed to contribute to any hematopoietic tissues, whereas Runx1$^{+/-}$ ES cells could contribute to bone marrow and peripheral blood, among others [66]. Studies using Runx1$^{+/LacZ}$ mice revealed that Runx1 expression marks HSCs [22, 68]. That is, LacZ expression is observed in the dorsal aorta, vitelline, and umbilical arteries where long-term repopulating HSCs reside. Rux1-LacZ cells localize in ECs of the dorsal aorta, HPCs budding from ECs, and in some mesenchymal cells surrounding the aorta. Recently, Runx1-LacZ cells in the yolk sac at E7.5 were demonstrated to contribute to definitive hematopoiesis [25]. These observations suggest that Runx1 is expressed in hemogenic ECs or their precursors and regulates the differentiation of HPCs from ECs. Runx1 directly suppresses Flk1 expression at the transcriptional level during ES cell differentiation [69]. Runx1 should be directly involved in EC-HPC transition by endowing the hematopoietic potential to their precursor cells.

Others

A Myb family transcription factor, c-myb is expressed in hemogenic ECs and is involved in definitive hematopoiesis. Rescue of c-myb expression in c-myb$^{-/-}$ ES cell-derived ECs restores the generation and proliferation of definitive HPCs, indicating that c-myb endows hemogenic properties to early ECs [70]. Ets transcription factors are expressed in hemangioblast, and four of the Ets genes are demonstrated to be essential for early EC specification and differentiation in zebrafish [71]. A homeobox transcription factor, HoxA9, was reported to act as a master switch to regulate the expression of prototypical endothelial-committed genes such as endothelial nitric oxide synthase, VEGFR2, and VE-cadherin in adult EC progenitor cells [72].

Conclusion

Recent studies on EC and HPC development uncovered various cellular and molecular mechanisms in the differentiation processes of these cells. Nevertheless, these findings do not efficiently contribute to the better understanding of EC and HPC development. Current trends in this research area are mainly accumulating evidence of the existence of the "heterogeneity" in the developmental processes. That is, multiple origins (with early segregation), multiple pathways, and multiple molecular functions exist and complicatedly interact with each other to form the functional circulation system. Cell tracing studies in vivo demonstrated multiple origins of and their early segregation to ECs and HPCs [24, 25]. Novel hemangioblastic cell populations are still being reported, such as Oct3/4$^+$/CD31$^+$ cells [73] and GATA1$^+$/VE-cadherin$^+$ cells [63].

   In the future, dual directions of studies will be required for a complete understanding of EC and HPC development. One direction is to extend the studies in various ways to accumulate novel results. In vivo animal models for cell tracing, in vitro analyses for the cell differentiation process, various large-scale biology, called "omics", for gene and protein expressions, epigenetic analyses, and so on—all of these strategies would provide novel outcomes from various points of view. The other direction is to make them relevant to each other and converge them to reconstitute a novel concept and understanding (although it is not easy). By repeating these processes, we should be able to approach a better or true "stereoscopic" understanding of the developmental processes.

References

1. Sabin FR (1920) Studies on the origin of blood-vessels and of red blood corpuscules as seen in the living blastoderm of chicks during the second day of incubation. Carnegie Contrib Embryol 272:214–262
2. Risaw W (1997) Mechanisms of angiogenesis. Nature 386:671–674
3. Carmeliet O (2000) Mechanisms of angiogenesis and arteriogenesis. Nat Med 6:389–395
4. Eichmann A, Yuan L, Moyon D, et al (2005) Vascular development: from precursor cells to branched arterial and venous networks. Int J Dev Biol 49:259–267
5. Coultas L, Chawengsaksophak K, Rossant J (2005) Endothelial cells and VEGF in vascular development. Nature 438:937–945
6. Alitalo K, Tammela T, Petrova T (2005) Lymphangiogenesis in development and human disease. Nature 438:946–953
7. Yamashita JK (2007) Differentiation of arterial, venous, and lymphatic endothelial cells from vascular progenitors. Trends Cardiovasc Med 17:59–63
8. Yamashita J, Itoh H, Hirashima M, et al (2000) Flk1 positive cells derived from embryonic stem cells serve as vascular progenitors. Nature 408:92–96
9. Yurugi-Kobayashi T, Itoh H, Schroeder T, et al (2006) Adrenomedullin/cyclic AMP pathway induces Notch activation and differentiation of arterial endothelial cells from vascular progenitors. Arterioscler Thromb Vasc Biol 26:1977–1984
10. Kono T, Kubo H, Shimazu C, et al (2006). Differentiation of lymphatic endothelial cells from embryonic stem cells on OP9 stromal cells. Arterioscler Thromb Vasc Biol 26:2070–2076
11. Liersch R, Nay F, Lu L, et al (2006) Induction of lymphatic endothelial cell differentiation in embryoid bodies. Blood 107:1214–1216
12. Kreuger J, Nilsson I, Kerjaschki D, et al (2006) Early lymph vessel development from embryonic stem cells. Arterioscler Thromb Vasc Biol 26:1073–1078
13. Cumano A, Godin I (2001) Pluripotent hematopoietic stem cell development during embryogenesis. Curr Opin Immunol 13:166–171
14. Ueno H, Weissman IL (2007) Blood lines from embryo to adult. Nature 446:996–997
15. Nishikawa SI (2001) A complex linkage in the developmental pathway of endothelial and hematopoietic cells. Curr Opin Cell Biol 13:673–678
16. Jaffredo T, Nottingham W, Liddiard K, et al (2005) From hemangioblast to hematopoietic stem cell: an endothelial connection? Exp Hematol 33:1029–1040
17. Mikkola HKA, Orkin SH (2006) The journey of developing hematopoietic stem cells. Development 133:3733–3744
18. de Bruijn MF, Speck NA, Peeters MC, et al (2000) Definitive hematopoietic stem cells first develop within the major arterial regions of the mouse embryo. EMBO J 19:2465–2474
19. Alvarez-Silva M, Belo-Diabangouaya P, Salaun J, et al (2003) Mouse placenta is a major hematopoietic organ. Development 130:5437–5444
20. Gekas C, Dieterlen-Lievre F, Orkin SH, et al (2005) The placenta is a niche for hematopoietic stem cells. Dev Cell 8:365–375
21. Ottersbach K, Dzierzak E (2005) The murine placenta contains hematopoietic stem cells within the vascular labyrinth region. Dev Cell 8:377–387
22. North TE, de Bruijn MF, Stacy T, et al (2002) Runx1 expression marks long-term repopulating hematopoietic stem cells in the midgestation mouse embryo. Immunity 16:661–672
23. Bertrand JY, Giroux S, Golub R, et al (2005) Characterization of purified intraembryonic hematopoietic stem cells as a tool to define their site of origin. Proc Natl Acad Sci U S A 102:134–139

24. Ueno H, Weissmann IL (2006) Clonal analysis of mouse development reveals a poly-clonal origin for yolk sac blood islands. Dev Cell 11:519–533
25. Samokhvalov IM, Samokhvalov NI, Nishikawa SI (2007) Cell tracing shows the contri-bution of the yolk sac to adult haematopoiesis. Nature 446:1056–1061
26. Yoder MC, Hiatt K, Mukherjee P (1997) In vivo repopulating hematopoietic stem cells are present in the murine yolk sac at day 9.0 postcoitus. Proc Natl Acad Sci U S A 94:6776–6780
27. Sabin F (1917) Origin and development of the primitive vessels of the chick and of the pig. Contrib Embyol 226:61–124
28. Choi K, Kennedy M, Kazarov A, et al (1998) A common precursor for hematopoietic and endothelial cells. Development 125:725–732
29. Huber TL, Kouskoff V, Fehling HJ, et al (2004) Haemangioblast commitment is initiated in the primitive streak of the mouse embryo. Nature 432:625–630
30. Smith RA, Glomski CA (1982) "Hemogenic endothelium" of the embryonic aorta: does it exist? Dev Comp Immunol 6:359–368
31. Nishikawa SI, Nishikawa S, Hirashima M, et al (1998) Progressive lineage analysis by cell sorting and culture identifies FLK1+VE-cadherin+ cells at a diverging point of endothelial and hemopoietic lineages. Development 125:1747–1757
32. Nishikawa SI, Nishikawa S, Kawamoto H, et al (1998) In vitro generation of lympho-hematopoietic cells from endothelial cells purified from murine embryos. Immunity 8:761–769
33. Jaffredo T, Gautier R, Eichmann A, et al (1998) Intraaortic hemopoietic cells are derived from endothelial cells during ontogeny. Development 125:4575–4583
34. Sugiyama D, Ogawa M, Hirose I, et al (2003) Erythropoiesis from acetyl LDL incorpo-rating endothelial cells at the preliver stage. Blood 101:4733–4738
35. Hirai H, Ogawa M, Suzuki N, et al (2003) Hemogenic and nonhemogenic endothelium can be distinguished by the activity of fetal liver kinase (Flk)-1 promoter/enhancer during mouse embryogenesis. Blood 101:886–893
36. Ogawa M, Kizumoto M, Nishikawa S, et al (1999) Expression of alpha4-integrin defines the earliest precursor of hematopoietic cell lineage diverged from endothelial cells. Blood 93:1168–1177
37. Shinoda G, Umeda K, Heike T, et al (2007) Alpha4-Integrin(+) endothelium derived from primate embryonic stem cells generates primitive and definitive hematopoietic cells. Blood 109:2406–2415
38. Fraser ST, Ogawa M, Yokomizo T, et al (2003) Putative intermediate precursor between hematogenic endothelial cells and blood cells in the developing embryo. Dev Growth Differ 45:63–75
39. Yamashita JK (2004) Differentiation and diversification of vascular cells from ES cells. Int J Hematol 80:1–6
40. Shalaby F, Rossant J, Yamaguchi TP, et al (1995) Failure of blood-island formation and vasculogenesis in Flk-1-deficient mice. Nature 376:62–66
41. Kataoka H, Takakura N, Nishikawa S, et al (1997) Expressions of PDGF receptor alpha, c-Kit and Flk1 genes clustering in mouse chromosome 5 define distinct subsets of nascent mesodermal cells. Dev Growth Differ 39:729–740
42. Yamaguchi TP, Dumont DJ, Conlon RA, et al (1993) Flk-1, an flt-related receptor tyro-sine kinase is an early marker for endothelial cell precursors. Development 118:489–498
43. Yamashita JK, Takano M, Hiraoka-Kanie M, et al (2005) Prospective identification of cardiac progenitor potentials by a novel single cell-based cardiomyocyte induction. FASEB J 19:1534–1536
44. Motoike T, Markham DW, Rossant J, et al (2003) Evidence for novel fate of Flk1+ pro-genitor: contribution to muscle lineage. Genesis 35:153–159
45. Eichmann A, Corbel C, Nataf V, et al (1997) Ligand-dependent development of the endothelial and hemopoietic lineages from embryonic mesodermal cells expressing

vascular endothelial growth factor receptor 2. Proc Natl Acad Sci U S A 94:5141–5146

46. Sakurai Y, Ohgimoto K, Kataoka Y, et al (2005) Essential role of Flk-1 (VEGF receptor 2) tyrosine residue 1173 in vasculogenesis in mice. Proc Natl Acad Sci U S A 102:1076–1081

47. Ema M, Faloon P, Zhang WJ, et al (2003) Combinatorial effects of Flk1 and Tal1 on vascular and hematopoietic development in the mouse. Genes Dev 17:380–393

48. Moretti A, Caron L, Nakano A, et al (2006) Multipotent embryonic isl1+ progenitor cells lead to cardiac, smooth muscle, and endothelial cell diversification. Cell 127:1151–1165

49. Green AR, Salvaris E, Begley CG (1991) Erythroid expression of the "helix-loop-helix" gene, SCL. Oncogene 6:475–479

50. Robb L, Lyons I, Li R, et al (1995) Absence of yolk sac hematopoiesis from mice with a targeted disruption of the scl gene. Proc Natl Acad Sci USA 92:7075–7079

51. Shivdasani RA, Mayer EL, Orkin SH (1995) Absence of blood formation in mice lacking the T-cell leukemia oncoprotein tal-1/SCL. Nature 373:432–434

52. Visvader JE, Fujiwara Y, Orkin SH (1998) Unsuspected role for the T-cell leukemia protein SCL/tal-1 in vascular development. Genes Dev 12:473–479

53. Endoh M, Ogawa M, Orkin S, et al (2002) SCL/tal-1-dependent process determines a competence to select the definitive hematopoietic lineage prior to endothelial differentiation. EMBO J 21:6700–6708

54. Warren AJ, Colledge WH, Carlton MB, et al (1994) The oncogene cysteine-rich LIM domain protein rbtn2 is essential for erythroid development. Cell 78:45–57

55. Pevny L, Simon MC, Robertson E, et al (1991) Erythroid differentiation in chimaeric mice blocked by a targeted mutation in the gene for transcription factor GATA-1. Nature 349:257–260

56. Wadman IA, Osada H, Grutz GG, et al (1997) The LIM-only protein Lmo2 is a bridging molecule assembling an erythroid, DNA-binding complex which includes the TAL1, E47, GATA1, and Ldb1/NL1 proteins. EMBO J 16:3145–3157

57. Cantor AB, Orkin SH (2002) Transcriptional regulation of erythropoiesis: an affair involving multiple partners. Oncogene 21:3368–3376

58. Orkin SH (1992) GATA-binding transcription factors in hematopoietic cells. Blood 80:575–581

59. Tsai FY, Keller G, Kuo FC, et al (1994) An early hematopoietic defect in mice lacking the transcription factor GATA-2. Nature 371:221–226

60. Ling KW, Ottersbach K, van Hamburg JP, et al (2004) GATA-2 plays two functionally distinct roles during the ontogeny of hematopoietic stem cells. J Exp Med 200:871–882

61. Lugus JJ, Chung YS, Mills JC, et al (2007) GATA2 functions at multiple steps in hemangioblast development and differentiation. Development 134:393–405

62. Fujimoto T, Ogawa M, Minegishi N, et al (2001) Step-wise divergence of primitive and definitive haematopoietic and endothelial cell lineages during embryonic stem cell differentiation. Genes Cells 6:1113–1127

63. Yokomizo T, Takahashi S, Mochizuki N, et al (2007) Characterization of GATA-1(+) hemangioblastic cells in the mouse embryo. EMBO J 26:184–196

64. Zheng W, Flavell RA. (1997) The transcription factor GATA-3 is necessary and sufficient for Th2 cytokine gene expression in CD4 T cells. Cell 89:587–596

65. Manaia A, Lemarchandel V, Klaine M, et al (2000) Lmo2 and GATA-3 associated expression in intraembryonic hemogenic sites. Development 127:643–653

66. Okuda T, van Deursen J, Hiebert SW, et al (1996) AML1, the target of multiple chromosomal translocations in human leukemia, is essential for normal fetal liver hematopoiesis. Cell 84:321–330

67. Wang Q, Stacy T, Binder M (1996) Disruption of the Cbfa2 gene causes necrosis and hemorrhaging in the central nervous system and blocks definitive hematopoiesis. Proc Natl Acad Sci U S A 93:3444–3449

68. North TE, Stacy T, Matheny CJ, et al (2004) Runx1 is expressed in adult mouse hematopoietic stem cells and differentiating myeloid and lymphoid cells, but not in maturing erythroid cells. Stem Cells 22:158–168
69. Hirai H, Samokhvalov IM, Fujimoto T, et al (2005) Involvement of Runx1 in the down-regulation of fetal liver kinase-1 expression during transition of endothelial cells to hematopoietic cells. Blood 106:1948–1955
70. Sakamoto H, Dai G, Tsujino K, et al (2006) Proper levels of c-Myb are discretely defined at distinct steps of hematopoietic cell development. Blood 108:896–903
71. Pham VN, Lawson ND, Mugford JW, et al (2007) Combinatorial function of ETS transcription factors in the developing vasculature. Dev Biol 303:772–783
72. Rossig L, Urbich C, Bruhl T, et al (2005) Histone deacetylase activity is essential for the expression of HoxA9 and for endothelial commitment of progenitor cells. J Exp Med 201:1825–1835
73. Furuta C, Ema H, Takayanagi S, et al (2006) Discordant developmental waves of angioblasts and hemangioblasts in the early gastrulating mouse embryo. Development 133:2771–2779

Atherosclerosis and Angiogenesis: Double Face of Neovascularization in Atherosclerotic Intima and Collateral Vessels in Ischemic Organs

Katsuo Sueishi, Mitsuho Onimaru, and Yutaka Nakashima

Summary. Angiogenesis is an essential process not only physiologically in terms of organ development and regeneration but also pathologically in regard to acute and chronic inflammation and cancer. Newly formed microvessels, mainly derived from adventitial blood vessels, are ubiquitously distributed in atherosclerotic plaque of human coronary arteries, and this angiogenesis correlates well with the inflammatory-repair reaction in plaques and luminal stenosis. Plaque angiogenesis is assumed to participate intimately in the progression of atherosclerosis partly through vascular endothelial growth factors (VEGFs)–VEGF receptors interactions, resulting in plaque destabilization, making it vulnerable to rupture. On the other hand, therapeutic angiogenesis for atherosclerosis-related ischemic diseases should be clinically effective in improving blood perfusion. Examining animal models of acute and severe limb ischemia indicated that fibroblast growth factor-2 (FGF-2) gene transfer, mediated by the Sendai virus vector (SeV) selectively in ischemic striated muscular tissue, promotes not only early induction of such angiogenic factors as VEGF-A, hepatocyte growth factor and platelet-derived growth factor-A (PDGF-A) (which relate mainly to angiogenic initiation) but also consecutive induction of PDGF-BB and VEGF-C participating in vascular maturation, thereby leading to functional rescue from limb ischemia. It is critical for therapeutic angiogenesis that several endogenous angiogenic factors temporally and spatially interact with each other in ischemic organs.

In this chapter, we review recent advances made in clarifying the pathophysiological role of angiogenesis in both atherosclerotic progression and atherosclerosis-induced organ ischemia—i.e., the "double face of angiogenesis in atherosclerosis"—as excessive and insufficient, respectively. Based on the emerging evidence, we discuss potent and novel therapeutic strategies for aiming to exploit "integrated and functional angiogenesis" in atherosclerosis-related organ ischemia using FGF-2 gene transduction.

Key words. Atherosclerosis · Angiogenesis · Vascular endothelial growth factors (VEGFs) · Fibroblast growth factor-2 (FGF-2) · Gene therapy

Introduction

Although angiogenesis is well known to distribute ubiquitously in advanced plaque of human atherosclerotic lesions [1–5], the pathophysiological significance of intraplaque angiogenesis in atherogenesis is still controversial. Intimal angiogenesis has been hitherto assumed to participate intimately in the promotion of atheroma growth and intraplaque edema and hemorrhage, possibly resulting in plaque rupture. Attention has recently refocused on intimal angiogenesis in atherosclerotic plaque. We are not only looking at the clinicopathological evaluation of vulnerable plaque to rupture relative to the recent advances in imaging inflammation and angiogenesis in atherosclerotic plaques and the accumulated comprehension of molecular mechanisms of angiogenesis and atherogenesis but are now focusing also on the availability of pharmacological and genetic interventions to manipulate angiogenesis-related diseases by antiangiogenic therapy as well as organ ischemia by therapeutic angiogenesis. In addition, recent emerging evidence supports the hypothesis that atherosclerosis is a chronic inflammatory disease [6, 7]. Angiogenesis is an essential vital response in various physiological and pathological conditions, such as embryonic development, the acute and chronic inflammatory-repair process including wound repair, rheumatoid arthritis, atherosclerosis, proliferative retinopathies caused by diabetes mellitus, senile macular degeneration and neonatal retinal prematurity, growth of solid cancers, and others [8–12]. On the contrary, it is hoped that clinical application of therapeutic angiogenesis for ischemic organs, such as the heart and lower extremities, in patients with atherosclerosis will not be far behind. Therefore, angiogenesis has a double face in patients with atherosclerosis: On one hand it is a possible promoter of atherogenesis when there is excessive angiogenesis, whereas on the other hand it is a determining factor in the fate of ischemic organs when there is insufficient angiogenesis, such as in the heart, brain, and lower extremities, resulting from atherosclerosis of the arteries in these organs.

In this chapter, we address the pathophysiological significance of plaque angiogenesis in atherogenesis and novel angiogenic gene therapy for atherosclerosis-related ischemia using fibroblast growth factor-2 (FGF-2) as a target gene, to localize the therapeutic effect and the induction of integrated and harmonized angiogenesis in ischemic organs.

Angiogenesis in Atherosclerotic Plaque

Histopathological Characteristics of Plaque Angiogenesis of Human Coronary Arteries

Newly formed blood vessels distribute irregularly but ubiquitously in atherosclerotic intima, particularly in the plaque shoulder and deeper part of the atheroma (Fig. 1a,c). Such vessels are thin-walled with variously sized diameters. Occasionally they are associated with fresh or old hemorrhage (Fig. 1c), extravascular deposition of fibrin, and rarely thrombi. Therefore, these vessels may be fragile to mechanistic stimuli. Morphometric analysis of inflammatory parameters to clarify the pathophysiological significance of angiogenesis in atheosclerosis has shown that the density of these

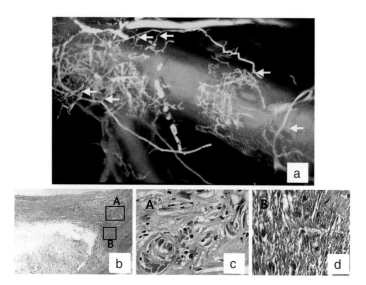

FIG. 1. Newly formed blood vessels of human atherosclerotic coronary artery. **a** Postmortem angiography of a methylsalicylate-cleared heart with silicone polymer, Microfil reveals the delicate and complicated vascular networks segmentally and circumferentially around narrowed coronary lumen, which connect directly with adventitial small arteries (*arrows*). **b–d** Histopathology of atheromatous intima of a human coronary artery (AHA lesion type IV). **b, c** Pathological findings of *squares A* and *B* of **a**, respectively. Note the many microvessels associated with fresh hemorrhage and lymphocytic infiltration (**c**) and an accumulation of foamy macrophages (**d**). **b, c** H&E. **d** Masson's trichrome stain

vessels is well correlated with the degree of luminal stenosis and the extent of leukocytic infiltration, which mainly comprises macrophages (Fig. 1d) and lymphocytes. These vessels originate largely from adventitial blood vessels and, rarely, from luminal endothelial overgrowth [13]. Although no vessel formation is characteristically noted in diffuse intimal thickening (DIT), which is considered adaptive intimal thickening, the more severe the atherosclerotic lesion type [as classified by the American Heart Association (AHA)], the more often the lesion contains intimal blood vessels—reaching more than 70% of advanced lesions (i.e., types IV–VI) [14, 15]. On the other hand, in early arteriosclerotic intima (AHA lesion type I and II) of human coronary arteries, intimal angiogenesis was associated in less than 40%. These findings strongly suggest that the degree of plaque angiogenesis correlates well with progression of atherosclerosis. Recent research obtained with micro-computed tomography images also supports this correlation of plaque angiogenesis and atherosclerotic progression and the origin of newly formed vessels in apolipoprotein E/low-density lipoprotein-negative (apoE$^{-/-}$/LDL$^{-/-}$) double knockout (KO) mice [16].

Despite the recent advances in vascular biology, including our knowledge of angiogenesis and vasculogenesis, little is known about target molecules, modulating mechanisms, and the fate of plaque angiogenesis. The expression of several angiogenic factors, including fibroblast growth factor (FGF) family members [17], vascular endothelial growth factor-A (VEGF-A) [14, 18], placental growth factor (PlGF) [19], and others have been suggested to participate in plaque angiogenesis; however, the spatial and temporal relation between their expression patterns and vascular distribution is not yet fully clarified. Both VEGF-A and VEGF-C, however, are expressed mainly by macrophages scattered around newly formed blood vessels in human coronary atherosclerotic plaques, and these positive cell densities increase as the atherosclerotic lesion advances [14, 15], suggesting that VEGF-A and VEGF-C act as local, endogenous regulators of intimal angiogenesis and contribute to plaque progression. In addition, despite overexpression of VEGF-C in atherosclerotic plaque, lymphatic vessels are rarely formed [15]. Thus, the imbalance of angiogenesis and lymphangiogenesis may be a factor promoting sustained inflammatory reaction during human atherogenesis.

Pathophysiological Significance of Plaque Angiogenesis in Atherogenesis

Together with histopathological characteristics of plaque angiogenesis in the human coronary artery, plaque angiogenesis associated with macrophagic and lymphocytic recruitment reflects the activity of the inflammatory-repair process in atherogenesis, possibly indicating a histopathological marker of an active atherosclerotic lesion leading to plaque instability. Macrophage recruitment is closely modulated by inflammatory function and partly mediated by VEGF–VEGFR interaction. Recent evidence obtained from several studies using experimental animal models of atherosclerosis strongly suggest an essential contribution of angiogenic factors—particularly via the linkage between VEGF-A and VEGFR-1 (Flt-1)—to the progression of arteriosclerosis in aspects of not only angiogenic and/or vasculogenic activation through VEGF-A and VEGFR-2 (Flk-1) interaction [20–23] but also induction of chronic inflammation. This is mainly due to recruitment of monocytic lineage cells [22, 24–26]. Similarly, PlGF (which is a ligand of VEGFR-1) and FGF family members has been recently reported to participate in atherogenesis using PlGF-deficient mice or adenovirus-mediated gene transfer of dominant-negative soluble FGFR-1, respectively [19, 25].

Based on these observations, intimal angiogenesis can be assumed to participate not only in the nutritional supply promoting plaque growth but also in sustaining the inflammatory process in situ. This supposition is indeed supported by the finding that administration of antiangiogenic agents suppresses plaque growth and progression [21, 22]. In addition, plaque angiogenesis may sustain the inflammation-repair process by activation of the coagulation-fibrinolysis system following hyperpermeability and hemorrhage as well as by serving the pathway of leukocytic recruitment by which overexpression of proinflammatory cytokines and leukocytic chemoattractants is induced [24, 26]. Furthermore, intimal microvessels can take part in the enlargement of not only the atheroma (due to increased intraplaque pressure caused by hemorrhage and edema) but also the necrotic core due to thrombosis in these blood vessels. Angiogenesis and the inflammatory process induced by VEGF-A and other angiogenesis-related factors promote atherogenesis in vivo, but the major cell species expressing these factors is macrophages. Therefore, it is more likely that angiogenesis and angiogenesis-related inflammation play an important role at the promotion phase but not at the initiation phase of atherogenesis.

Experimental animal models of atherosclerosis, however, are far from having the heterogeneity seen in human atherosclerotic morphology and pathophysiology. In fact, in rodent and rabbit models of arteriosclerosis, angiogenesis is scarce in intimal lesions and is mainly induced in injured arterial adventitia. Therefore, further studies on human plaque angiogenesis in vivo may constitute a more suitable "stage" on which to elucidate this issue and to apply clinical interventions, especially with the advent of potent imaging technology.

In addition, we examined the temporal and spatial relation between the accumulation of lipid and macrophage infiltration in the thickened intima of early human atherosclerotic coronary arteries, mainly atherosclerotic lesions AHA type I and partly type II (Fig. 2) [27]. We demonstrated that extracellular lipid deposition is first accumulated in the outer layer of preexisting DIT (Fig. 2b,e,h), where specific proteoglycans such as biglycan and decorin are co-localized; macrophages then infiltrate toward the deposited lipid to form foam cell lesions as the amount of deposited lipid increases (Fig. 2c,f,i). These findings are key to settling the long-term discussion of atherogenesis: Does lipid arrive first or do macrophages (the chicken versus egg conundrum), and are their origins the inner luminal or outer adventitial vascular wall? As few newly formed blood vessels were apparent in these early thickening intimas, it is

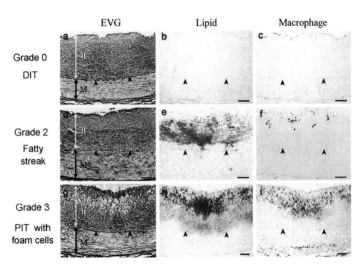

Fig. 2. Representative histologies of deposited lipid and infiltrating macrophages in human coronary arteries with early intimal thickening. a–c Grade 0 [equivalent to diffuse intimal thickening (DIT); a concentric intimal thickening composed of smooth muscle cells, elastin, and proteoglycans but devoid of lipid deposition]. The patient was a 36-year-old man. d–f Grade 2 (equivalent to fatty streak; a nonraised sudanophilic lesion as the earliest atherosclerotic lesion) from a 44-year-od man. g–i Grade 3 [equivalent to pathological intimal thickening (PIT) with foam cells].This is a preatheromatous lesion composed of extracellular lipid pools with an overlying layer of smooth muscle cells and lipid-laden macrophages, as defined by Virmani et al. [57]). The patient was a 29-year-old man. *Arrowheads* indicate internal elastic lamina. a, d, g Elastica van Gieson (*EVG*) stain. b, e, h Sudan IV stain. c, f, i Immunohistochemistry for a macrophage marker with anti-CD-68 antibody. *Bars* 100 μm. (From Nakashima et al. [27], with permission)

reasonable to assume that lipid imbibition and macrophage infiltration are from the luminal side. On the contrary, in advanced lesions, although advanced human athero- sclerotic lesions are morphologically heterogeneous and complicated, mononuclear cell infiltration and foam cell accumulation are usually pronounced around an athero- matous necrotic core or foam cell lesion, particularly in the atheroma shoulder of eccentrically thickening intimas, where newly formed blood vessels are frequently distributed in topographically close vicinity to intracellular or matrix-associated lipid deposition and macrophage infiltration (Fig. 1).

Together with these findings, preatheromatous or atheromatous lesions develop intimal angiogenesis; and newly formed microvessels can serve not only in inflam- matory cell recruitment, supporting and sustaining a chronic inflammatory-repair process, but also supply nutrition, thereby forming a vicious circle in relation to ath- eroma growth and instability. Current trials may successfully image macrophage functions by targeting either the metabolic activity using ^{18}F-fluorodeoxyglucose and positron emission tomography [28] or the phagocytic activity using ultra-small par- ticles in magnetic resonance imaging [29]. The approach recently reported by Hyafil et al. [30] using crystalline iodinated particles (N1177) solubilized with a biocompat- ible detergent could also demonstrate macrophages in atherosclerotic plaque in vivo. These approaches can provide functional as well as morphological information on atherosclerotic plaques and thus in the near future may offer a clear advantage for clinical evaluation and detection of culprit lesions vulnerable to plaque rupture in human atherosclerosis.

In addition of the contribution of plaque angiogenesis to atherosclerotic progres- sion, the hypothesis that plaque angiogenesis can promote plaque vulnerability to rupture has become a hot discussion point. Regarding the pathophysiological poten- tial of angiogenesis in plaque rupture, Virmani et al. [31] suggested the possibility that intraplaque hemorrhage from newly formed microvessels causes accumulation of erythrocyte-derived phospholipids and free cholesterol and may promote lesion instability as it relates to plaque rupture. Moreno et al. [32] examined histopathologi- cal characteristics including plaque angiogenesis in ruptured atherosclerotic plaques in human aorta and suggested that plaque angiogenesis plays a contributory role in the process of plaque rupture as well as in a thin fibrous cap (<60 μm), rupture of internal elastic lamina, inflammation in the fibrous cap consisting of varying degrees of macrophage and T-lymphocyte infiltration, and the intimal lipid area. Together with these findings, plaque angiogenesis is assumed to play an important role not only in atherosclerotic plaque progression but also in destabilization, leading to plaque rupture. Therefore, further comprehension of the mechanisms by which plaque angio- genesis occurs may contribute to exploiting novel therapeutic strategies to prevent the transition from a stable to an unstable atherosclerotic plaque.

Angiogenesis in Ischemic Organs

Adaptive Induction of Angiogenic Factors in Acute Ischemia

Blood vessels newly formed during the angiogenic process are microvessels consisting of endothelial cells (ECs) and pericytes. Spatial interactions mediated by angiogenic factors between ECs and surrounding nonendothelial mesenchymal cells (MCs) including pericytes are considered critical for functional angiogenesis. Pericytic damage or insufficient recruitment of pericytes strongly affects the structure, mechanical strength, and physiological function of the vascular microvessels [33]. In atherosclerosis-related ischemic organs, reactive/adaptive induction of endogenous angiogenic factors and subsequent stimulation of angiogenesis is a physiologically built-in essential response to recover from an ischemic condition.

Although it is generally well known that cells expressing the VEGF gene are equipped with reasonable intracellular signaling systems, such as hypoxia-inducible factor-1α (HIF-1α), to enhance VEGF expression in response to hypoxic stimulation, the temporal behavior of several angiogenesis-related factors in response to ischemia has been sparsely reported. Therefore, to clarify the temporal relation of angiogenesis-related factors during angiogenesis in vivo, we examined time-dependent gene expression levels of several representative angiogenesis-related factors including VEGFs using a murine model of acute and severe limb ischemia [34]. The principal feature of this model (limb salvage model) is that approximately 80% reduction of blood flow immediately following ischemic surgery is gradually restored to approximately a 40% reduction level; it attains a plateau by 7–10 days after surgery without loss of a limb probably owing to the action of the endogenous angiogenic system in response to ischemia [34–36]. Angiogenic initiation- or stimulation-related ligands—VEGF-A, FGF-2, hepatocyte growth factor (HGF), platelet-derived growth factor-A (PDGF-A), angiopoietin-2 [37–39]—were up-regulated compared to those in normal, nonischemic limbs; and the peak expression of these factors was during the early phase (1 day after the ischemia operation) in this murine model [35, 36]. Angiogenic maturation-related ligands—PDGF-BB and VEGF-C [40–42]—were also up-regulated, and their peak expression occurred during the late phase (7–10 days after the ischemia surgery). Interestingly, although HGF is an angiogenic factor down-regulated by hypoxia-related cAMP signaling in vitro [43], up-regulation of HGF was observed in this model and partly depended on up-regulation of endogenous FGF-2, which can stimulate the HGF gene in nonendothelial MCs in vitro, as evidenced by a study using FGF-2 KO mice [35]. Together with these findings, not only VEGF-A but also several angiogenic factors are directly or indirectly up-regulated in ischemic organs, and the main action of each factor is temporally different during angiogenesis in response to ischemia.

FGF-2 Gene Transfer Accelerates the Integrated and Functional Angiogenesis

"Therapeutic angiogenesis" is a clinical concept for atherosclerosis-related ischemic diseases such as myocardial and brain infarctions and arteriosclerosis obliterans. The end goal of this therapy is clinically improved blood perfusion through focal, active induction of angiogenesis and arteriogenesis in ischemic organs. Focal high-dose administration of an angiogenic factor into the ischemic organ is one of the methods to achieve this aim [34, 37, 43–47]. To establish an effective strategy it is important to explore scientifically which angiogenic factor is available.

A current report demonstrated that adenovirous- or myoblast-mediated focal administration of exogenous VEGF-A, a specific mitogenic stimulator to vascular endothelial cells, induced angioma-like, robust, structurally aberrant, newly formed blood vessels in murine ischemic myocardium [48, 49]. This suggested that a vital response induced by integration and harmonization of each function of several angiogenesis-related factors is essential for the induction of "functional" angiogenesis. In addition, our hitherto existing evidence demonstrated that focal, high-dose administration of VEGF-A mediated by Sendai virus vector (SeV) [34–36, 50, 51] showed no therapeutic effects, including restoration of blood flow, in murine ischemic limb, although the number of microvessels in treated thigh muscles were immunohistochemically increased fivefold over those in the control limb. On the other hand, SeV-mediated FGF-2 gene therapy showed a highly effective therapeutic response (Fig. 3a), and the number of microvessels were equally increased compared to those in the ischemic limb treated with SeV-VEGF-A [34]. These findings suggest that it is not necessarily the case that an increased number of vessels is correlated with a therapeutic effect. In our laboratory, an additional histological examination clarified that the number of newly formed vessels with pericyte recruitment was significantly increased in ischemic limbs treated by SeV-FGF-2 compared to those in ischemic limbs treated by SeV-VEGF-A [34] (Fig. 4). Taken together, structurally mature newly formed blood vessels are critical for a functional (i.e., therapeutic) effect.

Why can FGF-2 but not VEGF-A induce structurally mature newly formed blood vessels? Our previously reported evidence for a molecular mechanism related to the availability of FGF-2 demonstrated that SeV-FGF-2 gene transfer augmented approximately 20-fold the amount of exogenous FGF-2 protein in ischemic limb compared to that of the endogenous level. Moreover, time-dependent exogenous FGF-2 gene expression and FGF-2-mediated blood recovery showed that the peak expression of exogenous FGF-2 mRNA occurred 1–3 days after gene transfer, and restoration of blood flow gradually improved to a lower level, 20% reduction, and attained a plateau

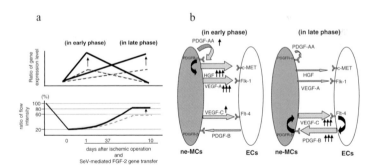

FIG. 3. **a** Time-dependent up-regulation (*upper graph*) of endogenous angiogenic factors and recovery of blood flow (*lower graph*) in response to ischemia (*dotted line*) and to Sendai virus (SeV)-mediated fibroblast growth factor-2 (FGF-2) gene transfer (*solid line*). Behaviors of endogenous factors; early responsive factors such as vascular endothelial growth factor-A (VEGF-A), hepatocyte growth factor (HGF), FGF-2, platelet-derived growth factor-AA (PDGF-AA), and angiopoietin-2; and later inducible factors including PDGF-BB and VEGF-C are shown with *thick* and *thin lines*, respectively. Temporal up-regulation of these factors and recovery of blood flow in response to FGF-2 gene therapy is similar to that seen in response to ischemia (*arrow*). **b** Exogenous FGF-2-dependent temporal and spatial hierarchical regulation of endogenous angiogenic factors. *ECs*, endothelial cells; *ne-MCs*; nonendothelial mesenchymal cells

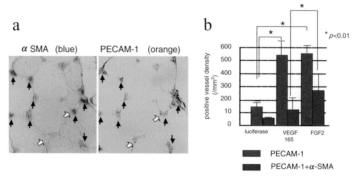

FIG. 4. **a** Typical immunohistochemical labeling of α-smooth muscle actin (αSMA)-positive cells (*left*, blue-labeled perivascular cells) and platelet endothelial cell adhesion molecule (PECAM-1)-positive cells with circumference (*right*, orange-labeled endothelial cells) in serial sections of thigh muscles treated with SeV-FGF-2. Some PECAM-1-positive vessels are apparently lined with blue-labeled cells (*arrows*), suggesting mature capillaries. Some other PECAM-1-positive vessels are without a blue cell lining (*open arrows*), suggesting immature vessels. **b** FGF-2 gene transfer effectively promotes blood vessel maturation. Densities of PECAM-1-positive vessels (*red bars*) and PECAM-1/SMA double-positive vessels (*blue bars*) were counted and quantified using serial sections of thigh muscles with SeV-mediated treatment of the indicated factors. (From Masaki et al. [34], with permission)

by 7–10 days after gene transfer in murine ischemic limbs. It should be noted that FGF-2 gene transfer is accompanied by up-regulation of endogenous angiogenesis-related factors such as VEGF-A, HGF, PDGF-A, PDGF-B, and VEGF-C, compared to their control levels; the peak expression of VEGF-A, HGF, and PDGF-A was during the early phase (1 day after FGF-2 gene transfer), and that of PDGF-BB and VEGF-C was during the late phase (7 days after FGF-2 gene transfer) [34–36]. Interestingly, the temporal expression pattern of these endogenous genes in response to FGF-2 is similar to that in murine limbs in response to ischemia, as described above (Fig. 3a). In addition, the functional blockade of endogenous VEGF-A, HGF, PDGF-A, or PDGF-BB by systemic administration of respective neutralizing antibodies was partially or completely diminished by the FGF-2-mediated therapeutic effects [34–36], suggesting that each endogenous factor plays an essential role and that they function with each other in FGF-2-induced rescue from ischemia. Moreover, our in vitro and in vivo examinations as well as other reported studies [52, 53] revealed that FGF-2 was an inducer for several angiogenesis-related factors—VEGF-A, HGF, PDGF-AA, VEGF-C—in regard to targeting such nonendothelial MCs as pericytes, vascular smooth muscle cells, and fibroblasts/myofibroblasts [35, 36]. In addition, SeV-FGF-2 gene transfer-dependent endogenous VEGF and HGF inductions were largely dependent on an autocrine system through PDGF-AA/PDGFRα in nonendothelial MCs, and these functions induced by FGF-2 comprise a key system during the early phase of the FGF-2-mediated angiogenesis in vivo [35, 36] (Fig. 3b). In contrast, although late-phase up-regulation of PDGF-B mRNA in response to SeV-FGF-2 transfer is observed in murine ischemic limbs, FGF-2 stimulation did not affect the PDGF-B gene in ECs in vitro. Further examination to resolve the discrepancy demonstrated that the PDGF-B gene was indirectly up-regulated via SeV-FGF-2-mediated activation of the VEGF-C/FLT-4 system in ischemic limbs, which was able to stimulate the PDGF-B gene in ECs directly in vitro. In addition, PDGF-BB up-regulated by the VEGF-C/Flt-4 system enhanced VEGF-C gene expression via the PDGF-BB/PDGFRβ system in murine ischemic limbs, which could directly and strongly stimulate the VEGF-C gene in nonendotheliol MCs in vitro. These findings demonstrate that this spatial paracrine linkage between the VEGF-C/FLT-4 system in ECs and the PDGF-BB/PDGFRβ system in nonendothelial MCs is one of the key systems during the late phase of FGF-2-mediated angiogenesis in vivo (Fig. 3). This positive-loop linkage can induce sustained and amplified up-regulation of PDGF-BB and VEGF-C and play an important role particularly in FGF-2-mediated mechanisms to promote mature blood vessel formation that was evidenced by electron microscopic observation of pericytes dropping off the dilated microvessels in thigh muscles of SeV-FGF-2-transferred murine ischemic limbs with blockade of endogenous function of PDGF-BB.

In summary, as angiogenesis-related factors play an essential role in the induction of "functional" new blood vessels, an agent for "therapeutic angiogenesis" should be an enhancer that is able to hierarchically induce spatially and temporally well-balanced activation of an endogenous angiogenic system. Based on our data, FGF-2 is an available, expedient therapeutic agent. It seems necessary to advance a novel concept—"integrated therapeutic angiogenesis"-based fundamental and clinical studies including combinatorial therapy [54, 55]—to establish more effective therapeutic strategies for ischemic diseases.

Conclusion

Atherosclerosis is a highly prevalent disease that results from a complex interaction between endogenous and exogenous atherogenic factors and the vascular wall and inflammatory cells by its link to the "inflammatory-repair process." Intraplaque angiogenesis is thought to be partly responsible for the promotion of atherosclerotic lesions [13–15, 20, 26], leading to organ ischemia such as heart and brain infarcts and arteriosclerosis obliterans, particularly at the occurrence of atherothrombosis. On the other hand, these ischemic organs may be functionally rescued by "therapeutic angiogenesis" [34–36, 51]. Thus, angiogenesis can participate in reverse two actions: promotion of not only atherogenesis but also organ ischemia in patients with advanced atherosclerotic lesions [56]. Further defining these pathophysiological mechanisms is necessary for the ultimate success of therapeutic modulation of angiogenesis selectively for atherosclerosis and atherosclerosis-related organ ischemia.

Acknowledgments. The authors thank Mr. Hiroshi Fujii for light microscopic and immunohistochemical examinations; Dr. Yoshikazu Yonemitsu, Department of Gene Therapy, Chiba University Graduate School of Medicine, Chiba, for providing excellent scientific comments; and Dr. Mamoru Hasegawa, DNAVEC Corporation, Ibaragi, for kindly supplying recombinant Sendai virus vectors. This work was supported in part by a Grant in Aid from the Japanese Ministry of Education, Culture, Sports, Science, and Technology, Tokyo, Japan (19209012, 16209012, and 13307009) and by a Grant of Promotion of Basic Science Research in Medical Frontier from the Organization for Pharmaceutical Safety and Research, Tokyo, Japan (project MF-21).

References

1. Winternitz MCR, Thomas RM, LeCompete PM (1938) The biology of arteriosclerosis. Charles C Thomas, Springerfield, IL
2. Paterson JC (1936) Vascularization and hemorrhage of the intima of arteriosclerotic arteries. Arch Pathol 22:312–324
3. Paterson JC (1938) Capillary rupture with intimal hemorrhage as a causative factor in coronary thrombosis. Arch Pathol 25:474–487
4. Geiringer E (1951) Intimal vascularization and atherosclerosis. J Pathol Bacteriol 63:201–211
5. Barger AC, Beeuwkes R III, Lainey LL, et al (1984) Hypothesis: vasa vasorum and neo-vascularization of human coronary arteries: a possible role in the pathology of atherosclerosis. N Engl J Med 310:175–177
6. Ross R (1999) Atherosclerosis—an inflammatory diseases. N Engl J Med 340:115–126
7. Libby P (2002) Inflammation in atherosclerosis. Nature 420:868–874
8. Murata T, Nakagawa K, Khalil A, et al (1996) The temporal and spatial vascular endothelial growth factor expression in retinal vasculogenesis of rat neonates. Lab Invest 74:68–77
9. Murata T, Nakagawa K, Khalil A, et al (1996) The relation between expression of vascular endothelial growth factor and breakdown of the blood-retinal barrier in diabetic rat retinas. Lab Invest 74:819–825
10. Shiraishi S, Nakagawa K, Kinugasa N, et al (1997) The immunohistochemical localization of vascular endothelial growth factor in human placenta associated with hydrops fetalis. Pediatr Pathol Lab Med 17:65–81
11. Carmeliet P, Jain RK (2003) Angiogenesis in cancer and other diseases. Nature 407:249–257
12. Carmeliet P (2003) Angiogenesis in health and disease. Nat Med 9:653–660
13. Kumamoto M, Nakashima Y, Sueishi K (1995) Intimal neovascularization in human coronary atherosclerosis: its origin and pathophysiological significance. Hum Pathol 26:450–456
14. Chen Y-X, Nakashima Y, Tanaka K, et al (1999) Immunohistochemical expression of vascular endothelial growth factor/vascular permeability factor in atherosclerotic intimas of human coronary arteries. Arterioscler Thromb Vasc Biol 19:131–139
15. Nakano T, Nakashima Y, Yonemitsu Y, et al (2005) Angiogenesis and lymphangiogenesis and expression of lymphangiogenic factors in the atherosclerotic intima of human coronary arteries. Hum Pathol 36:330–340
16. Langheinrich AC, Michniewicz A, Sedding DG, et al (2006) Correlation of vasa vasorum neovascularization and plaque progression in aortas of apolipoprotein E-/-/low-density lipoprotein-/- double knockout mice. Arterioscler Thromb Vasc Biol 26:347–352
17. Brogi E, Winkles JA, Underwood R, et al (1993) Distinct patterns of expression of fibroblast growth factors and their receptors in human atheroma and nonatherosclerotic arteries: association of acidic FGF with plaque microvessels and macrophages. J Clin Invest 92:2408–2418
18. Inoue M, Itoh H, Ueda M, et al (1998) Vascular endothelial growth factor (VEGF) expression in human coronary atherosclerotic lesions: possible pathophysiological significance of VEGF in progression of atherosclerosis. Circulation 98:2108–2116
19. Khurana R, Moons L, Shafi S, et al (2005) Placental growth factor promotes atherosclerotic intimal thickening and macrophage accumulation. Circulation 111:2828–2836
20. Yonemitsu Y, Kaneda Y, Morishita R, et al (1996) Characterization of in vivo gene transfer into the arterial wall mediated by the Sendai virus (hemagglutinating virus of Japan) liposomes: an effective tool for the in vivo study of arterial diseases. Lab Invest 75:313–323

21. Moulton KS, Heller E, Konerding MA, et al (1999) Angiogenesis inhibitors endostatin or TNP-470 reduce intimal neovascularization and plaque growth in apolipoprotein E-deficient mice. Circulation 99:726–732

22. Moulton KS, Vakili K, Zurakowski D, et al (2003) Inhibition of plaque neovascularization reduces macrophage accumulation and progression of advanced atherosclerosis. Proc Natl Acad Sci USA 100:1736–1741

23. Celletti FL, Waugh JM, Amabile PG, et al (2001) Vascular endothelial growth factor enhances atherosclerotic plaque progression. Nat Med 7:425–429

24. Zhao Q, Egashira K, Inoue S, et al (2002) Vascular endothelial growth factor is necessary in the development of arteriosclerosis by recruiting/activating monocytes in a rat model of long-term inhibition of nitric oxide synthesis. Circulation 105:1110–1115

25. Khurana R, Zhuang Z, Bhardwaj S, et al (2004) Angiogenesis-dependent and independent phases of intimal hyperplasia. Circulation 110:2436–2443

26. Ohtani K, Egashira K, Hiasa K, et al (2004) Blockade of vascular endothelial growth factor suppresses experimental restenosis after intraluminal injury by inhibiting recruitment of monocyte lineage cells. Circulation 110:2444–2452

27. Nakashima Y, Fujii H, Sumiyoshi S, et al (2007) Early human atherosclerosis: accumulation of lipid and proteoglycans in intimal thickening followed by macrophage infiltration. Arterioscler Thromb Vasc Biol 27:1159–1165

28. Tahara N, Kai H, Ishibashi M, et al (2006) Simbastatic attenuates plaque inflammation: evaluation by fluorodeoxyglucose positron emission tomography. J Am Coll Cardiol 48:1825–1831

29. Davies JR, Rudd JH, Weissberg PL, et al (2006) Radionuclide imaging for the detection of inflammation in vulnerable plaques. J Am Coll Cardiol 47:C57–C68

30. Hyafil F, Cornily J-C, Feig JE, et al (2007) Noninvasive detection of macrophages using a nanoparticle contrast agent for computed tomography. Nat Med 13:636–641

31. Virmani R, Kolodgie FD, Burke AP, et al (2005) Atherosclerotic plaque progression and vulnerability to rupture: angiogenesis as a source of intraplaque hemorrhage. Arterioscler Thromb Vasc Biol 25:2054–2061

32. Moreno PR, Purushothaman KR, Fuster V, et al (2004) Plaque neovascularization is increased in ruptured atherosclerotic lesions of human aorta: implications for plaque vulnerability. Circulation 110:2032–2038

33. Armulik A, Abramsson A, Betsholtz C (2005) Endothelial/pericyte interactions. Circ Res 97:512–523

34. Masaki I, Yonemitsu Y, Yamashita A, et al (2002) Angiogenic gene therapy for experimental critical limb ischemia; acceleration of limb loss by overexpression of vascular endothelial growth factor 165 but not fibroblast growth factor-2. Circ Res 90:966–973

35. Onimaru M, Yonemitsu Y, Tanii M, et al (2002) Fibroblast growth factor-2 gene transfer can stimulate hepatocyte growth factor expression irrespective of hypoxia-mediated down regulation in ischemic limb. Circ Res 91:923–930

36. Tsutsumi N, Yonemitsu Y, Shikada Y, et al (2004) Mechanism of angiogenic actions of rapamycin: an essential role of PDGFRa signaling in mesenchymal cells during in vivo angiogenesis. Circ Res 94:1186–1191

37. Losordo DW, Dimmeler S (2004) Therapeutic angiogenesis and vasculogenesis for ischemic disease. Part I. Angiogenic cytokines. Circulation 109:2487–2491

38. Maisonpierre PC, Suri C, Jones PF, et al (1997) Angiopoietin-2, a natural antagonist for Tie2 that disrupts in vivo angiogenesis. Science 277:48–50

39. Shikada Y, Yonemitsu Y, Koga T, et al (2005) Platelet-derived growth factor-AA is an essential and autocrine regulator of vascular endothelial growth factor expression in non-small cell lung carcinomas. Cancer Res 65:7241–7248

40. Kubo H, Fujiwara T, Jussila L, et al (2000) Involvement of vascular endothelial growth factor receptor-3 in maintenance of integrity of endothelial cell lining during tumor angiogenesis. Blood 96:546–553

41. Jain RK (2003) Molecular regulation of vessel maturation. Nat Med 9:685–693
42. Lindahl P, Johansson BR, Leveen P, et al (1997) Pericyte loss and microaneurysm formation in PDGF-B-deficient Mice. Science 277:242–245
43. Hayashi S, Morishita R, Nakamura S (1999) Potential role of hepatocyte growth factor, a novel angiogenic growth factor, in peripheral arterial disease: downregulation of HGF in response to hypoxia in vascular cells. Circulation 100(suppl II):II-301–II-308
44. Takeshita S, Pu LQ, Stein LA, et al (1994) Intramuscular administration of vascular endothelial growth factor induces dose-dependent collateral artery augmentation in a rabbit model of chronic limb ischemia. Circulation 90:II228–II234
45. Baumgartner I, Pieczek A, Manor O, et al (1998) Constitutive expression of phVEGF165 after intramuscular gene transfer promotes collateral vessel development in patients with critical limb ischemia. Circulation 97:1114–1123
46. Folkman J (1998) Therapeutic angiogenesis in ischemic limbs. Circulation 97:1108–1110
47. Losordo DW, Dimmeler S (2004) Vasculogenesis for ischemic disease. Part II. Cell-based therapies. Circulation 109:2344–2350
48. Carmeliet P (2000) VEGF gene therapy: stimulating angiogenesis or angioma-genesis? Nat Med 6:1102–1103
49. Lee RJ, Springer ML, Blanco-Bose WE, et al (2000) VEGF Gene delivery to myocardium: deleterious effects of unregulated expression. Circulation 102:898–901
50. Yonemitsu, Y, Kitson C, Ferrari S, et al (2000) Efficient gene transfer to the airway epithelium using recombinant Sendai virus. Nat Biotech 18:970–973
51. Masaki I, Yonemitsu Y, Komori K, et al (2001) Recombinant Sendai virus-mediated gene transfer to vasculature: a new class of efficient gene transfer vector to the vascular system. FASEB J 15:1294–1296
52. Stavri GT, Zachary IC, Baskerville PA, et al (1995) Basic fibroblast growth factor upregulates the expression of vascular endothelial growth factor in vascular smooth muscle cells: synergistic interaction with hypoxia. Circulation 92:11–14
53. Kubo H, Cao R, Brakkenhielm E, et al (2002) Blockade of vascular endothelial growth factor receptor-3 signaling inhibits fibroblast growth factor-2-induced lymphangiogenesis in mouse cornea. Proc Natl Acad Sci U S A 99:8868–8873
54. Cao R, Bråkenhielm E, Pawliuk R, et al (2003) Angiogenic synergism, vascular stability and improvement of hind-limb ischemia by a combination of PDGF-BB and FGF-2. Nat Med 9:604–613
55. Lu H, Xu X, Zhang M, et al (2007) Combinatorial protein therapy of angiogenic and arteriogenic factors remarkably improves collaterogenesis and cardiac function in pigs. Proc Natl Acad Sci U S A 104:12140–12145
56. Epstein SE, Stabile E, Kinnaird T, et al (2004) Janus phenomenon: the interrelated tradeoffs inherent in therapies designed to enhance collateral formation and those designed to inhibit atherogenesis. Circulation 109:2826–2831
57. Virmani R, Kolodgie FD, Burke AP, et al (2000) Lessons from sudden coronary death: a comprehensive morphological classification scheme for atherosclerotic lesions. Arterioscler Thromb Vasc Biol 20:1262–1275

Part 6   Hemorrhagic Diathesis

Hemophilia A and Inhibitors

MIDORI SHIMA AND AKIRA YOSHIOKA

Summary. Hemophilia A is the most common congenital coagulation disorder characterized by recurrent bleeding into various tissues. The clinical severity is associated with the level of factor VIII activity (FVIII:C). Therefore, precise measurement of FVIII:C is essential for diagnosis of hemophilia A. Recent advance in the laboratory method such as clot waveform analysis made possible to assess the whole clotting function and demonstrated various heterogeneities in the clotting function even in the same severe category due to the presence of very low level of FVIII activity and other plasma factors. Several types of FVIII gene mutations, including point mutations, deletions, insertions and intragene inversion have been identified, although inversion seems to be the most characteristic and frequent. The principle of hemostatic treatment for hemophilia A is a replacement therapy using FVIII concentrates. The center of the treatment is turning from an on-demand treatment into a regular prophylaxis. One of the main problems in the clinics of hemophilia A is the development of inhibitor, since it diminish the hemostatic effect of the replacement therapy. FVIII inhibitor is IgG and reacted with either or both A2 and C2 domain. Various inhibitory mechanisms such as inhibition on VWF, PL, thrombin and FXa binding have been reported. The main hemostatic treatment is a bypassing therapy with aPCC or rFVIIa except for low responder cases who can be treated by FVIII concentrates. The eradication of inhibitor by an immune tolerance induction (ITI) therapy by frequent regular infusion of FVIII is most important treatment.

Key words. hemophilia A · factor VIII · inhibitor · intragene inversion · immune tolerance induction

Introduction

Hemophilia A is the most common coagulation disorder and, in its severe form, is characterized by recurrent bleeding into various tissues, organs, and joints from early childhood. The most typical hemorrhagic symptom is hemarthrosis, eventually resulting in limitation of joint movement due to progressive degenerative and inflammatory changes. Hemophilia A is identified by a quantitative or qualitative deficiency of factor VIII (FVIII) caused by a mutation of the FVIII gene. This gene is located on the X chromosome, and hence the disease has an X-linked recessive mode of inheritance.

Approximately 30% of patients, however, are identified as sporadic cases with no familial history of bleeding. The basic treatment for acute bleeding is replacement therapy with FVIII concentrates. In recent years, the introduction of highly purified concentrates has considerably improved the management of hemophilia, and home-based treatment by regular infusions largely prevents spontaneous bleeding. Approximately 30% of patients with severe hemophilia develop FVIII inhibitors, and this has had a major impact on replacement therapy. The complication represents one of the most important current issues for the management of hemophilia. Several new therapeutic advances have been made however, including immune tolerance induction therapy and the use of hemostatic "bypassing" products, and they are providing encouraging progress for the treatment of patients with inhibitors.

Hemophilia A

Classification and Phenotype of Hemophilia A

Conventionally, hemophilia is divided into three types—severe (<1 IU/dl), moderate (1–5 IU/dl), mild (>5–40 IU/dl)—based on the plasma level of FVIII activity (FVIII: C) [1]. This classification follows the general understanding that the clinical severity of hemophilia correlates well with FVIII:C. Most severely affected patients have frequent spontaneous bleeds unless they receive regular prophylactic factor replacement therapy. In contrast, individuals in the moderate and mild categories bleed infrequently and present clinically only after trauma. Clinical heterogeneity is a feature in some patients, however, and some patients with very low FVIII:C exhibit a moderate clinical phenotype. Similarly, occasional patients in the moderate category appear to have clinically severe disease with frequent episodes of spontaneous bleeding. It seems likely, therefore, that several physiological parameters other than low FVIII:C, such as levels of other clotting/anticlotting factors and vascular integrity, may contribute to the clinical heterogeneity.

Measurements of FVIII:C and New Methods for Hemostatic Evaluation

Measurements of FVIII:C are generally performed by one-stage clotting assays [2] or amidolytic assays using a factor Xa (FXa) chromogenic substrate [3]. The lower limit of these assays is generally 1.0 IU/dl. This limit is often affected by a number of analytical variables, especially the quality of the known deficient plasma utilized in the assay procedures. It might be particularly pertinent that the presence of some FVIII, less than the 1 IU/dl detected in conventional assays, critically affects clinical outcome. Recent automated measuring systems have focused on patients of this nature and have led to an improvement in the reproducibility and sensitivity of the FVIII:C assays.

One such technique utilizes an automated photo-optical detection system to evaluate global clotting function [4]. The method provides the facility to observe and quantify the changes in light transmission that occur during the performance of routine clotting assays, such as those based on the activated partial thromboplastin time (aPTT). This technique, which has become known as aPTT waveform analysis, has identified significant differences in a number of patients with severe hemophilia A despite the fact that all were defined as severe by a conventional one-stage FVIII:C assay (<1.0 IU/dl) (Fig. 1) [5]. Further studies in this group of patients indicated that levels of FVIII:C < 1 IU/dl are likely to contribute to phenotypic heterogeneity (Fig. 2).

In addition to classification by procoagulant activity, hemophilia A (and hemophilia B, factor IX deficiency) can be subgrouped on the basis of clotting factor antigen. Usually factor VIII antigen (FVIII:Ag) and factor IX antigen (FIX:Ag) correlate well with FVIII:C and FIX:C, respectively, and patients with nondetectable antigen are classified as cross-reacting material negative (CRM⁻). In contrast, some severe, moder-

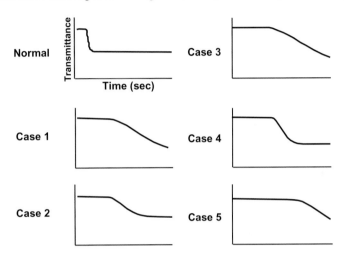

FIG. 1. Activated partial thromboplastin time (aPTT) clot waveform in normal individuals and in five patients with severe hemophilia A [5]. FVIII:C levels in all five patients with hemophilia A were measured as <1 IU/dl by conventional clotting assay. Light transmission is shown on the vertical axis and time in seconds on the horizontal axis

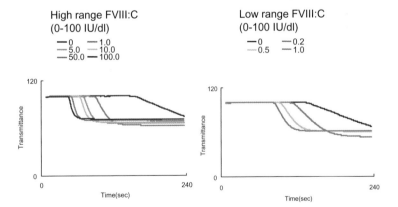

FIG. 2. Dose-dependent waveform changes in FVIII-deficient plasmas with various concentrations of FVIII:C. **Left** Dose-dependent pattern of clot waveform in FVIII-deficient plasma in the presence of added FVIII (0, 1.0, 5.0, 10.0, 50.0, 100.0 IU/dl). **Right** Dose-dependent pattern of native plasma samples with various levels of FVIII (0, 0.2, 0.5, 1.0 IU/dl)

ate, and mild hemophilia patients have normal levels of antigen and are termed cross-reacting material-positive (CRM⁺) [6, 7]. Alternatively, some patients have lower than normal antigen levels in excess of the respective clotting factor activity and are classified as cross-reacting material reduced (CRM^R). It is evident that CRM⁺ and CRM^R patients synthesize dysfunctional protein, and close analyses of such patients provide important clues for understanding the molecular pathogenesis of hemophilia.

Advances in the Molecular Biology of Hemophilia A

Structure of the FVIII Gene

The FVIII gene spans 186 kb and consists of 26 exons and 25 introns. It is located on the distal part of the X chromosome at Xq28 [8, 9]. Intron 22 is the largest intervening sequence, spanning 32.4 kb, containing a unique 9.5 kb designated the intron 22 homologous region (int22h-1) of which two copies (int22h-2, int22h-3) are present approximately 497 kb distal to the 5′ end of the FVIII gene. In the int22h-1 region, there are two intrachromosomal genes, F8A and F8B. F8A is an intron-less gene with approximately 2 kb transcribing in the opposite direction to the FVIII gene [10]. The F8B gene is transcribed in the same direction as exon 23–26 [11]. The F8A gene has been reported to code for a Huntington-associated protein, although the significance of this protein is not clear [12]. In hemophilia A patients with gross gene deletions including the F8B gene, symptoms other than bleeding are not seen; and the function of this gene is not thought to be significant.

Synthesis and Structure of FVIII Protein

Patients with hemophilia A recover FVIII activity to normal levels after liver transplantation, and there is little doubt that FVIII is synthesized in the liver. The precise cellular origin of FVIII remains controversial, although at present sinusoidal endothelial cells appear to be the likeliest site of synthesis. The mRNA is approximately 9 kb in size and codes a FVIII precursor protein with a single-chain peptide containing 2351 amino acid residues. This is processed into the mature FVIII protein with 2332 amino acid residues and a molecular mass of ~300 kDa. The FVIII molecule can be divided into three domains, arranged in the order A1-A2-B-A3-C1-C2 [8, 9, 13] based on amino acid homology (Fig. 3). It is processed into a series of metal ion-dependent heterodimers by cleavage at the B-A3 junction, generating a heavy chain of the A1 and A2 domains, together with heterogeneous fragments of a partially proteolyzed B domain, linked to a light chain consisting of the A3, C1, and C2 domains.

Cofactor Activity of FVIII

Factor VIII is transformed into its active form, FVIIIa, via limited proteolysis by thrombin or activated factor X (FXa) [14, 15]. The catalytic efficacy of the FX activating enzyme complex, tenase, is enhanced more than 10^5 times in the presence of FVIIIa [16]. Hence, a defect in FVIII severely impairs the clotting cascade. Thrombin and FXa cleave FVIII at Arg³⁷² and Arg⁷⁴⁰ of the heavy chain generating 50-kDa A1 and 40-kDa A2 subunits. The 80-kDa light chain is cleaved at Arg¹⁶⁸⁹, producing a 72-kDa A3-C1-C2 subunit. In addition, FXa generates a 67-kDa A3-C1-C2 subunit by cleavage at Arg¹⁷²¹. Proteolysis at Arg³⁷² and Arg¹⁶⁸⁹ is essential for FVIII cofactor function [16–18]. FVIIIa is inactivated by serine proteases such as activated protein C (APC), FXa, and FIXa

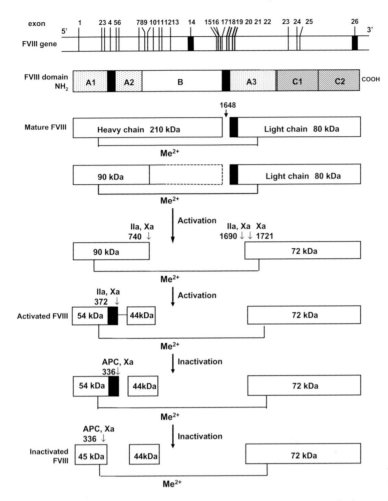

FIG. 3. Structure of FVIII gene and FVIII protein. Mature FVIII is composed of a metal ion-linked heterodimer containing a heavy chain, consisting of A1 and A2 domains together with partially proteolyzed B domains, combined with a light chain of A3, C1, and C2 domains. The three A domains are flanked by regions a1, a2, and a3 containing acidic amino acid residues. The reported epitope regions and the inhibitory effects on binding of FVIII ligands are shown

predominantly by proteolytic cleavage at Arg[336] [15, 19]. In addition, plasmin appears to inactivate FVIII by limited proteolysis at Arg[336] [20].

Molecular Aspects of Hemophilia A

Several types of FVIII gene mutations, including point mutations, deletions, insertions, and intragene inversions, have been identified in hemophilia A, although inversions seem to be the most characteristic and frequent. Mutations of this nature are found in approximately 40%–50% of severe patients. The discovery of inversions followed the decisive report of Naylor et al., who showed that no amplified mRNA product was obtained by the reverse transcription polymerase chain reaction (RT-

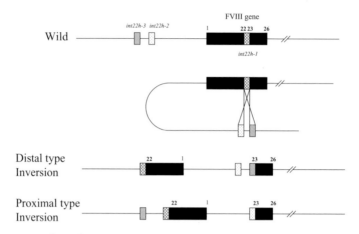

FIG. 4. Intron 22-derived inversion [22]. Two copies of an intron 22-related gene (int22h-1), int22h-3, and int22h-2 locate upstream of the FVIII gene. Intrachromosomal homologous recombination studies have identified inversions between int22–1 and extragenic either int22h-2 or int22h-3

PCR) beyond exon 22 and 23 in patients with no mutation evident by classic gene analysis [21]. Later studies demonstrated the presence of inversions between intragenic int22h-1 and extragenic int22h-2 or int22h-3 by intrachromosomal homologous recombination (Fig. 4) [22]. The int22h-2 inversion is termed the distal type, and the int22h-3 is called the proximal type. More recently another inversion between int1h-1 and int1h-2 in the intron 1 region, 125 kb upstream of exon 1, has been reported [23].

The second most common type of gene defect is represented by specific point mutations, with a frequency of approximately 40%–50%. These mutations are distributed across the whole FVIII gene, and no hot spots are evident. The database for hemophilia A mutations [24] (http://europium.csc.mrc.ac.uk/, October 2006), reports that 687 point mutations have been registered. In addition, 91 large gene deletions (>50 bp), 127 small gene deletions (<50 bp), and 31 insertions have been registered.

Advances in Therapy for Hemophilia A

Hemostatic treatment for acute bleeding in hemophilia A depends on replacement therapy with FVIII concentrates. On-demand FVIII replacement therapy requires prompt infusion of optimum doses for maximum effect. The trend in treatment for hemophilia A has shifted, however, from on-demand therapy for acute bleeding to regular prophylactic infusions. Treatment of this nature was developed on the basis that bleeding episodes, including life-threatening hemorrhage, are seen much less frequently in moderate hemophilia with FVIII levels >1 IU/dl. Subsequently, many studies have confirmed that the frequency and severity of bleeding episodes are dramatically decreased by regular FVIII infusions of 25–40 IU/kg three to three and a half times a week, maintaining FVIII:C levels above 1 IU/dl.

In addition, it is now clear that adequate prophylactic treatment prevents progressive joint damage. For example, Nilsson's group reported excellent results during long-term follow-up of prophylactic therapy in young patients with hemophilia A.

Orthopedic and radiological joint scores were found to have remained unchanged during follow-up in almost all patients, and the data supported the international consensus that hemophilic arthropathy can be prevented by administrating early, high-dose prophylaxis [25]. Many subsequent investigations have confirmed the efficacy of early prophylactic therapy [26–28], although most reports depended on observational clinical studies and were not based on substantive controlled studies. Manco-Johnson et al., however, commenced the Joint Outcome Study (JOS) in 1996 to evaluate prophylactic therapy using a prospective, randomized, controlled study [29].

Their 10-year study led to several important conclusions. First, the prognosis for joint abnormalities in patients given regular prophylaxis was better than in those who received on-demand therapy. Second, there was no strong relation between overt joint bleeding and magnetic resonance imaging (MRI) findings, especially in the on-demand treatment group. Abnormal MRI findings were demonstrated in patients with no or few reported bleeding episodes, suggesting persistent, although minimum, bleeding into the joints. During the 5-year follow up, joint damage developed in the on-demand group who had minimum joint bleeding. In contrast, no joint damage was evident in patients in the regular prophylaxis group even in the presence of one to five joint hemorrhages. On the basis of these and many other findings, therefore, primary prophylaxis therapy, starting in early childhood before the onset of joint damage, is now recommended internationally.

FVIII Inhibitors

Factor VIII inhibitors develop as alloantibodies in patients with congenital hemophilia A treated with extrinsic FVIII concentrates. In these circumstances FVIII replacement therapy becomes ineffective or markedly impaired. The development of an inhibitor, therefore, is one of the most serious complications seen in the clinical management of hemophilia A. FVIII inhibitors specifically bind to highly antigenic regions of the FVIII molecule, and the resulting antigen–antibody complexes are rapidly cleared from the circulation. Some inhibitors depress FVIII activity by interfering with functional interactions dependent on FVIII structure.

Immunological and Biochemical Characterization

Immunoglobulin Class and Subclass

Factor VIII inhibitors are oligoclonal in origin, with a mixed composition of immunoglobulin G (IgG) subtypes, including IgG1, IgG2, IgG4, and more rarely IgG3. Most FVIII inhibitors have restricted heterogeneity, however, and IgG4 antibody appears to be the dominant subtype [30–32].

Epitope localization of FVIII allo-inhibitors provides important information for understanding the characteristics inhibitor reactions and for investigating structure–function relations of the FVIII molecule. Immunoblot studies using purified FVIII fragments revealed specificity of FVIII inhibitors almost two decades ago [33]. Since that time, numerous studies have attempted to localize epitopes to more defined regions using various strategies, including immune precipitation assays, site-directed mutagenesis, chimeric FVIII with porcine sequences, synthetic peptides, phage display,

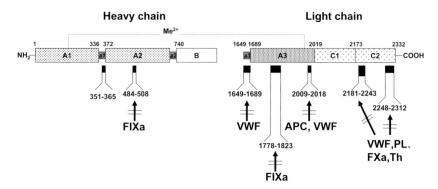

FIG. 5. Structure of factor VIII (FVIII) and epitopes of FVIII inhibitors. The reported epitope regions and the inhibitory effects on binding of FVIII ligands are shown. *FIXa*, activated factor IX; *FXa*, activated factor X; *VWF*, von Willebrand factor; *APC*, activated protein C; *PL*, phospholipid; *Th*, thrombin

and cloning of human hybridomas. Epitopes have been identified in all FVIII domains except for the B domain, although the principal epitopes appear to be located in the A2 and C2 domains [34, 35] (Fig. 5).

Functional Characterization

FVIII Inactivation Mechanism

Inhibitory Effects on FVIII Binding to VWF and Phospholipid. Association between negatively charged phospholipids (PLs) and von Willebrand factor (VWF) is essential for FVIII cofactor activity in the tenase enzyme complex [36] and for stabilization of FVIII:C, by protecting it from APC-catalyzed inactivation [37, 38]. VWF inhibits FVIII binding to PLs; and inhibition of FVIII binding to VWF and PLs is likely, therefore, to be a major FVIII inhibitory mechanism. Many anti-C2 antibodies inhibit both PL and VWF binding. The epitopes for these antibodies have been localized to both amino- and carboxy-terminal portions and to exposed positively charged hydrophobic residues constituting the membrane-binding motif.

When associated with VWF, FVIII exerts limited cofactor activity in the tenase complex. After thrombin activation, however, the active form of FVIII dissociates from VWF and is immediately incorporated into the tenase complex by binding to negative-charged PL micelles. Hence, if FVIII does not dissociate from VWF, coagulation function is heavily impaired. In our studies using a monoclonal antibody, ESH8 (epitope 2248–2285), and a hemophilia A alloantibody recognizing the C2 domain (epitope 2218–2307), the initial rate of release of thrombin-activated FVIII (FVIIIa) from VWF was reduced [39, 40]. These rare antibodies were shown to bind to an epitope situated more toward the amino-terminus than in other common anti-C2 inhibitors, inhibiting PL binding but not VWF binding. The studies identified a possible additional inhibitory mechanism for the anti-C2 antibodies. Most other anti-C2 antibodies inhibit both VWF and PL binding, and it is therefore evident that VWF- and PL-binding sites in the C2 domain do not overlap completely.

The amino-terminal acidic region of the FVIII light chain, spanning residues 1649–1689, has also been reported to be important for the binding of FVIII to VWF [41–43].

Data in these experiments were obtained, however, by epitope analysis of monoclonal antibodies that inhibit FVIII binding to VWF. Human FVIII inhibitors recognizing this region have not been described.

Type 2 FVIII autoantibodies recognizing the APC binding region in the A3 domain and spanning His^{2009}-Val^{2018}, have been shown to inhibit VWF binding and APC binding as described above [44]. Natural FVIII inhibitors recognizing the C1 domain have not been reported, although an anti-FVIII, IgG4k antibody, LE2E9, obtained from an immortalized B cell from a mild hemophilia A patient with a mutation at Arg^{2150}, inhibited VWF binding and recognized the C1 domain [45].

Inhibitory Effects on Factor IXa Binding. The active form of FVIII (FVIIIa) enhances factor Xa (FXa) generation by FIXa in the tenase complex mainly by increasing the k_{cat} by several orders of magnitude [36]. Inhibition of the FVIIIa–IXa association results in loss of FVIII cofactor function. Two FIXa interactive sites, residues 1811–1818 in the A3 domain [46] and residues 558–565 in the A2 domain [47], have been identified in the FVIII molecule. The A3 site showed a much higher affinity (Kd ~ 14 nM) than the A2 site (Kd ~ 300 nM), indicating that the A3 region was probably the dominant binding site. Inhibitor IgG, complexed with A3-C1 fragments, prevented binding of FIXa to the FVIII light chain, and binding was restricted in the presence of a peptide corresponding to residues 1804–1819 [48]. In addition, antibodies that reacted with residues 1778–1823 containing FIXa binding region inhibited binding of FIXa to immobilized FVIII light chain [49]. The affinity of the A2 reactive site for FIXa was much lower than that of the A3 site, however, and the role of the A2 site in the physiological interaction between FVIII and FIXa remains to be fully explored. Nevertheless, isolated A2 subunits were shown to increase the k_{cat} of FIXa-induced generation of FXa by approximately 100-fold, suggesting that the A2 domain plays an important role in this association [50]. This was further supported by finding that a panel of anti-A2 inhibitors and a monoclonal antibody inhibited A2 subunit-dependent FXa generation, and that the inhibitory effects correlated well with their neutralizing activity as measured in the Bethesda assay [51]. These results confirmed that disrupting the interaction between FVIIIa and FIXa represents a significant inhibitory mechanism in some FVIII inhibitors.

Inhibitory Effects on FXa and Thrombin. Factor VIII is transformed into an active form by limited proteolysis by two essential serine proteases, thrombin and FXa. The procoagulant activity is more pronounced after thrombin activation than after FXa activation [52]. Thrombin cleaves FVIII at Arg^{1689}, resulting in dissociation of FVIII from its complex with VWF, and promoting tenase activity on PLs by free FVIII. Thrombin-binding sites have been localized in both A2 and C2 domains, although the relative importance of the different determinants has not been established [53]. A C2-specific, affinity-purified FVIII alloantibody inhibited thrombin cleavage at Arg^{1689}, suggesting that the C2 domain contained the thrombin-binding site responsible for the cleavage at Arg^{1689}. The C2 epitope of the FVIII inhibitor was localized to the common C2 epitope region, residues 2248–2312, indicating that the carboxy-terminal region of the C2 domain may be involved in thrombin binding as well as VWF and PL binding.

Factor Xa cleaves only the free form of FVIII, and FXa-catalyzed FVIII activation is completely prevented by the formation of the FVIII–VWF complex. A anti-C2 monoclonal antibody, ESH8, inhibited FXa-catalyzed FVIII activation, indicating that the FXa-binding site was located within residues 2253–2270 of FVIII [54]. Four FVIII alloantibodies with C2 epitopes moderated FXa-catalyzed FVIII activation and inhibited proteolytic cleavage of FVIII by FXa. The antibodies inhibited Xa binding independent of PLs [55]. Three antibodies had the common C2 epitope in residues 2248–2312, and one antibody recognized a more amino-terminal region similar to the ESH8 epitope.

C2-specific antibodies are heterogeneous in their inhibitory effect on VWF and PL binding, and the relative importance of inhibiting FXa-mediated FVIII activation is difficult to determine. However, a synthetic peptide corresponding to residues 2253–2270 that selectively and specifically inhibited FVIII binding to FXa had profound anticoagulant activity [56]. It seems highly likely, therefore, that inhibition of this mechanism by C2 inhibitors would be manifested in a potent anti-FVIII effect.

Recently, Nogami et al. reported that a FVIII monoclonal antibody recognizing a FX(a) binding site within residues 351–365 in the A1 region inhibited both FX and FXa binding [57, 58].

Advances in the Treatment of Hemophilia A Patients with Inhibitor

Hemostatic Treatment

Mechanism of Hemostatic "Bypassing" Agents

The main commercially available bypassing agents are activated prothrombin complex (aPCC) (e.g., factor VIII bypassing agent, or FEIBA) and recombinant FVIIa (e.g., rFVIIa, NovoSeven). The mode of actions of FEIBA and NovoSeven are different, although both agents facilitate thrombin generation at the site of bleeding. Turecek et al. demonstrated that the hemostatic effect of FEIBA was similar to that of prothrombin and FXa in an antibody-induced rabbit FVIII inhibitor model [59]. Furthermore, factor V (FV) appears to be crucial for FEIBA-induced thrombin generation in vitro, and it is therefore reasonable to speculate that the key component of FEIBA facilitates thrombin generation through acceleration of the prothrombinase complex.

The rationale for the use of rFVIIa in hemophilia A with inhibitor was explained by Hoffman et al. using a cell-based model of coagulation [60]. Following vessel wall injury, tissue factor (TF) is exposed to blood and the TF–FVIIa complex formed on TF-bearing cells activates FX and results in conversion of prothrombin to thrombin. Initially, only trace amounts of thrombin are generated, which are not sufficient for stable fibrin formation. The effect on hemostasis in vivo at this time is minimal. The thrombin produced, however, stimulates platelets and activates FV, FVIII, and FXI by positive feedback mechanisms, resulting in the generation of relatively large quantities of thrombin on the activated platelets [61]. This platelet-dependent, TF-independent burst of thrombin promotes full fibrin plug formation and contributes to the inhibition of fibrinolysis by activating thrombin-activatable fibrinolysis inhibitor (TAFI) [62]. The effect of rFVIIa is dose-dependent and higher doses mediate the thrombin cycle even in the absence of FVIII and FIX [63].

Practical Hemostatic Therapy

The choice of hemostatic therapy in patients with FVIII inhibitors is selected on the basis of several clinical features, including the severity of bleeding, recent inhibitor titer, type of responder, previous hemostatic effect, and any potential surgical intervention. In addition, safety and economic factors might also need to be considered. Modern therapeutic guidelines specify that regular checks of inhibitor titer and confirmation or otherwise of an anamnestic response are essential in the decision-making process [64–67] (Fig. 6).

Low Responder (<5 BU/ml). The therapy of choice in low responders is continuation of FVIII replacement. The amount of FVIII required for neutralizing the inhibitor can be calculated using the formula $40 \times (100 - Ht)/100 \times$ body weight, where Ht is hematcrit (%). This amount is added to the amount that would be used to achieve the desired FVIII level in the absence of inhibitor. It is important to recognize, however, that the dose for neutralization is dependent on individual circumstances, such as the epitope and FVIII inactivation mode of the inhibitor. Recent observations have shown that the recovery of FVIII may differ between FVIII and FVIII–VWF concentrates, especially in cases where the antibody recognizes the light chain and inhibits FVIII–VWF binding [68]. The differences are probably due to competitive inhibition of VWF binding to FVIII. If FVIII replacement therapy is not effective, the activated clotting concentrates, rFVIIa or aPCC, are indicated.

High Responder (5–10 BU/ml). In high responders, bypassing therapy is the first choice. If bleeding is severe or in the event of major surgery, high doses of FVIII concentrates may be considered. Considerable caution is required with the use of FVIII products, however; and it is important to monitor for an increasing inhibitor titer that may occur generally 5–7 days after infusion. If the anamnestic response is high, it may be desirable to revert to bypassing therapy. The activated clotting factors are also the

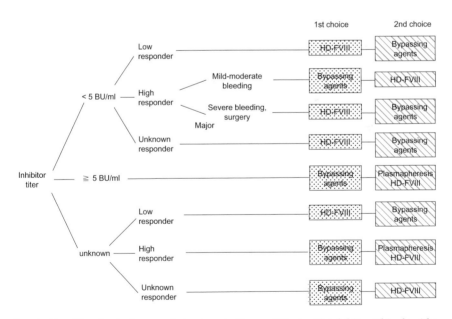

FIG. 6. Algorithm for the hemostatic treatment of hemophilia A with inhibitor. This algorithm is based on the measurement of recent inhibitor titer and responder type (low responder or high responder). Severe bleeding means life-threatening bleeding, such as intracranial bleeding. *HD-FVIII*, high-dose of FVIII concentrates

treatment of choice for acute bleeding during immune tolerance induction (ITI) therapy.

High Responder (>10 BU/ml/ml). In cases where the inhibitor response is very high, bypassing therapy is mainly advised. In the presence of the life-threatening bleeding or major surgery, however, when the inhibitor titer is close to 10 BU/ml FVIII replacement therapy may be considered.

Selection of Bypassing Therapeutic Concentrates

There is little evidence in favor of the FVIII bypassing agent, aPCC and rFVIIa. A recent study, however, called the FEIBA NovoSeven Comparative (FENOC) Study, was established for a randomized comparison of bypassing agents [69]. The study was designed to assess statistically the equivalence of these products in the treatment of typical joint bleeds, such as in the ankle, knee, or elbow. The response to either aPCC or rFVIIa was examined in 96 bleeding episodes in 48 participants. The main conclusion was that FEIBA and NovoSeven appeared to exhibit a similar effect on joint bleeds, although the efficacy between products was rated differently by a substantial proportion of patients, especially during the first 12 h after the start of symptoms. It may be particularly important, therefore, to clarify interindividual, pathophysiological, and pharmacological features during this time period, as early resolution of hemorrhagic symptoms might be critical for preventing cartilage destruction. The authors of this report emphasized that it is also imperative to develop improved treatment protocols as well as algorithms for the nonresponding patient.

In conclusion, specific bypassing agents should be selected on the basis of several factors, including previous hemostatic effect, interval after the onset of bleeding, difference in the disappearance half-time, whether plasma-derived or recombinant, anamnestic response due to the infusion of aPCC [70], patients' or parents' request, and economics.

A change from one type of concentrate to another may be a useful option in the event of a poor initial hemostatic response. Furthermore, it is important to recognize that long-term use of high-dose bypassing agents can result in unresponsiveness in some cases. For example, Hayashi et al. reported that improvements in thromboelastograph (TEG) parameters diminished after 10 days of daily infusions of aPCC in one patient. Changing the therapy to rFVIIa resulted in recovery of the improved TEG waveform. Interestingly, this patient also became unresponsive to rFVIIa as evidenced by TEG [71].

New Regimen of Bypassing Therapy

High-Doses of rFVIIa

The development of rFVIIa was a substantial advance for bypassing therapy. Frequent infusions are required, however, owing to the short disappearance half-time of FVIIa. On average, each bleeding episode requires more than two infusions with a 2.5-h interval. Moreover, the half-life of rFVIIa in pediatric patients appears to be shorter than that in adults [72]. This requirement for frequent infusions is one of its main disadvantages. As discussed above, the efficacy of rFVIIa depends on cell-based thrombin generation, and this is dose-dependent. In this respect, Kenet et al. reported that one 270 µg/kg infusion of rFVIIa had the same hemostatic effect as three infusions of 90 μg/kg, without any side effects [73]. More recently, a randomized controlled study confirmed that a single 270 μg/kg dose was at least as efficacious and safe as the 90 μg/kg × 3 regimen for the treatment of hemarthroses [74].

Prophylactic Treatment Using Bypassing Agents

The shorter half-life, higher cost, and risk of thrombotic symptoms have mainly precluded the use of bypassing agents for prophylactic treatment in patients with FVIII inhibitors. On the grounds that the dose required for prophylaxis is much lower than that needed for episodic treatment, however, several prophylactic trials using aPCC or rFVIIa have been undertaken.

Leissinger et al. studied five patients who received aPCC (FEIBA) prophylactically for ≥6 months to prevent postsurgical bleeding. The doses used ranged from 50 to 75 U/kg three times per week in four patients and 100 U/kg daily in one patient. The orthopedic status in these cases was maintained or even improved, and there were no thrombotic complications or other adverse events [75].

Another recent European study assessed prophylactic treatment with rFVIIa. Regular prophylactic infusions of rFVIIa reduced the number of bleeding episodes compared with previous treatments in 12 of 13 patients with inhibitor [76]. The doses of rFVIIa ranged from 200 to 250 μg/kg daily.

Immune Tolerance Induction Therapy

Immune tolerance induction (ITI) therapy has been widely adopted since first reported by Blackmann et al. in 1977 [77]. The main goal of treatment is to eradicate the inhibitor and to recover the response to conventional therapy. Three major protocols have been described. First, the German high-dose protocol proposed FVIII infusions of 200 IU/kg twice daily in high responder patients [78]. Second, the Malmö protocol included extracorporal immunoadsorption using a protein A column if needed, neutralization of the inhibitor, and replacement with factor concentrates; cyclophosphamide intravenously for 2 days and then orally for an additional 8–10 days; and intravenous γ-globulin daily for 5 days [79]. Third, a low-dose protocol was proposed in which infusions of 25–50 IU/kg are administered every other day [80]. The success rate with each protocol has been reported to be approximately 73%–83%. Statistically significant parameters affecting the success rate are reported to be lower inhibitor titer at initiation of ITI, historical peak titer, and peak titer during ITI [81, 82]. Other parameters, including dose, dosing interval, and type of FVIII concentrate may be important but remain controversial. International randomized, controlled studies are in progress to determine optimum regimens for ITI.

References

1. White GC II, Rosendaal F, Aledort LM, et al (2001) Definitions in hemophilia. Recommendation of the scientific subcommittee on factor VIII and factor IX of the Scientific and Standardization Committee of the International Society on Thrombosis and Haemostasis. Thromb Haemost 85:560
2. Hardisty RM (1962) A naturally occurring inhibitor of Christmas factor (factor IX). Thromb Diath Haemorrh 8:67–81

3. Rosén S, Andersson M, Blombäck M, et al (1985) Clinical application of a chromogenic substrate method for determination of factor VIII activity. Thromb Haemost 54:818–823

4. Braun PJ, Givens TB, Stead AG, et al (1997) Properties of optical data from activated partial thromboplastin time and prothrombin time assays. Thromb Haemost 78:1079–1087

5. Shima M, Matsumoto T, Fukuda K, et al (2002) The utility of activated partial thromboplastin time (aPTT) clot waveform analysis in the investigation of hemophilia A patients with very low levels of factor VIII activity (FVIII:C). Thromb Haemost 87:436–441

6. Denson KWE, Briggs R, Haddon ME, et al (1969) Two types of haemophilia (A+ and A–): a study of 48 cases. Br J Hamatol 17:163–171

7. Hoyer LW, Breckenridge RT (1968) Immunologic studies of antihemophilic factor (AHF, factor VIII): cross-reacting material in a genetic variant of hemophilia A. Blood 32:962–971

8. Wood WI, Capon DJ, Simonsen CC, et al (1984) Expression of active human factor VIII from recombinant DNA clones. Nature 312:330–337

9. Toole JJ, Knopf JL, Wozney JM, et al (1984) Molecular cloning of a cDNA encoding human antihaemophilic factor. Nature 312:342–347

10. Levinson B, Kenwrick S, Lakich D, et al (1990) A transcribed gene in an intron of the human factor VIII gene. Genomics 7:1–11

11. Levinson B, Kenwrick S, Gamel P, et al (1992) Evidence for a third transcript from the human factor VIII gene. Genomics 14:585–589

12. Peters MF, Ross CA (2001) Isolation of a 40-kDa Huntington-associated protein. J Biol Chem 276:3188–3194

13. Vehar GA, Keyt B, Eaton D, et al (1984) Structure of human factor VIII. Nature 312:337–342

14. Lollar P, Knutson GJ, Fass DN (1985) Activation of porcine factor VIII:C by thrombin and factor Xa. Biochemistry 24:8056–8064

15. Eaton D, Rodriguez H, Vehar GA (1986) Proteolytic processing of human factor VIII: correlation of specific cleavages by thrombin, factor Xa, and activated protein C with activation and inactivation of factor VIII coagulant activity. Biochemistry 25:505–512

16. Shima M, Ware J, Yoshioka A, et al (1989) An arginine to cysteine amino acid substitution at a critical thrombin cleavage site in a dysfunctional factor VIII molecule. Blood 74:1612–1617

17. Kamisue S, Shima M, Nishimura T, et al (1994) Abnormal factor VIII Hiroshima: defect in crucial proteolytic cleavage by thrombin at Arg1689 detected by a novel ELISA. Br J Haematol 86:106–111

18. Tuddenham EG, Schwaab R, Seehafer J, et al (1994) Haemophilia A: database of nucleotide substitutions, deletions, insertions and rearrangements of the factor VIII gene, second edition. Nucleic Acids Res 22:3511–3533

19. O'Brien DP, Johnson D, Byfield P, et al (1992) Inactivation of factor VIII by factor IXa. Biochemistry 31:2805–2812

20. Nogami K, Shima M, Matsumoto T, et al (2007) Mechanisms of plasmin-catalyzed inactivation of factor VIII: a crucial role for proteolytic cleavage at Arg336 responsible for plasmin-catalyzed factor VIII inactivation. J Biol Chem 282:5287–5295

21. Naylor JA, Green PM, Rizza CR, et al (1992) Factor VIII gene explains all cases of haemophilia A. Lancet 340:1066–1067

22. Lakich D, Kazazian HH Jr, Antonarakis SE, et al (1993) Inversions disrupting the factor VIII gene are a common cause of severe haemophilia A. Nat Genet 5:236–241

23. Bagnall RD, Waseem N, Green PM, et al (2002) Recurrent inversion breaking intron 1 of the factor VIII gene is a frequent cause of severe hemophilia A. Blood 99:168–174

24. Kemball-Cook G, Tuddenham EG (1997) The factor VIII mutation database on the world wide web: the haemophilia A mutation, search, test and resource site. HAMSTeRS update (version 3.0)

25. Löfqvist T, Nilsson IM, Berntorp E, et al (1997) Haemophilia prophylaxis in young patients—a long-term follow-up. J Intern Med 241:395–400

26. Manco-Johnson MJ, Nuss R, Geraghty S, et al (1994) Results of secondary prophylaxis in children with severe hemophilia. Am J Hematol 47:113–117

27. Liesner RJ, Khair K, Hann IM (1996) The impact of prophylactic treatment on children with severe haemophilia. Br J Haematol 92:973–978

28. Aledort LM, Haschmeyer RH, Pettersson H (1994) A longitudinal study of orthopaedic outcomes for severe factor-VIII-deficient haemophiliacs: the Orthopaedic Outcome Study Group. J Intern Med 236:391–399

29. Manco-Johnson MJ, Abshire TC, Shapiro AD, et al (2007) Prophylaxis versus episodic treatment to prevent joint disease in boys with severe hemophilia. N Engl Med 357:535–544

30. Hoyer LW, Gawryl MS, de la Fuente B (1981) Immunochemical characterization of human antibodies to factor VIII by immunoaffinity chromatography. Thromb Hemost 45:60

31. Fulcher CA, de Graaf Mahoney S, Zimmerman TS (1987) FVIII inhibitor IgG subclass and FVIII polypeptide specificity determined by immunoblotting. Blood 69:1475–1480

32. Matsumoto T, Shima M, Fukuda K, et al (2001) Immunological characterization of factor VIII autoantibodies in patients with acquired hemophilia A in the presence or absence of underlying disease. Thromb Res 104:381–388

33. Fulcher CA, de Graaf Mahoney S, Roberts JR, et al (1985) Localization of human factor FVIII inhibitor epitopes to two polypeptide fragments. Proc Natl Acad Sci U S A 82:7728–7732

34. Scandella D, deGraaf Mahoney S, Mattingly M, et al (1988) Epitope mapping of human factor VIII inhibitor antibodies by deletion analysis of factor VIII fragments expressed in Escherichia coli. Proc Natl Acad Sci U S A 85:6152–6156

35. Scandella D, Mattingly M, de Graaf Mahoney S, et al (1989) Localization of epitopes for human factor VIII inhibitor antibodies by immunoblotting and antibody neutralization. Blood 74:1618–1626

36. Van Dieijen G, Tans G, Rosing J, et al (1981) The role of phospholipid and factor VIIIa in the activation of bovine factor X. J Biol Chem 256:3433–3442

37. Nogami K, Shima M, Nishiya K, et al (2002) A novel mechanism of factor VIII protection by von Willebrand factor from activated protein C-catalyzed inactivation. Blood 99:3993–3998

38. Koedam JA, Meijers JC, Sixma JJ, et al (1988) Inactivation of human factor VIII by activated protein C: cofactor activity of protein S and protective effect of von Willebrand factor. J Clin Invest 82:1236–1243

39. Shima M, Nakai H, Scandella D, et al (1995) Common inhibitory effects of human anti-C2 domain inhibitor alloantibodies on factor VIII binding to von Willebrand factor. Br J Haematol 91:714–721

40. Saenko EL, Shima M, Gilbert GE, et al (1996) Slowed release of thrombin-cleaved factor VIII from von Willebrand factor by a monoclonal and a human antibody is a novel mechanism for factor VIII inhibition. J Biol Chem 271:27424–27431

41. Foster PA, Fulcher CA, Houghten RA, et al (1988) An immunogenic region within residues Val1670-Glu1684 of the factor VIII light chain induces antibodies which inhibit binding of factor VIII to von Willebrand factor. J Biol Chem 263:5230–5234

42. Leyte A, Mertens K, Distel B, et al (1989) Inhibition of human coagulation factor VIII by monoclonal antibodies: mapping of functional epitopes with the use of recombinant factor VIII fragments. Biochem J 263:187–194

43. Shima M, Yoshioka A, Nakai H, et al (1991) Epitope localization of monoclonal antibodies against factor VIII light chain which inhibit complex formation by factor VIII with von Willebrand factor. Int J Hematol 54:515–522

44. Nogami K, Shima M, Giddings JC, et al (2001) Circulating factor VIII immune complexes in patients with type 2 acquired hemophilia A and protection from activated protein C-mediated proteolysis. Blood 97:669–677

45. Jacquemin M, Benhida A, Peerlinck K, et al (2000) A human antibody directed to the factor VIII C1 domain inhibits factor VIII cofactor activity and binding to von Willebrand factor. Blood 95:156–163

46. Lenting PJ, van de Loo JW, Donath MJ, et al (1996) The sequence Glu1811-Lys1818 of human blood coagulation factor VIII comprises a binding site for activated factor IX. J Biol Chem 271:1935–1940

47. Fay PJ, Beattie T, Huggins CF, et al (1994) Factor VIIIa A2 subunit residues 558–565 represent a factor IXa interactive site. J Biol Chem 269:20522–20527

48. Zhong D, Saenko EL, Shima M, et al (1998) Some human inhibitor antibodies interfere with factor VIII binding to factor IX. Blood 92:136–142

49. Fijnvandraat K, Celie PH, Turenhout EA, et al (1998) A human alloantibody interferes with binding of factor IXa to the factor VIII light chain. Blood 91:2347–2352

50. Fay PJ, Koshibu K (1998) The A2 subunit of factor VIIIa modulates the active site of factor IXa. J Biol Chem 273:19049–19054

51. Fay PJ, Scandella D (1999) Human inhibitor antibodies specific for the factor VIII A2 domain disrupt the interaction between the subunit and factor IXa. J Biol Chem 274:29826–29830

52. Parker ET, Pohl J, Blackburn MN, et al (1997) Subunit structure and function of porcine factor Xa-activated factor VIII. Biochemistry 36:9365–9373

53. Nogami K, Shima M, Hosokawa K, et al (2000) Factor VIII C2 domain contains the thrombin-binding site responsible for thrombin-catalyzed cleavage at Arg1689. J Biol Chem 275:25774–25780

54. Nogami K, Shima M, Hosokawa K, et al (1999) Role of factor VIII C2 domain in factor VIII binding to factor Xa. J Biol Chem 274:31000–31007

55. Nogami K, Shima M, Nishiya K, et al (2002) Human factor VIII inhibitor alloantibodies with a C2 epitope inhibit factor Xa-catalyzed factor VIII activation: a new anti-factor VIII inhibitory mechanism. Thromb Haemost 87:459–465

56. Nogami K, Shima M, Nishiya K, et al (2002) Anticoagulant effects of a synthetic peptide containing residues Thr-2253-Gln-2270 within factor VIII C2 domain that selectively inhibits factor Xa-catalysed factor VIII activation. Br J Haematol 116:868–874

57. Nogami K, Lapan KA, Zhou Q, et al (2004) Identification of a factor Xa-interactive site within residues 337–372 of the factor VIII heavy chain. J Biol Chem 279:15763–15771

58. Nogami K, Freas J, Manithody C, et al (2004) Mechanisms of interactions of factor X and factor Xa with the acidic region in the factor VIII A1 domain. J Biol Chem 279:33104–33113

59. Turecek PL, Váradi K, Gritsch H, et al (2004) FEIBA: mode of action. Haemophilia 10(suppl 2):3–9

60. Hoffman M, Monroe DM (2001) A cell-based model of hemostasis. Thromb Haemost 85:958–965

61. Monroe DM, Hoffman M, Oliver JA, et al (1997) Platelet activity of high-dose factor VIIa is independent of tissue factor. Br J Haematol 99:542–547

62. Bajzar L, Manuel R, Nesheim ME (1995) Purification and characterization of TAFI, a thrombin-activable fibrinolysis inhibitor. J Biol Chem 270:14477–14484

63. Roberts HR, Monroe DM, White GC (2004) The use of recombinant factor VIIa in the treatment of bleeding disorders. Blood 104:3858–3864

64. Rubinger M, Rivard GE, Teitel J, et al (2000) Suggestions for the management of factor VIII inhibitors. Haemophilia 6:52–59

65. Gringeri A, Mannucci PM (2005) Italian guidelines for the diagnosis and treatment of patients with haemophilia and inhibitors. Haemophilia 11:611–619
66. Hay CR, Brown S, Collins PW, et al (2006) The diagnosis and management of factor VIII and IX inhibitors: a guideline from the United Kingdom Haemophilia Centre Doctors Organisation. Br J Haematol 133:591–605
67. Australian Health Ministers' Advisory Council (AHMAC) (2006) Evidence-based clinical practice guidelines for the use of recombinant and plasma-derived FVIII and FIX products. http://www.nba.gov.au
68. Inoue T, Shima M, Takeyama M, et al (2006) Higher recovery of factor VIII (FVIII) with intermediate FVIII/von Willebrand factor concentrate than with recombinant FVIII in a haemophilia A patient with an inhibitor. Haemophilia 12:110–113
69. Astermark J, Donfield SM, DiMichele DM, et al (2007) A randomized comparison of bypassing agents in hemophilia complicated by an inhibitor: the FEIBA NovoSeven Comparative (FENOC) study. Blood 109:546–551
70. Yoshioka A, Kamisue S, Tanaka I, et al (1991) Anamnestic response following infusion of prothrombin complex concentrates and activated prothrombin complex concentrates in haemophilia A patients with inhibitors. Blood Coagul Fibrinolysis 2:51–58
71. Hayashi T, Tanaka I, Shima M, et al (2004) Unresponsiveness to factor VIII inhibitor bypassing agents during haemostatic treatment for life-threatening massive bleeding in a patient with haemophilia A and a high responding inhibitor. Haemophilia 10:397–400
72. Villar A, Aronis S, Morfini M, et al (2004) Pharmacokinetics of activated recombinant coagulation factor VII (NovoSeven) in children vs. adults with haemophilia A. Haemophilia 10:352–359
73. Kenet G, Lubetsky A, Luboshitz J, et al (2003) A new approach to treatment of bleeding episodes in young hemophilia patients: a single bolus megadose of recombinant activated factor VII (NovoSeven). J Thromb Haemost 1:450–455
74. Kavakli K, Makris M, Zulfikar B, et al (2006) Home treatment of haemarthroses using a single dose regimen of recombinant activated factor VII in patients with haemophilia and inhibitors: a multi-centre, randomised, double-blind, cross-over trial. Thromb Haemost 95:600–605
75. Leissinger CA, Becton DL, Ewing NP, et al (2007) Prophylactic treatment with activated prothrombin complex concentrate (FEIBA) reduces the frequency of bleeding episodes in paediatric patients with haemophilia A and inhibitors. Haemophilia 13:249–255
76. Morfini M, Auerswald G, Kobelt RA, et al (2007) Prophylactic treatment of haemophilia patients with inhibitors: clinical experience with recombinant factor VIIa in European Haemophilia Centres. Haemophilia, OnlineEarly Articles
77. Brackmann HH, Gormsen J (1977) Massive factor-VIII infusion in haemophiliac with factor-VIII inhibitor, high responder. Lancet 29:933
78. Brackmann HH, Lenk H, Scharrer I, et al (1999) German recommendations for immune tolerance therapy in type A haemophiliacs with antibodies. Haemophilia 5:203–206
79. Freiburghaus C, Berntorp E, Ekman M, et al (1999) Tolerance induction using the Malmö treatment model 1982–1995. Haemophilia 5:32–39
80. Mauser-Bunschoten EP, Nieuwenhuis HK, Roosendaal G, et al (1995) Low-dose immune tolerance induction in hemophilia A patients with inhibitors. Blood 86:983–988
81. Mariani G, Scheibel E, Nogao T, et al (1994) Immune tolerance as treatment of alloantibodies to factor VIII in hemophilia: the International Registry of Immunotolerance Protocols. Semin Hematol 31:62–64
82. DiMichele DM, Kroner BL, North American Immune Tolerance Study Group (2002) The North American Immune Tolerance Registry: practices, outcomes, outcome predictors. Thromb Haemost 87:52–57

von Willebrand Disease: Diagnosis and Treatment

MASATO NISHINO

Summary. von Willebrand disease (VWD) is caused by quantitative or qualitative defects of von Willebrand factor (VWF), associated with mucocutaneous bleeding symptoms. VWD is classified into three major groups and four subcategories. The classification is intended to be simple, to rely on widely available laboratory tests, and to correlate with important clinical characteristics. It is meant to facilitate the diagnosis, treatment, and counseling of patients with VWD. The International Society on Thrombosis and Haemostasis/Scientific and Standardization Committee (ISTH/SSC) reviewed the classification and published update information on the diagnosis and treatment of VWD in 2006 based on the most recent pathophysiology data for VWF. Notably, Vicenza type has been classified into type 1 from type 2M. In epidemiology, there may be small differences of frequency in VWD subcategories (type 2) between Europe and Japan. For treatment, factor VIII/VWF concentrate is commonly employed in Japan, although DDAVP has been the first choice for VWD in Europe.

Key words. Classification of von Willebrand disease (VWD) · Bleeding symptoms · Laboratory tests · DDAVP · FVIII/VWF concentrate

Introduction

Von Willebrand disease (VWD) is a bleeding disorder caused by inherited defects in the concentration, structure, or function of von Willebrand factor (VWF) [1]. VWF mediates platelet adhesion at sites of vascular injury by binding to endothelial connective tissue and platelets. VWF also stabilizes blood clotting factor VIII (FVIII) in plasma. Therefore, impaired VWF may induce not only dysfunction of platelet plaque formation in injured vessels but also defective coagulation in hemostasis. The synthesis, assembly, and catabolism of VWF have been reviewed by the VWF Subcommittee of the ISTH/SSC (International Society on Thrombosis and Haemostasis/Scientific and Standardization Committee). The current diagnosis of and treatment for VWD are based on the update regarding the pathophysiology and classification of VWD [2].

Structure of VWF

The VWF is a multimeric plasma glycoprotein composed of subunits of molecular weight (MW) 250 kDa, multimerized through the N-N and C-C terminals with disulfide bonds. The size of multimers may vary from MW 500 kDa (dimers) to >10 000 kDa (>40 subunits). The VWF multimers can be estimated by the multimer bands: large multimers have >10 bands of > MW 10 000 kDa; intermediate multimers have 6–10 bands of MW 5000 kDa of ~10 000 kDa; and small multimers have ≤ 5 bands of < MW 5000 kDa. Several binding functions have been localized to discrete sites in the VWF subunit; for instance, the A1 domain interacts with platelet GPIb; domain A3 with fibrillar collagens; domains D′ and D3 with FVIII; and domain C1 with integrin $\alpha_{IIb}\beta_3$. High-molecular-weight (HMW) VWF multimers (large VWF to upper part of intermediate VWF) with enough lengths of the molecule may be effective hemostatically and can connect platelets or injured endothelial tissue (collagen) and platelets [3]. On the other hand, each size VWF multimer evenly binds and stabilizes FVIII in plasma.

Classification of VWD

Von Willebrand disease is classified into three primary categories (Table 1). Type 1 includes a partial quantitative deficiency, type 2 includes qualitative defects, and type 3 includes virtually complete deficiency of VWF. VWD type 2 is divided into four secondary categories. Type 2A includes variants with decreased VWF-dependent platelet adhesion caused by selective deficiency of HMW VWF multimers. Type 2B includes variants with markedly increased VWF-dependent platelet adhesion due to increased affinity of VWF for platelet glycoprotein Ib. Type 2M includes variants with decreased VWF-dependent platelet adhesion despite a relatively normal size distribution of VWF multimers. Type 2N includes variants with defective binding of VWF to FVIII. These six categories of VWD correlate with clinical features and therapeutic requirements [1]. In addition, the ISTH/SSC presented a database on mutations and gene abnormalities in VWD (http://www.shef.ac.uk/vwf/).

VWD Type 1

VWD type 1 includes a partial quantitative deficiency; the bleeding is attributed to a decrease in VWF concentration, not to a selective decrease in HMW multimers or to specific abnormalities in ligand-binding sites. The key findings are that plasma VWF may contain mutant subunits but has normal functional activities relative to the VWF antigen (VWF:Ag) level, and the proportion of HMW multimers is not significantly decreased [2, 4, 5]. Thus, the definition of VWD type 1 is now broader than that proposed in the 1994 classification [1]. Since the first half of 1990, the reduced secretion in type 1 has been thought to be caused by VWF mutations affecting gene expression. However, heterozygotes with VWD type 3 or normal persons with blood type O and a low VWF level have not been distinguishable from VWD type 1 patients with few bleeding episodes or in unrelated family [6]. Hence, the mechanism has been difficult to demonstrate consistently. VWF levels in plasma vary widely and continuously,

TABLE 1. Classification of von Willebrand disease (ISTH/SSC, 2006)

Type	VWF abnormality	VWF-dependent platelet adhesion	FVIII binding	VWF multimers	Heredity	Laboratoy findings (not definite)
1	Quantitatively deficient	Normal	Normal	HMW(+)	AD	VWF:RCo/VWF:Ag ratio >0.7; VWF:RCo,VWF:CB,VWF:Ag <30% or −2 SD
2A	Qualitative defect	Decreased	Normal	HMW(−)	AD (AR)	VWF:RCo (VWF:CB) <30% or −2 SD VWF:RCo/VWF:Ag ratio (mostly) <0.3
2B		Increased	Normal	HMW(−)	AD	Increased RIPA, thrombocytopenia VWF:RCo, VWF:CB <30% or −2 SD VWF:RCo/VWF:Ag ratio (mostly) 0.3–0.7
2M		Decreased	Normal	HMW(+)	AD	VWF:RCo <30% or −2 SD VWF:RCo/VWF:Ag ratio <0.7
2N		Normal	Decreased	HMW(+)	AR	Decreased FVIII:C <50% FVIII:C/VWF:Ag ratio <0.5
3	Completely deficient	~	~	~	AR	VWF:RCo, VWF:CB, VWF:Ag not detected

ISTH/SSC, International Society on Thrombosis and Haemostasis/Scientific & Standardization Committee; HMW, higher-molecular-weight multimers; AD, autosomal dominant; AR, autosomal recessive; VWF, von Willebrand factor; RCo, ristocetin cofactor; Ag, antigen; CB, collagen-binding capacity; RIPA, ristocetin-induced platelet aggregation; FVIII:C, factor VIII coagulant activity

associated with bleeding risks, so thresholds for abnormal values for plasma VWF among laboratories vary from 30 to 50 IU/dl or less than −2 SD below the mean [7–9].

Recent advances in research on VWF have indicated that patients with VWD type 1 can probably be sorted roughly into three groups. One is a group of patients with very low VWF levels in plasma (e.g., 5–20 IU/dl) and who are highly likely to have inheritable disease associated with bleeding symptoms caused by apparently dominant VWF mutations that induce intracellular transportation and secretion of VWF subunits or cause increased clearance and proteolysis of plasma VWF [10–13]. Another is a "suspicious" group of patients with the VWF level at the low end of the population distribution (e.g., 35–50 IU/dl), which expresses low hereditability; these patients rarely exhibit bleeding symptoms and rarely show linkage to the VWF locus. They could be managed as having a biomarker for an increased risk of bleeding instead of with a diagnosis of VWD [7, 14–18]. The others with an intermediate level of VWF (e.g., 20–35 IU/dl) might be classified as having probable VWD; their problem can apparently be diagnosed as VWD but is not often accompanied by severe bleeding symptoms. On the other hand, a recent multicenter survey, the European MCMDM-1 VWD survey of 143 families, showed that linkage of the VWD type 1 phenotype to the VWF gene depended on the severity of VWF deficiency. When the plasma VWF: Ag level was <30 IU/dl in the index case, linkage to the VWF gene was always observed. In contrast, if the plasma VWF:Ag level was >30 IU/dl, the proportion of linkage was markedly reduced. Furthermore, regarding the bleeding symptoms, there was a trend toward increased linkage for subjects with VWF:Ag <30 IU/dl [5, 19]. Thus, the group with VWF levels of 35–50 IU/dl should be well considered to have bleeding symptoms, family trait, and blood type, among other traits. Laboratory findings have indicated that most VWF in the patients with VWD type 1 has normal functional activity relative to the antigen level: a >0.7 ratio of VWF:ristocetin cofactor (RCo)/VWF:Ag [5, 20, 21]. Notably, VWD Vicenza type, with markedly increased clearance of VWF in plasma and with ultra-large VWF multimers, has been classified as the VWD type 1 from type 2M in the 2006 classification of VWD by the ISTH/SSC [2, 22–24].

Type 2

Type 2A

VWD type 2A includes qualitative variants with decreased VWF-dependent platelet adhesion and selective deficiency of HMW VWF multimers (large to intermediate), which is essential to establish linkages between platelets or between platelets and collagen of injured vessels. A significant or relative deficiency of HMW multimers caused by (1) defects in multimer assembly in endothelial cells and megakaryocytes, or (2) intrinsically increased sensitivity (proteolysis) to ADAMTS 13 in plasma. The former includes a variant with mutations in the propeptide of the VWF protomer in the D2 domain (previously named type IIC) [25–27], the CK domain (type IID) [28, 29], or the D3 domain (type IIE) [30, 31]. The latter (2) shows mutations that induce some fragile changes in the A2 domain, containing the VWF-cleaving site (Y1605–M1606) against ADAMTS13 [32]. Most of the type 2A group show autosomal dominant inheritance except those with type IIC. The key laboratory finding is that the

VWF:RCo (or VWF:CB;Collagen binding capacity)/VWF:Ag ratios in plasma are usually <0.3 [20, 33]. On other hand, the FVIII binding capacity (VWF:FVIIIB) does not show any difference between small and intermediate or large multimers [34].

Type 2B

VWD type 2B includes qualitative variants with increased affinity for platelet GPIb. The enhanced interaction of mutant VWF with platelet GPIb in vitro causes markedly increased ristocetin-induced platelet aggregation (RIPA), which can be observed at very low concentrations of ristocetin (0.3–0.7 mg/ml) [35], and in vivo shows spontaneous platelet aggregation in blood vessels and variable thrombocytopenia that can be exacerbated by stress or by desmopressin [36]. In type 2B, most subjects often show decreased to absent large multimers, resulting in a low VWF:RCo (VWF:CB) in plasma [30]. The defect of large multimers in plasma that may be caused by the increased affinity for platelet GPIb induces increased opportunity to be exposed to ADAMTS 13 [37]. The heterozygous mutations clustered within or near the A1 domain in patients change the conformation of A1 domain to have increased affinity for GPIb [38]. Exceptionally, type 2B Malmo and New York with the mutation of Pro1266Leu without thrombocytopenia and bleeding symptoms has been characterized as being associated with increased affinity for ristocetin and increased RIPA, but not increased affinity for GPIb [39].

Type 2M

VWD type 2M includes variants with decreased VWF-dependent platelet adhesion without a selective deficiency of HMW VWF multimers. The assembly and secretion of VWF multimers is almost normal, but in some cases (e.g., the previously named type IC) ultra-large multimers are observed owing to decreased affinity for GPIb, which consequently induces the decreased opportunity to be exposed to ADAMTS 13 [40]. The mutations in the patients have been shown mainly in the A1 domain, and most subjects may have plasma VWF:RCo/VWF:Ag ratios of 0.3–0.7. However, VWF:CB is insensitive to these mutations that impair platelet binding and decreased affinity for GPIb [41]. Recently, one family has been reported to have normal VWF:RCo but low VWF:CB, which is disproportionate to VWF:Ag in plasma, associated with a mutation in A3 domain [42]. This collagen-binding defect has remained uncommon and more data are required.

Type 2N

VWD type 2N includes variants with markedly decreased binding affinity for FVIII (VWF:FVIIIB) caused by homozygous or compound heterozygous VWF mutations in the D′ domain and a part of the D3 domain of the VWF subunit [43, 44]. In some compound heterozygous cases, one allele has an FVIII-binding mutation and the other expresses little or no VWF (type 2N/1' or type 2N/3') [45]. The plasma FVIII level correlates with specific type 2N mutations. One group shows severely impaired FVIII binding with a plasma FVIII level of 8.4 ± 5.2 IU/dl, and another group with a relatively common mutation (Arg854Gln) had a less severe FVIII level with 21.8 ± 9.8 IU/dl. In the FVIII-binding assay for diagnosis, VWF:FVIIIB values are sorted into a severe type

(<0.1) and a mild type (~0.3)[44]. Phenotypically, the patients appear to have mild hemophilia A with decreased FVIII:C, normal VWF activity, and bleeding symptoms. Hence, VWD type 2N must be carefully diagnosed in a female patient with hemophilia A (without compelling evidence for X-linked inheritance) versus VWD type 1 with mild decreased VWF and markedly decreased FVIII [43, 46].

Type 3

VWD type 3 includes virtually complete deficiency of VWF caused by usually nonsense mutations, frame shifts because of small insertion or deletion. Large deletions, splice site mutations, and missense mutations are less common. Type 3 is inherited as an autosomal recessive trait, and heterozygous relatives show mild or no bleeding episodes, with VWF activity at ~50 IU/dl. In most type 3 cases, VWF:Ag, VWF:RCo, and VWF:CB are <5 IU/dl, and FVIII levels are usually <10 IU/dl. When severe type 1 VWD with very low VWF cannot be distinguishable from type 3, a desmopressin (DDAVP) trial test should be undertaken for the diagnosis. Patients with VWD type 3 seldom show a measurable response to DDAVP, whereas in severe type 1 patients some increases are observed after DDAVP infusion.

Compound Heterozygosity

Co-inheritance of VWD type 2N and a nonexpressing or null VWF allele has often been reported to be phenotype with VWD type 1 (decreased VWF:Ag, VWF:RCo) or type 2N (decreased FVIII:C/VWF:Ag ratio). It should be described as "VWD type 2N/3" [45].

Epidemiology

Von Willebrand disease encompasses a wide spectrum of disease severity, ranging mild bleeding symptoms to severe hemorrhagic episodes that are similar to those of severe hemophilia. Because of this variation, the prevalence of VWD depends strongly on the criteria by which patients are identified. Therefore, estimates of VWD prevalence range from $0.56/10^5$ persons in Japan to $9.3/10^5$ in the United Kingdom. The prevalence of the various types of VWD tend to have certain distributions: type 1 60%–80%; type 2 7%–30%; type 3 5%–20%. Furthermore, the prevalence of subtypes may show variable distribution, for example: type 2A 10–15%, type 2B 2–7%, type 2M 1–7%, and type 2N 1–10% [47].

Clinical Manifestations

The most common bleeding symptoms in VWD reflect VWF-dependent platelet adhesion and include mucocutaneous bleeding (e.g., nosebleeds), subcutaneous hematoma, menorrhagia, and so on (Table 2). Life-threatening bleeding may especially occur in patients with VWD type 3 and certain cases of type 2, although most bleeding in patients with VWD is of mild to moderate severity compared with that in hemophiliacs. The symptoms vary among VWD types (including subtypes) in affected

TABLE 2. Bleeding symptoms in von Willebrand disease (%)

Bleeding symptomss	Japanese (n = 154)[a]				Italian (n = 1286)[b]			Scandinavian[b]	
	Type 1 (n = 83)	Type 2 (n = 44)	Type 3 (n = 11)	Total VWD	Type 1 (n = 944)	Type 2 (n = 268)	Type 3 (n = 74)	VWD (n = 264)	Normal (n = 500)
Nosebleed	81.9	70.4	63.6	74.8	56.3	62.6	74.3	62.5	4.6
Subcutaneous hematoma	66.3	77.3	81.8	71.4	14.4	18.6	31.1	49.2	11.8
Menorrhagia	43.7	66.6	100	58.4	30.7	31.8	32.4	60.1	25.3
Postpartum bleeding	14.6	25.9	33.3	23.2	16.6	18.5	26.1	23.3	19.5
Intraoral bleeding	36.1	34.1	54.5	34.5					
Gum bleeding	30.1	36.4	54.5	31.8	30.2	36.7	48.4	34.8	7.4
Bleeding after extraction	25.3	34.1	45.5	26.6	31.1	38.9	52.8	51.5	4.8
Hemarthrosis	7.2	11.4	36.4	10.2	2.4	4.7	41.9	8.3	0
Intestinal bleeding	3.6	18.2	0	7.5	5.1	10.9	17.6	14	0.6
Intracranial hemorrhage	1.2	4.5	0	2.7	0.5	2.3	8.1	—	—
Renal bleeding	1.2	2.2	0	1.4	2.1	3.9	10.9	6.8	0.6
Bleeding after operation	8.4	27.3	36.4	14.9	20.3	23.5	40.6	28	1.4

[a]Japanese Association of Thrombosis and Hemostasis, 2007
[b]Sadler and Rodeghiero [52]

family members and also over time in a single individual. In VWD type 3 and type 2N, the factor VIII level may decrease significantly, as in mild hemophiliacs, so bleeding and hematoma can be observed in joints and muscles, and hematuria may be evident.

The frequency of bleeding symptoms in VWD patients have been reported for Italian and Scandinavian populations [3, 48, 49]; and JTH/SSC also surveyed bleeding symptoms in 154 Japanese patients with VWD (Table 2). Nose bleeds, subcutaneous bleeding, and menorrhagia were the most common symptoms in VWD. Especially, menorrhagia was observed in 100% of VWD type 3 patients in Japan. Subsequently, intraoral bleeding and postoperative bleeding occurred frequently, as much as 4–10 times more often than in normal subjects. Some adult patients were diagnosed as having VWD based on prolonged bleeding after tooth extraction. Intestinal bleeding was relatively infrequent; but serious bleeding, such as gastrointestinal bleeding, may be associated with angiodysplasia [50, 51]. The ISTH/SSC reported guidelines to estimate abnormal bleeding symptoms for VWD type 1, some of which follow [53]:

- *Nose bleeding* more than twice without a history of trauma not stopped by short compression of <10 min, or one episode requiring blood transfusion. Nosebleeds are common during childhood in those with VWD but decline in frequency with increasing age.
- *Cutaneous hemorrhage and bruisability* with minimal or no apparent trauma, as a presenting symptom or requiring medical treatment. Small hematomas or ecchymoses are often observed in the lower extremities.
- *Prolonged bleeding* from trivial wounds, lasting >15 min or recurring spontaneously during the 7 days after the wound occurred.
- *Oral cavity bleeding* that requires medical attention, such as gingival bleeding, or bleeding with tooth eruption or bites to lips and tongue.
- *Prolonged recurrent bleeding after tooth extraction* or other oral surgery (e.g., tonsillectomy and adenoidectomy) requiring medical attention.
- *Menorrhagia* resulting in acute or chronic anemia or requiring medical treatment.

Menorrhagia is a frequent complaint and may be associated with iron deficiency anemia. The prevalence of menorrhagia in VWD is not clear because few studies have used quantitative measures of blood loss and there is bias in patient selection. With these limitations, approximately 60% of women with VWD appear to have menorrhagia (>80 ml of blood loss per one menstrual cycle) compared to 4%–9% of normal controls. Among Japanese women with VWD type 3, 100% of patients showed menorrhagia.

Pregnancy

Pregnancy usually is tolerated well by patients with VWD, although certain hemorrhagic complications are relatively common. The spontaneous miscarriage rate has been reported to be 10%–20% in VWD [48,53]. Delivery in women with VWD may be associated with substantial blood loss in severe-type VWD. In most patients except those with type 3 and certain patients with type 2, the levels of VWF and FVIII rise during pregnancy to >50 IU/dl, so the risk of hemorrhage during labor and delivery does not appear to be increased. After parturition, however, VWF and FVIII fall rapidly to their prepregnancy levels within a few days. Therefore, patients should be monitored for 1 week after delivery. In most patients with VWD type 2B, thrombocytopenia may occur during pregnancy and proceed gradually, and there should be concern about not having misdiagnosed the problem as autoimmune thrombocytopenia.

Laboratory Findings and Diagnosis of VWD

Laboratory tests relevant to the diagnosis of VWD can be grouped into screening tests that are part of the initial evaluation of any bleeding patient, specific tests that are necessary to confirm the diagnosis of VWD, and discriminating tests that are necessary to distinguish among the subtypes of VWD.

Screening Tests

Bleeding Time

The bleeding time (BT) had been essential test for diagnosis of bleeding disorders. However, there may not be apparent evidence that BT increases concordantly with VWF activity, especially with mild VWD. For example, some patients show a normal BT even if they have around 10%–20% of VWF:RCo. However, in the severe type of VWD (type 3 or some type 2) markedly prolonged values of BT usually are observed.

Coagulation Tests

Because plasma FVIII levels generally depend on the VWF:Ag levels, the activated partial thromboplastin time (APTT) may be prolonged in VWD patients. Especially with VWD types 3 and 2N, the plasma FVIII level is constantly decreased, and the APTT is markedly prolonged. However, most patients with mild VWD have normal or near-normal levels of FVIII and APTT.

Platelet Count

The platelet count may be decreased in patients with VWD type 2B, particularly in association with surgery or pregnancy. For the other VWD types, the platelet count is not affected directly.

Specific Tests

VWF:Ag, VWF:RCo, VWF:CB, FVIII:C

The distinction between primary categories of VWD can usually be made by measuring VWF:Ag, VWF:RCo (or VWF:CB), and FVIII:C. Concordant decreases in all levels suggest VWD type 1, disproportionate decreases in VWF:RCo (or VWF:CB), VWF:Ag, and FVIII suggest a form of VWD type 2. Virtual absence of VWF:Ag suggests VWD type 3.

TABLE 3. Plasma VWF:Ag levels in various blood groups

Blood type	VWF:Ag (IU/dl)	
	Geometric mean ± 2 SD[a]	Mean ± 2 SD[b]
OO	35.6–157.0 (74.8)	80.9 ± 28.9
AA	48.0–233.9 (105.9)	113.3 ± 46.6
AO		103.7 ± 33.8
BB	56.8–241.0 (116.9)	114.5 ± 37.6
BO		100.5 ± 28.8
AB	63.8–238.2 (123.3)	113.8 ± 34.4

[a]Gill et al., 1987 [6]
[b]Shima et al., 1995 [54]

Blood Type

Both plasma VWF and FVIII levels are dependent on the ABO blood type in normal subjects (Table 3). Blood type O subjects have significantly lower VWF and FVIII levels than subjects with other blood types (average <20–30 IU/dl) [6, 54]. For this reason, some laboratory employ two blood type-specific normal ranges: one for blood type O and one for non-type-O patients.

VWD Type 1

Patients with a proportional decrease of VWF:Ag and VWF:RCo (or VWF:CB) <30 IU/dl could be easily diagnosed as VWD type 1. Patients with values at 30–50 IU/dl or under ~2 SD to 30 IU/dl may be considered borderline for the diagnosis. For a diagnosis of VWD in these patients, they would be required to have inherited VWD, a non-O blood type, or significant bleeding episodes relevant to VWD. The guidelines by the ISTH/SSC recommended a VWF value of <~2 SD of the mean and significant bleeding symptoms for the diagnosis [52]. Normal and abnormal VWF function could be distinguished by comparisons of VWF activity and VWF:Ag. For example, the VWF:RCo/VWF:Ag ratio has been defined as >0.7 to distinguish between VWD types 1 and 2 [5, 20, 21]. Additionally, there might be some difficulty to discriminate normal but low-activity VWF subjects from those with VWD type 1. One report suggested that most patients with type 1 have an FVIII:C/VWF:Ag ratio >1.6, but normal subjects usually show a concordant decrease in plasma FVIII:C and VWF:Ag [55]. On other hand, the FVIII:C/VWF:Ag ratio should be >0.5 to distinguish VWD type 1 from type 2N and hemophilia A.

VWD Type 2

Decreased VWF:RCo or VWF:CB levels (mostly <30 IU/dl) and a disproportionate decrease between VWF:RCo (VWF:CB) and VWF:Ag suggests VWD type 2A when the VWF:RCo (VWF:CB)/VWF:Ag ratio is <0.3. Most subjects with VWD type 2B or 2M have a VWF:RCo/VWF:Ag ratio of 0.3 to <0.7. Especially, patients with type 2M have normal VWF:CB, although recently a family case with decreased affinity of VWF for collagen was reported in France [42]. Most subjects with VWD type 2N have markedly decreased FVIII and the FVIII:C/VWF:Ag ratio is usually <0.5 with normal or mildly decreased VWF. Further discrimination among type 2 variants often requires special tests: VWF multimers analysis for types 2A and 2M; RIPA for type 2B; and the FVIII-binding assay for type 2N.

Discrimination Tests

Ristocetin-Induced Platelet Aggregation

The RIPA test depends on both the concentration of VWF and the affinity of VWF for platelet GPIb. Increased affinity of VWF for GPIb can be estimated with RIPA for diagnosis of VWD type 2B; RIPA may be observed at a very low concentration of ristocetin (f.c. 0.3–0.7 mg/ml). In normal subjects, RIPA can usually be seen at f.c. 1.0–1.5 mg/ml and not be observed under f.c. 1.0 mg/ml. In contrast, most subjects from VWD type 1 and type 2M show decreased RIPA, and those from VWD type 3 and most VWD type 2A show completely deficient RIPA at f.c. 1.0–1.5 mg/ml of Ristocetine.

VWF Multimeric Analysis

The presence or absence of HMW multimers can first be judged conveniently in gels of 1.0%–1.5% agarose to distinguish between VWD types 2A, 2B, 2M, and 1. In the next step, VWF multimer distribution and satellite bands have been evaluated quantitatively by densitometric scanning to obtain information about multimer assembly (previously IIA, IIC, IID, IIE, and so on). However, the most important point is to distinguish normal or only subtly abnormal multimer distributions from a significant decrease in HMW multimers. Among VWD type 1 subjects, most have nearly normal VWF multimers, although some have ultra-large VWF multimers, such as the Vicenza type. In type 2A, an apparent defect of HMW multimers is observed with subtle abnormalities in the satellite bands caused by assembly defects (previously, IIC, IID, IIE) and accelerated proteolysis by ADAMTS13 (previously type IIA). Most patients with type 2B may have decreased or absent large multimers caused by accelerated GPIb binding that induces more exposure of VWF to ADAMTS13 and consequently increased proteolysis. On the other hand, most patients with VWD type 2M show normal VWF multimers, although some have ultra-large VWF multimers caused by decreased affinity for GPIb that induces less exposure of VWF to ADAMTS13 and decreased proteolysis. Patients with VWD type 2N show normal VWF multimers or an almost normal distribution of multimers with subtle abnormal satellite bands [43].

FVIII-Binding Assay

The affinity of VWF for FVIII from a small volume of plasma can be measured with a solid-phase FVIII-binding assay using monoclonal antibody to VWF and purified FVIII [43]. Most of VWD type 2N show VWF:FVIIIB titer to be under 0.5 with the FVIII binding assay [62]. Subjects with type 2N may be divided into two groups: those with the severe type (VWF:FVIIIB <0.1) and those with the mild type (VWF:FVIIIB ~0.3). The VWF:FVIIIB titer of carriers of type 2N usually varies from 0.5 to 1.0.

Diagnostic Chart for VWD (Nara)

In our laboratory, when patients have significant bleeding episodes or hereditary traits, the diagnostic chart may be available at the starting point of: VWF:RCo or VWF:Ag <30 IU/dl for VWD. A value of 30–50 IU/dl is considered as possible VWD when the patient's family has VWD traits. A decreased FVIII:C level is suspicious for VWD type 2N if the FVIII:C/VWF:Ag ratio is <0.5 without any hemophilia A traits in the family (Fig. 1).

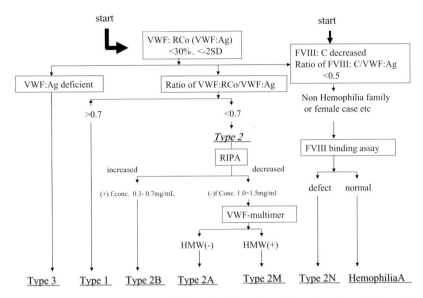

FIG. 1. Diagnostic chart for von Willebrand's disease (VWD) employed at the Nara Medical College, starting with VWF:RCo or VWF:Ag <30 IU/dl and FVIII:C <50 IU/dl. *RIPA*, ristocetin-induced platelet aggregation; *HMW*, high-molecular-weight von Willebrand factor (*VWF*) multimers; *RCo*, ristocetin cofactor; *Ag*, antigen; *FVIII:C*, factor VIII coagulant activity

Treatment of VWD

The two main therapeutic modalities used to treat VWD are DDAVP and transfusion with blood or recombinant products that contain factor VIII, factor VIII/VWF complex, or VWF. In Europe, DDAVP has been employed for hemostatic treatment of VWD type 1, whereas in Japan FVIII/VWF has been considered more effective in all types of VWD, as surveyed by the JTH. Other treatment includes antifibrinolytic drugs and hormonal therapy.

Desmopressin

Desmopressin (DDAVP), a synthetic analogue of vasopressin, increases factor VIII and VWF in plasma by releasing stored FVIII and VWF from liver cells and endothelial cells (or platelets), respectively. DDAVP given intravenously at a dose of 0.3 µg/kg usually increases these substances three- to sixfold above the basal levels. Patients with VWD type 1 have a high probability of responding to DDAVP, whereas those with VWD type 2 or 3 usually do not respond. Administration of DDAVP to patients with VWD type 2B is contraindicated because of worsened thrombocytopenia [56, 57]. In VWD type 1, the response to DDAVP correlates with the initial VWF:Ag level, so patients with VWF:Ag <10 IU/dl often do not have a useful increment in VWF and FVIII levels [58, 59]. However, in type 1 (Vicenza type) VWF increases markedly within a very short period [23, 60]. Patients with VWD type 2N with a mild defect in the FVIII-binding capacity often have a good response to DDAVP within a short time, whereas those with markedly decreased FVIII-binding capacity do not [61, 62].

A test dose of DDAVP at the time of diagnosis may be useful to facilitate the classification of some types of VWD and also to confirm the hemostatic effects beforehand. Additionally, infusion can be repeated every 12–24h, but the response to the increment level may gradually decrease.

The side effects of DDAVP include mild tachycardia, headache, and facial flushing. Hyponatremia and volume overload due to the antidiuretic effects of DDAVP are rare and occur mostly in young children after multiple doses. However, DDAVP should be used with caution in the elderly and patients with cardiac disorders. Additionally, DDAVP releases not only FVIII and VWF but also tissue plasmin activator (t-PA), which induces relatively fibrinolysis. Hence, antifibrinolytic agents may be given as adjunt to DDAVP. In the case of renal bleeding, DDAVP and antifibrinolytics should not be employed at the same time so as to prevent thrombosis in the kidney and anuria [59].

Replacement Therapy

FVIII/VWF Concentrate

Most FVIII/VWF concentrates contain HMW multimers of VWF and FVIII in a VWF:RCo/FVIII:C ratio of 2.0–2.5:1.0. Thus, all FVIII/VWF concentrates tend to correct the FVIII and VWF activity deficiencies. When a FVIII/VWF concentrate at a VWF:RCo concentration of 1.0 IU/kg is infused intravenously, the VWF:RCo in plasma increases 1.5–2.0 IU/dl in 30 min and decreases gradually with a half-life of 14–16 h. In patients with VWD type 3, the half-life in plasma may be shortened to 5–6 h at the first replacement therapy; and in the case of severe bleeding, the half-life may also be shortened unexpectedly in all types of VWD because of severe consumption of the VWF. The average loading dose of FVIII/VWF concentrate ranges from 40 to 80 IU of VWF:RCo per kilogram, and the dose and frequency correspond to the bleeding symptoms. The average dose of FVIII/VWF concentrates with VWF:RCo in our laboratory is shown for hemostatic management for VWD (Table 4).

Regarding the side effects, the viral inactivation/elimination in those plasma-derived concentrates could be well performed with several manufacturing steps [63]. Another concern with these products is that frequent infusions of FVIII/VWF may induce very high levels of FVIII in plasma, especially in VWD patient with capably normal production of FVIII. High exogenous and endogenous plasma FVIII levels of >400 IU/dl have been identified as a risk factor for venous thromboembolism (VTE). The mechanism and frequency on the thrombosis in VWD patients has not been clarified yet [64, 65].

FVIII Concentrates

In patients with VWD type 2N, the plasma FVIII levels may be significantly decreased but the VWF:RCo, VWF:Ag, and VWF:CB levels are normal. Therefore, FVIII recombinant products may be employed for short-time elevation of FVIII in plasma [62].

VWF Concentrates

Factor VIII levels in the plasma of patients with VWD are often decreased secondarily, caused by a defect or a decrease in VWF as the carrier protein for FVIII. Therefore,

TABLE 4. Guidelines for hemostatic management with FVIII/VWF concentrate for VWD (Nara)

Bleeding symptoms		Target (maintain level) VWF:RCo (U/dl)	Dose of VWF: RCo (U/kg)	No. of infusions/ day	Period (days)
Subcutaneous hematoma(severe)		50–100	60–80	2	Several
Nose bleeding	Moderate	30–50	40–60	1–2	1
	Severe	50–100	60–80	2	Several
Menorrhagia		50–100	60–80	2	Several
Pregnancy	Delivery	50–100	60–80	2	2
	Postpartum	30–50	40–60	1–2	4–5
Tooth extraction	At operation	50–100	60–80	2	1–2
	Post operation	30–50	40–60	1–2	7
Operation	At operation to ~2nd day	>100	100	2	2–3
	~7th Day	50–100	60–80	2	5–7
	~14th Day	30–50	40–60	1–2	7

From Nishino [47], with permission

replacement of normal VWF without FVIII may induce an elevation in the plasma FVIII level, which would be synthesized gradually within at least 5–6h. VWF product may have a benefit in regard to preventing thrombosis (VTE) due to a marked increase in FVIII after frequent infusion of FVIII/VWF concentrates to VWD patients [66, 67]. Unfortunately, this product is available only in Europe and is not available in Japan as of this writing.

Treatment for Women's Bleeding and Pregnancy

Menorrhagia in VWD patients may be treated successfully with hormonal therapy instead of, or with, DDAVP and FVIII/VWF concentrate. During pregnancy, VWF and FVIII levels tend to rise in type 1 and some type 2 cases, but no increase of VWF:RCo is observed in type 3 and some type 2 cases. Because the increase in VWF and FVIII levels is variable, patients should be monitored during pregnancy. When the VWF: RCo and FVIII:C levels increase to >50IU/dl (50%) at delivery in patients with type 1, no replacement therapy may be required at delivery with careful observation [53]. After delivery, the increased levels fall rapidly to baseline levels within 1–2 days and may cause late bleeding. In patients with VWD type 2 or 3, FVIII/VWF concentrates should be employed at delivery and for 4–5 days during the postpartum period [68, 69]. When the VWF levels do not increase enough (<50IU/dl), DDAVP or FVIII/VWF concentrate can be used in VWD type 1, but a test dose before pregnancy should be required. Most patients with VWD type 2 can be treated with FVIII/VWF concentrate at delivery and postpartum. Furthermore, in patients with VWD type 2B, the platelet count may be decreased gradually during pregnancy and sometimes to the minimum (to around $10 \times 10^3 \, \mu l$, so FVIII/VWF concentrate (never DDAVP) may be used at delivery and postpartum.

Prognosis

There have been no reports that mortality among patients with VWD was higher or lower than that of the normal population. However, with VWD type 3, intracranial bleeding and severe intestinal bleeding are life-threatening symptoms. On the other hand, patients with VWD unresponsive to DDAVP have been treated with plasma-derived FVIII/VWF concentrates, which has resulted in the patient having an unnecessarily high level of FVIII in the plasma. Thus, a small number of high-risk patients receiving repeated but limited infusions, particularly for major surgery, have had thrombotic complications such as VTE. Those cases would be analyzed worldwide, considering changes of ADAMTS13 values in plasma.

Note This work has been continuously discussed in the ISTH/SSC VWF Subcommittee and the Working Group for VWD Classification (Chairman: J.E. Sadler) since 1993.

References

1. Sadler JE (1994) A revised classification of von Willebrand disease. Thromb Haemost 71:520–525
2. Sadler JE, Budde U, Eikenboom JCJ, et al (2006) The working party on von Willebrand disease classification: update on the pathophysiology and classification of VWD; a report of the Subcommittee on VWF. J Thromb Haemost 4:2103–2114
3. Sadler JE, Mannucci PM, Berntorp E, et al (2000) Impact, diagnosis and treatment of von Willebrand disease. Thromb Haemost 84:160–174
4. James PD, Paterson AD, Notley C, et al (2006) Genetic linkage and association analysis in type 1 VWD; result from the Canadian type 1 VWD study. J Thromb Haemost 4:783–792
5. Eikenboom JC, Van Marion V, Putter H, et al (2006) Linkage analysis in families diagnosed with type 1 VWD in the European study, molecular and clinical markers for the diagnosis and management of type 1 VWD. J Thromb Haemost 4:774–782
6. Gill JC, Endre-Brooks J, Bauer PJ, et al (1987) The effect of ABO blood group on the diagnosis of VWD. Blood 69:1691–1695
7. Sadler JE (2003) VWD type 1; diagnosis in search of a disease. Blood 101:2089–2093
8. Rodeghiero F, Castaman G (2001) Congenital VWD type 1 definition, phenotype, clinical and laboratory assessment. Best Pract Res Clin Haematol 14:321–335
9. Sadler JE (2004) Slippery criteria for VWD type 1. J Thromb Haemost 2:1720–1723
10. Eikenboom JCJ, Matsushita T, Reitsma PH, et al (1996) Dominant type 1 VWD caused by mutated cysteine residues in the D3 domain of VWF. Blood 88:2433–2441
11. Bodo I, Katsumi A, Tulley EA, et al (2001) Type 1 VWD mutation Cys1149Arg causes intracellular retention and degradation of heterodimers; a possible general mechanism for dominant mutations of oligomeric proteins. Blood 98:2973–2979
12. Tjernberg P, Vos HL, Castaman G, et al (2004) Dimerization and multimerization defects of VWF due to mutated cysteine residue. J Thromb Haemost 2:257–265
13. Lethagen S, Isaksson C, Schaedel C, et al (2002) VWD caused by compound heterozygosity for a substitution mutation (T1156M) in the D3 domain of VWF and a stop mutation (Q2470X). Thromb Haemost 88:421–426
14. Souto JC, Almasy L, Borrell M, et al (2000) Genetic determinants of hemostasis phenotypes in Spanish families. Circulation 101:1546–1551
15. Vossen CY, Hasstedt SJ, Rosendaal FR, et al (2004) Heritability of plasma concentrations of clotting factor and measures of a prethrombotic state in a protein C deficient family. J Thromb Haemost 2:242–247

16. Miller CH, Graham JB, Goldin LR, et al (1979) Genetics of classic VWD; phenotypic variations within families. Blood 54:117–136
17. Castaman G, Eikenboom JC, Bertina RM, et al (1999) Inconsistency of association between type 1 VWD phenotype and genotype in families identified in an epidemiological investigation. Thromb Haemost 82:1065–1070
18. Casafia P, Martinez F, Haya S, et al (2001) Significant linkage and non-linkage of type 1 VWD to the VWF gene. Br J Hematol 115:692–700
19. Tosetto A, Rodeghiero F, Castaman G, et al (2007) Impact of plasma VWF levels in the diagnosis of type 1 VWD: results from a multicenter European study (MCMDM-1VWD). Thromb Haemost 5:715–721
20. Federici AB, Canciani MT, Forza I, et al (2000) Ristocetin cofactor and collagen binding activities normalized to antigen levels for a rapid diagnosis of type 2 VWD single center comparison of four different assays. Thromb Haemost 84:1127–1128
21. Laffan M, Brown SA, Collins PW, et al (2004) The diagnosis VWD; a guideline from the UK haemophilia center doctor's organization. Hemophilia 10:199–217
22. Holmberg L, Dent JA, Schneppenheim R, et al (1993) VWF mutation enhancing interaction with platelets in patients with normal multimeric structure. J Clin Invest 91:2169–2177
23. Mannucci PM, Lombardi R, Castaman G, et al (1988) VWD "Vicenza" with lager-than-normal (supranormal) von Willebrand factor multimers. Blood 71:65–70
24. Casonato A, Pontara E, Sartorello F, et al (2002) VWF survival in type Vicenza VWD. Blood 99:180–184
25. Ruggeri ZM, Nilsson IM, Lombardi R, et al (1982) Aberrant multimeric structure of VWF in a new variant of VWD (type IIC) J Clin Invest 70:1124–1127
26. Gaucher C, Dieval J, Mazurier C (1994) Characterization of VWF gene defects in two unrelated patients with type IIC VWD. Blood 84:1024–1030
27. Schneppenheim R, Thomas KB, Krey S, et al (1995) Identification of a candidate missense mutation in a family with VWD type IIC. Hum Genet 95:681–686
28. Schneppenheim R, Brassard J, Krey S (1996) Defective dimerization of VWF subunits due to a Cys~Arg mutation in type IID VWD. Proc Natl Acad Sci U S A 93:3581–3586
29. Kinoshita S, Harrison J, Lazerson J, et al (1984) A new variant of dominant type II VWD with aberrant multimeric pattern of FVIII-related antigen (type IID). Blood 63:1369–1371
30. Zimmerman TS, Dent JA, Ruggeri ZM, et al (1986) Subunit composition of plasma VWF: cleavage is present in normal individuals, increased in type IIA and IIB VWD, but minimal in variants with aberrant structure of individual oligomers (IIC, IID, and IIE). J Clin Invest 77:947–951
31. Schneppenheim R, Budde U, Ruggerri ZM (2001) A molecular approach to the classification of VWD. Best Pract Res Clin Haematol 14:281–298
32. Furlan M, Robles R, Lammle B (1996) Partial purification and characterization of a protease from human plasma cleaving VWF fragments produced by in vivo proteolysis. Blood 87:4223–4234
33. Casonato A, Pontara E, Bertomoro A, et al (2001) VWF collagen binding activity in the diagnosis of VWD; an alternative to ristocetin cofactor activity? Br J Haematol 112:578–583
34. Nishino M, Miura S, Yamamoto K, et al (1991) Factor VIII binding ability of VWF in several fractions of normal plasma and plasma from several types of VWD. Jpn J Thromb Hemost 2:152–158
35. Ruggeri ZM, Pareti FI, Mannucci PM, et al (1980) Heightened interaction between platelets and factor VIII/VWF in a new subtype of VWD. N Engl J Med 302:1047–1051
36. Holmberg L, Nilsson IM, Borge L, et al (1983) Platelet aggregation induced by DDAVP in type IIB VWD. N Engl J Med 309:816–821

37. Lankhof H, Damas C, Schiphorst ME, et al (1997) Functional studies on platelet adhesion with recombinant VWF type 2B mutants R543Q and R543W under conditions of flow. Blood 89:2766–2772

38. Huizinga EG, Tsuji S, Romijn RAP, et al (2002) Structure of glycoprotein Ibα and its complex with VWF A1 domain. Science 297:1176–1179

39. Holmberg L, Dent JA, Schneppenheim R, et al (1993) VWF mutation enhancing interaction with platelets in patients with normal multimeric structure. J Clin Invest 91:2169–2177

40. Ciavarella G, Ciavarella N, Antoncecchi S, et al (1985) High resolution of VWF multimeric composition defines a new variant of type 1 VWD with aberrant structure but presence of all size multimers (type IC) Blood 66:1423–1429

41. Rabinowiz I, Tuley EA, Mancuso DJ, et al (1992) VWD type B, a missense mutation selectively abolishes ristocetin induced VWF binding to platelet GPIb. Proc Natl Acad Sci U S A 89:9846–9849

42. Ribba AS, Loisel I, Lavergne JM, et al (2001) Ser968Thr mutation within A3 domain of VWF in two related patients leads to a defective binding of VWF to collagen. Thromb Haemost 86:848–854

43. Nishino M, Girma JP, Rothchild C, et al (1959) New variant of VWD with defective binding to FVIII. Blood 74:1951–1959

44. Mazurier C, Meyer D (1996) Factor VIII binding assay of VWF and the diagnosis of type 2N VWD: results of an international survey. Thromb Haemost 76:270–274

45. Eikenboom JC, Reistima P, Peerlinck KM, et al (1993) Recessive inheritance of VWD type 1. Lancet 341:982–986

46. Nishino M, Miura S, Yoshioka A, et al (1993) Variant VWD with defective binding to FVIII; the first case from Japan. Int J Hematol 57:163–173

47. Nishino M (1999) Consensus diagnosis and treatment of VWD. Jpn J Pediatr Hematol 13:410–420

48. Silwer J (1973) VWD in Sweden. Acta Pediatr Scand 238:1–159

49. Lak M, Peyvandi F, Mannucci PM (2000) Prevalence of bleeding manifestations hepatitis and alloantibodies to VWF in 348 Iranian patients with type 3 VWD. Br J Haematol 111:1236–1239

50. Ramsay DM, Buist TA, Macleod DA, et al (1976) Persistent gastrointestinal bleeding due to angiodysplasia of the gut in VWD. Lancet 2:275–278

51. Fressinaud E (1993) International survey of patients with VWD and angiodysplasia. Thromb Haemost 70:546

52. Sadler JE, Rodeghiero F (2005) Provisional criteria for diagnosis of type 1. J Thromb Haemost 3:775–777

53. Kadir RA, Lee CA, Sabin CA, et al (1998) Pregnancy in women with VWD or factor IX deficiency. Br J Obstet Gynaecol 105:314–321

54. Shima M, Fujimura Y, Nishiyama T, et al (1995) ABO blood group genotypes and plasma VWF in normal individuals. Vox Sang 68:236–240

55. Eikenboom JC, Castaman G, Kamphuisen PW, et al (2002) The FVIII/VWF ratio discriminates between reduced synthesis and increased clearance of VWF. Thromb Haemost 87:252–257

56. Ruggerri ZM, Mannucci PM, Lombardi R, et al (1982) Multimeric composition of FVIII/VWF following administration of DDAVP; implications for pathophysiology and therapy of VWD subtypes. Blood 59:1272–1278

57. Mannucci PM (2001) How I treat patients with VWD. Blood 97:1915–1919

58. Federici AB, Mazurier C, Berntorp E, et al (2004) Biologic response to DDAVP in patients with severe type 1 and type 2 VWD; results of a multicenter European study. Blood 103:2032–2038

59. Mannucci PM (2004) Treatment of VWD. N Engl J Med 351:683–694

60. Rodeghiero F, Castaman G, Di Bona E, et al (1988) Hyperresponsiveness to DDAVP for patients with type 1 VWD and normal intra platelet VWF. Eur J Haematol 40:163–167
61. Mazurier C, Gaucher C, Jorieux S, et al (1994) Biological effect of DDAVP in eight patients with type 2N VWD. Br J Haematol 88:849–854
62. Nishino M, Nishino S, Sugimoto M, et al (1996) Changes in FVIII binding capacity of VWF and FVIII:C in two patients with type 2N VWD after hemostatic treatment and during pregnancy. Int J Hematol 64:127–134
63. Mannucci PM, Chediak J, Hanna W, et al (2002) Treatment of VWD with a high purity FVIII/VWF concentrate: a new prospective multicenter study. Blood 99:450–456
64. Martineli I (2005) VWF and FVIII as risk factors for arterial and venous thrombosis. Semin Hematol 42:49–55
65. Mannucci PM (2002) Venous thromboembolism in VWD. Thromb Haemost 88:378–379
66. Konkle BA (2007) VWD: treatment with or without FVIII. J Thromb Haemost 5:1113–1114
67. Borel-Deron A, Fererici AB, Roussel-Robert V, et al (2007) Treatment of severe VWD with a high purity VWF concentrate: a prospective study of 50 patients. J Thromb Haemost 5:1115–1124
68. Boklage CE (1990) Survival probability of human conception from fertilization to term. Int J Fertil 35:75–94
69. Ramsahye BH, Davies SV, Dasani H, et al (1995) Obstetric management in VWD; a report of 24 pregnancies and a review of the literature. Hemophilia 1:140–144

Part 7 Other Topics of Thrombosis and Hemostasis

Age-Related Homeostasis and Hemostatic System

Kotoku Kurachi, Sumiko Kurachi, Toshiyuki Hamada, Emi Suenaga, Tatiana Bolotova, and Elena Solovieva

Summary. Aging is one of the inevitable aspects of life, affecting homeostatic states of every physiological system including that of hemostasis. Through systematic transgenic studies of age-related dynamic changes in expression of genes for hemostatic factors, we found the very first molecular mechanism of age-related homeostasis, the ASE/AIE-mediated genetic mechanism for age-related regulation of gene expression. In this mechanism, together with other essential elements for the promoter activity, two genetic elements—age-related stability element (ASE) and age-related increase element (AIE)—play critical roles, producing four age-related gene expression patterns. This mechanism is also found to be the first puberty-onset gene switch mechanism identified. This ASE/AIE-mediated regulatory mechanism has universal functionality across different genes and animal species. In addition to the age-related regulatory activity, ASE also has a unique tissue-specificity regulatory activity. Global analyses of lifetime age-related expression profiles of mouse liver genes and proteins were carried out, and the results support that there exists a small number of fundamental age-related regulatory mechanisms of genes and proteins that govern complex age-related homeostatic regulations. A large body of information obtained from the global analyses has been constructed into a versatile database, which is a platform resource for studying age-related homeostasis.

Key words. Age-related homeostasis · Aging · Molecular mechanisms · Age-dimension technology · Blood coagulation

Age-Related Homeostasis and Physiological Systems

Aging is the essential aspect of life, and aging itself is an undisputable risk factor for many diseases such as thrombosis/cardiovascular diseases, diabetes, cerebral vascular disease, geriatric diseases, cancers, and Alzheimer's disease. Despite its importance, literally nothing was known for a long time about how aging affects normal physiological homeostasis [1–3]. It was only recently that a possible existence of any molecular mechanism was investigated, resulting in the discovery of the first molecular mechanism of age-related homeostasis. This delay was due to the abstractive nature of studying age combined with the difficulty of setting up durable assay systems.

427

All physiological systems are under the control of homeostatic mechanisms and maintaining them within tolerable fluctuation ranges caused by various internal and external stresses, insults, and pathological challenges. If changes caused are outside of recoverable threshold ranges, a healthy state fails to be restored, and in some cases the body is rendered into a diseased state.

A healthy body with normally functioning physiological systems gradually changes along the age axis throughout the organism's lifespan. This process, aging, may synergistically play a critical role together with environmental conditions, affecting developmental processes of many age-associated diseases.

It is therefore of critical importance to understand precisely age-related homeostasis and its molecular/atomic mechanisms. This is a fledgling field of research, and its pioneering studies began during the mid-1990s. This new field of research has great potential for expanded research, providing novel targets for drug development with regard to various age-related diseases and other clinical applications.

Hemostatic System and Age-Related Homeostasis

We originally explored the little understood age-related homeostasis at the molecular level by taking the hemostatic system as our physiological model system for intensive study.

Activation of the blood coagulation cascade (Fig. 1) is triggered by exposure of tissue factor due to tissue injury, which through making a complex with circulating factor VII (FVII), induces activation of FVII to FVIIa. In turn, FVIIa triggers activation of extrinsic and intrinsic pathways of blood coagulation, eventually causing an avalanche of blood coagulation reactions involving nearly 20 procoagulant and anticoagulant factors. These cascade reactions are finely regulated by specific known (and yet unidentified) inhibitor/regulatory systems, positive and negative feedback reactions, resulting in a necessary and appropriate amount of stable fibrin clot at the required place within a matter of minutes. Most cascade steps of the blood coagulation system are of proteolytic reactions. In addition, procoagulant factors have higher plasma concentrations as their positions get lower and closer to the fibrin formation end step in the cascade, thus enabling efficient exponential blood clot formation. With all these reactions, the blood coagulation system functions efficiently.

How does the overall age-related homeostatic control of this system work? The blood coagulation activity in normal individuals makes a rapid increase during the perinatal stage and, at around the time of weaning, reaches a level similar to that of young adults, followed by gradual and continuous elevations with age [1–3]. The plasma concentration and/or activity of most known procoagulant factors (Fig. 1, arrows) increases with age, whereas anticoagulant factors such as antithrombin (ATIII), tissue factor pathway inhibitor (TFPI), and protein C (PC) and profibrinolytic factors including plasminogen and tissue plasminogen activator (t-PA) do not significantly increase with age, and some even show slight decreases with age. These epidemiological signals support the fact that the imbalance between procoagulant activity and anticoagulant and fibrinolytic activity increases in an age-dependent manner. Interestingly, plasminogen activator inhibitor-1 (PAI-1) increases its plasma concentration with aging in a manner similar to that found for procoagulant factors, resulting

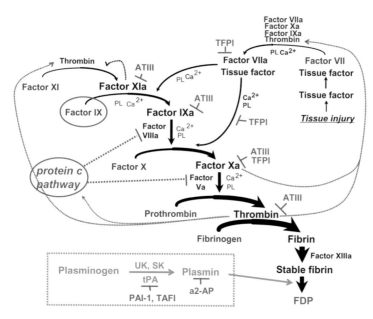

FIG. 1. Simplified basic cascades for blood coagulation and fibrinolysis. Blood coagulation and fibrinolysis cascades are shown with only the known major factors without the cellular information involved. Procoagulation factors are depicted with *connecting arrows*, and anticoagulation factors are shown with *T symbols* indicating the steps they inhibit. Fibrinolytic pathway factors are shown in the *dotted line box. Dotted thin lines with arrowheads* indicate positive feedback reactions. Factor IX and protein C (*circled*) were intensively investigated

in enhancement of the procoagulant tendency of the hemostatic system. As age advances, the overall intrinsic balance between procoagulant and anticoagulation activities is tipped toward procoagulant activity. This balance state may be easily affected by environmental (epigenetic) conditions such as consumed foods and drugs, in some cases enhancing the imbalance toward a more procoagulant state, thus augmenting the thrombogenic tendency, particularly in the elderly [4–7].

Molecular Mechanisms of Age-Related Homeostasis

Assay Methods

In the early quest for finding possible molecular mechanisms of age-related homeostasis, we first focused our efforts on factor IX (FIX) for establishing the optimal assay methods, and we completed its age-related gene expression profiles. The human FIX circulation level is known to increase significantly along the age axis [3, 8–10] (Fig. 2). We then applied similar analysis methods to establish the age-related expression profile for the human PC (hPC) gene. Unlike the hFIX gene, the gene for hPC, a factor in the potent anticoagulant PC pathway, shows an age-stable pattern with marginal age-associated fluctuations [3, 9, 11] (Fig. 2). Therefore, in our studies, hFIX and hPC

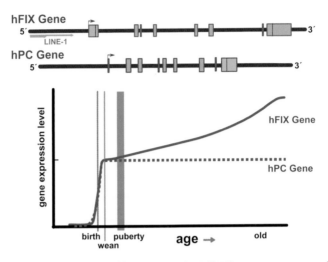

Fig. 2. Human factor IX (*hFIX*) and human protein C (*hPC*) gene structures and their age-related expression patterns. At the *top*, the hFIX gene is represented by *horizontal lines* with eight exons shown by *rectangles* [12]. Its age-related increase expression pattern is shown in the lower panel by a *thick line*. Relative positions of LINE-1 (retrotransposable element)-derived sequences in the 5′ upstream are shown with *overlapped horizontal arrows*. The hPC gene is composed of nine exons [13, 14]. Unlike the hFIX gene, this gene does not have LINE-1-derived sequences in its 5′ upstream. Its age-related expression shows a relatively stable age-related expression pattern as shown by the *thick dotted line* in the graph at the *bottom*

properly represented two types of age-related regulatory profile: age-related increase and age-related stable patterns.

For determining the molecular mechanisms underlying age-related homeostasis, age-axis longitudinal analysis of hFIX or hPC expression in transgenic mice carrying mini-gene vectors of hFIX gene or hPC gene was found to be the most suitable in vivo assay system (Fig. 3). A series of mini-gene expression vectors of hFIX and hPC genes, which are of approximately 40 and 13 kb pairs, respectively [12–14], were constructed. Importantly, these mini-gene vectors must be composed only of autologous components of the test gene, not of any heterologous reporter gene or parts of unrelated genes, which have been commonly used for gene expression analyses. These mini-genes were composed of variously truncated 5′ flanking sequences, first exon, first intron, a region spanning from the second exon through the last exon (taken from the corresponding cDNAs), and the 3′ flanking sequence derived from the corresponding gene [15–17].

After their expression activities were confirmed by assaying with HepG2 cells, mini-genes were subjected to construction of transgenic mice. Approximately 100 μl of blood samples were collected from snipped tail tips as scheduled for various age points, and circulating (serum) levels of hFIX or hPC produced in these transgenic mice were individually monitored for their entire lifespan (at least 2 years) by using hFIX- or hPC-specific enzyme-linked immunosorbent assay (ELISA). Importantly, mice have a blood coagulation system similar to that of humans, and age-related

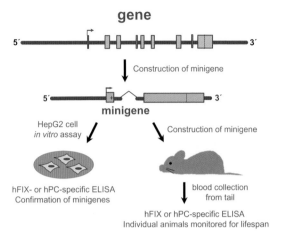

gene

minigene

FIG. 3. Methods used for analysis of age-related homeostatic mechanisms of hFIX and hPC. Mini-genes used for analyses are composed of only autologous components of either the hFIX gene or the hPC gene. Mini-genes are first tested with HepG2 cells to verify their correct construction and basic expression activities, after which they were used for construction of transgenic mice. Individual animals were then subjected to longitudinal monitoring of transgene expression by an enzyme-linked immunosorbent assay (ELISA) for at least 2 years

expression patterns of mouse FIX and PC genes, age-related increases or stable conditions, respectively, are also similar to those of their human counterparts [16, 17]. It is also noteworthy that hFIX and hPC are both circulating proteins, which can be specifically detected by a highly dependable ELISA established in our laboratory, allowing us to make reliable, reproducible assays without sacrificing animals [16, 17]. This approach also allowed us to avoid another difficult problem with regard to handling the differences of individual animals.

Discovery of the First Age-Related Regulatory Mechanism of Gene Expression

After systematic analyses of a series of transgenic mouse lines carrying hFIX minigenes for their entire lifespan (at least 2 years), we discovered the first molecular mechanism of age-related homeostasis (16) (Fig. 4), later designated the age-related stability element/age-related increase element (ASE/AIE)-mediated genetic mechanism for age-related regulation of gene expression [17]. With this mechanism, ASE and AIE represent critical genetic elements producing age-related stability and age-related increase in gene expression, respectively. ASE and AIE correspond to AE5′ and AE3′ in the original naming, respectively [16]. Together with the genetic elements for basal promoter activity, these two genetic elements play essential roles for generating four age-related patterns [16] (Fig. 4). In the absence of ASE, mini-genes with or without AIE show an age-unstable hFIX gene expression pattern with a rapid decline in expression over the puberty period, followed by a further decrease to low, but more or less stable, levels or to very low basal levels, respectively, during the subsequent 3–4

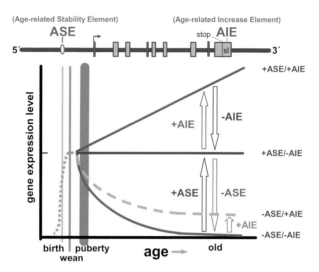

Fig. 4. Age-related stability element/age-related increase element (ASE/AIE)-mediated genetic mechanism for age-related regulation of gene expression. Relative positions of ASE and AIE in the gene are shown at the *top*. Four age-related mini-gene expression patterns generated by various combinations of ASE and AIE, depicted by − or + for their absence or presence, respectively, are shown. Age-related expression patterns of hFIX or hPC mini-genes determined by their specific ELISAs as circulatory protein levels parallel their mRNA levels in the liver. (Modified from Zhang et al. [17])

months. In the presence of ASE and the absence of AIE, gene expression showed an age-related stable pattern, whereas in the presence of both ASE and AIE, the age-related increase pattern, which mimics that of the wild-type hFIX gene, was reproduced. The ASE/AIE-mediated genetic mechanism for age-related regulation of gene expression is actually the puberty-onset gene switch mechanism, the first regulatory mechanism of that kind discovered to date.

ASE is located 5′ upstream to the basal promoter of the hFIX gene in a small region nucleotide (nt) −802 through nt −784 of the hFIX gene. AIE corresponds to an area of about 100 bp in the middle of the 3′ untranslated region (UTR) of the hFIX gene; it is composed of dinucleotide repeats with a potential to form RNA stem-loop structures after transcription.

Through footprinting analyses, an electrophoretic mobility shift assay (EMSA), and an in vivo function assay in transgenic mice, ASE was defined as having a nucleotide sequence, GAGGAAG, containing a minimal Ets consensus sequence (GGAA/T) [18, 19]. On the other hand, AIE has a stretch of dinucleotide repeat sequences (mixed with mostly AT, GT, and CA) with a potential to form three distinct stem loop structures when it is transcribed [12].

Similar to hFIX, mouse FIX (mFIX) also shows an age-dependent increase pattern for its circulating level and its gene expression (mRNA level) in the liver [20, 21]. The mFIX gene also contains an ASE sequence identical to the hFIX ASE as well as AIE, a stretch of 106-bp dinucleotide repeats in the middle of the 3′-UTR, the sequence of which is somewhat different from that of the hFIX gene [22] but having a stem-loop forming potential. Through testing in transgenic mice, the mFIX AIE element was also proved to be functional in a fashion similar to that of the hFIX gene, indicating that the ASE/AIE-mediated regulatory mechanism has universal functionality across species [17].

Our subsequent systematic studies on the transgenic mice carrying hPC mini-genes derived from the hPC gene completely confirmed the results obtained with the hFIX gene. Through these studies, we found that the hPC gene also has a functional ASE in the 5' region (again in the area of nt −832 through nt −826) upstream to the basal promoter region but lacks AIE [17] (Fig. 4). This explained why the hPC gene shows an age-stable expression pattern. The functional hPC ASE has a nucleotide sequence CAGGAAG with one base difference (G → C) in comparison to that of the hFIX gene, GAGGAAG. Interestingly, hPC and hFIX share significant similarities in the protein structure (corresponding to coding regions of their genes), but 5'-flanking regions and 3'-UTRs of their genes are grossly dissimilar, indicating unrelated evolutional origins [12, 14, 23, 24]. We showed that the 5'-flanking upstream region of the hFIX gene beyond approximately nt −350, including the region containing ASE, was derived from the retrotransposable element LINE-1 [24] (Fig. 2). In contrast, the hPC gene does not have LINE-1 or its remnant sequences in the 5'-flanking region, indicating no LINE-1 retrotranspositional event took place with this gene (Fig. 3). Furthermore, 3'-UTR sequences of hFIX and hPC genes share no similarity at all except the minimal local sequences at the very 3' end regions required for polyadenylation. The 3'-UTR sequence of the hFIX gene is about 1.4 kb in length, whereas that of the hPC gene is only 295 bp containing no AIE or AIE-like element [12, 14]. Dissimilarities in age-related expression patterns, as well as tissue specific expression patterns of hFIX and hPC genes, are consistent with their structural differences.

Importantly, both functional ASEs of the hFIX and hPC genes, GAGGAAG and CAGGAAG (together shown as G/CAGGAAG), bind the same liver nuclear protein as determined by competition EMSAs and functional assays in transgenic mice. Similar heptaoligonucleotide sequences with single base differences from G/CAGGAAG do not bind this protein and do not function as ASE [17]. GAGGAAA (a pseudo-ASE), present in the 5' upstream close to the functional hPC ASE, also binds a nuclear protein different from that which binds the functional ASEs, and it does not function as ASE [17]. Known Ets consensus elements, one in the first intron (GAGGATG) and the other in the last exon (CAGGATG), bind a similar nuclear protein; but it is again different from that which binds to functional ASEs. These elements also fail to function as ASE, suggesting that different Ets motif elements bind different nuclear proteins for different functions. As tested in transgenic mice, hFIX ASE can functionally substitute hPC ASE in regulating hPC gene expression [17].

It is important to note that hFIX ASE and hPC ASE have unrelated evolutionary origins. This clearly indicates a case of function-driven convergent evolution resulting in two closely related, but not identical, genetic elements that have identical homeostatic regulatory function.

The hPC gene does not have any AIE-like element. By introducing a unit of hFIX AIE into a hPC mini-gene (−1462hPCm1/AIE), however, its age-stable expression pattern is dramatically converted to an age-related increase pattern, similar to that of the hFIX gene [17]. This proves that ASE and AIE function universally across different species (at least humans and the mouse).

One of our unexpected observations was that there are no significant differences in the protein clearance rate of circulatory hFIX and hPC between young and old animals. This further supports the idea that age-related regulation is primarily done at transcription and RNA clearance steps [16, 17].

The ASE/AIE-mediated mechanism is presumably utilized by many other physiological systems [25]. In addition to this mechanism, yet unknown fundamental mechanisms likely exist and are involved in regulation of genes that encode not only factors in blood coagulation and fibrinolytic systems but also many genes involved in other physiological systems. Our finding of the ASE/AIE-mediated regulatory mechanism laid the foundation for extending our studies to find them and to prove the new hypothesis.

Toward an Integrated Understanding of Age-Related Homeostasis

Our new hypothesis regarding age-related homeostasis and its regulation is represented in Fig. 5. Age-related homeostasis of physiological reactions may be produced by highly dynamic and complex regulations of genes, proteins, sugars, lipid, and many other body components. If we can quantitatively define their dynamic lifetime changes along the age axis, we may be a step closer to understanding the complex regulatory mechanisms.

To test the durability of this idea, we first carried out global analyses of age-related expression profiles of mouse liver genes and proteins by taking quantitative microarray DNA chip assays and proteomic assays using two-dimensional gel electrophoresis (2DE) analyses combined with MALDI-TOF/MS PDF analyses. These analyses, which

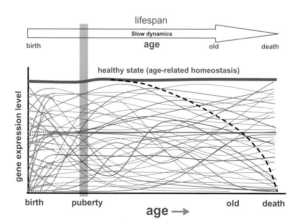

FIG. 5. Hypothetical representation of an age-related homeostatic state. A lifespan is shown at the *top* as an *open arrow* labeled *Slow dynamics*. A healthy age-related state (homeostatic state) is represented by a *thick horizontal line*, and imaginary age-related expression profiles of genes (could be proteins or any other parameters) are shown by the many *curved lines*. The *thick dotted line* represents a hypothetical age-related change in vitality

are still in progress, clearly indicated that we can identify about a dozen fundamental age-related expression profiles for both genes and proteins, including those that fit the four specific profiles produced by the ASE/AIE-mediated age-related regulatory mechanism of gene expression. There are a group of genes and proteins that show age-stable lifetime expression profiles (with fluctuation margins of less than 30% of the expression level at 1 month of age, approximately the weaning time). Expression profiles were further clustered into specifically unique groups, such as groups showing puberty-specific transient increase or decrease patterns, puberty-onset increase or decrease followed by their stabilization, old age-specific increase and decrease patterns, and other highly complex patterns. These strongly agree with the hypothesis mentioned above.

Age-related expression profiles obtained from gene expression analyses do not necessarily match those obtained from protein analysis, suggesting the importance of performing both gene and protein analyses. These differences are likely due to alternative splicings of precursor mRNAs produced from genes and largely to posttranslational modifications of proteins producing isomers with different functions.

Age Dimension Homeostasis, Diseases, and Epigenetics

When a large number of transgenic mice overexpressing hFIX at various levels were longitudinally monitored, we found that some animals died much earlier than their expected lifespans. These animals overexpressed hFIX at levels >1500 ng/ml serum, dying as young as at 3–5 months of age, whereas nontransgenic animals and transgenic animals expressing hFIX only at low levels (<200 ng/ml) lived a normal lifespan over 2 years [26]. These animals produced hFIX, which can normally function in mice in addition to their own inherent FIX, resulting in variously elevated levels of circulatory FIX. An age–survival–hFIX expression level plot (Kaplan-Meier analysis) clearly showed the close relation among them. Assuming that the plasma mFIX concentration in mice is similar to that of humans, any higher levels of total FIX in the circulation, even as low as 115% of the median, may work as a significant risk factor.

Histochemical analyses of the animals dying at younger ages than expected showed formation of thrombi in blood vessels of various organ tissues, including the brain, lungs, and heart. Furthermore, animals with high-level expression of hFIX were found to develop myocardial fibrosis in the left ventricle, presumably induced by thrombotic occlusions of blood vessels, mimicking a type of human myocardial infarction. These findings indicate that even relatively small elevations in the circulatory level of FIX, which occupies a position in the middle phase of the blood coagulation cascade, significantly shift the blood coagulation–anticoagulation balance toward the prothrombotic state. These analyses showed that there is a clear reciprocal relation between the duration of chronic maintenance of the elevated prothrombogenic state and survival. Interestingly, however, in our animal study, relatively short durations of high circulatory FIX concentration levels did not result in premature death. This represents a case for considering the relation of age, genetics, and epigenetic effectors.

Together, these observations suggest that any chronic environmental conditions such as consumption of prothrombogenic foods in combination with an elevated chronic pro-blood coagulation condition, such as an elevated circulatory level of pro-blood coagulation factor, may lead to a significantly enhanced thrombogenic state.

Age-Dimension Technology

The discovery of the ASE/AIE-mediated regulatory mechanism gene expression described here laid the foundation for developing age-dimension technology, a new field of research. ASE and AIE have a basic universal functionality, and therefore these elements can be utilized for designing various manipulations and interruptions of age-related expression of different genes.

A possible application is to utilize ASE for developing optimal gene delivery vector systems for durable gene therapy, ensuring long-term stable transgene expression. This may overcome two major problems encountered when developing gene therapy: (1) short duration of effective transgene expression in vivo even after efficient gene delivery; and (2) serious immunological side effects due to provoked immune reactions by repeated gene delivery procedures. Through studies of the hFIX mini-gene with and without ASE, we proved that ASE ensures age-stable, prolonged expression of transgenes, whereas mini-genes without ASE fail to do so.

For developing durable gene delivery systems, ASE provides an additional intriguing property: tissue-specific regulation of gene expression [27]. The presence of ASE in hFIX mini-gene expression vectors made their expression in transgenic mice strictly restricted to the liver, similar to that of the wild-type hFIX gene. When ASE is absent, strict liver-specific expression of the mini-gene cannot be maintained, allowing substantial expression in other tissues including the kidney and lungs. This unique spatial regulatory activity of ASE is not restricted to the liver. Transgenic mice carrying hPC mini-genes with ASE also showed precise tissue-specific expression in the liver and kidney, almost precisely reproducing the tissue specificity of the wild-type hPC gene. In the absence of ASE in the mini-genes, however, their expression failed to maintain the tissue specificity of the wild-type gene, showing expression in other tissues such as the stomach. This is the first time that a known transcriptional element was shown to have both temporal and spatial regulatory activities.

Age-dimension technology may also be useful for developing drugs capable of specific and safe modification of ASE and/or AIE activity through which we can interfere with and prevent age-related diseases, including cardiovascular diseases, diabetes, and cancer.

Conclusion

The first molecular mechanism of age-related homeostasis—ASE/AIE-mediated genetic mechanism for age-related regulation of gene expression—was successfully determined [16, 17, 26]. It has opened up a new research field for molecular mechanisms of age-related homeostasis and for age-dimension technology. Together, these studies have laid a solid foundation for establishing molecular regulatory mechanisms of age-related homeostasis, a new field of research with many possibilities. We look toward completing a truly integrated understanding of age-related homeostasis and its regulation, through which global analyses of age-related expression profiles of liver genes and proteins will play a critical role.

References

1. Andrew M, Vegh P, Johnston M, et al (1992) Maturation of the hemostatic system during childhood. Blood 80:998–2005
2. Marie D, Mannucci PM, Coppola R, et al (1995) Hypercoagulability in centenarians: the paradox of successful aging. Blood 85:3144–3149
3. Lowe GOD, Rumley A, Woodward M, et al (1997) Epidemiology of coagulation factors, inhibitors and activation markers: the Third Glasgow MONICA Survey. I. Illustrative reference ranges by age, sex and hormone use. Br J Haematol 97:775–784
4. Woodward M., Lowe GOD, Rumley A, et al (1997) Epidemiology of coagulation factors, inhibitors and activation markers: the Third Glasgow MONICA Survey. II. Relationships to cardiovascular risk factors and prevalent cardiovascular disease. Br J Haematol 97:785–797
5. Conlan MG, Folsom AR, Finch A, et al (1993) Associations of factor VIII and von Willebrand factor with age, race, sex, and risk factors for atherosclerosis: the Atherosclerosis Risk in Communities (ARIC) Study. Thromb Haemost 70:380–385
6. Conlan MG, Folsom AR, Finch A, et al (1994) Antithrombin III: associations with age, race, sex and cardiovascular disease risk factors: the Atherosclerosis Risk in Communities (ARIC) Study Investigators. Thromb Haemost 72:551–556
7. Van Hylckama Vlieg A, van der Linden IK, Bertina RM, et al (2000) High levels of factor IX increase the risk of venous thrombosis. Blood 95:3678–3682
8. Esmon CT (2000) Regulation of blood coagulation. Biochim Biophys Acta 1477:349–360
9. Dahlback B (2000) Blood coagulation. Lancet 355:1627–1632
10. Sweeney JD, Hoernig LA (1993) Age-dependent effect on the level of factor IX. Am J Clin Pathol 99:687–688
11. Bauer KA, Weiss LM, Sparrow D, et al (1987) Aging-associated changes in indices of thrombin generation and protein C activation in humans: Normative Aging Study. J Clin Invest 80:1527–1534
12. Yoshitake S, Schach BG, Foster DC, et al (1985) Nucleotide sequence of the gene for human factor IX (antihemophilic factor B). Biochemistry 24:3736–3750
13. Miao CH, Ho WT, Greenberg DL, et al (1996) Transcriptional regulation of the gene coding for human protein C. J Biol Chem 271:9587–9594
14. Foster DC, Yoshitake S, Davie EW (1985) The nucleotide sequence of the gene for human protein C. Proc Natl Acad Sci U S A 82:4673–4677
15. Kurachi S, Hitomi Y, Furukawa M, et al (1995) Role of intron I in expression of the human factor IX gene. J Biol Chem 270:5276–5281
16. Kurachi S, Deyashiki Y, Takeshita J, et al (1999) Genetic mechanisms of age regulation of human blood coagulation factor IX. Science 285:739–743
17. Zhang K, Kurachi S, Kurachi K (2002). Genetic mechanisms of age regulation of protein C and blood coagulation. J Biol Chem 277:4532–4540
18. Xin J-H, Cowie A, Lachance P, et al (1992) Molecular cloning and characterization of PEA3, a new member of the Ets oncogene family that is differentially expressed in mouse embryonic cells. Genes Dev 6:481–496
19. Chotteau-Lelievre A, Desbiens X, Pelczar H, et al (1997) Differential expression patterns of the PEA3 group transcription factors through murine embryonic development. Oncogene 15:937–952
20. Yao SN, DeSilva AH, Kurachi S, et al (1991) Characterization of a mouse factor IX cDNA and developmental regulation of the factor IX gene expression in liver. Thromb Haemost 65:52–58
21. Kurachi S, Hitomi E, Kurachi K (1996) Age and sex dependent regulation of the factor IX gene in mice. Thromb Haemost 76:965–969
22. Wu S-M, Stafford DW, Ware J (1990) Deduced amino acid sequence of mouse blood-coagulation factor IX. Gene 86:275–278

23. Salier J-P, Hirosawa S, and Kurachi K (1990) Functional characterization of the 5′-regulatory region of human factor IX gene. J Biol Chem 265:7062–7068
24. Hsu W, Kawamura S, Fountaine J-M, et al (1999) Organization and significance of LINE-1-derived sequences in the 5′ flanking region of the factor IX gene. Thromb Haemost 82:1782–1783
25. Kurachi K, Kurachi S (2005) Molecular mechanisms of age-related regulation of genes. J Thromb Haemost 3:909–914
26. Ameri A, Kurachi S, Sueishi K, et al (2003) Myocardial fibrosis in mice with overexpression of human blood coagulation factor IX. Blood 101:1871–1873
27. Zhang K, Kurachi S, Kurachi K (2002) New function for age-related stability element in conferring strict tissue-specific expression of human factor IX and protein C genes. Thromb Haemost 88:537–538

Molecular Evolution of Blood Clotting Factors with Special Reference to Fibrinogen and von Willebrand Factor

Sadaaki Iwanaga, Soutaro Gokudan, and Jun Mizuguchi

Summary. Comparative studies on "fibrinogen-like domains" and von Willebrand factor (vWF) domains found in various proteins have been performed. The "fibrinogen-like domain" is contained in more than 70 functional proteins, which are involved in innate immunity, angiogenesis, development, and differentiation, such as ficolins, tenascins, angiopoietins, and tachylectins. On the other hand, vWF A and D domains are identified extensively in integrins, extracellular matrix proteins, and secreted mucins. These vWF domains, in addition to "fibrinogen-like domains," are found throughout the vertebrates and in several invertebrate species, indicating that these domains have deep evolutionary roots associated with vertebrate circulatory systems and their roles in homeostasis. These facts suggest that a mechanism of hemostasis is evolutionarily conserved from arthropods to mammals.

Key words. Fibrinogen · Vitellogenin-related protein · Coagulogen · von Willebrand factor · Molecular evolution

Introduction

Blood circulating in all living organisms carries with it physiological functions that maintain a constant flow; once a lesion occurs in the tissue, blood thickens and ceases to flow. This phenomenon, known as blood coagulation, is a primary event in one of the biological defense systems of an organism. Clotting is thus critical in limiting loss of blood/hemolymph and initiating wound healing in all animals. It is also an important immune defense, quickly forming a secondary barrier to infection, thereby immobilizing and killing bacteria directly. No animal species has yet been described in which plasma/hemolymph coagulates without some contribution from cells: hemocytes, including platelets, thrombocytes, and amebocytes. Endothelial cells also participate in regulating the rate of the coagulation reaction and in lysing the fibrin clot.

Hemostatic reactions can be roughly divided into two stages: one in which the cellular components take part (primary hemostasis) and the other in which extracellular plasma factors mainly participate (secondary hemostasis). Cell adhesion and agglutination and the release of cellular components are biological reactions indispensable for primary hemostasis as well as for triggering secondary hemostasis [1]. In vertebrate animals, numerous plasma factors and tissue-derived protein factors, cofactors, serine proteases, phospholipids, and calcium ions participate in the process of secondary hemostasis (Fig. 1). One of the principal differences between vertebrate and invertebrate animals is the fact that body fluids in the former are mostly confined to blood and lymphatic vessels, whereas the latter has an open circulatory system. Therefore, after injury, invertebrates have efficient mechanisms that not only quickly prevent blood loss but also help in keeping pathogens from entering and spreading.

In 1961, Heilbrunn [2] suggested that plasma clotting in higher animals is the result of a progressive evolutionary extension into the plasma of the cytoplasmic sol-gel mechanism that is present in cells of all living organisms and underlies such basic reactions as protoplasmic streaming, ameboid movement, phagocytosis, muscle contraction, and cell division. However, our knowledge of the mechanisms of blood clotting and wound sealing in primitive animals was mostly morphological [3], and little was known of the biochemistry at that time. Nevertheless, currently available knowledge permits selection of specific examples that support Heilbrunn's idea [2] as observed in arthropods [4–7]. For instance, in horseshoe crabs, *Limulus polyphemus* and *Tachypleus tridentatus*, the clottable protein exists in a specific granule of amebocytes [7], as shown in Fig. 1. When these cells contact an activating surface, the cellular clotting factors are released into the surrounding fluid where they form an insoluble clot. Thus,

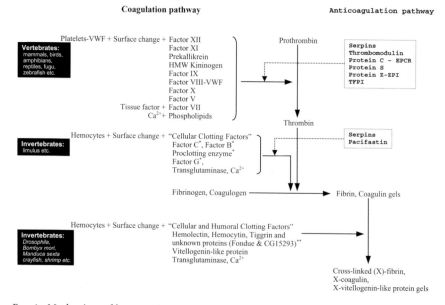

FIG. 1. Mechanism of hemostasis in various animal kingdoms [38–40, 70–74]. *Serpin*, serine proteinase inhibitor; *EPCR*, endothelial protein C receptor; *TFPI*, tissue factor pathway inhibitor. *These clotting factors have been identified in horseshoe crabs (*Limulus*). **Hemolectin, hemocytin, triggrin, and two unknown clotting factors (fondue and CG 15293) have recently been found in insect larvae hemolymph [70]. Vitellogenin-related protein and pacifastin are coagulable protein and proteinase inhibitor, respectively, identified in crayfish hemolymph [75, 76]. In addition to these coagulant factors, a number of anticoagulants (e.g., protein C, thrombomodulin, several serpins) participate in regulating the hemostatic reaction. Especially, the protein C anticoagulant pathway provides a unique anticoagulant, activated protein C, in response to thrombin generation. The system has multiple components, many of which are modulated by inflammatory mediators [1]

a hemolymph plasma clot is formed without the participation of plasma factors, unlike events in mammalian clotting systems. In contrast, in Crustacea, such as lobsters, shrimp, and crayfish, a clottable protein known as vitellogenin-related protein is present in the plasma [8] and is converted to its insoluble form by cellular transglutaminase released from certain hemocytes when activated by a foreign surface. Therefore, hemocytes in circulating blood always have a role in plasma coagulation, and there is an apparent evolutionary increase in the number of plasma factors involved in the hemostatic reaction (Fig. 1). These extra clotting factors amplify the reaction, thereby increasing its potential activity to meet the demand imposed by the increased circulatory efficiency and higher blood pressure in more advanced vertebrates.

This chapter focuses on the evolution of clotting factors at the molecular level, with a special reference to fibrinogen and von Willebrand factor (vWF). With respect to the evolution of other clotting factors, five γ-carboxyglutamic acid containing serine proteases (factors VII, IX, X, protein C, prothrombin) and five protein cofactors (tissue factor, factors V and VIII, thrombomodulin, protein S, protein Z) have been getting attention [9–11].

Structure of Fibrinogen

Fibrinogen, a glycoprotein of approximately 340 kDa, is usually present at a concentration of 2.5–3.0 mg/ml in human plasma. It is converted to an insoluble fibrin clot, thus playing an important role in blood coagulation [12].

Fibrinogen is composed of three subunits—Aα, Bβ, and γ chains—each of which is present in pairs. In its whole structure, three spheres are connected by two rods [13, 14], as shown in Fig. 2. The central nodule is called the E domain, and those at each end are called D domain. The rods are called coiled-coil regions. In D domain, mainly the Bβ and γ chains are folded up independently, and there are amino acid residues that are crosslinked by factor XIIIa [14]. The details are referred to in chapters by M.W. Mosesson and A. Ichinose elsewhere in this book. The COOH-terminal regions of the Bβ and γ chains that form the D domain have high homology with respect to primary and tertiary structures and are called "fibrinogen-like domains." It has been clarified recently that the fibrinogen-like domain is contained in a large number of functional proteins that are involved in innate immunity, angiogenesis, development, and differentiation.

Fibrinogen-Like Domains

Many proteins containing the fibrinogen-like domain have been recently identified [15–30], as shown in Table 1. Ficolins are composed of the fibrinogen-like and collagen-like domains. Several tissue-specific ficolins have been identified in various species ranging from mammals to ascidians (see squirts) [15, 24]. Human and ascidian ficolins show lectin activity, and their relation to the innate immune system has been suggested. Furthermore, the fibrinogen-like domain has been detected in some of the plasma lectins in horseshoe crabs, which are invertebrates and evolutionarily far from mammals; this domain plays an important role in the defense against infection [29]. The fibrinogen-like domain is also found in tenascins with one of the extracellular matrix (ECM) proteins known to have various physiological functions during the periods of

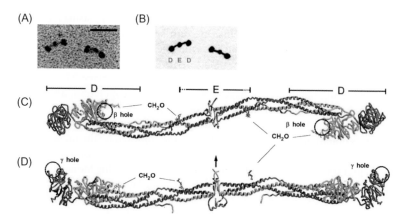

FIG. 2. Electron microscopic image and crystal structure of fibrinogen. **A** Electron microscopic image of human fibrinogen (analysis of a rotary projection image). Bar 50 nm. **B** The contour of the electron microscopic image in **A** is traced to create an image. *D* and *E* indicate the D and E domains, respectively. **C, D** Crystal structure of fibrinogen (bovine). Aα, Bβ, and γ chains are shown in blue, green, and red, respectively. In central domain E, monomers (Aα-Bβ-γ) in which the Aα, Bβ, and γ chains are combined by disulfide bonding form a dimmer (Aα-Bβ-γ)$_2$, and their NH$_2$-terminals are combined. *CH$_2$O* is a sugar chain. β *hole* is a space that is present at the COOH-terminal of the β chain. It is also called "pocket b." The structure of the b knob (B knob: Gly-His-Arg) binds to this site, which appears newly after the limited proteolysis of fibrinopeptide B. The γ *hole* is a space that is present at the COOH-terminal of the γ chain. The structure of the a knob (A knob: Gly-Pro-Arg) binds to this site, which appears newly after the limited proteolysis of fibrinopeptide A. The γ *hole* is also called "polymerization pocket a" or "pocket a." In this text, "pocket a" is used. In C and D, the angle of observation is rotated by 90°. (**A, B:** from Weisel et al. [13], with permission. **C, D:** By permission from Brown et al. [14], © 2007 National Academy of Science, USA)

development, differentiation, tissue repair, and regeneration [16,17]. It is also contained in angiopoietins, which participate in angiogenesis [18]; fibrinogen-related proteins (FREPs) contained in freshwater snail [26]; and scabrous, related to the development of ommatidium of *Drosophila* [28]. It has been reported recently that in *Suberites domuncula*, which is classified as Polifera, the fibrinogen-like domain is noted in some of the acute proteins that develop on exposure to (1→3)-β-D-glucan, and this molecule has been assumed to have the potential to act on self or nonself recognition [30].

Although these components are different from each other in terms of their physiological functions, all of them are considered to recognize the ligands via the fibrinogen-like domain to perform their functions. That is, the fibrinogen-like domain seems to be involved in intermolecular recognition.

Although most of these proteins have the fibrinogen-like domain in the COOH-terminal region, the whole structure of this molecule is highly diversified (Fig. 3). In the structure of the ficolin family, the NH$_2$-terminal short segment containing cysteine is connected to the collagen-like domain containing the repetitive structure of Gly-Xaa-Yaa, then to the short segment called "neck," and finally to the COOH-terminal fibrinogen-like domain [15]. In the structure of the tenascin family, the NH$_2$-terminal Cys-rich region is connected to the epidermal growth factor (EGF)-like repeats, then

TABLE 1. Various proteins containing "fibrinogen-like domain"

Species	Proteins	Functions
Vertebrates		
Human (*Homo sapiens*)	Ficolins	Recognition of bacterial components (the lectin-dependent complement activation pathway)
	Tenascins	Organization of extracellular matrix
	Angiopoietins	Angiogenesis
	Angiopoietin-like proteins	Angiogenesis
	MFAP-4 (mocrofibril-associated protein-4)	Unknown (cell adhesion?)
	Fgl-2/Fibroleukin	Unknown (immunomodulator?)
	Intelectin	Recognition of bacterial components
	Caspr (contactin-associated transmembrane receptor)	Developing nervous system (cell–cell comunication between glial cells and neurons)
Hedgehog (*Erinaceus europaeus*)	Erinacin	Metalloprotease inhibitor
Invertebrates		
Ascidian (*Halocynthia rorezi*)	Ficolins	Recognition of bacterial components
Slug (*Limax flavus*)	*L. flavus* agglutinin	Recognition of bacterial components
Freshwater snail (*Biomphalaria glabrata*)	FREPs (fibrinogen-related proteins)	Recognition of bacterial components
Sea cucumber (*Parastichopus parvimensis*)	FRePs (fibrinogen-related proteins)	Unknown
Fruitfly (*Drosophila melanogaster*)	Scabrous	Ommatidial differentiation
Horseshoe crab (*Tachypleus tridentatus*)	Tachylectin-5s	Recognition of bacterial components
Demosponge (*Suberites domuncula*)	FIB1_SUBDO	Host defense?

to the repetitive sequence of the fibronectin III-type structure, and finally to the COOH-terminal fibrinogen-like domain [17]. In the structure of the angiopoietin family, the portion corresponding to the collagen-like domain of ficolins is replaced by a coiled-coil region [18]. Erinacin, isolated from the muscle tissue of hedgehogs, is a metalloprotease inhibitor that inhibits the hemorrhagic activity of *Bothrops jararaca* snake venom [23]. This protein is composed of two subunits—i.e., α and β ($\alpha_{10}2\beta B_{10}$)— and each is composed of two domains, collagen-like and fibrinogen-like domains, thus resembling ficolins. The structure of FREPs isolated from the freshwater snail is characterized by the presence of the NH_2-terminal immunoglobulin (Ig)-like domain and that of the COOH-terminal fibrinogen-like domain [26]. In the structure of tachylectin-5A and tachylectin-5B (TLs-5), a plasma lectin in horseshoe crabs, the fibrinogen-like domain is also found on the COOH-terminal region, but there is no generally known domain structure in the NH_2-terminal region [29].

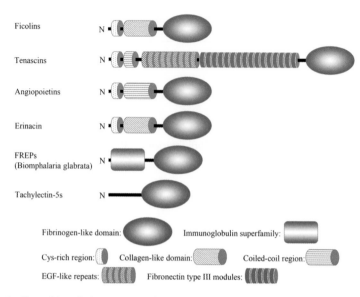

FIG. 3. Outline of the whole structure of proteins with the fibrinogen-like domain. Details are in the text

As mentioned above, the fibrinogen-like domain is contained in various functional proteins, and its primary structure is highly conserved. Of these proteins, the tertiary structure of tachylectin-5A (TL5A), one of the horseshoe crab plasma lectin TLs was elucidated for the first time. TLs are lectins that are responsible for most of the strong erythrocyte aggregating activity. Two kinds of TLs-5 (i.e., TL5A and TL5B) have been identified [29]. TLs-5 show unique binding specificity. That is, all the lectins bind to sugar chains by recognizing their structures, whereas TLs-5 recognize only *O*- and *N*-acetyl groups, which are simpler and have more universal structures than the sugar chains and bind to them [29]. Both show strong hemagglutination and bacterial agglutination activities in the presence of Ca^{2+}, and these activities are inhibited reversibly by EDTA [29]. TL5A and TL5B are glycoproteins composed of 269 and 289 amino acid residues, respectively. There is approximately 45% sequence similarity for the two in a whole molecule. In particular, approximately 200 COOH-terminal residues have the fibrinogen-like structure, showing high sequence homology (approximately 40%–50%) to the COOH-terminal of human L-ficolin [29].

Kairies et al. [31] succeeded in the crystallization of TL5A, forming a molecular complex with *N*-acetylglucosamine (GlcNAc) and clarified its tertiary structure by X-ray analysis (Fig. 4). The TL5A molecule can roughly be divided into three domains (A, B, P), as shown in Fig. 4B. Domain A is composed of two short α-helixes and a small reverse-parallel β-sheet, whereas domain B is composed of seven reverse-parallel β-strands. Domain P contains three short α-helixes and a small reverse-parallel β-sheet, but the random secondary structure accounts for most of it. Domain P is a functionally important region because Ca^{2+}- and ligand-binding sites exist. It is worthy of special attention that the tertiary structure of TL5A strongly resembles the structure of the fibrinogen γ chain (Fig. 4E). That is, the position of the Ca^{2+}- binding

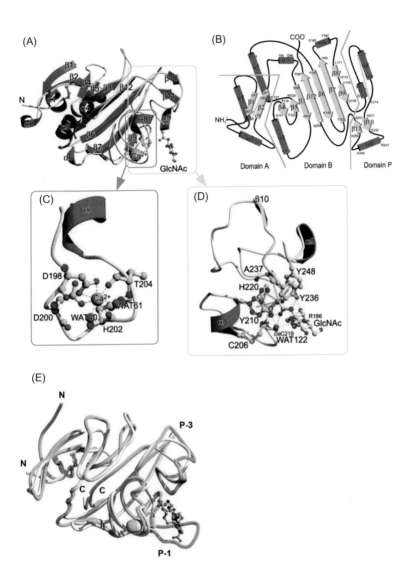

FIG. 4. Crystal structure of tachylectin-5A (TL5A), noting the similarity to the fibrinogen γ chain in three-dimensional structure. **A** Three-dimensional image of TL5A shown by a ribbon model. N-Acetylglucosamine (*GlcNAc*) is shown by a ball and stick model, and calcium ions are indicated by gold spheres. **B** Sequence of the secondary structure of TL5A shown by a topology diagram. The α-helix and β-strand are shown in red and blue, respectively. TL5A is roughly divided into three domains (A, B, P). Each domain was named based on its similarity to the domain structure of the fibrinogen γ chain. **C** Magnification of the calcium–binding site. Oxygen and nitrogen atoms are shown by red and blue balls and sticks, respectively. Water molecules and calcium ions are indicated by red and gold spheres, respectively. The calcium-binding site is present on the COOH-terminal side of the eighth α-helix (red). **D** Magnification of the GlcNAc-binding site. Hydrogen bonding, hydrophobic interaction, and water molecules are shown by a yellow broken line, a green broken line, and red spheres (*WAT*), respectively. It should be noted that -NH of Cys219 is bound to the oxygen atom at the acetoamide site of GlcNAc by hydrogen bonding. The funnel-shaped space made by Ala237 as the base and Tyr210, Tyr 236, Tyr248, and His220 forms a ligand-binding site. **E** The tertiary structure of TL5A (gray) and that of the fibrinogen γ chain (yellow) are superimposed. Their calcium ions are shown by gray and yellow spheres, respectively. The calcium-binding site is almost identical in the two structures. The balls and sticks in the figure indicate the superimposition of GlcNAc and GPRG peptide (A knob), which binds to the fibrinogen chain. The ligand-binding site is almost identical in the two structures. (Adapted from Kairies et al. [31], © 2007 National Academy of Science, USA)

site, which is conserved well in the primary structure, is almost identical [31]. Moreover, the position of "polymerization pocket a" on the γ chain at the time of fibrin formation is the ligand-binding site in TL5A, and five of the seven amino acid residues forming "pocket a" on the γ chain are also conserved in TL5A. Furthermore, TL5A has a *cis*-peptide bond at the ligand-binding site, and the fibrinogen γ chain also has it at Lys338-Cys339. This structure is necessary for the function of "pocket a." A clear difference between the two is that two loop structures, called P-1 and P-3, which are present at both ends of the Ca^{2+}- and ligand-binding sites on the γ chain, are much shorter in TL5A (Fig. 4E).

The tertiary structure of the fibrinogen-like domain of ficolins has recently been clarified [32, 33]. As members of the ficolin family, three ficolins are known in humans: L-ficolin and H-ficolin occurring mainly in plasma and M-ficolin as a secretory protein contained in the lungs and some leukocytes [15]. Although these three kinds of ficolin are different from each other with respect to binding specificity, it is known that each of them binds to the surface layer of an invading microorganism and activates a complement via the complement-lectin pathway [15]. The whole structure of each of these three ficolins is similar to that of TL5A, being composed of the three subdomains of A, B, and P (Fig. 5). The Ca^{2+}-binding site present in domain P of TL5A is also homologous in ficolins. The position of "pocket a" on the fibrinogen γ chain is almost identical with that of the ligand-binding pocket, which is present in domain P of TL5A, and one of the ligand-binding sites of ficolins is also present in the same position [32, 33]. On the other hand, the amino acid residues (Q and K), which are used for crosslinking by activated XIIIa, are present in the γ chain of fibrinogen, whereas such amino acid residues are not found in TLs-5 or ficolin. Furthermore, the binding sites for integrin $α_V β_3$ are present in the γ chain of fibrinogen; however, some of these sites are lost in TLs-5 and ficolin. Additionally, the "WXXW" sequence, a potential mannosylation site, is present in the fibrinogen γ chain and TLs-5A but is not found in human L-ficolin [29].

The angiopoietin family is a group of molecules identified as the ligands for Tie2, which is receptor-type tyrosine kinase on endothelial cells [18]. It is known to play an important role in the formation of blood and lymph vessels in mammals. Angiopoieitin-1 and angiopoietin-2 are considered, respectively, to function as an agonist and an antagonist on Tie2. The site where they bind to Tie2 is the COOH-terminal fibrinogen-like domain [18]. Barton et al. [34] examined the fibrinogen-like domain of recombinant angiopoieitin-2 and elucidated its three-dimensional structure by X-ray analysis. The whole structure of this fibrinogen-like domain as well as those of ficolins closely resembled that of TL5A, and the Ca^{2+}-binding site is homologous to those in TL5A and the fibrinogen γ chain (Fig. 5). Angiopoietin-2 recognizes Tie2 protein and binds to it; but unlike the fibrinogen γ chain, ficolins, and TL5A, it does not require Ca^{2+} for binding to the ligand. Ca^{2+} is required for the stability of angiopoietin [34]. Because the regions corresponding to "pocket a" on the fibrinogen γ chain and the ligand-binding pocket in TL5A are also highly conserved in angiopoietin-2, these regions may be involved in binding to Tie2. Thus, the functions of proteins with the fibrinogen-like domain are highly diversified, but their three-dimensional structures are extremely conserved [34].

Owing to rapid progress in the field of molecular biology in recent years, a large number of protein groups with the fibrinogen-like domain have been detected. In

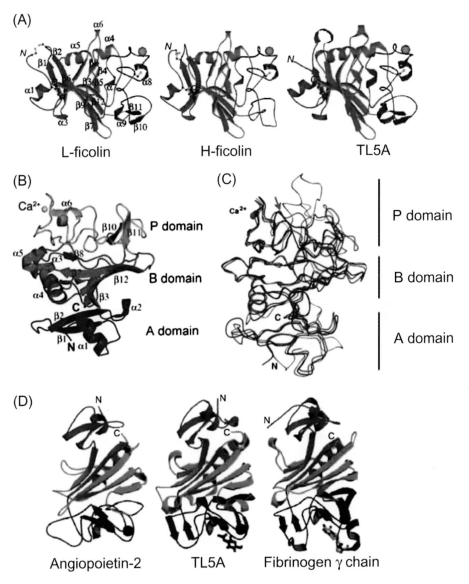

Fɪɢ. 5. Crystal structures of the fibrinogen-like domains of ficolins and angiopoietin-2. **A** Comparative views of the L-ficolin, H-ficolin, and TL5A monomers in the same orientation. Like TL5A, L-ficolin and M-ficolin are divided into three domains. Domains A, B, and P are shown in red, green, and blue, respectively. Calcium ions are shown by yellow spheres. The secondary structure of domain P is poorer in H-ficolin than in L-ficolin, but the whole structures of H-ficolin and L-ficolin closely resemble that of TL5A. **B** Three-dimensional image of the M-ficolin monomer shown by a ribbon diagram. M-ficolin is also divided into three domains. Domains A, B, and P are shown in blue, red, and cyan, respectively. Calcium ions are indicated by yellow spheres. **C** Superimposition of the tertiary structures of M-ficolin (magenta), TL5A (blue), and the γ chain (slate). The calcium ions binding to them are shown by spheres of the same colors as the tertiary structures. The site of binding is almost identical. **D** Side-by-side alignment of angiopoietin-2, TL5A, and the fibrinogen γ chain. TL5A is bound to GlcNAc, whereas the fibrinogen γ chain is bound to GPRP peptide. The ligand for each is shown by a ball and stick model. Angiopoietin-2 has three domains, like those of TL5A. Domains A, B, and P are shown in red, green, and blue, respectively. Calcium ions are indicated by black spheres. The calcium ion-binding site is almost identical in them. The secondary structure of domain P is poorer in angiopoietin-2 than in the other two, but its whole structure is similar to them. (**A**: From Kairies et al. [31], by permission from Macmillan Publishers Ltd. © 2007. **B, C**: From Barton et al. [34], with permission. **D**: From Barton et al. [34], with permission. © Elsevier 2007)

various proteins containing the fibrinogen-like domain, the structure of the function-ing site is considered to be similar, although their ligands are different. Therefore, the basic structure of the fibrinogen-like domain may be regarded as an important struc-tural motif for intermolecular recognition. There is a common point that most of the proteins with the fibrinogen-like domain, including fibrinogen, are molecules that function by forming homopolymers or heteropolymers with relative molecules (Table 2). The basic unit of ficolins is a trimer in which trimolecular protomers are combined by disulfide bonding, and the trimers are further combined by disulfide bonding to form a polymer [15]. The structure of erinacin may closely resemble that of ficolins, but it is structurally different from the ficolin family because it forms a heteropolymer [23]. Tenascin forms homopolymers that range from dimers to hexamers [17]. Angi-opoietin-1 and angiopoeitin-2 function by forming mainly hexamers and dimers, respectively [35]. TLs-5, plasma lectins of horseshoe crabs, have no typical multimer-forming motif; but like ficolins and angiopoietins, they form polymers in which two to four dimers combined by disulfide bonding are bound noncovalently [29].

Genome analysis has been completed for several species including humans. Accord-ing to the results, there are 26 genes containing the fibrinogen-like sequence in humans [36]. In recent years, genome information about organisms other than humans has been clarified one after another, and it has been gradually clarified that the genetic information about this fibrinogen-like domain is present in sponges, archaebacteria, and some viruses in addition to *Drosophila* and *Nematoda*, although its physiological significance is unknown in most of them. Taking these findings into account, the fibrinogen gene is considered to be present widely in primitive to higher animals, and its products possibly function mainly to recognize other molecules. Although there is a remarkable sequence homology between Bβ- and γ-chains, the Aα-chain is much less similar to those of other chains. In fact, when the Aα-chain is aligned with either the Bβ- or γ-chain, only 12% of the positions show identity, even when regions with differences in chain length are disregarded [37]. Therefore, the ancestor of Aα-chain might have separated from the common ancestor of the Bβ- and γ-chains. At present, there is no information about possible ancestral protein of the Aα-chain.

TABLE 2. Higher-order structural pattern of fibrinogen-related proteins

Proteins	Higher-order structural pattern
Fibrinogen	6-mer (hetero) $(A\alpha - B\beta - \gamma)_2$
Ficolins	
L-ficolin	12-mer (homo)
H-ficolin	18-mer (homo)
Tenascins	
Tenascin C	6-mer (homo)
Tenascin R	2,4,6-mer (homo)
Angiopoietins	
Angiopoietin-1	6-mer (homo)
Angiopoietin-2	dimer (homo)
Erinacin	30-mer (hetero) $(\alpha_{10}2\beta_{10})$
Tachylectin-5s	
Tachylectin-5A	6, 8-mer (homo)
Tachylectin-5B	4-mer (homo)

Vitellogenin-Related Proteins

In this and the next sections, body fluid coagulation proteins in invertebrate animals with an open blood vascular system are discussed. As stated briefly earlier in the chapter, although the body fluid coagulation system of invertebrate animals is much more diversified than that of vertebrate animals, little has been published concerning this system in invertebrate animals except for horseshoe crabs [38–40].

In the body fluid coagulation systems of tiger shrimp (*Penaeus mondon*) and crayfish (*Pacifastacus leniusculus*), "vitellogenin-related protein" (their coagulation protein) is cross-linked by transglutaminase (TGase) in the presence of calcium ions to form an insoluble gel [8, 41]. The coagulation protein and TGase are usually present separately in the body fluid and hemocytes, so circulation of the body fluid can be maintained. However, if the hemocytes show a lytic response to invading foreign substances, such as bacteria, TGase is released into the body fluid to cause covalent crosslinking of the coagulation protein, resulting in formation of an insoluble gel. These vitellogenin-related proteins are homodimeric glycoproteins composed of 1670 amino acid residues (molecular mass is approximately 380 kDa) in tiger shrimps and 1721 amino acid residues (molecular mass is approximately 420 kDa) in crayfish; and there is 57% sequence homology between the two glycoproteins [8, 41]. Their whole structures are roughly divided into NH₂-terminal vitellogenin-like domain 1 composed of approximately 80 amino acid residues (a Lys-rich domain) and COOH-terminal long vitellogenin-like domain 2 accounting for approximately two-thirds of the sequence. In addition, the Gln-rich domain and the vWF D-domain are present in the vitellogenin-like domain 2. The Lys-rich domain and the Gln-rich domain are crosslinked by TGase to form an insoluble gel. As a common characteristic in tiger shrimp and crayfish, the Lys-rich domain is composed of the structure in which the sequence "S/T–K–T–S/T" is repeated five times [8, 38]. On the other hand, in the Gln-rich domain, there is no such clear common characteristic of sequence and one additional Gln-rich domain that is present only in tiger shrimp [41].

As indicated by its name, vitellogenin was originally found and defined as a lipo-protein involved in vitellogenesis and egg laying in female insects [42]. However, at that time, its amino acid sequence and structural characteristics were not taken into consideration. Subsequently, it has been found that this substance has homology, although weak, to mammalian apolipoprotein B, microsomal triglyceride transfer protein, and vWF D-domain. As mentioned in the preceding section, fibrinogen was historically defined first as a coagulation protein on the basis of its function. However, as genome analysis has been performed in various organisms, it is being elucidated that "coagulation function" is only one of a great number of functions of fibrinogen. The same thing may be in the process of being clarified for vitellogenin-related protein. Briefly, vitellogenin-related protein is assumed to play an important role not only in vitellogenesis, which was found and defined first, but also in sex difference-unrelated phenomena, including lipid metabolism, coagulation, and immunity [42]. In tiger shrimp and crayfish, vitellogenin-related clotting protein is present in both males and females, playing a central role in the coagulation of the body fluid. Vitellogenin-related protein shows no similarity in sequence or structure to coagulogen or fibrinogen. Although horseshoe crabs are also arthropods, the molecular weight of coagulogen as their coagulation protein is lower, which is approximately 20 kDa; moreover, it develops and is stored in hemocytes.

In tiger shrimp, vitellogenin-related protein is expressed in most tissues excluding hemocytes, with its expression at the highest level in the gills and heart. The mean concentration of this protein in the body fluid is approximately 3 mg/ml, but it shows an approximately twofold increase immediately before molting, followed by a return to the original level after molting. It also shows a fourfold increase in the event of injury. These findings are highly interesting when it is taken into consideration that fibrinogen in vertebrate animals is an acute-phase protein. Such findings have not been obtained in crayfish. Furthermore, the Arg-Gly-Asp (RGD) sequence is present at two sites in the vitellogenin-related protein of tiger shrimp, whereas no such sequence is found in the coagulation protein of crayfish. On the other hand, RGD-recognizing receptors are assumed to be present in the hemocytes of crayfish, but its presence is unclear in tiger shrimps [8, 41]. In addition, vitellogenin-related protein contains 20% sugar (six potential Asn sites) in crayfish, whereas its content is low (3.8%, with four potential Asn sites) in tiger shrimp [8, 41].

Coagulogen

Coagulogen is a clotting protein localized in large granules contained in the hemocytes of invertebrate horseshoe crabs. It is functionally similar to the fibrinogen noted in vertebrate animals. When the living body of a horseshoe crab is injured, circulating hemocytes migrate to the site of injury and are brought into contact with foreign materials, resulting in degranulation of the hemocytes and activation of the body fluid coagulation system; finally, coagulogen is converted to coagulin by clotting enzyme to form an insoluble gel, which prevents loss of the body fluid (Fig. 6) [43, 44]. Coagulogen is a single-strand strongly basic simple protein composed of 175 amino acid residues. It is cleaved by the activated clotting enzyme of horseshoe crabs at two NH$_2$-terminal sites of the molecule. As a result, peptide C with 28 amino acid residues is released, and coagulogen is converted to a coagulin monomer composed of chain A (18 residues) and chain B (129 residues) connected by two disulfide bonds [44].

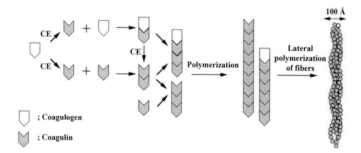

FIG. 6. Hypothetical scheme of coagulin polymerization [7, 43]. Coagulogen is converted to coagulin by clotting enzyme (CE) to form an insoluble aggregate. At this time, not only the coagulin monomer but also coagulogen is incorporated into the polymer. The resultant coagulin polymers swarm owing to interaction of the lateral surfaces to form thick fibers of approximately 100 Å. This interaction in the lateral direction is said to be ascribable to hydrophobic affinity. (From Kawasaki et al. [44], with permission)

Figure 7A shows the three-dimensional structure of coagulogen. The whole molecule resembles a football-like elongated rotary ellipsoid [45]. Because the A region corresponding to the NH$_2$-terminal side fluctuates, its three-dimensional structure has not yet been identified. This region is connected to the peptide C region composed of an α-helix making 4.5 revolutions, then to the B region mainly composed of β structures, and finally to COOH-terminal B6 [45]. Two Cys residues present in the A region are bound by S-S bonding to two Cys residues, which are present at B1 and B4, respectively, in the B region. Six β structures from B1 to B6, which comprise the B region, form a characteristic shape. B6 and B5 form a long, twisted reverse-parallel β-sheet, which is turned just one revolution at both ends. Similar reverse-parallel β-sheets are formed between B1 and B4 and also between B2 and B3 [45]. As mentioned below, these β structures are also present in nerve growth factor (NGF) and spätzle, which is known as a ligand for Toll receptors in *Drosophila* [46]. Peptide C, which is present on top of the coagulogen molecule and perpendicular to the paper surface, is cut off at the time of coagulin monomer formation. Just below the peptide C, there is a cluster of hydrophobic amino acids (Phe48, ILe50, Phe51, Phe57, Val109, Val111, Phe118, Tyr152, Leu154, Phe159). This site plays an important role in the formation of coagulin polymers [44, 45].

Comparison of coagulogen with other proteins shows that nearly half of the three-dimensional structure of the whole molecule of coagulogen strongly resembles that of NGF. The three-dimensional structure of spätzle in *Drosophila* has also been elucidated, and the structure of this molecule resembles that of coagulogen [44]. Figure 7B shows the three-dimensional models of these three substances. The topologies of the above-mentioned three pairs of reverse-parallel β-sheets resemble each other, and the positions of the three S-S bonds fixing their structures are almost identical among the three. On the basis of these topologies, comparison of the amino acid sequence of coagulogen with that of NGF shows 21% homology on the COOH-terminal sides of both molecules [44]. Therefore, coagulogen is assumed to be classified as the superfamily, including spätzle, transforming growth factor-β (TGFβ), and platelet-derived growth factor (PDGF) [46]. Although coagulogen, which is gelatinized to form a fibrillar structure, is functionally similar to vertebrate fibrinogen, its three-dimensional structure is quite different from that of fibrinogen.

von Willebrand Factor

Plasma vWF is a multimeric glycoprotein synthesized in endothelial cells and megakaryocytes. vWF synthesized by megakaryocytes is ultimately encountered in the α-granules of platelets. Secretion from platelets and endothelial cells results in the appearance of vWF in plasma (~10 μg/ml, ~50 nmol/l, as a monomer) and the subendothelium. Congenital deficiencies may lead to a mild, moderate, or severe bleeding diathesis, denoted von Willebrand disease (vWD) [3, 47].

The vWF is well known as a protein essential for the formation of the hemostatic plug. It acts as a molecular bridge that connects the platelets to the subendothelium that is exposed upon injury of the vessel wall. Moreover, vWF may promote platelet–platelet interaction together with other platelet-derived membrane glycoproteins, such as GPIb and GPIIb-IIIa ($\alpha_{IIb}\beta_3$), resulting in aggregation of platelets. These physiological functions have been studied extensively and reviewed in the recent literature [47, 48].

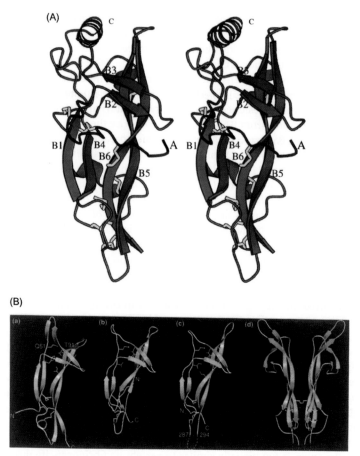

FIG. 7. Crystal structure of the coagulogen monomer [42]. **A** Stereo view of the coagulogen monomer. A and B1–B6 polypeptide chains correspond to the A (purple) and B (blue and green) chains, respectively. Because the A chain is fluctuating, its three-dimensional structure has not yet been identified. α-Helix is shown in reddish orange, and the main ones are found at three sites. In particular, it is found 4.5 times in the C chain (corresponding to peptide C). Disulfide bonding is shown in yellow, which is noted at eight sites. The three-dimensional structure of the whole molecule resembles that of cylindrical sugar (peptide C) picked with "sugar tweezers." Below peptide C, there is a "cove" where a cluster of hydrophobic amino acids is noted. On the other hand, there is also a cluster of hydrophobic amino acids in the lower most region of the molecule (corresponding to B5 and B6). If peptide C is cut off, this area probably shows interaction with the "cove" region to form a "head to tail" aggregate [44]. **B** Three-dimensional models of nerve growth factor (NGF) (*a*), coagulogen (*b*), and spätzle (*c*), and its dimer model (*d*). Although these three substances show approximately 15%–22% identity on the primary sequence, their tertiary structures resemble each other. In particular, the three-dimensional structure of cystine-knot (shown in yellow) strongly resembles the topology between them, and the Gln -51 (blue) and Thr -91 (red) topologies of NGF are also found at the same positions in coagulogen. (**B:** From Mizuguchi et al. [46], reprinted with permission. © Elsevier 2007)

Structure and Functional Domains of vWF

The vWF isolated from plasma, platelets, or endothelium consists of a series of multimers, ranging from dimers to high-molecular-weight proteins composed of 50 to about 100 vWF monomers. The largest vWF multimers have a molecular mass of ~20 000 kDa. vWF dimers are assembled from pairs of ~250-kDa polypeptide subunits in the endoplasmic reticulum via disulfide bridges between cysteine residues located in the COOH-terminal region [47].

The primary structure of human vWF has been determined by both protein sequencing and inference from the corresponding cDNA [49, 50]. Its gene lies at position p12-pter of the chromosome, covers approximately 178 kbp genomic DNA, and has been characterized [51]. The gene has 52 exons between 40 and 1379 bp in length; the intron lengths vary from 97 to approximately 19.9 kbp. The complete cDNA sequence of vWF shows that the 2050-residue-long protein is synthesized with both a 22-residue signal peptide and a 77-residue propeptide, giving pro-vWF a total length of 2791 residues (Fig. 8).

The 2050-amino-acid residues of the mature vWF monomer give a molecular mass of 226 kDa for the polypeptide, to which both N-bound and O-bound carbohydrate chains must be added. There are 13 potential sites for N-glycosylation and 10 potential sites for O-glycosylation. Thus, the glycosylated vWF polypeptide has an estimated molecular mass of about 280 kDa. The structures of these oligosaccharide chains contained in human vWF are described in detail in the chapter by K. Titani and T. Matsui in this book. Human vWF contains the sequence RGD, which may be expected to mediate binding to GPIIb-IIIa on platelets. The RGD sequence has also been found in the propeptide region of pro-vWF, but its significance is not known.

The whole amino acid sequence of pro-vWF allows the definition of four types of homologous domains, denoted A to D. Pro-vWF can thus be ordered D1-D2-D'-D3-A1-A2-A3-D4-B1-B2-C1-C2, where D' is a partial D-domain and where D1 and D2 together make up the propeptide region, as shown in Fig. 8 [47–49].

The propeptide part of pro-vWF is cleaved off within the cell, giving mature vWF directly; and the released peptide is necessary for the formation of multimers from the dimer subunit but not necessary for the formation of vWF dimers from the monomeric polypeptide chains. The dimerization of vWF monomer occurs by the formation of a disulfide bond between cysteine residues in the COOH-terminal region, and the union of dimers to give the complete multimer involves a disulfide bond in the NH$_2$-terminal region.

The vWF has two main functions in the bloodstream. The first is protection of factor VIII, and the second is mediating the adhesion of platelets to thrombogenic surfaces in connection with hemostasis, indicating that vWF contains various structural and functional domains. Up to now, the following functional domains have been identified (Fig. 8), the details of which can be found in the literature [47, 48]: (1) interaction of vWF with GPIb [49]; (2) interaction of vWF with GPIIb/IIIa (integrin $\alpha_{IIb}\beta_3$) [52–54]; (3) interactions with the subendothelium and basolateral endothelial cells [48]; (4) Interaction of vWF with factor VIII [47, 48]; (5) interaction of vWF with the components of the ECM, such as collagens VI, VII, XII, and XIV, and heparin sulfatides, and its related proteoglycans [47, 55, 56]; (6) interaction of vWF with snake venom components, such as botrocetin and its analogues [57, 58].

Prepro VWF

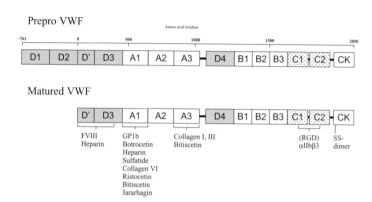

FIG. 8. Gross structures of prepro-vWF, mature vWF, and major ligand-binding sites with other components. The locations of five conserved structural domains (A, B, C, D, CK) are indicated. Intersubunit disulfide bonds are formed near the COOH-termini of pro-vWF dimers (CK) in the endoplasmic reticulum. Additional intersubunit disulfide bonds are formed near the NH$_2$-terminus of the mature subunits to assemble multimers in the Golgi apparatus [47, 48]. The major ligand-binding sites in the subunit have been localized as indicated. Activated platelet integrin $\alpha_{IIb}\beta_3$ binds vWF through a segment that includes the tripeptide sequence RGD in vWF domain C1 [48]

Various Proteins Containing vWF Domains

As expected from the multiple interactions of vWF with many proteins, homologue domains of vWF occur in unrelated proteins, indicating that they have been dispersed throughout the genome by duplication, exon shuffling, and insertion. For instance, the vWF-A domains, consisting of about 220–amino-acid modules are found extensively in proteins of vertebrate and even invertebrate animals (Fig. 9). More than 55 vWF-A domains are known to be distributed among 27 distinct genes [53]. Such A domains are found in the α subunits of certain integrins, including five leukocyte adhesion receptors and two collagen receptors. vWF-A domains also occur in several nonfibrillar collagens, cartilage matrix protein, and the closely related matrilins-2, -3, and -4, the noninhibitory $\alpha2$ and $\alpha3$ subunits of an inter-α-trypsin Kunitz-type protease inhibitor, the $\alpha2$ subunit of a dihydropyridine-sensitive calcium channel, and complements factor B, factor B-like proteins, C2, and C3-like protein [47, 53, 59–61].

Many of these proteins contain several A domains in tandem. For example, the collagen $\alpha1$-V1 and $\alpha2$-V1 chains each has three A domains, and the $\alpha3$-VI chain and Hydra (*Hydra vulgaris*) collagen VI contain 12 and 5 A domains, respectively [59]. The vWF-A domains of complement factors B and C2 and several integrin α chains have been shown to be macromolecular ligands, and their binding functions appear to be regulated [60]. These A domains also have divalent metal ion (Mg^{2+})-binding sites that must be occupied to bind ligands [61].

The B, C, D, and CK domains near the COOH-terminal end of the vWF subunit are shared with certain epithelial mucins, often in association with vWF-D domains and sometimes, vWF-B domains [55, 56]. These proteins form homodimers or oligomers. As already mentioned, phylogenetic analysis has permitted the proposal of an evolutionary history of the four human mucin genes located on chromosome 11 from an ancestor gene common to the vWF gene [56]. cDNA sequences flanking the central part of the subclass of human mucins (MUC2, MUC5AC, MUC5B) and animal mucins (RMUC2, FIM-B, 1, PSM) code for cysteine-rich domains that are similar to the cysteine-rich domains that flank the three consecutive A (A1, A2, A3) domains of vWF (Fig. 9). These cysteine-rich domains are named D1-D2-D'-D3 upstream of the central part in mucins and upstream of the A1-A2-A3 domains followed by B, C, and CK domains in the vWF molecule [56]. Dimerization of porcine submaxillary mucin is mediated by disulfide bonds between COOH-terminal CK domains, and this mechanism may apply to other structurally similar mucins. The vWF-CK domains are homologous to the neutrophin family of growth factors and are particularly similar to TGFβ [47]. Like mucins that contain CK domains, growth factors of the neutrophin family are usually dimeric [55], suggesting that CK domains can function as dimerization motifs in several contexts.

On the other hand, the several structural domains found in human vWF are distributed even more broadly in the animal kingdom. Five vWF-D domains are found in zonadhesin, a mammalian sperm membrane glycoprotein that appears to function in species-specific recognition of egg extracellular matrix. Moreover, vWF-A domains occur in protozan parasite and nematode proteins (Fig. 9), including *Plasmodium falciparum* malarial thrombospondin-related anomymous protein (TRAP), and circumsporozoate TRAP-related protein (CTRP), *Eimeria tenella* microsome protein Etp100, and *Caenorhabditis elegans* 338-kDa (partial) hypothetical protein T20G5.3 [47, 59].

Insect-Derived Proteins Containing vWF Domains

Recently, Kotani et al. [62] reported that silkworm (*Bombyx mori*) hemolymph contains a lectin-like protein, named hemocytin, which shows characteristic features of the carbohydrate recognition domain of C-type animal lectin (Fig. 9). The total amino acid residues of hemocytin predicted from the cDNA sequence (10477 bp) is 3133, and it shows significant sequence homology with human vWF. The amino acid sequences of domains D1, D2, D3, D', D'', and D''' of hemocytin has about 14.2%–30.0% identity with those of domains D1, D2, D3, D4, and D' of human vWF. In the hemocytin sequence, repetitive B1 and B2 and C1 and C2 domains have also about 29.0%–52.3% and 26.2%–29.3% identity with those domains of vWF, respectively. More than 40% of the whole precursor is constituted by the repetitive region homologous to the repetitive D, D', B, and C domains of vWF, suggesting that hemocytin and vWF have diverged from the same ancestral origin. Moreover, there are also 20.5%–34.2% identical residues in a repetitive region, E1 and E2, of hemocytin with phospholipids-binding domains of clotting factors V and VIII.

Based on these homology analyses, the hemocytin preprosequence is presumed to consist of duplication and rearrangements of the small sequences, which are evolutionary conserved in mammalian adhesive proteins. Probably the hemocytin molecule is related to hemostasis or encapsulation of foreign substances for self-defense in *Bombyx mori* [62].

In addition to *Bombyx* hemocytin, Goto et al. [63, 64] identified a novel *Drosophila* protein of >400kDa, named hemolectin, which is secreted from the *Drosophila* hemocyte-derived cell line Kc167. Its 11.7-kbp cDNA contains an open reading frame of 3843 amino acid residues, with conserved domains in human vWF, clotting factors V and VIII, and complement factors. The hemolectin gene is located in the third

| **Proteins** | **Schematic structure of domains** |

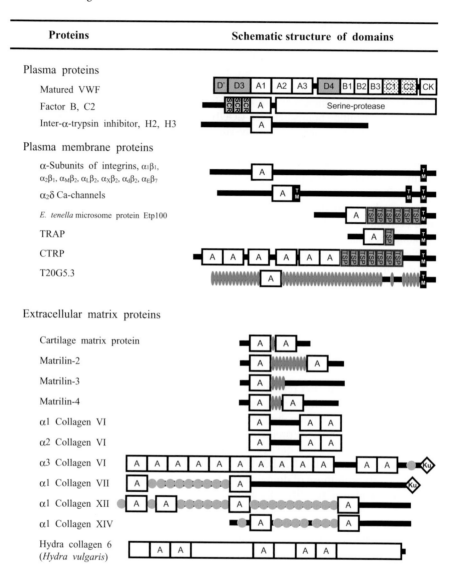

FIG. 9. Structures of various proteins with conserved vWF domains. Plasma proteins include vWF, complement factors B and C2, and complement B-like and C-like proteins from invertebrates. Plasma membrane proteins include the α subunits of integrins, the α₂ subunit of a dihydropyridine-sensitive calcium channel (Ca channel), *E. tenella* membrane protein Ep100, *P. falciparum* thrombospondin-related anonymous protein (TRAP), circumsporozite TRAP-related protein (CTRP), and *C. elegans* hypothetical protein T20G5. Extracellular matrix proteins include cartilage matrix protein (CMP), the related matrilin-2, -3, and -4, and collagens VI, VII, XII, and XIV. The colored symbols for structural motifs are shown and include vWF A, B, C, D, and CK domains, the mammalian epithelial mucins, bovine and porcine submaxillary mucins, *X. laevis* integumentary mucin FIM-B1, and the large family of neutrophins represented here by transforming growth factor-β (TGFβ). See text for other references

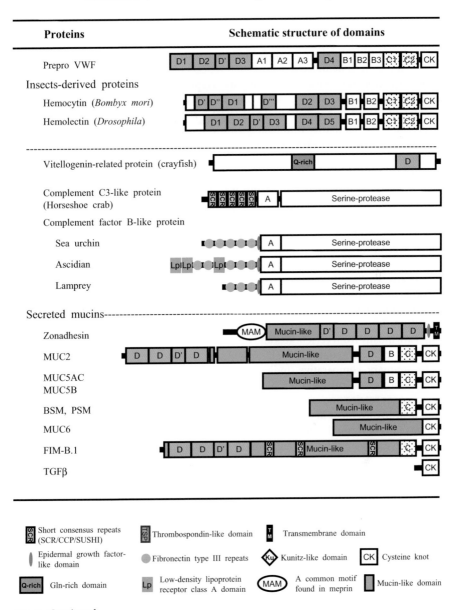

FIG. 9. *Continued*

chromosome, position 70CI -5, and consists of 25 exons. As shown in Fig. 9, the organi-zation of D1-D2-D′-D3 essential for vWF to be processed by furin, to bind to factor VIII, and to form interchain disulfide linkages is mostly conserved. Moreover, the NH_2-terminal region of hemolectin has mutual repeats of three complement control pro-teins (CCPs), also known as a small consensus repeat (SCR) and sushi domain, and two EGF-like domains, both of which are contained in complement factors [61], and other proteins involved in hemostasis. The COOH-terminal two-thirds of the hemolectin sequence shares similarity with *Bombyx mori*-derived hemocytin, mentioned above.

Representatives of all vWF domains are thus found throughout the vertebrates and in several invertebrate species, indicating that vWF has deep evolutionary roots asso-ciated with vertebrate circulatory systems and its role in hemostasis [65–68]. These facts strongly suggest that a mechanism of hemostasis is evolutionarily conserved from insects to mammals. Because *Drosophila* has no platelets, thrombin, fibrinogen, or typical fibrillar collagens, the mechanism of hemostasis may be quite different from those of vertebrate animals. In fact, hemolectin, in addition to hemocytin (*Bombyx mori*), lacks vWF-A domains, which are critical for vWF to bind to fibrillar collagens, platelet-derived GPIb, and botrocetin. This suggests that hemolectin functionally differs from vWF, although it is involved in hemostasis/coagulation in *Drosophila* larval hemolymph [69].

According to the essay by Whittaker and Hynes [59], the phylogenetic distribution of vWF-A domains is listed in more than 500 proteins. Although the vWF domains in Fig. 9 are mostly from vertebrates and invertebrates, there are also plant, fungal, bacte-rial, and archaean sequences [53].

In summary, comparative studies have been performed on vWF domains found in various proteins. Consideration of the known properties of vWF domains in plasma membrane proteins, such as integrins, ECM proteins, and secreted mucins, allows insights into their involvement in protein–protein interactions and the roles of bound divalent cations and conformational changes. Further functional, structural, and genomic studies should give a clearer understanding of the vWF domains as a functional unit.

Acknowledgment. The authors thank Tamami Sonoda for assisting in typing the manuscript.

References

1. High K, Roberts HR (eds) (1995) Molecular basis of thrombosis and hemostasis. Marcel Dekker, New York
2. Heibrunn LV (1961) The evolution of the hemostatic mechanism. In: Macfarlane RG, Robb -Smith ATH (eds) Functions of blood. Academic, London
3. Sparkling NW (1981) Comparative physiology of blood clotting. Comp Biochim Physiol 68A:541–548
4. Gregoire C (1974) Hemolymph coagulation. In: Rockstein M (ed) The Physiology of Insects, vol 5. Academic, San Diego, pp 309–360
5. Doolittle RF (1993) The evolution of vertebrate blood coagulation: a case of Yin and Yang. Thromb Haemost 70:24–28
6. Pathy L (1990) Evolutionary assembly of blood coagulation. Semin Thromb Hemost 16:245–259
7. Iwanaga S (1993) Primitive coagulation systems and their message to modem biology. Thromb Haemost 70:48–55

8. Hall M, Wang R, Antwerpen R, et al (1999) The crayfish plasma clotting protein: a vitellogenin -related protein responsible for clot formation in crustacean blood. Proc Natl Acad Sci U S A 96:1965–1970

9. Davidson CJ, Tunddenham EG, Mcvey JH (2003) 450 Million years of hemostasis. J Thromb Haemost 1:1487–1494

10. Rowley AF, Hill DJ, Ray CE, et al (1997) Haemostasis in fish: an evolutionary perspective. Thromb Haemost 77:227–233

11. Davidson CJ, Hirt RP, Lal K, et al (2003) Molecular evolution of the vertebrate blood coagulation network. Thromb Haemost 89:420–428

12. Blombäck B (1996) Fibrinogen and fibrin: proteins with complex roles in hemostasis and thrombosis. Thromb Res 83:1–75

13. Weisel JW, Francis CW, Nagaswami C, et al (1993) Determination of the topology of factor XIIIa-induced fibrin γ-chain cross-links by electron microscopy of ligated fragments. J Biol Chem 268:26618–26624

14. Brown JH, Volkmann N, Jun G, et al (2000) The crystal structure of modified bovine fibrinogen. Proc Natl Acad Sci U S A 97:85–90

15. Fujita T, Matsushita M, Endo Y (2004) The lectin-complement pathway: its role in innate immunity and evolution. Immunol Rev 198:185–202

16. Hsia HC, Schwarzbauer JE (2005) Meet the tenascins: multifunctional and mysterious. J Biol Chem 280:26641–26644

17. Chiquet-Ehrismann R (2004) Tenascins. Int J Biochem Cell Biol 36:986–990

18. Morisada T, Kubota Y, Urano T, et al (2006) Angiopoietins and angiopoietin-like proteins in angiogenesis. Endothelium 13:71–79

19. Zhao Z, Lee CC, Jiralerspong S, et al (1995) The gene for a human microfibril-associated glycoprotein is commonly deleted in Smith-Magenis syndrome patients. Hum Mol Genet 4:589–597

20. Rüegg C, Pytela R (1995) Sequence of a human transcript expressed in T-lymphocytes and encoding a fibrinogen-like protein. Gene 160:257–262

21. Tsuji S, Uehori J, Matsumoto M, et al (2001) Human intelectin is a novel soluble lectin that recognizes galactofuranose in carbohydrate chains of bacterial cell wall. J Biol Chem 276:23456–23463

22. Peles E, Nativ M, Lustig M, et al (1997) Identification of a novel contactin-associated transmembrane receptor with multiple domains implicated in protein-protein interactions. EMBO J 16:978–988

23. Omori-Satoh T, Yamakawa Y, Mebs D (2000) The antihemorrhagic factor, erinacin, from the European hedgehog (Erinaceus europaeus), a metalloprotease inhibitor of large molecular size possessing ficolin/opsonin P35 lectin domains. Toxicon 38:1561–1580

24. Kenjo A, Takahashi M, Matsushita M, at al (2001) Cloning and characterization of novel ficolins from the solitary ascidian, Halocynthia roretzi. J Biol Chem 276:19959–19965

25. Kurachi S, Song Z, Takagaki M, et al (1998) Sialic-acid-binding lectin from the slug Limax flavus cloning, expression of the polypeptide, and tissue localization. Eur J Biochem 254:217–222

26. Léonard PM, Adema CM, Zhang SM, et al (2001) Structure of two FREP genes that combine IgSF and fibrinogen domains, with comments on diversity of the FREP gene family in the snail Biomphalaria glabrata. Gene 269:155–165

27. Xu X, Doolittle RF (1990) Presence of a vertebrate fibrinogen-like sequence in an echinoderm. Proc Natl Acad Sci U S A 87:2097–2101

28. Baker NE, Mlodzik M, Rubin GM (1990) Spacing differentiation in the developing Drosophila eye: a fibrinogen-related lateral inhibitor encoded by scabrous. Science 250:1370–1377

29. Gokudan S, Muta T, Tsuda R, et al (1999) Horseshoe crab acetyl group-recognizing lectins involved in innate immunity are structurally related to fibrinogen. Proc Natl Acad Sci U S A 96:10086–10091

30. Perović-Ottstadt S, Adell T, Proksch P, et al (2004) A (1→3)-β-D-glucan recognition protein from the sponge Suberites domuncula. Eur J Biochem 271:1924–1937

31. Kairies N, Beisel HG, Fuentes-Prior P, et al (2001) The 2.0-Å crystal structure of tachylectin 5A provides evidence for the common origin of the innate immunity and the blood coagulation systems. Proc Natl Acad Sci U S A 98:13519–13524
32. Garlatti V, Belloy N, Martin L, et al (2007) Structural insights into the innate immune recognition specificities of L- and H-ficolins. EMBO J 26:623–633
33. Tanio M, Kondo S, Sugio S, et al (2007) Trivalent recognition unit of innate immunity system. J Biol Chem 282:3889–3895
34. Barton WA, Tzvetkova D, Nikolov DB (2005) Structure of the angiopoietin-2 receptor binding domain and identification of surfaces involved in Tie2 recognition. Structure 13:825–832
35. Davis S, Papadopoulos N, Aldrich TH, et al (2003) Angiopoietins have distinct modular domains essential for receptor binding, dimerization and superclustering. Nat Struct Biol 10:38–44
36. Venter JC, Adams MD, Myers EW, et al (2001) The sequence of the human genome. Science 291:1304–1351
37. Doolittle RF, Spraggon G, Everse SJ (1997) Evolution of vertebrate fibrin formation and the process of its dissolution. Ciba Found Symp 212:4–23
38. Iwanaga S, Kawabata S (1998) Evolution and phylogeny of defense molecules associated with innate immunity in horseshoe crab. Front Biosci 3:973–984
39. Muta T, Iwanaga S (1996) Clotting and immune defense in Limulidae. Prog Mol Subcell Biol 15:154–189
40. Muta T, Iwanaga S (1996) The role of hemolymph coagulation in innate immunity. Curr Opin Immunol 8:41–47
41. Yeh MS, Huang CJ, Leu JH, et al (1999) Molecular cloning and characterization of a hemolymph clottable protein from tiger shrimp (Penaeus monodon). Eur J Biochem 266:624–633
42. Avarre JC, Lubzens E, Babin PJ (2007) Apolipocrustacein, formerly vitellogenin, is the major egg yolk precursor protein in decapod crustaceans and is homologous to insect apolipophorin II/I and vertebrate apolipoprotein B. BMC Evol Biol 7: http://www.biomedcentral.com/1471-2148/7/3
43. Iwanaga S, Kawabata S, Muta T (1998) New types of clotting factors and defense molecules found in horseshoe crab hemolymph: their structures and functions. J Biochem (JB review) 123:1–15
44. Kawasaki H, Nose T, Muta T, et al (2000) Head-to-tail polymerization of coagulin, a clottable protein of the horseshoe crab. J Biol Chem 275:35297–35301
45. Bergner A, Oganessyan V, Muta T, et al (1996) Crystal structure of coagulogen, the clotting protein from horseshoe crab: a structural homologue of nerve growth factor. EMBO J 15:6789–6797
46. Mizuguchi K, Parker JS, Blundell TL, et al (1998) Getting knotted: a model for the structure and activation of Spätzle. Trends Biochem Sci 23:239–242
47. Sadler JE (1998) Biochemistry and genetics of von Willebrand factor. Annu Rev Biochem 67:395–424
48. Ruggeri ZM (1999) Structure and function of von Willebrand factor. Thromb Haemost 82:576–584
49. Sadler JE, Shelton Inloes BB, Sorace JM, et al (1985) Cloning and characterization of two cDNAs coding for human von Willebrand factor. Proc Natl Acad Sci U S A 82:6394–6398
50. Titani K, Kumar S, Takio K, et al (1986) Amino acid sequence of human von Willebrand factor. Biochemistry 25:3171–3186
51. Ginsberg D (1999) Molecular genetics of von Willebrand disease. Thromb Haemost 82:585–591
52. Maita N, Nishio K, Nishimoto E, et al (2003) Crystal structure of von Willebrand factor A1 domain complexed with snake venom, Bitiscetin: insight into glycoprotein Ibα binding mechanism induced by snake venom proteins. J Biol Chem 278:37777–37781

53. Tuckwell D (1999) Evolution of von Willebrand factor A (VWA) domains. Biochem Soc Transact 27:835–840
54. Ulrichts H, Udvardy M, Lenting PJ, et al (2006) Shielding of the A1 domain by the D′ D3 domains of von Willebrand factor modulates its interaction 1b-IX-V. J Biol Chem 281:4699–4707
55. Perez-Vilar J, Hill RL (1999) The structure and assembly of secreted mucins. J Biol Chem 274:31751–31754
56. Desseyn JL, Aubert JP, Porchet N, et al (2000) Evolution of the large secreted gel-forming mucins. Mol Biol Evol 17:1175–1184
57. Matsui T, Hamako J (2005) Structure and function of snake venom toxins interacting with von Willebrand factor. Toxicon 45:1075–1087
58. Obert B, Romijn RA, Houllier A, et al (2006) Characterization of bitiscetin-2, a second form of the bitiscetin from the venom of *Bitis arietans*: comparison of its binding site with the collagen-binding site on the von Willebrand factor A3-domain. J Thromb Haemost 4:1596–1601
59. Whitaker CA, Hynes RO (2002) Distribution and evolution of von Willebrand/integrin A domains: widely dispersed domains with roles in cell adhesion and elsewhere. Mol Biol Cell 13:3369–3387
60. Zhu Y, Thangamani S, He B, Ding JL (2005) The origin of the complement system. EMBO J 24:382–394
61. Nonaka M (2001) Evolution of the complement system. Curr Opin Immunol 13:69–73
62. Kotani E, Yamakawa M, Iwamoto S, et al (1995) Cloning and expression of the gene of hemocytin, an insect humoral lectin which is homologous with the mammalian von Willebrand factor. Biochim Biophys Acta 1260:245–258
63. Goto A, Kadowaki T, Kitagawa Y (2003) *Drosophila* hemolectin gene is expressed in embryonic and larval hemocytes and its knock down causes bleeding defects. Dev Biol 264:582–591
64. Goto A, Kumagai T, Kumagai C, et al (2001) A *Drosophila* hemocyte-specific protein, hemolectin, similar to human von Willebrand factor. Biochem J 359:99–108
65. Theopold U, Fabbri DLM, Scherfer C, et al (2002) The coagulation of insect hemolymph. CMLS Cell Mol Life Sci 59:363–372
66. Krem MM, Dicera E (2002) Evolution of enzyme cascades from embryonic development to blood coagulation. Trends Biochem Sci 27:67–74
67. Theopold U, Schmidt O, Söderhäll K, et al (2004) Coagulation in arthropods: defense, wound closure and healing. Trends Immunol 25:289–294
68. Bohn H, Barwig B (1984) Hemolymph clotting in the cockroach *Leucophaea maderae* (Blattaria). J Comp Physiol B 154:457–467
69. Bidla G, Lindgren M, Theopold U, et al (2005) Hemolymph coagulation and phenoloxidase in *Drosophila* larvae. Dev Comp Immunol 29:668–679
70. Iwanaga S (1993b) The limulus clotting reaction. Curr Opin Immunol 5:74–82
71. Iwanaga S (2002) The molecular basis of innate immunity in the horseshoe crab. Curr Opin Immunol 14:87–95
72. Iwanaga S, Miyata T, Tokunaga F, et al (1992) Molecular mechanism of hemolymph clotting systems in *Limulus*. Thromb Res 68:1–32
73. Kawabata S, Muta T, Iawnaga S (1996) Clotting cascade and defense molecules found in the hemolymph of the horseshoe crab. In: Söderhäll K, Iwanaga S, Vasta G (eds) New directions in invertebrate immunology. SOS Publication, Fair Haven, CT, pp 255–283
74. Tokunaga F, Iwanaga S (1993) Horseshoe crab transglutaminase. Methods Enzymol 223:378–388
75. Jiravanichpaisal P, Lee BL, Söderhäll K (2006) Cell-mediated immunity in arthropods: hematopoiesis, coagulation, melanization and opsonization. Immunobiology 211:213–236
76. Hall M, Heusden MC, Söderhäll K (1995) Identification of the major lipoproteins in crayfish hemolymph as proteins involved in immunorecognition and clotting. Biochem Biophys Res Commun 216:939–946

Snake Venoms and Other Toxic Components Affecting Thrombosis and Hemostasis

Yasuo Yamazaki and Takashi Morita

Summary. Blood coagulation and platelet aggregation are the primary host defense systems against invasion by nonself factors such as viruses and pathogens. Some exogenous proteins/peptides from snake venoms and the saliva of hematophagous organisms target blood coagulation and platelet aggregation systems. Many unique proteins/peptides that do not exist in mammals have been isolated from various snake venoms and the salivary glands of leeches and ticks. These compounds specifically attack critical molecules of prey and hosts to disrupt and destroy their homeostatic systems. Such potent and highly specific exogenous proteins/peptides have thus been indispensable in elucidating the complex physiology of mammalian blood coagulation and platelet aggregation. This chapter provides an overview of the structures and functions of snake venom toxins, salivary proteins of hematophagous organisms, and also some bacteria-derived proteins that affect blood coagulation and platelet aggregation.

Key words. Blood coagulation · Platelet aggregation · Leech · Tick · *Staphylococcus aureus*

Introduction

The coagulation cascade is outlined in Fig. 1. Coagulation factors are generally circulating in the bloodstream as inactive precursors, and a small amount of factor VII (~1% of the total factor VII) is known to exist in its active form (factor VIIa) (see Fig. 1). When tissue factor (TF) is exposed to blood as a result of vascular wall injury, it immediately forms a complex with factor VIIa, after which the complex directly activates factor IX to produce factor IXa. Factor X and prothrombin are then sequentially activated to produce factor Xa and α-thrombin, respectively. The coagulation factors VII/VIIa, IX/IXa, X/Xa, and prothrombin are vitamin K-dependent and contain 10–13 γ-carboxyglutamic acid (Gla) residues in their N-terminal small domains (Gla domain). The Gla domains are essential for Ca^{2+}-dependent binding of these coagulation factors to platelet and endothelial cell phospholipids, resulting in efficient localized fibrin clot formation.

Platelet receptors and endogenous ligands (agonists) are summarized in Fig. 2. Platelet activation/aggregation is elicited by von Willebrand factor (vWF), collagen, adenosine diphosphate (ADP), and α-thrombin through several distinct platelet

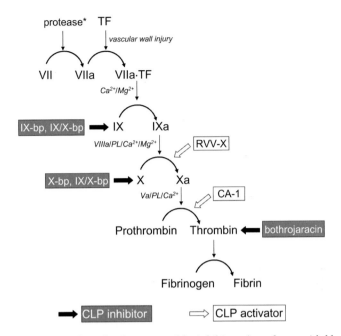

Fig. 1. Coagulation cascade and points targeted by inhibitors (*gray boxes with black arrows*) and activators (*white boxes with white arrows*) of venom C-type lectin-like proteins (*CLP*). *Although physiological activating enzyme for factor VII is unclear, factor VII-activating protease/plasma hyaluronan-binding protein (FSAP/PHBP) [149] and hepsin [150, 151], a type II transmembrane serine protease, are reported to activate factor VII in vitro. FSAP/PHBP is a secretory serine protease that is homologous with hepatocyte growth factor activator [152]

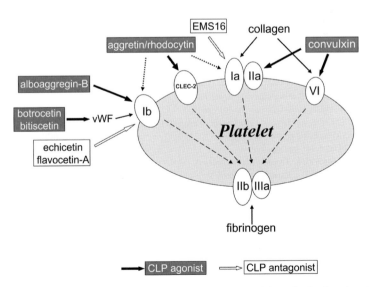

Fig. 2. C-type lectin-like proteins (*CLP*) of snake venom that modulate platelet function. Agonist molecules are depicted as *gray boxes* with *black arrows*; antagonist molecules are depicted as *white boxes* with *white arrows*

receptors, including glycoprotein Ib (GPIb), GPIa/IIa (also known as integrin $\alpha_2\beta_1$), GPVI, and protease-activated receptor-1 (Fig. 2). Receptor-mediated platelet activation ultimately results in platelet aggregation due to an interaction between fibrinogen and GPIIb/IIIa (integrin $\alpha_{IIb}\beta_3$) (Fig. 2).

Snake Venom Proteins Affecting Blood Coagulation and Platelet Aggregation

Snake venom proteins/peptides have evolved to produce effective toxins for capturing prey [1]. Snake venom proteins/peptides can be classified into the following structural groups: serine proteases, Kunitz-type protease inhibitors, metalloproteinases, phospholipases A_2, L-amino acid oxidases, C-type lectin(-like) proteins, disintegrins, bradykinin-potentiating peptides, atrial natriuretic peptides, nerve growth factors, vascular endothelial growth factors, cystatin-like proteins, sarafotoxins, three-finger toxins, and cysteine-rich secretory proteins (Table 1) [2]. Members of the C-type lectin-like protein (CLP) family comprise most major venom proteins that affect blood coagulation and platelet aggregation (Figs. 1, 2; Table 2). A number of discoveries have resulted from functional and structural studies of CLPs as follows. Functional studies of factors IX- and X-binding protein (IX/X-bp) have revealed that Mg^{2+} ions are crucial for the blood coagulation cascade [3, 4]. Crystal structures of the Gla domains of coagulation factors IX and X have been determined from structural studies of complexes between the Gla domain of factor X and factor X-binding protein (X-bp, a venom CLP) [5] and the Gla domain of factor IX and factor IX-binding protein (IX-bp, also a venom CLP) [6]. A three-dimensional (3D) domain-swapping phenomenon was discovered in the first study to identify IX/X-bp, the first known CLP [7]. The discovery of a calcium-dependent prothrombin activator with C-type lectin-like domains, carinactivase (a Ca^{2+}-dependent prothrombin activator) [8], has led to its clinical application in the measurement of plasma prothrombin levels [9, 10].

TABLE 1. Toxins affecting blood coagulation factors and platelets

Toxin (molecular weight, derivation)	Target molecule
Coagulation factor-activating/cleaving proteins	
Ancrod[b] (35.4 kDa, *Calloselasma rhodostoma*)	Fibrinogen
Fibrolase[c] (23 kDa, *Agkistrodon contortrix contortrix*)	Fibrinogen
Ecarin[c] (55 kDa, *Echis sochureki*)	Prothrombin
CA-1[c] (87 kDa, *Echis leucogaster*)	Prothrombin
Trocarin D[b] (46.5 kDa, *Tropidechis carinatus*)	Prothrombin
Pseutarin C[b] (250 kDa, *Pseudonaja textiles*)	Prothrombin
Staphylocoagulase (64 kDa, *Staphylococcus aureus*[a])	Prothrombin
RVV-X[c] (79 kDa, *Vipera russelli russelli*)	Factor X
Gingipain R1[h] (110 kDa, *Porphyromonas gingivalis*[a])	Factor X, protein C
RVV-V[b] (29 kDa, *Vipera russelli russelli*)	Factor V
Agkicontin[b] (37 kDa, *Agkistrodon contortrix contortrix*)	Protein C
TSV-PA[b] (33 kDa, *Trimeresurus stejnegeri*)	Plasminogen
Streptokinase (47 kDa, *Streptococcus equisimilis*[a])	Plasminogen
Bradykinin-releasing enzyme[b] (30 kDa, *Gloydius blomhoffi*)	Kininogen
Hemorrhagic factor[c] (25 kDa, *Trimeresurus flavoviridis*)	Basement membrane

Table 1. *Continued*

Toxin (molecular weight, derivation)	Target molecule
Coagulation factor-inhibiting proteins/peptides	
Hirudin (7 kDa, medicinal leech, *Hirudo medicinalis*[a])	α-Thrombin
Bothrojaracin[d] (27 kDa, *Bothrops jararaca*)	α-Thrombin
Ornithodorin[j] (12.6 kDa, soft tick, *Ornithodoros moubata*[a])	α-Thrombin
Rhodniin (11 kDa, insect, *Rhodnius prolixus*[a])	α-Thrombin
Haemadin (4.3 kDa, Indian leech, *Haemadipsa sylvestris*[a])	α-Thrombin
X-bp[d] (27 kDa, *Deinagkistrodon acutus*)	Factor X/Xa
Ammodytoxin A[i] (13.8 kDa, *Vipera ammodytes ammodytes*)	Factor Xa
TAP[e] (7 kDa, soft tick, *Ornithodoros moubata*[a])	Factor Xa
Ixolaris[e] (15.7 kDa, black-legged tick, *Ixodes scapularis*[a])	Factor X/Xa, factor VIIa/TF
Penthalaris[e] (35 kDa, black-legged tick, *Ixodes scapularis*[a])	Factor X/Xa, factor VIIa/TF
Antistasin (15 kDa, Mexican leech, *Haementeria officinalis*[a])	Factor Xa
IX/X-bp[d] (27 kDa, *Trimeresurus flavoviridis*)	Factors IX/IXa, X/Xa
IX-bp[d] (27.5 kDa, *Trimeresurus flavoviridis*)	Factor IX/IXa
M-LAO (60 kDa, *Gloydius blomhoffi*)	Factor IX
Prolixin-S (20 kDa, insect, *Rhodnius prolixus*[a])	Factor IX/IXa
Hemextin AB complex[g] (14 kDa, *Hemachatus haemachatus*)	Factor VIIa
NAP c2 (9.6 kDa, dog hookworm, *Ancylostoma caninum*[a])	Factor VIIa/TF
Haemaphysalin[j] (16.2 kDa, tick, *Haemaphysalis longicornis*[a])	Factor XII/XIIa, HK
Platelet aggregation inducers	
Botrocetin[d] (25 kDa, *Bothrops jararaca*)	vWF/GPIb
Bitiscetin[d] (29 kDa, *Bitis arietans*)	vWF/GPIb
Alboaggregin-B[d] (23 kDa, *Trimeresurus albolabris*)	GPIb
Aggretin/rhodocytin[d] (29 kDa, *Calloselasma rhodostoma*)	CLEC-2
Convulxin[d] (72 kDa, *Crotalus durissus terrificus*)	GPVI
Alborhagin[c] (60 kDa, *Trimeresurus albolabris*)	GPVI
Platelet aggregation inhibitors	
Echicetin[d] (26 kDa, *Echis carinatus*)	GPIb
Agkicetin/agkistin[d] (29 kDa, *Deinagkistrodon acutus*)	GPIb
Flavocetin-A[d] (149 kDa, *Trimeresurus flavoviridis*)	GPIb
Mocarhagin[c] (55 kDa, *Naja mocambique mocambique*)	GPIb
Jararhagin[c] (52 kDa, *Bothrops jararaca*)	vWF, GPIa/IIa
EMS16[d] (33 kDa, *Echis multisquamatus*)	GPIa/IIa
Rhodocetin[d] (35 kDa, *Calloselasma rhodostoma*)	GPIa/IIa
Trigramin[f] (8 kDa, *Trimeresurus gramineus*)	GPIIb/IIIa
Piscivostatin[f] (14.2 kDa, *Agkistrodon piscivorus piscivorus*)	GPIIb/IIIa
Contortrostatin[f] (15 kDa, *Agkistrodon contortrix contortrix*)	GPIIb/IIIa
Mambin[g] (6.8 kDa, *Dendroaspis jamesonii*)	GPIIb/IIIa
Decorsin (4.4 kDa, North American leech, *Macrobdella decora*[a])	GPIIb/IIIa
Disagregin[j] (7 kDa, soft tick, *Ornithodoros moubata*[a])	GPIIb/IIIa
Savignygrin[j] (7 kDa, *Ornithodoros savignyi*[a])	GPIIb/IIIa

[a]Organisms other than snakes
[b]Serine protease
[c]Metalloproteinase
[d]C-type lectin-like protein
[e]Kunitz-type inhibitor
[f]Disintegrin
[g]Three-finger toxin homologue
[h]Cysteine protease
[i]Phospholipase A$_2$ homologue
[j]Kunitz-type inhibitor homologue

TABLE 2. C-type lectin-like proteins of snake venom that modulate hemostasis

CLP	Structure	Ligand/substrate
Anticoagulants		
IX/X-bp, IX-bp	$\alpha\beta$	Factor IX
IX/X-bp, X-bp	$\alpha\beta$	Factor X
Bothrojaracin	$\alpha\beta$	α-Thrombin
Coagulants		
RVV-X	$\alpha\beta$/metalloproteinase	Factor X
Carinactivase-1 (CA-1)	$\alpha\beta$/metalloproteinase	Prothrombin
Platelet agonists		
Botrocetin, bitiscetin	$\alpha\beta$	vWF
Alboaggregin-B	$\alpha\beta$	GPIbα
Agglucetin	Tetrameric structure[a]	GPIbα
Aggretin/rhodocytin	$\alpha\beta$	CLEC-2
Convulxin	$(\alpha\beta)_4$	GPVI
Platelet antagonists		
Echicetin	$\alpha\beta$	GPIbα
Flavocetin-A (FL-A)	$(\alpha\beta)_4$	GPIbα
EMS16	$\alpha\beta$	GPIa/IIa

CLP, C-type lectin-like proteins
[a]Speculated to be a tetramer composed of $\alpha1$-$\beta2$-$\alpha2$-$\beta1$

Many CLPs are used as critical reagents in studies of platelet activation [11]. Recently, a novel platelet receptor, CLEC-2, has been identified using a CLP-type platelet aggregation inducer, aggretin/rhodocytin (Tables 1, 2) [12].

C-Type Lectin-Like Proteins

Structure of CLPs

Despite their similarity in name, the C-type lectin-like proteins (CLPs) of snake venom differ from C-type lectins. CLPs demonstrate a range of biological activities, including anticoagulant- and platelet-modulating activities, but they do not possess lectin activity, unlike C-type lectins [13, 14]. The amino acid sequences of CLP subunits share 15%–40% homology with the carbohydrate recognition domains (CRDs) of C-type lectins, including mammalian mannose-binding protein (MBP) [15, 16], as well as snake venom galactose-specific lectin [17]. A prominent feature of CLPs is that they are heterodimers or oligomeric complexes of heterodimers. In contrast, C-type lectins are Ca^{2+}-dependent lectins of snake venom, which are exclusively homodimers or homooligomers. Table 2 lists various CLPs that affect hemostasis, including CLPs that function as anticoagulants, coagulants, and platelet modulators (agonists and antagonists), all of which lack lectin activity.

Snake Venom CLP Anticoagulants

Both CLP activators and inhibitors in snake venom target intermediate stages of the clotting cascade (Fig. 1) [18]. The venom CLP anticoagulants IX/X-bp [19, 20], IX-bp

[21, 22], and X-bp [23], inhibit coagulation factors IX and X (Fig. 1, Table 1) by binding to their Gla domains [24]. Crystal structures of the Gla domains of factors IX and X have been determined by studying the complexes that form between coagulation factors IX and X and various CLP anticoagulant proteins [5, 6]. Mg^{2+} ions also bind to the Gla domain of factor IX and are critical to the blood coagulation cascade [4, 6]. Another venom CLP anticoagulant, bothrojaracin, has been identified and purified from the venom of *Bothrops jararaca*, the most common venomous snake of South America (Fig. 1, Table 1) [25, 26]. Bothrojaracin forms a noncovalent complex with α-thrombin without altering its ability to catalyze the activity of small synthetic peptide substrates. This CLP binds with high affinity ($K_d = 0.6$ nM) to thrombin exosite 1, indicating that binding at this site is a major determinant of thrombin-bothrojaracin specificity [25, 26].

Snake Venom CLP Procoagulants

The CLP coagulants directly reduce plasma levels of factor X and prothrombin or indirectly convert fibrinogen to fibrin in plasma, resulting in serious hemostatic disturbances, including disseminated intravascular coagulation (DIC) [18]. Factor X activator, RVV-X, isolated from *Daboia russelli* (formerly *Vipera russelli*; Russell's viper) venom, is one of the most potent procoagulants identified to date (Fig. 1, Table 1). RVV-X is composed a 92 880-dalton glycoprotein with a heavy chain (the catalytic subunit) and two distinct light chains [27]. The heavy chain contains three distinct domains: a metalloproteinase domain, a disintegrin (a platelet aggregation inhibitor, see Disintegrins)-like domain, and a Cys-rich domain of unknown function, which are commonly found in mammalian ADAMs (a disintegrin and metalloproteinase) [28–30]. The two light chains of RVV-X are thought to be regulatory subunits with C-type lectin-like domains that recognize the Gla domain of the factor X light chain [18] (our unpublished observation). Activation of factor X by RVV-X is inhibited by flooding with the isolated Gla domain (amino acids 1–44) of factor X [18] (our unpublished observation), suggesting that RVV-X light chains bind to the factor X Gla domain. Further understanding of the role of Ca^{2+} in this process comes from studies of the Ca^{2+}-dependent prothrombin activator, CA-1, from *Echis carinatus leucogaster* venom (Fig. 1, Table 1). CA-1 has a structure similar to that of RVV-X. Like RVV-X, CA-1 has three polypeptide chains of 62 kDa, 17 kDa, and 14 kDa [8, 18]. The isolated regulatory subunits of CA-1 (two light chains homologous to CLPs) recognize the Ca^{2+}-bound conformation of the Gla domain in prothrombin [8]. These observations suggest that RVV-X primarily recognizes the Ca^{2+}-bound conformation of the Gla domain of factor X via its CLP-type regulatory subunit, with subsequent conversion of factor X to factor Xa, which is then catalyzed by its catalytic metalloproteinase subunit [18]. RVV-X can also activate factor IX and protein C through a similar catalytic mechanism, although activation activity against factor IX and protein C is significantly weaker than that for factor X.

vWF/GPIb-Binding CLPs

Botrocetin, isolated from the venom of *B. jararaca*, was the first identified CLP to induce platelet agglutination (Fig. 2, Table 1) [31, 32]. This protein first binds to the

A1 domain of vWF to form an active complex, which then induces platelet agglutination (Fig. 2). The crystal structure of a complex formed between botrocetin and a "gain-of-function" mutant A1 domain (I546V) of vWF demonstrates that botrocetin can switch the conformation adopted by the mutant A1 domain back toward the wild-type conformation, suggesting that the affinity of botrocetin for vWF is enhanced by exposing the GPIb-binding surface [33]. Bitiscetin, isolated from the venom of *Bitis arietans*, is another CLP that induces platelet agglutination (Fig. 2, Table 1) [34]. Several biochemical studies of bitiscetin indicate that the vWF A1 domain, which binds GPIb, is essential for its biological activity [35]. The crystal structure of bitiscetin suggests that a negatively charged region on its surface may bind to a positively charged region of the vWF A1 domain [36]. A crystallographic study of the vWF A1 domain complexed with bitiscetin has further demonstrated that the vWF A1 domain contacts a concave depression on bitiscetin [37].

GPIb-Binding CLPs

Alboaggregin-B, a member of the CLP family isolated from *Trimeresurus albolabris* venom, aggregates platelets directly, without the need for any cofactors (Fig. 2, Table 1) [38, 39]. It competes with vWF in binding to platelet GPIb [39]. In contrast, echicetin, a CLP from the venom of *Echis carinatus sochureki*, specifically inhibits the aggregation of fixed platelets induced by various platelet GPIb agonists, including bovine vWF, alboaggregin-B, and human vWF in the presence of botrocetin (Fig. 2, Table 1) [40, 41]. In addition to alboaggregin-B and echicetin, a high-molecular-mass GPIb-binding protein known as flavocetin-A has been isolated from the habu snake, *T. flavoviridis* (Fig. 2, Table 1) [42]. This protein has a molecular mass of 149 kDa and binds with the greatest affinity to platelet GPIb. The K_d value for flavocetin-A (0.35 nM) is 10- to 100-fold less than that for echicetin. In addition, flavocetin-A strongly inhibits vWF-dependent aggregation of fixed platelets and inhibits shear-induced platelet aggregation at high shear stress. The X-ray structure of flavocetin-A demonstrates a cyclic tetramer made up of four identical $\alpha\beta$-heterodimers arranged head to tail (Fig. 3).

Platelet Collagen Receptor (GPIa/IIa and GPVI)-Binding CLPs

Platelets adhere to collagen fibers exposed at sites of damage to the endothelial lining through specific membrane receptors, resulting in platelet activation. Platelet GPVI and GPIa/IIa (integrin $\alpha_2\beta_1$) are collagen receptors that participate in platelet activation. Convulxin, a CLP isolated from the venom of the tropical rattlesnake *Crotalus durissus terrificus*, is a platelet agonist that acts on GPVI (Fig. 2, Table 1) [43]. Although previous studies indicate that convulxin is an $\alpha_3\beta_3$ heterotrimer, its crystal structure demonstrates a disulfide-linked cyclic $\alpha_4\beta_4$ heterotetramer, similar to flavocetin-A (Fig. 3) [44, 45].

Recently, EMS16, a CLP isolated from the venom of *E. multisquamatus*, has been identified as the first selective GPIa/IIa antagonist to be isolated from snake venom

FIG. 3. Crystal structure of flavocetin-A, a GPIb antagonistic CLP from the venom of the Habu snake *Trimeresurus flavoviridis*. Flavocetin-A consists of a cyclic tetramer of dimers linked by eight disulfide bridges. Subunits α and β are shown in magenta and green, respectively. Interchain disulfide bridges within and between the αβ-heterodimers are shown as ball-and-stick representations (*yellow and black*)

(Fig. 2, Table 1) [46]. It binds to the integrin α_2-I domain, which is the collagen-binding site of integrin α_2, inhibiting collagen-induced platelet aggregation and cell migration. Amino acid sequencing and crystallographic studies have been performed to clarify how the integrin α_2-I domain–EMS16 complex is formed [47–49].

Initially, aggretin/rhodocytin, a CLP from *Calloselasma rhodostoma* venom, was thought to induce platelet aggregation by acting as a GPIa/IIa and GPIb agonist (Fig. 2) [50–54]. However, it has recently been shown that aggretin/rhodocytin induces platelet aggregation via a novel platelet receptor, CLEC-2 (Fig. 2, Table 1) [12]. The extracellular domain of CLEC-2 demonstrates structural similarities to members of the CLP family. A single tyrosine residue is found in its cytoplasmic tail and undergoes tyrosine phosphorylation upon platelet activation by aggretin/rhodocytin but not collagen, thrombin receptor agonists, or convulxin [12]. Activation of platelets by aggretin/rhodocytin is abolished in either Syk- or PLCg2-deficient mice and is partially reduced in the absence of other signaling molecules, such as LAT, SLP-2, or Vav1/Vav3 [12]. These findings indicate the existence of a novel platelet activation pathway distinct from other known pathways. Quite recently, podoplanin, a transmembrane sialoglycoprotein, has shown to act as an endogenous ligand for CLEC-2 [153] . Crystal structural study has recently revealed that an extracellular domain of CLEC-2 has a fold similar to that of C-type lectin [55].

Disintegrins

Disintegrins are a family of small, naturally occurring proteins (49–84 amino acid residues long) found in snake venoms [56]. Disintegrins inhibit platelet aggregation by binding to the fibrinogen receptor, which is the major integrin on the surface of platelets, referred to as GPIIb/IIIa (also known as integrin $\alpha_{IIb}\beta_3$). Upon activation of platelets by various physiological agonists, GPIIb/IIIa undergoes a conformational change, allowing it to bind to fibrinogen [57, 58]. Binding of fibrinogen to the receptor mediates platelet aggregation. Disintegrins prevent fibrinogen binding to GPIIb/IIIa by binding directly with the receptor. Venom disintegrins contain an RGD sequence (Arg-Gly-Asp) that is known to function as a universal cell recognition sequence, and disintegrins inhibit platelet aggregation by binding specific integrins, such as vitronectin and fibronectin receptors, on the surface of endothelial cells and fibroblasts.

Disintegrins have been divided into three groups based on their length and number of cysteine residues: (1) short disintegrins, composed of 49–51 amino acid residues and 8 cysteine residues; (2) medium-size disintegrins, composed of about 70 amino acid residues and 12 cysteine residues; and (3) long disintegrins, composed of 84 amino acid residues and 14 cysteine residues. In addition to these monomeric disintegrins, dimeric disintegrins (13–15 kDa), such as contortrostatin [59], EMF10 [60], piscivostatin [61], acostatin [62], and schistatin [63], have been discovered more recently. To date, more than 40 disintegrins have been identified. All disintegrins show a high degree of homology with each other with regard to the arrangement of cysteine residues in each group. Recently, the crystal structures of three disintegrin structures—monomeric [64], homodimeric [63], and heterodimeric disintegrins [65]— have been elucidated. In each, cysteine residues are involved in disulfide bond formation with an integrin-binding loop protruding from one end of the long axis of each elongated molecule [63–65]. The C-terminal region is located spatially near the integrin-binding loop, where it is readily exposed, suggesting that it might mediate binding to the receptor [63–65]. The monomers of dimeric disintegrins are structurally similar to monomeric disintegrins and are firmly linked through two intermolecular disulfide bridges at their N-terminals [63, 65]. The two integrin-binding loops of the dimer are divergent [63, 65]. Although it has been speculated that the disintegrin-like domains of mammalian ADAMs might be structurally distinct from snake venom disintegrins, the disintegrin-like domain of ADAM10 has been shown to adopt a conformation similar to that of snake venom proteins [66]. The recently reported crystal structure of snake venom ADAM (VAP1) has revealed that ADAM has a C-shaped arm-like structure composed of an N-terminal metalloproteinase domain following a disintegrin-like domain and a C-terminal cysteine-rich domain [67]. The integrin-binding loop is packed by the cysteine-rich domain and is inaccessible for protein binding [67].

It is believed that disintegrin arises from the same precursor as metalloproteinases [68]. However, recent findings indicate that some disintegrin precursors, including one of the heterodimer chains, are encoded on short mRNA sequences lacking almost all propeptide and metalloproteinase domains [62, 69]. Several unique disintegrins, in which integrin-binding tripeptide motifs are replaced by non-RGD sequences, are found in a number of viper venoms [69–72]. These molecules demonstrate considerable divergence in their integrin-binding selectivity (summarized by Calvete et al.) [73].

Other Toxic Components Affecting Blood Coagulation and Platelet Aggregation

Additional Snake Venom-Derived Mediators of Blood Coagulation

The venoms of some Australian elapid snakes contain other classes of prothrombin activators, distinct from the metalloproteinases discussed above, such as CA-1. Trocarin D from the Australian rough-scaled snake (*Tropidechis carinatus*) is structurally similar to human coagulation factor Xa with an identical domain architecture, including an N-terminal Gla domain followed by two epidermal growth factor (EGF)-like domains and a serine protease domain (Table 1) [74]. The prothrombinase activity of trocarin D is potentiated by Ca^{2+}, factor Va, and phospholipids (also cofactors of human prothrombinase), suggesting a mechanism of activation similar to that of human factor Xa. Pseutarin C from *Pseudonaja textilis* venom is a large protein complex consisting of catalytic and nonenzymatic subunits, which are functionally similar to the mammalian factor Xa–factor Va complex (Table 1) [75, 76]. Both proteases cleave prothrombin at the same loci as human factor Xa. Two distinct classes of fibrin(ogen)olytic enzymes have been identified in various snake venoms, including serine and metalloproteinases (Table 1). These fibrin(ogen)olytic proteases preferentially cleave either the α or β chains of fibrinogen in the absence of any cofactors but have little to no effect on fibrin clot formation [77]. Fibrin(ogen)olytic and hemorrhagic responses are closely related but are generally caused by different proteases (Table 1) [78, 79]; however, enzymes from some snake venom species mediate both fibrin(ogen)olytic and hemorrhagic activity. Russell's viper (*Vipera russelli russelli*, also known as *Daboia r. russelli*) venom contains a serine protease RVV-V that increases the coagulant activity of coagulation factor V (Table 1) [80]. Human thrombin cleaves factor V at three loci: Arg 709, Arg1018, and Arg 1545. Cleavage at the former two sites by thrombin is insufficient for activation but necessary for rapid cleavage at Arg1545, which results in full activation of factor V [81]. In contrast to thrombin, RVV-V directly cleaves factor V at Arg1545, and this single cleavage at Arg1545 is sufficient for complete activation of factor V [81].

A serine protease that specifically activates protein C is found in the venom of the southern copperhead snake *Agkistrodon contortrix contortrix* (Table 1) [82, 83]. Activated protein C is an anticoagulant protease that inactivates coagulation factors Va and VIIIa, resulting in inhibition of blood coagulation [84]. Physiologically, protein C is activated by meizothrombin, an active intermediate of prothrombin, in the presence of phospholipids and thrombomodulin, a transmembrane protein that acts as a cofactor for protein C activation by meizothrombin [85–87]. Snake venom protein C activator directly and efficiently activates protein C in the absence of these cofactors. As such, this enzyme is now used in several clinical assays for protein C concentration and protein C resistance [88]. A specific plasminogen activator from *Trimeresurus stejnegeri* (also known as *Viridovipera stejnegeri*) venom (TSV-PA) is a serine protease that cleaves plasminogen at Arg561 to generate two-chain plasmin, a key enzyme in fibrinolysis (Table 1) [89, 90]. The sequence of TSV-PA exhibits marked homology with RVV-V and protein C activator despite considerable differences in their substrate specificities.

In addition to the above-listed coagulation factor-cleaving proteases, coagulation factor-inhibiting toxins are also found in various snake venoms. Most of the anticoagulants in snake venoms are CLPs [14, 91, 92]. Phospholipase A$_2$ (PLA$_2$) enzymes and their homologues are found in most snake venoms. They induce widespread pharmacological effects, such as anticoagulation [93, 94], as well as myotoxic [95, 96] and neurotoxic [97, 98] effects despite similarities in their structure. Snake venoms contain two classes of PLA$_2$ enzymes that exhibit anticoagulant effects. The first class weakly inhibits the extrinsic tenase complex by hydrolyzing and destroying phospholipids, essential cofactors in the coagulation response [93, 94]. The second class of anticoagulant PLA$_2$ enzymes, including CM-IV from *Naja nigricollis* venom and ammodytoxin A from *Vipera ammodytes ammodytes* venom, inhibit the activities of both tenase and prothrombinase complexes through combined enzymatic and nonenzymatic effects (Table 1) [93, 94, 99]. These PLA$_2$ enzymes have been shown to bind directly to factor Xa and block formation of the prothrombinase complex by competing with factor Va [93, 94, 99]. Recently, recombinant proteins have been used to show that the anticoagulation site of ammodytoxin A is located on its C-terminal loop and β-wing regions [99].

L-Amino acid oxidases (LAAOs) are present in several snake venoms. LAAOs are generally homodimeric flavin adenine dinucleotide (FAD)-binding glycoproteins that catalyze the oxidative deamination of L-amino acids to α-keto acids, thereby releasing ammonia and hydrogen peroxide [100]. L-Amino acid oxidase (M-LAO), an LAAO from the venom of *Gloydius blomhoffi*, acts as a specific inhibitor of coagulation factor IX (Table 1) [101]. M-LAO selectively prolongs factor IX-mediated plasma clotting time through an indirect mechanism but does not affect the clotting time mediated by other factors, such as factor X, prothrombin, and fibrinogen [101]. A hydrogen peroxide-mediated indirect mechanism is thought to be responsible for M-LAO-mediated prolongation of factor IX-mediated plasma clotting time, but the exact mechanism remains unclear.

Recently, two synergistically acting anticoagulant proteins, hemextin A and hemextin B, were isolated from the venom of *Hemachatus haemachatus* (Table 1) [102]. Hemextin A and hemextin B form a 2:2 complex and inhibit the coagulant activity of factor VIIa in a noncompetitive manner [102]. On its own, hemextin A is a weak and partial inhibitor of factor VIIa that exhibits mild anticoagulant activity; hemextin B has no effect at all [102]. The N-terminal sequences of both proteins are highly homologous and suggest that hemextins belong to the three-finger toxin family [103] that is commonly observed in snake venom-derived nicotinic acetylcholine receptor blockers.

Other Components of Snake Venom Affecting Platelet Aggregation

The CLPs and disintegrins are the major components of snake venoms that modulate platelet aggregation [11, 14, 56, 91, 104, 105]. Some snake venoms contain metalloproteinases that affect platelet membrane receptors, such as GPIb, GPIa/IIa, and GPVI. Mocarhagin, a metalloproteinase from *Naja mocambique mocambique* venom, cleaves at Glu282 in the anionic sulfated region of the GPIb α chain, resulting in inhibition of vWF-mediated platelet aggregation (Table 1) [106]. Jararhagin was first isolated as a hemorrhagic metalloproteinase from the venom of *Bothrops jararaca* (Table 1) [107, 108]. This enzyme also inhibits platelet aggregation by targeting two distinct molecules, vWF and GPIa/IIa [107, 108]. Jararhagin cleaves vWF at several sites in the A1 domain, the GPIb-binding site, and the GPIIa (β_1 integrin) subunit of the platelet surface GPIa/IIa receptor ($\alpha_2\beta_1$ integrin); it also acts as a binding protein to the I domain of GPIa (α_2 integrin), resulting in inhibition of collagen-induced platelet activation [107, 108]. In contrast to these aggregation inhibitory enzymes, a unique metalloproteinase, alborhagin, is found in the venom of *Trimeresurus albolabris* (Table 1) [109]. This enzyme activates platelets via the GPVI-mediated pathway in a manner similar to that of convulxin (see C-Type Lectin-Like Proteins) [11, 110]. Alborhagin partially blocks convulxin binding to the cell surface-expressed GPVI (40% at maximum inhibition), indicating that each protein recognizes similar but distinct sites on the GPVI molecule [109].

In addition to the well-known disintegrins (see Disintegrins), unique disintegrin-like molecules are found in the venom of snakes belonging to the genus *Dendroaspis*. These proteins, including mambin from *Dendroaspis jamesonii* venom, are not homologous with disintegrins despite having an RGD motif; however, they exhibit significant homology with short-chain postsynaptic neurotoxins belonging to the three-finger toxin family (Table 1) [111].

Toxins Affecting Blood Coagulation and Platelet Aggregation from Other Organisms

Bacteria and hematophagous organisms such as leeches, ticks, and hookworms—are plentiful sources of a number of substances that prevent blood coagulation. Staphylocoagulase from *Staphylococcus aureus* [112–122] and streptokinase from *Streptococcus equisimilisa* [123–125] are activators of prothrombin and plasminogen, respectively, each with a unique catalytic mechanism (Table 1). Both proteins form stable complexes with their substrate proteins in a 1:1 molar ratio and subsequently convert zymogens to activated forms without proteolysis [122, 123]. Gingipains are secretory trypsin-like cysteine proteases derived from *Porphyromonas gingivalis*, a gram-negative anaerobic bacterium; they are strongly associated with human periodontal disease [126]. Gingipains, such as gingipain R1, are effective activators of several coagulation factors, such as factor X and protein C (Table 1) [127, 128].

Hirudin is a potent, highly selective inhibitor of α-thrombin derived from the salivary glands of *Hirudo medicinalis* leeches. The N-terminal region of hirudin directly associates with an area near the active site cleft of α-thrombin, and the carboxyl tail wraps around the α-thrombin overlying the fibrinogen-recognition site (anion-binding exosite 1) [129]. Two double-headed thrombin inhibitors, rhodniin, a Kazal-type inhibitor from the bug *Rhodnius prolixus* [130, 131], and ornithodorin, a Kunitz-type inhibitor from the soft tick *Ornithodoros moubata* [132], have also been shown to contact both the active site cleft and the anion-binding exosite 1 of α-thrombin (Table 1). Ornithodorin interacts with the protease via distinct regions from the reactive loops of two Kunitz-type domains [132]. Haemadin, isolated from an Indian leech *Haemadipsa sylvestris* [133], also interacts with thrombin at two sites: Its N-terminal portion binds to the active site similar to hirudin, and its C-terminal acidic tail interacts with the heparin-binding site of thrombin (anion-binding exosite 2) rather than exosite 1 (Table 1) [134].

Several factor Xa inhibitors are found in the saliva of ticks. Tick anticoagulant peptide (TAP), isolated from the soft tick *Ornithodoros moubata*, is a potent and specific inhibitor of factor Xa, which belongs to the Kunitz-type protease inhibitor family (Table 1) [135]. Recently, two additional Kunitz-type Xa inhibitors, ixolaris [136] and penthalaris [137], have been cloned following isolation from the saliva of the tick *Ixodes scapularis* (Table 1). Ixolaris and penthalaris have two and five tandem Kunitz domains, respectively, and directly bind to and inhibit factor X/Xa. They also exhibit factor Xa-dependent factor VIIa/tissue factor inhibitory activity similar to that of the human tissue factor pathway inhibitor (TFPI) molecule. Nematode anticoagulant protein c2 (NAPc2) is also a TFPI-like factor Xa-dependent inhibitor of factor VIIa/tissue factor; however, binding to factor Xa does not efficiently attenuate its amidolytic activity [138].

Antistasin from the Mexican leech *Haementeria officinalis* is a 15-kDa protein that inhibits factor Xa (Table 1) [139, 140]. It is composed of two homologous tandem domains and a basic C-terminal tail. Its N-terminal domain binds directly to and inhibits factor Xa, and its C-terminal tail disrupts interactions with phospholipids, factor Va, and prothrombin in the prothrombinase complex [141, 142]. A hemoprotein prolixin-S from *Rhodnius prolixus* blocks the coagulant activity of factors IX and IXa by binding to its Gla domain in a manner similar to that of snake venom IX/X-bp (Table 1) [143]. Prolixin-S also functions as a nitric oxide transporter, causing vasodilation and inhibition of platelet aggregation. Haemaphysalin from the hard tick *Haemaphysalis longicornis* directly binds factor XII/XIIa and high-molecular-weight kininogen (HK), and it interferes with the association of these molecules with biological activating surface (Table 1) [144].

Several platelet aggregation inhibitors have been isolated from leeches and ticks. A GPIIb/IIIa antagonist, decorsin, is found in the saliva of the North American leech *Macrobdella decora* (Table 1) [145]. Crystal structure analysis reveals that decorsin folds in a manner similar to that of the potent thrombin inhibitor hirudin [146]. Disagregin [147] and savignygrin [148] also show disintegrin-like properties despite their structural similarity to members of the Kunitz-type protease inhibitor family (Table 1). Savignygrin has an RGD motif that corresponds with the reactive center of a protease inhibitor [148], whereas the corresponding sequence of disagregin replaces to RED [147].

Conclusion

Exogenous organisms are abundant sources of unique bioactive molecules. Proteins/peptides from the venoms of snakes and the saliva of leeches and ticks specifically target blood coagulation and platelet aggregation systems of their prey and hosts. That is, exogenous organisms are valuable sources of proteins that might aid in furthering our understanding of mammalian coagulation and platelet aggregation systems. Furthermore, studies of exogenous factors can lead to the design and development of novel drugs.

References

1. Fry BG, Vidal N, Norman JA, et al (2006) Early evolution of the venom system in lizards and snakes. Nature 439:584–588
2. Yamazaki Y, Morita T (2007) Snake venom components affecting blood coagulation and the vascular system: structural similarities and marked diversity. Curr Pharm Des 13:2872–2886
3. Sekiya F, Yamashita T, Atoda H, et al (1995) Regulation of the tertiary structure and function of coagulation factor IX by magnesium (II) ions. J Biol Chem 270: 14325–14331
4. Sekiya F, Yoshida M, Yamashita T, et al (1996) Magnesium (II) is a crucial constituent of the blood coagulation cascade: potentiation of coagulant activities of factor IX by Mg^{2+} ions. J Biol Chem 271:8541–8544
5. Mizuno H, Fujimoto Z, Atoda H, et al (2001) Crystal structure of an anticoagulant protein in complex with the Gla domain of factor X. Proc Natl Acad Sci U S A 98:7230–7234
6. Shikamoto Y, Morita T, Fujimoto Z, et al (2003) Crystal structure of Mg^{2+}- and Ca^{2+}-bound Gla domain of factor IX complexed with binding protein. J Biol Chem 278:24090–24094
7. Mizuno H, Fujimoto Z, Koizumi M, et al (1997) Structure of coagulation factors IX/X-binding protein, a heterodimer of C-type lectin domains. Nat Struct Biol 4:438–441
8. Yamada D, Sekiya F, Morita T (1996) Isolation and characterization of carinactivase, a novel prothrombin activator in *Echis carinatus* venom with a unique catalytic mechanism. J Biol Chem 271:5200–5207
9. Yamada D, Morita T (1999) CA-1 method, a novel assay for quantification of normal prothrombin using a Ca^{2+}-dependent prothrombin activator, carinactivase-1. Thromb Res 94:221–226
10. Iwahashi H, Kimura M, Nakajima K, et al (2001) Determination of plasma prothrombin level by Ca^{2+}-dependent prothrombin activator (CA-1) during warfarin anticoagulation. J Heart Valve Dis 10:388–392
11. Clemetson KJ, Navdaev A, Dormann D, et al (2001) Multifunctional snake C-type lectins affecting platelets. Haemostasis 31:148–154
12. Suzuki-Inoue K, Fuller GL, Garcia A, et al (2006) A novel Syk-dependent mechanism of platelet activation by the C-type lectin receptor CLEC-2. Blood 107:542–549
13. Drickamer K (1999) C-type lectin-like domains. Curr Opin Struct Biol 9:585–590
14. Morita T (2004) C-type lectin-related proteins from snake venoms. Curr Drug Targets Cardiovasc Haematol Disord 4:357–373
15. Weis WI, Drickamer K, Hendrickson WA (1992) Structure of a C-type mannose-binding protein complexed with an oligosaccharide. Nature 360:127–134
16. Weis WI, Kahn R, Fourme R, et al (1991) Structure of the calcium-dependent lectin domain from a rat mannose-binding protein determined by MAD phasing. Science 254:1608–1615
17. Hirabayashi J, Kusunoki T, Kasai K (1991) Complete primary structure of a galactose-specific lectin from the venom of the rattlesnake *Crotalus atrox*: homologies with Ca^{2+}-dependent-type lectins. J Biol Chem 266:2320–2326
18. Morita T (1998) Proteases which activate factor X. In: Bailey GS (ed) Snake venom enzymes. Alaken, Ft. Collins CO, pp 179–208.
19. Atoda H, Morita T (1989) A novel blood coagulation factor IX/factor X-binding protein with anticoagulant activity from the venom of *Trimeresurus flavoviridis* (Habu snake): isolation and characterization. J Biochem (Tokyo) 106:808–813
20. Atoda H, Hyuga M, Morita T (1991) The primary structure of coagulation factor IX/factor X-binding protein isolated from the venom of *Trimeresurus flavoviridis*: homology with asialoglycoprotein receptors, proteoglycan core protein, tetranectin,

and lymphocyte Fc epsilon receptor for immunoglobulin E. J Biol Chem 266:14903–14911

21. Atoda H, Ishikawa M, Yoshihara E, et al (1995) Blood coagulation factor IX-binding protein from the venom of *Trimeresurus flavoviridis*: purification and characterization. J Biochem (Tokyo) 118:965–973

22. Mizuno H, Fujimoto Z, Koizumi M, et al (1999) Crystal structure of coagulation factor IX-binding protein from habu snake venom at 2.6 A: implication of central loop swapping based on deletion in the linker region. J Mol Biol 289:103–112

23. Atoda H, Ishikawa M, Mizuno H, et al (1998) Coagulation factor X-binding protein from *Deinagkistrodon acutus* venom is a Gla domain-binding protein. Biochemistry 37:17361–17370

24. Atoda H, Yoshida N, Ishikawa M, et al (1994) Binding properties of the coagulation factor IX/factor X-binding protein isolated from the venom of *Trimeresurus flavoviridis*. Eur J Biochem 224:703–708

25. Arocas V, Zingali RB, Guillin MC, et al (1996) Bothrojaracin: a potent two-site-directed thrombin inhibitor. Biochemistry 35:9083–9089

26. Zingali RB, Jandrot-Perrus M, Guillin MC, et al (1993) Bothrojaracin, a new thrombin inhibitor isolated from *Bothrops jararaca* venom: characterization and mechanism of thrombin inhibition. Biochemistry 32:10794–10802

27. Gowda DC, Jackson CM, Hensley P, et al (1994) Factor X-activating glycoprotein of Russell's viper venom: polypeptide composition and characterization of the carbohydrate moieties. J Biol Chem 269:10644–10650

28. Paine MJ, Desmond HP, Theakston RD, et al (1992) Purification, cloning, and molecular characterization of a high molecular weight hemorrhagic metalloprotease, jararhagin, from *Bothrops jararaca* venom: insights into the disintegrin gene family. J Biol Chem 267:22869–22876

29. Perry AC, Barker HL, Jones R, et al (1994) Genetic evidence for an additional member of the metalloproteinase-like, disintegrin-like, cysteine-rich (MDC) family of mammalian proteins and its abundant expression in the testis. Biochim Biophys Acta 1207:134–137

30. Takeya H, Nishida S, Miyata T, et al (1992) Coagulation factor X activating enzyme from Russell's viper venom (RVV-X): a novel metalloproteinase with disintegrin (platelet aggregation inhibitor)-like and C-type lectin-like domains. J Biol Chem 267:14109–14117

31. Brinkhous KM, Barnes DS, Potter JY, et al (1981) Von Willebrand syndrome induced by a *Bothrops* venom factor: bioassay for venom coagglutinin. Proc Natl Acad Sci U S A 78:3230–3234

32. Brinkhous KM, Read MS, Fricke WA, et al (1983) Botrocetin (venom coagglutinin): reaction with a broad spectrum of multimeric forms of factor VIII macromolecular complex. Proc Natl Acad Sci U S A 80:1463–1466

33. Fukuda K, Doggett TA, Bankston LA, et al (2002) Structural basis of von Willebrand factor activation by the snake toxin botrocetin. Structure 10:943–950

34. Hamako J, Matsui T, Suzuki M, et al (1996) Purification and characterization of bitiscetin, a novel von Willebrand factor modulator protein from *Bitis arietans* snake venom. Biochem Biophys Res Commun 226:273–279

35. Matsui T, Hamako J, Matsushita T, et al (2002) Binding site on human von Willebrand factor of bitiscetin, a snake venom-derived platelet aggregation inducer. Biochemistry 41:7939–7946

36. Hirotsu S, Mizuno H, Fukuda K, et al (2001) Crystal structure of bitiscetin, a von Willebrand factor-dependent platelet aggregation inducer. Biochemistry 40:13592–13597

37. Maita N, Nishio K, Nishimoto E, et al (2003) Crystal structure of von Willebrand factor A1 domain complexed with snake venom, bitiscetin: insight into glycoprotein Ibalpha binding mechanism induced by snake venom proteins. J Biol Chem 278:37777–37781

38. Peng M, Lu W, Kirby EP (1991) Alboaggregin-B: a new platelet agonist that binds to platelet membrane glycoprotein Ib. Biochemistry 30:11529–11536

39. Yoshida E, Fujimura Y, Miura S, et al (1993) Alboaggregin-B and botrocetin, two snake venom proteins with highly homologous amino acid sequences but totally distinct functions on von Willebrand factor binding to platelets. Biochem Biophys Res Commun 191:1386–1392

40. Peng M, Emig FA, Mao A, et al (1995) Interaction of echicetin with a high affinity thrombin binding site on platelet glycoprotein GPIb. Thromb Haemost 74:954–957

41. Polgar J, Magnenat EM, Peitsch MC, et al (1997) Amino acid sequence of the alpha subunit and computer modelling of the alpha and beta subunits of echicetin from the venom of Echis carinatus (saw-scaled viper). Biochem J 323 (Pt 2):533–537

42. Taniuchi Y, Kawasaki T, Fujimura Y, et al (1995) Flavocetin-A and -B, two high molecular mass glycoprotein Ib binding proteins with high affinity purified from Trimeresurus flavoviridis venom, inhibit platelet aggregation at high shear stress. Biochim Biophys Acta 1244:331–338

43. Clemetson JM, Polgar J, Magnenat E, et al (1999) The platelet collagen receptor glycoprotein VI is a member of the immunoglobulin superfamily closely related to FcalphaR and the natural killer receptors. J Biol Chem 274:29019–29024

44. Batuwangala T, Leduc M, Gibbins JM, et al (2004) Structure of the snake-venom toxin convulxin. Acta Crystallogr D Biol Crystallogr 60:46–53

45. Murakami MT, Zela SP, Gava LM, et al (2003) Crystal structure of the platelet activator convulxin, a disulfide-linked alpha4beta4 cyclic tetramer from the venom of Crotalus durissus terrificus. Biochem Biophys Res Commun 310:478–482

46. Marcinkiewicz C, Lobb RR, Marcinkiewicz MM, et al (2000) Isolation and characterization of EMS16, a C-lectin type protein from Echis multisquamatus venom, a potent and selective inhibitor of the alpha2beta1 integrin. Biochemistry 39:9859–9867

47. Horii K, Okuda D, Morita T, et al (2004) Crystal structure of EMS16 in complex with the integrin alpha2-I domain. J Mol Biol 341:519–527

48. Horii K, Okuda D, Morita T, et al (2003) Structural characterization of EMS16, an antagonist of collagen receptor (GPIa/IIa) from the venom of Echis multisquamatus. Biochemistry 42:12497–12502

49. Okuda D, Horii K, Mizuno H, et al (2003) Characterization and preliminary crystallographic studies of EMS16, an antagonist of collagen receptor (GPIa/IIa) from the venom of Echis multisquamatus. J Biochem (Tokyo) 134:19–23

50. Bergmeier W, Bouvard D, Eble JA, et al (2001) Rhodocytin (aggretin) activates platelets lacking alpha(2)beta(1) integrin, glycoprotein VI, and the ligand-binding domain of glycoprotein Ibalpha. J Biol Chem 276:25121–25126

51. Chung CH, Peng HC, Huang TF (2001) Aggretin, a C-type lectin protein, induces platelet aggregation via integrin alpha(2)beta(1) and GPIb in a phosphatidylinositol 3-kinase independent pathway. Biochem Biophys Res Commun 285:689–695

52. Huang TF, Liu CZ, Yang SH (1995) Aggretin, a novel platelet-aggregation inducer from snake (Calloselasma rhodostoma) venom, activates phospholipase C by acting as a glycoprotein Ia/IIa agonist. Biochem J 309(Pt 3):1021–1027

53. Navdaev A, Clemetson JM, Polgar J, et al (2001) Aggretin, a heterodimeric C-type lectin from Calloselasma rhodostoma (malayan pit viper), stimulates platelets by binding to alpha 2beta 1 integrin and glycoprotein Ib, activating Syk and phospholipase Cgamma 2, but does not involve the glycoprotein VI/Fc receptor gamma chain collagen receptor. J Biol Chem 276:20882–20889

54. Shin Y, Morita T (1998) Rhodocytin, a functional novel platelet agonist belonging to the heterodimeric C-type lectin family, induces platelet aggregation independently of glycoprotein Ib. Biochem Biophys Res Commun 245:741–745

55. Watson AA, Brown J, Harlos K, et al (2007) The crystal structure and mutational binding analysis of the extracellular domain of the platelet-activating receptor CLEC-2. J Biol Chem 282:3165–3172

56. McLane MA, Sanchez EE, Wong A, et al (2004) Disintegrins. Curr Drug Targets Cardiovasc Haematol Disord 4:327–355
57. Phillips DR, Charo IF, Parise LV, et al (1988) The platelet membrane glycoprotein IIb-IIIa complex. Blood 71:831–843
58. Xiao T, Takagi J, Coller BS, et al (2004) Structural basis for allostery in integrins and binding to fibrinogen-mimetic therapeutics. Nature 432:59–67
59. Zhou Q, Hu P, Ritter MR, et al (2000) Molecular cloning and functional expression of contortrostatin, a homodimeric disintegrin from southern copperhead snake venom. Arch Biochem Biophys 375:278–288
60. Marcinkiewicz C, Calvete JJ, Vijay-Kumar S, et al (1999) Structural and functional characterization of EMF10, a heterodimeric disintegrin from *Eristocophis macmahoni* venom that selectively inhibits alpha 5 beta 1 integrin. Biochemistry 38:13302–13309
61. Okuda D, Morita T (2001) Purification and characterization of a new RGD/KGD-containing dimeric disintegrin, piscivostatin, from the venom of *Agkistrodon piscivorus piscivorus*: the unique effect of piscivostatin on platelet aggregation. J Biochem (Tokyo) 130:407–415
62. Okuda D, Koike H, Morita T (2002) A new gene structure of the disintegrin family: a subunit of dimeric disintegrin has a short coding region. Biochemistry 41:14248–14254
63. Bilgrami S, Tomar S, Yadav S, et al (2004) Crystal structure of schistatin, a disintegrin homodimer from saw-scaled viper (*Echis carinatus*) at 2.5 A resolution. J Mol Biol 341:829–837
64. Fujii Y, Okuda D, Fujimoto Z, et al (2003) Crystal structure of trimestatin, a disintegrin containing a cell adhesion recognition motif RGD. J Mol Biol 332:1115–1122
65. Bilgrami S, Yadav S, Kaur P, et al (2005) Crystal structure of the disintegrin heterodimer from saw-scaled viper (*Echis carinatus*) at 1.9 A resolution. Biochemistry 44:11058–11066
66. Janes PW, Saha N, Barton WA, et al (2005) Adam meets Eph: an ADAM substrate recognition module acts as a molecular switch for ephrin cleavage in trans. Cell 123:291–304
67. Takeda S, Igarashi T, Mori H, et al (2006) Crystal structures of VAP1 reveal ADAMs' MDC domain architecture and its unique C-shaped scaffold. EMBO J 25:2388–2396
68. Yamada D, Shin Y, Morita T (1999) Nucleotide sequence of a cDNA encoding a common precursor of disintegrin flavostatin and hemorrhagic factor HR2a from the venom of *Trimeresurus flavoviridis*. FEBS Lett 451:299–302
69. Sanz L, Chen RQ, Perez A, et al (2005) cDNA cloning and functional expression of jerdostatin, a novel RTS-disintegrin from *Trimeresurus jerdonii* and a specific antagonist of the alpha1beta1 integrin. J Biol Chem 280:40714–40722
70. Marcinkiewicz C, Weinreb PH, Calvete JJ, et al (2003) Obtustatin: a potent selective inhibitor of alpha1beta1 integrin in vitro and angiogenesis in vivo. Cancer Res 63:2020–2023
71. Eble JA, Bruckner P, Mayer U (2003) *Vipera lebetina* venom contains two disintegrins inhibiting laminin-binding beta1 integrins. J Biol Chem 278:26488–26496
72. Bazan-Socha S, Kisiel DG, Young B, et al (2004) Structural requirements of MLD-containing disintegrins for functional interaction with alpha 4 beta 1 and alpha 9 beta1 integrins. Biochemistry 43:1639–1647
73. Calvete JJ, Marcinkiewicz C, Monleon D, et al (2005) Snake venom disintegrins: evolution of structure and function. Toxicon 45:1063–1074
74. Joseph JS, Chung MC, Jeyaseelan K, et al (1999) Amino acid sequence of trocarin, a prothrombin activator from *Tropidechis carinatus* venom: its structural similarity to coagulation factor Xa. Blood 94:621–631
75. Rao VS, Kini RM (2002) Pseutarin C, a prothrombin activator from *Pseudonaja textilis* venom: its structural and functional similarity to mammalian coagulation factor Xa-Va complex. Thromb Haemost 88:611–619

76. Rao VS, Swarup S, Kini RM (2003) The nonenzymatic subunit of pseutarin C, a prothrombin activator from eastern brown snake (*Pseudonaja textilis*) venom, shows structural similarity to mammalian coagulation factor V. Blood 102:1347–1354

77. Swenson S, Markland FS Jr (2005) Snake venom fibrin(ogen)olytic enzymes. Toxicon 45:1021–1039

78. Markland FS Jr, Swenson S (2004) Venombin AB. In: Barrett AJ, Rawlings N, Woessner JF (ed) Handbook of proteolytic enzymes. Elsevier, Amsterdam, pp 1715–1723

79. Swenson S, Markland FS Jr, Toombs C (2004) Fibrolase. In: Barrett AJ, Rawlings N, Woessner JF (ed) Handbook of proteolytic enzymes. Elsevier, Amsterday, pp 640–643

80. Swenson S, Markland FS Jr (2004) Russell's viper venom factor V activator. In: Barrett AJ, Rawlings N, Woessner JF (ed) Handbook of proteolytic enzymes. Elsevier, Amsterdam, pp 1724–1726

81. Keller FG, Ortel TL, Quinn-Allen MA, et al (1995) Thrombin-catalyzed activation of recombinant human factor V. Biochemistry 34:4118–4124

82. Kisiel W, Kondo S, Smith KJ, et al (1987) Characterization of a protein C activator from *Agkistrodon contortrix contortrix* venom. J Biol Chem 262:12607–12613

83. McMullen BA, Fujikawa K, Kisiel W (1989) Primary structure of a protein C activator from *Agkistrodon contortrix contortrix* venom. Biochemistry 28:674–679

84. Esmon CT (2003) The protein C pathway. Chest 124:26S–32S

85. Cote HC, Bajzar L, Stevens WK, et al (1997) Functional characterization of recombinant human meizothrombin and Meizothrombin(desF1): thrombomodulin-dependent activation of protein C and thrombin-activatable fibrinolysis inhibitor (TAFI), platelet aggregation, antithrombin-III inhibition. J Biol Chem 272:6194–6200

86. Hackeng TM, Tans G, Koppelman SJ, et al (1996) Protein C activation on endothelial cells by prothrombin activation products generated in situ: meizothrombin is a better protein C activator than alpha-thrombin. Biochem J 319(Pt 2):399–405

87. Koike H, Okuda D, Morita T (2003) Mutations in autolytic loop-2 and at Asp^{554} of human prothrombin that enhance protein C activation by meizothrombin. J Biol Chem 278:15015–15022

88. Gempeler-Messina PM, Volz K, Buhler B, et al (2001) Protein C activators from snake venoms and their diagnostic use. Haemostasis 31:266–272

89. Zhang Y, Wisner A, Maroun RC, et al (1997) *Trimeresurus stejnegeri* snake venom plasminogen activator: site-directed mutagenesis and molecular modeling. J Biol Chem 272:20531–20537

90. Zhang Y, Wisner A, Xiong Y, et al (1995) A novel plasminogen activator from snake venom: purification, characterization, and molecular cloning. J Biol Chem 270:10246–10255

91. Morita T (2005) Structures and functions of snake venom CLPs (C-type lectin-like proteins) with anticoagulant-, procoagulant-, and platelet-modulating activities. Toxicon 45:1099–1114

92. Morita T (2004) Use of snake venom inhibitors in studies of the function and tertiary structure of coagulation factors. Int J Hematol 79:123–129

93. Kini RM (2005) Structure-function relationships and mechanism of anticoagulant phospholipase A_2 enzymes from snake venoms. Toxicon 45:1147–1161

94. Mounier CM, Bon C, Kini RM (2001) Anticoagulant venom and mammalian secreted phospholipases A_2: protein- versus phospholipid-dependent mechanism of action. Haemostasis 31:279–287

95. Lomonte B, Angulo Y, Calderon L (2003) An overview of lysine-49 phospholipase A_2 myotoxins from crotalid snake venoms and their structural determinants of myotoxic action. Toxicon 42:885–901

96. Gutierrez JM, Ownby CL (2003) Skeletal muscle degeneration induced by venom phospholipases A_2: insights into the mechanisms of local and systemic myotoxicity. Toxicon 42:915–931

97. Rigoni M, Caccin P, Gschmeissner S, et al (2005) Equivalent effects of snake PLA$_2$ neurotoxins and lysophospholipid-fatty acid mixtures. Science 310:1678–1680

98. Schiavo G, Matteoli M, Montecucco C (2000) Neurotoxins affecting neuroexocytosis. Physiol Rev 80:717–766

99. Prijatelj P, Charnay M, Ivanovski G, et al (2006) The C-terminal and beta-wing regions of ammodytoxin A, a neurotoxic phospholipase A$_2$ from *Vipera ammodytes ammodytes*, are critical for binding to factor Xa and for anticoagulant effect. Biochimie 88:69–76

100. Du XY, Clemetson KJ (2002) Snake venom L-amino acid oxidases. Toxicon 40:659–665

101. Sakurai Y, Shima M, Matsumoto T, et al (2003) Anticoagulant activity of M-LAO, L-amino acid oxidase purified from *Agkistrodon halys blomhoffii*, through selective inhibition of factor IX. Biochim Biophys Acta 1649:51–57

102. Banerjee Y, Mizuguchi J, Iwanaga S, et al (2005) Hemextin AB complex, a unique anticoagulant protein complex from *Hemachatus haemachatus* (African Ringhals cobra) venom that inhibits clot initiation and factor VIIa activity. J Biol Chem 280:42601–42611

103. Tsetlin V (1999) Snake venom alpha-neurotoxins and other "three-finger" proteins. Eur J Biochem 264:281–286

104. Lu Q, Navdaev A, Clemetson JM, et al (2005) Snake venom C-type lectins interacting with platelet receptors: structure-function relationships and effects on haemostasis. Toxicon 45:1089–1098

105. Lu Q, Clemetson JM, Clemetson KJ (2005) Snake venoms and hemostasis. J Thromb Haemost 3:1791–1799

106. Ward CM, Andrews RK, Smith AI, et al (1996) Mocarhagin, a novel cobra venom metalloproteinase, cleaves the platelet von Willebrand factor receptor glycoprotein Ibalpha: identification of the sulfated tyrosine/anionic sequence Tyr-276–Glu-282 of glycoprotein Ibalpha as a binding site for von Willebrand factor and alpha-thrombin. Biochemistry 35:4929–4938

107. Laing GD, Moura-da-Silva AM (2005) Jararhagin and its multiple effects on hemostasis. Toxicon 45:987–996

108. Laing GD, Paine MJI (2004) Jararhagin. In: Barrett AJ, Rawlings N, Woessner JF (ed) Handbook of proteolytic enzymes. Elsevier, Amsterdam, pp 654–656

109. Andrews RK, Gardiner EE, Asazuma N, et al (2001) A novel viper venom metalloproteinase, alborhagin, is an agonist at the platelet collagen receptor GPVI. J Biol Chem 276:28092–28097

110. Polgar J, Clemetson JM, Kehrel BE, et al (1997) Platelet activation and signal transduction by convulxin, a C-type lectin from *Crotalus durissus terrificus* (tropical rattlesnake) venom via the p62/GPVI collagen receptor. J Biol Chem 272:13576–13583

111. McDowell RS, Dennis MS, Louie A, et al (1992) Mambin, a potent glycoprotein IIb-IIIa antagonist and platelet aggregation inhibitor structurally related to the short neurotoxins. Biochemistry 31:4766–4772

112. Igarashi H, Morita T, Iwanaga S (1979) A new method for purification of staphylocoagulase by a bovine prothrombin-Sepharose column. J Biochem (Tokyo) 86:1615–1618

113. Morita T, Igarashi H, Iwanaga S (1981) Staphylocoagulase. Methods Enzymol 80:311–319

114. Kawabata S, Morita T, Iwanaga S, et al (1985) Staphylocoagulase-binding region in human prothrombin. J Biochem (Tokyo) 97:325–331

115. Kawabata S, Morita T, Iwanaga S, et al (1985) Difference in enzymatic properties between alpha-thrombin-staphylocoagulase complex and free alpha-thrombin. J Biochem (Tokyo) 97:1073–1078

116. Kawabata S, Morita T, Iwanaga S, et al (1985) Enzymatic properties of staphylothrombin, an active molecular complex formed between staphylocoagulase and human prothrombin. J Biochem (Tokyo) 98:1603–1614

117. Kawabata S, Miyata T, Morita T, et al (1986) The amino acid sequence of the procoagulant- and prothrombin-binding domain isolated from staphylocoagulase. J Biol Chem 261:527–531

118. Kawabata S, Morita T, Miyata T, et al (1986) Isolation and characterization of staphylocoagulase chymotryptic fragment: localization of the procoagulant- and prothrombin-binding domain of this protein. J Biol Chem 261:1427–1433

119. Kaida S, Miyata T, Yoshizawa Y, et al (1987) Nucleotide sequence of the staphylocoagulase gene: its unique COOH-terminal 8 tandem repeats. J Biochem (Tokyo) 102:1177–1186

120. Kaida S, Miyata T, Yoshizawa Y, et al (1989) Nucleotide and deduced amino acid7 sequences of staphylocoagulase gene from *Staphylococcus aureus* strain 213. Nucleic Acids Res 17:8871

121. Kawabata S, Morita T, Miyata T, et al (1987) Structure and function relationship of staphylocoagulase. J Protein Chem 6:17–32

122. Friedrich R, Panizzi P, Fuentes-Prior P, et al (2003) Staphylocoagulase is a prototype for the mechanism of cofactor-induced zymogen activation. Nature 425:535–539

123. Wang X, Lin X, Loy JA, et al (1998) Crystal structure of the catalytic domain of human plasmin complexed with streptokinase. Science 281:1662–1665

124. Boxrud PD, Verhamme IM, Fay WP, et al (2001) Streptokinase triggers conformational activation of plasminogen through specific interactions of the amino-terminal sequence and stabilizes the active zymogen conformation. J Biol Chem 276:26084–26089

125. Sun H, Ringdahl U, Homeister JW, et al (2004) Plasminogen is a critical host pathogenicity factor for group A streptococcal infection. Science 305:1283–1286

126. Imamura T, Pike RN, Potempa J, et al (1994) Pathogenesis of periodontitis: a major arginine-specific cysteine proteinase from *Porphyromonas gingivalis* induces vascular permeability enhancement through activation of the kallikrein/kinin pathway. J Clin Invest 94:361–367

127. Imamura T, Potempa J, Tanase S, et al (1997) Activation of blood coagulation factor X by arginine-specific cysteine proteinases (gingipain-Rs) from *Porphyromonas gingivalis*. J Biol Chem 272:16062–16067

128. Hosotaki K, Imamura T, Potempa J, et al (1999) Activation of protein C by arginine-specific cysteine proteinases (gingipains-R) from *Porphyromonas gingivalis*. Biol Chem 380:75–80

129. Grutter MG, Priestle JP, Rahuel J, et al (1990) Crystal structure of the thrombin-hirudin complex: a novel mode of serine protease inhibition. EMBO J 9:2361–2365

130. Van de Locht A, Lamba D, Bauer M, et al (1995) Two heads are better than one: crystal structure of the insect derived double domain Kazal inhibitor rhodniin in complex with thrombin. EMBO J 14:5149–5157

131. Friedrich T, Kroger B, Bialojan S, et al (1993) A Kazal-type inhibitor with thrombin specificity from *Rhodnius prolixus*. J Biol Chem 268:16216–16222

132. Van de Locht A, Stubbs MT, Bode W, et al (1996) The ornithodorin-thrombin crystal structure: a key to the TAP enigma? EMBO J 15:6011–6017

133. Strube KH, Kroger B, Bialojan S, et al (1993) Isolation, sequence analysis, and cloning of haemadin: an anticoagulant peptide from the Indian leech. J Biol Chem 268:8590–8595

134. Richardson JL, Kroger B, Hoeffken W, et al (2000) Crystal structure of the human alpha-thrombin-haemadin complex: an exosite II-binding inhibitor. EMBO J 19:5650–5660

135. Waxman L, Smith DE, Arcuri KE, et al (1990) Tick anticoagulant peptide (TAP) is a novel inhibitor of blood coagulation factor Xa. Science 248:593–596

136. Francischetti IM, Valenzuela JG, Andersen JF, et al (2002) Ixolaris, a novel recombinant tissue factor pathway inhibitor (TFPI) from the salivary gland of the tick, *Ixodes scapularis*: identification of factor X and factor Xa as scaffolds for the inhibition of factor VIIa/tissue factor complex. Blood 99:3602–3612

137. Francischetti IM, Mather TN, Ribeiro JM (2004) Penthalaris, a novel recombinant five-Kunitz tissue factor pathway inhibitor (TFPI) from the salivary gland of the tick vector of Lyme disease, *Ixodes scapularis*. Thromb Haemost 91:886–898

138. Stassens P, Bergum PW, Gansemans Y, et al (1996) Anticoagulant repertoire of the hookworm *Ancylostoma caninum*. Proc Natl Acad Sci U S A 93:2149–2154

139. Tuszynski GP, Gasic TB, Gasic GJ (1987) Isolation and characterization of antistasin: an inhibitor of metastasis and coagulation. J Biol Chem 262:9718–9723

140. Nutt E, Gasic T, Rodkey J, et al (1988) The amino acid sequence of antistasin: a potent inhibitor of factor Xa reveals a repeated internal structure. J Biol Chem 263:10162–10167

141. Lapatto R, Krengel U, Schreuder HA, et al (1997) X-ray structure of antistasin at 1.9 Å resolution and its modeled complex with blood coagulation factor Xa. EMBO J 16:5151–5161

142. Mao SS, Przysiecki CT, Krueger JA, et al (1998) Selective inhibition of factor Xa in the prothrombinase complex by the carboxyl-terminal domain of antistasin. J Biol Chem 273:30086–30091

143. Isawa H, Yuda M, Yoneda K, et al (2000) The insect salivary protein, prolixin-S, inhibits factor IXa generation and Xase complex formation in the blood coagulation pathway. J Biol Chem 275:6636–6641

144. Kato N, Iwanaga S, Okayama T, et al (2005) Identification and characterization of the plasma kallikrein-kinin system inhibitor, haemaphysalin, from hard tick, *Haemaphysalis longicornis*. Thromb Haemost 93:359–367

145. Seymour JL, Henzel WJ, Nevins B, et al (1990) Decorsin: a potent glycoprotein IIb-IIIa antagonist and platelet aggregation inhibitor from the leech *Macrobdella decora*. J Biol Chem 265:10143–10147

146. Krezel AM, Wagner G, Seymour-Ulmer J, et al (1994) Structure of the RGD protein decorsin: conserved motif and distinct function in leech proteins that affect blood clotting. Science 264:1944–1947

147. Karczewski J, Endris R, Connolly TM (1994) Disagregin is a fibrinogen receptor antagonist lacking the Arg-Gly-Asp sequence from the tick, *Ornithodoros moubata*. J Biol Chem 269:6702–6708

148. Mans BJ, Louw AI, Neitz AW (2002) Savignygrin, a platelet aggregation inhibitor from the soft tick *Ornithodoros savignyi*, presents the RGD integrin recognition motif on the Kunitz-BPTI fold. J Biol Chem 277:21371–21378

149. Romisch J, Feussner A, Vermohlen S, et al (1999) A protease isolated from human plasma activating factor VII independent of tissue factor. Blood Coagul Fibrinolysis 10:471–479

150. Leytus SP, Loeb KR, Hagen FS, et al (1988) A novel trypsin-like serine protease (hepsin) with a putative transmembrane domain expressed by human liver and hepatoma cells. Biochemistry 27:1067–1074

151. Kazama Y, Hamamoto T, Foster DC, et al (1995) Hepsin, a putative membrane-associated serine protease, activates human factor VII and initiates a pathway of blood coagulation on the cell surface leading to thrombin formation. J Biol Chem 270:66–72

152. Choi-Miura NH, Tobe T, Sumiya J, et al (1996) Purification and characterization of a novel hyaluronan-binding protein (PHBP) from human plasma: it has three EGF, a kringle and a serine protease domain, similar to hepatocyte growth factor activator. J Biochem (Tokyo) 119:1157–1165

153. Suzuki-Inoue K, Kato, Y, Inoue O, et al (2007) Involvement of the snake toxin receptor CLEC-2, in podoplanin-mediated platelet activation, by cancer cells. J Biol Chem 282:25993–26001

Structures and Potential Roles of Oligosaccharides in Human Coagulation Factors and von Willebrand Factor

Taei Matsui, Jiharu Hamako, and Koiti Titani

Summary. Almost all plasma proteins are glycosylated to a degree with either Asn-linked or Ser/Thr-linked sugar chains. Oligosaccharides on these glycoproteins have a supporting role in eliciting biological activity and/or a protective role from enzymic cleavage, as well as acting as a molecular tag for clearance by hepatic lectins. The structure and physiological function of oligosaccharides in plasma glycoproteins participating in thrombosis and hemostasis have been reviewed. Because glycosylation is strictly regulated at limited sites in the molecule, the extra glycosylation sites found in some fibrinogen abnormalities decrease fibrinogen activity. Although typical Ser/Thr-linked sugar chains have N-acetylgalactosamine as a linking sugar, coagulation factors VII, IX, and XII have been found to possess glucose and/or fucose residues for the linking. The ABO(H) blood group antigens are atypically associated with the Asn-linked oligosaccharides of coagulation factor VIII (FVIII) and von Willebrand factor (VWF). The ABO blood group has been known to affect plasma concentrations of FVIII/VWF. Blood group sugar chains on VWF appear to be correlated with an ABO genotype and susceptibility to ADAMTS13, a plasma metalloproteinase specific for VWF, which may provide the missing link between ABO(H) antigens on VWF and clinical evidence that blood group O donors have a reduced concentration of VWF when compared with non-O donors.

Key words. Oligosaccharide structure · Coagulation factors · von Willebrand factor · ABO(H) blood group antigens · Fuc (Glc)-O-linked sugar chains

Introduction

Plasma contains more than 100 species of protein, almost all of which are glycosylated. These sugar chain moieties have various functions, such as acting as molecular guards against proteolysis, molecular tags for lysosome targeting, molecular guides for proper assembly and secretion, clearance markers for asialoglycoprotein receptor (ASGPR) [1], and/or molecular props for maintaining structural conformation [2].

When compared to protein and DNA, the oligosaccharide structures of glycoproteins are not homogeneous and are highly complex. Glycosylation appears to be a nonspecific event, but their attachment sites are strictly limited to the Asn, Ser, and

Thr residues in the protein folds facing the molecular surface. Extra glycosylation sites, if arising from amino acid substitutions forming a consensus sequence for N-glycosylation, sometimes affect protein function. Thus, glycosylation is strictly regulated, and proper glycosylation is essential for eliciting suitable biological activities.

In this chapter, the structure and function of oligosaccharides on plasma glycoproteins, particularly human coagulation factors and von Willebrand factor (VWF), are discussed. Some coagulation factors have been found to possess unusual Ser/Thr glycosylation, and ABO(H) blood group carbohydrate antigen structures have been found in limited species of human plasma glycoproteins, including coagulation factor VIII (FVIII) and VWF. The potential roles of these oligosaccharides are also discussed.

Glycosylation of Proteins

The glycosylation patterns of glycoproteins are broadly classified into two groups: those covalently linked to carbohydrate chains through Asn or Ser/Thr residues, forming amide-linked (N-linked) or O-glycosidically linked (O-linked) sugar chains. N-linked sugar chains are transferred to accessible Asn residues as a preformed core set at the endoplasmic reticulum and are then processed successively by glycosidases and glycosyltransferases to a mature form while passing through the Golgi apparatus. The N-linked sugar chains are composed of a trimannosyl core structure with an oligomannose structure (high-mannose type), N-acetyllactosamine repeats (complex-type), or both (hybrid type) (Fig. 1a). Complex-type oligosaccharides are processed from the high-mannose type via the hybrid type structure. Nonreducing termini of the mature complex-type sugar chains are often capped with N-acetylneuraminic acid (NeuNAc: sialic acid) penultimate to β-galactose (Gal) residues.

O-Linked sugar chains are synthesized directly by the action of specific glycosyltransferases. Typical O-linked sugar chains are linked to Ser or Thr residues through N-acetylgalactosamine (GalNAc) residues (Fig. 1b) [3], but some glycoproteins have recently been found to be linked to Ser/Thr through xylose (Xyl), fucose (Fuc), glucose (Glc), N-acetylglucosamine (GlcNAc), or mannose (Man) residues [4]. The nonreducing ends of the O-linked sugar chains are also frequently terminated with sialic acid residues. In the case of O-linked sugar chains, they often exist as clusters in the sequence, resulting in the formation of a highly negatively charged field.

In addition to above the examples of glycosylation, some sugar chains (e.g., Le antigens) are noncovalently associated with proteins as glycolipids.

Oligosaccharides of Coagulation Factors

Typical complex-type N-linked sugar chains have been proposed among plasma proteins and are traditionally referred to as "serum-type" oligosaccharides. The glycosylation sites and oligosaccharide types of coagulant proteins are summarized in Table 1. Like other plasma glycoproteins, almost all coagulation factors are synthesized in the liver.

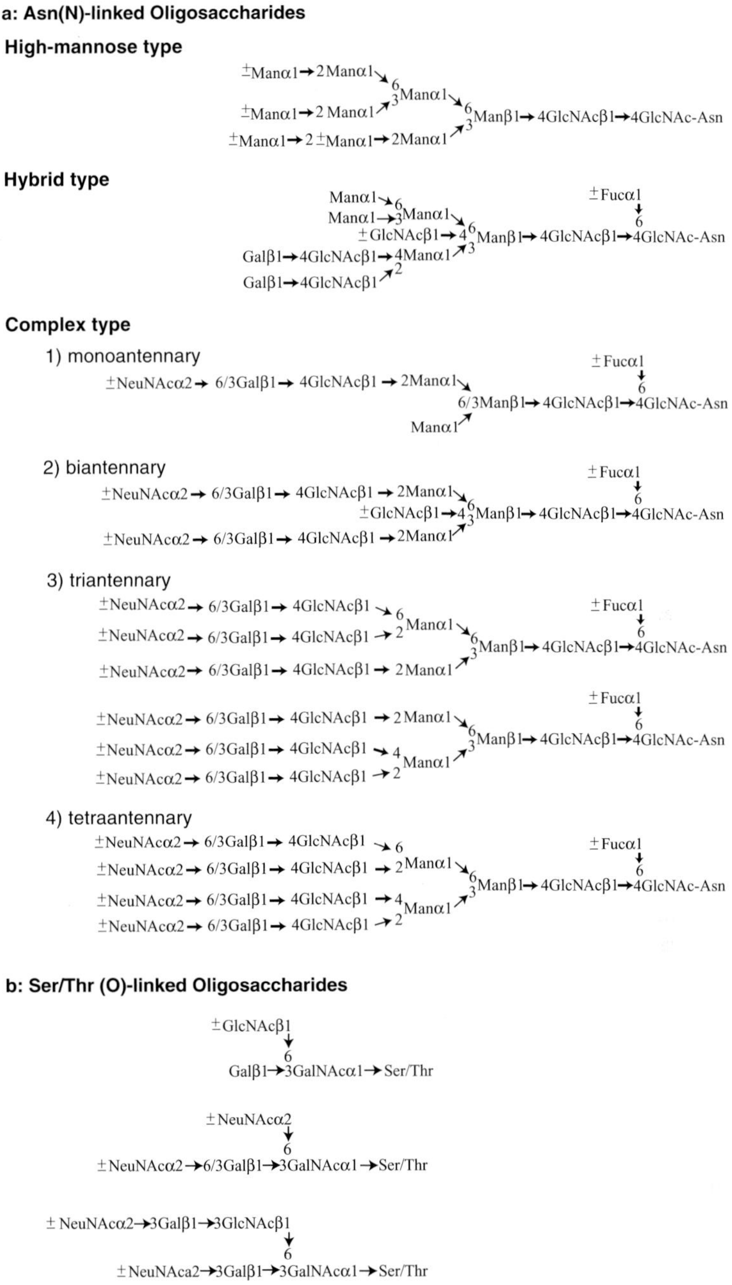

a: Asn(N)-linked Oligosaccharides

High-mannose type

$\pm$Manα1→2Manα1

$\pm$Manα1→2 Manα1 $\nearrow^{3}$Manα1

$\pm$Manα1→2 $\pm$Manα1→2Manα1 $\nearrow$

Manα1$\searrow^{6}$$_{3}$Manβ1 → 4GlcNAcβ1→4GlcNAc-Asn

Hybrid type

Manα1$\searrow_{6}$
Manα1→3Manα1$\searrow$
$\pm$GlcNAcβ1→4^{6}Manβ1→ 4GlcNAcβ1→4GlcNAc-Asn
Galβ1→4GlcNAcβ1→4Manα1$\nearrow_{3}$
Galβ1→4GlcNAcβ1$\nearrow^{2}$

$\pm$Fucα1
↓
6

Complex type

1) monoantennary

$\pm$NeuNAcα2→ 6/3Galβ1→ 4GlcNAcβ1 → 2Manα1$\searrow$
6/3Manβ1→ 4GlcNAcβ1→4GlcNAc-Asn
Manα1$\nearrow$

$\pm$Fucα1
↓
6

2) biantennary

$\pm$NeuNAcα2→ 6/3Galβ1→ 4GlcNAcβ1 → 2Manα1$\searrow$
$\pm$GlcNAcβ1→4$_{3}^{6}$Manβ1→ 4GlcNAcβ1→4GlcNAc-Asn
$\pm$NeuNAcα2 → 6/3Galβ1→ 4GlcNAcβ1 → 2Manα1$\nearrow$

$\pm$Fucα1
↓
6

3) triantennary

$\pm$NeuNAcα2 → 6/3Galβ1→ 4GlcNAcβ1 $\searrow_{6}$
$\pm$NeuNAcα2→ 6/3Galβ1→ 4GlcNAcβ1 → 2 Manα1$\searrow$
$\pm$NeuNAcα2 → 6/3Galβ1→ 4GlcNAcβ1 → 2Manα1$\nearrow$
$^{6}_{3}$Manβ1→ 4GlcNAcβ1→4GlcNAc-Asn

$\pm$Fucα1
↓
6

$\pm$NeuNAcα2 → 6/3Galβ1→ 4GlcNAcβ1 → 2Manα1$\searrow$
$\pm$NeuNAcα2→ 6/3Galβ1→ 4GlcNAcβ1 $\searrow_{4}$ Manα1
$\pm$NeuNAcα2→ 6/3Galβ1→ 4GlcNAcβ1 $\nearrow^{2}$
$^{6}_{3}$Manβ1→ 4GlcNAcβ1→4GlcNAc-Asn

$\pm$Fucα1
↓
6

4) tetraantennary

$\pm$NeuNAcα2→ 6/3Galβ1→ 4GlcNAcβ1 $\searrow_{6}$
$\pm$NeuNAcα2→ 6/3Galβ1→ 4GlcNAcβ1 → 2Manα1$\searrow$
$\pm$NeuNAcα2→ 6/3Galβ1→ 4GlcNAcβ1 $\searrow_{4}$ Manα1
$\pm$NeuNAcα2→ 6/3Galβ1→ 4GlcNAcβ1 $\nearrow^{2}$
$^{6}_{3}$Manβ1→ 4GlcNAcβ1→4GlcNAc-Asn

$\pm$Fucα1
↓
6

b: Ser/Thr (O)-linked Oligosaccharides

$\pm$GlcNAcβ1
↓
6
Galβ1→3GalNAcα1→ Ser/Thr

$\pm$NeuNAcα2
↓
6
$\pm$NeuNAcα2→6/3Galβ1→3GalNAcα1→Ser/Thr

$\pm$ NeuNAcα2→3Galβ1→3GlcNAcβ1
↓
6
$\pm$ NeuNAcα2→3Galβ1→3GalNAcα1→Ser/Thr

Fig. 1. Typical structures of Asn-linked and Ser/Thr-linked oligosaccharides. a Asn-linked oligosaccharides are classified into three types: high-mannose type, hybrid type, and complex-type. b Typical Ser/Thr-linked oligosaccharides. Sugars are abbreviated as follows: *NeuNAc*, N-acetylneuraminic acid; *Gal*, galactose; *GlcNAc*, N-acetylglucosamine; *GalNAc*, N-acetylgalactosamine; *Man*, mannose; *Fuc*, fucose

TABLE 1. Glycosylation type and species in coagulation factors, protein C, and von Willebrand factor

Glycoprotein	Mr (kDa) (subunit)	CHO content (%)	N-linked		O-linked		ABO(H) antigen
			No.[a]	Type[b]	No.[c]	Type[d]	
FI (Fbg)	340 $[A(\alpha)B(\beta)\gamma]_2$	3	4	Bi	0		No
FII	72	12	3	Mo, Bi	0		No
FV	330	13–25	26		n.d.		No
FVII	50	12	2		2	Glc-O Fuc-O	n.d.
FVIII	330	24	25	HM, Bi, Tri, Tetra	n.d.		Yes
FIX	56	17	2		5	Glc-O Fuc-O GalNAc-O	No
FX	59	15	2	Bi, Tri, Tetra	2	GalNAc-O	No
FXI	160 (dimer)	12	5		n.d.		n.d.
FXII	80	17	2		7	Fuc-O GalNAc-O	n.d.
FXIII	320 (A_2B_2)	4	1 (A chain) 2 (B chain)		0		n.d.
PC	62	14–19	4		0		n.d.
VWF	270	12–15	12	Mo, Bi, Tri, Tetra	10	GalNAc-O	Yes

CHO, carbohydrate; F, coagulation factor; Fbg, fibrinogen; PC, protein C; VWF, von Willebrand factor; n.d., not determined

[a]Number of N-glycosylation sites (including potential sites) in mature subunit

[b]HM, high-mannose type; Mo, monoantennary; Bi, biantennary; Tri, triantennary; Tetra, tetraantennary complex-type

[c]Number of O-glycosylation sites in mature subunit

[d]Glc-O, O-linked glucose; Fuc-O, O-linked fucose type; GalNAc-O, O-linked N-acetylgalactosamine type

Fibrinogen (Factor I)

Fibrinogen (Fbg), or factor I (FI), is a terminal protein in the coagulation cascade and is transformed to fibrin by limited proteolysis mediated by thrombin. Fbg has a relative molecular mass of 340 kDa comprised of disulfide-linked dimers of three subunits, $[A(\alpha)B(\beta)\gamma]_2$. It contains approximately 3% carbohydrate with four N-linked sugar chains per molecule. Biantennary complex-type oligosaccharides with or without sialic acid residues have been reported [5].

Abnormal Fbg with an extra glycosylation site caused by the formation of an additional Asn-X-Ser/Thr consensus sequence due to amino acid substitution has also been reported. Fbg Lima, in which Aα Arg141 is substituted with Ser, has an extra N-glycosylation site at Asn139 [6]. Terminal sialic acids of the extra oligosaccharides appear to have contributed to impaired fibrin polymerization, as desialylation corrects both the coagulation time and fibrin polymerization. The extra carbohydrate moiety, located in the a coiled-coil region, impairs the protofibril lateral association process, giving rise to thinner, more curved fibers [7]. Fbg Caracas II also has an extra glycosylation site at A(α) Asn434 caused by the substitution of Ser434, and both the thrombin time and polymerization are corrected by desialylation [8, 9]. Fbg Kaiserslautern has a novel sialylated N-glycosylation site caused by γ chain Lys380→Asn substitution, which results in electrostatic repulsion between condensing protofibrils, leading to impaired polymerization [10]. Fbg Asahi contains Met310→Thr substitution in the γ chain, producing a novel N-glycosylation site at Asn308. Desialylation has no correcting effects in the case of Fbg Asahi, suggesting that the additional glycosylation affects the conformation required for fibrin monomer polymerization [11]. Fbg Niigata shows B(β) Asn160→Ser substitution, which is associated with extra glycosylation at Asn 158. Enzymatic deglycosylation of the N-linked sugar chains of Fbg Niigata accelerates fibrin monomer polymerization [12]. The N-linked sugar chains thus appear to be necessary to build less porous fibrin networks.

Prothrombin (Factor II)

Human prothrombin (FII) has approximately 12% carbohydrate with three N-glycosylation sites, of which two are present in the first activation peptide portion and one exists in the thrombin heavy-chain portion. Human prothrombin [13] and thrombin [14] have mono- or disialylated biantennary complex-type oligosaccharides (Fig. 2), whereas bovine prothrombin contains unique NeuNAcα2–3Galβ1–3(NeuNAcα2–6) GlcNAc chains (Fig. 2) [13]. Although the function of oligosaccharides in the activation peptide has not been studied, partial deglycosylation of human thrombin using

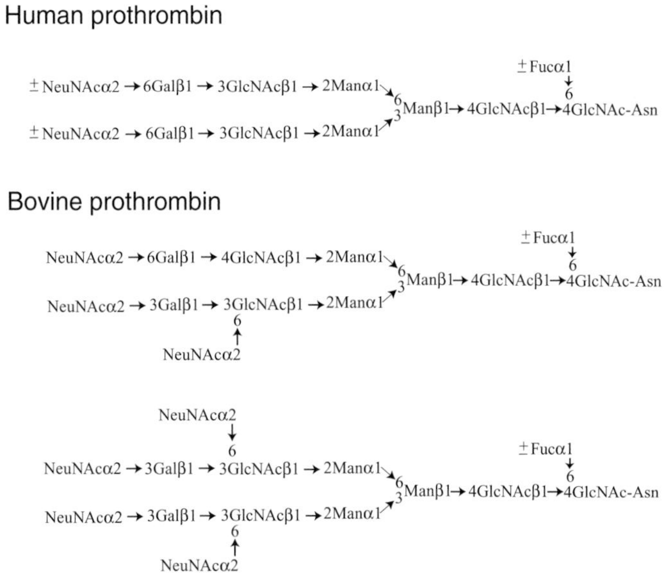

FIG. 2. Oligosaccharide structures of Asn-linked sugar chains on human and bovine prothrombin. Human prothrombin (FII) contains sialylated biantennary complex type oligosaccharides. Bovine prothrombin contains a biantennary chain with NueNAcα2–6GlcNAc branch structure

neuraminidase, β-galactosidase and β-*N*-acetylglucosaminidase had no apparent effect on biological activities of thrombin (such as Fbg clotting, FVIII coagulant activation), inhibition by antithrombin III or α_2-macroglobulin, binding to polymerized fibrin, or stimulation of platelet release and aggregation [15].

Factor V

Coagulation factor V (FV) is highly sialylated, as N-linked and O-linked oligosaccharides account for 13%–25% of its mass. A total of 26 potential N-glycosylation sites are distributed in the heavy chain (8 sites), the activation peptide (15 sites), and the light-chain portion (3 sites). FV is mainly produced by the liver but is also partially synthesized in endothelial cells, leukocytes, macrophages, and megakaryocytes. The domain structure of FV resembles that of FVIII. This suggests that FV has oligosaccharides similar to those of FVIII, but the ABO(H) blood group antigens found in FVIII have not been found in FV (Table 1) [16].

Factor V is activated by thrombin and inactivated by activated protein C (APC). Although structural analysis of the oligosaccharides of FV is not complete, the effects of desialylation or deglycosylation on the functional activity of FV have been studied [17]. Neither the procoagulant activity of FV nor its activation by thrombin was affected by mild deglycosylation with neuraminidase and *N*-glycanase. The *N*-glycanase (but not the *O*-glycanase) digestion of FV increased its susceptibility to inactivation by APC, which suggests that N-linked sugar chains influence the APC resistance of FV.

Factor VII

Factor VII (FVII) is a single-chain vitamin K-dependent serine protease precursor in plasma that participates in the extrinsic blood coagulation pathway. FVII is converted to FVIIa after cleavage between Arg152 and Ile153 by FIXa, FXa, or thrombin. Although FVIIa shows no apparent proteolytic activity, after forming a complex with tissue factor and Ca^{2+} FVIIa converts FIX and FX to FIXa and FXa by limited proteolysis. In plasma, small amounts of FVIIa (~1%) circulate and initiate the extrinsic pathway when complexed with tissue factor. The positive feedback of FIXa, Xa, and thrombin increase the amounts of FVIIa.

Human FVII contains two N-linked (Asn145 and Asn322) and two O-linked (Ser52 and Ser60) sugar chains. Glycosylation of FVII may be important for its function and its plasma half-life. Unique oligosaccharides have been identified in the O-linked sugar chains of the first epidermal growth factor (EGF)-like domain of FVII. FVII has Glc, Xylα1–3Glc, Xylα1–3Xylα1–3Glc structures linked to Ser52 [18] and a fucose residue linked to Ser60 [19]. These O-linked (Xyl)n-Glc-Ser structures have been also found in factor IX, protein Z, and thrombospondin [20] (Fig. 3). Site-directed mutagenesis of Ser52 to Ala showed that mutant recombinant FVIIa exhibited ~60% of the coagulant activity of wild-type FVIIa, but there was no significant loss of affinity for tissue factor or the expression of proteolytic activity toward factor X or factor IX following complex formation with tissue factor [21]. Nuclear magnetic resonance (NMR) studies of the O-fucosylated and non-O-fucosylated EGF-like domain of FVII indicated that the overall structures of both forms are almost the same, but the Ca^{2+}

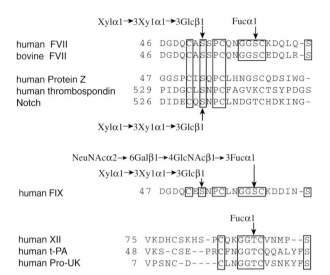

FIG. 3. Glc-O- and Fuc-O-linked Ser/Thr sugar chain structures and consensus amino acid sequences of factors VII, IX, and XII. *t-PA*, human tissue-type plasminogen activator; *Pro-UK*, human urokinase-type plasminogen. Consensus amino acid residues are in *boxes*

dissociation constant is twofold higher in the non-O-fucosylated form [22]. Because the first EGF-like domain is a potential Ca^{2+}-binding site, O-fucosylation may influence the Ca^{2+}-binding affinity of FVIIa.

In a recombinant human FVIIa, the glycosylation profile differed between Asn145 and Asn322. The degree of sialylation at Asn22 was lower than that at Asn145 [23]. N-Glycosylation is normally a co-translational process in endoplasmic reticulum; but of the two N-glycosylation sites in FVII, posttranslational glycosylation occurs at Asn322, which is required for posttranslational folding of the nascent protein [24].

Factor VIII

Factor VIII (FVIII) is a factor IXa cofactor that converts factor X to the activated form (Xa) in the presence of Ca^{2+} and a phospholipid surface. FVIII circulates in blood as a complex with VWF. FVIII consists of domains A1-A2-B-A3-C1-C2, in which the B domain is a heavily glycosylated region that is essential for procoagulant activity. FVIII is processed to a noncovalently associated heterodimer of the heavy chain (domains A1-A2-B) and the light chain (domains A3-C1-C2). Human FVIII contains 19 N-linked sugar chain-binding sites in the B domain out of a potential 25 N-glycosylation sites. Structures of the N-linked oligosaccharides in human plasma FVIII, recombinant human FVIII produced in baby hamster kidney (BHK) cells [25] or Chinese hamster ovary (CHO) cells [26], and porcine FVIII have been elucidated [27]. All the N-linked oligosaccharides contained mainly high-mannose type, bi-, tri- and tetraantennary complex-type sugar chains with or without sialic acid, whereas a Galα1-3Gal structure was found in the N-linked sugar chains of recombinant FVIII

(3%) and porcine FVIII (67%). Because humans and Old World monkeys typically contain natural antibodies against the Galα1–3Gal group, these saccharide antigens are potently pathogenic in humans when introduced as recombinant drugs. Infusion experiments of recombinant FVIII into baboons showed that its half-life in the blood circulation is similar to that of plasma-derived FVIII [25]. In in vitro experiments using VWF and porcine FVIII, VWF prevented the binding of anti-Galα1–3Gal antibody to the Galα1–3Gal group of porcine FVIII [27]. These results suggest that the effects of Galα1–3Gal antigens on recombinant FVIII are minor, as VWF acts like a steric barrier to protect FVIII against the natural antibodies. The safety of recombinant FVIII (recombinate, kognate) produced by BHK or CHO cells has been widely evaluated [28, 29]. It is interesting that, like VWF, human FVIII has ABO(H) blood group antigens (discussed below). These ABO(H) antigens are not attached to recombinant or porcine FVIII [25]. Oligosaccharides in the B domain of FVIII also have important roles in the secretion process of FVIII [30] and in ASGPR recognition in vivo, similar to other plasma glycoproteins [31].

Factor IX

Factor IX (FIX, Christmas factor) is a single-chain precursor of vitamin K-dependent serine protease (FIXa) in plasma that participates in the intrinsic coagulation pathway by converting FX to the active FXa in the presence of Ca^{2+}, phospholipids, and FVIIIa. Human FIX contains two N-linked and four O-linked sugar chains, of which two O-linked oligosaccharides are attached to Ser53 and Ser61 through glucose and fucose residues, respectively (Fig. 3), in the L-chain of FIXa (active form). The former sugar chain is $(Xyl\alpha1-3)_{0-2}Glc\beta1$-Ser53 as found in FVII, but the latter structure is a unique tetrasaccharide: NeuNAcα2–6Galβ1–4GlcNAcβ1–3Fucα1-Ser61 [19, 32, 33]. This type of O-linked fucose oligosaccharide attached to Thr/Ser residues in a consensus sequence of Cys-X-X-Gly-Gly-Thr/Ser-Cys in the first EGF-like domain of FIX has been also identified in urokinase-type plasminogen activator (Thr18), tissue-type plasminogen activator (Thr61), FVII (Ser60), factor XII (Thr90), and the recombinant Notch 1 [34] (Fig. 3). The isolated EGF-like domain of urokinase-type plasminogen activator with Fuc-O-Thr had no stimulative effects on cell proliferation after defucosylation without influencing receptor-binding activity, suggesting that fucose residues are essential for eliciting this proliferative response [35]. The function of Fuc-O-Ser oligosaccharides of FIX has not yet been elucidated, but they may function as a molecular trigger of signal transduction for FIX.

The N-linked sugar chains of FIX are sialylated bi- (10%), tri- (20%), and tetraantennary (70%) complex-type oligosaccharides bound to Asn157 and Asn167 in the activation peptide region [36], and 27% of the oligosaccharides have the (sialyl) Le^x structure (±NeuNAc-Galβ1–4(Fucα1–3)GlcNAc). Thr159 and Thr169 in the activation peptide are partially glycosylated, and the ± NeuNAc-Gal-GalNAc-Ser and NeuNAc-(Gal)-GalNAc-Ser structures have been elucidated [37]. Because FIX is highly sialylated, the negatively charged group of sugar chain moieties has an important role in the conformation of FIX, particularly in the activation peptide region. Enzymatic removal of sialic acid residues from human FIX eliminates its clotting activity without affecting activation of FIX to FIXa [38], but the apparently conflicting observation that neuraminidase-treated FIX retains full enzymatic activity has also been reported [39]. The sugar chain moiety may contribute to the in vivo survival of FIX in circulation. Arg 94 → Ser mutation in FIX caused a naturally occurring hemophilia B defect, as it introduces a new O-linked glycosylation site in the second EGF-like domain, which markedly impairs activation by FXIa and turnover of FX [40].

Factor X

Factor X (FX, Stuart factor) is synthesized as a single chain and is cleaved into L- and H-chains linked together by an S-S linkage. This processed mature FX is converted to active FXa by the FIXa–FVIII complex (intrinsic pathway), the FVIIa–TF complex (extrinsic pathway), or Russell's viper venom, which activates prothrombin (FII) into thrombin (FIIa, active form) in the presence of phospholipid and Ca^{2+}. Human FX has two N-linked sugar chains at Asn181 and Asn191 and two O-linked sugar chains at Thr171 and Thr159 in the activation peptide region, which is cleaved off during activation by FIXa. The major N-linked oligosaccharides of human FX are neutral bi-, tri-, and tetraantennary complex-type without fucose residues, followed by mono-sialylated biantennary complex-type with NeuNAcα2–6Gal group [41]. The O-linked sugar chains of human FX are thought to be disialylated Galβ1–3GalNAc sequences [41].

A lectin from *Sambucus nigra* specific for the terminal sialic acid α2–6-linked galactose or N-acetylgalactosamine inhibits the activation of human FX in both pathways [42]. Neuraminidase digestion of FX also reduces the activation of zymogen by these complexes or clotting activity [42], although other experiments have shown that neuraminidase treatment of FX has no significant effect on clotting activity or cleavage susceptibility by either intrinsic complex or Russell's viper venom [39]. These conflicting results may be caused by small amounts of contaminating protease(s) in the neuraminidase used or in the FX preparation.

A deletion mutagenesis study on the activation peptide region of FX indicated that deletion of neither the activation peptide region nor the N-linked oligosaccharide (by Asn191 → Ala mutation) affect the catalytic activity of the resultant FXa, but activation of the mutants by FVIIa, FIXa, or thrombin was accelerated compared to wild-type FX. Furthermore, the rate of activation was faster for mutants without Asn191, suggesting that the sugar chain at Asn191 (and a portion of the activation peptide region) serves primarily as a negative autoregulation mechanism to prevent spurious activation of FX by FVIIa, FIXa, and thrombin [43].

Factor XII

Factor XII (FXII, Hageman factor) is the zymogen of a serine protease that participates in the initial phase of the intrinsic coagulation cascade, fibrinolysis, and the generation of bradykinin and angiotensin. Prekallikrein is cleaved by FXII to kallikrein, which then cleaves FXII first to α-FXIIa and then to β-FXIIa. α-FXIIa activates FXI to form FXIa. Human FXII is a single-chain glycoprotein (596 amino acid (aa) residues) containing 16.8% carbohydrate, which is distributes in two N-linked and seven O-linked tentative glycosylation sites. Six proposed O-linked sugar chains are present in the connecting region between the kringle domain and the catalytic region corresponding to the activation peptide area of other coagulation factors. Among coagulation factors, these areas are generally rich in carbohydrates. Structural data for the sugar chain moieties of FXII have not yet been reported, except for the O-linked sugar chain at Thr90 [44]. Human FXII is fully fucosylated at Thr90, as found in FVII and FIX, where the consensus amino acid sequence for O-fucosylation is conserved in the first EGF-like domain (Fig. 3).

Protein C

Protein C (PC) is an antithrombotic serine protease precursor that circulates mostly as disulfide-linked L- and H-chains of 62 kDa. When PC is activated by forming a complex with thrombin and thrombomodulin, the resulting APC (active form) strongly inactivates FVa and FVIIIa in the presence of Ca^{2+} and protein S. Human PC (419 aa) has no O-linked oligosaccharide but does have four N-glycosylation sites at Asn97, Asn248, Asn313, and Asn329. The glycosylation site at Asn329 has an unusual consensus sequence, Asn-X-Cys, which was discovered first in bovine PC [45] followed by human VWF [46]. The same N-glycosylation at Asn-X-Cys has been reported in human plasma α_1T-glycoprotein [47] and leukocyte CD69 [48]. Site-directed mutagenesis (Asn→Gln) at each glycosylation site of PC indicated that glycosylation at Asn97 in the second EGF-like domain is critical for efficient secretion of PC and affects the degree of core glycosylation at Asn329 [49]. Glycosylation at Asn248 has effects on intracellular processing into L- and H-chains. Furthermore, alteration of glycosylation sites in the serine-protease domain at Asn248, Asn313, and Asn329 (Asn→Gln) increased the anticoagulant activity of PC, and the accelerated activation observed in a mutant PC with Asn313→Gln occurred because of enhanced affinity for the thrombin–thrombomodulin complex. PC has two EGF-like domains but has no consensus sequence for the O-glycosylation with Glc or Fuc-O-Ser/Thr linkages observed in the first EGF-like domains of FVII, FIX, and FXII [44].

Oligosaccharides on von Willebrand Factor

Human VWF is a highly multimeric plasma glycoprotein composed of disulfide-linked homopolymers of 2050-aa subunits. VWF has two major functions: performing as a carrier and protector of FVIII and acting as molecular glue for platelets against subendothelium at the sites of vascular injury [50]. Mature VWF subunits contain functional domains designated A–D, which align as D'-D3-A1-A2-A3-D4-B-C1-C2-CK (cystine knot) from the N-terminus (Fig. 4) [46, 50]. FVIII binds noncovalently with the N-terminal D'-D3 domains. The A domains play crucial roles as binding sites for platelet membrane glycoprotein Ib (GPIb) (A1 domain) and for subendothelial collagen (A3 domain) or as a cleavage site for ADAMTS13 (a disintegrin-like and metalloprotease with thrombospondin type 1 motif 13) (A2 domain). VWF modulators such as the antibiotic ristocetin and snake venoms botrocetin and bitiscetin also bind to the A1 domain to induce platelet agglutination in vitro [51]. Biosynthesis of VWF is limited to endothelial cells and megakaryocytes, in contrast to other plasma glycoproteins produced almost exclusively by the liver. VWF is constitutively secreted into plasma from endothelial cells, which is responsible for >95% of plasma VWF. The VWF is partially stored in Weibel-Palade bodies of endothelial cells or in α granules of platelets. Quantitative deficiencies of VWF in plasma or functional inefficiencies of VWF caused by aberrant structure results in the congenital hemostatic disorder von Willebrand disease (VWD), the most prevalent bleeding disorder in humans.

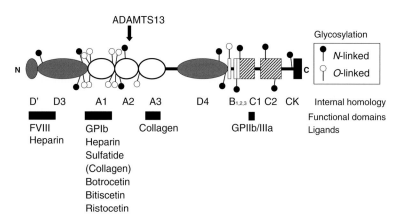

FIG. 4. Structure of human von Willebrand factor (VWF). The VWF subunit is composed of *A–D* domains and C-terminal cystine knot (*CK*) domains. Distribution of N-linked and O-linked sugar chains is indicated. *Bars* indicate the binding sites for each ligand, and an *arrow* indicates the cleaving site for ADAMTS13

Human VWF subunits have 12%–15% carbohydrates distributed as 12 N-linked and 10 O-linked sugar chains. The N-linked sugar chains are attached relatively evenly along the subunit, except for the A1 and A3 domains, whereas the O-linked sugar chains form clusters, particularly at both sides of the A1 domain (Fig. 4). The N-linked glycosylation site at Asn1147 (numbered from the N-terminal of the prepro-VWF) in the D3 domain does not have the Asn-X-Ser/Thr consensus sequence but does have the Asn-X-Cys sequence [46], similar to PC. The major component of the N-linked sugar chains of human plasma VWF is the α2–6 linked monosialylated biantennary complex-type oligosaccharide chain with a fucosylated core structure [52, 53]. VWF also contains a biantennary structure with a bisecting *N*-acetylglucosamine residue or 2–4 and 2–6 branched tri- and tetraantennary complex-type oligosaccharides (Fig. 1) [53, 54]. More than 86% of the N-linked oligosaccharides of human VWF are mono- or disialylated. The major O-linked oligosaccharides were elucidated to be tetrasaccharides of NeuNAcα2–6 (NeuNAcα2–3Galβ1–3)GalNAc linked to Ser/Thr [55]. Almost all of the O-linked oligosaccharides of VWF are sialylated, while the N-linked oligosaccharides partially contain unsialylated Galβ1–4GlcNAc group at the non-reducing termini [56]. These oligosaccharides are rather common structures observed with many other plasma glycoproteins, but VWF a typically expresses ABO(H) blood group antigen structures in contrast to other plasma proteins.

The ABO(H) blood group antigens are oligosaccharides attached to glycoproteins or glycolipids. The H-antigen (Fucα1-2Galβ1-4/3GlcNAc-) is a basic structure found in all humans, except for the Bombay blood group, in which the AB(H) blood group is deficient owing to the absence of the H-enzyme (FUT 1, α1–2fucosyltransferase). The A and B enzymes, which transfer α-*N*-acetylgalactosamine and α-galactose to the terminal β-galactose of H-antigen with α1–2 linkages, respectively, successively produce blood group A and B antigens. The genes for the H-enzyme and A/B-enzymes are located at chromosomes 19q13.3 and 9q34.1–34.2, respectively.

In 1979, Sodetz et al. [57] reported that the human VWF–FVIII complex showed ABO(H) blood group antigenicity on the hemaglutination-inhibiting activity assay, but sugar composition analysis of VWF could not identify the presence of the nonre-ducing terminal GalNAc needed for the blood group A antigen [58]. In 1992, structural analysis of oligosaccharides of human VWF showed that ABO(H) blood group anti-gens are attached to the nonreducing terminal of the N-linked sugar chains [53] (Fig. 5). Because *N*-glycanase digestion deprives VWF of anti-blood-group antigenicity, these structures are localized in the N-linked sugar chains [59], although most blood group antigens in body fluids (e.g., gastrointestinal mucin) are found in the O-linked sugar chains [60, 61]. The blood group antigens found in VWF have the type 2 (Galβ1–4GlcNAc) core structures but not the type 1 (Galβ1–3GlcNAc) structure. Expression of blood group antigens on VWF is not influenced by the secretor status of the donor [59], but the plasma concentration of VWF shows a weak effect based on secretor status [62]. The occurrence of blood group antigens in human plasma proteins is limited to FVIII and some α_2-macroglobulins, in addition to VWF, among 35 plasma proteins surveyed [16, 59]. FVIII also has blood group antigens in the N-linked oligosaccharides, but it is unknown whether these antigens are included in the O-linked sugar chains [25].

The synthetic pathways of ABO(H) blood group antigens on VWF, FVIII, and α_2-macroglobulin remain unclear. Although plasma contains some glycosyltransferases,

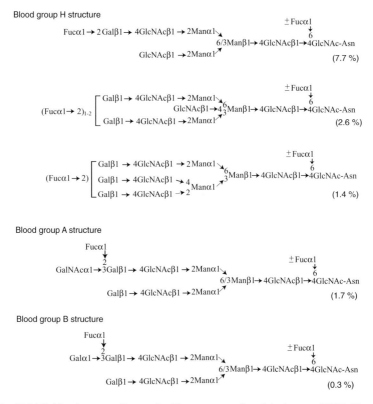

FIG. 5. ABO(H) blood group oligosaccharide structures found in human VWF. *Numbers in parentheses* indicate the percent ratio in the Asn-linked sugar chains of VWF

it is uncertain whether it contains sugar nucleotide donors, suggesting that blood group antigens of these plasma proteins might be processed intracellularly rather than extracellularly. VWF from platelets has no blood group antigens despite their presence among platelet membrane glycoproteins [63, 64]. Blood group antigens on VWF after ABO-mismatched bone marrow transplantation are not altered; the recipient's blood type continues to be expressed in contrast to erythrocytes, which gradually convert to express the donor's type [63]. These observations suggest that the ABO(H) blood group antigens of VWF are not synthesized by hematopoietic cells but are synthesized exclusively by endothelial cells. In primary culture experiments using endothelial cells, secreted VWF possessed blood group antigens [64], whereas human umbilical vein endothelial cells (HUVECs) did not produce VWF carrying blood group antigens [64, 65]. The reason for the lack of blood group antigens on VWF from HUVEC is not obvious, but it is possible that physiological differences between HUVEC and other endothelial cells are responsible or that neonatal blood group expression is weaker when compared to that of adults. Human endothelial cells contain α1-3fucosyl transferase VI in Weibel-Palade bodies, suggesting the presence of glycosylation pathways other than Golgi-associated glycosylation [66]. In contrast to endothelial VWF, platelet-derived VWF has not been studied in detail. The size and species of sugar chains appear to differ from those of endothelial VWF [67].

Factor VIII is reportedly synthesized by hepatic sinusoidal endothelial cells [68], lung microvascular endothelial cells [69], and other tissues [70]. Recombinant human FVIII produced in CHO cells exhibited oligosaccharides different from those carried by plasma FVIII and did not contain the blood group structure [25]. Human α_2-macroglobulin is a tetramer of identical subunits (1451 aa) containing eight N-linked sugar chains that are 70% sialylated (exclusively NeuAcα2–6Gal group). α_2-Macroglobulin is synthesized by various organs, including the liver and macrophages, but the origin of the blood group antigens has not yet been investigated. Recently, hepatocytes were found to synthesize α_2-macroglobulin under co-culture conditions with endothelial cells, suggesting that endothelial cells modulate hepatocyte gene expression by direct cellular interaction [71]. These accumulating results suggest that blood group antigen synthesis might be directly or indirectly regulated by endothelial cells.

Function of Sugar Chains in VWF

In VWF, the carbohydrate moiety had evoked numerous incompatible results until some time ago. Carbohydrate deficiency of FVIII–VWF has been reported in a variant of VWD [72, 73]. The type 2B VWD variant, in which loss of a high-molecular-weight multimer resulted from increased affinity for GPIb, was thought to be caused by increased exposure of terminal galactose residues [74]. To date, numerous point mutations have been found in the A1 domain, which may influence the conformation of VWF to facilitate GPIb accession [75]. Desialylation of VWF has also shown conflicting results [76]; some reports have found that asialo-VWF exhibits lower collagen-binding activity and platelet agglutinability by ristocetin [77], whereas others have reported that asialo-VWF exhibits increased activity [78, 79]. Some consensus has been reached based on cooperative experiments using the same materials that found that VWF becomes sensitive to proteolysis after desialylation [80]. Desialylation facilitates the cleavage of VWF by small contaminating proteases in the glycosidases used or the

VWF preparation itself in vitro, resulting in the loss of high-molecular-weight multimers that mediate the platelet agglutinability or collagen-binding activity of VWF. Thus, terminal sialic acid residues are important in protecting VWF from proteolytic degradation, as reported for other glycoproteins. Furthermore, the established effects of desialylation of VWF without proteolysis on platelet interactions are that asialo-VWF spontaneously augments platelet agglutination in the absence of modulators [81]. The mechanisms underlying this autoinduction are not yet clearly resolved. However, desialylation of the O-linked sugar chain cluster around the A1 domain may increase the accessibility of the negatively charged GPIb on the platelets or directly affect the conformation of the A1 domain. Recombinant VWF that has been selectively deprived of O-linked sugar chains shows normal multimer assembly, secretion, and binding to collagen and heparin but decreased platelet agglutinability induced by ristocetin [82]. Because botrocetin-induced platelet agglutination was not affected by O-glycosylation deficiency, O-linked sugar chains appear to be necessary to maintain conformations essential for ristocetin interaction.

In the intact VWF, however, all the N-linked sugar chains are not always concealed by sialic acid residues. Although insufficient for recognition by ASGPR, lectin blotting analysis using RCA_{120} has suggested the presence of terminal β1–4-linked galactose residues in intact VWF [56]. Clearance of VWF by ASGPR is also increased by desialylation of other plasma glycoproteins [83]. VWF has two types of sialyl linkage— NeuAcα2–6 and α2–3—up to the penultimate galactose, and these linkages are synthesized by sialyltransferases ST6Gal-I and ST3Gal-IV, respectively. ST3Gal-IV-deficient mice exhibited reductions in plasma VWF promoted by ASGPR [84]. As NeuAcα2–3 linkages are the major structure in O-linked sugar chains (but the minor structure in N-linked sugar chains) in human VWF, sialylation by ST3Gal-IV sialyltransferase plays a key role in modifying plasma levels of VWF. Further glycosylation modifications of VWF have been reported using RIIIS/J mice, in which VWF levels are reduced to 5% when compared to normal mice; hence, this strain is useful as a model for investigating type 1 VWD [85]. The responsible gene in RIIIS/J was revealed to be the *N*-acetylgalactosaminyl transferase gene (Galgt2), which is specifically expressed in gastrointestinal epithelial cells in normal mice, but in RIIIS/J Galgt2 gene expression is moved from epithelial cells to vascular endothelial cells. The resultant VWF carrying terminal GalNAc residues shows high affinity toward ASGPR and is rapidly cleared from the circulation. The same phenomenon has not been seen in humans, but the altered glycosylation of VWF due to changes in gene expression suggest that the presence of other genes can also modify the lifespan of VWF.

Blood Group and VWD

A significant correlation has been observed between the blood group and the plasma concentration of VWF. Blood group O donors tend to have lower concentrations of VWF when compared with non-O groups [86–88]. Conversely, blood group AB donors show relatively higher concentrations of VWF. These differences also reflect the incidence of VWD. Type 1 VWD, which is the major (~70%) VWD in partial quantitative VWF deficiency, is predominant in blood group O [86]. Thus, it is necessary to consider the blood group when type 1 VWD is detected.

Translation, synthesis, and pooled VWF are not significantly different among blood groups [89]. Biological activities of VWF, such as platelet agglutinability and collagen binding, are not altered by the blood group. Expression of ABO(H) blood group antigens are not directly linked to those of VWF, as the VWF gene is mapped to chromosome 12. These results suggest that the differences in plasma concentrations of VWF between the blood groups are based on a posttranslational process. As VWF itself possesses ABO(H) blood group antigens, it is possible that blood group antigens might play a role in plasma concentrations of human VWF.

Blood Group and ADAMTS13

The VWF circulates as a series of multimers ranging from 500 kDa to more than 10 000 kDa; and higher-molecular-weight VWF elicits higher thrombotic activity. This heterogeneity in molecular size is derived from gradual degradation by plasma proteases or disulfatase. ADAMTS13 has been identified as a physiological metalloprotease specific for VWF that regulates the effective plasma concentrations of VWF (higher-molecular-weight VWF) by cleaving it between Tyr1605 and Met1606 residues in the A2 domain under high-shear stress [90]. ADAMTS13 is a 193 kDa glycoprotein having ten N-linked and seven O-linked sugar chains. Recently, Glcβ1–3Fucα1-O-Ser structures were found in the seven thrombospondin type 1 repeats (TSRs) of human ADAMTS13 [91]. Mutation of Ser to Ala in TSRs reduced the secretion of ADAMTS13 suggesting that O-fucosylation in TSRs is functionally significant for secretion of ADAMTS13. The relation between ADAMTS13, blood group, and plasma VWF levels has been investigated. ADAMTS13 activity is about 10% higher in blood group O than in other groups, and the susceptibility of VWF to ADAMTS13 is significantly higher in blood group O [92]. VWF levels are higher in A_1A_1 donors than A_1O_1, followed by A_2O_1 or OO individuals, and the amount of blood group antigens on VWF also correlates with blood group genotype [93, 94]. VWF from the Bombay phenotype, in which all ABO(H) blood group antigens are absent, showed lower VWF levels and marked susceptibility to ADAMTS13 cleavage compared to blood group O [95, 96]. These results indicate that the ABO(H) blood group antigens on VWF function as a molecular guard that protects VWF from ADAMTS13 proteolysis [76, 89]. Only the A2 domain has two N-linked sugar chains (Asn1515 and Asn1574) near the ADAMTS13 cleaving site (Fig. 4). Deglycosylation or mutation of Asn1574 increased the susceptibility of VWF to ADAMTS13 proteolysis, indicating a critical role for the glycan at Asn1574 in modulating the VWF-ADAMTS13 interaction [97]. It is likely that ADAMTS13 has relatively good access to the cleaving site on VWF in blood group O (and Bombay type), as the H structure has one fewer sugar than A and B structures, which would lessen the steric hindrance at the cleaving site.

Gastrointestinal mucins are known to contain numerous sialic acid residues, but they also abundantly express blood group antigens [61]. Although the biological significance of blood group antigens on mucin is unknown, they may protect the intestinal surface from the excess hydrolases in the digestive juice, similar to negatively charged sialic acid residues. Human parotid α-amylase has a Galβ1–4(Fucα1–3)GlcNAc (X-antigenic determinant) structure on its N-linked sugar chains in addition to sialic acid [98], suggesting that blood group-related sugar chains also protect enzymes against hydrolysis. Secretary immunoglobulins are also present in secreted body fluids. Thus, it would be beneficial for anti-blood-group antibodies to form immune complexes with alloantigens but not with digestive enzymes or epidermal surfaces bearing the same blood group antigens.

Alternatively, the blood group sugar chains on VWF (and FVIII) are also recognized for clearance by hepatic receptors. It is possible that fucose-exposed glycoproteins (H-structure) tend to be cleared faster from the circulation by fucose-binding lectins [99, 100]. Further studies exploring novel lectins in the liver, which selectively bind the H-structure on oligosaccharides, would provide better insight into why blood group O donors have relatively low concentrations of VWF.

Conclusion

Post-genome interests have again returned to protein molecules, particularly with regard to 3D structure elucidation and functional analysis. Glycosylation of proteins is also an important co-translational or posttranslational modification that influences biological activity. Structural analysis of oligosaccharides on glycoproteins has gradually become more sensitive and more rapid with high-throughput techniques owing to recent technical innovations [101, 102]. Although their flexibility makes 3D structural data for sugar chains elusive, 3D structural databases of NMR data for oligosaccharides are gradually being compiled. Site-specific mutagenesis of certain glycosylation sites should also provide insight into the function of sugar chains in glycoproteins involved in thrombosis and hemostasis in conjunction with the functional abnormalities found in specific thrombotic diseases. However, O-glycosylation sites and their oligosaccharide structures on coagulation factors have not yet been fully determined compared to N-linked sugar chains. More efficient endo-O-glycanases would facilitate the structural and functional analysis of O-linked oligosaccharides. As is the case for FVII, FIX, and FXII, unusual glycosylation that cannot be predicted from DNA analysis alone might also be found. In VWF, structural analysis of the A2 domain has apparently elucidated the cleaving mechanism by ADAMTS13 as well as the role of N-linked sugar chains in the A2 domain, particularly the relation between blood group sugar chains and susceptibility to ADAMTS13. Synthetic pathways for these blood group-bearing plasma glycoproteins are slowly being clarified at the molecular level, and the biological significance of ABO blood group antigens is becoming apparent.

References

1. Ashwell G, Morell AG (1974) The role of surface carbohydrates in the hepatic recognition and transport of circulating glycoproteins. Adv Enzymol Relat Areas Mol Biol 41:99–128
2. Dwek RA (1995) Glycobiology: "towards understanding the function of sugars." Biochem Soc Transact 23:1–25
3. Perez-Vilar J, Hill RL (1999) The structure and assembly of secreted mucins. J Biol Chem 274:31751–31754
4. Van den Steen P, Rudd PM, Dwek RA, et al (1998) Concepts and principles of O-linked glycosylation. Crit Rev Biochem Mol Biol 33:151–208
5. Mizuochi T, Taniguchi T, Asami Y, et al (1982) Comparative studies on the structures of the carbohydrate moieties of human fibrinogen and abnormal fibrinogen Nagoya. J Biochem 92:283–293
6. Maekawa H, Yamazumi K, Muramatsu S, et al (1992) Fibrinogen Lima: a homozygous dysfibrinogen with an A alpha-arginine-141 to serine substitution associated with extra N-glycosylation at A alpha-asparagine-139: impaired fibrin gel formation but

normal fibrin-facilitated plasminogen activation catalyzed by tissue-type plasminogen activator. J Clin Invest 90:67–76

7. Marchi R, Arocha-Pinango CL, Nagy H, et al (2004) The effects of additional carbohydrate in the coiled-coil region of fibrinogen on polymerization and clot structure and properties: characterization of the homozygous and heterozygous forms of fibrinogen Lima (Aalpha Arg141 → Ser with extra glycosylation). J Thromb Haemost 2:940–948

8. Maekawa H, Yamazumi K, Muramatsu S, et al (1991) An A alpha Ser-434 to N-glycosylated Asn substitution in a dysfibrinogen, fibrinogen Caracas II, characterized by impaired fibrin gel formation. J Biol Chem 266:11575–11581

9. Woodhead JL, Nagaswami C, Matsuda M, et al (1996) The ultrastructure of fibrinogen Caracas II molecules, fibers, and clots. J Biol Chem 271:4946–4953

10. Brennan SO, Loreth RM, George PM (1998) Oligosaccharide configuration of fibrinogen Kaiserslautern: electrospray ionisation analysis of intact gamma chains. Thromb Haemost 80:263–265

11. Yamazumi K, Shimura K, Terukina S, et al (1989) A gamma methionine-310 to threonine substitution and consequent N-glycosylation at gamma asparagine-308 identified in a congenital dysfibrinogenemia associated with posttraumatic bleeding, fibrinogen Asahi. J Clin Invest 83:1590–1597

12. Sugo T, Nakamikawa C, Takano H, et al (1999) Fibrinogen Niigata with impaired fibrin assembly: an inherited dysfibrinogen with a Bbeta Asn-160 to Ser substitution associated with extra glycosylation at Bbeta Asn-158. Blood 94:3806–3813

13. Mizuochi T, Yamashita K, Fujikawa K, et al (1979) The carbohydrate of bovine prothrombin: occurrence of Gal beta 1 leads to 3GlcNAc grouping in asparagine-linked sugar chains. J Biol Chem 254:6419–6425

14. Nilsson B, Horne MK, Gralnick HR (1983) The carbohydrate of human thrombin: structural analysis of glycoprotein oligosaccharides by mass spectrometry. Arch Biochem Biophys 224:127–133

15. Horne MK, Gralnick HR (1984) The oligosaccharide of human thrombin: investigations of functional significance. Blood 63:188–194

16. Matui T (1995) Structure and function of sugar chains of human von Willebrand factor. Jpn J Thromb Hemost 6:421–432 (in Japanese)

17. Fernandez JA, Hackeng TM, Kojima K, et al (1997) The carbohydrate moiety of factor V modulates inactivation by activated protein C. Blood 89:4348–4354

18. Nishimura H, Kawabata S, Kisiel W, et al (1989) Identification of a disaccharide (Xyl-Glc) and a trisaccharide (Xyl2-Glc) O-glycosidically linked to a serine residue in the first epidermal growth factor-like domain of human factors VII and IX and protein Z and bovine protein Z. J Biol Chem 264:20320–20325

19. Nishimura H, Takao T, Hase S, et al (1992) Human factor IX has a tetrasaccharide O-glycosidically linked to serine 61 through the fucose residue. J Biol Chem 267:17520–17525

20. Iwanaga S, Nishimura H, Kawabata S, et al (1990) A new trisaccharide sugar chain linked to a serine residue in the first EGF-like domain of clotting factors VII and IX and protein Z. Adv Exp Med Biol 281:121–131

21. Bjoern S, Foster DC, Thim L, et al (1991) Human plasma and recombinant factor VII: characterization of O-glycosylations at serine residues 52 and 60 and effects of site-directed mutagenesis of serine 52 to alanine. J Biol Chem 266:11051–11057

22. Kao YH, Lee GF, Wang Y, et al (1999) The effect of O-fucosylation on the first EGF-like domain from human blood coagulation factor VII. Biochemistry 38:7097–7110

23. Klausen NK, Bayne S, Palm L (1998) Analysis of the site-specific asparagine-linked glycosylation of recombinant human coagulation factor VIIa by glycosidase digestions, liquid chromatography, and mass spectrometry. Mol Biotechnol 9:195–204

24. Bolt G, Kristensen C, Steenstrup TD (2005) Posttranslational N-glycosylation takes place during the normal processing of human coagulation factor VII. Glycobiology 15:541–547

25. Hironaka T, Furukawa K, Esmon PC, et al (1992) Comparative study of the sugar chains of factor VIII purified from human plasma and from the culture media of recombinant baby hamster kidney cells. J Biol Chem 267:8012–8020

26. Medzihradszky KF, Besman MJ, Burlingame AL (1997) Structural characterization of site-specific N-glycosylation of recombinant human factor VIII by reversed-phase high-performance liquid chromatography-electrospray ionization mass spectrometry. Anal Chem 69:3986–3994

27. Hironaka T, Furukawa K, Esmon PC, et al (1993) Structural study of the sugar chains of porcine factor VIII–tissue- and species-specific glycosylation of factor VIII. Arch Biochem Biophys 307:316–330

28. Bray GL, Gomperts ED, Courter S, et al (1994) A multicenter study of recombinant factor VIII (recombinate): safety, efficacy, and inhibitor risk in previously untreated patients with hemophilia A: the Recombinate Study Group. Blood 83:2428–2435

29. Yoshioka A, Fukutake K, Takamatsu J, et al (2006) Clinical evaluation of recombinant factor VIII preparation (Kogenate) in previously treated patients with hemophilia A: descriptive meta-analysis of post-marketing study data. Int J Hematol 84:158–165

30. Kaufman RJ, Pipe SW, Tagliavacca L, et al (1997) Biosynthesis, assembly and secretion of coagulation factor VIII. Blood Coagul Fibrinolysis 8(suppl 2):S3–S14

31. Bovenschen N, Rijken DC, Havekes LM, et al (2005) The B domain of coagulation factor VIII interacts with the asialoglycoprotein receptor. J Thromb Haemost 3:1257–1265

32. Harris RJ, van Halbeek H, Glushka J, et al (1993) Identification and structural analysis of the tetrasaccharide NeuAc alpha(2→6)Gal beta(1→4)GlcNAc beta(1→3)Fuc alpha 1→O-linked to serine 61 of human factor IX. Biochemistry 32:6539–6547

33. Kuraya N, Omichi K, Nishimura H, et al (1993) Structural analysis of O-linked sugar chains in human blood clotting factor IX. J Biochem 114:763–765

34. Moloney DJ, Shair LH, Lu FM, et al (2000) Mammalian Notch1 is modified with two unusual forms of O-linked glycosylation found on epidermal growth factor-like modules. J Biol Chem 275:9604–9611

35. Rabbani SA, Mazar AP, Bernier SM, et al (1992) Structural requirements for the growth factor activity of the amino-terminal domain of urokinase. J Biol Chem 267:14151–14156

36. Makino Y, Omichi K, Kuraya N, et al (2000) Structural analysis of N-linked sugar chains of human blood clotting factor IX. J Biochem 128:175–180

37. Agarwala KL, Kawabata S, Takao T, et al (1994) Activation peptide of human factor IX has oligosaccharides O-glycosidically linked to threonine residues at 159 and 169. Biochemistry 33:5167–5171

38. Chavin SI, Weidner SM (1984) Blood clotting factor IX. Loss of activity after cleavage of sialic acid residues. J Biol Chem 259:3387–3390

39. Bharadwaj D, Harris RJ, Kisiel W, et al (1995) Enzymatic removal of sialic acid from human factor IX and factor X has no effect on their coagulant activity. J Biol Chem 270:6537–6542

40. Hertzberg MS, Facey SL, Hogg PJ (1999) An Arg/Ser substitution in the second epidermal growth factor-like module of factor IX introduces an O-linked carbohydrate and markedly impairs activation by factor XIa and factor VIIa/tissue factor and catalytic efficiency of factor IXa. Blood 94:156–163

41. Nakagawa H, Takahashi N, Fujikawa K, et al (1995) Identification of the oligosaccharide structures of human coagulation factor X activation peptide at each glycosylation site. Glycoconj J 12:173–181

42. Sinha U, Wolf DL (1993) Carbohydrate residues modulate the activation of coagulation factor X. J Biol Chem 268:3048–3051

43. Rudolph AE, Mullane MP, Porche-Sorbet R, et al (2002) The role of the factor X activation peptide: a deletion mutagenesis approach. Thromb Haemost 88:756–762

44. Harris RJ, Ling VT, Spellman MW (1992) O-linked fucose is present in the first epidermal growth factor domain of factor XII but not protein C. J Biol Chem 267:5102–5107

45. Stenflo J, Fernlund P (1982) Amino acid sequence of the heavy chain of bovine protein C. J Biol Chem 257:12180–12190

46. Titani K, Walsh KA (1988) Human von Willebrand factor: the molecular glue of platelet plugs. Trends Biochem Sci 13:94–97

47. Araki T, Haupt H, Hermentin P, et al (1998) Preparation and partial structural characterization of alpha1T-glycoprotein from normal human plasma. Arch Biochem Biophys 351:250–256

48. Vance BA, Wu W, Ribaudo RK, et al (1997) Multiple dimeric forms of human CD69 result from differential addition of N-glycans to typical (Asn-X-Ser/Thr) and atypical (Asn-X-cys) glycosylation motifs. J Biol Chem 272:23117–23122

49. Grinnell BW, Walls JD, Gerlitz B (1991) Glycosylation of human protein C affects its secretion, processing, functional activities, and activation by thrombin. J Biol Chem 266:9778–9785

50. Sadler JE (2005) von Willebrand factor: two sides of a coin. J Thromb Haemost 3:1702–1709

51. Matsui T, Hamako J (2005) Structure and function of snake venom toxins interacting with human von Willebrand factor. Toxicon 45:1075–1087

52. Debeire P, Montreuil J, Samor B, et al (1983) Structure determination of the major asparagine-linked sugar chain of human factor VIII—von Willebrand factor. FEBS Lett 151:22–26

53. Matsui T, Titani K, Mizuochi T (1992) Structures of the asparagine-linked oligosaccharide chains of human von Willebrand factor: occurrence of blood group A, B, and H(O) structures. J Biol Chem 267:8723–8731

54. Samor B, Michalski JC, Debray H, et al (1986) Primary structure of a new tetraantennary glycan of the N-acetyllactosaminic type isolated from human factor VIII/von Willebrand factor. Eur J Biochem 158:295–298

55. Samor B, Michalski JC, Mazurier C, et al (1989) Primary structure of the major O-glycosidically linked carbohydrate unit of human von Willebrand factor. Glycoconj J 6:263–270

56. Matsui T, Kihara C, Fujimura Y, et al (1991) Carbohydrate analysis of human von Willebrand factor with horseradish peroxidase-conjugated lectins. Biochem Biophys Res Commun 178:1253–1259

57. Sodetz JM, Paulson JC, McKee PA (1979) Carbohydrate composition and identification of blood group A, B, and H oligosaccharide structures on human factor VIII/von Willebrand factor. J Biol Chem 254:10754–10760

58. Samor B, Mazurier C, Goudemand M, et al (1982) Preliminary results on the carbohydrate moiety of factor VIII/von Willebrand factor (FVIII/vWf). Thromb Res 25:81–89

59. Matsui T, Fujimura Y, Nishida S, et al (1993) Human plasma alpha 2-macroglobulin and von Willebrand factor possess covalently linked ABO(H) blood group antigens in subjects with corresponding ABO phenotype. Blood 82:663–668

60. Hounsell EF, Feizi T (1982) Gastrointestinal mucins: structures and antigenicities of their carbohydrate chains in health and disease. Med Biol 60:227–236

61. Watkins WM (2001) The ABO blood group system: historical background. Transfus Med 11:243–265

62. O'Donnell J, Boulton FE, Manning RA, et al (2002) Genotype at the secretor blood group locus is a determinant of plasma von Willebrand factor level. Br J Haematol 116:350–356

63. Matsui T, Shimoyama T, Matsumoto M, et al (1999) ABO blood group antigens on human plasma von Willebrand factor after ABO-mismatched bone marrow transplantation. Blood 94:2895–2900

64. Brown SA, Collins PW, Bowen DJ (2002) Heterogeneous detection of A-antigen on von Willebrand factor derived from platelets, endothelial cells and plasma. Thromb Haemost 87:990–996

65. O'Donnell J, Mille-Baker B, Laffan M (2000) Human umbilical vein endothelial cells differ from other endothelial cells in failing to express ABO blood group antigens. J Vasc Res 37:540–547

66. Schnyder-Candrian S, Borsig L, Moser R, et al (2000) Localization of alpha 1,3-fucosyltransferase VI in Weibel-Palade bodies of human endothelial cells. Proc Natl Acad Sci U S A 97:8369–8374

67. Kagami K, Williams S, Horne M, et al (2000) A preliminary analysis of platelet von Willebrand factor oligosaccharides. Nagoya J Med Sci 63:51–56

68. Do H, Healey JF, Waller EK, et al (1999) Expression of factor VIII by murine liver sinusoidal endothelial cells. J Biol Chem 274:19587–19592

69. Jacquemin M, Neyrinck A, Hermanns MI, et al (2006) FVIII production by human lung microvascular endothelial cells. Blood 108:515–517

70. Wion KL, Kelly D, Summerfield JA, et al (1985) Distribution of factor VIII mRNA and antigen in human liver and other tissues. Nature 317:726–729

71. Talamini MA, McCluskey MP, Buchman TG, et al (1998) Expression of alpha2-macroglobulin by the interaction between hepatocytes and endothelial cells in coculture. Am J Physiol 275:R203–R211

72. Gralnick HR, Coller BS, Sultan Y (1976) Carbohydrate deficiency of the factor VIII/von Willebrand factor Protein in von Willebrand's disease variants. Science 192:56–59

73. Gralnick HR, Cregger MC, Williams SB (1982) Characterization of the defect of the factor VIII/von Willebrand factor protein in von Willebrand's disease. Blood 59:542–548

74. Grainick HR, Williams SB, McKeown LP, et al (1985) Von Willebrand's disease with spontaneous platelet aggregation induced by an abnormal plasma von Willebrand factor. J Clin Invest 76:1522–1529

75. Sadler JE, Budde U, Eikenboom JC, et al (2006) Update on the pathophysiology and classification of von Willebrand disease: a report of the Subcommittee on von Willebrand Factor. J Thromb Haemost 4:2103–2114

76. Millar CM, Brown SA (2006) Oligosaccharide structures of von Willebrand factor and their potential role in von Willebrand disease. Blood Rev 20:83–92

77. Kao KJ, Pizzo SV, McKee PA (1980) Factor VIII/von Willebrand protein: modification of its carbohydrate causes reduced binding to platelets. J Biol Chem 255:10134–10139

78. De Marco L, Shapiro SS (1981) Properties of human asialo-factor VIII: a ristocetin-independent platelet-aggregating agent. J Clin Invest 68:321–328

79. Grainick HR, Williams SB, Coller BS (1985) Asialo von Willebrand factor interactions with platelets: interdependence of glycoproteins Ib and IIb/IIIa for binding and aggregation. J Clin Invest 75:19–25

80. Federici AB, De Romeuf C, De Groot PG, et al (1988) Adhesive properties of the carbohydrate-modified von Willebrand factor (CHO-vWF). Blood 71:947–952

81. De Marco L, Girolami A, Russell S, et al (1985) Interaction of asialo von Willebrand factor with glycoprotein Ib induces fibrinogen binding to the glycoprotein IIb/IIIa complex and mediates platelet aggregation. J Clin Invest 75:1198–1203

82. Carew JA, Quinn SM, Stoddart JH, et al (1992) O-linked carbohydrate of recombinant von Willebrand factor influences ristocetin-induced binding to platelet glycoprotein Ib. J Clin Invest 90:2258–2267

83. Sodetz JM, Pizzo SV, McKee PA (1977) Relationship of sialic acid to function and in vivo survival of human factor VIII/von Willebrand factor protein. J Biol Chem 252:5538–5546
84. Ellies LG, Ditto D, Levy GG, et al (2002) Sialyltransferase ST3Gal-IV operates as a dominant modifier of hemostasis by concealing asialoglycoprotein receptor ligands. Proc Natl Acad Sci U S A 99:10042–10047
85. Mohlke KL, Purkayastha AA, Westrick RJ, et al (1999) Mvwf, a dominant modifier of murine von Willebrand factor, results from altered lineage-specific expression of a glycosyltransferase. Cell 96:111–120
86. Gill JC, Endres-Brooks J, Bauer PJ, et al (1987) The effect of ABO blood group on the diagnosis of von Willebrand disease. Blood 69:1691–1695
87. Shima M, Fujimura Y, Nishiyama T, et al (1995) ABO blood group genotype and plasma von Willebrand factor in normal individuals. Vox Sang 68:236–240
88. Souto JC, Almasy L, Muniz-Diaz E, et al (2000) Functional effects of the ABO locus polymorphism on plasma levels of von Willebrand factor, factor VIII, and activated partial thromboplastin time. Arterioscler Thromb Vasc Biol 20:2024–2028
89. Jenkins PV, O'Donnell JS (2006) ABO blood group determines plasma von Willebrand factor levels: a biologic function after all? Transfusion 46:1836–1844
90. Dong JF (2005) Cleavage of ultra-large von Willebrand factor by ADAMTS-13 under flow conditions. J Thromb Haemost 3:1710–1716
91. Ricketts LM, Dlugosz M, Luther KB, Haltiwanger RS, Majerus EM (2007) O-fucosylation is required for ADAMTS13 secretion. J Biol Chem 282:17014–17023
92. Bowen DJ, Collins PW (2006) Insights into von Willebrand factor proteolysis: clinical implications. Br J Haematol 133:457–467
93. Morelli VM, de Visser MC, van Tilburg NH, et al (2007) ABO blood group genotypes, plasma von Willebrand factor levels and loading of von Willebrand factor with A and B antigens. Thromb Haemost 97:534–541
94. O'Donnell J, Boulton FE, Manning RA, et al (2002) Amount of H antigen expressed on circulating von Willebrand factor is modified by ABO blood group genotype and is a major determinant of plasma von Willebrand factor antigen levels. Arterioscler Thromb Vasc Biol 22:335–341
95. Bowen DJ (2003) An influence of ABO blood group on the rate of proteolysis of von Willebrand factor by ADAMTS13. J Thromb Haemost 1:33–40
96. O'Donnell JS, McKinnon TA, Crawley JT, et al (2005) Bombay phenotype is associated with reduced plasma-VWF levels and an increased susceptibility to ADAMTS13 proteolysis. Blood 106:1988–1991
97. McKinnon TA, Chion AC, Millington AJ, Lane DA, Laffan MA (2007) N-linked glycosylation of VWF modulates its interaction with ADAMTS13. Blood Nov. 1
98. Yamashita K, Tachibana Y, Nakayama T, et al (1980) Structural studies of the sugar chains of human parotid alpha-amylase. J Biol Chem 255:5635–5642
99. Sarkar K, Sarkar HS, Kole L, et al (1996) Receptor-mediated endocytosis of fucosylated neoglycoprotein by macrophages. Mol Cell Biochem 156:109–116
100. Lehrman MA, Pizzo SV, Imber MJ, et al (1986) The binding of fucose-containing glycoproteins by hepatic lectins: re-examination of the clearance from blood and the binding to membrane receptors and pure lectins. J Biol Chem 261:7412–7418
101. Bunkenborg J, Pilch BJ, Podtelejnikov AV, et al (2004) Screening for N-glycosylated proteins by liquid chromatography mass spectrometry. Proteomics 4:454–465
102. Liu T, Qian WJ, Gritsenko MA, et al (2005) Human plasma N-glycoproteome analysis by immunoaffinity subtraction, hydrazide chemistry, and mass spectrometry. J Proteome Res 4:2070–2080

Part 8   Clinicals: Thrombosis-related Disorders and Recent Topics

Hypercoagulable States

Tetsuhito Kojima and Hidehiko Saito

Summary. Hypercoagulable states are clinical conditions of patients who are unusually predisposed to venous or arterial thromboembolism. They are also called thrombophilias or prothrombotic disorders. Numerous congenital or acquired risk factors for hypercoagulable states were identified over the last three decades. Patients with inherited thrombotic disorders, including deficiencies of antithrombin, protein C, or protein S, are referred to as having a congenital hypercoagulable state. These deficiencies are uncommon but strong risk factors for thromboembolism, whereas the more recently discovered genetic variants, such as factor V Leiden and prothrombin variant, are common and weak risk factors, causing disease only in the presence of other factors. Patients with an increased risk of developing thrombotic complications because they are of advanced age, are immobilized, are in a pregnancy or puerperium, are undergoing surgery, are having cancer, and/or are using oral contraceptives or hormone replacement therapy are referred to as having an acquired hypercoagulable state. In these patients, the cause of thrombosis is frequently multifactorial and complex. Identification of such conditions may indicate a need for aggressive prophylaxis during high-risk periods, a need for prolonged treatment after an initial episode of thromboembolism, avoidance of oral contraceptives, and investigation of asymptomatic family members when a familial disorder is identified.

Key words. Hypercoagulable state · Congenital · Secondary · Multifactorial · Thromboembolism

Introduction

The hypercoagulable states are clinical conditions of patients who are unusually predisposed to venous or arterial thromboembolism; they are also referred to as thrombophilias or prothrombotic disorders. They may be classified as congenital (inherited), acquired (secondary), or both (mixed) conditions (Table 1) [1]. Congenital hypercoagulable states are caused by inherited thrombotic disorders due to mutations in genes encoding plasma proteins involved in coagulation mechanisms. Acquired hypercoagulable states are due to the many heterogeneous disorders that have been associated with an increased risk of thrombotic complications.

TABLE 1. Risk factors and conditions for hypercoagulable state

Acquired	Inherited	Mixed/Unknown
Bed rest	Antithrombin deficiency	High levels of factor VIII
Plaster cast	Protein C deficiency	High levels of factor IX
Trauma	Protein S deficiency	High levels of factor XI
Major surgery	Factor V Leiden (FVL)	High levels of fibrinogen
Orthopedic surgery	Prothrombin 20210A	High levels of TAFI
Malignancy	Dysfibrinogenemia	Low levels of TFPI
Pregnancy	Factor XIII 34Val	APC resistance in the absence of FVL
Oral contraceptives		
Hormonal replacement therapy		Hyperhomocysteinemia
Antiphospholipid syndrome		High levels of PCI (PAI-3)
Myeloproliferative disorders		
Polycythemia vera		
Central venous catheters		
Age		
Obesity		

Modified from Rosendaal [1]
TAFI, thrombin-activatable fibrinolysis inhibitor; TFPI, tissue factor pathway inhibitor; PCI, protein C inhibitor; PAI-3, plasminogen activator inhibitor-3; APC, activated protein C

Pathophysiology

Pathogenesis of venous and arterial thrombosis is complex and multifactorial. The classic Virchow's triad includes (1) vascular endothelial injury, (2) reduced blood flow, and (3) alterations in the constitution of the blood [2]. The interactions among these factors play a role in activation of the hemostatic system and thrombus formation. The pathogenesis of venous thrombosis appears to be different from that of arterial thrombosis: reduced blood flow and activation of blood coagulation play a major role in the former, whereas activation of platelets under high shear flow is responsible for the latter.

Thrombosis is considered to be hemostasis in the wrong place at the wrong time. Normally, blood fluidity is maintained by intact intimal endothelial cells and physiological inhibitors of thrombin generation. Comprehension of the mechanism of pathological thrombus formation by the hypercoagulable states requires an understanding of normal hemostasis, which consists of two important processes: the platelet plug formation and fibrin clot formation that occur concurrently at a site of vascular injury.

The formation of a platelet plug results from three processes: adhesion, activation, and aggregation of platelets. Platelets adhere to the vascular subendothelium exposed collagen fibers directly via collagen receptors and indirectly by attaching to collagen-bound von Willebrand factor (VWF) molecules at a site of vascular injury. Once adherent, platelets are activated by a number of agonists including thrombin, collagen, epinephrine, and thromboxane A_2 and release their α- and dense-granule contents, promoting further platelet recruitment, activation, and aggregation to form the platelet plug.

The fibrin clot formation is induced by a series of reactions, called the coagulation cascade, in which the plasma coagulation proteins are activated sequentially leading to thrombin generation [3]. Thrombin, in turn, converts fibrinogen to fibrin, which is finally stabilized by activated factor XIII. In response to vascular injury, in vivo coagulation is triggered by exposure of tissue factor (TF). Trace amounts of activated factor VII (factor VIIa), which is present in the circulation of normal individuals, complex with TF exposed on cellular surfaces at sites of vascular injury and then activates factor VII to factor VIIa. The factor VIIa–TF complex is able to activate factor X (factor Xa), which in turn catalyzes factor Va-mediated thrombin generation. The factor VIIa–TF complex is also able to activate factor IX (factor IXa), which interacts with activated factor VIII (factor VIIIa) on activated platelets bound to the extracellular matrix (ECM). This complex generates additional amounts of factor Xa, thereby magnifying the process of thrombin generation. Thrombin also promotes ongoing coagulation by activation of factors VIII, V, XI, and XIII. The end result of these sequential reactions is conversion of fibrinogen to fibrin monomers; and factor XIIIa crosslinks fibrin to promote stabilized hemostatic thrombus formation. In normal hemostasis, thrombin generation is tightly controlled by the anticoagulant system, which includes antithrombin (AT), protein C (PC), protein S (PS), and tissue factor pathway inhibitor (TFPI).

Congenital Hypercoagulable States

Almost all currently recognized congenital hypercoagulable states involve defects in the proteins of the coagulation or fibrinolytic system, rather than platelet abnormalities [4]. Congenital abnormalities of procoagulant or anticoagulant proteins result in an increased risk for venous thromboembolism (VTE) as well as arterial thrombosis [5]. Typical clinical presentations of congenital hypercoagulable states are idiopathic VTE in patients of relatively young age (<50 years) with recurrence events, thrombosis at unusual sites, and a positive family history of VTE [6, 7].

Numerous risk factors for hypercoagulable states were identified during the last three decades. Patients with inherited thrombotic disorders, such as AT deficiency, PC deficiency, or PS deficiency, are referred to as having a congenital hypercoagulable state. These were uncommon but strong risk factors for VTE, whereas the more recently discovered genetic variants, including factor V Leiden or prothrombin gene variant are common but weak, causing disease only in the presence of other factors [1]. The prevalences of genetic and mixed risk factors in the general population and in patients with VTE are listed in Table 2 [8]. The prevalence of genetic risk factors varies among different ethnic groups (discussed later). It should be also noted that VTE in patients with genetic risk factors usually occurs when they are exposed to some acquired (environmental) risk factors.

Antithrombin Deficiency

Antithrombin is a plasma protease inhibitor that inhibits thrombin and other activated coagulation factors, such as factors IXa, Xa, XIa, and XIIa [9], thereby contributing to the maintenance of blood fluidity. AT deficiency, identified by Egeberg [10] in

TABLE 2. Prevalence of genetic and mixed risk factors in the general population and in patients with VTE

Risk factor	General population (%)	Patients with VTE (%)
AT deficiency	0.02	1–3
PC deficiency	0.2–0.4	3–5
PS deficiency	0.03–0.13	1–5
Factor V Leiden[a]	1–15	10–50
Factor II G20210A[a]	2–5	6–18
Hyperhomocysteinemia	~5	~10
High plasma factor VIII level[b]	11	25
High plasma factor IX level[c]	3	7.5
High plasma factor XI level[d]	10	19

Modified from Franco and Reitsma [8]

VTE, venous thromboembolism; AT, antithrombin; PC, protein C; PS, protein S

[a]Common in Caucasians but not in Africans or Asians [48]

[a]Factor VIII level ≥150 IU/dl [57]

[b]Factor IX level >90% (i.e., 129.0 IU/dl) [58]

[c]Factor XI level >90% (i.e., 120.8 IU/dl) [59]

1965 in a Norwegian family, was the first inheritable hypercoagulable state associated with familial thrombosis. Since then, numerous studies have described similar clinical and laboratory findings in additional families, establishing the concept of AT deficiency as a risk factor for thrombophilia [9]. Women and men are equally affected, and congenital AT deficiency is usually heterozygous. Rare cases of homozygous qualitative AT deficiency have been reported [11–13], but homozygous quantitative AT deficiency has not. Thus, complete deficiency of AT is thought to be lethal at the embryonic stage as is suggested in AT null mice [14].

Antithrombin deficiency is classified into two types: type I (low plasma levels of both functional and immunological AT) and type II (variant AT in plasma). Type II is further subclassified into RS (defective reactive site), HBS (defective heparin-binding site), and PE (pleiotropic, i.e. multiple effects on function) [15]. The gene coding for AT (*SERPINC1*) is localized on chromosome 1q23-25, spans 13.4 kb of DNA, and has seven exons [16]. The current database contains 172 mutations (http://www. hgmd.cf.ac.uk/ac/gene.php?gene=SERPINC1), revealing that the molecular basis of AT deficiency is highly heterogeneous [17, 18]. Missense mutations are the most frequent genetic defects found in AT deficiency, but other gene lesions (e.g., nonsense mutations, splice site mutations, deletions, insertions) have also been reported.

The prevalence of AT deficiency is 0.02% in the general population, whereas it is more than 50 times higher in patients with VTE (Table 2). The level of AT in heterozygotes is usually 40%–70% of normal. The risk of thrombosis increases as the functional AT activity decreases, with the highest risk occurring when AT levels are <60% of normal [19]. The initial thrombotic event in most affected individuals is deep venous thrombosis (DVT) in the legs [20]. Women with AT deficiency appear to develop thrombosis earlier in life [21].

Management for patients who develop acute thrombotic complications due to AT deficiency is heparin administration in conjunction with exogenous AT administration [5]. The AT concentration should be maintained at >80% of normal during management of an acute thrombotic event with heparin and before surgery in a patient with an AT deficiency. This is achieved by administering fresh frozen plasma or AT concentrate, with the concentrate preferred. Long-term warfarin therapy is recommended for affected patients who have suffered a thrombotic event. All affected women should receive heparin and AT therapy during pregnancy. Patients from families with AT deficiencies should be studied and if they have an AT deficiency should be protected with heparin or warfarin during times of increased risk (i.e., surgery, trauma, pregnancy).

Protein C or Protein S Deficiency

Proteins C and protein S are vitamin K-dependent natural anticoagulants synthesized by the liver. PC, a zymogen of serine protease, is activated after thrombin binds to its endothelial receptor (thrombomodulin). Activated PC (APC) cleaves and inactivates factors Va and VIIIa rapidly, thereby inhibiting clot formation. APC also decreases plasminogen activator inhibitor-1 (PAI-1) activity, increasing fibrinolytic potential by reducing inhibition of the conversion of plasminogen to plasmin. PS acts as a nonenzymatic cofactor for APC, promoting the efficiency of these reactions. PC and PS deficiencies result in defects in the APC anticoagulant system and lead to a hypercoagulable state with increasing risk for VTE [22–26].

Congenital PC deficiency is transmitted as an autosomal dominant trait, with a prevalence of approximately 1/300 [27]. PC deficiency is classified into type I, in which both the protein and functional levels of PC are low, and type II, in which only the functional level is low. The gene encoding for PC is located on chromosome 2q13-14, and it spans approximately 10 kb containing 9 exons. Like AT deficiency, the molecular defects underlying PC deficiency have been elucidated in many families and are highly heterogeneous [28–31]. A total of 236 PC gene mutations have been reported in the database, most of them being missense mutations (http://www.hgmd.cf.ac.uk/ac/gene.php?gene=PROC). Reported defects include promoter mutations, splice site abnormalities, in-frame and frameshift deletions, insertions, nonsense and silent mutations.

Protein S, in the presence of phospholipid, greatly enhances the rate of inactivation of coagulation factor Va by APC. It also has an APC-independent ability to reduce directly the degradation of prothrombin and factor X through the binding of factors Va, VIIIa, and Xa [32, 33]. In vivo, 60%–70% of PS is bound to C4-binding protein. There are two PS genes in the human genome, *PROS1* and *PROS2*, which have been mapped to 3p11.1- q11.2. *PROS1* is the active gene responsible for the production of PS, whereas *PROS2* is a pseudogene. *PROS1* spans 80 kb genomic DNA and contains 15 exons. Loss-of-function mutations in *PROS1* lead to a deficiency of PS, an established inherited cause of venous thrombotic disease [8].

The inheritance of familial PS deficiency is usually autosomal dominant with a prevalence of 1:500 in the general population [5]. Based on plasma measurements, PS deficiency is classified into type I (a quantitative deficiency with a reduction of both total and free PS), type II (a qualitative deficiency characterized by decreased activity and normal total and free PS antigen levels), and type III (normal levels of total PS and low levels of free PS). However, the demonstrated coexistence of type I and type III in several PS-deficient families suggests that the two types are phenotypic variants of the same genetic disease [34]. Like PC deficiency, homozygosity is associated with severe neonatal purpura fulminans.

The identification of DNA abnormalities underlying PS deficiency has revealed its highly heterogeneous molecular basis [29, 35, 36]. A database compiling the defects identified in the PS gene reported 182 mutations (http://www.hgmd.cf.ac.uk/ac/gene.php?gene=PROS1). Missense mutations account for approximately 60% of the gene defects; nonsense mutations, splice site mutations, and large and small deletions and insertions have been detected in the remaining cases [35]. Recently, Kimura et al. reported that PS-K196E mutation is a genetic risk factor for DVT in Japanese patients, with the allele frequency estimated at 0.009 by genotyping the Japanese general population [37]. In many of the patients with PS deficiency studied, however, DNA abnormalities have not been identified.

There is little evidence to support prophylactic anticoagulation in asymptomatic individuals with a PC or PS deficiency. However, following a thrombotic episode, heparin therapy should be initiated followed by conversion to warfarin.

Factor V Leiden and APCR

Factor V Leiden (FVL) is a single point mutation in the factor V (FV) molecule in which glutamine is substituted for arginine at position 506 (Arg506Gln). This change makes factor Va extremely resistant to proteolytic cleavage by APC (APC resistance, or APCR) [38]. As factor Va functions as a cofactor in the activation of prothrombin to thrombin, the mutation leads to larger amounts of factor Va being available for coagulation reactions, shifting the hemostatic balance toward greater thrombin generation [39]. Therefore, those carrying this autosomal dominant trait are at markedly increased risk for venous thrombosis.

The extent of the increased thrombotic risk with the FVL mutation is probably less than that associated with AT, PC, and PS deficiencies [40]. The data obtained in case–control and cohort studies suggest that heterozygosity for the FVL mutation increases the risk of venous thrombosis three- to eightfold [41–43], although higher risks have also been claimed [44]. Homozygosity increases the thrombotic risk 50- to 100-fold [45].

Screening for the FVL mutation involves measuring the activated partial thromboplastin time (aPTT) in the presence and the absence of APC, determining the resistance to APC. This screening test, however, has low sensitivity and specificity because it is influenced by a variety of factors including, sex, age, and the presence of lupus anticoagulant. It is also unreliable in patients on oral anticoagulants. Genetic screening of DNA or RNA obtained from white blood cells for the FVL mutation is the most direct test available.

Factor V Leiden is highly prevalent in Caucasians, with carrier frequencies in the population ranging from 1% to 15% [46, 47] (Table 2), but it has not been found in Africans or Asians. FVL, present in 10%–50% of patients with a VTE, is considered the most common genetic defect involved in the etiology of venous thrombosis in Caucasians. The FVL abnormality originated from a single mutational event that occurred about 21 000–30 000 years ago (i.e., after the divergence of Africans from non-Africans and of Caucasoids from Mongoloids) [48]. The fact that the incidence of venous thrombosis is much higher in Caucasoids than in Mongoloids may be partly explained by the presence of FVL. The high prevalence of the mutation in the general population suggests that there is also a positive selection pressure for FVL. Evidence to support this notion has been provided by a study that revealed a significantly reduced risk of intrapartum bleeding complications in women carrying FVL in comparison with noncarriers [49].

Prothrombin Gene Variant (G20210A)

Prothrombin, a vitamin K-dependent zymogen, plays a key role in its activated form (thrombin) in the conversion of fibrinogen to fibrin. In 1996, a novel genetic factor involved in the etiology of VTE was described: a G→A transition at nucleotide position 20210 (G20210A) in the 3′-untranslated region of the coagulation prothrombin gene [50]. This mutation, found in 1%–3% of subjects in the general Caucasian population and in 6%–18% of patients with VTE, is associated with a two- to fivefold increased risk of VTE [8]. The prothrombin G20210A mutation results in elevated concentrations of plasma prothrombin and a tendency to hypercoagulability due to the greater availability of prothrombin for conversion to thrombin.

Patients with G20210A mutation have been demonstrated to have an increased risk of both coronary and cerebral arterial thromboses. It has been shown that in the presence of traditional risk factors for coronary disease the prothrombin mutation acts synergistically to increase the risk of myocardial infarction [51]. Prothrombin G20210A mutation can be diagnosed only by gene analysis. Prothrombin G20210 A is considered the second most prevalent genetic abnormality linked to thrombophilia in Caucasians.

Hyperhomocysteinemia

The pathogenesis of thrombosis in hyperhomocysteinemia is unclear. Proposed mechanisms include direct endothelial injury, increased TF activity, inhibition of PC activation, increased platelet activation and aggregation, suppression of thrombomodulin expression, and impaired fibrinolysis by inhibition of tissue plasminogen activator (t-PA) binding to its endothelial cell receptor [52, 53].

Inherited severe hyperhomocysteinemia, as seen in classic homocystinuria, may result from homozygous methylene tetrahydrofolate reductase (MTHFR) or cystathionine β-synthase (CBS) deficiency and, more rarely, from inherited errors of cobalamin (vitamin B$_{12}$) metabolism [54]. Inherited mild to moderate hyperhomocysteinemia may result from heterozygous MTHFR and CBS deficiencies but most commonly results from the C677T gene polymorphism, which is the most common mutation in the gene that codes for the MTHFR enzyme [53, 54]. This single-point mutation (C677T) in the coding region for the MTHFR-binding site (exon 4) is autosomal recessive, leads to substitution of a valine for an alanine, and results in a thermolabile variant of MTHFR deficiency [53]. Recent studies showed, however, that *MTHFRC667T* homozygosity with hyperhomocysteinemia is not associated with VTE [55, 56].

Aquired Hypercoagulable States

The acquired or secondary hypercoagulable states include a variety of clinical conditions that are known to have an increased risk for developing thrombotic complications [4, 5]. Transient or reversible conditions include pregnancy, use of oral contraceptives, hormone replacement therapy, immobilization, trauma or major surgery, and prolonged travel (economy class syndrome). Conditions that are generally irreversible include malignancy, myeloproliferative disorders, nephrotic syndrome, antiphospholipid syndrome, and paroxysmal nocturnal hemoglobinuria. Certain acquired conditions (e.g., elevated factor VIII, IX, and XI levels [57–59], lupus anticoagulant, pregnancy, oral contraceptive use) could result in the laboratory phenotype of APC resistance. This type of APC resistance is not associated with FVL. The origin and molecular basis of APCR in the absence of FVL is not well known and is likely to be of mixed genetic and acquired origin. It should be also noted that AT, PC, and PS deficiencies may occur in acquired hypercoagulable states, such as nephrotic syndrome, pregnancy, or disseminated intravascular coagulation [60].

Pregnancy and Puerperium

Pregnancy and puerperium are well-recognized hypercoagulable states. About 70% of DVTs in pregnancy are iliofemoral, with 80% of these persisting on the left side [61]. The risk of thromboembolism associated with pregnancy increases with age, the presence of hypertension, and the mode of delivery. It is particularly high in women who are confined to bed because of preeclampsia or eclampsia and who undergo a cesarean section [4].

Various anatomical, physiological, and biochemical changes during pregnancy and the postpartum period may predispose a patient to thromboembolism [4]. Pregnancy is associated with a hypercoagulable state due to increased concentrations of coagulation factors I, VII, VIII, IX, X, XI, and XII. There is also an elevated platelet count, and concentrations of PS [62] and AT decrease. Hypercoagulability is further enhanced by a marked depression of fibrinolytic activity during pregnancy, which is promptly restored within an hour of delivering the placenta [63]. These changes in the coagulation system are compounded by stasis generated by the gravid uterus compressing venous return from the lower extremities.

The risk of thrombosis during pregnancy is further increased if a woman has a concomitant genetic risk for thrombosis. Most notably is the FVL mutation that confers a threefold higher risk of thrombosis in pregnant women than in pregnant women without the mutation. The prothrombin gene variant (G20210A) is associated with a twofold increase in pregnancy-related thrombosis [64].

Warfarin crosses the placenta, is possibly teratogenic, and presents a hazard of hemorrhage to the fetus. Thus, heparin, which does not traverse the placenta, is the anticoagulant of choice during the first trimester and during late pregnancy. The hypercoagulable state persists up to 2 months following delivery of the fetus, and consideration should be given to continuing therapy after delivery [5].

Oral Contraceptives and Hormone Replacement Therapy

Venous thromboembolic events are serious complications of oral contraceptives and hormone replacement therapy (HRT). The administration of estrogen has been associated with a two- to sixfold increased relative risk of VTEs with either therapy[65]. Use of oral contraceptives is associated with increased levels of activated factor VII and decreased activities of PS and circulating thrombomodulin [66]. Oral contraceptives and HRT are associated with exponentially higher VTE relative risks when used by women with an inherited hypercoagulable state [8].

It remains impractical to screen for hypercoagulable disorders in every woman currently taking HRT or women who wish to begin therapy with oral contraceptives. If there is a family history of thrombosis or recurrent venous thrombosis, screening for a congenital hypercoagulable state is reasonable. Women who develop thrombosis while taking oral contraceptives should be treated like any patient with an acute thrombotic event and counseled regarding alternate methods of contraception. In addition, they should be tested for other causes of the hypercoagulable state [5].

Polycythemia Vera

Polycythemia vera (PV), a myeloproliferative disorder, carries with it an increased risk of thrombotic events. In a large series of 1213 patients with PV followed for 20 years, 634 fatal and nonfatal arterial and venous thromboses were recorded in 485 patients (41%) [67]. In PV, the prothrombotic effect of an elevated hematocrit is well established. The high hematocrit causes blood hyperviscosity, which plays a major role in the pathogenesis of both microcirculatory disturbances and arterial and venous thromboses. The multivariable, time-dependent analysis suggested that a high white blood cell count was also an independent predictor of vascular risk [68]. The predisposition for arterial thrombosis is reduced by cytoreductive therapy. However, aggressive treatment with cytoreductive agents carries an increased risk of malignancy [67].

Recently, a somatic mutation of the Janus kinase 2 (*JAK2*) gene resulting in constitutive activation of tyrosine kinase was identified with high frequency in patients with PV [69–71]. It is worth noting that the *JAK2* V617F mutation has been reported in a large proportion of patients with Budd-Chiari syndrome [72], in those with portal and mesenteric venous thrombosis [73, 74], and in a small but significant number of patients with cerebral vein thrombosis [75].

Laboratory Diagnosis

Table 3 shows the primary and secondary methods employed in the investigation of inherited thrombophilias [8, 76]. Diagnosis of AT, PC, and PS deficiency is established by plasma measurements of each protein using functional and immunological methods. APCR may be diagnosed by the aPTT modified assay or by identification of the FVL mutation with gene analysis techniques. Prothrombin G20210A is detected by gene analysis only. The usefulness of measuring plasma levels of coagulation factors in patients with VTE remains to be demonstrated [8].

Laboratory testing for hypercoagulable states is costly; and it rarely influences acute VTE management. It should be performed in selected patients with an objectively diagnosed venous thrombotic event under the following circumstances: relatively young patients (<50 years), recurrence of VTE, thrombosis at unusual sites, a positive family history of venous thrombotic disease [6, 7]. It may also be considered in selected asymptomatic individuals, particularly women who are relatives of patients with known inherited hypercoagulability, if the results may affect their decision to begin oral contraceptives use or HRT.

One should keep in mind the following points during the workup of patients with hypercoagulable states. Assays performed during acute illness or while a patient is

TABLE 3. Laboratory diagnosis of inherited hypercoagulable states

Risk factor	Primary method	Secondary method
AT deficiency	Plasma measurement[a]	DNA analysis
PC deficiency	Plasma measurement[a]	DNA analysis
PS deficiency	Plasma measurement[a]	DNA analysis
APCR	aPTT-based test	—
FVL	DNA analysis	—
Factor II G20210A	DNA analysis	—
Hyperhomocysteneia	Plasma measurement	DNA analysis[b]

Modified from Franco and Reitsma [8]

APCR, activated protein C resistance; FVL, factor V Leiden; aPTT, activated partial thromboplastin time

[a] Functional method for AT and PC measurements and immunological method for (free and total) PS measurement. Immunological methods are useful for further characterization of cases of AT and PC deficiency

[b] No mutation was unequivocally associated with risk for VTE

anticoagulated may be unreliable, leading to misdiagnosis. For example, anticoagulant therapy with warfarin may influence the PC and PS levels; and in some tests for APC resistance, heparin treatment may influence the measurement of AT. Recent thrombosis, inflammatory disease, and pregnancy may also affect some of these tests [77]. Abnormal results for inherited hypercoagulable states should, in general, be confirmed by a second measurement obtained under ideal circumstances.

Management

Anticoagulant therapy is a mainstay of management. The timing, type, duration, and intensity of anticoagulation depend on the clinical situation. The decision to extend therapy beyond 6–12 months after an incident thrombotic event must be made on an individual basis [61]. Patients with acquired hypercoagulable states who develop VTE during transient high-risk clinical settings—such as oral contraceptive use, in a post-surgical state, or with limb immobilization—has a low risk of developing recurrent thrombosis after 3 months of treatment. They should, however, have appropriate thromboprophylaxis for future high-risk situations; and if they fall into a category where a prothrombotic disorder is likely, they may warrant further investigation. Patients who develop VTE with ongoing malignancy, lupus anticoagulant paroxysmal nocturnal hemoglobinuria (PNH) or other continued risk factors are at high risk of having a recurrent thrombosis. For most of these patients, long-term anticoagulation should be considered, and expert advice is recommended.

Patients with inherited hypercoagulable states require appropriate counseling, including the advisability of testing first-degree relatives [61]. They should be considered for prolonged anticoagulation after a first episode of spontaneous VTE. However, long-term management should be individualized and depends on such factors as the precise nature of the disorder or disorders, circumstances of thrombosis, anticoagulant risk, and patient preference. Expert advice is recommended for all patients.

Recommendations may change as new knowledge related to natural history and results of clinical trials become available. Patients with congenital and some acquired thrombophilias are at increased, but variable, risk of VTE with pregnancy or use of oral contraceptives or HRT. Management in these patients should be guided by expert advice and an informed patient's preferences.

References

1. Rosendaal FR (2005) Venous thrombosis: the role of genes, environment, and behavior. Hematology [Am Soc Hematol Educ Program] 1–12
2. Kitchens C (1985) Concept of hypercoagulability: a review of its development, clinical application, and recent progress. Semin Thromb Hemost 11:293–315
3. Tripodi A, Mannucci PM (2007) Abnormalities of hemostasis in chronic liver disease: reappraisal of their clinical significance and need for clinical and laboratory research. J Hepatol 46:727–733
4. Schafer AI (1985) The hypercoagulable states. Ann Intern Med 102:814
5. Johnson CM, Mureebe L, Silver D (2005) Hypercoagulable states: a review. Vasc Endovasc Surg 39:123–133
6. Bauer KA (2001) The thrombophilias: well-defined risk factors with uncertain therapeutic implications. Ann Intern Med 135:367–373
7. Cushman M (2005) Inherited risk factors for venous thrombosis. Hematology 2005:452–457
8. Franco R, Reitsma P (2001) Genetic risk factors of venous thrombosis. Hum Genet 109:369–384
9. Van Boven HH, Lane DA (1997) Antithrombin and its inherited deficiency states. Semin Hematol 34:118–204
10. Egeberg O (1965) Inherited antithrombin deficiency causing thrombophilia. Thromb Diath Haemorrh 13:516–530
11. Chowdhury V, Lane D, Mille B, et al (1994) Homozygous antithrombin deficiency: report of two new cases (99 Leu to Phe) associated with arterial and venous thrombosis. Thromb Haemost 72:198–202
12. Okajima K, Ueyama H, Hashimoto Y, et al (1989) Homozygous variant of antithrombin III that lacks affinity for heparin, AT III Kumamoto. Thromb Haemost 61:20–24
13. Boyer C, Wolf M, Vedrenne J, et al (1986) Homozygous variant of antithrombin III: AT III Fontainebleau. Thromb Haemost 56:250–255
14. Ishiguro K, Kojima T, Kadomatsu K, et al (2000) Complete antithrombin deficiency in mice results in embryonic lethality. J Clin Invest 106:873–878
15. Lane D, Bayston T, Olds R, et al (1997) Antithrombin mutation database: 2nd (1997) update: For the Plasma Coagulation Inhibitors Subcommittee of the Scientific and Standardisation Committee of the International Society on Thrombosis and Haemostasis. Thromb Haemost 77:197–211
16. Olds RJ, Lane DA, Chowdhury V, et al (1993) Complete nucleotide sequence of the antithrombin gene: evidence for homologous recombination causing thrombophilia. Biochemistry 32:4216–4224
17. Bayston T, Lane D (1997) Antithrombin: molecular basis of deficiency. Thromb Haemost 78:339–343
18. Van Boven HH, Olds RJ, Thein SL, et al (1994) Hereditary antithrombin deficiency: heterogeneity of the molecular basis and mortality in Dutch families. Blood 84:4209–4213
19. Sagar S, Stamatakis JD, Higgins AF, et al (1976) Efficacy of low-dose heparin in prevention of extensive deep-vein thrombosis in patients undergoing total-hip replacement. Lancet 307:1151–1154

20. Bucciarelli P, Rosendaal FR, Tripodi A, et al (1999) Risk of venous thromboembolism and clinical manifestations in carriers of antithrombin, protein C, protein S deficiency, or activated protein C resistance: a multicenter collaborative family study. Arterioscler Thromb Vasc Biol 19:1026–1033

21. Pabinger I, Schneider B (1996) Thrombotic risk in hereditary antithrombin III, protein C, or protein S deficiency: a cooperative, retrospective study. Arterioscler Thromb Vasc Biol 16:742–748

22. Griffin J, Evatt, B, Zimmerman, TS, et al (1981) Deficiency of protein C in congenital thrombotic disease. J Clin Invest 68:1370–1373

23. Broekmans A, Veltkamp, JJ, Bertina, RM (1983) Congenital protein C deficiency and venous thromboembolism: a study of three Dutch families. N Engl J Med 309:340–344

24. Comp P, Nixon RR, Cooper DW, et al (1984) Familial protein S deficiency is associated with recurrent thrombosis. J Clin Invest 74:2082–2088

25. Schwarz HP, Fischer M, Hopmeier P, et al (1984) Plasma protein S deficiency in familial thrombotic disease. Blood 64:1297–1300

26. Kamiya T, Sugihara T, Ogata K, et al (1986) Inherited deficiency of protein S in a Japanese family with recurrent venous thrombosis: a study of three generations. Blood 67:406–410

27. Miletich J, Sherman L, Broze G (1987) Absence of thrombosis in subjects with heterozygous protein C deficiency. N Engl J Med 317:991–996

28. Reitsma PH, Poort SR, Allaart CF, et al (1991) The spectrum of genetic defects in a panel of 40 Dutch families with symptomatic protein C deficiency type I: heterogeneity and founder effects. Blood 78:890–894

29. Aiach M, Gandrille S, Emmerich J (1995) A review of mutations causing deficiencies of antithrombin, protein C and protein S. Thromb Haemost 74:81–89

30. Reitsma P (1997) Protein C deficiency: from gene defects to disease. Thromb Haemost 78:344–350

31. Millar D, Johansen B, Berntorp E, et al (2000) Molecular genetic analysis of severe protein C deficiency. Hum Genet 106:646–653

32. Van Wijnen M, Stam J, van't Veer C, et al (1996) The interaction of protein S with the phospholipid surface is essential for the activated protein C-independent activity of protein S. Thromb Haemost 76:397–403

33. Koppelman SJ, Hackeng TM, Sixma JJ, et al (1995) Inhibition of the intrinsic factor X activating complex by protein S: evidence for a specific binding of protein S to factor VIII. Blood 86:1062–1071

34. Zoller B, Garcia de Frutos P, Dahlback B (1995) Evaluation of the relationship between protein S and C4b-binding protein isoforms in hereditary protein S deficiency demonstrating type I and type III deficiencies to be phenotypic variants of the same genetic disease. Blood 85:3524–3531

35. Gandrille S, Borgel D, Sala N, et al (2000) Protein S deficiency: a database of mutations—summary of the first update; for the Plasma Coagulation Inhibitors Subcommittee of the Scientific and Standardization Committee of the International Society on Thrombosis and Haemostasis. Thromb Haemost 84:918

36. Makris M, Leach M, Beauchamp NJ, et al (2000) Genetic analysis, phenotypic diagnosis, and risk of venous thrombosis in families with inherited deficiencies of protein S. Blood 95:1935–1941

37. Kimura R, Honda S, Kawasaki T, et al (2006) Protein S-K196E mutation as a genetic risk factor for deep vein thrombosis in Japanese patients. Blood 107:1737–1738

38. Bertina RM, Koeleman BPC, Koster T, et al (1994) Mutation in blood coagulation factor V associated with resistance to activated protein C. Nature 369:64–67

39. Seligsohn U, Lubetsky A (2001) Genetic susceptibility to venous thrombosis. N Engl J Med 344:1222–1231

40. Martinelli I, Mannucci PM, De Stefano V, et al (1998) Different risks of thrombosis in four coagulation defects associated with inherited thrombophilia: a study of 150 families. Blood 92:2353–2358
41. Koster T, Rosendaal FR (1993) Venous thrombosis due to poor anticoagulant response to activated protein C: Leiden Thrombophilia Study. Lancet 342:1503–1506
42. Svensson PJ, Dahlback B (1994) Resistance to activated protein C as a basis for venous thrombosis. N Engl J Med 330:517–522
43. Ridker PM, Hennekens CH, Lindpaintner K, et al (1995) Mutation in the gene coding for coagulation factor V and the risk of myocardial infarction, stroke, and venous thrombosis in apparently healthy men. N Engl J Med 332:912–917
44. Salomon O, Steinberg DM, Zivelin A, et al (1999) Single and combined prothrombotic factors in patients with idiopathic venous thromboembolism: prevalence and risk assessment. Arterioscler Thromb Vasc Biol 19:511–518
45. Rosendaal FR, Koster T, Vandenbroucke JP, et al (1995) High risk of thrombosis in patients homozygous for factor V Leiden (activated protein C resistance). Blood 85:1504–1508
46. Rees DC, Cox M (1995) World distribution of factor V Leiden. Lancet 346:1133–1134
47. Ridker P, Miletich JP, Hennekens CH, et al (1997) Ethnic distribution of factor V Leiden in 4047 men and women: implications for venous thromboembolism screening. JAMA 277:1305–1307
48. Zivelin A, Griffin JH, Xu X, et al (1997) A single genetic origin for a common caucasian risk factor for venous thrombosis. Blood 89:397–402
49. Lindqvist P, Svensson PJ, Dahlback B, et al (1998) Factor V Q506 mutation (activated protein C resistance) associated with reduced intrapartum blood loss: a possible evolutionary selection mechanism. Thromb Haemost 79:69–73
50. Poort SR, Rosendaal FR, Reitsma PH, et al (1996) A common genetic variation in the 3′-untranslated region of the prothrombin gene is associated with elevated plasma prothrombin levels and an increase in venous thrombosis. Blood 88:3698–3703
51. Doggen CJM, Cats VM, Bertina RM, et al (1998) Interaction of coagulation defects and cardiovascular risk factors: increased risk of myocardial infarction associated with factor V Leiden or prothrombin 20210A. Circulation 97:1037–1041
52. Welch GN, Loscalzo J (1998) Homocysteine and atherothrombosis. N Engl J Med 338:1042–1050
53. Guba S, Fonseca V, Fink L (1999) Hyperhomocysteinemia and thrombosis. Semin Thromb Hemost 25:291–309
54. De Stefano V, Casorelli I, Rossi E, et al (2000) Interaction between hyperhomocysteinemia and inherited thrombophilic factors in venous thromboembolism. Semin Thromb Hemost 26:305–311
55. Frederiksen J, Juul K, Grande P, et al (2004) Methylenetetrahydrofolate reductase polymorphism (C677T), hyperhomocysteinemia, and risk of ischemic cardiovascular disease and venous thromboembolism: prospective and case-control studies from the Copenhagen City Heart Study. Blood 104:3046–3051
56. Mansilha A, Araujo F, Severo M, et al (2005) Genetic polymorphisms and risk of recurrent deep venous thrombosis in young people: prospective cohort study. Eur J Vasc Endovasc Surg 30:545–549
57. Rosendaal FR (1999) Venous thrombosis: a multicausal disease. Lancet 353:1167–1173
58. Vlieg AvH, van der Linden IK, Bertina RM, et al (2000) High levels of factor IX increase the risk of venous thrombosis. Blood 95:3678–3682
59. Meijers JCM, Tekelenburg WLH, Bouma BN, et al (2000) High levels of coagulation factor XI as a risk factor for venous thrombosis. N Engl J Med 342:696–701
60. Joist J (1990) Hypercoagulability: introduction and perspective. Semin Thromb Hemost 16:151–157

61. Bockenstedt PL (2006) Management of hereditary hypercoagulable disorders. Hematology 2006:444–449
62. Horinaga H, Otsuka H, Ishizuka B (2005) Changes in protein S activities and its significance in the coagulating and fibrinolytic system during normal pregnancy. J Obstet Gynecol Neonatal Hematol 14:36–42
63. Bonnar J, McNicol GP, Douglas AS (1969) Fibrinolytic enzyme system and pregnancy. BMJ 3:387–389
64. Dilley A, Austin H, El-Jamil M, et al (2000) Genetic factors associated with thrombosis in pregnancy in a United States population. Am J Obstet Gynecol 183:1271–1277
65. Gomes MPV, Deitcher SR (2004) Risk of venous thromboembolic disease associated with hormonal contraceptives and hormone replacement therapy: a clinical review. Arch Intern Med 164:1965–1976
66. Quehenberger P, Loner U KS, Handler S, et al (1996.) Increased levels of activated factor VII and decreased plasma protein S activity and circulating thrombomodulin during use of oral contraceptives. Thromb Haemost 76:729–734
67. Gruppo Italiano Studio P (1995) Polycythemia vera: the natural history of 1213 patients followed for 20 years. Ann Intern Med 123:656–664
68. Landolfi R, Di Gennaro L, Barbui T, et al (2007) Leukocytosis as a major thrombotic risk factor in patients with polycythemia vera. Blood 109:2446–2452
69. Kralovics R, Passamonti F, Buser AS, et al (2005) A gain-of-function mutation of JAK2 in myeloproliferative disorders. N Engl J Med 352:1779–1790
70. Baxter EJ, Scott LM, Campbell PJ, et al (2005) Acquired mutation of the tyrosine kinase JAK2 in human myeloproliferative disorders. Lancet 365:1054–1061
71. James C, Ugo V, Le Couedic J-P, et al (2005) A unique clonal JAK2 mutation leading to constitutive signalling causes polycythaemia vera. Nature 434:1144–1148
72. Patel RK, Lea NC, Heneghan MA, et al (2006) Prevalence of the activating JAK2 tyrosine kinase mutation V617F in the Budd-Chiari syndrome. Gastroenterology 130: 2031–2038
73. Primignani M, Barosi G, Bergamaschi G, et al (2006) Role of the JAK2 mutation in the diagnosis of chronic myeloproliferative disorders in splanchnic vein thrombosis. Hepatology 44:1528–1534
74. Colaizzo D, Amitrano L, Tiscia G, et al (2007) The JAK2 V617F mutation frequently occurs in patients with portal and mesenteric venous thrombosis. J Thromb Haemost 5:55–61
75. De Stefano V, Fiorini A, Rossi E, et al (2007) Incidence of the JAK2 V617F mutation among patients with splanchnic or cerebral venous thrombosis and without overt chronic myeloproliferative disorders. J Thromb Haemost 5:708–714
76. De Stefano V, Finazzi G, Mannucci PM (1996) Inherited thrombophilia: pathogenesis, clinical syndromes, and management. Blood 87:3531–3544
77. Tripodi A, Mannucci PM (2001) Laboratory investigation of thrombophilia. Clin Chem 47:1597–1606

Etiopathology of the Antiphospholipid Syndrome

Tatsuya Atsumi, Olga Amengual, and Takao Koike

Summary. The antiphospholipid syndrome (APS) is an autoimmune disorder in which vascular thrombosis and pregnancy morbidity occur in patients with laboratory evidence of antiphospholipid antibodies (aPLs). Phospholipid-binding plasma proteins, β_2-glycoprotein I (β2GPI) and prothrombin, are the dominant antigenic targets in APS. The aPLs affect the normal coagulation or anticoagulation reactions occurring on cell membranes and also interact with certain cells, altering the expression and secretion of procoagulant substances. Despite the clear association between aPLs and thrombotic events, the precise underlying disease mechanisms in APS remain unclear. Recently, great interest has focused on the cell membrane receptors and on the signal transduction pathways involved in the procoagulant cell activation mediated by aPLs. The roles of cell surface receptors for β2GPI, including those of the apolipoprotein E receptor 2′ and annexin II, in aPL-mediated cell activation have been investigated. Moreover, the crucial role of p38 mitogen-activated protein kinase on aPL-mediated cell activation has been demonstrated. Those extensive researches are helping to unveil the etiopathology of APS, ultimately leading to specific treatment for the affected patients.

Keywords. Antiphospholipid antibodies · Anti-β_2-glycoprotein I antibodies · Antiprothombin antibodies · Thrombosis · Complement

Introduction

The antiphospholipid syndrome (APS) is a clinical disorder characterized by thrombosis and pregnancy morbidity associated with the persistent presence of anti-phospholipid antibodies (aPLs), anti-cardiolipin antibodies (aCLs), and lupus anticoagulant (LA). The syndrome was first proposed to be a distinct entity—the anticardiolipin syndrome—in 1985, and it was later renamed the antiphospholipid syndrome, nowadays also recognized as Hughes syndrome [1].

Although aCLs and LA are of considerable clinical importance in APS, evidence shows that the dominant antigenic targets in APS are phospholipid-binding plasma proteins, β_2-glycoprotein I (β2GPI), and prothrombin [2, 3]. Other antigenic targets have also been identified in patients with APS, including tissue plasminogen activator (t-PA), plasmin, and thrombin.

521

Despite the clear association between aPLs and the thrombotic tendency, the precise underlying disease mechanisms in APS remain unclear. During the last few years, the binding of aPLs to procoagulant cells and how this binding mediates cell dysfunctions that potentially induce the clinical manifestations of the APS have been the focus of interest for many researchers. It is now recognized that the p38 mitogen-activated protein kinase (MAPK) pathway plays an important role in aPL-mediated cell activation [4, 5].

In this chapter, we review the etiopathology of APS and the aPL–cell interactions involved in the development of thrombotic complications in patients with APS.

Clinical Manifestations of Antiphospholipid Syndrome

The first evidence for APS was the description by Wassermann, in 1906, of complement-fixing antibodies that react with extracts from bovine hearts. In 1941, the relevant antigen was identified as cardiolipin, a phospholipid that is the basis for the biological false-positive serological test for syphilis (BFP-syphilis) [6]. Blood screening for this disease revealed that BFP-syphilis tests were associated with systemic lupus erythematosus (SLE) with no evidence of syphilis.

During the early 1950s, the development of a partial thromboplastin time coagulation test led to the recognition of anticoagulant activity in patients with SLE. This phenomenon called "lupus anticoagulant" was only observed in vitro and was not associated with bleeding disorders, although paradoxically related to thrombotic events. A major step toward the recognition of the APS was the description, in 1983, of a solid-phase immunoassay for detection of aCLs [7], leading to the development of the current assays used to detect aPLs.

Antiphospholipid syndrome can occur as an isolated disease or in association with SLE. Venous thrombosis is the most common manifestation, particularly deep venous thrombosis (DVT) in the lower extremities occasionally complicated by pulmonary embolism. Thrombosis in other venous territories, including axillary, central retinal, glomerular, and adrenal veins, has been described in APS. In addition, APS represents one of the most common causes of Budd-Chiari syndrome associated with hepatic vein thrombosis. In the arterial circulation, the involvement of intracranial arteries, manifesting as cerebral infarction and transient ischemic attacks, is the most frequent characteristic and APS is considered one of the most important risk factors for ischemic cerebral events. Other manifestations of arterial occlusion include ischemic heart disease, peripheral arterial occlusion, acute abdomen, and ischemic colitis due to mesenteric thrombosis [8].

Recurrent fetal losses comprise one of the most consistent complications of the APS, and aPLs may be found in up to 20% of woman with recurrent pregnancy losses [9]. aPL-related fetal losses are strikingly frequent during the second and third trimesters, although they can occur at any stage of pregnancy. APS patients are susceptible to early onset of pregnancy complications, such as severe preeclampsia and hemolysis, elevated liver enzymes, and low platelet count (HELLP) syndrome [10].

Since the first description of APS, the range of abnormalities associated with aPL has considerably broadened with recognition of other manifestations, such as livedo reticularis, nonthrombotic neurological syndromes, psychiatric manifestations, skin

TABLE 1. Revised classification criteria for the antiphospholipid syndrome

Clinical criteria

Vascular thrombosis

One or more clinical episodes of arterial, venous, or small vessel thrombosis in any tissue or organ confirmed by objective validated criteria by imaging or histopathology in the absence of significant evidence of inflammation in the vessel wall

Pregnancy morbidity

One or more unexplained deaths of a morphologically normal fetus at or beyond the 10th week of gestation, *or*

One or more premature deaths of a morphologically normal neonate before the 34th week of gestation due to eclampsia, severe preeclampsia, or placental insufficiency, *or*

Three or more unexplained consecutive spontaneous abortions before the 10th week of gestation (maternal anatomical or hormonal abnormalities and paternal and maternal chromosomal causes excluded)

Laboratory criteria

Lupus anticoagulant present in plasma on two or more occasions at least 12 weeks apart detected according to the guidelines of the International Society on Thrombosis and Haemostasis [69, 70]

IgG and/or IgM anti-cardiolipin antibodies present in medium or high titer in serum or plasma on two or more occasions at least 12 weeks apart, measured by a standardized ELISA [71]

IgG and/or IgM anti-β2 glycoprotein I antibodies present in >99th percentile titer in serum or plasma on two or more occasions at least 12 weeks apart, measured by a standardized ELISA [72]

From Pruemer [10]

Antiphospholipid syndrome is present if at least one of the clinical criteria and one of the laboratory criteria are met

Ig, immunoglobulin; ELISA, enzyme-linked immunosorbent assay

ulcers, hemolytic anemia, thrombocytopenia, pulmonary hypertension, nephropathy, heart valve abnormalities, and atherosclerosis [8].

In 1992, the term "catastrophic" was added to the designation APS to highlight an accelerated form of this syndrome with multiorgan failure, severe thrombocytopenia, and adult respiratory distress syndrome. Catastrophic APS is a rare but life-threatening condition with a poor prognosis and a high rate of mortality [11].

APS is diagnosed when venous and/or arterial thrombosis and/or pregnancy morbidity occurs in a patient whose laboratory tests for aPLs are positive. An international consensus statement on classification criteria for definite APS was published in 1999 [12] and revised in 2006 [10] (Table 1).

Autoantibodies Associated with Antiphospholipid Syndrome

Antigenic Targets

Two phospholipid-binding plasma proteins, β_2-glycoprotein I (β2GPI) and prothrombin, are the dominant antigenic targets recognized by aPLs in patients with APS [2, 3]. β2GPI, also known as apolipoprotein H, was first described in 1961 as a component of the β-globulin fraction of human serum. β2GPI is a single-chain polypeptide of 326 amino acid residues with a molecular mass of 50 kDa; it is present in normal human plasma at approximately 200 μg/ml. It contains a high proportion of proline and cysteine residues and is heavily glycosylated [13].

β2GPI, a member of the complement control protein repeat or short consensus repeat (SCR) superfamily, is composed of five homologous motifs of approximately 60 amino acids designated as SCR, or sushi, domains. Each motif contains four conserved half cysteine residues related to the formation of two internal disulfide bridges. Although the first four domains are typical, the fifth domain of β2GPI is a modified form containing 82 amino acid residues and six half cysteines. The tertiary structure of β2GPI revealed a highly glycosylated protein with an elongated fishhook-like arrangement of the globular SCR domains (Fig. 1) [14, 15]. β2GPI binds to solid-phase phospholipids through a major phospholipid binding site located in the fifth domain, $C^{281}KNKEKKC^{288}$ close to the hydrophobic loop [16], and the phospholipids-binding property is significantly reduced by cleavage of one particular site (K317-T318) on the domain V of β2GPI, nicked β2GPI.

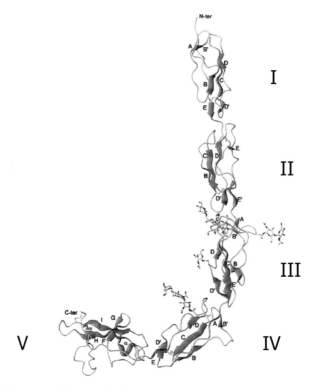

FIG. 1. Cristal structure of human β₂-glycoprotein I. Structural representation of human blood β₂-glycoprotein I reveals the extended chain of the five short consensus domains. The β-strands are shown in a darker color. (From Schwarzenbacher et al. [15], with permission)

Prothrombin is a vitamin K-dependent single-chain glycoprotein of 579 amino acid residues with a molecular weight of 72 kDa; it is present at a concentration of approximately 100 µg/ml in normal plasma [17]. Prothrombin undergoes γ-carboxylation during its liver biosynthesis. These γ-carboxyglutamic residues, known as the Gla domain, are located on fragment 1 of prothrombin. The Gla domain is essential for calcium-dependent binding of phospholipid to prothrombin. A kringle domain containing two kringle structures and a carboxyl-terminal serine protease follows the Gla domain.

Epitopes Recognized by Antiphospholipid Antibodies

It is widely accepted that aCLs found in patients with APS recognize the cardiolipin–β2GPI complex, and that the epitope locates on the β2GPI molecule. Therefore, both aCLs and anti-β2GPI antibodies represent the autoantibodies that bind to the phospholipids–β2GPI complex. However, the mechanisms by which anti-β2GPI antibodies bind to β2GPI are still a matter of debate. It has been proposed that binding of β2GPI to anionic phospholipids increases the local concentration of β2GPI, thus promoting an increase in the intrinsic affinity and binding of autoantibodies to β2GPI [18]. Other theory is based on the recognition of cryptic epitope(s) on β2GPI by anti-β2GPI antibodies. Those cryptic epitope(s) are exposed only when β2GPI interacts with negatively charged phospholipids, such as cardiolipin, or when adsorbed on a polyoxygenated polystyrene surface [19].

Another focus of intensive discussion has been the location of the epitope(s) for anti-β2GPI antibodies on the β2GPI molecule. In fact, anti-β2GPI antibodies recognize different epitopes located in all five domains of β2GPI. Domains IV and I were reported as candidates for epitopic location by using a series of deletion mutant proteins of β2GPI [20]. Wang et al. [21] showed anti-β2GPI antibodies directed to domain V, and domain IV was revealed as one of the major epitopic locations for monoclonal aCLs raised from APS patients [22].

In contrast, Iverson et al. [23] demonstrated that anti-β2GPI antibodies in the major population of APS patients bind to a particular structure in domain I of β2GP I, and that single amino acid substitution of full-length β2GPI disrupts antibody binding to β2GPI. However, some particular mutations made in domain IV also affected antibody binding to β2GPI in the anti-β2GPI antibody enzyme-linked immunosorbent assay (ELISA) [20].

Recently, de Laat et al. showed that pathogenic aCLs bind a cryptic epitope on domain I of β2GPI (Gly40-Arg43), which is accessible for aCLs only after conformational change and is induced by the binding of β2GPI to a negatively charged surface via a positive-charge patch in domain V [24]. Moreover, our group demonstrated that epitopic structures recognized by anti-β2GPI antibodies are cryptic and that three electrostatic interactions between domain IV and V are involved in their exposure [20]. Binding of anti-β2GPI antibodies to solid-phase β2GPI was significantly diminished by L replacement of W^{235}, an amino acid residue partly located in an inner region on domain IV and commonly found in the epitopic structures. This hypothesis is also supported by our previous data showing that replacement of a single amino acid at position 247 of β2GPI, which is important for the interaction between domains IV and V, can alter the antigenicity of β2GPI for pathogenic autoantibodies [25, 26].

Autoreactive CD4+ T cells against β2GPI have been identified in patients with APS [27]. These β2GPI-specific CD4+ T cells recognize a cryptic peptide encompassing amino acid residues 276–290 in the context of DRB4 *0103 (DR53); this epitope is located in domain V and contains the major phospholipid-binding site. The binding of β2GPI to anionic phospholipids facilitates processing and presentation of a cryptic epitope that activates pathogenic autoreactive T cells [28]. The activation of β2GPI-reactive CD4+ T cells subsequently stimulates B cells to produce pathogenic anti-β2GPI antibodies through interleukin 6 (IL-6) expression and CD40-CD40 ligand engagement [29].

Unlike β2GPI, prothrombin requires calcium ions for its binding to anionic phospholipids. Anti-prothrombin antibodies may be directed against cryptic or neoepitopes exposed when prothrombin binds to those phospholipids. In the presence of calcium, the N-terminal region of prothrombin undergoes a conformational change that results in exposure of a hydrophobic patch thought to be crucial for functional phospholipid binding [30]. In addition, it has been proposed that a second conformational change creates a surface-exposed hydrophilic cleft that may be complementary in shape and charge to that of the polar group [31].

On the other hand, anti-prothrombin antibodies may be low-affinity antibodies recognized more efficiently when the prothrombin is bound to phosphatidylserine coated on ELISA plates [32], or they may bind bivalently to immobilized prothrombin [33]. Thus, prothrombin complexed with phosphatidylserine could allow clustering and better orientation of the antigen, offering optimal conditions for antibody recognition.

Clinical Significance of Antiphospholipid Antibodies

Anti-β2GPI antibodies represent an independent risk factor for thrombosis and pregnancy morbidity. The detection of IgG and IgM anti-β2GPI antibodies are listed in the criteria for the classification of APS with a threshold for positive anti-β2GPI antibodies > 99th percentile of controls. The interpretation of anti-β2GPI antibody results has some limitations mainly due to the difference of methodology and the lack of standardization. Furthermore, the possible interference of cryoglobulins and rheumatoid factor should be considered in the interpretation of IgM anti-β2GPI antibodies [10].

Anti-prothrombin antibodies detected by ELISA are a heterogeneous population including antibodies against prothrombin alone coated on gamma irradiated or activated polyvinylchloride ELISA plates (aPT-A) [32, 34], and antibodies to the phosphatidylserine–prothrombin complex (aPS/PT)[32, 35]. Data on the clinical significance of aPT-A are contradictory, implying low sensitivity of these antibodies for the diagnosis of APS. APT-As are found in 60% of APS patients, but no more than two-third of those antibodies had LA activity [32, 34]. In contrast, aPS/PT strongly correlated with the presence of LA and showed higher sensitivity and specificity for the diagnosis of APS than aPT-A. Therefore, aPS/PT may serve as a confirmatory assay for LA and APS [35, 36].

Antiphospholipid Antibody–Procoagulant Cell Interaction

Platelet Interactions

Activated platelets are present in patients with APS, and aPLs can stimulate platelet agglutination and aggregation [37]. One in vivo study supported the importance of platelet activation in the pathogenesis of thrombosis, showing that the thrombi resulting from the infusion of anti-β2GPI antibodies in hamsters, in which carotid arteries were primed with a photochemical injury, were rich in platelets [38].

Membranes of activated platelets are an important source of negatively charged phospholipids, which provides a catalytic surface for blood coagulation. β2GPI is able to bind to surface membranes of activated platelets. β2GPI inhibits the generation of activated factor X by activated platelets, and anti-β2GPI antibodies interfered with this inhibition [39].

Monocytes and Endothelial Cell Interactions

Injured and/or activated monocytes or endothelial cells may be a predominant target of aPLs [40–42]. Cultured endothelial cells incubated with aPLs express increased levels of adhesion molecules, such as intercellular cell adhesion molecule-1, vascular cell adhesion molecule-1, and P-selectin [43]. This effect is mediated by β2GPI and may promote inflammation and thrombosis [42]. aPLs also induce tissue factor (TF) activity, antigens, or mRNA on endothelial cells and monocytes [40, 41, 44] Moreover, markedly elevated plasma levels of endothelin-1 (ET-1), the most potent endothelium-derived contracting factor modulating vascular smooth muscle tone, were found in patients with APS with arterial thrombosis. Human monoclonal aCL induced prepro ET-1 mRNA levels, confirming the direct effect of aPLs on endothelium regarding ET-1 production. Thus, the production of ET-1 induced by aPLs may play an important role in altering arterial tone and probably contributing to arterial occlusion [45].

Microparticle production is a hallmark of cell activation, and aPLs have been shown to stimulate the release of microparticles from endothelial cells. This may represent a new pathogenic mechanism for the thrombotic complications of APS [46].

Cell Receptors for Antiphospholipid Antibody Interactions

Anti-β2GPI antibodies activate platelets, monocytes, and endothelial cells in a β2GPI-dependent manner. This cell activation might require an interaction between β2GPI and a specific cell receptor.

The apolipoprotein E receptor 2 (ApoER2′), the only member of the low-density lipoprotein receptor present on platelets, has been suggested to mediate a role in the activation of platelets. Blockage of ApoER2′ on platelets using a receptor-associated protein leads to a loss of the increased adhesion of platelets to collagen induce by the β2GPI–anti-β2GPI antibodies complex. Moreover, dimeric β2GPI interacts with ApoER2′, and it has been hypothesized that the β2GPI–anti-β2GPI antibodies complex initially has to form on exposed phosphatidylserine on the cell surface before interacting with ApoER2' [47]. This procedure requires an initial independent priming stimulus to lead the platelets to the phosphatidylserine exposure on the cell surface. The binding site in the β2GPI molecule for ApoER2' on platelets has been reported to be located in domain V, which does not overlap with the phospholipid-binding site [48].

β2GPI has been also shown to bind directly to the glycoprotein (GP) Ibα subunit of the platelet adhesion receptor GPIb/IX/V in vitro [49, 50]. GPIbα subunit is located on platelets, with von Willebrand factor being the most important ligand, and it also serves to localize factor XI and thrombin on platelet surfaces. Binding of β2GPI to GPIbα enables anti-β2GPI antibodies directed against domain I to activate platelets, leading to thromboxane production and to activation of the phosphoinositol-3 kinase (PI3-kinase)/Akt pathway [49], which is involved in the signaling of GPIbα downstream, contributing to platelet adhesion and aggregation.

Annexin II, also known as annexin A2, is an endothelial cell receptor for t-PA and plasminogen, which has been suggested to interact with the β2GPI–anti-β2GPI antibodies complex on the endothelial cell surface, mediating cell activation [51]. However, it is still unclear whether such a putative receptor is actually involved in cell activation because annexin II does not span the cell membrane, and the presence of an unknown "adaptor" was suggested to induce activation. This receptor can also be directly targeted on endothelial cells activated by anti-annexin II antibodies [52].

The Toll-like receptor (TLR) family may also play a role in the interaction of the β2GPI–anti-β2GPI antibodies complex on the endothelial cell surface. Anti-β2GPI antibodies might crosslink β2GPI molecules, likely together with TLR, eventually favoring receptor polymerization and signaling cascade activation. TLR2 and TLR4 may be the responsible receptors in this family [53, 54].

The involvement of Fcγ receptor in cell activation has been investigated in vivo [38] and in in vitro studies on platelets [47], endothelial cells [51] and monocytes [55]. Results suggest that this receptor is not strictly necessary for cellular activation.

Signaling Pathways of Cell Activation

Great interest has arisen regarding the signal transduction mechanisms implicated in the increased expression of procoagulants in response to aPLs. The adapter molecule myeloid differentiation protein (MyD88)-dependent signaling pathway and nuclear factor kappa B (NFκB) have been involved in endothelial cell activation by aPLs [56–59]. IgG purified from APS patients induced nuclear translocation of NFκB, leading to the transcription of a large number of genes that have a NFκB-responsive element in their promoter. This nuclear translocation of NFκB mediates, at least in part, the elevated expression of TF and adhesion molecules by endothelial cells.

Several groups have demonstrated the crucial role of p38 MAPK pathway in aPL-mediated cell activation. Our group reported that monocytes stimulated by monoclonal anti-β2GPI antibodies derived from APS patients induce phosphorylation of p38MAPK, a locational shift of NFκB into the nucleus, and up-regulation of TF expression [4]. TF expression occurs only in the presence of β2GPI, suggesting that perturbation of monocytes by anti-β2GPI antibodies is initiated by interaction between the cell and the autoantibody-bound β2GPI (Fig. 2).

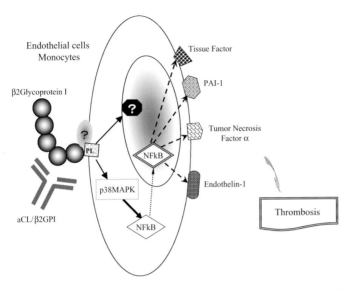

FIG. 2. Proposed functions of anti-phospholipid antibodies on monocytes or endothelial cells. Stimulation of procoagulant cells by aCL/β2GPI induce p38 MAPK phosphorylation, which leads to nuclear translocation of NFκB, transcription of procoagulant substances, and subsequently thrombus formation. *aCL/β2GPI*, β$_2$-glycoprotein I-dependent anti-cardiolipin antibodies; *p38MAPK*, p38 mitogen-activated protein kinase; *NFκB*, nuclear factor kappa B; *PAI-1*, plasminogen activator inhibitor-1

Almost simultaneously, Vega-Ostertag et al demonstrated that phosphorylation of p38 MAPK is involved in aPL-mediated production of thromboxane B$_2$ by platelets [60]. Pretreatment of platelets with SB203580, a p38 MAPK specific inhibitor, completely abrogated aPL-mediated platelet aggregation. The same group also showed the involvement of p38 MAPK in the up-regulation of TF in endothelial cells [5]. Activation of the p38 MAPK pathway increases the activity of inflammatory cytokines such as tumor necrosis factor-α (TNFα) and IL-1β. We also showed up-regulation of TNFα, IL-1β, and macrophage inflammatory protein 3β (MIP3β) on monocytes treated with aPLs [4]. In addition, the p38 MAPK pathway had been involved in the regulation of TF expression in monocytes, endothelial cells, and smooth muscle cells.

Complement and Antiphospholipid Antibodies

Complement activation has been associated with the pathogenesis of pregnancy loss in patients with APS [61, 62]. Elevated levels of C3 and C4 predict subsequent miscarriages in patients with unexplained recurrent miscarriages [63]. Moreover, higher levels of complement activation products were found in the plasma of APS patients with cerebral ischemic events than in patients with non-APS-related cerebral ischemia [64].

Pierangeli et al. [65], using a murine model of surgically induced thrombosis, showed that activation of complements C3 and C5 by aPLs mediates the activation of endothelial cells and the induction of thrombosis. Furthermore, Fischetti et al. [66]

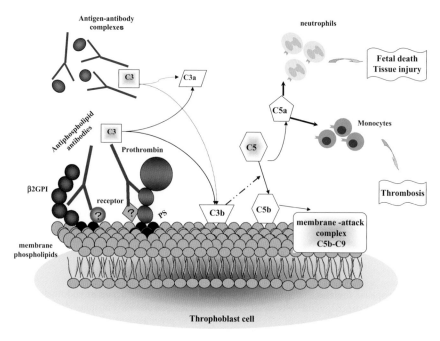

FIG. 3. Potential mechanisms for the pathogenicity of anti-phospholipid antibodies on fetal and placental injury. Placenta trophoblast cells are targeted by phospholipid-binding proteins/ antiphospholipid antibodies complexes, leading to activation of the complement cascade through the classic pathway. Complements C3 and subsequently C5 are activated. The generated C5a recruits and activates polymorphonuclear leukocytes and monocytes and stimulates the release of mediators of inflammation, which ultimately results in thrombosis and fetal injury. $\beta2GPI$, β_2-glycoprotein I; PS, phosphatidylserine

demonstrated the involvement of complement activation in the pathogenesis of thrombosis in an animal model in which polyconal IgG aPLs from APS patients were transfer to rats pretreated with lipopolysaccharide. Rats receiving IgG aPLs developed thrombus, whereas those receiving polyclonal IgG antibodies from healthy volunteers did not.

The IgG isotype of aPLs is the one most frequently found in patients with APS, with IgG2 subclass being the most prevalent [67, 68]. IgG2 and IgG4 subclasses have relatively weak ability to fix complement via the classic pathways, suggesting that an additional mechanism may be involved in the enhancement of complement activation in patients with aPLs.

Figure 3 shows the potential mechanisms for the pathogenicity of aPLs upon fetal and placental injury.

Conclusion

Thrombotic complications in patients with aPLs are unpredictable, and the mechanisms triggering thrombosis in susceptible individuals are not clearly defined. During the last few years, the identification of cell receptors and signaling pathways of aPL-mediated cell activation have allowed new insights into understanding the hypercoagulable state characteristic of APS patients. Recognition of the crucial role of p38 MAPK in the intercellular activation mediated by aPLs may translate into a new therapeutic approach in affected patients leading to the investigation of target therapy to down-regulate the specific pathway of cell activation.

Acknowledgments. This work was supported by the Japanese Ministry of Health, Labor, and Welfare and the Japanese Ministry of Education, Culture, Sports, Science, and Technology.

References

1. Khamashta MA (2006) Hughes syndrome: history. In: Khamashta MA (eds) Hughes syndrome: antiphospholipid syndrome. Springer, London, pp 3–8
2. Matsuura E, Igarashi Y, Fujimoto M, et al (1990) Anticardiolipin cofactor(s) and differential diagnosis of autoimmune disease. Lancet 336:177–178
3. Bevers EM, Galli M, Barbui T, et al (1991) Lupus anticoagulant IgG's (LA) are not directed to phospholipids only, but to a complex of lipid-bound human prothrombin. Thromb Haemost 66:629–632
4. Bohgaki M, Atsumi T, Yamashita Y, et al (2004) The p38 mitogen-activated protein kinase (MAPK) pathway mediates induction of the tissue factor gene in monocytes stimulated with human monoclonal anti-beta2Glycoprotein I antibodies. Int Immunol 16:1633–1641
5. Vega-Ostertag M, Casper K, Swerlick R, et al (2005) Involvement of p38 MAPK in the up-regulation of tissue factor on endothelial cells by antiphospholipid antibodies. Arthritis Rheum 52:1545–1554
6. Pangborn MC (1941) A new serologically active phospholipid from beef heart. Proc Soc Exp Biol Med 48:745
7. Harris EN, Gharavi AE, Boey ML, et al (1983) Anti-cardiolipin antibodies: detection by radioimmunoassay and association with thrombosis in systemic lupus erythematosus. Lancet 2:1211–1214
8. Cervera R, Piette JC, Font J, et al (2002) Antiphospholipid syndrome: clinical and immunologic manifestations and patterns of disease expression in a cohort of 1,000 patients. Arthritis Rheum 46:1019–1027
9. Branch DW, Silver R, Pierangeli S, et al (1997) Antiphospholipid antibodies other than lupus anticoagulant and anticardiolipin antibodies in women with recurrent pregnancy loss, fertile controls, and antiphospholipid syndrome. Obstet Gynecol 89:549–555
10. Miyakis S, Lockshin MD, Atsumi T, et al (2006) International consensus statement on an update of the classification criteria for definite antiphospholipid syndrome (APS). J Thromb Haemost 4:295–306
11. Asherson RA (1992) The catastrophic antiphospholipid syndrome. J Rheumatol 19:508–512
12. Wilson WA, Gharavi AE, Koike T, et al (1999) International consensus statement on preliminary classification criteria for definite antiphospholipid syndrome: report of an international workshop. Arthritis Rheum 42:1309–1311
13. Lozier J, Takahashi N, Putnam FW (1984) Complete amino acid sequence of human plasma β2-glycoprotein I. Proc Natl Acad Sci U S A 81:3640–3644
14. Bouma B, de Groot PG, van den Elsen JM, et al (1999) Adhesion mechanism of human β(2)-glycoprotein I to phospholipids based on its crystal structure. EMBO J 18:5166–5174

15. Schwarzenbacher R, Zeth K, Diederichs K, et al (1999) Crystal structure of human beta2-glycoprotein I: implications for phospholipid binding and the antiphospholipid syndrome. EMBO J 18:6228–6239
16. Hunt J, Krilis S (1994) The fifth domain of β2-glycoprotein I contains a phospholipid binding site (Cys281-Cys288) and a region recognized by anticardiolipin antibodies. J Immunol 152:653–659
17. Chow BK, Ting V, Tufaro F, et al (1991) Characterization of a novel liver-specific enhancer in the human prothrombin gene. J Biol Chem 1991:18927–18933
18. Sheng Y, Kandiah DA, Krilis SA (1998) Anti-β2-glycoprotein I autoantibodies from patients with the "antiphospholipid" syndrome bind to β2-glycoprotein I with low affinity: dimerization of β2-glycoprotein I induces a significant increase in anti-β2-glycoprotein I antibody affinity. J Immunol 161:2038–2043
19. Matsuura E, Igarashi Y, Yasuda T, et al (1994) Anticardiolipin antibodies recognize β2-glycoprotein I structure altered by interacting with an oxygen modified solid phase surface. J Exp Med 179:457–462
20. Kasahara H, Matsuura E, Kaihara K, et al (2005) Antigenic structures recognized by anti-beta2-glycoprotein I auto-antibodies. Int Immunol 17:1533–1542
21. Wang MX, Kandiah DA, Ichikawa K, et al (1995) Epitope specificity of monoclonal anti β2-glycoprotein I antibodies derived from patients with the antiphospholipid syndrome. J Immunol 155:1629–1636
22. George J, Gilburd B, Hojnik M, et al (1998) Target recognition of β2-glycoprotein I (β2GPI)-dependent anticardiolipin antibodies: evidence for involvement of the fourth domain of β2GPI in antibody binding. J Immunol 150:3917–3923
23. Iverson GM, Reddel S, Victoria EJ, et al (2002) Use of single point mutations in domain I of beta 2-glycoprotein I to determine fine antigenic specificity of antiphospholipid autoantibodies. J Immunol 169:7097–7103
24. De Laat B, Derksen RH, van Lummel M, et al (2006) Pathogenic anti-beta2-glycoprotein I antibodies recognize domain I of beta2-glycoprotein I only after a conformational change. Blood 107:1916–1924
25. Atsumi T, Tsutsumi A, Amengual O, et al (1999) Correlation between beta2-glycoprotein I valine/leucine247 polymorphism and anti-beta2-glycoprotein I antibodies in patients with primary antiphospholipid syndrome. Rheumatology 38:721–723
26. Yasuda S, Atsumi T, Matsuura E, et al (2005) Significance of valine/leucine247 polymorphism of beta2-glycoprotein I in antiphospholipid syndrome: increased reactivity of anti-beta2-glycoprotein I autoantibodies to the valine247 beta2-glycoprotein I variant. Arthritis Rheum 52:212–218
27. Hattori N, Kuwana M, Kaburaki J, et al (2000) T cells that are autoreactive to beta2-glycoprotein I in patients with antiphospholipid syndrome and healthy individuals. Arthritis Rheum 43:65–75
28. Kuwana M, Matsuura E, Kobayashi K, et al (2005) Binding of beta 2-glycoprotein I to anionic phospholipids facilitates processing and presentation of a cryptic epitope that activates pathogenic autoreactive T cells. Blood 105:1552–1557
29. Arai T, Yoshida K, Kaburaki J, et al (2001) Autoreactive CD4(+) T-cell clones to beta2-glycoprotein I in patients with antiphospholipid syndrome: preferential recognition of the major phospholipid-binding site. Blood 98:1889–1896
30. Wu JR, Lentz BR (1994) Phospholipid-specific conformational changes in human prothrombin upon binding to procoagulant acidic lipid membranes. Thromb Haemost 71:596–604
31. McDonald JF, Shah AM, Schwalbe RA, et al (1997) Comparison of naturally occurring vitamin K-dependent proteins: correlation of amino acid sequences and membrane binding properties suggests a membrane contact site. Biochemistry 36:5120–5127
32. Galli M, Beretta G, Daldossi M, et al (1997) Different anticoagulant and immunological properties of anti-prothrombin antibodies in patients with antiphospholipid antibodies. Thromb Haemost 77:486–491

33. Bevers EM, Zwaal RF, Willems GM (2004) The effect of phospholipids on the formation of immune complexes between autoantibodies and beta2-glycoprotein I or prothrombin. Clin Immunol 112:150–160
34. Arvieux J, Darnige L, Caron C, et al (1995) Development of an ELISA for autoantibodies to prothrombin showing their prevalence in patients with lupus anticoagulant. Thromb Haemost 74:1120–1125
35. Atsumi T, Ieko M, Bertolaccini ML, et al (2000) Association of autoantibodies against the phosphatidylserine-prothrombin complex with manifestations of the antiphospholipid syndrome and with the presence of lupus anticoagulant. Arthritis Rheum 43:1982–1993
36. Amengual O, Atsumi T, Koike T (2003) Specificities, properties, and clinical significance of antiprothrombin antibodies. Arthritis Rheum 48:886–895
37. Wiener MH, Burke M, Fried M, et al (2001) Thromboagglutination by anticardiolipin antibody complex in the antiphospholipid syndrome: a possible mechanism of immune-mediated thrombosis. Thromb Res 103:193–199
38. Jankowski M, Vreys I, Wittevrongel C, et al (2003) Thrombogenicity of beta 2-glycoprotein I-dependent antiphospholipid antibodies in a photochemically induced thrombosis model in the hamster. Blood 101:157–162
39. Shi W, Chong BH, Chesterman CN (1993) β2-Glycoprotein I is a requirement for anticardiolipin antibodies binding to activated platelets: differences with lupus anticoagulants. Blood 81:1255–1262
40. Kornberg A, Blank M, Kaufman S, et al (1994) Induction of tissue factor-like activity in monocytes by anti-cardiolipin antibodies. J Immunol 153:1328–1332
41. Amengual O, Atsumi T, Khamashta MA, et al (1998) The role of the tissue factor pathway in the hypercoagulable state in patients with the antiphospholipid syndrome. Thromb Haemost 79:276–281
42. Del Papa N, Sheng YH, Raschi E, et al (1998) Human beta 2-glycoprotein I binds to endothelial cells through a cluster of lysine residues that are critical for anionic phospholipid binding and offers epitopes for anti-beta 2-glycoprotein I antibodies. J Immunol 160:5572–5578
43. Pierangeli SS, Espinola RG, Liu X, et al (2001) Thrombogenic effects of antiphospholipid antibodies are mediated by intercellular cell adhesion molecule-1, vascular cell adhesion molecule-1, and P-selectin. Circ Res 88:245–250
44. Reverter JC, Tassies D, Font J, et al (1998) Effects of human monoclonal anticardiolipin antibodies on platelet function and on tissue factor expression on monocytes. Arthritis Rheum 41:1420–1427
45. Atsumi T, Khamashta MA, Haworth RS, et al (1998) Arterial disease and thrombosis in the antiphospholipid syndrome: a pathogenic role for endothelin 1. Arthritis Rheum 41:800–807
46. Dignat-George F, Camoin-Jau L, Sabatier F, et al (2004) Endothelial microparticles: a potential contribution to the thrombotic complications of the antiphospholipid syndrome. Thromb Haemost 91:667–673
47. Lutters BC, Derksen RH, Tekelenburg WL, et al (2003) Dimers of beta 2-glycoprotein I increase platelet deposition to collagen via interaction with phospholipids and the apolipoprotein E receptor 2′. J Biol Chem 278:33831–33838
48. Van Lummel M, Pennings MT, Derksen RH, et al (2005) The binding site in β2-glycoprotein I for ApoER2′ on platelets is located in domain V. J Biol Chem 280:36729–36736
49. Shi T, Giannakopoulos B, Yan X, et al (2006) Anti-beta2-glycoprotein I antibodies in complex with beta2-glycoprotein I can activate platelets in a dysregulated manner via glycoprotein Ib-IX-V. Arthritis Rheum 54:2558–2567
50. Pennings MT, Derksen RH, van Lummel M, et al (2007) Platelet adhesion to dimeric beta-glycoprotein I under conditions of flow is mediated by at least two receptors: glycoprotein Ibalpha and apolipoprotein E receptor 2′. J Thromb Haemost 5:369–377

51. Zhang J, McCrae KR (2005) Annexin A2 mediates endothelial cell activation by antiphospholipid/anti-beta2 glycoprotein I antibodies. Blood 105:1964–1969
52. Cesarman-Maus G, Rios-Luna NP, Deora AB, et al (2006) Autoantibodies against the fibrinolytic receptor, annexin 2, in antiphospholipid syndrome. Blood 107: 4375–4382
53. Satta N, Dunoyer-Geindre S, Reber G, et al (2007) The role of TLR2 in the inflammatory activation of mouse fibroblasts by human antiphospholipid antibodies. Blood 109: 1507–1514
54. Meroni PL, Raschi E, Testoni C, et al (2004) Innate immunity in the antiphospholipid syndrome: role of toll-like receptors in endothelial cell activation by antiphospholipid antibodies. Autoimmun Rev 3:510–515
55. Zhou H, Wolberg AS, Roubey RA (2004) Characterization of monocyte tissue factor activity induced by IgG antiphospholipid antibodies and inhibition by dilazep. Blood 104:2353–2358
56. Raschi E, Testoni C, Bosisio D, et al (2003) Role of the MyD88 transduction signaling pathway in endothelial activation by antiphospholipid antibodies. Blood 101: 3495–3500
57. Meroni PL, Raschi E, Testoni C, et al (2001) Statins prevent endothelial cell activation induced by antiphospholipid (anti-beta2-glycoprotein I) antibodies: effect on the pro-adhesive and proinflammatory phenotype. Arthritis Rheum 44:2870–2878
58. Dunoyer-Geindre S, De Moerloose P, Galve-De Rochemonteix B, et al (2002) NFkappaB is an essential intermediate in the activation of endothelial cells by anti-beta(2) glyco-protein 1 antibodies. Thromb Haemost 88:851–857
59. Espinola RG, Liu X, Colden-Stanfield M, et al (2003) E-Selectin mediates pathogenic effects of antiphospholipid antibodies. J Thromb Haemost 1:843–848
60. Vega-Ostertag M, Harris EN, Pierangeli SS (2004) Intracellular events in platelet activation induced by antiphospholipid antibodies in the presence of low doses of thrombin. Arthritis Rheum 50:2911–2919
61. Girardi G, Redecha P, Salmon JE (2004) Heparin prevents antiphospholipid antibody-induced fetal loss by inhibiting complement activation. Nat Med 10:1222–1226
62. Salmon JE, Girardi G, Lockshin MD (2007) The antiphospholipid syndrome as a dis-order initiated by inflammation: implications for the therapy of pregnant patients. Nat Clin Pract Rheumatol 3:140–147
63. Sugiura-Ogasawara M, Nozawa K, Nakanishi T, et al (2006) Complement as a predictor of further miscarriage in couples with recurrent miscarriages. Hum Reprod 21: 2711–2714
64. Davis WD, Brey RL (1992) Antiphospholipid antibodies and complement activation in patients with cerebral ischemia. Clin Exp Rheumatol 10:455–460
65. Pierangeli SS, Girardi G, Vega-Ostertag M, et al (2005) Requirement of activation of complement C3 and C5 for antiphospholipid antibody-mediated thrombophilia. Arthritis Rheum 52:2120–2124
66. Fischetti F, Durigutto P, Pellis V, et al (2005) Thrombus formation induced by antibod-ies to beta2-glycoprotein I is complement dependent and requires a priming factor. Blood 106:2340–2346
67. Sammaritano LR, Ng S, Sobel R, et al (1997) Anticardiolipin IgG subclasses: association of IgG2 with arterial and/or venous thrombosis. Arthritis Rheum 40:1998–2006
68. Amengual O, Atsumi T, Khamashta MA, et al (1998) IgG2 restriction of anti-beta2-glycoprotein I as the basis for the association between IgG2 anticardiolipin antibodies and thrombosis in the antiphospholipid syndrome: comment on the article by Sam-maritano et al. Arthritis Rheum 41:1513–1515
69. Brandt JT, Triplett DA, Alving B, et al (1995) Criteria for the diagnosis of lupus antico-agulants: an update. On behalf of the Subcommittee on Lupus Anticoagulant/Antiphos-pholipid Antibody of the Scientific and Standardization Committee of the ISTH. Thromb Haemost 74:1185–1190

70. Wisloff F, Jacobsen EM, Liestol S (2002) Laboratory diagnosis of the antiphospholipid syndrome. Thromb Res 108:263–271
71. Harris EN, Pierangeli SS (2002) Revisiting the anticardiolipin test and its standardization. Lupus 11:269–275
72. Reber G, Tincani A, Sanmarco M, et al (2004) Proposals for the measurement of anti-beta2-glycoprotein I antibodies: standardization group of the European Forum on Antiphospholipid Antibodies. J Thromb Haemost 2:1860–1862

Malignancy and Thrombosis

Hiroshi Kobayashi, Ryuji Kawaguchi, Yoriko Tsuji, Yoshihiko Yamada, Mariko Sakata, Seiji Kanayama, Shoji Haruta, and Hidekazu Oi

The following members and their institutions participated in this review article: M Ikeda, Department of Surgery, Graduate School of Medicine, Osaka University, Japan; N Yamada, Department of Cardiology, Mie University Graduate School of Medicine, Japan; K Sueishi (a director), Division of Pathophysiological and Experimental Pathology, Department of Pathology, Graduate School of Medical Sciences, Kyushu University, Japan; K Susuki (a chair), Department of Molecular Pathobiology, Mie University Graduate School of Medicine, Japan, who belong to Scientific Promotion Committee (SPC) of The Japanese Society on Thrombosis and Hemostasis (JSTH).

Summary. The association between cancer and an increased incidence of venous thromboembolism (VTE; Trousseau syndrome) is well characterized. This chapter reviews the pertinent literature related to VTE in patients with malignancy and the impact and scope of cancer-associated thrombosis, possible risk factors, etiology, pathogenesis, current practice patterns, and future directions.

Key words. Venous thromboembolism · Pulmonary embolism · Cancer · Prophylaxis · Etiology

Introduction

An estimated incidence of venous thromboembolism (VTE) is 1 per 1000 person-years for the general population [1, 2]. In Japan, the overall incidence of pulmonary embolism (PE) after general surgery is 0.33% [3]. Fatal PE was reported in 0.08% of the surgical population, and the mortality rate of patients with PE was about 30% [3]. Despite advances in diagnostic techniques and improved anticoagulation prophylaxis, the incidence of VTE has not changed significantly over the past four decades. The mortality rate directly ascribed to PE has ranged from 45% [4] to 91% [5] in large studies worldwide. Thus, PE remains a commonly underestimated and potentially lethal condition. Hospitalized patients with acute medical conditions are at significant risk of VTE, as approximately 10%–30% of general medical patients may develop deep vein thrombosis (DVT) or PE [6].

Venous thromboembolism has been estimated to occur in up to 30% of patients with hematological and solid tumor malignancies [7, 8], with 15% of those developing disseminated intravascular coagulation (DIC) [9]. Despite advances in the prophylaxis and treatment of VTE in the cancer population, cancer-associated thrombosis remains a serious and potentially life-threatening disease [10]. Although most thrombotic events occur postoperatively, VTE occurs commonly throughout the clinical course of patients with cancer. In an autopsy study of patients with intracranial neoplasms, 40% of VTE occurred outside the postoperative period [11]. In a study screening for asymptomatic DVT in a postoperative population, the rate rose to 60% [11].

Thrombotic events are identified as the second leading cause of mortality in cancer patients following the disease itself [10]. Among patients with cancer, the risk of recurrence is about threefold higher in those with metastatic disease [10, 12]. The risk of recurrent VTE and all-cause death is threefold higher in patients with concurrent VTE and malignancy compared to noncancer patients with VTE [10]. Thus, cancer patients are considered to be at increased risk of both initial and recurrent VTE events [13].

Etiology and Pathogenesis

The risk factors for VTE can be divided into inherited and acquired factors [14]. An inherited deficiency may be due to a lack of proteins and/or their activity that normally protect against VTE. A genetic predisposition to thrombophilia (hereditary risk factors) may be due to factor V Leiden mutation, G20210A prothrombin gene mutation, deficiencies of protein C, protein S, and antithrombin, or hyperhomocysteinemia [15]. Ethnic differences in the genetic background of thrombophilia have been reported [16]: No patients with factor V Leiden were detected in a Japanese population, whereas in contrast protein S mutation K196E is a common genetic risk factor among the Japanese [17]. More than 190 mutations have been identified in the protein S gene [17].

Acquired risk factors include medical conditions, including malignancy, cardiac disease, respiratory disease, inflammatory bowel disease, infectious diseases, hospitalization, prolonged immobility, surgery, venous trauma, estrogen therapy, pregnancy, anti-phospholipid antibodies, a history of VTE, increasing age, acquired thrombophilia, and obesity [6, 15]. Malignancy is one of the most important acquired homeostatic factors. An association between malignancy and VTE was first recognized in 1864 by Armand Trousseau in patients with occult carcinoma (Trousseau syndrome) [18]. There are many hypotheses concerning the causal relationship between cancer and the development of VTE and PE. An idiopathic VTE can be a predictor of occult malignancy, with one study suggesting that individuals who present with an unprovoked episode of VTE have a 10% frequency of subsequent cancer [10]. Supporting this concept, sera of patients with cancer showed decreased clotting times with increased circulating levels of plasminogen activator inhibitor type-1 (PAI-1) and elevated levels of markers of activated blood coagulation, including D-dimer, thrombin–antithrombin complexes, and soluble fibrin monomers, indicative of intravascular coagulation and fibrinolysis/DIC [11]. There are several risk factors (biochemical and clinical factors) for the development of VTE in cancer patients that are well described in the literature.

Biochemical Factors

The pathogenesis of cancer-associated VTE is complex, involving multiple interactions between tumors and various components of hemostasis [13]. Malignancy is associated with a hypercoagulable state, which is attributed to tumor expression of tissue factor (TF) and other procoagulants, activation of vascular endothelial cells by tumor-derived proinflammatory cytokines, and adhesive interactions between tumor cells and host cells [13, 19]. As shown in Fig. 1, prothrombotic factors in malignancy

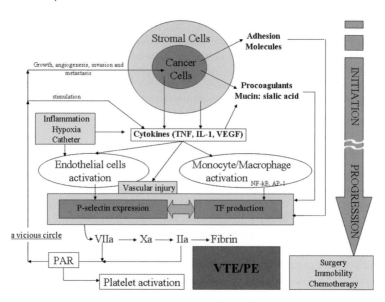

FIG. 1. Pathogenesis of the hypercoagulable state in cancer: mechanism of venous thromboembolism (*VTE*) in patients with cancer. Tumor-derived proinflammatory cytokines stimulate expression of tissue factor (*TF*) and cancer procoagulant by monocytes and endothelial cells and promote the development of a prothrombotic endothelial cell surface. *TNF*, tumor necrosis factor; *IL-1*, interleukin-1; *VEGF*, vascular endothelial growth factor; *NFκB*, nuclear factor κB; *AP-1*, activating protein-1; *PAR*, protease-activated receptor; *PE*, pulmonary embolism

include tumor production of procoagulants—i.e., TF, cancer procoagulant (CP), hepsin [20, 21]—and inflammatory cytokines and the interaction between tumor cells and host cells including blood cells (i.e., monocytes, macrophages, neutrophils, platelets) and vascular endothelial cells [22].

Tissue Factor and Thrombin

Evidence has accumulated indicating that TF circulates in normal plasma [23], both associated with cell-derived membrane microvesicles [24] and as a soluble, alternatively spliced form [25]. Tumor-derived TF may gain access to the systemic circulation and activate the coagulation cascade at distant sites. For example, ovarian cancer (clear cell carcinoma in particular) reportedly increases the risk of VTE. Clear cell carcinoma of the ovary produces excessive levels of TF, and the patient is more likely to develop VTE [26]. TF is the cell surface receptor of the serine protease factor VII/activated factor VII (VIIa) that plays a crucial role in the initiation of coagulation. Formation of complexes between TF and factor VIIa initiates the extrinsic blood coagulation cascade by activating factors IX and X [27], which in turn causes thrombin activation, platelet aggregation, fibrin deposition, and local hypercoagulable hemostasis.

Tissue factor might be derived from cancer cells themselves and activated host cells, including monocytes, macrophages, fibroblasts, and endothelial cells. Regulation of TF expression is controlled by the transcription factors nuclear factor-κB (NFκB) and activating protein-1 (AP-1). TF is commonly not expressed on fibroblasts and endothelial cells but is rapidly induced in response to not only thrombin [28, 29] but also inflammatory stimuli including endotoxin, proinflammatory cytokines, and growth factors such as vascular endothelial growth factor (VEGF), platelet-derived growth factor (PDGF), fibroblast growth factor (FGF), epidermal growth factor (EGF), tumor necrosis factor-α (TNFα), interleukin-1β (IL-1β), and transforming growth factor-β (TGFβ) produced by inflammatory cells and malignant cells, rendering the endothelial cell surface of blood vessels thrombogenic [30]. These mediators also up-regulate plasminogen activator inhibitor-1 (PAI-1), thereby decreasing fibrinolytic activity and down-regulating both thrombomodulin and the endothelial cell protein C receptor, which are both required for optimal protein C activation [31]. Increased expression of TF in tumors may contribute to angiogenesis by increasing VEGF expression [32] and down-regulating the expression of thrombospondin, one of the antiangiogenic proteins [33]. Factor VIIa, factor Xa, and thrombin act via protease-activated receptors (PARs) and TF to stimulate signal transduction and alter gene expression in both cancer cells and host cells [20]. The up-regulation of TF under these pathophysiological conditions may be responsible for thrombotic complications. The expression of TF contributes to not only thrombus formation but also chronic inflammation [34]. The association of coagulation factors thrombin and TF with proinflammatory cytokines resembles a vicious circle, where coagulation seems to cause a predisposition to more coagulation.

Thrombin interacts with PAR-1, PAR-3, and PAR-4 [35]. Activated PAR is coupled via several members of the heterotrimeric G-proteins $G\alpha_{o/i}$, $G\alpha_{12/13}$, and $G\alpha_q$ to transduce a substantial network of signaling pathways [35]. PAR activates multiple intracellular signaling pathways, including the p44/42 mitogen-activated protein (MAP) kinase and phosphatidylinositol-3 (PI3) kinase [13]. These signals also act as key pathways regulating tumor cell angiogenesis, growth, invasion and metastasis, and survival [13]. Loss of PTEN, a negative regulator of PI3-kinase, and hypoxia in glioma cells up-regulate TF expression and promote plasma clotting [36]. Furthermore, the small GTPases RhoA and RhoD act as a molecular switch controlling the negative and positive invasive pathways triggered by PARs [37].

Thrombin up-regulates expression of proinflammatory cytokines [38]. In response to thrombin-induced proinflammatory cytokines, induction of adhesion molecules and the expression of endothelial cell surface procoagulant proteins can also occur. Thrombin also activates tumor cell adhesion to platelets, endothelial cells, and matrix proteins. Angiogenesis, adhesion, migration, and cell survival induced via coagulation factors such as thrombin and TF are mediated via PARs [13, 39] and are independent of fibrin formation, which can be modulated without interfering with blood coagulation [40].

P-Selectin

Microvesicles bearing TF appear to participate in VTE by binding activated endothelial cells, a process that depends on the interaction between P-selectin glycoprotein ligand-1 (PSGL-1, CD162) on microvesicles and P-selectin on activated endothelial cells [41]. Like platelets, endothelial cells contain large amounts of P-selectin stored in their intracellular granules and express it on their surface upon activation by several cytokines [41]. Furthermore, P-selectin mediates heterotopic aggregation of activated platelets to cancer cells and adhesion of cancer cells to stimulated endothelial cells. The biological importance of P-selectin mediates cell adhesive interactions in the pathogenesis of inflammation, thrombosis, and cancer progression [42].

Cancer Procoagulant

Not only coagulation proteins and inflammatory cytokines but also cancer procoagulant (CP) may be elevated in patients with malignancy [22]. Although the protein sequence and cDNA encoding CP is not clear, CP is likely a cysteine protease that may be produced by malignant tissue. The possible role of CP in the pathogenesis of cancer-related thrombosis has been suggested [43]. CP is a direct activator of factor X, without contribution of factor VII or any other cofactors. CP differs from other procoagulants in regard to its physical, chemical, and enzymatic activity [43]. It can also induce platelet activation [44]. Tumor cell release of proinflammatory cytokines enhances CP activity (44). The CP activity is elevated in serum and neoplastic tissue of women with genital carcinoma [45]. These patients have decreased coagulation times and thus are likely to develop coagulation disturbances during the course of their cancer.

Hepsin

Hepsin, a putative membrane-associated serine protease, activates human factor VII and initiates a pathway of blood coagulation on the cell surface leading to thrombin formation [21]. This enzyme activates human factor VII by limited proteolytic cleavage of the Arg-Ile peptide bond [21]. Decreased expression of hepsin was observed in human hepatocellular carcinomas [46]. The incidence of VTE and PE in patients with hepatocellular carcinoma is relatively low, possibly due to decreased expression of hepsin and decreased production of coagulation factors in an affected liver.

Proinflammatory Cytokines

Other mechanisms of thrombus promotion include some general responses of the host to the tumor (i.e., acute phase, inflammation, adhesion, angiogenesis), decreased levels of coagulation inhibitors, and impaired fibrinolysis. The inflammatory response is an important characteristic of cancer-associated thrombosis and aggressiveness. A linkage between the coagulation cascade and the proinflammatory cytokines/chemokines involved in cancer has been reported. All of the inflammatory events tend to shift the hemostatic balance in favor of clot formation [31]. A number of cytokines potentially involved in the pathogenesis of VTE, including monocyte chemoattractant protein-1 (MCP-1), macrophage inflammatory protein-1 (MIP-1), epithelial/neutrophil-activating protein-78 (ENA-78), TNFα, and IL-6 [2]. These cytokines enhance vessel wall neutrophil extravasation early after thrombus induction. IL-6 can increase both the platelet count and their responsiveness to agonists such as thrombin. An increasing body of evidence indicates that cytokine-induced activation of hemostasis in malignant disease contributes to tumor growth and progression [13]. IL-8 is particularly considered to be associated with tumor progression.

Clinical Factors

Clinicopathological Characteristics

Clinical risk factors for cancer-associated thrombosis include tumor size, tumor grade, tumor pathology, mode of treatment including anticancer therapy, surgical procedures, longer anesthesia times, inherited and acquired thrombophilia, catheter use, and immobility [9, 10]. In women with cancer, the clinical course is often complicated by VTE episodes [9], although some recent studies have reported a substantially higher risk of recurrence among males [47]. Age and ABO blood group status have also been linked with increased VTE risk in patients with cancer [11]. Large tumors may have a greater impact on ambulation or result in more TF production. High-grade tumors have been shown to express higher levels of TF, which may precipitate greater activation of coagulation [11]. Therefore, like ovarian neoplasms, larger and higher-grade tumors have been associated with a greater risk of VTE. This risk is further complicated in patients undergoing cancer-related surgery due to immobility and biological changes associated with surgery [19]. Various oncological treatments can also further increase the thrombotic risk [48]. There was no significant difference in the incidence of PE after laparoscopic and open abdominal surgery [3].

The risk of VTE among cancer patients has been observed to vary by tumor type. The malignancies associated with the highest incidence risk of DVT and PE included cancers of the brain, pancreas, lung, colon, stomach, or kidney; gynecological cancers; and lymphoma [7, 44, 49]. The incidence of PE after cancer surgery ranges from 0.57% after colon malignancy to 3.85% after pancreatic cancer surgery and is reportedly significantly higher than that after surgery for noncancerous conditions (0.20%) [3]. The level of this risk is also influenced by the presence or absence of metastases [48].

Anticancer Therapy (Chemotherapy)

The prothrombotic tendency in cancer patients is enhanced by anticancer therapy, including surgery, chemotherapy, hormone therapy, and radiotherapy [50, 51]. Indeed, 70% of patients with VTE are estimated to have one or more risk factors in addition to their malignancy: chemotherapy (25%), radiotherapy (21%), hormonal therapy (10%) [51]. At present, several original basic science studies and clinical trials are underway in an effort to enhance our understanding of the mechanisms by which various chemotherapeutic agents can generate a prothrombotic state [52]. Induction of thrombosis by drugs involves a variety of mechanisms: vascular (endothelial cell) damage, enhancement of procoagulant activity, reduction in anticoagulants synthesis, and stimulation of platelet aggregation [44, 53].

Numerous drugs are associated with thrombotic microangiopathy, including thalidomide, tamoxifen, cisplatin, bleomycin, gemcitabine, L-asparaginase, corticosteroids, cyclosporine A, tacrolimus, granulocyte colony-stimulating factor, macrophage-granulocyte colony-stimulating factor, and erythropoietin [55]. L-Asparaginase is associated with reduced hepatic synthesis of antithrombin [44]. Significant activation of the coagulation cascade (elevation of prothrombin activation fragments 1 and 2) was observed in the patients treated with cisplatin [54]. Clinical trials have suggested that the combination of VEGF inhibitor (prinomastat) with chemotherapy (gemcitabine/cisplatin regimen or paclitaxel/carboplatin regimen) approximately doubles the risk of a VTE among patients with advanced non-small-cell lung cancer [55]. Thalidomide given in combination with multiagent chemotherapy and dexamethasone is associated with a significantly increased risk of DVT [56]. A high rate of VTE was observed in metastatic renal cell carcinoma patients treated with the combination of weekly intravenous gemcitabine with continuous-infusion fluorouracil (5-FU) and daily oral thalidomide. The addition of thalidomide to gemcitabine and 5-FU added significant vascular toxicity [57]. All agents, including tamoxifen, used for breast carcinoma chemoprevention and adjuvant therapy appear to increase the risk of VTE [58].

In addition to chemotherapy, many supportive therapies are also associated with an increased risk for the development of VTE [52]. Corticosteroids, commonly used to control chemotherapy-induced vomiting in individuals with cancer, also appear to elevate VTE risk, an effect attributed to increases in the levels of factors VII, VIII, and XI and fibrinogen levels [11]. Treatment with erythropoietin increased the risk of thromboembolic events [59]; use of this drug results in increased reticulocytosis, increased platelet reactivity, and endothelial activation [44]. Thrombotic risk may be also increased by thrombogenic catheter material, a large catheter diameter and greater number of lumens, catheter tip malposition or preexisting venous obstruction, prothrombotic therapeutic agents, catheter-associated infections, and fibrinous catheter lumen occlusion [8].

Treatment

Prophylaxis

The incidence of VTE and PE in Japanese surgical patients is not as low as previously thought [3]. Recently, many patients with cancer may receive appropriate thromboprophylaxis due to consensus-group recommendations [6]. Thromboprophylaxis significantly reduces the risk of VTE, the clinical care and quality of life in these patients [6, 60]. Any fatal PE, which included sudden death from possible fatal PE, are also potentially preventable. Thus, thromboprophylaxis is a well-established therapy worldwide and may be appropriate and safe for selected high-risk patients. Patients with cancer undergoing abdominal surgery are at substantially higher risk for VTE than patients without cancer [60]. The routine use of thromboprophylaxis should be recommended in patients with cancer who are undergoing surgical procedures [60].

Mechanical Approaches

The mechanical approaches, which include early and frequent ambulation, graduated compression stockings, electrical calf muscle stimulation, and intermittent external pneumatic compression devices, are partially effective. The incidence of PE was four to six times lower in patients who had received mechanical prophylaxis, indicating that perioperative prophylaxis against VTE is important [3]. Although these devices decrease VTE rates when compared with no prophylaxis, significant differences in efficacy have been noted between research settings and routine medical care [61].

Mechanical prophylaxis is recommended for patients who have a contraindication to anticoagulant prophylaxis or are at high risk for bleeding [19]. Recent systemic bleeding, severe hypertension, and endocarditis are relative contraindications to anticoagulation [11].

Anticoagulation

Anticoagulation can be considered early. The thromboprophylaxis regimens consist of a single preoperative dose of unfractionated heparin (UFH) (5000 IU two or three times daily) or low-molecular-weight heparin (LMWH), such as enoxaparin, dalteparin, nadroparin, tinzaparin, or reviparin. Heparin thromboprophylaxis in cancer surgery patients is initiated 2 h preoperatively with a subcutaneous dose appropriate to the level of risk [19]. Current practice involves giving thromboprophylaxis for 7–10 days. However, late thrombotic events can occur after the procedure has been performed [19]. Therefore, prolonged thromboprophylaxis for up to 4 weeks is more effective than short-term administration in the high-risk patients [60]. Fondaparinux is a selective inhibitor of activated factor X that was recently approved for thromboprophylaxis. Postoperative fondaparinux was at least as effective as perioperative LMWH (dalteparin) in patients undergoing high-risk abdominal surgery [62]. However, recent prospective studies demonstrated that prophylaxis with low-dose LMWH or low-dose warfarin was not effective in reducing symptomatic events [63]. Furthermore, approaches to preventing or managing catheter-associated thrombosis, including the use of thrombolytic agents, are guided by limited experience [8]. Thromboprophylaxis may be appropriate and safe for selected high-risk patients, including those with known thrombophilia, prior DVT history, and/or anatomical or technical risk factors [11]. Primary and secondary thromboprophylaxis in patients with cancer should be considered part of any integrated oncological treatment [48].

Acute Management

In patients with cancer, VTE portends a poor prognosis; in fact, only 12% of those who suffer an event survive beyond 1 year [52]. Appropriate management of thromboembolism in women with cancer has the potential to reduce the negative clinical outcomes related to these complications [9]. Roentgenographic and perfusion scanning studies have demonstrated that if treated properly perfusion of primary affected areas normalizes rapidly during the first few days after PE [11]. The duration of anticoagulation depends on the underlying VTE risk. Cancer patients with VTE should be treated for as long as their disease is active to minimize the chance of a recurrent VTE [64]. Therefore, anticoagulation is continued indefinitely in most of these patients. Long-term treatment options for cancer patients who experience VTE include vitamin K antagonists (VKAs) (e.g., warfarin), UFH, LMWHs, and inferior vena cava (IVC) filters (64). Progression and recurrence of VTE can be prevented by therapy with UFH or LMWH followed by warfarin for at least 3 months [9]. In cases of cure, such as entirely resected tumors, anticoagulation can be discontinued after 6–12 months [65]. The inability to resect most primary tumors completely and to cure most metastatic systemic cancers leads to a persistent hypercoagulable state [11].

Mechanical Approaches

The concurrent use of graduated compression stockings has a synergistic effect on the reduction in VTE risk [60]. The literature supports early use of compression stockings to reduce postthrombotic syndrome [66]. Although indications for its use are not established, the strategy for preventing PE is roughly as follows [67]: A permanent IVC filter can be implanted in patients with failure of, or a contraindication for, anticoagulation and those with a residual proximal DVT who have a permanent risk factor. A temporary filter can be used in DVT patients without a PE who are scheduled for surgery, those with a floating thrombus, and those with a residual proximal DVT who have a transient risk factor. In cases of transient contraindications, an IVC filter followed by anticoagulation at a later date may limit future thrombotic complications. Concern over the possibility of intracranial bleeding has limited the use of anticoagulation in patients with intracranial malignancies [11]. In this hypercoagulable population, however, IVC filters frequently lead to complications [11]. The filter complication rate is reported to be 62% in patients with brain tumors after the filter placement, including 26% filter thrombosis, 12% recurrent PE, and 10% postphlebitic syndrome [1]). Recurrent VTE was observed in 40% of patients with brain metastases who had IVC filters. It is of note that in one study the IVC filter was an independent risk factor for recurrent DVT but not a risk factor for PE [24]. Thus, IVC filters may be modestly efficacious for preventing PE, but there is little evidence to support their use [66].

Pharmacologic Approaches

Several new anticoagulants are being investigated that promise greater therapeutic choices and potentially better outcomes for cancer patients with VTE [64]. Reassuring safety data and superior efficacy when compared with IVC filters have led to the increasing use of anticoagulation to treat VTE in most patients with cancer. Traditional agents such as heparin and warfarin are generally preferred.

Unfractionated Heparin

Intravenous UFH has a well-documented safety record. The short half-life and easy reversibility of UFH make it an ideal initial treatment in patients deemed candidates for anticoagulation. However, this drug requires close monitoring because hemorrhagic complications occur most commonly in the context of overanticoagulation. The use of an initial bolus, routine in most clinical contexts, should be considered carefully in this population [11]. Although a high loading dose provides therapeutic levels of anticoagulation immediately (compared with approximately 6 h when heparin is administered without a bolus), it does so at considerable risk for transient overanticoagulation [68]. Standard management involves the initial administration of UFH by intravenous injection or infusion for about 7 days [69]. The dose of UFH administered is adjusted to maintain an activated partial thromboplastin time (aPTT) of approximately 1.5–2.5 times normal.

Heparin-induced thrombocytopenia (HIT) is a potentially life-threatening adverse effect of heparin treatment, yet it is a treatable prothrombotic disease that develops in approximately 0.5%–5.0% of heparin-treated patients and dramatically increases their risk of thrombosis [70]. HIT is caused by immunoglobulin G (IgG) antibodies that bind to epitopes on platelet factor 4 (PF4) released from activated platelets that develop when it forms complexes with heparin [71]. HIT is more often caused by UFH than by LMWH and is more common in postsurgical patients than in medical patients.

Low-Molecular-Weight Heparin

Standard management involves initial administration of weight-adjusted LMWH by subcutaneous injection [69]. The role of LMWH—dalteparin (once daily), tinzaparin (once daily), enoxaparin (twice daily)—in the treatment of VTE in patients with cancer is evolving. It is recommended that treatment with LMWH for at least the first 3–6 months of long-term treatment [72]. Recent clinical trials showed that for initial therapy LMWH is at least as efficacious as UFH in reducing recurrent thrombosis. Prolonged treatment with LMWH may be as effective as (CANTHANOX study) or more effective than (CLOT trial) warfarin therapy [41]. Cancer patients with VTE who were given long-term dalteparin were 50% less likely to develop recurrent VTE than patients treated with warfarin. Therefore, in general patient populations, these agents have been used effectively as initial therapy for DVT and PE [41, 64]. Although LMWH is used routinely and successfully as initial anticoagulation in patients with cancer, its longer half-life and less complete protamine reversibility sometimes make it less appealing as initial therapy in patients in whom bleeding can be life-threatening. Individuals with a current episode or recent history of HIT should not receive heparin products, including LMWH. In their place, direct thrombin inhibitors (lepirudin, Argatroban) could be considered, although few data exist to guide their use in patients with cancer. Instead, LMWH might best be used after a patient has tolerated a successful trial of UFH, as a bridge to therapeutic warfarin use. One factor potentially limiting the broader use of LMWH for chronic therapy is its higher cost [64]. The more recent evidence strongly supports that use of LMWH is cost-saving or, at a minimum, cost-effective compared with use of UFH for VTE, generally regardless of the treatment setting [66]. In particular, use of LMWH in outpatients was less costly than hospitalization for UFH administration [73].

Anticoagulant drugs appear to be an attractive strategy in cancer therapy, with an effect that would surpass the benefit of preventing thrombosis [74]. Recent studies have suggested that LMWH may also influence overall survival in patients with cancer [48, 75]. The effects seem to be apparent only in limited cases (patients with a low cancer burden) [48]. Thus, results from individual studies and some meta-analyses show that selected cancer patients with VTE treated with LMWH have a better survival than patients treated with UFH or warfarin [75]. These findings raise the likelihood that the use of LMWH in selected oncological patients may increase significantly in the near future. At present, however, there are no data regarding whether LMWH prolongs survival of patients with minimal-disease cancer or those with all types of cancer [48, 72, 75].

The proposed mechanisms of this possibility are as follows. First, LMWH prevents fatal thromboembolic disease. Second, LMWH shows coagulation-independent antineoplastic properties. Suppression of coagulation proteases may be related to inhibition of cell proliferation, adhesion, invasion, and metastasis (see Etiology and Pathogenesis). It can influence proliferation of cancer cells through down-regulation of proto-oncogenes, kinase phosphorylation, and TF [74]. Finally, direct antitumor effects include inhibition of heparanase and angiogenic activity and induction of tumor apoptosis [75]. The inhibitory effect of LMWH on endothelial tube formation is associated with the release of TF pathway inhibitor [76]. In summary, evidence is ample that LMWH is superior to UFH for the treatment of DVT, particularly for reducing mortality and the risk of major bleeding during the initial therapy [66].

Warfarin

Oral agents are potentially the most attractive because of the route of administration. Patients with a provoked episode of VTE may be well served with just 3 months of anticoagulation [66]. Extended-duration conventional-intensity oral anticoagulation may be optimal treatment for patients with unprovoked VTE or following a second episode of VTE [66]. As with UFH, hemorrhagic complications associated with warfarin use frequently occur in the context of overanticoagulation [77]. Cancer patients have more episodes of major bleeding during chronic warfarin therapy than do patients without cancer [64]. Weekly monitoring of such patients improves the outcome [77]. Patients have an increased chance of bleeding during the initial month of therapy. With close clinical follow-up and regular laboratory monitoring, patients with cancer have undergone warfarin therapy for months to years without bleeding complications.

Anticoagulation intensity can fluctuate significantly because many of these patients take medications that interact with warfarin [11]. Common examples include cimetidine and omeprazole (both prescribed for peptic ulcer prophylaxis in patients on corticosteroids), trimethoprim-sulfamethoxazole (*Pneumocystis pneumoniae* prophylaxis for patients undergoing chemotherapy and irradiation), and the anticonvulsant phenytoin [11, 78]. Notwithstanding these limitations, warfarin remains the standard long-term anticoagulant used in patients with malignancies. Efficacy, cost, drug availability, patient co-morbidities, and concomitant medications all need to be considered when selecting chronic VTE therapy.

Plasminogen Activators

Primary therapy with fibrinolysis (urokinase-type or tissue-type plasminogen activator) or embolectomy is generally considered for patients presenting with a massive PE [79]. However, there is no evidence that fibrinolytics reduce the mortality rate [79]. Surgical or catheter-based embolectomy has been demonstrated to be a safe and effective treatment option.

Conclusion

Anticoagulation should be considered the principal modality to prevent (prophylaxis) and treat (management) VTE in most patients with cancer. Optimizing methods for the treatment and prevention of thrombosis is of particular importance in this population. Cancer patients with VTE risk factors should receive appropriate thromboprophylaxis as well as acute management of VTE and long-term secondary prophylaxis. Recent randomized trials [66] support the idea that LMWH is modestly superior to UFH for initial treatment of DVT and is at least as effective as UFH for treatment of PE. Although oral anticoagulation is the optimal treatment for patients with unprovoked VTE in the general population, the use of LMWH instead of oral anticoagulation may be recommended, particularly in patients with malignant conditions. LMWH use is cost-saving or cost-effective compared with UFH for treatment of VTE [66]. In patients with cancer, optimal anticoagulant regimens for either prevention or treatment have not been established. Further studies are needed to determine when anticoagulant therapy can be reasonably discontinued. High-quality prospective, randomized, controlled trials are needed to identify the safest, most effective anticoagulant agents and treatment strategies for patients with or without massive cancer.

Notes: Supported in part by the Grant No. 18591842 from the Japan Ministry of Health, Labour and Welfare.

References

1. Clagett GP, Anderson FA Jr, Heit J (1995) Prevention of venous thromboembolism. Chest 108:312S–334S
2. Kroegel C, Reissig A (2003) Principle mechanisms underlying venous thromboembolism: epidemiology, risk factors, pathophysiology and pathogenesis. Respiration 70:7–30
3. Sakon M, Kakkar AK, Ikeda M, et al (2004) Current status of pulmonary embolism in general surgery in Japan. Surg Today 34:805–810
4. Goldhaber SZ, Visani L, De Rosa M (1999) Acute pulmonary embolism: clinical outcomes in the International Cooperative Pulmonary Embolism Registry (ICOPER). Lancet 353:1386–1389
5. Kasper W, Konstantinides S, Geibel A (1997) Management strategies and determinants of outcome in acute major pulmonary embolism: results of a multicenter registry. J Am Coll Cardiol 30:1165–1171
6. Cohen AT, Alikhan R, Arcelus JI, et al (2005) Assessment of venous thromboembolism risk and the benefits of thromboprophylaxis in medical patients. Thromb Haemost 94:750–759
7. Sorensen HT, Mellemkjaer L, Olsen JH, et al (2000) Prognosis of cancers associated with venous thromboembolism. N Engl J Med 343:1846–1850
8. Linenberger ML (2006) Catheter-related thrombosis: risks, diagnosis, and management. J Natl Compr Canc Netw 4:889–901
9. Prandoni P (2005) Venous thromboembolism risk and management in women with cancer and thrombophilia. Gend Med 2:S28–S34
10. Pruemer J (2005) Prevalence, causes, and impact of cancer-associated thrombosis. Am J Health Syst Pharm 62:S4–S6
11. Gerber DE, Grossman SA, Streiff MB (2006) Management of venous thromboembolism in patients with primary and metastatic brain tumors. J Clin Oncol 24:1310–1318
12. Drake TA, Cheng J, Chang A, et al (1993) Expression of tissue factor, thrombomodulin, and E-selectin in baboons with lethal *Escherichia coli* sepsis. Am J Pathol 142:1458–1470
13. Winter PC (2006) The pathogenesis of venous thromboembolism in cancer: emerging links with tumour biology. Hematol Oncol 24:126–133
14. PIOPED Investigators (1990) Value of the ventilation/perfusion scan in acute pulmonary embolism: results of the prospective investigation of pulmonary embolism (PIOPED). JAMA 263:2753–2759
15. Kearon C (2001) Epidemiology of venous thromboembolism. Semin Vasc Med 1:7–26
16. Miyata T, Kimura R, Kokubo Y, et al (2006) Genetic risk factors for deep vein thrombosis among Japanese: importance of protein S K196E mutation. Int J Hematol 83:217–223
17. Mizukami K, Nakabayashi T, Naitoh S, et al (2006) One novel and one recurrent mutation in the PROS1 gene cause type I protein S deficiency in patients with pulmonary embolism associated with deep vein thrombosis. Am J Hematol 81:787–797

18. Trousseau A (1865) Phlegmasia alba dolens. Clinique Médicale de l'Hôtel-Dieu de Paris, the New Sydenham Society, London 3:94–111
19. Bergqvist D (2007) Risk of venous thromboembolism in patients undergoing cancer surgery and options for thromboprophylaxis. J Surg Oncol 95:167–174
20. Sampson MT, Kakkar AK (2002) Coagulation proteases and human cancer. Biochem Soc Trans 30:201–207
21. Kazama Y, Hamamoto T, Foster DC, et al (1995) Hepsin, a putative membrane-associated serine protease, activates human factor VII and initiates a pathway of blood coagulation on the cell surface leading to thrombin formation. J Biol Chem 270:66–72
22. De Cicco M (2004) The prothrombotic state in cancer: pathogenic mechanisms. Crit Rev Oncol Hematol 50:187–196
23. Koyama T, Nishida K, Ohdama S, et al (1994) Determination of plasma tissue factor antigen and its clinical significance. Br J Haematol 87:343–347
24. Giesen PL, Rauch U, Bohrmann B, et al (1999) Blood-borne tissue factor: another view of thrombosis. Proc Natl Acad Sci U S A 96:2311–2315
25. Bogdanov VY, Balasubramanian V, Hathcock J, et al (2003) Alternatively spliced human tissue factor: a circulating, soluble, thrombogenic protein. Nat Med 9:458–462
26. Uno K, Homma S, Satoh T, et al (2007) Tissue factor expression as a possible determinant of thromboembolism in ovarian cancer. Br J Cancer 96:290–295
27. Guba M, Yezhelyev M, Eichhorn ME, et al (2005) Rapamycin induces tumor-specific thrombosis via tissue factor in the presence of VEGF. Blood 105:4463–4469
28. Wang X, Wang E, Kavanagh JJ, et al (2005) Ovarian cancer, the coagulation pathway, and inflammation. J Transl Med 3:25
29. Stahli BE, Camici GG, Steffel J, et al (2006) Paclitaxel enhances thrombin-induced endothelial tissue factor expression via c-Jun terminal NH_2 kinase activation. Circ Res 99:149–155
30. Shen BQ, Lee DY, Cortopassi KM, et al (2001) Vascular endothelial growth factor KDR receptor signaling potentiates tumor necrosis factor-induced tissue factor expression in endothelial cells. J Biol Chem 276:5281–5286
31. Esmon CT (2004) The impact of the inflammatory response on coagulation. Thromb Res 114:321–327
32. Cao R, Bjorndahl MA, Religa P, et al (2004) PDGF-BB induces intratumoral lymphangiogenesis and promotes lymphatic metastasis. Cancer Cell 6:333–345
33. Rickles FR, Levine M, Edwards RL (1992) Hemostatic alterations in cancer patients. Cancer Metastasis Rev 11:237–248
34. Osterud B (1998) Tissue factor expression by monocytes: regulation and pathophysiological roles. Blood Coagul Fibrinolysis 9(suppl 1):S9–S14
35. Coughlin SR (2000) Thrombin signalling and protease-activated receptors. Nature 407:258–264
36. Rong Y, Post DE, Pieper RO, et al (2005) PTEN and hypoxia regulate tissue factor expression and plasma coagulation by glioblastoma. Cancer Res 65:1406–1413
37. Nguyen QD, De Wever O, Bruyneel E, et al (2005) Commutators of PAR-1 signaling in cancer cell invasion reveal an essential role of the Rho-Rho kinase axis and tumor microenvironment. Oncogene 24:8240–8251
38. Fan Y, Zhang W, Mulholland M (2005) Thrombin and PAR-1-AP increase proinflammatory cytokine expression in C6 cells. J Surg Res 129:196–201
39. Nierodzik ML, Karpatkin S (2006) Thrombin induces tumor growth, metastasis, and angiogenesis: evidence for a thrombin-regulated dormant tumor phenotype. Cancer Cell 10:355–362
40. Mohle R, Green D, Moore MA, et al (1997) Constitutive production and thrombin-induced release of vascular endothelial growth factor by human megakaryocytes and platelets. Proc Natl Acad Sci U S A 94:663–668

41. Lopez JA, Kearon C, Lee AY (2004) Deep venous thrombosis. Hematology (Am Soc Hematol Educ Program) 439–456
42. Chen M, Geng JG (2006) P-selectin mediates adhesion of leukocytes, platelets, and cancer cells in inflammation, thrombosis, and cancer growth and metastasis. Arch Immunol Ther Exp (Warsz) 54:75–84
43. Gordon SG, Mielicki WP (1997) Cancer procoagulant: a factor X activator, tumor marker and growth factor from malignant tissue. Blood Coagul Fibrinolysis 8:73–86
44. Heit JA (2005) Cancer and venous thromboembolism: scope of the problem. Cancer Control 12(suppl 1):5–10
45. Szajda SD, Jozwik M, Jozwik M, et al (2004) Cancer procoagulant activity in serum and neoplastic tissue in cases of cervical and uterine carcinoma. Ginekol Pol 75:705–712
46. Chen CH, Su KY, Tao MH, et al (2006) Decreased expressions of hepsin in human hepatocellular carcinomas. Liver Int 26:774–780
47. Kyrle PA, Minar E, Bialonczyk C, et al (2004) The risk of recurrent venous thromboembolism in men and women. N Engl J Med 350:2558–2563
48. Cunningham MS, White B, O'Donnell J (2006) Prevention and management of venous thromboembolism in people with cancer: a review of the evidence. Clin Oncol (R Coll Radiol) 18:145–151
49. Levitan N, Dowlati A, Remick SC, et al (1999) Rates of initial and recurrent thromboembolic disease among patients with malignancy versus those without malignancy: risk analysis using Medicare claims data. Medicine (Baltimore) 78:285–291
50. Tateo S, Mereu L, Salamano S, et al (2005) Ovarian cancer and venous thromboembolic risk. Gynecol Oncol 99:119–125
51. Joung S, Robinson B (2002) Venous thromboembolism in cancer patients in Christchurch, 1995–1999. N Z Med J 115:257–260
52. Haddad TC, Greeno EW (2006) Chemotherapy-induced thrombosis. Thromb Res 118:555–568
53. Nadir Y, Hoffman R, Brenner B (2004) Drug-related thrombosis in hematologic malignancies. Rev Clin Exp Hematol 8:E4
54. Kuenen BC, Levi M, Meijers JC, et al (2003) Potential role of platelets in endothelial damage observed during treatment with cisplatin, gemcitabine, and the angiogenesis inhibitor SU5416. J Clin Oncol 21:2192–2198
55. Behrendt CE, Ruiz RB (2003) Venous thromboembolism among patients with advanced lung cancer randomized to prinomastat or placebo, plus chemotherapy. Thromb Haemost 90:734–737
56. Zangari M, Anaissie E, Barlogie B, et al (2001) Increased risk of deep-vein thrombosis in patients with multiple myeloma receiving thalidomide and chemotherapy. Blood 98:1614–1615
57. Desai AA, Vogelzang NJ, Rini BI, et al (2002) A high rate of venous thromboembolism in a multi-institutional phase II trial of weekly intravenous gemcitabine with continuous infusion fluorouracil and daily thalidomide in patients with metastatic renal cell carcinoma. Cancer 95:1629–1636
58. Deitcher SR, Gomes MP (2004) The risk of venous thromboembolic disease associated with adjuvant hormone therapy for breast carcinoma: a systematic review. Cancer 101:439–449
59. Bohlius J, Wilson J, Seidenfeld J, et al (2006) Recombinant human erythropoietins and cancer patients: updated meta-analysis of 57 studies including 9353 patients. J Natl Cancer Inst 98:708–714
60. Negus JJ, Gardner JJ, Tann O, et al (2006) Thromboprophylaxis in major abdominal surgery for cancer. Eur J Surg Oncol 32:911–916
61. Cornwell EE 3rd, Chang D, Velmahos G (2002) Compliance with sequential compression device prophylaxis in at-risk trauma patients: a prospective analysis. Am Surg 68:470–473

62. Agnelli G, Bergqvist D, Cohen AT, et al (2005) Randomized clinical trial of postoperative fondaparinux versus perioperative dalteparin for prevention of venous thromboembolism in high-risk abdominal surgery. Br J Surg 92:1212–1220

63. Geerts WH, Pineo GF, Heit JA, et al (2004) Prevention of venous thromboembolism: the Seventh ACCP Conference on Antithrombotic and Thrombolytic Therapy. Chest 126:338S–400S

64. Streiff MB (2006) Long-term therapy of venous thromboembolism in cancer patients. J Natl Compr Canc Netw 4:903–910

65. Bates SM, Ginsberg JS (2004) Clinical practice: treatment of deep-vein thrombosis. N Engl J Med 351:268–277

66. Segal JB, Streiff MB, Hoffman LV, et al (2007) Management of venous thromboembolism: a systematic review for a practice guideline. Ann Intern Med. 146:211–222

67. Kai R, Imamura H, Kumazaki S, et al (2006) Temporary inferior vena cava filter for deep vein thrombosis and acute pulmonary thromboembolism: effectiveness and indication. Heart Vessels 21:221–225

68. Heres EK, Speight K, Benckart D (2001) The clinical onset of heparin is rapid. Anesth Analg 92:1391–1395

69. Falanga A, Zacharski L (2005) Deep vein thrombosis in cancer: the scale of the problem and approaches to management. Ann Oncol 16:696–701

70. Levy JH, Hursting MJ (2007) Heparin-induced thrombocytopenia, a prothrombotic disease. Hematol Oncol Clin North Am 21:65–88

71. Cines DB, Rauova L, Arepally G, et al (2007) Heparin-induced thrombocytopenia: an autoimmune disorder regulated through dynamic autoantigen assembly/disassembly. J Clin Apher 22:31–36

72. Horton J (2005) Venous thrombotic events in cancer: the bottom line. Cancer Control 12:31–37

73. Hull RD, Raskob GE, Rosenbloom D, et al (1997) Treatment of proximal vein thrombosis with subcutaneous low-molecular-weight heparin vs intravenous heparin: an economic perspective. Arch Intern Med 157:289–294

74. Lecumberri R, Paramo JA, Rocha E (2005) Anticoagulant treatment and survival in cancer patients: the evidence from clinical studies. Haematologica 90:1258–1266

75. Kakkar A (2005) Low-molecular-weight heparin and survival in patients with malignant disease. Cancer Control 12:22–30

76. Mousa SA, Mohamed S (2004) Inhibition of endothelial cell tube formation by the low molecular weight heparin, tinzaparin, is mediated by tissue factor pathway inhibitor. Thromb Haemost 92:627–633

77. Hylek EM, Singer DE (1994) Risk factors for intracranial hemorrhage in outpatients taking warfarin. Ann Intern Med 120:897–902

78. Holbrook AM, Pereira JA, Labiris R (2005) Systematic overview of warfarin and its drug and food interactions. Arch Intern Med 165:1095–1106

79. Piazza G, Goldhaber SZ (2006) Acute pulmonary embolism. II. Treatment and prophylaxis. Circulation 114:e42–e47

Disseminated Intravascular Coagulation

Hideo Wada

Summary. Disseminated intravascular coagulation (DIC) is an acquired syndrome that induces extensive intravascular coagulation due to various causes. Microthrombus formation and endothelial cell injuries are generated mainly in the small veins and arteries, and organ failure can even occur in patients demonstrating severe DIC. DIC is characterized by an elevated generation of fibrin-related products and the presence of hemostatic abnormalities due to either inflammation (with vascular endothelial cell injury) or noninflammatory causes (without vascular endothelial cell injury). This disease is divided into two types: overt DIC and nonovert DIC. The frequency of DIC was reported to be about 1% among hospitalized patients in a 1992 investigation. The outcome of DIC might also be poorer in patients with infectious diseases than in those with noninfectious diseases. Although three diagnostic criteria for DIC have been established, more sensitive and specific criteria for DIC are required for patients with infectious diseases. The treatment of underlying diseases is essential for DIC. Anticoagulant therapy is also essential, but heparin/heparinoids should be carefully administered in patients with marked bleeding. As a result, antifibrinolytic therapy needs to be carefully monitored using hemostatic molecular markers. The use of physiological protease inhibitor might be an effective treatment for DIC. Early diagnosis and treatment are therefore required to improve the outcome of patients with DIC.

Key words. DIC · Infectious diseases · Poor outcome · Hemostatic molecular marker · Protease inhibitor

Introduction

Although many definitions of disseminated intravascular coagulation (DIC) have been advocated up to now, the definition and concept of DIC have not yet been formally established [1]. DIC is generally considered to result from disseminated fibrin formation, which mainly occurs in the small veins and arteries owing to various causes, and most of the fibrin in such cases tends to be simultaneously dissolved by secondary fibrinolysis [2]. The definition and concept of DIC were proposed by the International Society of Thrombosis and Haemostasis (ISTH)/Scientific Standardization Committee (SSC) of 2001 [3] (Table 1). Two mechanisms have been proposed to play a role in the onset of DIC: noninflammatory pathogenesis (without vascular

TABLE 1. Definition and concept of DIC

Definition—DIC is an acquired syndrome that causes extensive intravascular coagulation due to various causes. Microthrombus formation and endothelial cell injuries are mainly generated in small veins and arteries. Organ failure can occur in severe cases of DIC.
Concept—DIC is characterized by increased generation of fibrin-related products and hemostatic abnormalities due to inflammation (with vascular endothelial cell injury) or noninflammatory causes (without vascular endothelial cell injury).
Stage of disease—DIC is classified into two types: overt DIC (noncompensatory DIC) and nonovert DIC (compensatory DIC).

The definition and concept are cited from Taylor et al. [3]
DIC, disseminated intravascular coagulation

endothelial cell injury) and inflammatory pathogenesis (with vascular endothelial cell injuries). Noninflammatory DIC tends to occur in patients with acute leukemia and aortic aneurysms, among other conditions, whereas inflammatory DIC tends to occur in patients with, for example, severe sepsis, trauma, or burns. The importance of elevated fibrin-related products [4] has also been previously proposed to play a role in the pathological state of DIC. ISTH proposed that DIC should be divided into overt DIC (noncompensation stage) and nonovert DIC (compensation stage) similar to pre-DIC [5] owing to the fact that greater efficacy was achieved in the treatment of pre-DIC than in the treatment of DIC (overt DIC) in a retrospective study [6]. In addition, in overt DIC, comparatively controllable DIC (e.g., obstetrics DIC) should also be distinguished from uncontrollable DIC (e.g., severe sepsis).

Pathogenesis

The pathogenesis for DIC differs greatly between noninflammatory DIC and inflammatory DIC. Noninflammatory DIC is caused by a markedly elevated expression or release of tissue factor (TF) [7], plasminogen activator (PA) [8], or anexin II on tumor cells in patients with acute leukemia or a solid cancer and due to a blood flow abnormality in aneurysms (Fig. 1). TF can trigger the extrinsic pathway of coagulation. High TF activity results in the activation of prothrombin to thrombin to form a fibrin thrombus. Increased TF production is considered to be the most important factor regarding the onset of DIC. TF is significantly high in both leukemic cells [7] and the plasma of patients with DIC, suggesting that DIC in leukemia is caused by elevated TF from leukemic cells. In hemostatic abnormalities, hyperfibrinolysis and hypercoagulability are marked, but the plasma levels of antithrombin (AT) and thrombomodulin (TM), which are vascular endothelial cell injury markers, are often normal. The degree of vascular endothelial cell injury tends to be slight, and few clinical manifestations occur except for the hemorrhaging. In contrast, the leukocytes are markedly activated, and the expression of TF mRNA [9] in leukocytes is reported to be significantly high. High plasma levels of inflammatory cytokines and PA inhibitor I (PAI-I) have been observed in inflammatory DIC, which is frequently accompanied by vascular endothelial cell injury. Therefore, fluid, albumins, and AT, among others, leak out of blood vessels; thereafter, edema and clinical manifestations such as shock are observed (capillary leak syndrome). In addition, AT in the blood and TM in injured vascular endothelial

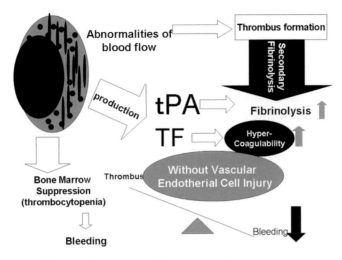

FIG. 1. Pathogenesis of noninflammatory disseminated intravascular coagulation (DIC). *TF*, tissue factor; *tPA*, tissue-type plasminogen activator

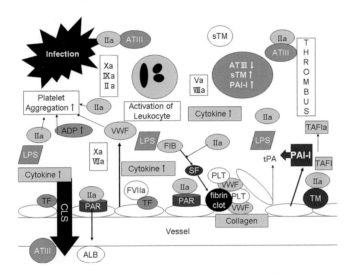

FIG. 2. Pathogenesis of inflammatory DIC. *ALB*, albumin; *CLS*, capillary leak syndrome; *PAR*, protease-activated receptor; *VWF*, von Willebrand factor; *PLT*, platelet; *TAFI*, thrombin-activatable fibrinolysis inhibitor; *FIB*, fibrinogen; *sF*, soluble fibrin; *ATIII*, antithrombin III; *TM*, thrombomodulin; *PAI*, plasminogen activator inhibitor

cells tend to decrease, and the expression of TF and PAR (protease activated receptors, or PARs) [10] increases, leading to the formation of microthrombus. Therefore, although various organ symptoms have been observed, no fluctuation in the fibrinolytic markers tend to be obvious. Both AT and protein C (PC), which are the vascular endothelial cell injury markers, decrease, and soluble TM increases [11] (Fig. 2).

Diagnosis

During the early stages of DIC, microthrombus formation begins in the microvasculature, but secondary fibrinolysis and the consumption of coagulation factors are not usually significant factors. Therefore, fluctuation of the prothrombin time (PT) and platelet counts, fibrinogen, fibrin, and fibrinogen degradation products (FDP) values, which are used as the general coagulation tests, tend to be slight. Only hemostatic molecular markers such as prothrombin fragments F1 + 2 (F1 + 2), soluble fibrin (SF), and the thrombin–AT complex (TAT) are increased in this stage. During the second stage of inflammatory DIC, both AT and PC are lower, whereas the PT becomes prolonged, and the platelet counts decrease. In noninflammatory DIC, the fibrinogen levels decrease, whereas the FDP, D-dimer, and plasmin–plasmin inhibitor complex (PPIC) dramatically increase (Fig. 3). In final stages of DIC, the symptoms of organ failure mainly appear in inflammatory DIC, whereas bleeding symptoms tend to mainly appear in noninflammatory DIC. These findings suggest that hemostatic molecular markers such as TAT, SF, and F1 + 2 are useful for the early diagnosis of DIC and that the general coagulation tests are useful for the diagnosis of overt-DIC but not for the early diagnosis of DIC. Up to now, three diagnostic criteria of DIC have been established: the DIC diagnostic criteria established by the Japanese Ministry of Health and Welfare (JMHW) [12], the ISTH overt DIC diagnostic criteria [3], and the new Japanese diagnostic criteria for DIC [13]. Three DIC diagnostic criteria in patients with infectious diseases are shown in Table 2. Regarding the diagnostic criteria for DIC based on general coagulation tests, the ISTH overt DIC diagnostic criteria

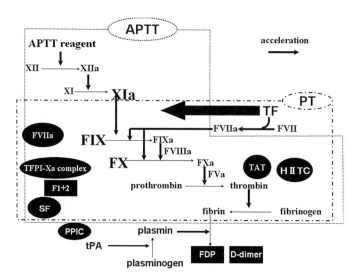

FIG. 3. Hemostatic molecular markers in DIC. *APTT*; activated partial thromboplatin time, *PT*; prothrombin time, *TF*; tissue factor, *TFPI*; tissue factor pathway inhibitor, *SF*; soluble fibirin, *TAT*; thrombin antithrombin complex, *PPIC*; plasmin plasmin inhibitor complex, *tPA*; tissue type plasminogen activator, *F1+2*; prothrombin fragment 1 + 2, *HIITC*; heparin cofactor II thrombin complex, *FDP*; fibrinogen and fibrin degradation products

TABLE 2. Comparison of three diagnostic criteria [3, 13, 13] for DIC in patients with infectious diseases (modified version)

Parameter	JMHW DIC criteria	ISTH overt DIC criteria	New Japanese DIC criteria
Underlying disease	Positive: 1 p	Necessary	Necessary
Clinical symptoms	Organ failure: 1 p	No points	SIRS (>3 items): 1 p
	Bleeding: 1 p	No points	
Platelet count ($\times 10^3/\mu l$))	80–120: 1 p	50–100: 1 p	80–120 or >30%
	50–80: 2 p	<50: 2 p	reduction/24 h: 1 p
	<50: 3 p		<80 or >50% reduction/ 24 h: 3 p
Fibrin-related markers	FDP ($\mu g/ml$)	FDP, D-dimer, SF	FDP ($\mu g/ml$)
	10–20: 1 p	Moderate increase: 2 p	10–25: 1 p
	20–40: 2 p	Marked increase: 3 p	<25: 3 p
	<40: 3 p		
Fibrinogen	100–150 mg/dl: 1p	<100 mg/dl: 1 p	
	<100 mg/dl: 2 p		
PT	PT ratio	PT (seconds)	PT ratio
	1.25–1.67: 1 p	3–6 s prolongation: 1 p	>1.2: 1 p
	>1.67: 2 p	>6 s prolongation: 2 p	
DIC	≥6 p	≥5 p	≥4 p

JMHW, Japanese Ministry of Health and Welfare; ISTH, International Society of Thrombosis and Haemostasis; p, point; SIRS, systemic inflammatory response syndrome; FDP, fibrin degradation product; SF, soluble fibrin

demonstrate high specificity but low sensitivity [14]. In contrast, the new Japanese diagnostic criteria have high sensitivity and low specificity [13], suggesting that overdiagnosis for DIC might be caused by these diagnostic criteria. It is expected that the addition of an evaluation of F1 + 2, SF, and TAT increases the sensitivity and specificity for the diagnosis of DIC. In the ISTH overt DIC diagnostic criteria, there was no cutoff values of fibrin-related markers, that is, moderate increase (2 points) and marked increase (3 points). The hemostatic markers also have to be standardized [2].

Frequency of DIC and Underlying Diseases

In a national investigation regarding gynecology, pediatrics, surgery, and internal medicine conducted in 1992, a total of 1286 (about 1%) patients among 123 231 admitted to large hospitals were found to have some association with DIC, suggesting that DIC is not such a rare disease. The frequency was in the order of gynecology, pediatrics, surgery, and internal medicine, but these investigations were not carried out in emergency departments and intensive care units (ICUs) in which the frequency of DIC tend to be markedly higher than in the other departments. In an investigation conducted by the Japanese Ministry of Health and Welfare in 1998, the absolute number in the underlying diseases of DIC was high for sepsis, shock, and non-Hodgkin lymphoma, whereas the frequency of DIC was high in patients with acute promyelocytic leukemia, fulminant hepatitis, and placenta praevia (Table 3) [2].

TABLE 3. Underlying diseases associated with DIC based on a report by the Japanese Ministry of Health and Welfare in 1998

Rating	Underlying disease	DIC	Total	Frequency (%)
High absolute number				
1	Sepsis	166	410	40.5
2	Non Hodgkin's' lymphoma	154	777	19.8
3	Hepatoma	113	3545	3.2
4	Acute myeloblastic leukemia	91	288	31.6
5	Lung cancer	82	1026	8.0
6	Respiratory infections	78	1205	6.5
7	Liver cirrhosis	72	3335	2.2
8	Acute promyelocytic leukemia	71	91	78.0
9	Gastric cancer	46	1090	4.2
10	Acute lymphoblastic leukemia	45	151	29.8
High frequency				
1	Acute promyelocytic leukemia	71	91	78.0
2	Fulminant hepatitis	29	64	45.3
3	Sepsis	166	410	40.5
4	Breast cancer	7	19	36.8
5	Acute myeloblastic leukemia	91	288	31.6
6	Acute lymphoblastic leukemia	45	151	29.8
7	Acute myelomonoblastic leukemia	11	40	27.5
8	Others	33	124	26.6
9	Chronic myelocytic leukemia	22	84	26.2
10	Acute monoblastic leukemia	7	28	25.0

Regarding the underlying disease, five infectious diseases, including sepsis, shock, respiratory infection, biliary system infection, and acute respiratory distress syndrome (ARDS), had a high absolute number of DIC presentations, and the absolute number of DIC episodes associated with infectious disease was about 44% of all patients with DIC. Based on the findings of this investigation, it appears that it is necessary to establish new DIC diagnostic criteria for infectious diseases. The frequency of DIC in the KyberSept Trial [15] and the PROWESS study [16], which were Phase III clinical trials of severe sepsis, was 40.7% and 22.4%, respectively, suggesting that the frequency of DIC in severe sepsis might range from 20% to 40%.

Outcome of DIC

In a randomized controlled trial (RCT) of severe sepsis, the mortality rates for the placebo group of the 28 days in the KyberSept Trial [15], the PROWESS study [16] and the OPTIMIST Trial [17] were 38.7%, 30.8%, and 33.9%, respectively. The mortality on the 28th day in the unfractionated heparin (UFH) group was 34.6% in the Phase III trial of recombinant soluble TM [18], suggesting that the mortality of severe sepsis was about 30%–40%. The evaluation of the outcome on RCT of physiological protease inhibitors [15–17] is shown in Table 4. In the placebo group of the KyberSept Trial [15], the mortallity after 28th days was significantly higher in the patients with DIC (40.0%) than in those without DIC (22.2%), suggesting that the association with DIC

TABLE 4. Outcome of DIC and underlying diseases (mortality at 28 days)

Treatment	With infection (%)			Without infection (%)
	DIC	Without DIC	With and without DIC	DIC
Recombinant TM [15]	28.0			17.2
Plasma-derived APC [19]				20.4
Recombinant APC [16]			24.0	
AT [15]	25.4	22.1		
UFH [15]	34.6			18.0
UFH [19]				40.0
Heparin [15–17]			36.6, 28.0, 29.8	
Placebo [15]	40.0	22.2		
Placebo [16]	46.2	26.5	39.0	

TM, thrombomodulin; APC, activated protein C; AT, antithrombin; UFH, unfractionated heparin

results in a worse outcome in patients with sepsis. After the administration of recombinant TM [18] or AT [15], the mortality in DIC patients with infectious disease improved to 28.0% and 25.4%, respectively, suggesting that physiological protease inhibitors might be useful against DIC in patients with severe sepsis. On the other hand, in patients with noninfectious diseases, the mortality rates for DIC patients treated with APC [19] or TM [18] were 20.2% or 20.4%, respectively. These findings suggest that the outcome might be worse in patients with infectious diseases than in those with noninfectious diseases.

Treatment

Among patients with leukemia or lymphoma, advances in adjuvant therapies such as antibiotics, granulocyte-colony stimulating factor (G-CSF), and blood transfusion, coupled with the administration of all-*trans*-retinoic acid (ATRA) for APL, have decreased the frequency and mortality due to DIC. However, the frequency and mortality associated with DIC are still high in ICU patients, especially in those with sepsis. An examination on the correlation between the efficacy of treatment and the DIC score at the time of starting treatment has revealed a greater efficacy to be achieved in the treatment of pre-DIC than in the treatment of overt DIC. The outcome was also poorer in patients with a high DIC score than in those with a low DIC score, suggesting the importance of early diagnosis and treatment.

Treatment of Underlying Disease

There is no evidence regarding the relation between the treatment for underlying disease and the outcome of DIC. However, the administration of antibiotics for the treatment of infectious diseases and chemotherapy, including ATRA for leukemia/lymphoma, are widely accepted treatment; and surgical obliteration and drainage of the infectious focus also tend to be frequently carried out as soon as possible. Generally, DIC is often not cured when the underlying disease does not improve. On the other hand, it is better to treat DIC before treating the underlying diseases when the

TABLE 5. Treatment of DIC

Treatment of underlying diseases
Antibiotics, anticancer drugs including ATRA, surgery, drainage, among others
Anticoagulation therapy
Heparin/heparinoid—UFH, LMWH, danaparoid sodium
Synthetic protease inhibitors—gabexate mesilate, nafamstat mesilate
Physiological protease inhibitors—AT, APC,[a] TM[a]
Antifibrinolysis/fibrinolysis therapy—tranexamic acid or SPI
Transfusion—FFP, PC

ATRA, all-*trans*-retinoic acid; UFH, unfractionated heparin; LMWH, low-molecular-weight heparin; AT, antithrombin; APC, activated protein C; SPI, synthetic protease inhibitor; FFP, fresh frozen plasma; PC, platelet concentrate

[a]APC and TM have not yet been approved as treatment for DIC

risk of bleeding due to DIC is high. Regarding the treatment of sepsis, a report describing the Surviving Sepsis Campaign guidelines for management of severe sepsis and septic shock has recently been published [20].

Anticoagulant Therapy

As anticoagulant therapy is essential for DIC treatment, it is strongly recommended. There are many types of DIC, depending on the pathology and stage of DIC; in some cases, strong anticoagulant therapy is necessary, whereas in the other cases mild anticoagulant therapy is recommended. Regarding each anticoagulant drug for DIC treatment, there are not enough reports that have strongly recommended it for the treatment of DIC.

Heparin/Heparinoid

The heparin/heparinoid group includes UFH, low-molecular-weight heparin (LMWH), and danaparoid sodium (DS). Although heparin or heparinoid does not itself have any anticoagulant activity, it increases the activity of AT to suppress thrombin activity and thus alleviates the hemostatic abnormalities associated with DIC. The side effects include hemorrhaging, shock, heparin-induced thrombocytopenia (HIT), and deterioration of thrombotic thrombocytopenic purpura (TTP). Regarding the administration of heparin for marked bleeding, serious hepatic or renal failure is a fundamental contraindication. LMWH [21] and DS [22] both have a relatively strong anti-Xa activity but less AT activity, resulting in a decreased incidence of hemorrhage in comparison to UFH. In a retrospective analysis [23] on RCTs [15–17] looking at AT, activated protein C (APC), and tissue factor pathway inhibitor (TFPI) in regard to severe sepsis, the mortality was lower in the low-dose heparin group than in the placebo group. As this analysis was not an RCT for heparin use, physician bias may have favored the administration of heparin treatment.

Unfractionated Heparin

No systematic review, meta-analysis, or RCT exists regarding the treatment of DIC with UFH. In the ICU, most patients are treated with low-dose heparin to prevent thrombosis and embolism. In a comparison with synthesis protease inhibitor

(SPI), no significant difference was observed in either the survival rates or the usefulness. Even though there is still insufficient evidence of UFH on DIC treatment, it is nevertheless considered to be the standard drug for DIC. Although UFH is recommended in patients with thrombosis, it is not used when marked hemorrhaging occurs.

Low-Molecular-Weight Heparin

Dalteparin sodium in LMWH is the only JMHW-approved treatment for DIC. In a multicenter, cooperative, double-blind trial [24] comparing dalteparin sodium with UFH, dalteparin sodium significantly reduced organ failure ($P < 0.05$), reduced bleeding symptoms ($P < 0.1$), and showed a higher safety rate than UFH ($P < 0.05$). The treatment of DIC with LMWH is therefore recommended because the bleeding tendency was less for UFH.

Danaparoid Sodium

Danaparoid sodium is also approved for the treatment of DIC by the JMHW. In a double-blind comparative study [25], DS alleviated bleeding symptoms and organ failure, and it demonstrated a high safety rating ($P < 0.05$) than UFH. As DS does not cancel the antiinflammatory action of AT [26], the bleeding tendency after treatment with DS tends to be less than after UFH.

Synthetic or Purified Protease Inhibitors

Synthetic or purified protease inhibitors (SPIs) include gabexate mesilate (GM), nafamstat mesilate (NM), urinastatin, argatroban, and sivelestat sodium hydrate. GM, NM, and urinastatin were initially approved for the treatment of pancreatitis, and thereafter GM and NM were approved for the treatment of DIC by the JMHW. GM [27] mildly inhibits the activity of thrombin, FXa, plasmin, and plasma kallikrein. As these substances do not cause bleeding, they are frequently used in patients with DIC in Japan. Argatroban is a specific thrombin inhibitor that has strong anticoagulant activity, but it also poses a high risk for bleeding. Urinastatin and sivelestat sodium hydrate inhibit granulocyte elastase. Two RCTs have studied the usefulness of GM in DIC. These studies, which were small, did not demonstrate any significant difference in the outcome or in the alleviation of DIC between the patients treated with GM and those without.

Two randomized nonblind clinical trials evaluating the use of ethyl p-(6-guanidino-hexanoxybenzoate (FOY) [27] and NM [28] in the treatment of DIC have been performed over the past two decades. No significant difference was observed in the outcome or improvement of DIC between GM or NM and UFH. Because those older clinical trials were nonblinded, objectivity may have been reduced. GM and NM are recommended in cases in which hemorrhaging immediately after an operation, for example, tends to be marked and also in cases in which there is the high possibility of hemorrhaging due to marked decreases in the platelet counts.

Physiological Protease Inhibitors

Physiological protease inhibitors including AT, APC, TFPI, and TM have recently been evaluated for their efficacy in the treatment of severe sepsis and DIC. TFPI binds the TF–FVIIa complex and inhibits the extrinsic pathway (i.e., activation of factor X or IX). However, in the OPTIMIST study [17], the usefulness of TFPI for sepsis was not proven.

Antithrombin

Antithrombin inhibits factors IIa, VIIa, Xa, and XIa, and its activity is markedly increased by heparin. AT also demonstrates an antiinflammatory and antiorgan derangement, action but its effects tend to be canceled out by heparin. Many Phase II clinical trials [29, 30] of AT for the treatment of sepsis have been shown to improve the outcome. In one Phase III clinical trial (KyberSept trial) [15], no significant difference was observed in the overall survival rate between the AT-treated group and the placebo group. In a subgroup analysis of the KyberSept trial [15], high-dose AT demonstrated a significant improvement in the patients without heparin and in those with DIC [31]. These analyses suggest that AT is effective for the treatment of patients with severe sepsis who are at high risk for death and DIC.

Activated Protein C

Activated protein C has an anticoagulatory effect via the inactivation of factors VIIIa and Va, and it also activates protease-activated receptor-1 (PAR-1) to inhibit inflammation and apoptosis [10]. There are two types of APC: plasma-derived APC and recombinant APC. In the Prowess trial [16], recombinant APC significantly reduced the mortality among severely septic patients. In addition, plasma D-dimer and serum interleukin-6 (IL-6) levels were significantly lower in the APC-treated group than in the placebo group. The U.S. Food and Drug Administration (FDA) has approved APC for the treatment of severe sepsis.

Thrombomodulin

Thrombomodulin binds thrombin, and the thrombin–TM complex activates PC to APC. TM also binds high-mobility group-B1 (HMGB-1), thus inhibiting the inflammatory process. In Phase III clinical trials regarding the treatment of DIC, TM extracted from urine significantly improved the DIC score in comparison to UHF but not the mortality at 28 days (unpublished data). In a Phase III trial using recombinant TM [18], the mortality rate at 28 days did not differ significantly between TM and UFH, but DIC did show significant alleviation in the patients treated with TM. As a result, TM appears to be an effective drug for the treatment of DIC.

Antifibrinolysis and Fibrinolysis Therapy

Antifibrinolytic therapy including tranexamic acid or SPI for DIC has been shown to suppress dissolving of the microthrombus and to add to the deterioration of organ failure. Therefore, it seems to be contraindicated in DIC induced by sepsis. However, antifibrinolytic therapy is recommended markedly hyperfibrinolytic type of DIC under monitoring with hemostatic molecular markers.

As advanced DIC often demonstrates hyperfibrinolysis and consumption coagulopathy, fibrinolytic therapy is thus considered to increase the risk of hemorrhage.

Transfusion

Fresh Frozen Plasma

The main purpose of administring of fresh frozen plasma (FFP) is to replenish several clotting factors, and the activated plasma thromboplastin time (aPTT), prothrombin time (PT), and fibrinogen values should be monitored before administering FFP. The adaptation of FFP is more than 2.0 of the PT-INR, twofold prolongation of the aPTT, and the fibrinogen level at < 100 mg/dl.

Platelet Concentrates

Although there is usually no fear of hemorrhaging in patients with a platelet count of >50 000/μl, the platelet count dramatically changes in DIC. In patients in whom a bleeding tendency is obvious, as well as in those in whom the platelet count rapidly decreases to <50 000/μl, PC infusion is considered. However, platelet transfusion should be very carefully administered when organ failure is obvious.

References

1. Müllar-Berghaus G, Blombäck M, ten Cate JW (1993) Attempts to define disseminated intravascular coagulation. In: Müllar-Berghaus G, Madlener K, Blombäck M, ten Cate JW (eds) DIC—pathogenesis, diagnosis and therapy of disseminated intravascular coagulation. Excerpta Medica, Amsterdam, pp 1–8
2. Wada H (2004) Disseminated intravascular coagulation. Clin Chim Acta 344:13–21
3. Taylor FB Jr, Toh CH, Hoots WK, et al (2001) Towards definition, clinical and laboratory criteria, and a scoring system for disseminated intravascular coagulation—on behalf of the Scientific Subcommittee on disseminated intravascular coagulation (DIC) of the International Society on Thrombosis and Haemostasis (ISTH). Thromb Haemost 86:1327–1330
4. Müller-Berghaus G, ten Cate H, Levi M (1999) Disseminated intravascular coagulation: clinical spectrum and established as well as new diagnostic approaches. Thromb Haemost 82:706–712
5. Wada H, Minamikawa K, Wakita Y, et al (1993) Hemostatic study before onset of disseminated intravascular coagulation. Am J Hematol 43:190–194
6. Wada H, Wakita Y, Nakase T, et al (1995) Outcome of disseminated intravascular coagulation in relation to the score when treatment was begun. Thromb Haemost 74:848–852
7. Wada H, Nagano T, Tomeoku M, et al (1982) Coagulant and fibrinolytic activities in the leukemic cell lysates. Thromb Res 30:315–322
8. Wada H, Kumeda Y, Ogasawara Z, et al (1993) Plasminogen activators and their inhibitors in leukemic cell homogenates. Am J Hematol 42:166–170
9. Sase T, Wada H, Kamikura Y, et al (2004) Tissue factor messenger RNA levels in leukocytes compared with tissue factor antigens in plasma from patients in hypercoagulable state caused by various diseases. Thromb Haemost 92:132–139
10. Riewald M, Petrovan RJ, Donner A, et al (2002) Activation of endothelial cell protease activated receptor 1 by the protein C pathway. Science. 296:1880–1882
11. Wada H, Mori Y, Shimura M, et al (1998) Poor outcome in disseminated intravascular coagulation or thrombotic thrombocytopenic purpura patients with severe vascular endothelial cell injuries. Am J Hematol 58:189–194
12. Kobayashi N, Maegawa T, Takada M, et al (1983) Criteria for diagnosis of DIC based on the analysis of clinical and laboratory findings in 345 DIC patients collected by the Research Committee on DIC in Japan. Bibl Haematol 49:265–275
13. Gando S, Iba T, Eguchi Y, et al (2006) Japanese Association for Acute Medicine Disseminated Intravascular Coagulation (JAAM DIC) Study Group: A multicenter, prospective validation of disseminated intravascular coagulation diagnostic criteria for critically ill patients: comparing current criteria. Crit Care Med 34:625–631
14. Wada H, Gabazza EC, Asakura H, et al (2003) Comparison of diagnostic criteria for disseminated intravascular coagulation (DIC): diagnostic criteria of the International

Society of Thrombosis and Haemostasis (ISTH) and of the Japanese Ministry of Health and Welfare for overt DIC. Am J Hematol 74:17–22

15. Warren BL, Eid A, Singer P, et al (2001) High-dose antithrombin in severe sepsis: a randomized controlled trial. JAMA 286:1869–1878

16. Bernard GR, Vincent JL, Laterre PF, et al (2001) Efficacy and safety of recombinant human protein C for severe sepsis. N Engl J Med 8:699–709

17. Abraham E, Reinhart K, Opal S, et al (2003) Efficacy and safety of tifacogin (recombinant tissue factor pathway inhibitor) in severe sepsis: a randomized controlled trial. JAMA 290:238–247

18. Saito H, Maruyama I, Shimazaki S, et al (2007) Efficacy and safety of recombinant human soluble thrombomodulin (ART-123) in disseminated intravascular coagulation: results of phase III randomized, double blind, clinical trial. J Thromb Haemost 5:31–41

19. Aoki N, Matsuda T, Saito H, et al (2002) A comparative double blind randomized trial of activated protein C and unfractionated heparin in the treatment of disseminated intravascular coagulation. Int J Hematol 75:540–547

20. Dellinger RP, Carlet JM, Masur H, et al (2004) Surviving Sepsis Campaign Management Guidelines Committee: Surviving Sepsis Campaign guidelines for management of severe sepsis and septic shock. Crit Care Med 32:858–873

21. Weitz JL (1997) Low-molecular-weight heparins. N Engl J Med 337:688–698

22. Meuleman DG (1992) Orgaran (Org 10172); its pharmacological profile in experimental models. Haemostasis 22:58–65

23. Polderman KH, Girbes ARJ (2004) Drug intervention trials in sepsis: divergent results. Lancet 363:1721–1723

24. Sakuragawa N, Hasegawa H, Maki M, et al (1993) Clinical evaluation of low-molecular-weight heparin (FR-860) on disseminated intravascular coagulation (DIC): a multicenter co-operative double-blind trial in comparison with heparin. Thromb Res 72:475–500

25. Yasunaga K, Ogawa K, Mori K, et al (1995) Evaluation of clinical effect on danaparoid sodium (KB-101) on disseminated intravascular coagulation (DIC): double blind comparative study. Jpn Pharmacol Ther 23:2815–2834

26. Harada N, Okajima K, Kohmura H, et al (2007) Danaparoid sodium reduces ischemia/reperfusion-induced liver injury in rats by attenuating inflammatory responses. Thromb Haemost 97:81–87

27. Ohno H, Kambayashi J, Chang SW, et al (1981) FOY[ethyl p-(6-guanidino-hexanoxy) benzonate] methaneslfonate as a serine protease inhibitor. II. In vivo effect on coagulofibrinolytic system in comparison with heparin or aprotinin. Thromb Res 24: 452–455

28. Sibata S, Takahashi H, Aoki N, et al (1988) Evaluation of clinical effect on FUT-175 (nafamostat mesilate) on disseminated intravascular coagulation (DIC): multiple facilities controlled clinical trial. Clin Res 65:921–940

29. Albert J, Blomqvist H, Gardlund B, et al (1992) Effect of antithrombin concentrate on haemostatic variables in critically ill patients. Acta Anaesthesiol Scand 36:745–752

30. Fourrier F, Chopin C, Huart JJ, et al (1993) Double-blind, placebo controlled trial of antithrombin III concentrates in septic shock with disseminated intravascular coagulation. Chest 104:882–888

31. Kienast J, Juers M, Wiedermann CJ, et al (2006) Treatment effects of high-dose antithrombin without concomitant heparin in patients with severe sepsis with or without disseminated intravascular coagulation. J Thromb Haemost 4:90–97

Current Clinical Status of Venous Thromboembolism in Japan

MASHIO NAKAMURA, TAKESHI NAKANO, SATOSHI OTA,
NORIKAZU YAMADA, MASATOSHI MIYAHARA, NAOKI ISAKA, AND
MASAAKI ITO

Summary. Venous thromboembolism (VTE) is the third most common cardiovascular disease for which many clinical guidelines have been published in the Western world. In Japan, VTE has recently been recognized as a common disease and has received increased social as well as medical attention; however, the clinical diagnostic yield is still low. The most effective means of reducing unexpected death from VTE is to implement a comprehensive institutional policy of primary prophylaxis in patients at risk. Although it is difficult to prepare Japanese guidelines for prophylaxis of VTE based on reliable evidence, epidemiological information is available in the general surgical, orthopedic, and gynecological areas of practice. Finally, the first Japanese guidelines for prophylaxis of VTE were published in the spring of 2004. In addition, the diagnostic and management strategies have changed along with improvement in modalities and coverage by thrombolytic agents. The next goal is to gather evidence that provides the foundation for guidelines on treatment and prophylaxis. Large-scale research studies similar to those performed in the West should be conducted to elucidate the pathological mechanisms of VTE and improve its management.

Key words. Anticoagulation therapy · Prophylaxis · Thrombolytic therapy · Vena cava filters · Venous thromboembolism

Introduction

Venous thromboembolism (VTE), manifesting as either deep vein thrombosis (DVT) or pulmonary thromboembolism (PTE), is an extremely common medical problem in the Western world, occurring either in isolation or as a complication of other diseases or procedures. Many clinical guidelines for VTE have been published, such as "Guidelines on diagnosis and management of acute pulmonary embolism" (2000) by the European Society of Cardiology [1]; "Guidelines for the management of suspected acute pulmonary embolism" (2003) by the British Thoracic Society [2]; "Prevention of venous thromboembolism" (2004) by the American College of Chest Physicians [(3]; and "Prevention and treatment of venous thromboembolism" (2006) by the Internal Union of Angiology [4]. At the same time, the incidence of VTE in Japan was considered to be minimal, and it was not a focus of attention in most Japanese medical societies. Nonetheless, a few researchers suspected that VTE might be common in Japan and could be responsible for some unexplained sudden deaths. The lower diagnostic yield of VTE in Japan was also thought to be a contributor because the disorder is observed in multiple medical areas, including cardiovascular and respiratory medicine and surgery, and information on these cases was not disseminated systematically.

The Japanese Society of Pulmonary Embolism Research was established in 1994 [5], and international symposiums on pulmonary embolism in Japan took place in Mie in 1998 [6] and in Chiba and Sendai in 2003 and 2000, respectively [7]. These efforts were aimed at raising awareness of VTE and compiling data on Japanese cases. As a result, awareness in the medical profession has improved, diagnostic techniques have advanced, and the number of cases of VTE reported in Japan has increased rapidly. Furthermore, because of the increase in venous thromboembolic risk factors, such as continuing westernization of Japanese life and social aging, and of postoperative sudden death and travelers' thrombosis (so-called economy-class syndrome), as well as the high incidence of VTE noted following the recent mid-Niigata Prefecture earthquake, it has recently received a lot of attention, both medically and socially. In 2004, Japanese guidelines for VTE [8, 9] were published, and the increasing awareness of this disorder in Japan is changing the perception of VTE from that of a rare disease to a common one. However, large-scale studies that would help provide evidence of VTE have not been conducted in Japan. Moreover, Japan lags behind Western countries in clinical management of this disease.

Incidence and Prognosis of Venous Thromboembolism

Although VTE is diagnosed and treated in as many as 260 000 patients in the United States each year, it is estimated that more than half the cases that actually occur are never diagnosed. Thus, the true incidence may be as high as 600 000 cases [10, 11]. At the same time, the incidence of PTE in serial autopsy cases in Japan and the United States was reported as 0.8% and 23.8%, respectively, according to a previous international collaboration [12]. Moreover, the annual reports of the pathological autopsy cases in Japan noted a rate of 1.41%–1.72% [13, 14]. These results indicated that the incidence of VTE was extremely low in Japan. However, more recent reports have indicated that the incidence of DVT after joint replacement in Japan is the same as in Western countries [15, 16], and the incidence of PTE in serial autopsy cases ranged from 11% to 24% [17–20]. Consequently, the incidence of VTE in Japan is no longer considered low. Despite this evidence, data compiled from recent questionnaire surveys showed that the number of clinically diagnosed cases of PTE in Japan was only 28–32 per 1 million people [21, 22], which is 1/15th that of Western countries. We must therefore conclude that Japanese clinicians are still not entirely familiar with VTE.

Racial and environmental factors may contribute to the differences in incidence reported among Japanese and Westerners. Reports that show a higher incidence of PTE in black Americans than in black Africans [23] and no differences between white and nonwhite South Africans [24] suggest that an environmental factor is involved in the incidence of PTE. On the other hand, the factor V Leiden mutation with activated protein C resistance, which has drawn attention as a new coagulation abnormality, is a high-frequency gene mutant that appears in 30%–50% of Caucasians with VTE [25]. No Japanese patients were found with this mutation. Similarly, no Japanese patient has the prothrombin G20210A mutation, which is associated with VTE in Caucasians [26, 27]. These findings suggest that a racial factor is also involved in the incidence of PTE. Studies among races in the United States also show that the incidence of VTE among Asians is one-fifth to one-quarter that of whites [28, 29]. Meanwhile, in our study that investigates coagulopathy in detail, only 5 of 61 consecutive patients (8%) with PTE did not have an inherited coagulopathy or a secondary risk factor for VTE [30]. These results suggest that Japanese patients might have unknown inherited coagulopathies.

In the United States, widespread prevention of VTE in hospitals led to a significant decline in the incidence of PTE beginning during the mid-1980s, although the incidence had been increasing before that period [31]. On the other hand, deaths from PTE in Japan were projected to increase by 10 times over the 50 years from 1951 to 2000 according to the vital statistics of the Ministry of Health, Labor, and Welfare [32]. The PTE rate in autopsy cases from the annual reports of the pathological autopsy cases in Japan also increased 3.5-fold from 1958 to 1997 [13, 14, 33]. These increases in PTE reflect both an "increase in appearance" and a "true increase." The "increase in appearance" indicates improved accuracy of diagnosis, such as the rising awareness among health care professionals, increasing visibility of VTE, and advances in modalities. The recent rapid increase in the incidence of this disease stems largely from these factors, which has resulted from educational campaigns in various quarters. Among the factors contributing to the "true increase" of VTE are, first, the aging of the Japanese population, which itself confers a risk of VTE; and the elderly are likely to have other risks for VTE. The second factor is change in life style. In recent decades, eating habits in Japan have become westernized, and the percentage of the population that is overweight is growing. Although the relation between VTE and metabolic syndrome is unknown, it should be considered. The third factor is advances in medical care. The risk of VTE at an inpatient setup rises according to the complexity of treatment, and the number of patients at risk further increases because the prognosis has improved for conditions such as malignancy that increase the risk of VTE. In addition, many medicines, such as oral contraceptives and drugs used in hormone replacement therapy, may increase the risk of VTE (Fig. 1).

From the point of view of prognosis, the hospital mortality rate was 14% for 309 patients diagnosed with PTE from 1994 to 1997 according to a collaborative study conducted by the Japanese Society of Pulmonary Embolism Research [34]. Among those who died, mortality among patients with shock at presentation was 30%, and it was 20% except for cases diagnosed by an autopsy, and 6% for those who presented without shock. In addition, the mortality after discharge was 3%. These results indicate that patients with PTE who are correctly diagnosed have a relatively good prognosis. However, in the following decade, the prognosis of severe cases did not improve [35] (Table 1). In addition to underdiagnosis of PTE, the high frequency of sudden death, which afflicts 25% of severe cases, and death due to recurrent PTE affect the mortality rates [36] (Fig. 2). In particular, many patients with unexpected circulatory

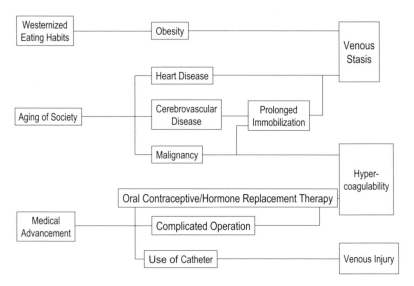

Fig. 1. Change in social structure and increasing risk of pulmonary thromboembolism

Table 1. Change in the mortality rate of pulmonary thromboembolism in Japan

Subjects	Change in morality (%)		
	1/1994–10/1997 (*n* = 309)	11/1997–10/2000 (*n* = 257)	11/2000–8/2003 (*n* = 461)
Total	14	12	8
Cases with arrest or shock	30	32	27
Other mild cases	6	3	2

Data are from Sakuma et al. [35]

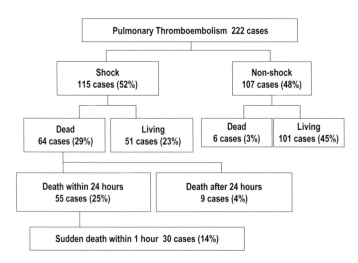

Fig. 2. In-hospital mortality of 222 patients with pulmonary thromboembolism. Sudden deaths within 1 h from onset account for 43% of total deaths with pulmonary thromboembolism

arrest succumb even with percutaneous cardiopulmonary support. Consequently, it is important to prevent PTE.

Development of Diagnostic Methods and Management of Pulmonary Thromboembolism

Many guidelines on the diagnosis and management of PTE have been published and reported at Western scientific meetings. Guidelines for the diagnosis and management of PTE have also been developed by the Japanese Society of Circulation in cooperation with other societies in Japan [9] (Fig. 3). Application of Western guidelines for PTE in Japan, as for other disorders, was uncertain because of unknown differences in the incidence of PTE in Caucasian and Japanese populations and the differences in medical care among countries. In addition, guidelines for diagnosis that are based on the situation in each country are necessary because of differences in the development of diagnostic modalities and cost.

Traditionally, the gold standard for diagnosis of PTE has been pulmonary angiography. However, efforts have been made to restrict the indications for use of pulmonary angiography, which is invasive; currently, the diagnosis of PTE tends to rely on noninvasive modalities [34, 35, 37] (Table 2). Although the main noninvasive diagnostic modality for PTE in Western countries is the perfusion lung scan, contrast-enhanced multislice computed tomography (CT) is useful in Japan. Current machines can evaluate a thromboembolism even in segmental arteries except in patients who cannot hold their breath for the necessary imaging interval. Multislice CT has the additional

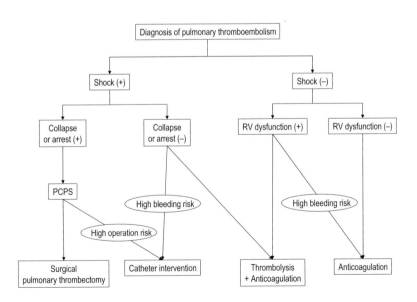

FIG. 3. Management strategy in pulmonary thromboembolism. *PCPS*, percutaneous cardiopulmonary support; *RV*, right ventricle

TABLE 2. Development of diagnostic methods of pulmonary thromboembolism in Japan

Diagnostic methods	Development of methods (%)		
	1/1994–10/1997 (n = 309)	11/1997–10/2000 (n = 257)	11/2000–8/2003 (n = 461)
Perfusion lung scan	74	77	62
Pulmonary angiography	45	57	37
Contrast-enhanced CT	14	58	62
MRI	2	6	2

Data are from Nakamura et al. [34] and Sakuma et al. [35, 37] CT, computed tomography; MRI, magnetic resonance imaging

TABLE 3. Development of diagnostic methods for DVT after the occurrence of pulmonary thromboembolism in Japan

Diagnostic method	Development (%)		
	1/1994–10/1997 (n = 309)	11/1997–10/2000 (n = 257)	11/2000–8/2003 (n = 461)
Operation rate of examination for DVT	61	65	83
Venous ultrasonography	18	24	55
Venography	81	73	41
Radio nuclear angiography	13	12	2

Operation rate indicates the ratio of patients examined for operation during each interval to the total number of patients diagnosed with DVT during that interval.
DVT, deep vein thrombosis
Data are from Nakamura [34] and Sakuma [35, 37]

advantage of being able to examine veins in the abdomen, pelvis, and legs after scanning the pulmonary arteries. The efficacy of multislice CT in PTE has been examined, and this modality is now included in the diagnostic algorithm for PTE. In addition, because of the importance of diagnosing DVT as a source of thromboembolism, venous ultrasonography, which is convenient and can be used noninvasively at the bedside, is recommended. Implementation of this modality has nevertheless been inadequate [34, 35, 37] (Table 3), and it is important to train sonographers and physicians to perform venous ultrasonography.

In the West, management guidelines for anticoagulation and thrombolysis based on a wealth of evidence are being released. No clinical trials are planned or underway to gather such evidence in Japan [38]. With regard to thrombolytic agents, which are used commonly for PTE, Monteplase, mutant tissue plasminogen activator (t-PA), has been permitted for use in patients by the Japanese government only since autumn 2005. At the same time, the effect of thrombolysis on the prognosis of patients with PTE is not clear, even in Western countries [39]. In many randomized controlled trials for PTE therapy, although thrombolysis has been shown to bring about more rapid thrombolytic and hemodynamic effects than anticoagulation, there have been no reports that thrombolysis reduces PTE-related mortality. Therefore, the ability of thrombolysis to improve the prognosis is unknown, and its indications have not been sufficiently established. In addition, it is arguable whether thrombolysis has an effect on prognosis in severe cases. Recently, right ventricular dysfunction was identified as an important factor linked to prognosis. In the most recent randomized controlled trial of PTE in normotensive patients with right ventricular dysfunction, t-PA was found to improve the clinical course and prevent clinical deterioration; it requires an escalation of treatment compared with heparin, although the differences in mortality were not significant [40]. Therefore, thrombolytic therapy might be recommended for PTE patients with right ventricular dysfunction; however, the choice of this therapy has little scientific basis and needs further research.

Over the past few years, the use of catheter intervention for various kinds of vascular diseases has progressed, and catheter thrombolysis and catheter fragmentation are now used in patients with VTE. These techniques are being used to dissolve a thromboembolus and reduce right ventricular overload in PTE, rapidly restore a venous valve, and decrease the frequency of postthrombotic syndrome. Although it is often difficult to dissolve an obstructive DVT using conventional pharmacotherapy, a recent Japanese study reported that catheter-directed thrombolysis achieves a high dissolution rate of DVT and a reduction in the incidence of postthrombotic syndrome [41]. Thus, this technique appears to be promising.

The development of inferior vena cava (IVC) filters constitute a marked advance in the prevention of primary and recurrent PTE with DVT. When the technique was initially introduced, only permanent filters were used, and the recurrence rate of PTE after IVC filter implantation was reported to be 2% [42]. However, results of a French prospective multicenter study indicated that the long-term prognosis was worse after IVC filter implantation than after anticoagulation [43]. As a result, prophylactic use of permanent IVC filters for stable patients is not recommended. On the other hand, when anticoagulation alone is performed as PTE management, it may lead to massive recurrent PTE, a critical complication. To avoid such cases of recurrent, severe PTE, nonpermanent IVC filters have recently come into use. These filters are implanted only during the acute phase when a venous thrombus can easily migrate and the pulmonary vascular bed is small. The filter can then be removed after thrombus dissolution, decreasing the risk of embolism. Temporary IVC filters [44] and retrievable IVC filters [45] can now be used as nonpermanent IVC filters. Although temporary filters carry a risk of infection and bleeding, because they are joined to an external catheter they fare better in younger subjects and women of child-bearing age because they can be readily removed. At the same time, retrievable IVC filters are implanted as permanent filters and can be retrieved with a catheter after use. They impose fewer limitations of patient activities, and the risk of infection and bleeding is small. Moreover, they can be implanted permanently. Although the indications for use of these IVC filters have not yet been determined, the decision to use these devices should be based on the severity of the right ventricular overload and the size and ease of migration of the DVT. The trend in Japan, more so than in the West, is toward increased use of IVC filters. Once the utility of anticoagulation has been confirmed, indications for use of IVC filters should be determined carefully, especially when considering the use of permanent IVC filters (Table 4).

TABLE 4. Development of methods of management of pulmonary thromboembolism in Japan

Methods of management	Development (%)		
	1/1994–10/1997 (n = 309)	11/1997–10/2000 (n = 257)	11/2000–8/2003 (n = 461)
Anticoagulation	74	82	92
Thrombolytic therapy	50	48	58
Catheter intervention	6	6	10
Surgical pulmonary thrombectomy	2	3	2
Vena cava filter	18	34	35

Data are from Nakamura [34] and Sakuma [35, 37]

Current Status of Prophylaxis of Venous Thromboembolism

In-hospital prophylaxis of VTE is important for the following reasons: (1) the incidence of VTE is remarkably high in hospitals; (2) early diagnosis of DVT is difficult because it produces few symptoms; (3) PTE has a high fatality rate; and (4) prophylaxis of VTE is cost-efficient. Many Western medical societies have developed guidelines for the prevention of VTE based on extensive evidence. Sudden death from PTE after surgery and in other conditions is an increasing problem in Japan, and medical malpractice litigation is becoming more frequent when an appropriate treatment for PTE has not been provided. Until now, it was difficult to estimate the incidence of PTE accurately because of the difficulty of its diagnosis. However, adequate diagnosis and management of PTE is currently regarded as a matter of course, and an inadequate response by health care providers could be judged a lack of due diligence. For these reasons, a guideline for prophylaxis of VTE was anxiously awaited in Japan. As previously noted, it was impossible to use Western guidelines without modification because of differences in the incidence and use of prophylactic agents between Japan and the West. In response to this situation, the "Japanese Guideline for Prevention of Venous Thromboembolism" [8] was published in the spring of 2004 by the Editorial Committee organized by the Japanese Society of Anesthesiologists, Japanese Society of Phlebology, Japanese Society of Pulmonary Embolism Research, Japan Society of Obstetrics and Gynecology, Japanese College of Cardiology, Japanese Orthopaedic Association, Japanese Society of Intensive Care Medicine, Japanese Society of Thrombosis and Homeostasis, Japanese Urological Association, and Japan Society of Obstetrical Gynecological and Neonatal Hematology. Because few clinical studies are available that provide evidence specific to Japanese patients, the guideline remains incomplete. Even so, the participation of numerous medical societies in the development of the Japanese guideline is a significant step that is unprecedented elsewhere in the world. At about the same time, a PTE prophylaxis management fee was listed in the revised health insurance as a part of medical treatment fees in the spring of 2004. As a result, in-hospital VTE prophylaxis is starting to be implemented in many institutions. In the future, collection of objective information from Japanese patients is urgently needed, and approval to use new anticoagulants is eagerly awaited. Fondaparinux, an anti-Xa agent, and enoxa-parin, a low molecular weight heparin, became available for use in Japan in the spring of 2007 and 2008, respectively. It is expected that in the near future pharmaceutical prophylaxis with new anticoagulants will become mainstream [46, 47].

Future Outlook

The most significant development in research and practice relating to VTE is that it has been recognized as a more common disease than was previously considered. However, it is still difficult to diagnose VTE unless suspected, although diagnostic modalities are improving. A diagnostic device that can be used at the bedside to easily identify VTE is needed. The management and prophylaxis of VTE will move forward with the development of new agents. Antithrombotic agents with a lower risk of bleed-ing are needed because VTE is a condition that is associated with easy bleeding. The quality of diagnosis and treatment in Japan has reached Western standards. The next goal is to gather evidence that provides the foundation for guidelines of treatment and prophylaxis. Full-scale research studies such as those performed in the West should be carried out to elucidate the pathological mechanism of VTE and improve its management.

References

1. Task Force on Pulmonary Embolism, European Society of Cardiology (2000) Guide-lines on diagnosis and management of acute pulmonary embolism. Eur Heart J 21:1301–1336
2. British Thoracic Society Standards of Care Committee Pulmonary Embolism Guide-line Development Group (2003) British Thoracic Society guidelines for the manage-ment of suspected acute pulmonary embolism. Thorax 58:470–483
3. Geerts WH, Pineo GF, Heit JA, et al (2004) Prevention of venous thromboembolism: the Seventh ACCP Conference on Antithrombotic and Thrombolytic Therapy. Chest 126:338S–400S
4. Nicolaides AN, Breddin HK, Fareed J, et al (2001) Prevention of venous thromboem-bolism: international consensus statement—guidelines compiled in accordance with the scientific evidence. Int Angiol 20:1–37
5. Japanese Society of Pulmonary Embolism Research: JaSPER. URL http://jasper.gr.jp/
6. Nakano T, Goldhaber SZ (eds) (1999) Pulmonary embolism. Springer, New York
7. Kunio S (ed) (2005) Venous thromboembolism. Springer, Tokyo
8. Editorial Committee on Japanese Guideline for Prevention of Venous Thromboembo-lism (2004) Japanese guideline for prevention of venous thromboembolism. Medical Front International, Tokyo
9. Guidelines for the diagnosis, treatment and prevention of pulmonary thromboembo-lism and deep vein thrombosis (2004) Jpn Circ J 69:1077–1126
10. Gillum RF (1987) Pulmonary embolism and thrombophlebitis in the United States, 1970–1985. Am Heart J 114:1262–1264
11. Bell WR, Simon TL (1982) Current status of pulmonary thromboembolic disease: pathophysiology, diagnosis, prevention, and treatment. Am Heart J 103:239–262
12. Gore I, Hirst AE, Tanaka K (1964) Myocardial infarction and thrombosis. Arch Intern Med 113:323–330

13. Hasegawa H, Nagata H, Yamauchi M, et al (1981) Statistical status of pulmonary embolism in Japan (II). Jpn J Chest Dis 40:677–681 (in Japanese)
14. Mieno T, Kitamura S (1989) Incidence of pulmonary thromboembolism in Japan. Kokyu to Junkan 37:923–927 (in Japanese)
15. Kitajima I, Tachibana S, Hirota Y, et al (1999) The incidence of pulmonary embolism following total hip arthroplasty. Seikei Geka 50:1287–1290 (in Japanese)
16. Fujita S, Fuji T, Mitsui T, et al (2000) Prospective multicenter study on prevalence of deep vein thrombosis after total hip or total knee arthroplasty. Seikei Geka 51:745–749 (in Japanese)
17. Mizukami Y, Murai Y, Fukushima Y, et al (1976) Pulmonary thromboembolism in a sequent autopsy series of 200 elders. Kokyu to Junkan 24:979–984 (in Japanese)
18. Ito S (1982) Clinico-pathological studies on pulmonary thromboembolism. Mie-Igaku 25:586–597 (in Japanese)
19. Ito M (1991) Pathology of pulmonary embolism. Kokyu to Junkan 39:567–572
20. Nakamura Y, Yutani C, Imakita M, et al (1996) Pathophysiology of clinicopathological aspect of venous thrombosis and pulmonary thromboembolism. Jpn J Phlebol 7:17–22 (in Japanese)
21. Kumasaka N, Sakuma M, Shirato K (1999) Incidence of pulmonary thromboembolism in Japan. Jpn Circ J 63:439–441
22. Kitamukai O, Sakuma M, Takahashi T, et al (2003) Incidence and characteristics of pulmonary thromboembolism in Japan 2000. Intern Med 42:1090–1094
23. Thomas WA, Davies JNP, O'Neal RM, et al (1960) Incidence of myocardial infarction correlated with venous and pulmonary thrombosis and embolism: a geographic study based on autopsies in Uganda, East Africa and St. Louis, USA. Am J Cardiol 5:4–47
24. Joffe SN (1974) Racial incidence of postoperative deep vein thrombosis in South Africa. Br J Surg 61:982–983
25. Dahlback B, Carlsson M, Svensson PJ (1993) Familial thrombophilia due to a previously unrecognized mechanism characterized by poor anticoagulant response to activated protein C: prediction of a cofactor to activated protein C. Proc Natl Acad Sci U S A 90:1004–1008
26. Seki T, Okayama H, Kumagai T, et al (1998) Arg506Gln mutation of the coagulation factor V gene not detected in Japanese pulmonary thromboembolism. Heart Vessels 13:195–198
27. Ro A, Hara M, Takada A (1999) The factor V Leiden mutation and the prothrombin G20210A mutation was not found in Japanese patients with pulmonary thromboembolism. Thromb Haemost 82:1769
28. White RH, Zhou H, Romano PS (1998) Incidence of idiopathic deep venous thrombosis and secondary thromboembolism among ethnic groups in California. Ann Intern Med 128:737–740
29. Klatsky AL, Armstrong MA, Poggi J (2000) Risk of pulmonary embolism and/or deep venous thrombosis in Asian-Americans. Am J Cardiol 85:1334–1337
30. Yamada N, Nakamura M, Ishikura K (2003) Epidemiological characteristics of acute pulmonary thromboembolism in Japan. Int Angiol 22:50–54
31. Lilienfeld DE (2000) Decreasing mortality from pulmonary embolism in the United States, 1979-1996. Int J Epidemiol 29:465–469
32. Sakuma M, Konno Y, Shirato K (2002) Increasing mortality from pulmonary embolism in Japan, 1951-2000. Circ J 66:1144–1149
33. Sakuma M, Takahashi T, Kitamukai O, et al (2002) Incidence of pulmonary embolism in Japan: analysis using "Annual of the pathological autopsy cases in Japan." Ther Res 23:632–634
34. Nakamura M, Fujioka H, Yamada N, et al (2001) Clinical characteristics of acute pulmonary thromboembolism in Japan: results of a multicenter registry in the Japanese Society of Pulmonary Embolism Research. Clin Cardiol 24:132–138

35. Sakuma M, Nakamura M, Nakanishi N, et al (2004) Inferior vena cava filter is a new additional therapeutic option to reduce mortality from acute pulmonary embolism. Circ J 68:816–821

36. Ota M, Nakamura M, Yamada N, et al (2002) Prognostic significance of early diagnosis in acute pulmonary thromboembolism with circulatory failure. Heart Vessels 17:7–11

37. Sakuma M, Okada O, Nakamura M, et al (2003) Recent developments in diagnostic imaging techniques and management for acute pulmonary embolism: multicenter registry by the Japanese Society of Pulmonary Embolism Research. Intern Med 42:470–476

38. Nakamura M, Nakanishi N, Yamada N, et al (2005) Effectiveness and safety of the thrombolytic therapy for acute pulmonary thromboembolism: results of a multicenter registry in the Japanese Society of Pulmonary Embolism Research. Int J Cardiol 99:83–89

39. Wan S, Quinlan DJ, Agnelli G, et al (2004) Thrombolysis compared with heparin for the initial treatment of pulmonary embolism: a meta-analysis of the randomized controlled trials. Circulation 110:744–749

40. Konstantinides S, Geibel A, Heusel G, et al (2002) Management strategies and prognosis of Pulmonary Embolism-3 Trial investigators: heparin plus alteplase compared with heparin alone in patients with submassive pulmonary embolism. N Engl J Med 347:1143–1150

41. Yamada N, Ishikura K, Ota S, et al (2006) Pulse-spray pharmacomechanical thrombolysis for proximal deep vein thrombosis. Eur J Vasc Endovasc Surg 31:204–211

42. Yazu T, Fujioka H, Nakamura M, et al (2000) Long-term results of inferior vena cava filters: experiences in a Japanese population. Intern Med 39:707–714

43. Decousus H, Leizorovicz A, Parent F, et al (1998) A clinical trial of vena cava filters in the prevention of pulmonary embolism in patients with proximal deep-vein thrombosis. N Engl J Med 338:409–415

44. Yamada N, Niwa A, Sakuma M, et al (2001) Status of use of temporary vena cava filters in Japan. Ther Res 22:1439–1441 (in Japanese)

45. Ishikura K, Yamada N, Oota M, et al (2002) Clinical experience with retrievable vena cava filters for prevention of pulmonary thromboembolism. J Cardiol 40:267–273 (in Japanese)

46. Samama MM, Cohen AT, Darmon JY, et al (1999) A comparison of enoxaparin with placebo for the prevention of venous thromboembolism in acutely ill medical patients: prophylaxis in medical patients with Enoxaparin Study Group. N Engl J Med 341:793–800

47. Eriksson BI, Bauer KA, Lassen MR, et al (2001) Steering Committee of the Pentasaccharide in Hip-Fracture Surgery Study: fondaparinux compared with enoxaparin for the prevention of venous thromboembolism after hip-fracture surgery. N Engl J Med 345:1298–1304

Platelet Activity and Antiplatelet Therapy in Patients with Ischemic Stroke and Transient Ischemic Attack

SHINICHIRO UCHIYAMA, TOMOMI NAKAMURA, YUMI KIMURA, AND MASAKO YAMAZAKI

Summary. Ischemic stroke and transient ischemic attack as well as coronary and peripheral artery diseases are platelet-dependent disease states that are categorized as atherothrombosis because they are mainly attributable to the obstruction of brain arteries by platelet-rich thrombi. Indeed, many platelet function tests demonstrate evidence of platelet activation in these patients.

The most widely used antiplatelet agent is aspirin. Aspirin can significantly reduce vascular events including stroke, myocardial infarction, and vascular death in patients at risk of atherothrombosis, although aspirin reduces only approximately one-fourth of total vascular events, which is not sufficient. This limited efficacy of aspirin has recently been much discussed in terms of "aspirin resistance." Strategies against aspirin resistance might include use of alternative antiplatelet agents or combined use of other antiplatelet agents with aspirin. Candidates for alternative antiplatelet agents would be thienopyridines or phosphodiesterase (PDE) inhibitors.

Dual antiplatelet therapy with aspirin and thienopyridine is a potent antiplatelet strategy, although such combination therapy produces further prolongation of the bleeding time and in reality increases hemorrhagic complications. By contrast, PDE inhibitors are associated with less bleeding risk. They do not prolong the bleeding time or increase hemorrhagic complications, which may be a great advantage for dual antiplatelet therapy. Recent evidence indicates that PDE inhibitors have not only antiplatelet effects but also endothelial protective effects, which may be related to a low bleeding risk.

Key words. Stroke · Platelet activation · Aspirin · Thienopyridine · Phosphodiesterase inhibitor

Introduction

Ischemic cerebrovascular disease (CVD), coronary artery disease (CAD), and peripheral artery disease (PAD) have recently been categorized into atherothrombosis because they share a common pathophysiological mechanism, that is, obstruction of arteries by platelet-rich thrombi formed during rupture of unstable plaques

Therefore, atherothrombosis including noncardioembolic ischemic stroke and transient ischemic attack (TIA) is a platelet-dependent disease state and thus an indication for treatment with antiplatelet therapy [1].

In reality, it has been reported that platelet activation is observed in many platelet function tests in CVD patients [2–4]. Unlike local intravascular intervention, because antiplatelet therapy may act on the whole body through the circulating blood (e.g., even when an antiplatelet agent is administered for the prevention of CVD), it may simultaneously prevent CAD and PAD. Therefore, the efficacy of antiplatelet therapy should be evaluated by risk reduction of total atherothrombotic events in all vascular beds [5].

Platelet Activity in Platelet Function Tests

Platelet Survival and Deposition

Platelet survival is a marker of in vivo platelet activation because the survival is shortened by platelet consumption during thrombogenesis. We measured the platelet survival time using platelet labeled with indium-111 tropolne in patients with ischemic stroke. Marked shortening of the platelet survival time was frequent in patients with atherothrombotic and cardioembolic stroke, indicating platelet consumption by thrombus formation in the arterial lumen and the heart chamber, respectively, whereas only minimal shortening of platelet survival was observed in the small proportion of patients with lacunal stroke (Fig. 1) [3, 4]. We also evaluated platelet deposition by platelet imaging at 48h after reinjection of autologous platelets labeled with indium-111 tropolone. Platelet deposition was often observed in patients with atherothrombotic stroke but never in those with cardioembolic or lacunar stroke (Fig. 2) [3, 4].

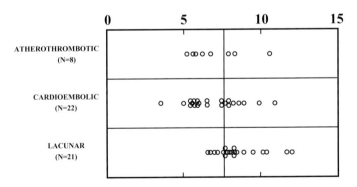

FIG. 1. Platelet survival time measured using platelets labeled with indium-111 tropolone. *Vertical bar* represents the mean minus 2 SD of the values for normal, healthy, nonsmoking adult male volunteers

T.O. 77 F
INFARCT IN R.MCA AREA

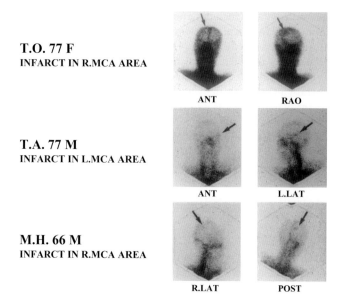

ANT   RAO

T.A. 77 M
INFARCT IN L.MCA AREA

ANT   L.LAT

M.H. 66 M
INFARCT IN R.MCA AREA

R.LAT   POST

FIG. 2. Platelet imaging visualized at 48 h after reinjection of autologous indium-labeled platelets. Platelet deposition was detected in the area surrounding the ischemic core, called the ischemic penumbra, in patients with atherothrombosis. *MCA*, middle cerebral artery; *L*, left; *R*, right; *ANT*, anterior; *RAO*, right anterior oblique; *LAT*, lateral; *POST*, posterior

Shear-Induced Platelet Aggregation

Shear-induced platelet aggregation (SIPA) plays an important role in thrombogenesis at sites of arterial bifurcation, stenosis, and spasm. We measured SIPA at a high shear stress of 108 dynes/cm^2 using a rotational cone-plate streaming chamber in patients with ischemic stroke [6]. SIPA was increased in patients with atherothrombotic stroke and TIA but was not increased in patients with cardioembolic or lacunar stroke (Table 1) [7]. The extent of SIPA was significantly correlated with the amount of large von Willebrand factor (vWF) multimers displayed by sodium dodecyl sulfate (SDS)-agarose gel electrophoresis and semiquantitated by densitometry (Fig. 3) [7]. As for the molecular mechanism of SIPA, it is known that vWF, especially its large multimers, as well as glycoproteins (GP) Ib/IX and IIb/IIIa are required at high shear [8]. Therefore, these results suggest that SIPA was increased by an increase in large vWF multimers released from injured or stimulated endothelial cells or a decrease in breakdown of the multimers in plasma.

Platelet Fibrinogen Binding and P-Selectin Expression

Measurement of binding to platelets or expression on platelets of adhesion molecules is a more sensitive and specific technique for detecting activated platelets than conventional platelet aggregometry. We measured platelet fibrinogen binding and P-selectin expression in whole blood using flow cytometry. Fibrinogen binding to

TABLE 1. Shear-induced platelet aggregation in subtypes of cerebral ischemia

Subgroup	No.	SIPA (%), (mean ± SD)	
Atherothrombotic stroke	21	57.9 ± 10.4	
Cardioembolic stroke	11	48.3 ± 11.8	
Lacunar stroke	31	49.2 ± 8.1	
Transient ischemic attack	12	57.5 ± 8.6	
Patient controls	12	46.3 ± 10.3	
Normal controls	14	44.9 ± 2.7	

SIPA, Shear-induced platelet aggregation
*$P < 0.05$
†$P < 0.01$
‡$P < 0.001$ (ANOVA)

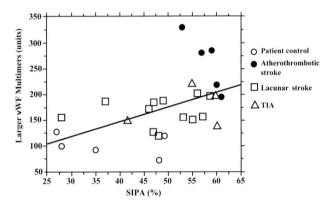

FIG. 3. Correlation between large von Willebrand factor (*vWF*) multimers and the extent of shear-induced platelet aggregation (*SIPA*). Extent of SIPA was significantly correlated with the amount of large von Willebrand factor multimers, which was analyzed and determined by agarose gel electrophoresis followed by densitometry. *TIA*, transient ischemic attack

platelets requires lower concentrations of activating stimuli than platelet secretion and thus an early phase of platelet activation. P-selectin is expressed only by the degranulated platelets, thereby reflecting a late phase of platelet activation. Patients with atherothrombotic stroke showed significant increases in both (Fig. 4) [9]. Patients with lacunar stroke also showed significant increase in both, but P-selectin expression was significantly less than that in those with atherothrombotic stroke (Fig. 4) [9]. These results indicate that stronger platelet activation occurs in the presence of atherothrombotic stroke than in lacunar stroke. Patients with cardioembolic stroke showed a significant increase in P-selectin expression without any increase in fibrinogen binding (Fig. 4) [9]. One possible explanation for this discrepancy might be an increase in "empty exhausted platelets," which are refractory platelets that have undergone a release reaction, in this subtype of ischemic stroke [3].

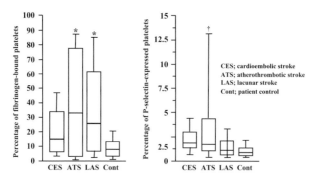

FIG. 4. Platelet fibrinogen binding and P-selectin expression in subgroups of ischemic stroke patients. The box plots display median values with 25th and 75th percentiles, and the bar chart shows 10th and 90th percentiles. *P < 0.05, compared with control. †P < 0.05, compared with lacunar stroke

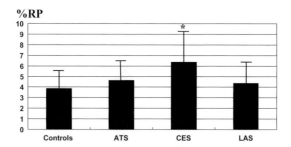

FIG. 5. Percentage of reticulated platelets in patients with various subtypes of ischemic stroke and control subjects. Data are presented as the mean ± SD. *P < 0.05 for atherothrombotic stroke and lacunar stroke patients compared to controls

Percentage of Reticulated Platelets

Circulating platelets usually do not have RNA because they do not have nuclei. However, young platelets newly produced by bone marrow have residual amounts of RNA. It is known that the fluorescent dye thiazol orange can stain residual RNA in platelets. These platelets are called "reticulated platelets." We measured the percentage of reticulated platelets (%RT) as a marker of platelet turnover using thiazol orange by flow cytometry. The %RT was significantly increased in patients with cardioembolic and atherothrombotic stroke but was not significantly increased in those with lacunar stroke (Fig. 5) [10]. In addition, patients with cardioembolic stroke had significantly higher %RT not only than controls but also than patients with atherothrombotic and lacunar stroke (Fig. 5) [10]. These results indicate that enhanced platelet turnover as a result of platelet consumption is most prominent in cardioembolic stroke among subtypes of ischemic stroke.

Antiplatelet Therapy for Stroke and Transient Ischemic Attack

Aspirin

Among antiplatelet agents, aspirin is most widely used in the world, and its use is supported by the most substantial evidence. Aspirin inhibits platelet aggregation by inhibiting thromboxane (TX) A_2 synthesis through blocking cyclooxygenase (COX) activity. According to the meta-analysis by the Antithrombotic Trialists' Collaboration (ATT), risk reduction of vascular events—stroke, myocardial infarction (MI), vascular death—by aspirin in high-risk patients with atherothrombotic disease was highly significant [5]. However, the total risk reduction of vascular events by aspirin is only 22% [5]. This limited effect of aspirin has recently been addressed in terms of "aspirin resistance" (AR). Definition of AR has not been established, although it is generally defined as appearing in patients in whom aspirin fails to prevent vascular events due to the failure of inhibitory effects by aspirin on platelet activation [11].

Mechanisms of AR may include (1) etiologies of vascular events other than atherothrombosis; (2) decreased bioavailability of aspirin; (3) activation of alternative pathways other than the COX–TXA_2 pathway; (4) increased production of platelets unexposed to aspirin as a result of increased platelet turnover during thrombogenesis; and (5) gene polymorphisms of platelet membrane GPs and agonist receptors as well as enzymes for the platelet arachidonic acid (AA) cascade [1, 11].

We studied AR in 857 consecutive patients with suspected stroke or TIA who had been given only aspirin and undergone examination by the platelet aggregation test using conventional platelet aggregometry in platelet-rich plasma. Our definition of AR (biological AR) was failure in the inhibition of platelet aggregation induced by AA, which is totally dependent on TXA_2. Complete resistance was defined as full aggregation induced by 1 mM AA despite taking aspirin; and partial resistance was defined as trace aggregation induced by 1 mM AA but no aggregation induced by <1 mM AA. Complete and partial AR was observed in 35 (4.1%) and 75 (8.8%) patients, respectively; thus, the total prevalence of AR was 13.1% [1]. We analyzed factors affecting AR on data available from these patients. Inadequate dose of aspirin (<80 mg), multiple vascular risk factors (hypertension, diabetes mellitus, and hypercholesterolemia), enhanced alternative pathway of platelet aggregation (every 10% increase in platelet aggregation induced by adenosine diphosphate and platelet-activating factor) were associated with biological AR [1].

We are investigating associations of gene polymorphisms with AR in 1000 Japanese patients with atherothrombosis as a nationwide multicenter cooperative study (Profile and Genetic Factors of Aspirin Resistance; ProGEAR) supported by a grant from the Japanese Ministry of Health, Labor, and Welfare (Principle Investigator: Toshiyuki Miyata, National Cardiovascular Center, Osaka). The subjects are patients with noncardioembolic stroke, TIA, or acute coronary syndrome who are given aspirin for secondary prevention. Platelet function tests include platelet aggregation induced by AA and collagen as well as serum thromboxane B_2, and urinary 11-dehydro-thromboxane B_2. Target genes are single nucleotide polymorphisms (SNPs) for GPIIb/IIIa, GPIV, TP (thromboxane A_2 receptor), COX-1, COX-2, and TXA_2 synthase. SNPs at 50 sites of the whole microsome area will be also investigated by DNA microarray.

The patients are followed up for 2 years to observe vascular events including stroke, TIA, MI, and vascular death.

Strategies against AR may include (1) increasing the aspirin dose; (2) alternative use of other antiplatelet agents; (3) combination of aspirin with other antiplatelet agents; (4) tailor-made medicine based on the results in genetic studies such as the ProGEAR study; and (5) investigation of molecular targets to develop novel antiplatelet agents [1].

Thienopyridines

According to the ATT's meta-analysis for direct comparison of aspirin with other antiplatelet agents, risk reductions of vascular events by ticlopidine and clopidogrel were 12% and 10% higher, respectively, than that by aspirin, although these percentages were not statistically significant [5]. However, when these data were combined, risk reduction of vascular events by thienopyridines became significantly higher than that by aspirin [12]. Thienopyridines such as ticlopidine and clopidogrel inhibit platelet aggregation by inhibiting the binding of adenosine diphosphate (ADP) to its receptor $P2Y_{12}$ [1]. Clopidogrel was proven to produce a significantly greater risk reduction of vascular events than aspirin in patients with atherothrombosis (ischemic stroke, MI, PAD) [12]. We have previously reported that SIPA was inhibited by ticlopidine but not by aspirin in patients with ischemic stroke or TIA (Fig. 6) [7]. This difference in the inhibition of SIPA between thienopyridine and aspirin may explain, at least in part, the difference in risk reduction of vascular events between the drugs.

It is well known that clopidogrel has less serious adverse effects, such as neutropenia, severe liver dysfunction, and thrombotic thrombocytopenic purpura (TTP), than ticlopidine [13]. However, there have been no clinical trials for direct comparison of clopidogrel versus ticlopidine in Western countries. Thus, we recently conducted a combined analysis of two Phase III multicenter, double-blind, placebo-controlled

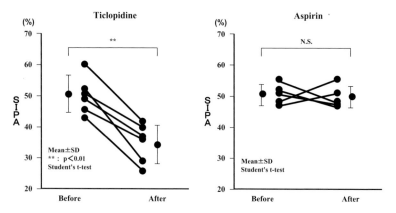

FIG. 6. Ex vivo effects of antiplatelet agents on SIRA in patients with ischemic stroke and TIA

trials to compare clopidogrel versus ticlopidine in 1921 Japanese patients with cerebral infarction. The safety endpoint was a composite of hematological changes, hepatic dysfunction, hemorrhagic complications, and other adverse drug reactions considered to be related to the drug. Safety events were fewer in patients on clopidogrel than in those on ticlopidine [hazard ratio 0.610, 95% confidence interval (CI) 0.529–0.703], whereas the vascular events (ischemic stroke, MI, vascular death) were similar in patients on clopidogrel and in those on ticlopidine (hazard ratio 0.918, 95% CI 0.518–1.626) [14].

Phosphodiesterase Inhibitors

The Cilostazol Stroke Prevention Study (CSPS), a randomized placebo-controlled double-blind trial, showed that cilostazol significantly reduced the risk of recurrent stroke in more than 1000 patients with cerebral infarction proven by brain magnetic resonance imaging (MRI) [15]. Cilostazol is a PDE inhibitor, which is the same as dipyridamole [16]. However, cilostazol inhibits PDE type 3, which is specific for cyclic adenosine monophosphate (cAMP), whereas dipyridamole inhibits mainly PDE type 5, which is specific for cyclic guanosine monophosphate (cGMP) [17]. Cilostazol as well as dipyridamole is known to have not only an antiplatelet effect [18, 19] but also a vasodilating effect [20, 21], an essential effect of PDE inhibitors. In addition, recent studies have reported that cilostazol has many pleiotropic effects, such as antiatherogenic effects [22, 23], protective effects on endothelial functions [24, 25], and antiinflammatory effects [26, 27].

We have reported that cilostazol, but not aspirin, inhibits enhancement of shear-induced platelet activation by remnant-like lipoprotein particles [28], which is an independent risk factor for symptomatic cerebral infarction as reported by us [29] and is an important biomarker for metabolic syndrome. We have also reported that cilostazol improves hemorrheology impaired by ADP, platelet-activating factor (PAF), and formyl-methionyl phenylalanine (FMLP) in a microcirculation model [30]. ADP activates only platelets, whereas PAF activates both platelets and leukocytes, and FMLP activates only leukocytes. Therefore, our results indicate that cilostazol can alleviate microcirculation derangement by inhibiting activation of not only platelets but also leukocytes.

We performed a meta-analysis on the data collected from 13 double-blind, placebo-controlled trials of cilostazol, which included nine PAD, two CVD, and one CAD trials—in total, 6194 patients with atherothrombosis [31]. The total vascular events, including cerebrovascular and cardiovascular events, were fewer in the cilostazol group than in the placebo group (risk ratio 0.84, 95% CI 0.73–0.96). In particular, cerebrovascular events were drastically fewer in the cilostazol group than in the placebo group (risk ratio 0.60, 95% CI 0.45–0.80), although cardiovascular events were similar in the two groups (risk ratio 0.95, 95% CI 0.81–1.12). Interestingly, serious bleeding complications were not different in the two groups (risk ratio 1.00, 95% CI 0.66–0.52).

Combination of Aspirin and Thienopyridine

We have previously reported that the combination of aspirin and ticlopidine can be a more potent antiplatelet strategy than either drug alone as the combination can block a multipathway of platelet aggregation [32]. We have also showed, using flow cytometry, that platelet fibrinogen binding is not inhibited in patients taking aspirin or ticlopidine alone but is inhibited in those taking aspirin plus ticlopidine; we also showed that P-selectin expression is not inhibited in patients taking aspirin alone, is significantly inhibited in those taking ticlopidine alone, and again is markedly inhibited in patients taking both aspirin and ticlopidine (Fig. 7) [9]. These results indicate that combination therapy with aspirin and ticlopidine is a potent antiplatelet strategy, at least more than monotherapy with aspirin.

Management of Atherothrombosis with Clopidogrel in High-Risk Patients (MATCH) study was conducted to compare the efficacy and safety of clopidogrel plus aspirin versus clopidogrel alone in 7600 patients with minor stroke or TIA with one or more risk factors [33]. Although there was no significant difference in ischemic events, including stroke, between the two groups, primary intracranial hemorrhage was significantly (two times) higher in the group of patients treated with clopidogrel plus aspirin than in the group of patients treated with clopidogrel alone.

Another large clinical trial (Clopidogrel for High Atherothrombotic Rsik and Ischemic Stabilization, Management, and Avoidance, or CHARISMA) also showed that the rate of vascular events, including stroke, MI, and vascular death, was not different between patients on aspirin plus clopidogrel and those on aspirin alone, although hemorrhagic complications were more frequent in patients on aspirin plus clopidogrel than in those on aspirin alone for patients with atherothrombosis or multiple vascular risk factors [34].

We have previously reported that the bleeding time, which is a marker for normal hemostasis, is significantly prolonged by aspirin, more prolonged by ticlopidine, and most prolonged by aspirin plus ticlopidine in patients with stroke and TIA [32]. Our results suggest that impairment of normal hemostasis leads to the increase of intracranial hemorrhage in patients allocated to the group of clopidogrel plus aspirin because clopidogrel may affect normal hemostasis similar to ticlopidine.

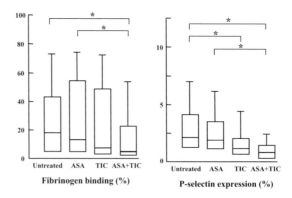

FIG. 7. Effects of aspirin alone (300 mg/day), ticlopidine alone (200 mg/day), and aspirin (81 mg/day) plus ticlopidine (100 mg/day) on the bleeding time in patients with ischemic stoke or transient ischemic attack

Combination of Aspirin and Phosphodiesterase Inhibitor

The European Stroke Prevention Study 2 (ESPS-2) showed that the combination of aspirin and dipyridamole has greater risk reduction of subsequent stroke than either drug alone [35]. Our in vitro experiment demonstrated that aspirin enhances the inhibitory effect of dipyridamole on SIPA despite the fact that aspirin alone never inhibits SIPA, which supports the results of the ESPS-2 (Fig. 8) [36].

According to the European/Australasian Stroke Prevention in Reversible Ischemia Trial (ESPRIT), the composite endpoint of ischemic events (stroke, MI, vascular death) and hemorrhagic events (serious bleeding) was significantly lower in patients on aspirin plus dipyridamole than in those on aspirin alone among 2739 patients with minor stroke of atherothrombotic origin (hazard ratio 0.80, 95% CI 0.66–0.98) [37]. Interestingly, in this trial, hemorrhagic events showed a strong trend toward reduction in patients on aspirin plus dipyridamole in comparison with those on aspirin alone (hazard ratio 0.67, 95% CI 0.44–1.03).

It is known that intracranial arterial stenosis (IAS) is more common in Asian than Caucasian populations [38], and such patients are at high risk of recurrence even with medical treatment [39]. According to the Warfarin-Aspirin Symptomatic Intracranial Disease (WASID) trial, warfarin (INR2-3) treatment is concerned with increases in deaths and intracranial hemorrhage, whereas aspirin treatment is not enough to prevent ischemic stroke [40]. Dual therapy with aspirin and another antiplatelet agent might be a new strategy for symptomatic IAS. Such dual therapy, however, has to reduce more the risk of ischemic stroke without an increased risk of hemorrhagic stroke.

A randomized control trial, or RCT (Treatment on Symptomatic Intracranial Stenosis, or TOSS), has been conducted to investigate the efficacy of cilostazol in patients with symptomatic intracranial arterial stenosis (IAS). In TOSS, 160 IAS patients with acute ischemic stroke were given aspirin 100 mg or placebo and followed up for 6

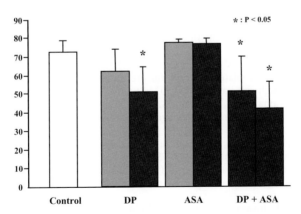

FIG. 8. In vitro effects of dipyridamole (*DP*) and/or aspirin (*ASA*) on SIPA in whole blood. Dipyridamole did not inhibit SIPA at 5 μM but significantly inhibited it at 20 μM. Aspirin did not inhibit SIPA at 5 or 20 μM. A combination of 20 μM dipyridamole and 20 μM aspirin inhibited SIPA more than 20 μM dipyridamole alone

months to study the progression of IAS using magnetic resonance angiography (MRA) and transcranial Doppler (TCD) imaging. The results showed that progression of IAS was significantly less frequent in patients treated with aspirin plus cilostazol than in those treated with aspirin plus placebo on both MRA and TCD [41].

We also started another RCT, named Cilostazol Aspirin Therapy in Recurrent Stroke in Patients with Intracranial Arterial Stenosis, or CATHARSIS to compare long-term efficacy and safety of aspirin plus cilostazol versus aspirin alone in 200 Japanese patients with symptomatic IAS (Principle Investigator: Shinichiro Uchiyama, Clinical Trials Gov. Identifier NCT0033164).

It has been reported that cilostazol does not produce further prolongation of bleeding time when combined with aspirin and/or clopidogrel in PAD patients [42]. This might be a great advantage of cilostazol in dual or triple antiplatelet therapy to avoid an increased bleeding risk. Recent evidence suggested that PDE inhibitors such as cilostazol [25–27] and dipyridamole [43–45] have not only antiplatelet effects but also endothelial protective effects, which may be related to less bleeding risk.

Acknowledgments. The studies described in this chapter have been supported in part by research grants for cardiovascular diseases from the Japanese Ministry of Health, Labor and Welfare (62-A-2, 9-A-8, and 12-A-2) and the Japanese Ministry of Education, Science, and Culture (0370425 and 11670645).

References

1. Uchiyama S, Nakamura T, Yamazaki M, et al (2006) New modalities and aspects of antiplatelet therapy for stroke prevention. Cerebrovasc Dis 21(suppl 1):7–16
2. Uchiyama S, Takeuchi M, Osawa M, et al (1983) Platelet function tests in thrombotic cerebrovascular disorders. Stroke 14:511–517
3. Uchiyama S, Yamazaki M, Hara Y, et al (1997) Alterations of platelet, coagulation, and fibrinolysis markers in patients with acute ischemic stroke. Semin Thromb Hemost 23:535–541
4. Uchiyama S, Nakamura T, Yamazaki M, et al (2003) Platelet function and antiplatelet therapy in patients with ischemic stroke. In: Satoh K, Suzuki S, Matsunaga M (eds) Advances in brain research. International Congress Series 1251. Elsevier Science, Amsterdam
5. Antithrombotic Trialists' Collaboration (2002) Collaborative meta-analysis of randomised trials of antiplatelet therapy for prevention of death, myocardial infarction, and stroke in high risk patients. BMJ 324:71–96
6. Uchiyama S, Yamazaki M, Maruyama S, et al (1993) Shear-induced platelet aggregation and its inhibition by antiplatelet agents in cerebral ischemia. Clin Hemorrheol 13:623–636
7. Uchiyama S, Yamazaki M, Maruyama S, et al (1994) Shear-induced platelet aggregation in cerebral ischemia. Stroke 25:1547–1551
8. Ikeda Y, Handa M, Kawano K, et al (1991) The role of von Willebrand factor and fibrinogen in platelet aggregation under varying shear stress. J Clin Invest 87:1234–1240
9. Yamazaki M, Uchiyama S, Iwata M, et al (2001) Measurement of platelet fibrinogen binding and P-selectin expression by flow cytometry in patients with cerebral infarction. Thromb Res 104:197–205
10. Nakamura T, Uchiyama S, Yamazaki M, et al (2002) Flow cytometric analysis of reticulated platelets in patients with ischemic stroke. Thromb Res 106:171–177

11. Hankey GJ, Eikelboom JW (2004) Aspirin resistance: may be a cause of recurrent ischaemic vascular events in patients taking aspirin. BMJ 328:477–479
12. Hankey GJ, Sudlow CLM, Danbabin DW (2000) Thienopyridines or aspirin to prevent stroke and other serious vascular events in patients at risk of vascular disease? A systematic review of the evidence from randomized trials. Stroke 31:1779–1784
13. CAPRIE Steering Committee (1996) A randomized, blinded, trial of clopidogrel versus aspirin in patients at risk of ischaemic events (CAPRIE). Lancet 348:1329–1339
14. Uchiyama S, Yamaguchi T, Fukuuchi Y (2007) The safety and efficacy of clopidogrel versus ticlopidine in Japanese stroke patients: combined results of two phase III multicentre randomised clinical trials. Cerebrovasc Dis 23(suppl 2):143
15. Gotoh F, Tohgi H, Hirai S, et al (2000) Cilostazol Stroke Prevention Study: a placebo-controlled double-blind trial for secondary prevention of cerebral infarction. J Stroke Cerebrovasc Dis 9:147–157
16. Sorkin EM, Markham A (1999) Cilostazol. Drugs Aging 14:63–73
17. Umekawa H, Tanaka T, Kimura Y, et al (1984) Purification of cyclic adenosine monophosphate phosphodiesterase from human platelets using new-inhibitor Sepharose chromatography. Biochem Pharmacol 33:3339–3344
18. Inoue T, Sohma R, Morooka S (1999) Cilostazol inhibits the expression of activation-dependent membrane surface glycoprotein on the surface of platelets stimulated in vitro. Thromb Res 93:137–143
19. Nomura S, Shouzu A, Omoto S, et al (1998) Effect of cilostazol on soluble adhesion molecules and platelet-derived microparticles in patients with diabetes. Thromb Haemost 80:388–392
20. Mochizuki Y, Oishi M, Mizutani T (2001) Effects of cilostazol on cerebral blood flow, P300, and serum lipid levels in the chronic stage of cerebral infarction. J Stroke Cerebrovasc Dis 10:63–69
21. Nakamura T, Houchi H, Minami A, et al (2001) Endothelial-dependent relaxation by cilostazol, a phosphodiesterase III inhibitor, on rat thoracic aorta. Life Sci 69:1709–1715
22. Takahashi S, Oida K, Fujiwara R, et al (1992) Effect of cilostazol, a cyclic AMP phosphodiesterase inhibitor, on the proliferation of rat aortic smooth muscle cells in culture. J Cardiovasc Pharmacol 20:900–906
23. Mizutani M, Okuda Y, Yamashita K (1996) Effect of cilostazol on the production of platelet-derived growth factor in cultured human vascular endothelial cells. Biochem Mol Med 57:156–158
24. Aoki M, Morishita R, Hayashi S, et al (2001) Inhibition of neointimal formation after balloon injury by cilostazol, accompanied by improvement of endothelial dysfunction and induction of hepatocyte growth factor in rat diabetes model. Diabetologia 44:1034–1042
25. Kim KY, Shin HK, Choi JM, et al (2002) Inhibition of lipopolysaccharide-induced apoptosis by cilostazol in human umbilical vein endothelial cells. J Pharmacol Exp Ther 300:709–715
26. Nishio Y, Kashiwagi A, Takahara N, et al (1997) Cilostazol, a cAMP phosphodiesterase inhibitor, attenuates the production of monocyte chemoattractant protein-1 in response to tumor necrosis factor-α in vascular endothelial cells. Horm Metab Res 29:491–495
27. Otsuki M, Saito H, Xu X, et al (2001) Cilostazol repress vascular cell adhesion molecule-1 gene transcription via inhibiting NF-κB binding to its recognition sequence. Atherosclerosis 158:121–128
28. Yamazaki M, Uchiyama S, Xiong Y, et al (2005) Effects of remnant-like particle on shear-induced platelet activation and its inhibition by antiplatelet agents. Thromb Res 115:211–218
29. Uchiyama S, Yamazaki M, Iwata M (2004) Remnant lipoprotein and lipoprotein (a) as risk factors for stroke and carotid disease. In: Matsuzawa Y, Kita T, Nagai R et al (eds)

Atherosclerosis XIII. International Congress Series 1262. Elsevier, Amsterdam, pp 478–481

30. Kimura Y, Uchiyama S, Iwata M (2004) Effects of antiplatelet agents on hemorheology in a microcirculation model. In: Abstracts, 5th World Stroke Congress, Vancouver

31. Uchiyama S, Goto S, Shinohara Y, et al (2007) Stroke prevention by cilostazol in patients with cerebrovascular disease, peripheral artery disease, and coronary stenting: a meta-analysis of clinical trials. Cerebrovasc Dis 23(suppl 2):12

32. Uchiyama S, Nagayama T, Sone R, et al (1989) Combination therapy with low-dose aspirin and ticlopidine in cerebral ischemia. Stroke 20:1643–1647

33. Diener HC, Bogouslavsky J, Brass LM, et al (2004) Aspirin and clopidogrel compared with clopidogrel alone after recent ischaemic stroke or transient ischaemic attack in high-risk patients (MATCH): randomised, double-blind, placebo-controlled trial. Lancet 364:331–337

34. Bhatt DL, Fox KAA, Hacke W, et al (2006) Clopidogrel and aspirin versus aspirin alone for the prevention of atherothrombotic events. N Engl J Med 354:1706–1717

35. Diener HC, Cunha L, Forbes C, et al (1996) European Stroke Prevention Study 2: dipyridamole and acetylsalicylic acid in the secondary prevention of stroke. J Neurol Sci 143:1–13

36. Nakamura T, Uchiyama S, Yamazaki M, et al (2002) Effects of dipyridamole and aspirin on shear-induced platelet aggregation in whole blood and platelet-rich plasma. Cerebrovasc Dis 14:234–238

37. ESPRIT Study Group (2006) Aspirin plus dipyridamole versus aspirin alone after cerebral ischaemia of arterial origin (ESPRIT): randomised controlled trial. Lancet 367:1665–1673

38. Sacco RL, Kargman DE, Gu Q, et al (1995) Race-ethnicity and determinants of intracranial atherosclerotic cerebral infarction: the Northern Manhattan Stroke Study. Stroke 26:14–20

39. Chimowitz MI, Kokkinos J, Strong J, et al (1995) The Warfarin-Aspirin Symptomatic Intracranial Disease Study. Neurology 45:1488–1493

40. Chimowitz MI, Lynn MJ, Howlett-Smith H, et al (2005) Comparison of warfarin and aspirin for symptomatic intracranial arterial stenosis. N Engl J Med 1305–1316

41. Kwon SU, Cho Y-J, Koo JS, et al (2005) Cilostazol prevents the progression of the symptomatic intracranial arterial stenosis: the multicenter double-blind placebo-controlled trial of cilostazol in symptomatic intracranial arterial stenosis. Stroke 36:782–786

42. Whilhite DB, Comerota AJ, Schmieder FA, et al (2003) Managing PAD with multiple platelet inhibitors: the effect of combination therapy on bleeding time. J Vasc Surg 38:710–713

43. Eisert WG, Muller TH (1990) Dipyridamole: evaluation of an established antithrombotic drug in view of modern concepts of blood vessel-wall interactions. Thromb Res 12(suppl):65–72

44. Eisert WG (2001) Near-field amplification of antithrombotic effects of dipyridamole through vessel wall cells. Neurology 57(suppl 2):20–23

45. Eisert WG (2007) Dipyridamole. In: Michelson AD (ed) Platelets, 2nd edn. Elsevier Science, Amsterdam

Clinical Role of Recombinant Factor VIIa in Bleeding Disorders

Harold R. Roberts, Dougald M. Monroe, and Nigel S. Key

Summary. Recombinant VIIa is approved in the United States and Europe for the treatment of bleeding episodes in patients with hemophilia A and B who have developed inhibitory antibodies to factors VIII and IX, respectively. Factor VIIa is also approved for the treatment of bleeding in patients with acquired hemophilia and factor VII deficiency. In Europe, factor VIIa is also approved for the control of bleeding in patients with Glanzmann's thrombasthenia. In addition to approved uses, factor VIIa is also used "off-label" to control bleeding in an astonishingly wide variety of acquired hemorrhagic conditions such as bleeding following trauma, intracranial bleeding, bleeding following surgery, pulmonary hemorrhage, and a host of other bleeding conditions that do not respond to the usual hemostatic agents. Although thromboembolic side effects have been reported, they are unusual and usually occur in patients with underlying conditions that predispose them to thrombosis. Thromboembolic side effects seem to be more likely to occur following prolonged use of the drug. All in all, however, factor VIIa is considered both safe and effective but should be reserved for approved uses or for "off-label" uses when conventional therapy fails to control hemorrhage.

Key words. Factor VIIa · Hemorrhage · Bleeding · Thrombotic side effects · Safety · Efficacy · Off-level use

Introduction

The development of activated factor VII (FVIIa) was stimulated by the advent of agents used to "bypass" inhibitor antibodies that arose in patients with classic hemophilia or hemophilia B. After the development of prothrombin complex concentrates (PCCs) from plasma, it was known that these products also contained small amounts of activated factors VII, IX, and X [1]. Later Shanbrom and Fekete (then at Hyland Laboratories, a division of Baxter Bioscience) noted that PCCs shortened the prothrombin and partial thromboplastin times of hemophilic plasma. Later, during the early 1970s, they noted that a PCC preparation was effective in controlling hemorrhage in a patient with acquired hemophilia [2]. They postulated that the effectiveness of PCC was due to the presence of activated factors in the PCC concentrates. Based on this hypothesis, two "activated" PCCs were developed, Autoplex by Baxter and FEIBA (factor VIII inhibitor bypassing activity) by Immuno [3, 4]. These products were prepared by controlled activation, and both contained prothrombin plus zymogen and activated forms of factors VII, IX, and X. However, they differed somewhat in their content of the various factors, and the precise concentration of the factors in the products was not known. Initially, both PCCs and activated PCCs were used to treat hemorrhagic episodes, but later the activated products were used almost exclusively. The early observations that components of PCCs were effective in "bypassing" factors VIII and IX raised the question as to which agent or agents in the PCCs were required for the so-called bypassing activity.

This question led Ulla Hedner and Walter Kisiel to manufacture factor VIIa from plasma; and in 1983 they published an article describing the effective use of plasma-derived FVIIa in controlling bleeding in two patients with classic hemophilia who had high titer anti-factor VIII inhibitors [5]. It is interesting to note that their first article was initially rejected by another journal. Following the production of FVIIa from plasma, Hedner later directed the production of a recombinant product by Novo Nordisk [6]. Since that time the recombinant product has been used extensively world-wide for an astonishing variety of hereditary and acquired bleeding disorders unresponsive to conventional replacement therapy [7].

Use of Factor VIIa in Hereditary Clotting Factor Deficiencies

Recombinant FVIIa has traditionally been used in patients with hemophilia A or B who have developed inhibitor antibodies against factor VIII or IX, respectively [7]. Extensive experience suggests that rFVIIa is highly effective in controlling hemorrhage in 80%–90% of patients. Recommended doses range from 90 to 120 µg/kg of body weight given every 2–3 h until bleeding stops [8]. For uncomplicated hemarthroses, this usually requires two to three doses of the product. Some physicians give a follow-up dose about 12–24 h after the last dose to prevent rebleeding [9]. When more than two to three doses of FVIIa are required, the interval between doses may be increased from every 2–3 h to every 4–6 h depending on the patient's response. When hemophiliacs with inhibitors require surgery, the duration of treatment requires several days of therapy, ranging from a week to 10 days or longer. An example of the regimen for the use of rFVIIa in surgery is that of Ingerslev and colleagues, who used factor VIIa every 2–3 h for the first 2 days but later increased the intervals between doses to every 3–6 h for 7–10 days [10].

Failure of rFVIIa to control hemorrhage has been reported in a small number of hemophilic patients with inhibitors [11]. In some but not all patients, failure may be attributed to inadequate dosing, and in such patients higher doses of FVIIa may be effective. Kenet and colleagues have used doses as high as 270 µg/kg in a small number of patients with good results and without serious adverse side effects [12, 13].

Anecdotal reports suggest that rFVIIa may be effective in patients with antibody inhibitors against other hereditary clotting factor deficiencies including factors XI, V, and X [14, 15]. In addition, the agent has been shown to be effective in patients with factor XI deficiency even in the absence of an inhibitor.

Factor VIIa is the treatment of choice in patients with factor VII deficiency. In these patients, lower doses of the product (i.e., 15–20 µg/kg) are effective in controlling bleeding episodes [16]. Although inhibitors against rFVIIa are not a problem in patients with clotting factor deficiencies other than factor VII, anti-factor VII antibodies have been reported after treatment of severely affected factor VII-deficient patients [17].

rFactor VIIa has also been an effective treatment for bleeding in patients with hereditary (and acquired) von Willebrand disease (VWD) complicated by antibodies to the von Willebrand factor (VWF) [18]. Most of these patient have type III VWD with virtual absence of VWF, but rFVIIa has been used in some patients with type I and IIA VWD. The doses of rFVIIa used in these patients are similar to those recommended for the treatment of hemophilic patients with inhibitors. It is interesting that rFVIIa does not appear to be effective in dog models of VWD, perhaps because canine platelets do not contain VWF, whereas VWF is present in human platelets [6].

Use of Factor VIIa in Hereditary Platelet Disorders

Patients with the Bernard-Soulier syndrome have a defect in platelet adhesion due to genetic mutations in the genes coding for the glycoprotein (GP)Ib–V–IX complex. The disease is inherited as an autosomal recessive disorder with clinical manifestations of mild thrombocytopenia characterized by very large platelets and easy bruising and bleeding from mucous membranes. The syndrome is rare and has traditionally been treated with platelet transfusions, with the result that patients may become refractory to such therapy. Therefore, a few patients have been treated with rFVIIa with good results [19]. Because the syndrome is so rare, it is unlikely that a clinical trial of rFVIIa for this syndrome will ever be possible, but anecdotal reports of the effectiveness of FVIIa for this condition makes it an attractive candidate for further use.

Patients with Glanzmann's thrombocytopathy have mutations in the genes coding for GPIIb–IIIa integrins that are necessary for normal platelet aggregation. The disease is inherited as a recessive trait with clinical manifestations of easy bruising, bleeding from mucous membranes, and menorrhagia. Bleeding episodes can be severe, especially during and after surgery. Although the platelet count is normal, the platelets fail to aggregate to form a normal platelet plug. Traditional treatment consists of platelet transfusions, but many patients develop antibodies against the missing integrin so the patients become refractory to further platelet transfusions. Poon and colleagues have established a registry of Glanzmann's thrombasthenic patients treated with rFVIIa and report that most patients respond, at least partially, to treatment with rFVIIa [20]. The registry is ongoing and can be found on the Internet [21]. It appears that doses of rFVIIa of up to 80 μg/kg give better results than lower doses [20]. Factor VIIa appears to act in platelet disorders presumably by increasing thrombin generation to the extent that more platelets are attracted to the site of bleeding and the fibrin clot structure is denser and more stable. Although failure to respond to rFVIIa by patients with this disorder has been reported, it is probably the most effective agent available when the patient is refractory to platelet transfusions. Thrombotic side effects have been reported after use of FVIIa in patients with Glanzmann's syndrome, but these cases are rare.

Patients with refractory thrombocytopenia may also respond to rFVIIa provided the platelet count is in the range of 10 000–20 000/μl [22]. Although several new agents are being developed to treat refractory thrombocytopenia, including oral and parenteral agonists for the thrombopoietin receptor, these agents are not yet available for routine clinical use.

Factor VIIa in Acquired Bleeding Disorders

Acquired inhibitors against factor VIII occur rarely in otherwise normal individuals, although many of these patients have a personal or family history of autoimmune disorders such as rheumatoid arthritis or ulcerative colitis, among others. Acquired factor VIII inhibitors also occur during pregnancy and the postpartum period, in those with underlying malignancies, and after ingestion of certain drugs. In other patients, particularly the elderly, the occurrence of inhibitors is entirely idiopathic. Bleeding in patients with acquired hemophilia can be severe and may be fatal.

Treatment of acute bleeding episodes in patients can be problematic as they do not respond to even high doses of factor rVIIa if the inhibitor titer is >5 Bethesda units/ml. The treatment of choice is use of one of the inhibitor bypassing agents. Recombinant factor VIIa has recently been approved in the United States for the treatment of hemorrhage in such patients. The recommended dose of factor VIII is 90–120 μg/kg, with a good to excellent response being seen in about 90% of patients [23]. Administration of rFVIIa is used for intercurrent bleeding episodes and does not raise the inhibitor titer. The long-term goal of treatment is to eradicate the inhibitor by administering appropriate immunosuppressive agents, such as prednisone/cytoxan, rutuximab, or other immunosuppressive agents, depending on the patient's response.

"Off-Label" Use of Factor VIIa

The use of rFVIIa for bleeding in other conditions is based on anecdotal reports and represents "off-label" use of the agent. In the following paragraphs, some of the more common off-label experience with rFVIIa is described.

Recombinant factor VIIa has been recommended for reversal of bleeding secondary to anticoagulant therapy. Bleeding following warfarin therapy has been reported to be controlled by giving one or two doses of rFVIIa [24]. It has also been used to stop bleeding following administration of heparin and low-molecular-weight heparin (LMWH) [25]. In these cases, response to the drug is more likely to occur provided that anticoagulation with heparin or heparin derivatives is not excessive. If factor Xa inhibition by LMWH is virtually complete, it is unlikely that FVIIa will be of benefit; but when even small amounts of factor X are available, FVIIa may be effective. There are limited data on how effective the agent is in staunching bleeding following the use of hirudin and its derivatives or other potent direct thrombin inhibitors such as argatroban, but there are occasional reports of successful use of recombinant (r)FVIIa in these situations [26]. In patients who are bleeding severely from the effects of these drugs, and when conventional therapy is not effective in controlling hemorrhage, a trial of rFVIIa is warranted.

The reputation of rFVIIa as a "general hemostatic agent" may have sprung from an initial report on the successful use of the product to control severe hemorrhage in an Israeli soldier undergoing major surgery for a gunshot wound to the abdomen [27]. The patient had received numerous transfusions of blood and blood products and appeared to be bleeding from "dilutional coagulopathy." Two doses of FVIIa rapidly effected hemostasis in this patient. As a result of this experience, two randomized trials of FVIIa in a relatively large number of patients with uncontrolled bleeding from blunt and penetrating trauma revealed that bleeding could be controlled in many patients; however, although there was a significant decrease in the requirement for red blood cells, there was no significant effect on mortality, the primary endpoint [28]. In these trials the number of serious adverse events was no greater than that observed among patients receiving placebo. Despite these trials, however, there has been a great deal of controversy over the use of rFVIIa after trauma. Some authors caution that thromboembolic events are much more common than initially reported. Others have urged caution in administering rFVIIa to patients with arterial injury or have warned that the use of FVIIa in elderly patients may be accompanied by a greater number of thrombotic side effects than in younger patients [29]. However, there seems to be general agreement that patients with severe, life-threatening hemorrhage not controlled by conventional therapy are candidates for rFVIIa therapy despite the potential risks involved. This is especially true on the field of combat where early treatment of life-threatening hemorrhage is essential [30]. It also appears that lower doses of rFVIIa may be effective in these patients, so doses of 35–60 μg/kg are often tried before giving higher doses of the agent.

One of the more promising clinical uses of FVIIa was to control expansion of the hematoma in patients with intracranial hemorrhage [31]. Based on previous observations that hematomas in patients with intracranial hemorrhage tend to expand, especially during the first few hours after the initial hemorrhage, it was postulated that early use of rFVIIa would be efficacious in preventing such expansion and could improve clinical outcomes. In an initial randomized trial, the data indicated that patients receiving rFVIIa had a better outcome with a significant reduction in mortality and improved neurological function. However, there was a 5% frequency of thrombotic events, most of which were not fatal [32]. A Phase III trial has now been completed, but the results have not yet been published.

Factor VIIa has been found to be useful for the control of bleeding due to liver disease but only in selected patients [33]. It has been used to treat bleeding from esophageal varices, but the response has been variable, as might be expected. If large veins are bleeding due to rupture of the vessel wall, one would not expect to stop the bleeding except by mechanical means. In general, it appears that rFVIIa is not effective in reducing mortality from bleeding in patients with liver disease, but more complete data are needed before final judgment is made on its therapeutic value [34].

Recombinant factor VIIa has also been used to control bleeding from the gastrointestinal tract; but again, responses have been variable and seem to be dependent on the type of bleeding. Mucosal bleeding from capillaries or small blood vessels tend to be more responsive than bleeding from large vascular lesions. For example, bleeding from intestinal angiodysplasia has been reported to be controlled by administration of rFVIIa [35]. The doses used are those recommended for hemophilic bleeding, but the dose response has not been carefully evaluated.

Factor VIIa has been used successfully to control pulmonary hemorrhage and has been particularly useful in patients who have failed to respond to conventional therapies [36]. Most patients have responded to one or two doses of FVIIa at the doses recommended for the treatment of hemophiliacs with inhibitors. Again, the use of rFVIIa in these conditions has not been approved by the U.S. Food and Drug Administration (FDA), and no large scale clinical trials have been possible; thus, the optimal dose needed to control bleeding in these patients is not known. Nevertheless, occasional dramatic responses occur, and given the high mortality associated with alveolar hemorrhage, use of rFVIIa in this condition seems reasonable.

The number of bleeding episodes in which FVIIa has been reported to be effective continues to expand, and anecdotal reports of its success in controlling hemorrhage complicating postpartum bleeding, hemorrhage following spinal surgery, bleeding following eye surgery or eye injury, the control of hemorrhage during surgery on hemangiomas, bleeding secondary to renal failure, and many other hemorrhagic conditions have been reported [37]. The widespread use of FVIIa for a wide variety of bleeding disorders has been described in case reports and case series too numerous to be listed here, but clinical trials to substantiate the safety and effectiveness of rFVIIa for all of the conditions that have been reported is not realistic. Nevertheless, it does appear that rFVIIa may well be indicated in patients who experience serious bleeding that is not responsive to conventional therapy.

Factors Influencing the Effectiveness of Factor VIIa

The mode of action of rFVIIa is to increase thrombin generation on the surface of activated platelets. The initial activation of platelets requires that the tissue factor pathway be intact, but subsequently the large pharmacologic doses of FVIIa result in binding of the agent to the surface of platelets such that direct activation factor X takes place independently of tissue factor [39]. Activated platelets are attracted to and localized to the site of vessel injury, so systemic activation of the clotting system does not occur. Although rFVIIa increases thrombin generation, complete normalization of thrombin generation, at least as measured by in vitro tests, does not occur. Nevertheless, increasing the dose of the factor has some effect up to a point, although even the highest doses (e.g., 270 μg/kg) do not completely normalize thrombin generation in hemophilic plasma. The reasons for failure of FVIIa to normalize thrombin generation under these circumstances is not known, but in vitro experiments suggest that increasing the prothrombin or factor X levels plays a role. If the concentrations of these factors are increased in the presence of pharmacologic doses of FVIIa, thrombin generation is significantly improved, sometimes to "normal" levels [41]. In addition, platelets from different individuals vary in their ability to generate thrombin. The reasons for this are not known, but the difference can be striking. Thus, failure of rFVIIa to control bleeding may not simply be due to dose effects but may depend on the patient's platelets and levels of certain procoagulants, such as prothrombin and factor X.

Safety of Factor VIIa

The effectiveness of rFVIIa in controlling hemorrhage in a variety of hereditary and acquired conditions is well established. The product has also been shown to be relatively safe, but the subject of the safety of rFVIIa has become controversial. The thrombotic side effects of rFVIIa have recently been highlighted by both the mass media and individual investigators to the extent that there seems to be a widespread impression that thrombosis occurs frequently following use of the agent. It has also been suggested that thrombotic events are relatively increased following use of rFVIIa when compared to FEIBA, but this suggestion has been challenged [42, 43]. Contrary to the above reports, there is evidence that when rFVIIa is administered to normal subjects no side effects are observed even though circulating levels of FVIIa are many-fold higher than the trace amounts of FVIIa that normally circulate in healthy humans [44]. This observation suggests that for FVIIa to cause thrombotic events it does so because of some underlying pathology that may or may not be clinically evident [45].

In an interesting article that described heart attacks and strokes in patients with classic hemophilia, Girolami and his colleagues found reports of 36 patients with classic hemophilia who had a myocardial infarction (MI) [46]. These cases were the result of searching the literature over a number of years, suggesting that the occurrence of MIs in hemophilic patients is rare. Of these 36 patients with an MI, 15 received either no treatment or treatment with factor VIII, suggesting that the cause of the infarct was due to underlying pathology. However, at least half of the 36 patients received a bypassing agent, either FEIBA, FVIIa, or both. Both agents are known to improve thrombin generation, so one might surmise that improving coagulation to control bleeding was also sufficient to result in coronary thrombosis perhaps due to undetected coronary atherosclerosis. The disturbing observation in this article was that some of the hemophilic patients who experienced an MI were young, so underlying risk factors for a thromboembolic event would not be suspected. This report is important because it shows that even when a congenital deficiency of a clotting factor exists, thrombotic complications can occur. Although bypassing agents may control bleeding, an agent that improves thrombin generation may also result in unwanted thrombosis.

There is no doubt that thromboembolic complications may accompany the use of rFVIIa as well as other bypassing agents, but these effects must be viewed in the perspective of the patients receiving the agent. In a recent review of the safety of rFVIIa written by knowledgeable representatives of the FDA, the records of more than 11 000 hospitalized patients who received FVIIa for various types of hemorrhage were analyzed with respect to thrombotic side effects [47]. Among these 11 000 patients, 431 adverse events were reported to the FDA. Of the 431 events, 185 thromboembolic complications were observed in 168 patients, only 12 of whom had congenital hemophilia. Thus, of the several thousand patients receiving FVIIa, roughly 1.5%–2.0% suffered a thrombotic event. Almost all patients had an underlying condition that predisposed them to thrombosis. Furthermore, not all the thrombotic events could be attributed to FVIIa. Not all events were fatal, even though thromboses included myocardial infarction, strokes, peripheral arterial and venous thromboses, and pulmonary emboli. The mortality rate among the patients with thromboses was less than 1%. Two major shortcomings of the study were noted: The analysis was retrospective in nature, and it was unlikely that all adverse events related to FVIIa were reported to the FDA. However, given the notoriety surrounding the potential thrombotic side effects of the product, one would assume that most of the adverse events would be known to the FDA or to Novo Nordisk, the manufacturer of the product. Despite flaws in the study, it seems likely that the adverse events that occur following administration of rFVIIa are low and that mortality from the agent is as low or lower than that seen with warfarin or heparin. Even if rFVIIa is considered to be generally safe, it is clear that it must be used with caution, especially in elderly patients or those with known conditions that predispose to thrombosis. When rFVIIa is used for bleeding in patients with conditions not approved by the FDA, care must be taken to ensure that the bleeding cannot be controlled by conventional therapy, that the hemorrhagic event is limb- or life-threatening, and that the risk/benefit ratio is judged to be favorable to the patient.

Conclusion

Currently, FVIIa is approved for the following: bleeding episodes in patients with hemophilia A or B who have inhibitors to factor VIII or IX, respectively; intercurrent hemorrhage in patients with acquired hemophilia; and treatment of bleeding in patients with hereditary factor VII deficiency. It is particularly useful in hemophilia B patients who have inhibitors as it does not cause an increase in inhibitor titer and has not been associated with anaphylaxis or the nephrotic syndrome, as has exposure to factor IX-containing products. In Europe, FVIIa is approved for the above conditions as well as for patients with Glanzmann's thrombasthenia. The product is generally considered to be both safe and effective for these conditions even though serious adverse events, including thromboembolic episodes, have been reported in about 1%–2% of patients. rFactor VIIa has also been shown to be highly effective in controlling bleeding in a wide variety of other conditions, although the FDA has not sanctioned its use for these disorders. Factor VIIa improves thrombin generation and can reverse hypocoagulability, so bleeding is controlled. Improvement in thrombin generation occasionally results in thrombotic phenomena that usually occur in patients with underlying conditions predisposing to thrombosis. Nevertheless, the approved, as well as off-label, uses of the product may prove life-saving in patients with life- or limb-threatening hemorrhagic episodes that do not respond to traditional therapy.

References

1. Gilchrist GS, Ekert H, Shanbrom E, et al (1969) Evaluation of a new concentrate for the treatment of factor IX deficiency. N Engl J Med 280:291–295
2. Shanbrom E, Fekete S. Hyland Laboratories, Division of Baxter Bioscience, personal communication.
3. Abildgaard CF, Penner JA, Watson-Williams EJ (1980) Anti-inhibitor coagulant complex (Autoplex) for treatment of factor VIII inhibitors in hemophilia. Blood 56:978–984
4. Thomas T, Williams H, Williams Y, et al (1977) FEIBA in haemophiliacs with factor VIII inhibitor. BMJ 1:52
5. Hedner U, Kisiel W (1983) Use of human factor VIIa in the treatment of two hemophilic patients with high titer inhibitors. J Clin Invest 21:1836–1841
6. Brinkhous KM, Hedner U, Garris JB, et al (1989) Effect of recombinant factor VIIa in dogs with hemophilia A, hemophilia B, and von Willebrand disease. Proc Nat Acad Sci USA 86:1382–1386
7. Roberts HR, Monroe DM, White GC (2004) The use of recombinant factor VIIa in the treatment of bleeding disorders. Blood 104:3858–3864
8. Roberts HR (2001) Recombinant factor VIIa and the safety of treatment. Semin Hematol 38:48–58

9. Hedner U (2007) Recombinant factor VIIa: its background, development and clinical use. Curr Opin Hematol 14:225–229
10. Ingerslev J, Friedman D, Gastineau D, et al (1996) Major surgery in hemophilic patients with inhibitors using recombinant factor VIIa. Haemostasis 26:118–123
11. Makris M, Hampton KK, Preston E (2001) Failure of recombinant factor VIIa in treatment of abdominal bleeding in acquired hemophilia. Am J Hematol 66:67–68
12. Kenet G (2006) High-dose factor VIIa therapy in hemophilic patients with inhibitors. Semin Hematol 43:5108–5110
13. Abshire T, Kenet G (2004) Recombinant factor VIIa: review of efficacy, dosing, regimens, and safety in patients with congenital and acquired factor VIII or IX inhibitors. J Thromb Haemost 2:899–909
14. Solomon O, Zivelin A, Livnat T, et al (2006) Inhibitors to factor XI in patients with severe factor XI deficiency. Semin Hematol 43:510–512
15. Boggio L, Green D (2001) Recombinant factor VIIa in the management of myeloid-associated factor X deficiency. Br J Haematol 112:1074–1075
16. Brummal ZK, Rivard GE, Poulist RL, et al (2004) Factor VIIa in factor VII deficiency. J Thromb Haemost 2:1735–1744
17. Mariani G, Testa MG, Di Pauloantonio T, et al (1999) Use of recombinant activated factor VII in congenital factor VII deficiencies. Vox Sang 77:131–136
18. Di Paola J, Nugent D, Young G (2001) Current therapy for rare factor deficiencies. Haemophilia 7:16–22
19. White GC (2006) Congenital and acquired platelet disorders: current dilemmas and treatment strategies. Semin Hematol 43:537–541
20. Poon MC, D'Orion R, Von Depka M, et al (2004) Prophylactic and therapeutic factor VIIa in patients with Glanzmann's thrombasthenia. J Thromb Haemost 2:1096–1103
21. Poon MC, Zotz R, Di Menno G, et al (2002) Glanzmann's thrombasthenia treatment: a prospective operational registry on the use of recombinant factor VIIa and other hemostatic agents. Semin Hematol 43:533–536
22. Aquilar C, Lucia J (2007) Successful control of severe post-operative bleeding in a case of refractory idiopathic thrombocytopenia purpura. Am J Hematol 82:246–247
23. Hedner U, Erhardtsen E (2003) Potential role of recombinant factor VIIa as a hemostatic agent. Clin Adv Hematol Oncol 2:112–119
24. Deveras RA, Kessler CW (2002) Reversal of warfarin-induced excessive anticoagulation with recombinant human factor VIIa. Ann Intern Med 637:884–888
25. Ingerslev J, Vanek T, Culic S (2007) Use of recombinant factor VIIa for emergency reversal of anticoagulation. J Postgrad Med 53:17–22
26. Oh JJ, Akers WS, Lewis D, et al (2006) Recombinant factor VIIa for refractory bleeding after cardiac surgery secondary to anticoagulation with the direct thrombin inhibitor lepisuden. Pharmacotherapy 26:569–577
27. Kenet G, Walden R, Eldad A, et al (1999) Treatment of traumatic bleeding with recombinant factor VIIa. Lancet 353:1879
28. Rizoli SB, Boffars KD, Riou B, et al (2006) Recombinant activated factor VII as an adjunctive therapy for bleeding control in severe trauma patients with coagulopathy: subgroup analysis from two randomized trials. Crit Care 10:R178
29. Thomas GO, Dutton RP, Hemlock B, et al (2007) Thromboembolic complications associated with factor VIIa administration. J Trauma 62:564–569
30. Martinowitz U, Michaelson M (2005) Guidelines for the use of recombinant activated factor VII (rFVIIa) in uncontrolled bleeding: a report by the Israeli Multidisciplinary rFVIIa Task Force. J Thromb Haemost 3:640–648
31. Mayer SA (2007) Recombinant activated factor VII for acute intracerebral hemorrhage. Stroke 38:763–767
32. Mayer SA, Brun NC, Begtrup K, et al (2005) Recombinant activated factor VII for acute intracerebral hemorrhage. N Engl J Med 352:777–785

33. Chino J, Paolini D, Tran A, et al (2005) Recombinant activated factor VII as an adjunct to packing for liver injury with hepatic venous disruption. Am Surg 71:595–597
34. Marti-Carvajal AJ, Salanti G, Marti-Carvajal PI (2007) Human recombinant activated factor VII for upper gastrointestinal bleeding in patients with liver diseases. Cochrane Database Syst Rev 24:CD004887
35. Meijer K, Peters FT, van der Meer J (2001) Recurrent severe bleeding from gastrointestinal angiodysplasia in a patient with von Willebrand's disease, controlled with recombinant factor VIIa. Blood Coagul Fibrinolysis 12:211–213
36. Henke D, Falk RJ, Gabriel DA (2004) Successful treatment of diffuse alveolar hemorrhage with activated factor VII. Ann Intern Med 140:493–494
37. Midathada MV, Mehta P, Waner M, et al (2004) Recombinant factor VIIa in the treatment of bleeding. Am J Clin Pathol 121:124–137
38. Key NS (2003) Recombinant FVIIa for intractable hemorrhage: more questions than answers. Transfusion 43:1649–1651
39. Monroe DM, Roberts HR (2003) Mechanism of action of high-dose factor VIIa: points of agreement and disagreement. Arterioscler Thromb Vasc Biol 23:8–9
40. Monroe DM, Hoffman M, Roberts HR (2002) Platelets and thrombin generation. Arterioscler Thromb Vasc Biol 22:1381–1389
41. Allen GA, Wolberg AS, Oliver JA, et al (2004) Impact of procoagulant concentration on rate, peak and total thrombin generation in a model system. J Thromb Haemost 2:402–413
42. Aledort LM (2004) Comparative thrombotic event incidence after infusion of recombinant factor VIIa versus factor VIII inhibitor bypass activity. J Thromb Haemost 2:1700–1708
43. Sallah S, Isaksen M, Seremetis S, et al (2005) Comparative thrombotic event incidence after infusion of recombinant factor VIIa vs. factor VIII inhibitor bypass activity—a rebuttal. J Thromb Haemost 3:820–822
44. Friederich PW, Levi M, Bauer KA, et al (2001) Ability of recombinant factor VIIa to generate thrombin during inhibition of tissue factor in human subjects. Circulation 103:2555–2559
45. Roberts HR, Monroe DM 3rd, Hoffman M (2004) Safety profile of recombinant factor VIIa. Semin Hematol 41:101–108
46. Girolami A, Ruzzon E, Fabris F, et al (2006) Myocardial infarction and other arterial occlusions in hemophilia a patients: a cardiological evaluation of all 42 cases reported in the literature. Acta Haematol 116:120–125
47. O'Connell KA, Wood JJ, Wise RP, et al (2006) Thromboembolic adverse events after use of recombinant human coagulation factor VIIa. JAMA 296:43–44

Clinical Role of Protein S Deficiency in Asian Population

Naotaka Hamasaki and Taisuke Kanaji

Summary. Thrombophilia is defined as an increased tendency to thrombosis and can be either inherited or acquired. It has been recognized that in Caucasians a polymorphism of coagulation factor V, factor V Leiden (R506Q), is a major risk factor for venous thrombosis. However, this does not appear to be the case for Asian thrombophilia. Accumulated studies from Asian and Caucasian countries indicate that the frequencies of protein S deficiency and of protein C deficiency were higher in Asian patients suffering from deep vein thrombosis than in Caucasian patients, indicating that deficiency of the activated protein C (APC) anticoagulant system is more common in Asian than in Caucasian patients. Our study indicates that the frequencies of protein S gene variants and of protein C gene variants in the Japanese patients were much higher than those of these gene variants in Caucasian patients, suggesting that these gene variants are responsible for the deficiency of the APC anticoagulant system and that the deficiency of the APC anticoagulant system is a major risk factor for venous thrombosis. Because factor V Leiden shows resistance to the APC anticoagulant system, coagulation regulation by the APC anticoagulant system profoundly fails in individuals having factor V Leiden, yielding the same phenotypes as those with reduced APC anticoagulation activity. In this chapter, we summarize studies from Asian and Caucasian countries and highlight the importance of the APC anticoagulant system in coagulation regulation in vivo.

Key words. Deep vein thrombosis · Protein S gene variant · Factor V Leiden · Japanese thrombophilia · Protein S Tokushima

Introduction

Protein S, a vitamin K-dependent protein with a molecular weight of approximately 75 000 daltons, plays an important role in the regulation of coagulation. Protein S is synthesized by hepatocytes, endothelial cells, megakaryocytes, human testis Leydig

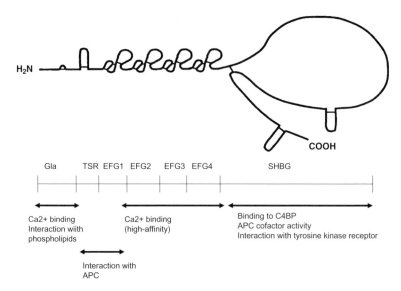

Fig. 1. Module structure of protein S protein. *Gla*, γ-carboxyglutamyl (Gla) module; *TSR*, thrombin-sensitive region; *EGF*, epidermal growth factor-like module; *SHBG*, a region homologous to sex-hormone-binding globulin; *APC*, activated protein C [2–4]

cells, and brain. Mature protein S, a single-chain protein of 635 amino acids, acts as a nonenzymic cofactor to activated protein C (APC) in the proteolytic degradation of factors Va and VIIIa. Protein S also displays an APC-independent inhibitory function in the prothrombinase complex [1]. In plasma, protein S circulates in two forms, free (approximately 30%–40%) and bound to a regulator of the classic complement pathway, C4b-binding protein (C4BP) [2]. Only free protein S has cofactor activity for APC. The main function of C4BP is to down-regulate the complement system, which is important for the inflammation response and host defense. The N-terminal of protein S consists of a Gla module in which glutamyl residues have been carboxylated to γ-carboxyglutamyl (Gla) residues in a vitamin K-dependent reaction. The Gla residues are crucial for binding Ca^{2+}. A thrombin-sensitive region (TSR) is followed by four epidermal growth factor (EGF)-like modules. TSR and the first EGF region are important for the interaction with APC [3]. The C-terminal part of protein S is homologous to sex-hormone-binding globulin (SHBG) [4]. It contains the binding site for C4BP (Fig. 1).

The protein S gene (*PROS1*) spans approximately 80 kb of DNA, is composed of 15 exons and 14 introns, and is located near the centromere of chromosome 3q11.1-11.2 [5]. This locus also contains a pseudogene, *PROS2*. *PROS2* lacks exon 1 and contains multiple frame shifts and stop codons. The two exhibit 97% identity between their exons and 95% between their introns, which complicates genotyping patients with protein S deficiency (Fig. 2).

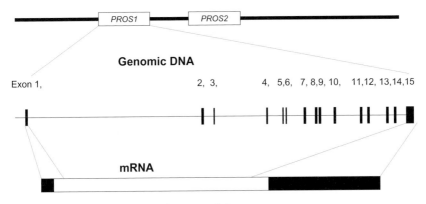

FIG. 2. Gene and mRNA structure of protein S [5]

Classification of Protein S Deficiency

Hereditary protein S deficiency is an autosomal dominant disorder that is associated with a risk of recurrent venous thrombosis. Protein S deficiency has been traditionally classified into type I (quantitative deficiency with low total protein S antigen, low free protein S antigen, and low APC cofactor activity), type II (qualitative deficiency with normal total protein S antigen, normal free protein S antigen levels, but low APC cofactor activity), and type III (normal total protein S antigen but low free protein S antigen and low APC cofactor activity). However, several reports have suggested that types I and III are two phenotypic expressions derived from the same genetic defect [6].

Analytical Considerations for Protein S Assays

Theoretically, a functional assay for protein S (APC cofactor activity) would be the best choice for diagnosing protein S deficiency, but it is not specific enough for detecting protein S deficiency. Patients with APC resistance, such as factor V Leiden, might be misdiagnosed as type II deficiency, especially in Caucasians [7]. Accordingly, laboratory diagnosis of protein S deficiency in Caucasians should be primarily based on the measurement of protein S antigen assay, because factor V Leiden is frequent in Caucasians. However, we could not detect qualitative abnormalities such as protein S Tokushima by the antigen measurement alone (discussed later). Thus, screening of protein S deficiency in Japanese should be based on the measurement of APC cofactor activity, because protein S Tokushima is frequent in Japanese. In general, the laboratory diagnosis of protein S deficiency should be based on the measurements of total protein S antigen and of APC cofactor activity. A pair of easy assay systems of total protein S antigen and of APC cofactor activity would be desired.

It is reported that free protein S levels increased with rising temperature, dilution, and length of preanalytical incubation of diluted samples [8, 9]. This effect was greater on the samples from the protein S-deficient patients than on those from healthy subjects [9]. The assays should be performed at room temperature (20°–25°C), and the diluted samples should be analyzed immediately.

Factors or Conditions that Affect Protein S Level

Various physiological and clinical conditions affect the free protein S level. Women have lower reference intervals than men. During pregnancy, the free protein S level falls during the second trimester. It is reported that protein S, protein C, and anti-thrombin levels may be transiently suppressed after acute ischemic stroke [10]. Other acquired protein S deficiencies occur with liver disease, varicella infection, extensive thrombosis, warfarin intake, disseminated intravascular coagulation (DIC), and systemic lupus erythematosus (SLE). Recently as a part of the Genetic Analysis of Idiopathic Thrombophilia (GAIT) project in Spain, a genome scan searching for quantitative trait loci (QTLs) affecting free protein S was investigated [11]. The C4BP gene region showed strong evidence of linkage with free protein S levels, suggesting that C4BP might affect the free protein S levels [12].

Two polymorphisms (Pro626 and/or nt 2698) of protein S have previously been reported to influence free protein S levels in healthy individuals in France and Italy, especially women [13–15]. However, there was no correlation to free protein S levels in Finland [16]. A racial difference in plasma protein S levels was also reported. The black African control group had significantly lower free protein S levels than either the black Caribbean group or the white Caucasian group [17].

Variants in the Protein S Gene (PROS1) in Japanese

Combined with our analysis [18], including unpublished data (updated on March 2006) and results of other studies [19–33c], a total of 51 variants (27 missense variants, 9 nonsense variants, 7 variants involving the splice sites, 1 duplication and 3 deletion mutations) were identified in a Japanese population (Table 1). These variants are distributed throughout the coding sequence except exon 1. Table 1 shows all the variants that were found in our study with their location in the PROS1, the predicted amino acid change, and levels of protein S antigen. In addition to deep vein thrombosis (DVT) and pulmonary embolism (PE), infarction, arterial thrombosis, and obstetrical complications such as intrauterine growth retardation (IUGR) and placental abruption were listed in our accumulated data of protein S deficiency. The close sequence homology between PROS1 and PROS2 suggests the possibility of gene rearrangements like vWF gene. However, no recombination between these two genes has been described. Large deletion of the PROS1 has been found in families where mutations in the PROS1 have not been deleted despite sequencing [34], suggesting that screening for large deletions in PROS1 may be useful for Protein S deficiency patients.

Eight variants (Lys9Glu [36], Glu26Ala [35, 36], Thr37Met [35], Val46Leu [37, 32], Trp342Arg, Arg474Cys [31, 38], Pro626Leu[39, 40], Lys633Glu) have been reported previously in Caucasians, and 31 variants are unique to Japanese (Table 1). Four variants (Val46Leu, Lys155Glu, Arg474Cys, Pro626Leu) were recurrent. Most of the variants were mutations which suggest quantitative deficiency (type I or type III). Two variants (Lys9Glu [36], Lys155Glu [29, 42–44]) were associated with qualitative deficiency (type II). Three patients were a homozygote of the protein S Tokushima (Lys155Glu) [18, 46] and 8 of the patients were compound heterozygote of the protein S

TABLE 1. A summary of PROS1 variants in Japanese population

Exon	Mutation	Predicted AA change	Total PS Ag (%)	Free PS Ag (%)	PS Ac (%)	Age	Sex	OAC	Thrombosis	Ref	Comments
Exon 1											
Exon 2	AAA → GAA	Lys9Glu	83	75	60	69	M		DVT	20	TypeII
	GAA → GCA	Glu26Ala + Lys155Glu	48	27	6	36	M	+	DVT,PE	18	ISTH 2007 P-T-074
	CAG → AG del	Gln10Arg . . . Stop								18	
Exon 3	ACG → ATG	Thr37Met	104	70	51	35	M	–	PE	20	PC deficiency
	GTT → CTT	Val46Leu	50	38	44	50	F		DVT	32	
	GTT → CTT	Val46Leu + Exon 15 – 1g → t; Asp583Val . . . 587Stop	20	ND	ND	28	M		DVT,PE	18	
	GTT → CTT	Val46Leu	43	24	5>	32	M	+		26	
	GTT → CTT	Val46Leu	ND	ND	46	5	M	–	PE	18,29	
Exon 4	GGG → AGG	Gly54Arg	74	25	34	37	M		DVT,PE	27	Expression study +/ normal secretion
Exon 5	TCA → TGA	Ser62stop	41	15	<5	63	M		DVT,PE,BI	26,33c	Exon 4 skipping
	TGT → TAT	Cys80Tyr	49	39	21	62	M	+	DVT	18	Expression study +/ impaired secretion
Exon 6	GAA → TAA	Glu119stop + Lys155Glu	103	36	3	34	F		DVT	18	
	AAG → GAG	Lys155Glu (Proten S Tokushima)								41	
Exon 7	GAA → CAA	Glu201Gln	98	56	49	26	F		DVT	18	Expression study +/ normal secretion
Exon 8	GAT → AAT	Asp202Asn	56.7	39.6	29.2	59	F	–	DVT	23	
	TGC → TTC	Cys206Phe	84	32	19		F		DVT	18	Expression study +/ impaired secretion
	del CTCTG	Cys206Stop + Lys155Glu	95	30	<10	11	M		DVT	26	
	GAG → TAG	Glu208Stop	60	26	ND	47	M	+	DVT,PE	26	
	CAG → TAG	Gln238Stop	29.8	7.4	16.5	22	M		DVT	23	
	TGT → TGG	Cys241Trp	42	17	6	48	M		DVT,PE	18	
	TGT → TGG	Cys241Trp + Lys155Glu	66	23	20		M	–	Double vision	18	Expression study +/ impaired secretion

TABLE 1. *Continued*

Exon	Mutation	Predicted AA change	Total PS Ag (%)	Free PS Ag (%)	PS Ac (%)	Age (year)	Sex	OAC	Thrombosis	Ref	Comments
Exon 9	CGT → TGT	Arg275Cys	63		38	34	F	+	BI	33c	Expression study +/ impaired secretion
	CGT → TGT	Arg275Cys								18	
Exon 10	GGC → AGC	Gly295Ser	67	24	<10	26	F	–	DVT,pregnancy	21	
	GGC → AGC	Gly295Ser	50	29	24	56	M	–		20	
	CTG → CCG	Leu298Pro	40	16.3	18.5				DVT	23	PC intron mutation +/ familial DVT
	GAA → TAA	Glu301Stop	31	14	21	26	M	+	DVT	18	
	CGT → CAT	Arg314His	77	49	36	31	F	–	DVT	26,33c	Expression study +/ impaired secretion
Exon 11	TGG → CGG	Trp342Arg	31	27	9	31	M		DVT,PE	24	
	CCG → CAG	Pro375Gln + Asn455Tyr(Exon12)	66	26	26	40	M	–	DVT	33c	Expression study +/ impaired secretion
	AAA → TAA	Lys392Stop	44.4	19.4	19				DVT	23	
Exon 12	CGA → TGA	Arg410Stop	35	10	11	16	M	+	DVT,PE	18	
	AGC → AGGC	Ser411Arg . . . Stop								18	ISTH 2007 P-T-074
	GCT → GAT	Ala450Asp	46	22	33	40	M		DVT	18	Expression study +/ impaired secretion
Exon 13	GAT → TAT	Asn455Tyr								33c	Expression study +/ impaired secretion
	CGT → TGT	Arg474Cys	56	6	13	47	M		DVT	31	
	CGT → TGT	Arg474Cys			22	33	M		DVT,PE	18	
	CGT → TGT	Arg474Cys		35	32	50	F		DVT,PE, UC	18	
	CGT → TGT	Arg474Cys	60			36	M	–	DVT	18	
	CGT → TGT	Arg474Cys				32	M	+	PE	18	
	CGT → TGT	Arg474Cys + Lys155Glu			5>	16	F	+		18	
Exon 14	TAT → TAA	Tyr519Stop									ISTH 2007 P-T-074
	CGG → TGG	Arg520Trp	63	38	34	44	F	+	DVT,PE	18	Expression study +/ impaired secretion
	CAG → TAG	Gln522Stop	31	14	<10	57	M		DVT,PE	30,33a	Reduced mutated mRNA, expression study

Location	Nucleotide change	Amino acid change					Sex		Clinical	Ref	Rinsho Ketsueki 44(8). OS-1-1
	ACA → GCA	Thr576Ala + Lys155Glu	49	27	10	69	M	–	DVT	18	Expression study +/ impaired secretion
	GGT → AGT	Gly579Ser	68	27	22	59	M	–	BI	18	Expression study +/ normal secretion
Exon 15	ACA → ATA	Thr589Ile	88	73	60	27	M	–	DVT,ITP	18,29	Expression study +/ ER retention
	TAT → TGT	Tyr595Cys	50	34	28	22	M	–	DVT	18,28,29	
	CCA → CTG	Pro626Leu	55	49	27	29	M		Symptom free	18	
	CCA → CTG	Pro626Leu	79	53	48	67	F	–	DVT	18	
	CCA → CTG	Pro626Leu	69	37	30	36	M	–	DVT	26	
	AAG → AA del	Lys633Glu ... Stop	37	20	12	11	M		DVT	18	
Large deletion	107 kb	at least the whole PROS1 gene			16	22	F	–	DVT	70	
Promoter	–190 c → g		101	85	52	33	F	–	abruptio placentae	18	
	–189 g → c		67	69	56	41	M	–	TTP	18	
	–189 g → c		60	39	35	32	F	–	IUGR	18	
	–168 c → t		68	31		32	M		Portal vein thrombosis	19	Expression study +/ luciferase assay
	–22 c → t		63	56	63	36	M	+	PE,AMI,BI	18	PC deficiency
Intron a	–1: g → c		67	37	<10	10	F	–	DVT	18,33b	Reduced mutated mRNA
Intron c	Exon 4 –1g → a						M				JJTH 2006 17(5) P-16; Reduced mutated mRNA
Intron f	Exon 7 –2a → t		51	19	13	25	M	+	DVT	24	Exon 7 skipping
Intron J	Exon 10 +5g → a		ND	ND	22	20	F		DVT	8	
Intron L	Exon 12 +2t → a		76	12	<10	29	M	–	DVT,PE, AO	18,22	Reduced mutated mRNA
Intron M	Exon 14 –2a → g		58	24	10	60	M		DVT	18,25	Reduced mutated mRNA
Intron N	Exon 15 –1g → t	Asp583Val ... 587Stop	50	17	28		M	–	Symptom free	32	
3' UTR	533 c → g		66	51	40	45	M	+	multiple BI	18	

M, male; F, female; PE, pulmonary embolism; DVT, deep vein thrombosis; BI, brain infraction; UC, ulcerative colitis; AMI, acute myocardial infarction; TTP, thrombotic thrombocytopenic purpura; AO, aortic obustruction; ITP, idiopathic thrombocytopenic purpura; OAC, oral anticoagulant therapy

S variants, including 6 Tokushima and other variants. Protein S Tokushima (Lys-155Glu) variant was originally identified in Japanese patients suffering from DVT, and the variant was found in 1.65% (3/182) of healthy Japanese volunteers [41]. We discuss this variant later in the chapter. Two patients carried variants in both the *PROS1* and *PROC* genes.

Expression studies were performed on several variants. Most of the variants (Arg474Cys [31], Arg275Cys [33c], Arg314His [33c], Pro375Glu + Asn455Tyr [33c], Gln522Stop [33a], Tyr595Cys [29], Cys206Phe, Cys241Trp, Ala450Asp, Arg520Trp, Gly579Ser (unpublished observations)) showed inefficient secretion. However, three variants (Gly54Arg [29], Glu201Gln (unpublished observations), Thr589Ile [29]) were efficiently secreted with an APC cofactor activity compatible with that of wild-type protein S and were inhibited by C4BP with a dose dependence similar to that of wild-type protein S. These variants were candidates for a neutral change, although it is possible that they are in linkage disequilibrium with a causative variant, such as a large deletion.

Protein S Tokushima (Lys155Glu) Variant as a Genetic Risk Factor for Deep Vein Thrombosis in Japanese Patients

Protein S Tokushima (Lys155Glu) variant was originally identified in 1993 in Japanese patients with DVT in two independent families [41, 42]. Yamazaki et al. originally reported that 1.65% (3/182) of the normal population was heterozygous for this variant [41]. In vitro studies showed that protein S Tokushima (Lys155Glu) has diminished capability to act as an APC cofactor [29, 42]. We observed four heterozygous and one homozygous carrier for protein S Tokushima (Lys155Glu) among 39 patients with reduced protein S activity in 85 Japanese deep vein thrombosis (DVT) patients [18]. The odds ratio (OR) for DVT associated with protein S Tokushima was 3.74 [95% confidence interval (CI) 1.06–13.2] [18], suggesting that protein S Tokushima is a risk factor for DVT. This result was confirmed by another group [46]; in that case, the OR was 5.58 (95% CI 3.11–10.01) (Table 2). Thus, we may conclude that protein S Tokushima is an established genetic risk factor for DVT in Japanese [18, 47]. We have analyzed the *PROS1* gene in 258 of 1604 cases. Thrombophilia testing was performed on these individuals from 1994 to March 2006, and 52 variants of protein S (20%) were found. Among them, 26 variants were protein S Tokushima (unpublished observation).

TABLE 2. Summary of studies on protein S Tokushima (K155E) and deep vein thrombosis

Study (year)	No.of cases/ controls	PS Tokushima: odds ratio (95% confidence interval)	Frequency of PS Tokushima (%)	Adjusted factors
Kinoshita et al. [18] (2005)	85/304	3.74 (1.06–13.23)	1.64	None
Kimura et al. [46] (2006)	161/3651	5.58 (3.11–10.01)	1.81	None

We measured total protein S, free protein S, and APC cofactor activity in 30 patients and their families with protein S Tokushima. Total and free protein S levels were usually within the normal range, but APC cofactor activity was reduced compared to total or free protein S, indicating type II deficiency. In some patients, however, APC cofactor activity was not reduced (e.g., 74% and 69%), which indicates that screening with the APC cofactor activity alone may not be a good tool for detecting protein S Tokushima. Thus, genetic typing is necessary to detect protein S Tokushima. Type II protein S deficiency was detected more frequently in Asian countries than Caucasian countries. Haplotype analysis will enable us to pinpoint when and where it occurred and how it spread to the Asians.

Caucasians, especially in Dutch and Swedish populations, with combined protein S deficiency and APC resistance (factor V Leiden) were thought to have higher risk for thrombosis than those carrying either of the two defects [48, 49]. Compound heterozygotes of protein S Tokushima and other variants may have a higher risk for thrombosis than heterozygotes of protein S deficiency.

Prevalence of Protein S Deficiency in the General Population

Protein S deficiency was a recognized risk factor for venous thrombosis, but assessment of the risk of thrombosis associated with protein S deficiency was difficult owing to the lack of sufficiently large studies examining the prevalence of deficiency in the general population. Table 3 shows the prevalence of protein S deficiency in the general populations of Caucasians (one large study in Scotland) and Asians (two studies in Japanese, one study in Thailand). Healthy blood donors (3788 individuals) in Scotland were studied [50, 51]. Eight (0.21%) had persistently low levels of protein S over a 4- to 7-year follow-up period. *PROS1* gene analysis identified at least one defect (intron K, exon 12:-3 g→t) in six donors, and five were heterozygous for the Heerlen polymorphism predicting a Ser460Pro substitution. The authors estimated the prevalence of heritable protein S deficiency in the Scottish population at 0.16%–0.21%. Protein S Heerlen existed in 0.8% of the populations studied, and this variant may be a potential risk factor for venous thrombosis in Caucasians [52]. However, there was no report that this variant is associated with an increased risk of venous thrombosis. In any case, the prevalence of protein S deficiency is speculated to be low (0.16%–0.21%) in healthy Caucasians, although type II deficiency could not be detected in this study because APC activity was not measured in either study.

Two studies were reported from Japan [53, 54]. Protein S deficiency was suspected when protein S activity was more than 2 SD below the mean for each sex. They excluded pseudo-protein S deficiency by checking the ratio of protein S and other coagulation factors such as antithrombin (AT), factor X, and factor II. The frequencies of protein S deficiency in two studies were 1.12% ($n = 2690$) [53] and 2.04% ($n = 392$), respectively [54]. Protein S deficiency was found in 3.7% of healthy Thais [55]. In contrast, the frequencies of protein C deficiency in healthy Asian populations (0.27%, 0.50%) were equivalent to that of Caucasians, and no antithrombin deficiency was detectable in the Asian population (Table 3). Thus, the frequency of protein S

TABLE 3. Frequency of protein S deficiency in the general population and in patients with DVT

Population	Definition	PS def. (%)	Type I, III (%)	Type II (%)	PC def. (%)	AT def. (%)	PS + PC def. (%)	No.	Ref.
Healthy population									
Scotland	tPS,fPS/gene analysis	0.16–0.21						3788	50, 51
Japanese	tPS,PS,PS/FII, PS/FX activity ratio	2.04	1.02	1.02	0.50	0.00		392	54
Japanese	ATIII/PS activity ratio	1.12						2690	53
Thai	tPS,fPS	3.70	1.99	1.71	0.27	0.00	0.00	352	55
Deep vein thrombosis (unselected)									
Caucasian	tPS,fPS,PS activity	2.3			3.70	1.90		2008	63
Japanese	tPS,fPS,PS/FII, PS/FX activity ratio	18.6	8.9	9.7	7.96	1.77	0.88	113	64
Japanese	tPS,fPS,PS activity,gene analysis	22.4			9.41	2.35	1.18	85	18
Hong Kong	tPS,fPS,PS activity	21.2			17.31	9.62	0.91	52	65
Taiwan	tPS,fPS,PS activity	26.7	20.8	5.9	17.24	5.17	2.35	116	66, 67

PS activity: APC cofactor activity

deficiency in the Asian population was 5–10 times higher than that in the Caucasian population. The definition of protein S deficiency was different in these four studies. International standardization is essential for proper comparison.

Protein S Deficiency: Recognized Risk Factor for Venous Thrombosis

In 1984, members of several kindreds who exhibited reduced levels of protein S were described who had a striking history of recurrent venous thrombosis disease [56, 57]. Subsequently, many additional families with this disorder have been reported. Although one large prospective case–control study failed to find an association between protein S deficiency and venous thrombosis [58], a 2.4-fold increased risk of thrombosis has been reported among protein S-deficient patients [59]; furthermore, family-based studies have reported a 5.0- to 8.5-fold higher risk of thrombosis among affected relatives of protein S-deficient patients compared with unaffected relatives [60, 61].

In addition to DVT and pulmonary embolism, cerebral vein thrombosis, mesenteric and portal venous thrombosis, subclavian vein thrombosis, and purpura fulminans have been reported in association with congenital protein S deficiency. It has also been reported that congenital protein S deficiency is responsible to obstetrical complications such as preeclampsia/eclampsia, recurrent fetal loss, and intrauterine fetal restriction [62]. Arterial thrombosis has been associated more often with protein S deficiency than with protein C deficiency.

Table 3 shows the frequency of protein S deficiency in unselected patients with venous thrombosis in Caucasians (cumulative analysis of 2008 cases) and Asians (two studies in Japanese, one study in Hong Kong, one study in Taiwan) [18, 63–67]. The frequency of protein S deficiency in Caucasians was 2.3% for unselected patients with venous thrombosis, a frequency roughly similar to that of protein C deficiency (3.7%) [63]. Two independent analyses were reported in a Japanese population. A total of 113 consecutive Japanese patients with DVT were studied, and 21 of the 113 (18.6%) patients were identified as protein S deficient. Among them, one patient with combined protein C deficiency (type I) and protein S deficiency (type II) was found (0.88%) [64]. Half of the patients (11/20) had type II deficiency (9.7%) (Table 3). Separately, we studied 85 unselected patients who were confirmed to have DVT [18]. Gene analysis was performed on factors related to the reduced activities in the affected patients. We found gene substitution in the *PROS1* gene in 19 of 39 DVT patients [18]. Five individuals had protein S Tokushima, and others had different variants (Table 1). The frequency of variants of *PROS1* was as much as 22.4% (19/85) (Table 3). The frequency of protein S abnormality in Japanese DVT patients in two studies showed similar percentages: 18.6% (21/113) [64] and 22.4% (19/85) [18].

A group of 52 unselected Chinese patients with documented venous thrombosis were studied in Hong Kong [65], 11 (21.2%) of whom were found to have protein S deficiency (Table 3). One of the eleven patients had combined protein S deficiency and protein C deficiency (0.91%) [65]. In Taiwan, 116 consecutive, unrelated patients with venous thrombosis were studied [66, 67]. In total, 31 patients had protein S deficiency (26.7%), and the frequency of type II deficiency was 5.9% (Table 3). It would be interesting to know whether protein S Tokushima can be detected in a Taiwanese population. The frequency of protein S deficiency in Asian patients was approximately 10 times higher than that of Caucasian patients (Table 3). In Japanese, more than 10% of protein S deficiency is caused by protein S Tokushima, but the prevalence of protein S deficiency is still high even after excluding protein S Tokushima. The variants of protein S are distributed throughout the coding sequence (Table 1). The frequency of protein C deficiency is also high in Asians. Some limited variants, such as protein C Arg147Trp (R147W), probably contributes to it because protein C R147W was frequently detected in Taiwan [68], Japan [18], and Thailand [69]. It would be interesting to compare racial differences of these deficient frequencies based on gene analysis relating to protein S/protein C gene variants and factor V Leiden.

The frequency of protein S deficiency was much higher in Asian DVT patients (18.6%–26.7%) than in Caucasian patients (2.3%) (Table 3). The frequency of protein C deficiency (8.0%–17.3%) in Asian patients was also higher than that of Caucasians patients (Table 3), indicating that abnormalities of APC anticoagulant proteins (proteins S and C) in patients suffering from DVT are more common in Asians than in Caucasians. These frequencies were 5–10 times higher than that of Caucasian DVT patients (Table 3). Although gene analysis was performed only in our study [18], we may conclude that gene abnormalities of protein S and protein C in Japanese DVT patients are quite high compared to those in Caucasian patients, suggesting that gene abnormalities of proteins S and C are the major risk factor for DVT in Japanese population.

Role of APC Anticoagulant System in Thrombophilia

Because factor V Leiden variant shows resistance to the protein S/protein C anticoagulant system, coagulation regulation by the protein S/protein C anticoagulation system profoundly fails in individuals having factor V Leiden variant (Fig. 3a,b), yielding the same phenotypes as those with reduced protein S/protein C anticoagulant activity in heterozygous deficient subjects (Fig. 3a,c). It is interesting that although factor V Leiden variant was not detected in Japanese DVT patients, the frequency of Japanese DVT patients having heterozygous *PROS1* or *PROC* gene variants (22% for *PROS1* and 9% for *PROC*) [18] is equivalent to the frequency of Caucasian DVT patients having factor V Leiden variant (20%–40%) [63]. In Caucasian thrombophilia coagulation regulation is disturbed most directly by factor V Leiden variant bypassing the protein S/protein C anticoagulant system (Fig. 3b), whereas in Japanese thrombophilia coagulation regulation is disturbed by weak protein S/protein C activity caused by *PROS1* and *PROC* gene mutations (Fig. 3c). Phenotypically, the relative hypofunction of the protein S/protein C (APC) anticoagulant system is an important risk factor for DVT in Japanese as well as in Caucasian individuals.

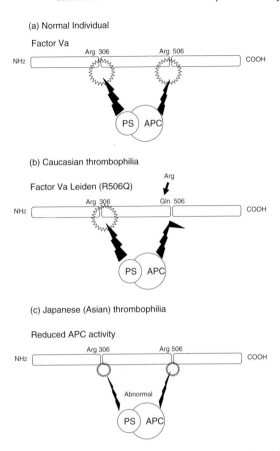

FIG. 3. **a** Coagulation regulation by protein S (*PS*)/protein C anticoagulation in normals. *APC*, activated protein C. **b** Coagulation dysregulation by factor V Leiden in Caucasian thrombophilia. **c** Coagulation dysregulation by protein S/protein C anticoagulation in Asian thrombophilia

Conclusion

We summarized the importance of the APC anticoagulant system in thrombophilia. As is already known, factor V Leiden (R506Q) is the major risk factor for Caucasian thrombophilia, but it does not appear to be the case for other races. However, accumulated studies indicate that for Asian populations the reduced activity of APC (protein S/protein C) anticoagulant system is a risk factor for Asian thrombophilia. Especially, the frequency of protein S gene variants in Japanese DVT patients is approximately 10 times higher than that of protein S gene variants in Caucasian DVT patients. Furthermore, the frequency of protein C deficiency in these patients is also much higher in Asian counties than in Caucasian countries. The frequency (20%–40%) of factor V Leiden in Caucasian DVT patients is similar to that (30%) of heterozygous gene variants of the APC anticoagulant system in Japanese DVT patients. As shown in Fig. 3, individuals with factor V Leiden and those with protein S (or protein

C) heterozygous gene abnormality suffer from relatively reduced APC activity and tend to have thrombophilia.

References

1. Dahlback B (1995) The protein C anticoagulant system: inherited defects as basis for venous thrombosis. Thromb Res 77:1–43
2. Dahlback B, Stenflo J (1981) High molecular weight complex in human plasma between vitamin K-dependent protein S and complement component C4b-binding protein. Proc Natl Acad Sci U S A 78:2512–2516
3. Suzuki K, Nishioka J, Hashimoto S (1983) Regulation of activated protein C by thrombin-modified protein S. J Biochem (Tokyo) 94:699–705
4. Hillarp A, Dahlback B (1988) Novel subunit in C4b-binding protein required for protein S binding. J Biol Chem 263:12759–12764
5. Schmidel DK, Tatro AV, Phelps LG, et al (1990) Organization of the human protein S genes. Biochemistry 29:7845–7852
6. Zoller B, Garcia de Frutos P, et al (1995) Evaluation of the relationship between protein S and C4b-binding protein isoforms in hereditary protein S deficiency demonstrating type I and type III deficiencies to be phenotypic variants of the same genetic disease. Blood 85:3524–3531
7. Deitcher SR, Kottke-Marchant K (2003) Pseudo-protein S deficiency due to activated protein C resistance. Thromb Res 112:349–353
8. Persson KE, Dahlback B, Hillarp A (2003) Diagnosing protein S deficiency: analytical considerations. Clin Lab 49:103–110
9. Tsuda T, Tsuda H, Yoshimura H, et al (2002) Dynamic equilibrium between protein S and C4b binding protein is important for accurate determination of free protein S antigen. Clin Chem Lab Med 40:563–567
10. Bushnell CD, Goldstein LB (2000) Diagnostic testing for coagulopathies in patients with ischemic stroke. Stroke 31:3067–3078
11. Almasy L, Soria JM, Souto JC, et al (2003) A quantitative trait locus influencing free plasma protein S levels on human chromosome 1q: results from the Genetic Analysis of Idiopathic Thrombophilia (GAIT) project. Arterioscler Thromb Vasc Biol 23:508–511
12. Esparza-Gordillo J, Soria JM, Buil A, et al (2003) Genetic determinants of variation in the plasma levels of the C4b-binding protein (C4BP) in Spanish families. Immunogenetics 54:862–866
13. Castaman G, Biguzzi E, Razzari C, et al (2007) Association of protein S p.Pro667Pro dimorphism with plasma protein S levels in normal individuals and patients with inherited protein S deficiency. Thromb Res 120:421–426
14. Leroy-Matheron C, Duchemin J, Levent M, et al (1999) Genetic modulation of plasma protein S levels by two frequent dimorphisms in the PROS1 gene. Thromb Haemost 82:1088–1092
15. Leroy-Matheron C, Duchemin J, Levent M, et al (2000) Influence of the nt 2148 A to G substitution (Pro 626 dimorphism) in the PROS1 gene on circulating free protein S levels in healthy volunteers—reappraisal of protein S normal ranges. Thromb Haemost 83:798–799
16. Heinikari T, Huoponen O, Partanen J, et al (2005) Protein S gene polymorphisms Pro626 and nt2698—no correlation to free protein S levels or protein S activities. Thromb Haemost 94:1340–1341
17. Jerrard-Dunne P, Evans A, McGovern R, et al (2003) Ethnic differences in markers of thrombophilia: implications for the investigation of ischemic stroke in multiethnic populations: the South London Ethnicity and Stroke Study. Stroke 34:1821–1826

18. Kinoshita S, Iida H, Inoue S, et al (2005) Protein S and protein C gene mutations in Japanese deep vein thrombosis patients. Clin Biochem 38:908–915
19. Sanda N, Fujimori Y, Kashiwagi T, et al (2007) An Sp1 binding site mutation of the PROS1 promoter in a patient with protein S deficiency. Br J Haematol 138:663–669
20. Fujimura H, Kambayashi J, Kato H, et al (1998) Three novel missense mutations in unrelated Japanese patients with type I and type II protein S deficiency and venous thrombosis. Thromb Res 89:151–160
21. Hirose M, Kimura F, Wang HQ, et al (2002). Protein S gene mutation in a young woman with type III protein S deficiency and venous thrombosis during pregnancy. J Thromb Thrombolysis 13:85–88
22. Iida H, Nakahara M, Komori K, et al (2001) Failure in the detection of aberrant mRNA from the heterozygotic splice site mutant allele for protein S in a patient with protein S deficiency. Thromb Res 102:187–196
23. Iwaki T, Mastushita T, Kobayashi T, et al (2001) DNA sequence analysis of protein S deficiency—identification of four point mutations in twelve Japanese subjects. Semin Thromb Hemost 27:155–160
24. Mizukami K, Nakabayashi T, Naitoh S, et al (2006) One novel and one recurrent mutation in the PROS1 gene cause type I protein S deficiency in patients with pulmonary embolism associated with deep vein thrombosis. Am J Hematol 81:787–797
25. Nakahara M, Iida H, Urata M, et al (2001) A novel splice acceptor site mutation of protein S gene in affected individuals with type I protein S deficiency: allelic exclusion of the mutant gene. Thromb Res 101:387–393
26. Okada H, Takagi A, Murate T, et al (2004) Identification of protein Salpha gene mutations including four novel mutations in eight unrelated patients with protein S deficiency. Br J Haematol 126:219–225
27. Okamoto Y, Yamazaki T, Katsumi A, et al (1996) A novel nonsense mutation associated with an exon skipping in a patient with hereditary protein S deficiency type I. Thromb Haemost 75:877–882
28. Tsuda H, Tokunaga F, Nagamitsu H, et al (2006) Characterization of endoplasmic reticulum-associated degradation of a protein S mutant identified in a family of quantitative protein S deficiency. Thromb Res 117:323–331
29. Tsuda H, Urata M, Tsuda T, et al (2002) Four missense mutations identified in the protein S gene of thrombosis patients with protein S deficiency: effects on secretion and anticoagulant activity of protein S. Thromb Res 105:233–239
30. Yamazaki T, Hamaguchi M, Katsumi A, et al (1995) A quantitative protein S deficiency associated with a novel nonsense mutation and markedly reduced levels of mutated mRNA. Thromb Haemost 74:590–595
31. Yamazaki T, Katsumi A, Kagami K, et al (1996) Molecular basis of a hereditary type I protein S deficiency caused by a substitution of Cys for Arg474. Blood 87:4643–4650
32. Yamazaki T, Katsumi A, Okamoto Y, et al (1997) Two distinct novel splice site mutations in a compound heterozygous patient with protein S deficiency. Thromb Haemost 77:14–20
33a. Yamazaki T, Saito H, Dahlback B (2002) Rapid intracellular degradation of a truncated mutant protein S (Q522X). Thromb Haemost 87:171–172
33b. Tatewaki H, Iida H, Nakahara M, et al (1999). A novel splice acceptor site mutation which produces multiple splicing abnormalities resulting in protein S deficiency type I. Thromb Haemost 82:65–71
33c. Okada H, Yamazaki T, Takagi A, et al (2006). In vitro characterization of missense mutations associated with quantitative protein S deficiency. J Thromb Haemost 4:2003–2009
34. Johansson AM, Hillarp A, Sall T, et al (2005) Large deletions of the PROS1 gene in a large fraction of mutation-negative patients with protein S deficiency. Thromb Haemost 94:951–957

35. Gandrille S, Borgel D, Eschwege-Gufflet V, et al (1995) Identification of 15 different candidate causal point mutations and three polymorphisms in 19 patients with protein S deficiency using a scanning method for the analysis of the protein S active gene. Blood 85:130–138

36. Simmonds RE, Ireland H, Kunz G, et al (1996) Identification of 19 protein S gene mutations in patients with phenotypic protein S deficiency and thrombosis: Protein S Study Group. Blood 88:4195–4204

37. Espinosa-Parrilla Y, Morell M, Souto JC, et al (1999) Protein S gene analysis reveals the presence of a cosegregating mutation in most pedigrees with type I but not type III PS deficiency. Hum Mutat 14:3039

38. Borgel D, Duchemin J, Alhenc-Gelas M, et al (1996) Molecular basis for protein S hereditary deficiency: genetic defects observed in 118 patients with type I and type IIa deficiencies: The French Network on Molecular Abnormalities Responsible for Protein C and Protein S Deficiencies. J Lab Clin Med 128:218–227

39. Espinosa-Parrilla Y, Morell M, Borrell M, et al (2000) Optimization of a simple and rapid single-strand conformation analysis for detection of mutations in the PROS1 gene: identification of seven novel mutations and three novel, apparently neutral, variants. Hum Mutat 15:463–473

40. Espinosa-Parrilla Y, Yamazaki T, Sala N, et al (2000) Protein S secretion differences of missense mutants account for phenotypic heterogeneity. Blood 95:173–179

41. Yamazaki T, Sugiura I, Matsushita T, et al (1993) A phenotypically neutral dimorphism of protein S: the substitution of Lys155 by Glu in the second EGF domain predicted by an A to G base exchange in the gene. Thromb Res 70:395–403

42. Hayashi T, Nishioka J, Shigekiyo T, et al (1994) Protein S Tokushima: abnormal molecule with a substitution of Glu for Lys-155 in the second epidermal growth factor-like domain of protein S. Blood 83:683–690

43. Hayashi T, Nishioka J, Suzuki K (1995) Molecular mechanism of the dysfunction of protein S(Tokushima) (Lys155→Glu) for the regulation of the blood coagulation system. Biochim Biophys Acta 1272:159–167

44. Hayashi T, Nishioka J, Suzuki K (1996) Characterization of dysfunctional protein S-Tokushima (K155→E) in relation to the molecular interactions required for the regulation of blood coagulation. Pol J Pharmacol 48:221–223

45. Tsuda T, Yoshimura H, Hamasaki N (2006) Effect of phosphatidylcholine, phosphatidylethanolamine and lysophosphatidylcholine on the protein C/protein S anticoagulation system. Blood Coagul Fibrinolysis 17:453–458

46. Kimura R, Honda S, Kawasaki T, et al (2006) Protein S-K196E mutation as a genetic risk factor for deep vein thrombosis in Japanese patients. Blood 107:1737–1738

47. Miyata T, Kimura R, Kokubo Y, et al (2006) Genetic risk factors for deep vein thrombosis among Japanese: importance of protein S K196E mutation. Int J Hematol 83:217–223

48. Koeleman BP, van Rumpt D, Hamulyak K, et al (1995) Factor V Leiden: an additional risk factor for thrombosis in protein S deficient families? Thromb Haemost 74:580–583

49. Zoller B, Berntsdotter A, Garcia de Frutos P, et al (1995) Resistance to activated protein C as an additional genetic risk factor in hereditary deficiency of protein S. Blood 85:3518–3523

50. Beauchamp NJ, Dykes AC, Parikh N, et al (2004) The prevalence of, and molecular defects underlying, inherited protein S deficiency in the general population. Br J Haematol 125:647–654

51. Dykes AC, Walker ID, McMahon AD, et al (2001) A study of protein S antigen levels in 3788 healthy volunteers: influence of age, sex and hormone use, and estimate for prevalence of deficiency state. Br J Haematol 113:636–641

52. Duchemin J, Gandrille S, Borgel D, et al (1995) The Ser 460 to Pro substitution of the protein S alpha (PROS1) gene is a frequent mutation associated with free protein S (type IIa) deficiency. Blood 86:3436–3443

53. Sakata T, Okamoto A, Mannami T, et al (2004) Prevalence of protein S deficiency in the Japanese general population: the Suita Study. J Thromb Haemost 2:1012–1013.
54. Nomura T, Suehisa E, Kawasaki T, et al (2000) Frequency of protein S deficiency in general Japanese population. Thromb Res 100:367–371
55. Akkawat B, Rojnuckarin P (2005) Protein S deficiency is common in a healthy Thai population. J Med Assoc Thai 88(suppl 4):S249–S254
56. Comp PC, Nixon RR, Cooper MR, et al (1984) Familial protein S deficiency is associated with recurrent thrombosis. J Clin Invest 74:2082–2088
57. Schwarz HP, Fischer M, Hopmeier P, et al (1984) Plasma protein S deficiency in familial thrombotic disease. Blood 64:1297–1300
58. Koster T, Rosendaal FR, Briet E, et al (1995) Protein C deficiency in a controlled series of unselected outpatients: an infrequent but clear risk factor for venous thrombosis (Leiden Thrombophilia Study). Blood 85:2756–2761
59. Faioni EM, Valsecchi C, Palla A, et al (1997) Free protein S deficiency is a risk factor for venous thrombosis. Thromb Haemost 78:1343–1346
60. Makris M, Leach M, Beauchamp NJ, et al (2000) Genetic analysis, phenotypic diagnosis, and risk of venous thrombosis in families with inherited deficiencies of protein S. Blood 95:1935–1941
61. Martinelli I, Mannucci PM, De Stefano V, et al (1998) Different risks of thrombosis in four coagulation defects associated with inherited thrombophilia: a study of 150 families. Blood 92:2353–2358
62. Adachi T (2005) Protein S and congenital protein S deficiency: the most frequent congenital thrombophilia in Japanese. Current Drug Targets 6:585–592
63. Seligsohn U, Lubetsky A (2001) Genetic susceptibility to venous thrombosis. N Engl J Med 344:1222–1231
64. Suehisa E, Nomura T, Kawasaki T, et al (2001) Frequency of natural coagulation inhibitor (antithrombin III, protein C and protein S) deficiencies in Japanese patients with spontaneous deep vein thrombosis. Blood Coagul Fibrinolysis 12:95–99
65. Liu HW, Kwong YL, Bourke C, et al (1994) High incidence of thrombophilia detected in Chinese patients with venous thrombosis. Thromb Haemost 71:416–419
66. Shen MC, Lin JS, Tsay W (1997) High prevalence of antithrombin III, protein C and protein S deficiency, but no factor V Leiden mutation in venous thrombophilic Chinese patients in Taiwan. Thromb Res 87:377–385
67. Shen MC, Lin JS, Tsay W (2000) Protein C and protein S deficiencies are the most important risk factors associated with thrombosis in Chinese venous thrombophilic patients in Taiwan. Thromb Res 99:447–452
68. Tsay W, Shen MC (2004) R147W mutation of PROC gene is common in venous thrombotic patients in Taiwanese Chinese. Am J Hematol 76:8–13
69. Chumpia W, Peerapittayamongkol C, Angchaisuksiri P, et al (2006) Single nucleotide polymorphisms and haplotypes of protein C and protein S genes in the Thai population. Blood Coagul Fibrinolysis 17:13–18
70. Yin T, Takeshita S, Sato Y, et al (2007). A large deletion of the PROS1 gene in a deep vein thrombosis patient with protein S deficiency. Thromb Haemost: 98:783–789

Atherothrombosis and Thrombus Propagation

Atsushi Yamashita and Yujiro Asada

Summary. The rapid closure of coronary arteries as a result of occlusive thrombi is the major cause of acute myocardial infarction, and disrupted (via rupture or erosion) coronary atherosclerotic plaques can trigger a coronary thrombosis. Accumulating evidence indicates that inflammation plays a key role in plaque instability and thrombus formation. Autopsy studies of acute myocardial infarction have revealed that thrombi on disrupted plaques are principally composed of aggregated platelets and a large amount of fibrin with many inflammatory cells. Thrombi from ruptured plaques are significantly richer in fibrin than those on eroded plaques, and more tissue factor and C-reactive protein are expressed in ruptured than eroded plaques. These compounds appear to contribute more to fibrin-rich thrombus formation upon plaque rupture than plaque erosion. On the other hand, plaque disruption does not always result in total thrombotic occlusion because microscopic coronary thrombi are frequent in autopsies of noncardiac death. However, the mechanisms of arterial thrombus propagation in vivo remain unclear. Recent studies have demonstrated that increased vascular wall thrombogenicity together with substantial blood flow alteration is crucial for occlusive thrombus formation. Furthermore, not only plasma von Willebrand factor and vascular wall tissue factor but also intrinsic coagulation factors play important roles in thrombus propagation in vivo.

Key words. Tissue factor · von Willebrand factor · Blood flow · Vascular wall thrombogenicity · Coagulation factors

Introduction

Acute cardiovascular events usually involve thrombus formation at disrupted atherosclerotic plaques, and they are the clinical manifestation of a generalized and progressive vascular disease that is currently called atherothrombosis. This term has emerged in place of atherosclerosis because it describes the pathophysiology of the disease more precisely. Although thrombosis is a major complication of atherosclerosis, it does not always result in complete thrombotic occlusion with subsequent acute symptomatic events [1]. Therefore, thrombus propagation is critical to the onset of clinical events. Thrombus formation is regulated by many factors, such as the thrombogenicity of exposed plaque constituents, local hemorheology, systemic thrombogenicity,

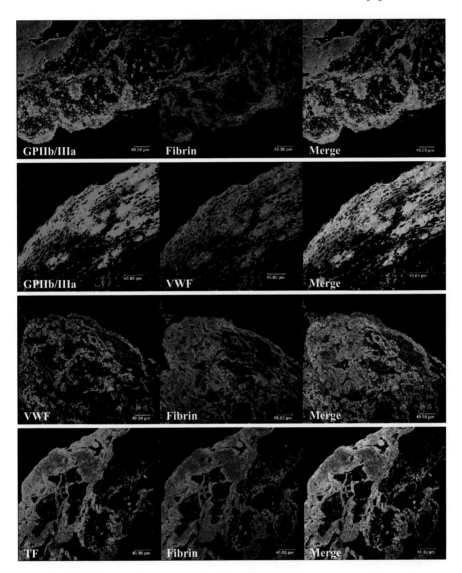

FIG. 1. Immunofluorescence micrographs of fresh coronary thrombi from patients with acute myocardial infarction. *Left* Staining with fluorescein isothiocyanate-labeled glycoprotein (GP) IIb/IIIa, von Willebrand factor (VWF), and tissue factor (TF) (green). *Center* Staining with Cy3-labeled with fibrin and VWF (red). *Right* Merged immunofluorescence images. Co-localized areas of each factor are stained yellow. (From Yamashita et al. [3], with permission. © Elsevier 2006)

and fibrinolytic activity. The mechanisms of thrombus formation have been intensively investigated, whereas those involved in thrombogenesis and thrombus propagation on atherosclerotic lesions remain obscure. This chapter addresses recent advances in the mechanisms of thrombogenesis and thrombus propagation on atherosclerotic lesions, especially in coronary atherothrombosis.

Pathology of Coronary Atherothrombosis

The traditional view of arterial thrombi is that they are mainly composed of aggregated platelets because the coagulation pathway is not effectively activated under conditions of rapid flow. However, recent studies have demonstrated that thrombi on disrupted plaques are composed of aggregated platelets, fibrin, and other blood cells [2–4]. Occlusive thrombi consistently comprise platelets and a large amount of fibrin. Figure 1 shows the immunofluorescence emitted by fresh coronary thrombi obtained from patients with acute myocardial infarction (AMI). The thrombi are composed of platelets [glycoprotein (GP) IIb/IIIa] intermingled with a large amount of fibrin, and the surface of the thrombi is mainly covered with GPIIb/IIIa and von Willebrand factor (VWF), a blood adhesion molecule. Tissue factor (TF), an initiator of the coagulation cascade, is closely associated with fibrin [3], and VWF and/or TF might contribute to thrombus propagation and obstructive thrombus formation on atherosclerotic lesions.

Coronary thrombus develops at the site of plaque rupture or plaque erosion. Plaque *rupture* is defined as fibrous cap disruption, allowing blood contact with a thrombogenic necrotized core, resulting in thrombus formation. Plaques with obvious macrophage and T-lymphocyte infiltration, a large necrotized core, and thin fibrous caps <65 µm thick tend to rupture [5]. On the other hand, *erosion* is defined as disruption or surface injury of plaques with abundant smooth muscle cells (SMCs) and proteoglycan matrix, especially versican and hyaluronan. A necrotized core is often absent. Plaque erosion is associated with relatively few or absent macrophages and T cells compared with plaque rupture [5]. Thrombi that develop during plaque rupture and erosion are consistently composed of platelets and fibrin but in different proportions. Thrombi on ruptured plaques are significantly richer in fibrin, whereas those on eroded plaque are relatively platelet-rich (Fig. 2). Both TF and C-reactive protein are abundant in ruptured plaque compared with eroded plaques [2]. The contents of disrupted plaques might contribute to the nature and mechanisms of thrombus formation after disruption.

Plaque Rupture

Plaque rupture probably involves thinning and disruption of the fibrous cap by metalloproteases together with local rheological forces and emotional status. Accumulating evidence indicates that inflammation plays a key role in the pathogenesis of plaque rupture. Inflammatory cells are quite numerous in rupture-prone atherosclerotic plaques and can produce enzymes that degrade the extracellular matrix of the fibrous cap. Macrophages in human atheroma overexpress interstitial collagenases, gelatinases, and elastolytic enzymes. Activated T lymphocytes and macrophages can secrete

Plaque rupture   Plaque erosion

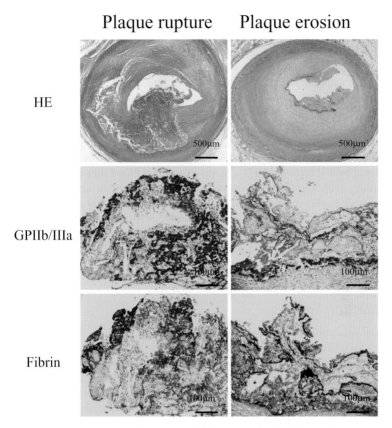

FIG. 2. Coronary plaque rupture and erosion with thrombi. Ruptured plaque has a large necrotic core and disrupted thin fibrous cap accompanied by thrombus formation. Eroded plaque has smooth muscle cell-rich atherosclerotic lesion with thrombus formation. Both thrombi comprise platelets and fibrin. *HE*, H&E stain. (From Sato et al. [2], with permission)

interferon γ (IFNγ), which inhibits collagen synthesis and induces the apoptotic death of SMCs [6]. Moreover, IFNγ can induce interleukin-18 (IL-18), which accelerates inflammation. IL-18 is expressed in macrophages and is up-regulated in plaques from patients with unstable angina. Furthermore, IL-18 expression is more increased in plaques with thrombus than without thrombus [7]. The important antiinflammatory cytokine IL-10 is also up-regulated in macrophages in atherosclerotic lesions from patients with unstable, compared with stable, angina [8]. This evidence indicates that an inflammatory state participates in plaque instability. Intraplaque hemorrhage could promote necrotic core expansion, and iron accumulation due to hemoglobin breakdown can act as a catalyst in free radical formation. Thus, intraplaque hemorrhage might contribute to coronary plaque instability in addition to progressive plaque expansion via the generation of reactive oxygen species (ROS) [9]. However, direct evidence has yet to support a correlation between intraplaque hemorrhage and plaque instability.

Plaque Erosion

The mechanisms of plaque erosion are poorly understood. About 80% of thrombi from eroded plaques are nonocclusive despite sudden coronary death [5]. Platelet-rich emboli are more frequent in patients who die suddenly with plaque erosion compared with those who have ruptured plaques. Because activated platelets release vasoconstrictive agents such as 5-hydroxytriptamine (5-HT, serotonin) and thromboxane A_2 (TXA_2), these molecules released from thrombi might induce distal microvascular constriction, which increases peripheral resistance leading to altered coronary blood flow. Our animal study showed that SMC-rich atherosclerotic vessels are hypervasoconstricted by 5-HT via the 5-HT_{2A} receptor and that blood flow is reduced after vascular injury [10]. Such hypervasoconstriction might play an important role in the pathogenesis of plaque erosion. On the other hand, Durand et al. reported that the induction of endothelial apoptosis in vivo leads to both endothelial denudation and vessel thrombosis in the rabbit femoral artery [11]. However, the endogenous triggers of endothelial apoptosis remain unknown.

Thrombus Propagation

Coagulation Factors

The most fundamental difference between normal and atherosclerotic arteries is the presence or absence of abundant active TF in the intima [12]. The TF-dependent extrinsic coagulation pathway produces fibrin, which is essential for thrombus propagation. Recent studies indicate that the small amount of TF detected in the circulating blood can support clot formation in vitro. Plasma TF levels are elevated in patients with unstable angina and AMI, and these levels correlate with adverse outcomes [13]. However, plasma TF has little procoagulant activity, and the TF-dependent coagulation pathway is rapidly shut down by TF pathway inhibitor [14]. Therefore, whether blood-derived TF contributes thrombus propagation in vivo remains controversial. Hematopoietic cell-derived, TF-positive microparticles (MPs) contribute to laser injury-induced thrombosis in the microvasculature of the mouse cremaster muscle [15]. In contrast, MPs do not contribute to photochemically induced thrombosis in the mouse carotid artery [16]. Using a rabbit model we demonstrated that injury to the neointima induces large fibrin-rich thrombi, whereas injury to a normal artery induces small platelet thrombi (Figure 3a–c,f,g,j); furthermore, we found that TF derived from the vascular wall rather than blood contributes to thrombus propagation on atherosclerotic lesions [17]. Leroyer et al. recently reported that MPs in human atherosclerotic plaques originate mainly from macrophages, erythrocytes, SMCs, and lymphocytes and that they are more abundant and have more procoagulant activity than those in plasma [18]. MPs from both plaques and plasma express TF and generate thrombin, but procoagulant activity is twice as high in MPs isolated from plaques [18]. Their results indicate that plaque-derived MPs play a more important role in thrombus formation and propagation after plaque disruption. Further studies are required to clarify the contribution of blood-derived TF to thrombus propagation on atherosclerotic lesions.

The intrinsic coagulation pathway might promote thrombus propagation in vivo. This pathway is initiated when coagulation factor XII (FXII) contacts negatively

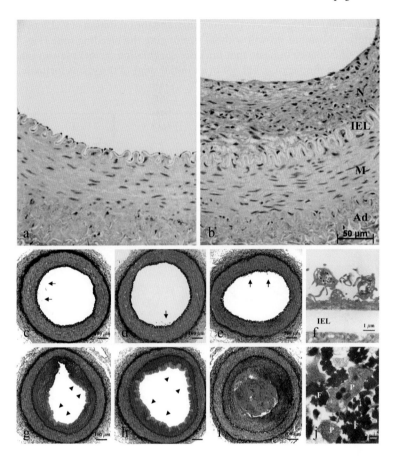

FIG. 3. Tissue factor (TF) expression, and thrombi 30 min after balloon injury in a rabbit femoral artery. Neointima and adventitia are immunoreactive for TF. **a** Normal femoral artery. **b** Injured femoral artery. *N*, neointima; *IEL*, internal elastic lamina; *M*, media; *Ad*, adventitia. Small thrombi (*arrows*) induced by balloon injury without (**c**) or with blood flow reduced to 50% (**d**) or 25% (**e**). Transmission electron micrograph (**f**) shows that thrombi consist of aggregated platelets. Fibrin-rich mural thrombi (*arrowheads*) arose due to balloon injury of the neointima (**g**) and neointimal injury with 50% blood flow (**h**). Moreover, neointimal injury with 25% blood flow caused occlusive thrombus formation (**i**). Transmission electron micrograph (**j**) shows that these thrombi are composed of abundant fibrin strands (*F*) admixed with platelets (*P*). (From Yamashita et al. [17], with permission)

charged surfaces in a reaction involving the plasma proteins, high-molecular-mass kininogen, and plasma kallikrein. In contrast to TF, FXII activation probably appears on the negatively charged surface of activated platelets. Factor XI (FXI) is activated by FXIIa, thrombin, and XIa. Feedback activation of FXI by thrombin promotes further thrombin generation in vitro [19]. Recent studies have revealed that the intrinsic coagulation pathway plays a significant role in thrombus formation in vivo. Initial platelet adhesion and aggregation are normal in mice lacking FXII and FXI, but thrombus propagation is suppressed [20]. Our animal study identified FXI in fibrin-rich thrombus on injured atherosclerotic neointima, and the inhibition of FXI activity obviously reduced thrombus growth without prolonging the bleeding time (Fig. 4) [21]. These lines of evidence indicate that intrinsic coagulation factors play an important role in thrombus propagation but not in an initial stage of thrombus formation.

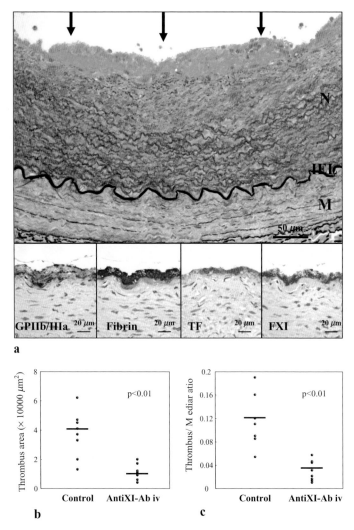

FIG. 4. Light and immunohistochemical microphotographs of a thrombus and the thrombus area 30 min after balloon injury in a rabbit atherosclerotic iliac artery. **a** Injured neointima is covered with fresh thrombi (*arrows*). Thrombi are immunopositive for GPIIb/IIIa, fibrin, TF, and factor XI. Graph shows area of the thrombus (**b**) and the thrombus/media ratio (**c**). Inhibition of plasma FXI by monoclonal antibody reduces thrombus growth. *Symbols* represent average areas and ratios in eight animals. *Horizontal bar* represents the median of the values. (From Yamashita et al. [21], with permission)

Blood Flow

Changes in blood flow play a crucial role in thrombus propagation. Some clinical studies have demonstrated marked changes in blood flow during the process of coronary thrombosis and after interventions. Marzilli et al. reported that coronary blood flow in patients with unstable angina is reduced by about 80% during ischemia [22]. Plaque rupture and coronary interventions can induce distal microvascular embolism and vasoconstriction, which reduce coronary blood flow via the rapid elevation of distal vascular resistance [23, 24]. However, how blood flow affects thrombus propagation on atherosclerotic lesions has not been examined in detail. Our animal study showed that injury with blood flow reduced by >75% promotes the propagation of fibrin-rich mural thrombi, resulting in vessel occlusion. In contrast, small platelet thrombi that developed on normal arteries did not grow even with a 90% reduction in blood flow (Fig. 3c–j, Table 1) [17]. This evidence indicates that increased vascular wall thrombogenicity together with a substantial reduction in blood flow is crucial for the formation of occlusive thrombus. Reduced blood flow at plaque disruption sites might contribute to thrombus propagation and lead to an acute coronary syndrome.

Platelet Adhesion Molecules and Their Receptors

Adhesion molecules and their receptors on platelets play pivotal roles in thrombus formation by supporting platelet tethering, firm adhesion, aggregation, and platelet recruitment to the thrombus surface. The essential roles of GPIbα and GPIIb/IIIa have been confirmed by many studies in gene-targeted mice. The absence of these factors leads to highly impaired thrombus formation, independently of the type of vascular injury [25]. von Willebrand factor is a large, multimeric, plasma protein that undergoes a conformational change that permits its binding to GPIbα. Platelet recruitment on the thrombus surface is primarily mediated by VWF and GPIbα on circulating

TABLE 1. Occlusive rate and time to occlusion after balloon injury of rabbit femoral artery

Artery condition	Femoral artery BF (%)			
	100%	50%	25%	10%
Normal artery ($n = 5$ each)	0	0	0	0
Atherosclerotic artery ($n = 5$ each)	0	0	80*	100**
Time to occlusion (s)			160 ± 18^a	71 ± 17
Atherosclerotic artery + anti-VWF antibody ($n = 5$ each)	ND	ND	0	0

From Yamashita et al. [17], with permission
BF, blood flow; ND, not done
[a]Data are from four occluded vessels
*$P < 0.05$ (vs. normal artery with 25% BF, atherosclerotic artery with 50% BF, atherosclerotic artery + anti-VWF antibody with 25% BF)
**$P < 0.01$ (vs. normal artery with 10% BF, atherosclerotic artery + anti-VWF antibody with 10% BF)

platelets [26, 27]. The interaction between VWF and GPIbα is crucial in both the initial step of platelet adhesion to an injured vessel wall and recruitment to the growing thrombus. Our animal studies demonstrated that inhibiting VWF–GPIbα interaction significantly suppresses fibrin-rich thrombus formation on injured atherosclerotic neointima [28] and prevents occlusive thrombus formation on injured neointima under disturbed flow conditions (Table 1) [17]. These results indicate that VWF plays a significant role in thrombus propagation on atherosclerotic lesions via platelet recruitment.

Ecto-NTPDase and Prostaglandin E_2

Activated platelets release adenosine diphosphate (ADP), 5-HT, and TXA_2, all of which promote further platelet activation. This self-amplifying process leads to thrombus propagation and stabilization. ADP receptors mediate the cyclic calcium signaling required to sustain GPIIb/IIIa activation and thrombus stabilization in vitro [29]. Ecto-nucleoside triphosphate diphosphohydrolase (E-NTPDase) is a cell membrane protein that rapidly hydrolyzes both ADP and adenosine triphosphate to adenosine monophosphate and thereby inhibits platelet aggregation on endothelial cells and SMCs [30]. The expression of E-NTPDase in atherosclerotic lesions is significantly down-regulated in patients with unstable angina compared to that in patients with stable angina [31]. Therefore, the increase of E-NTPDase in plaques might suppress thrombus formation after plaque disruption. Our animal study found that adenovirus-mediated gene transfer of E-NTPDase in arterial walls inhibited occlusive thrombus formation in rat carotid arteries [30]. These results suggest that E-NTPDase activity in atherosclerotic lesions modulates thrombus formation and the onset of vascular events.

Prostaglandin (PG) E_2 is a bioactive lipid derived from arachidonic acid that enhances the aggregation of submaximally stimulated platelets in a process called potentiation. An inflammatory condition in unstable plaque might up-regulate PGE_2 through cyclooxygenases and microsomal PGE synthase-1 in plaques [32]. Gross et al. demonstrated that PGE_2 produced in the vascular wall exacerbates atherothrombosis on carotid plaques of apolipoprotein E knockout mice [33]. Inhibiting PGE_2 release might help prevent the onset of acute coronary events.

References

1. Davies MJ, Thomas A (1984) Thrombosis and acute coronary-artery lesions in sudden cardiac ischemic death. N Engl J Med 310:1137–1140
2. Sato Y, Hatakeyama K, Yamashita A, et al (2005) Proportion of fibrin and platelets differs in thrombi on ruptured and eroded coronary atherosclerotic plaques in humans. Heart 91:526–530
3. Yamashita A, Sumi T, Goto S, et al (2006) Detection of von Willebrand factor and tissue factor in platelets-fibrin rich coronary thrombi in acute myocardial infarction. Am J Cardiol 97:26–28
4. Hoshiba Y, Hatakeyama K, Tanabe T, et al (2006) Co-localization of von Willebrand factor with platelet thrombi, tissue factor and platelets with fibrin, and consistent presence of inflammatory cells in coronary thrombi obtained by an aspiration device from patients with acute myocardial infarction. J Thromb Haemost 4:114–120

5. Virmani R, Kolodgie FD, Burke AP, et al (2000) Lessons from sudden coronary death: a comprehensive morphological classification scheme for atherosclerotic lesions. Arterioscler Thromb Vasc Biol 20:1262–1275

6. Shah PK (2003) Mechanisms of plaque vulnerability and rupture. J Am Coll Cardiol 41:15S–22S

7. Nishihira K, Imamura T, Hatakeyama K, et al. (2007) Expression of interleukin-18 in coronary plaque obtained by atherectomy from stable and unstable angina. Thromb Res 121:275–279

8. Nishihira K, Imamura T, Yamashita A, et al (2006) Increased expression of interleukin-10 in unstable plaque obtained by directional coronary atherectomy. Eur Heart J 27:1685–1689

9. Kolodgie FD, Gold HK, Burke AP, et al (2003) Intraplaque hemorrhage and progression of coronary atheroma. N Engl J Med 349:2316–2325

10. Nishihira K, Yamashita A, Tanaka N, et al (2006) Inhibition of 5-hydroxytryptamine receptor prevents occlusive thrombus formation on neointima of the rabbit femoral artery. J Thromb Haemost 4:247–255

11. Durand E, Scoazec A, Lafont A, et al (2004) In vivo induction of endothelial apoptosis leads to vessel thrombosis and endothelial denudation: a clue to the understanding of the mechanisms of thrombotic plaque erosion. Circulation 109:2503–2506

12. Hatakeyama K, Asada Y, Marutsuka K, et al (1997) Localization and activity of tissue factor in human aortic atherosclerotic lesions. Atherosclerosis 133:213–219

13. Mackman N (2004) Role of tissue factor in hemostasis, thrombosis, and vascular development. Arterioscler Thromb Vasc Biol 24:1015–1022

14. Monroe DM, Hoffman M, Roberts HR (2002) Platelets and thrombin generation. Arterioscler Thromb Vasc Biol 22:1381–1389

15. Chou J, Mackman N, Merrill-Skoloff G, et al (2004) Hematopoietic cell-derived microparticle tissue factor contributes to fibrin formation during thrombus propagation. Blood 104:3190–3197

16. Day SM, Reeve JL, Pedersen B, et al (2005) Macrovascular thrombosis is driven by tissue factor derived primarily from the blood vessel wall. Blood 105192–105198

17. Yamashita A, Furukoji E, Marutsuka K, et al (2004) Increased vascular wall thrombogenicity combined with reduced blood flow promotes occlusive thrombus formation in rabbit femoral artery. Arterioscler Thromb Vasc Biol 24:2420–2424

18. Leroyer AS, Isobe H, Leseche G, et al (2007) Cellular origins and thrombogenic activity of microparticles isolated from human atherosclerotic plaques. J Am Coll Cardiol 49:772–777

19. Gailani D, Broze GJ Jr (1991) Factor XI activation in a revised model of blood coagulation. Science 253:909–912

20. Renne T, Pozgajova M, Gruner S, et al (2005) Defective thrombus formation in mice lacking coagulation factor XII. J Exp Med 202:271–281

21. Yamashita A, Nishihira K, Kitazawa T, et al (2006) Factor XI contributes to thrombus propagation on injured neointima of the rabbit iliac artery. J Thromb Haemost 4:1496–1501

22. Marzilli M, Sambuceti G, Fedele S, et al (2000) Coronary microcirculatory vasoconstriction during ischemia in patients with unstable angina. Am J Coll Cardiol 35:327–334

23. Erbel R, Heusch G (2000) Coronary microembolization. J Am Coll Cardiol 36:22–24

24. Taylor AJ, Bobik A, Berndt MC, et al (2002) Experimental rupture of atherosclerotic lesions increases distal vascular resistance: a limiting factor to the success of infarct angioplasty. Arterioscler Thromb Vasc Biol 22:153–160

25. Denis CV, Wagner DD (2007) Platelet adhesion receptors and their ligands in mouse models of thrombosis. Arterioscler Thromb Vasc Biol 27:728–739

26. Kulkarni S, Dopheide SM, Yap CL, et al (2000) A revised model of platelet aggregation. J Clin Invest 105:783–791

27. Bergmeier W, Piffath CL, Goerge T, et al (2006) The role of platelet adhesion receptor GPIb alpha far exceeds that of its main ligand, von Willebrand factor, in arterial thrombosis. Proc Natl Acad Sci U S A 103:16900–16905
28. Yamashita A, Asada Y, Sugimura H, et al (2003) Contribution of von Willebrand factor to thrombus formation on neointima of rabbit stenotic iliac artery under high blood-flow velocity. Arterioscler Thromb Vasc Biol 23:1105–1110
29. Goto S, Tamura N, Ishida H, et al (2006) Dependence of platelet thrombus stability on sustained glycoprotein IIb/IIIa activation through adenosine 5'-diphosphate receptor stimulation and cyclic calcium signaling. J Am Coll Cardiol 47:155–162
30. Furukoji E, Matsumoto M, Yamashita A, et al (2005) Adenovirus-mediated transfer of human placental ectonucleoside triphosphate diphosphohydrolase to vascular smooth muscle cells suppresses platelet aggregation in vitro and arterial thrombus formation in vivo. Circulation 111:808–815
31. Hatakeyama K, Hao H, Imamura T, et al (2005) Relation of CD39 to plaque instability and thrombus formation in directional atherectomy specimens from patients with stable and unstable angina pectoris. Am J Cardiol 95:632–635
32. Cipollone F, Prontera C, Pini B, et al (2001) Overexpression of functionally coupled cyclooxygenase-2 and prostaglandin E synthase in symptomatic atherosclerotic plaques as a basis of prostaglandin E(2)-dependent plaque instability. Circulation 104: 921–927
33. Gross S, Tilly P, Hentsch D, et al (2007) Vascular wall-produced prostaglandin E_2 exacerbates arterial thrombosis and atherothrombosis through platelet EP3 receptors. J Exp Med 20:4311–4320

Thrombotic Microangiopathy

Yoshihiro Fujimura, Masanori Matsumoto, and Hideo Yagi

Summary. Thrombotic microangiopathy (TMA) is a pathological condition characterized by generalized microvascular occlusion by platelet thrombi, thrombocytopenia, and microangiopathic hemolytic anemia (MAHA). Thrombotic thrombocytopenic
purpura (TTP) and hemolytic-uremic syndrome (HUS) are both life-threatening diseases with TMA lesions. A "pentad" of thrombocytopenia, MAHA, fluctuating neurological signs, renal failure, and fever was classically a hallmark of TTP, whereas a
"triad" of thrombocytopenia, MAHA, and renal insufficiency was thought to be unique
to HUS. Thus, TTP has been characterized as neurotropic and HUS as nephrotropic
owing to the affected organs. However, the two conditions are often indistinguishable
in clinical practice owing to the similarity of symptoms. In 1996, the presence of von
Willebrand factor (VWF)-cleaving protease (VWF-CP) was identified in normal
plasma. This protease specifically cleaves unusually large VWF multimers released
from vascular endothelial cells into smaller multimers. Then, in 2001, VWF-CP was
found to be a new member of the metalloproteinase family and was termed ADAMTS13.
Following this discovery, it has been established that a deficiency of plasma ADAMTS13
activity due to genetic mutation causes a complex thrombohemorrhagic disease
termed Upshaw-Schulman syndrome (USS) and that a secondary deficiency of
ADAMTS13 activity due to autoantibodies causes acquired TTP. In contrast, typical
acquired HUS associated with enterohemorrhagic *Escherichia coli* O157-H7 infection
presents with almost normal ADAMTS13 activity. The TMAs also develop in association with various underlying diseases, such as connective tissue disease, stem cell
transplantation, malignancies, drug treatment, pregnancy, and infections. Recent
studies have indicated that atypical HUS can be induced by genetic mutations in
complement regulatory cofactors, including factor I, factor H, and membrane cofactor
protein or CD46.

Key words. TMA · ADAMTS13 · TTP · HUS · USS

Introduction

Thrombotic microangiopathy (TMA) is a pathological condition marked by vast
microvascular occlusions by platelet thrombi, thrombocytopenia, and microangiopathic hemolytic anemia (MAHA) [1]. Thrombotic thrombocytopenic purpura (TTP)
and hemolytic-uremic syndrome (HUS) are life-threatening diseases with features
of TMA.

First described by Moschcowitz in 1924 [2], TTP was documented a 16-year-old girl who died of multiorgan failure after a 1-week clinical course. At her autopsy, hyaline membrane thrombi were recognized in the small arteries of multiple organs, except for lung. In 1966, Amorosi and Ultmann [3] reported on 16 patients and reviewed 255 previously documented patients, establishing a "pentad" of clinical features: intravascular platelet clumping leading to severe thrombocytopenia, MAHA, fluctuating neurological signs, renal failure, and fever. TTP was initially thought to occur mostly in adults, but recent results indicate that this may not be true.

In 1955, Gasser et al. [4] described five children who died of acute renal insufficiency; their autopsies showed prominent necrosis of the renal cortex. This study established the "triad" of clinical features of HUS: renal insufficiency, thrombocytopenia, and MAHA. Furthermore, after discovery of a close relationship between HUS and enterohemorrhagic *Escherichia coli* infection, in particular by strain O157-H7, which produces a Shiga-like toxin, it is thought that this disease typically affects children [5]. Because of organ preference, TTP has been referred to as neurotropic and HUS as nephrotropic.

In 1996, the presence of von Willebrand factor (VWF)-cleaving protease (VWF-CP, later named ADAMTS13) was identified [6, 7]. It is now accepted that a deficiency of plasma VWF-CP/ADAMTS13 activity due to genetic mutations or autoantibodies causes TTP, but this is not the case for HUS [8, 9]. Unfortunately, it is often difficult to differentiate TTP from HUS by clinical signs alone because of similar symptoms. Furthermore, TMAs also occur secondarily, in association with various underlying diseases, such as connective tissue disease, stem cell transplantation, malignancies, drug treatment, pregnancy, and infections. Finally, recent studies have indicated that atypical HUS can be induced by genetic alterations in complement regulatory factors, such as factor H [10, 11], factor I [12], and membrane cofactor protein (MCP) or CD46 [13, 14].

Since 1998, our institution (Nara Medical University) has been collecting a large data set on 783 patients with TMA as a nationwide referral center (Table 1). In this chapter we review the current progress of TMA research together with an analysis of the data based on the Japanese experience.

TTP Pathogenesis

UL-VWFM

A macromolecular plasma glycoprotein, VWF is synthesized in vascular endothelial cells (ECs) and megakaryocytes [15]. EC-derived VWF bears ABO blood type structures in Asn-linked sugar chains [16], is stored in Weibel-Palade bodies and is constitutively secreted into either subendothelial matrices or the circulation upon stimulation. In contrast, megakaryocyte- and platelet-derived VWF contains no ABO sugar chains, is stored in the α-granules of platelets, and is released upon stimulation into the circulation [15]. The amount of platelet-derived VWF in the circulation is assumed to be less than 2% of the total. EC-VWF plays an essential role in primary hemostasis by anchoring platelets onto injured denuded subendothelial matrices under high shear stress, but the role of platelet-VWF remains to be investigated.

TABLE 1. ADAMTS 13 activity and its inhibitor titers in 783 Japanese patients with TMA (Nara Medical University from July 1997 to December 2006)

	Congenital TMA (n = 56)		Acquired TMA (n = 727)											Total
	USS	Etiology unknown	Idiopathic[a] (n = 338)		Drugs (n = 30)			Connective tissue disease	Malignancy	SCT	Pregnancy	E. coli O157-HUS	Others	
			TTP	HUS	TC/CL	MMC	Others							
	n = 33	n = 23	n = 244	n = 94	n = 21	n = 7	n = 2	n = 187	n = 54	n = 43	n = 13	n = 28	n = 34	n = 783
ADAMTS13 activity (%)														
<3	33	0	158	0	17	0	2	35	4	0	4	0	5	258
3 ≤ 25	0	4	73	19	2	1	0	62	19	18	4	3	14	219
25 ≤ 50	0	8	12	38	1	3	0	55	20	12	2	16	6	173
≥50	0	11	1	37	1	3	0	35	11	13	3	9	9	133
Inhibitor (Bethesda U/ml)	n = 33	n = 23	n = 209	n = 37	n = 21	n = 7	n = 2	n = 84	n = 22	n = 11	n = 6	n = 15	n = 10	n = 480
<0.5	33	23	22	37	2	7	0	34	11	9	1	15	3	197
0.5 ≤ 2.0	0	0	105	0	8	0	2	35	7	2	2	0	5	166
≥2.0	0	0	82	0	11	0	0	15	4	0	3	0	2	117

[a]Clinically diagnosed by physicians in the referring hospitals
TC, ticlopidine; CI, clopidogrel; MMC, mitomycine C; others, peginterferon and sildenafil; SCT, stem cell transplantation

Plasma VWF is composed of heterogeneous multimers with molecular masses ranging from 500 to 20 000 kDa [17]. These multimers are composed of VWF subunits of 2050 amino acid residues, linked together by disulfide bonds in a head-to-head and tail-to-tail fashion [18]. Functionally critical binding domains for factor VIII (FVIII), collagen, platelet glycoprotein (GP) Ib, and GPIIb/IIIa have been identified on VWF subunits [15].

The formation of VWF multimers (VWFMs) occurs in the endoplasmic reticulum. The initial dimer is assembled from a pair of 250-kDa VWF subunits via intersubunit disulfide bonds formed by cysteine residues located in the C-terminal region of each subunit. The subsequent polymerization is achieved by interdimer disulfide bond formation in the N-terminal region of each subunit; this is regulated by thiol-dependent complex formation of thrombospondin-1 properdin domains and VWF-A3 domains [19]. This explains the observation that plasma VWF in patients with Upshaw-Schulman syndrome consists of a series of VWFMs, including unusually large VWFMs (UL-VWFM), despite severe deficiency of ADAMTS13 activity (see below).

VWF-CP/ADAMTS13

Some patients with von Willebrand disease (VWD) have a congenital bleeding disorder caused by *VWF* mutations in which the larger VWFM is absent, presumably due to heightened proteolysis of the abnormal VWF molecule; type 2A VWD is such a case. Using immuno-purified type 2A-VWF from patient plasma, Dent et al. [20] showed that the heightened proteolysis of VWF subunit is caused by specific cleavage of the VWF Tyr1605-Met1606 bond. In 1996, Furlan et al. [6] and Tsai [7] demonstrated that a metalloproteinase termed VWF-CP was responsible for this cleavage. Subsequently, four groups of investigators cloned the VWF-CP cDNA and identified it as a member of the ADAMTS (a disintegrin-like and metalloproteinase with a thrombospondin type I motif) family, termed ADAMTS13 [21–24]. The predicted amino acid sequence of ADAMTS13 consists of 1427 residues, which include a signal peptide, a short propeptide terminating in the sequence RQRR, a reprolysin-like metalloproteinase domain, a disintegerin-like domain, a thrombospondin-1 repeat (Tsp-1), a Cys-rich domain, an ADAMTS spacer, 7 additional Tsp-1 repeats, and two CUB domains. The gene encoding ADAMTS13 is located on chromosome 9q34.

Northern blotting initially suggested that the ADAMTS13 mRNA is uniquely expressed in the liver. Furthermore, immunological studies together with in situ hybridization studies unambiguously indicated that ADAMTS13 is produced in hepatic stellate cells (HSCs), or Itoh cells [25]. Subsequently, platelets [26] and vascular ECs [27] have also been implicated as ADAMTS13-producing cells, although the amount produced by these cell types appears to be far less than that produced by HSCs. The mechanism of release and cell type of origin of circulating ADAMTS13 remain to be elucidated.

TTP

The mechanism underlying TMA in patients with congenital or acquired deficiency of ADAMTS13 activity has not yet been fully elucidated; however, one can be hypothesized based on previous reports, as illustrated in Fig. 1 [28]. UL-VWFM is released

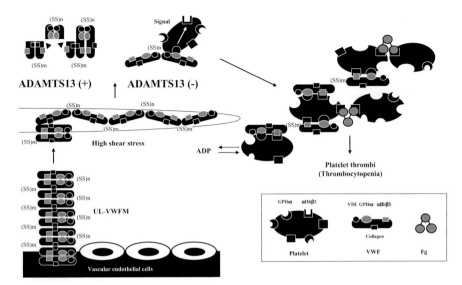

FIG. 1. Proposed mechanism of platelet thrombus formation, resulting in thrombocytopenia and organ damage in patients with thrombotic thrombocytopenic purpura (TTP). Unusually large von Willebrand factor multimers (*UL-VWFM*), held together by intersubunit disulfide bonds (*SS*), are produced in vascular endothelial cells and released into the circulation. They are transported to peripheral small arteries where high shear stress is generated. Under these circumstances, they change their molecular conformation from a globular form to an extended form, to which ADAMTS13 is accessible. However, when this enzyme activity is deficient with or without its inhibitors, UL-VWFMs interact with platelet glycoprotein (*GP*) Ibα, which generates signals to activate the self-platelets, including adenosine diphosphate (*ADP*) release from platelets. A series of these reactions leads to the formation of platelet aggregates with VWF and fibrinogen (*Fg*), resulting in a platelet thrombus

into the circulation by vascular ECs and is transported to peripheral small arteries, where high shear stress is continuously generated. Siedlecki et al. [29] showed that shear stress changes the three-dimensional structure of VWF from a globular inactive state to an extended activated state. The ADAMTS13 protease efficiently cleaves the extended, active form of UL-VWFM. When ADAMTS13 activity is reduced, UL-VWFM interacts more intensively with platelet GPIb and generates signals that result in further acceleration of platelet activation. A series of these reactions leads to the formation of platelet microaggregates and the generation of thrombocytopenia. Furthermore, microangiopathic changes in red blood cells, schistocytes, are thought to take place upon collision with fibrin strands, which are formed surrounding platelet thrombi in TTP [28]. More recently, it has been shown that plasma ADAMTS13 efficiently cleaves newly produced UL-VWFM when it binds to vascular EC surfaces via P-selectin [30]. However, the UL-VWFM preexisting in USS plasma disappears within 1 h after supplementation of ADAMTS13 by infusion of frozen plasma, suggesting that exogenous ADAMTS13 smoothly cleaves its substrate without its binding to vascular EC [28].

ADAMTS13 Assays

ADAMTS13 Activity

Classic analyses of ADAMTS13 activity were performed using purified VWF as a substrate under the following conditions: (1) the presence of divalent cations such as Ca^{2+} or Ba^{2+}; (2) the presence of a protein denaturant such as urea or guanidine HCl at a final concentration of 1.0–1.5 mol/l; (3) a low ionic strength buffer; (4) pH ~8.0; (5) the presence of a serine protease inhibitor such as Pefabloc; and (6) a long incubation period (~24 h) at 37°C followed by EDTA quenching. The products were then analyzed visually by sodium dodecyl sulfate (SDS)-agarose [8] or polyacrylamide [9] gel electrophoresis or functionally by ristocetin-induced platelet aggregation [31] or collagen binding [32]. All of these methods are time-consuming. The SDS-agarose gel method [8] was one of the most frequently used because of its excellent reproducibility and a lower detection limit of 3% of the normal control. Other methods have been reported to have lower detection limits of 5%–10% of the normal control. To determine ADAMTS13 activity more rapidly, Cruz et al. introduced a double-tagged recombinant VWF-A2 domain (residues 1481–1668) (6xHis tag-VWF187-Tag100) substrate [33]. Subsequently, Kokame et al. [34] reported a minimal 73-residue sequence of the VWF-A2 domain, expressed in *Escherichia coli* as a fusion protein tagged with N-terminal glutathione-*S*-transferase and C-terminal histidine tags (GST-VWF73-His) that was completely cleaved by ADAMTS13 within 1 h in the absence of protein denaturant (Fig. 2). Using these double-tagged recombinant VWF-A2 fragments as substrates, Western blot assay, and sandwich enzyme-linked immunosorbent assay (ELISA) were developed. The former remains time-consuming in practice, and the latter has a lower detection limit of approximately 3%–12% of the normal control. It is thought that these ELISAs have low sensitivity because they do not measure the amount of final product generated by the enzymatic digestion but, rather, determine the amount of substrate left undigested. Kokame et al. [35] developed a more convenient FRETS-VWF73 assay in which the D1599 residue of the VWF73 polypeptide is modified with a 2-*N*-methyamino-benzoyl group (NMA) and the N1610 residue with a 2,4-dinitrophenyl group (DNP). If the bond between Y1605 and M1606 is cleaved, the energy transfer that quenches the fluorescence does not occur, allowing fluorescent emission at 440 nm from NMA. This assay is measurable to 3% of the normal control and is now widely used in research laboratories. However, it has recently been shown that the presence of hemoglobin, bilirubin, or chylomicrons in samples significantly interferes with the FRETS-VWF73 assay [36].

Kato et al. [37] have recently produced a mouse monoclonal antibody (mAb) termed N10 that specifically recognizes the Y1605 residue of the VWF-A2 domain generated by ADAMTS13 cleavage. Using this mAb, a highly sensitive sandwich ELISA, termed ADAMTS13-act-ELISA, has been used to determine enzyme activity. The lower limit of this assay was 0.5% of the normal control. Because of its high sensitivity, easy handling, and lack of interference from plasma components, the ADAMTS13-act-ELISA would be a method recommended for routine laboratory use.

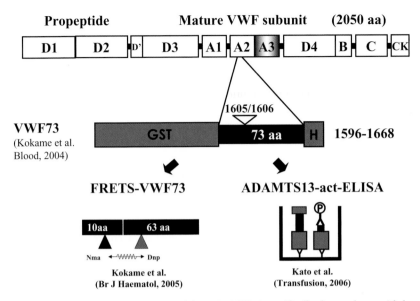

FIG. 2. Assays measuring ADAMTS13 activity. ADAMTS13 specifically cleaves the peptide bond between Tyr1605 and Met1606 of the VWF-A2 domain. Kokame et al. [34] identified a minimal 73-amino-acid residue sequence (VWF73) in the VWF-A2 domain that can be cleaved by ADAMTS13 in the absence of a protein denaturant in vitro. Kokame et al. [35] developed a FRETS-VWF73 assay using a fluorogenic substrate. Kato et al. [37] reported a novel, highly sensitive enzyme-linked immunosorbent assay measuring ADAMTS13 activity using a monoclonal antibody that recognizes a specific epitope on Tyr1605, a residue exposed only by ADAMTS13 cleavage

ADAMTS13 Inhibitors

ADAMTS13-neutralizing autoantibodies (mostly immunoglobulin G-type inhibitors) can be measured using the Bethesda method employed for determining factor VIII inhibitors [38]. Before assay, however, the tested samples must be heat-treated at 56°C for 60 min to eliminate endogenous enzyme activity. Briefly, the assay consists of two steps: First, the tested or control normal plasma is heat-inactivated, mixed with an equal volume of intact normal pooled plasma, and incubated for 2 h. Then, after incubation, the residual enzyme activity is measured. One Bethesda unit is defined as the amount of inhibitor that reduces activity by 50% of the control value. Values of >0.5 U/ml are significant.

ADAMTS13 Antigen

The ADAMTS13 antigen can be measured using polyclonal or monoclonal antibodies [39–41]. In USS patients, plasma ADAMTS13 antigen is deficient or severely decreased, so antigen levels parallel ADAMTS13 activity [39]. In acquired TTP, however, this is not the case. The reasons for this are not fully elucidated, but it is speculated that once immunoglobulin G (IgG)-neutralizing or IgM-nonneutralizing autoantibodies to

ADAMTS13 develop, the resulting immune complexes would be rapidly removed from the circulation by proteolysis or phagocytosis. Using sandwich ELISA with the inhibitory mAb coated on the wells, a concordant decrease of ADAMTS13 activity and antigen has been shown; this might be useful as a diagnostic tool for TTP [42].

Congenital TMA

Congenital TTP—Upshaw-Schulman Syndrome

History

Classic hallmarks of Upshaw-Schulman syndrome (USS) are repeated childhood episodes of chronic thrombocytopenia and hemolytic anemia that are reversed by infusions of fresh frozen plasma. The most striking clinical feature is severe neonatal jaundice with a negative Coombs' test; these symptoms were required for exchange blood transfusion. Today, however, USS is defined as a congenital deficiency of ADAMTS13 activity due to genetic mutations. The lengthy history leading to this conclusion has been described in detail in previous publications [28]. In fact, the term USS had almost been embedded in 1997, when the assay for VWF-CP (ADAMTS13) activity was established. This is because any candidate pathogenesis initially postulated for the disease, such as a defect of platelet-stimulating factor, decreased plasma fibronectin level, or lack of thrombopoietin, has been entirely excluded by subsequent investigations. Instead, the practical diagnostic term chronic relapsing TTP (CR-TTP) has long been used. This term was coined by Moake et al. [43], who found that UL-VWFM was present in the plasma of four CR-TTP patients during the remission phase but disappeared during the acute phase. In 1997, Furlan et al. [44] showed that VWF-CP (ADAMTS13) activity was deficient in four patients with CR-TTP, but they did not address ADAMTS13 inhibitors. Retrospectively, two CR-TTP patients in these studies had congenital TTP, and two had acquired TTP. The term USS was revisited by Kinoshita et al. [45], who reported three Japanese USS patients with severe deficiencies of VWF-CP activity and their asymptomatic parents, who had moderately decreased activity. This study suggested that USS is inherited in an autosomal recessive fashion. Soon after this, solid evidence of a link between congenital TTP or USS and *ADAMTS13* mutations was provided by Levy et al. [23], and approximately 80 mutations have been identified to date [46–48]. Notably, the term familial TTP has sometimes been used as a synonym for USS, but data on siblings of patients with acquired idiopathic TTP caused by ADAMTS13-inhibitors have been reported [49]. Thus, this term appears to be incorrect for USS. (See the home page of Online Mendelian Inheritance in Man (OMIM) Entry 274150 <http://www.genome.ad.jp/dbget-bin/ www_bget?mim: 274150>).

Clinical Manifestations of USS Versus Knockout Mice

Although USS patients consistently lack ADAMTS13 activity, they do not always have acute symptoms; symptoms often become evident only when the patients have infections or become pregnant. In both instances, vascular EC injuries might be involved, and elevated plasma levels of cytokines or soluble thrombomodulin have been indirectly shown in such cases. However, in studies of *ADAMTS13* knockout mice [50, 51],

UL-VWFM appeared in blood, but acute symptoms were not observed. Considering these results together, investigators have assumed that ADAMTS13 deficiency is prothrombotic but by itself is insufficient to provoke acute symptoms; therefore, second hits or triggers must exist. However, the lack of symptoms in the knockout mice are in sharp contrast with the clinical symptoms of USS. For example, acute clinical aggravation within 1 h of infusing 1-desamino-8-D-arginine vasopressin (DDAVP) into a USS patient was reported [52] and was confirmed in a different USS patient (Dr. Mutsuko Konno, Sapparo Kosei Hospital, personal communication). Furthermore, a striking difference between mice and humans was observed during pregnancy. In our studies of 33 USS patients, 9 women had a history of pregnancy and all had thrombocytopenia during the second to third trimester. When this thrombocytopenia was not well managed, they developed clinical signs of TTP, and death of the fetus occurred in many cases.

In the cases of DDAVP and pregnancy, a common feature was a marked increase in plasma VWF levels with the appearance of UL-VWFM. Additionally, the degree of platelet aggregation and/or thrombus size under high shear stress in vitro is dependent on the concentration of large VWFMs. In pregnant women with USS, thrombocytopenia inevitably occurs at the second to third trimester when plasma levels of VWF rapidly increase with an appearance of UL-VWFMs (unpublished observations). Thus, we assume that an important trigger of TTP development is elevated plasma UL-VWFM levels in USS patients. It is noteworthy that severe neonatal jaundice is a typical sign of early-onset bouts of USS; however, recent analysis suggests that such cases represent a relatively small number (13/33, 39%) of patients. In fact, most USS patients have childhood episodes of thrombocytopenia, but they may be overlooked or misdiagnosed as idiopathic thrombocytopenic purpura (ITP) or Evans syndrome.

Incidence and Inheritance

The incidence of USS is unknown, but women with USS are more likely to be diagnosed (on the occasion of pregnancy) than men.

The ADAMTS13 gene is located on chromosome 9q34; therefore, the disease is inherited in an autosomal recessive fashion, indicating that the parents of patients are asymptomatic carriers. Furthermore, a major population of the patients with unrelated parents is a compound heterozygote and in a minor population of the patients with related parents a homozygote. Most recently, a patient with a de novo mutation with a compound heterozygote has been found [53].

Congenital HUS

Because of predominant renal involvement, the patients with congenital HUS are often diagnosed as having atypical HUS (aHUS). A number of genetic mutations in complement regulatory factors (e.g., factor H [10, 11], factor I [12], MCP or CD46 [13, 14]) have been identified in aHUS patients. The description of these mutations is important but is beyond the scope of this chapter, so the reader is referred to information at the home page of University College London (UCL)–FH aHUS Mutation Database C (<http://www.biochem.ucl.ac.uk/~becky/FH/index.php>).

Acquired TMA

Idiopathic TMA

Patients with idiopathic TMA are defined as those who develop TMA without discernible underlying diseases or pathological conditions. A total of 338 patients of this type were included in our survey. Distinction between TTP and HUS based on clinical symptoms and routine laboratory data is usually difficult; thus, the initial differential diagnoses of these patients were made by physicians and included the predominance of neurotropic or nephrotropic clinical signs. However, after measurement of plasma ADAMTS13 activity, patients with severe deficiency were classified as having TTP.

TTP-HUS

From among the 338 idiopathic TTP-HUS patients, 244 were classified as having TTP. Among the 244, 158 (65%) had severe deficiency of ADAMTS13 activity (<3% of the normal control), 73 (30%) had moderately decreased activity (3% to <25%), 12 (5%) had mildly decreased activity (26% to <50%), and 1 patient had normal activity (>50%). ADAMTS13 inhibitors were detected in 187 (89%) of 209 patients tested, and 22 patients had no ADAMTS13 inhibitors (<0.5 Bethesda unit/ml).

On the other hand, 94 of the 338 TTP-HUS patients were classified as having idiopathic HUS. Of these 94 patients, 19 had moderately decreased ADAMTS13 activity, 38 had mildly decreased activity, and 37 had normal activity. None of these patients were positive for ADAMTS13 inhibitors.

Secondary TMA

The diseases in the category "secondary" TMA occur in association with another underlying disease or pathological condition. Thus, the differential diagnosis of TTP or HUS is not practical, and the term TMA is used to categorize these diseases.

Connective Tissue Disease

A total of 187 patients were identified in the connective tissue disease (CTD) category. These patients suffered from systemic lupus erythematosus (SLE, $n = 74$), systemic sclerosis (SSc, $n = 44$), polymyositis/dermatomyositis (PM/DM, $n = 12$), rheumatoid arthritis (RA, $n = 8$), mixed connective tissue disease (MCTD, n = 7), Sjögren's syndrome (SjS, $n = 7$), antiphosphoid syndrome (APS, $n = 6$), anti-neutrophil cytoplasmic antibody-related nephritis (ANCA-ARN, $n = 6$), microscopic polyangiitis (MPA, $n = 5$), overlap syndrome ($n = 3$), anti-glomerular basement membrane antibody-related nephritis (BGM-GRN, $n = 3$), and other CTDs ($n = 12$).

Of these patients, 35 (19%) had severe deficiency of ADAMTS13 activity. The proportion of severely decreased ADAMTS13 activity in each disease was SLE (18/74, 23%), SSc (2/44, 5.2%), PM/DM (2/12, 8.3%), RA (1/8, 17%), MCTD (5/7, 71%), SjS (3/7, 43%), APS (3/6, 50%), and ANCA-ARN (1/6, 17%). None of the remaining 23 patients with other CTDs had severely decreased ADAMTS13 activity.

Malignancy

In all, 54 patients with malignancy were surveyed. Among them, 25 patients suffered from hematological malignancies [non-Hodgkin's lymphoma (NHL, $n = 13$), myelo-dysplastic syndrome ($n = 5$), acute myeloid leukemia ($n = 2$), acute lymphoblastic leukemia ($n = 2$), chronic myeloid leukemia ($n = 1$), plasmacytoma ($n = 1$), or multiple myeloma ($n = 1$)]; 28 had carcinoma [gastric cancer ($n = 9$), pancreatic cancer ($n = 3$), colon cancer ($n = 3$), lung cancer ($n = 2$), ovarian cancer ($n = 2$), or other ($n = 9$)]; and one suffered from Ewing's sarcoma. Among these patients, severely decreased ADAMTS13 activity was found in three with NHL, including two with intravascular lymphoma (IVL) [54] and one with angioimmunoblastic T-cell lymphoma (AILT), and one patient with pancreatic cancer.

Stem Cell Transplantation

A total of 43 patients with stem cell transplantation (SCT) were enrolled in the survey. These patients were classified as undergoing bone marrow transplantation (BMT) ($n = 19$), peripheral blood SCT ($n = 16$), or cord blood SCT ($n = 8$). None of these patients had severe deficiency of ADAMTS13 activity. These results are in good accord with the reports of van der Plas et al. [55].

Pregnancy

Thirteen pregnant patients were surveyed. Among them, six developed TTP-HUS between weeks 10 and 40 of pregnancy, six developed TTP-HUS soon after delivery, and one developed symptoms 3 months after delivery. Four patients had severe deficiency of ADAMTS13 activity. Pregnancy-associated TMA must be carefully differentiated from USS, which initially presents as thrombocytopenia without other accompanying clinical symptoms during the second or third trimester but soon after develops into TTP-HUS, especially if inappropriate treatment is given.

Drugs

Thirty patients with drug treatment-associated TMA were identified. The drugs included ticlopidine (TC, $n = 20$), clopidogrel (CL, $n = 1$), mitomycin C (MMC, $n = 7$), PEG-interferon ($n = 1$), and sildenafil (Viagra®, $n = 1$). Both TC and CL are thieno-pyridine derivatives, and CL has been available in Japan since April 2006. Therefore, most thienopyridine-associated TMAs are associated with TC; 17 of these 20 patients showed severely decreased ADAMTS13 activity, 2 had moderately decreased ADAMTS13 activity, and 1 had normal activity. Altogether, 19 of the patients ($n = 19/20$, 95%) with TC-associated TMAs had ADAMTS13 inhibitors. The single patient with CL-associated TMA had mildly decreased ADAMTS13 activity and did not have ADAMTS13 inhibitors. These results, in part, reflect a recent report by Bennett et al. [56], who showed two mechanistic pathways for thienopyridine-associated TTPs in which TC-associated TTP is more likely to be ADAMTS13-dependent, whereas CL-associated TTP is less ADAMTS13-dependent.

No patients with MMC-associated TMA showed severely decreased ADAMTS13 activity. The patients with PEG-interferon- and sildenafil-associated TMA had severely decreased ADAMTS13 activity owing to the presence of inhibitors.

Escherichia Coli O157-H7

Escherichia coli-associated disease is often referred to as diarrhea-associated HUS. The clinical signs of this disease are usually indistinguishable from those of TTP because the patients sometimes present with encephalopathy induced by Shiga-like toxin (or verotoxin). Of 28 patients in this category, none had severe deficiency of ADAMTS13 activity, and ADAMTS13 inhibitors were not detected in any of the patients. An excellent review by Moake [57] discusses the mechanism of Shiga-toxin-induced TMA.

Other Disorders

Altogether, 34 TMA patients did not fit the aforementioned categories. Six of these patients had chronic liver disease, two with severely decreased ADAMTS13 activity, three with moderately decreased activity, and one with mildly decreased activity. An additional six patients developed TMA after living-related liver transplantation; among them, one had severely decreased ADAMTS13 activity, four had moderately decreased activity, and one had mildly decreased activity. ADAMTS13 is produced mainly in the liver and exclusively by stellate cells. Therefore, various types of liver damage may result in reduction of plasma ADAMTS13 activity and thereby contribute to the development of TMA. Seven additional patients had TMA associated with infectious diseases, five of which cases were related to bacterial and two to viral infections. Of these patients, two (including a patient with human immunodeficiency virus-associated TMA) had severely decreased ADAMTS13 activity, two had moderately decreased activity, and one had mildly decreased activity, and two had normal activity. In the remaining 15 patients, the underlying diseases varied significantly.

Conclusion

Diseases with common pathological features of TMA are often life-threatening. As discussed herein, TMA occurs both spontaneously and in association with a variety of underlying diseases. One-third of TMA patients have deficient ADAMTS13 activity due either to genetic mutations or the development of autoantibodies. For these patients, plasma exchange therapy, with or without accompanying immunosuppressive therapy, has become an efficacious treatment based on evidence-based medicine (EBM) results. However, the remaining two-thirds of TMA patients have normal or only slightly reduced ADAMTS13 activity. For this group of patients, an EBM appropriate treatment has not yet been established.

References

1. Warkentin T, Kelton J (1994) In: Bloom A, Forbes C, Thomas D, et al (eds) Haemostasis and thrombosis, vol 1.Churchill Livingstone, Edinburgh. pp 780–782
2. Moschcowitz E (1924) Hyaline thrombosis of the terminal arterioles and capillaries: a hitherto undescribed disease. Proc N Y Pathol Soc 24:21–24
3. Amorosi EL, Ultmann JE (1966) Thrombotic thrombocytopenic purpura: report of 16 cases and review of the literature. Medicine 45:139–159

4. Gasser C, Gautier E, Steck A, et al (1955) Hemolytic-uremic syndrome: bilateral necrosis of the renal cortex in acute acquired hemolytic anemia. Schweiz Med Wochenschr 85:905–909

5. Karmali MA, Petric M, Lim C, et al (1985) The association between idiopathic hemolytic uremic syndrome and infection by verotoxin-producing *Escherichia coli*. J Infect Dis 151:775–782

6. Furlan M, Robles R, Lämmle B (1996) Partial purification and characterization of a protease from human plasma cleaving von Willebrand factor to fragments produced by in vivo proteolysis. Blood 87:4223–4234

7. Tsai HM (1996) Physiologic cleavage of von Willebrand factor by a plasma protease is dependent on its conformation and requires calcium ion. Blood 87:4235–4244

8. Furlan M, Robles R, Galbusera M, et al (1998) von Willebrand factor-cleaving protease in thrombotic thrombocytopenic purpura and the hemolytic-uremic syndrome. N Engl J Med 339:1578–1584

9. Tsai HM, Lian EC (1998) Antibodies to von Willebrand factor-cleaving protease in aute thrombotic thrombocytopenic purpura. N Engl J Med 339:1585–1594

10. Rougier N, Kazatchkine MD, Rougier JP, et al (1998) Human complement factor H deficiency associated with hemolytic uremic syndrome. J Am Soc Nephrol 9: 2318–2326

11. Warwicker P, Goodship, TH, Donne RL, et al (1998) Genetic studies into inherited and sporadic hemolytic uremic syndrome. Kidney Int 53:836–844.

12. Fremeaux-Bacchi V, Dragon-Durey MA, Blouin J, et al (2004) Complement factor I: a susceptibility gene for atypical haemolytic uraemic syndrome. J Med Genet 41:e84

13. Noris M, Brioschi S, Caprioli J, et al (2003) Familial haemolytic uraemic syndrome and an MCP mutation. Lancet 362:1542–1547

14. Richards A, Kemp EJ, Liszewski MK, et al (2003) Mutations in human complement regulator, membrane cofactor protein (CD46), predispose to development of familial hemolytic uremic syndrome. Proc Natl Acad Sci U S A 100:12966–12971

15. Fujimura Y, Titani K (1994) In: Bloom A, Forbes C, Thomas D, et al (eds) Haemostasis and thrombosis, vol. 1. Churchill Livingstone, Edinburgh, pp 379–395

16. Matsui T, Shimoyama T, Matsumoto M, et al (1999) ABO blood group antigens on human plasma von Willebrand factor after ABO-mismatched bone marrow transplantation. Blood 94:2895–2900

17. Ruggeri ZM, Zimmerman TS (1981) The complex multimeric composition of factor VIII/von Willebrand factor. Blood 57:1140–1143

18. Titani K, Kumar S, Takio K, et al (1986) Amino acid sequence of human von Willebrand factor. Biochemistry 25:3171–3184

19. Xie L, Chesterman CN, Hogg PJ (2001) Control of von Willebrand factor multimer size by thrombospondine-1. J Exp Med 12:1341–1349

20. Dent JA, Berkowitz SD, Ware J, et al (1990) Identification of a cleavage site directing the immunochemical detection of molecular abnormalities in type IIA von Willebrand factor. Proc Natl Acad Sci U S A 87:6306–6310

21. Fujikawa K, Suzuki H, McMullen B, et al (2001) Purification of human von Willebrand factor-cleaving protease and its identification as a new member of the metalloproteinase family. Blood 98:1662–1666

22. Gerritsen HE, Robles R, Lämmle B, et al (2001) Partial amino acid sequence of purified von Willebrand factor-cleaving protease. Blood 98:1654–1661

23. Levy GG, Nichols, WC, Lian EC, et al (2001) Mutations in a member of the ADAMTS gene family cause thrombotic thrombocytopenic purpura. Nature 413:488–494

24. Soejima K, Mimura N, Hirashima M, et al (2001) A novel human metalloprotease synthesized in the liver and secreted into the blood: possibly, the von Willebrand factor-cleaving protease? J Biochem (Tokyo) 130:475–480

25. Uemura M, Tatsumi K, Matsumoto M, et al (2005) Localization of ADAMTS13 to the stellate cells of human liver. Blood 106:922–924

26. Suzuki M, Murata M, Matsubara Y, et al (2004) Detection of von Willebrand factor-cleaving protease (ADAMTS-13) in human platelets. Biochem Biophys Res Commun 313:212–216

27. Turner N, Nolasco L, Tao Z, et al (2006) Human endothelial cells synthesize and release ADAMTS-13. J Thromb Haemost 4:1396–1404.

28. Fujimura Y, Matsumoto M, Yagi H, et al (2002) Von Willebrand factor-cleaving protease and Upshaw-Schulman syndrome. Int J Hematol 75:25–34

29. Siedlecki CA, Lestini, BJ, Kottke-Marchant KK, et al (1996) Shear-dependent changes in the three-dimensional structure of human von Willebrand factor. Blood 88:2939–2950

30. Padilla A, Moake JL, Bernardo A, et al (2004) P-selectin anchors newly released ultra-large von Willebrand factor multimers to the endothelial cell surface. Blood 103:2150–2156

31. Bohm M, Vigh T, Scharrer I (2002) Evaluation and clinical application of a new method for measuring activity of von Willebrand factor-cleaving metalloprotease (ADAMTS13). Ann Hematol 81:430–435

32. Gerritsen HE, Turecek PL, Schwarz HP, et al (1999) Assay of von Willebrand factor (vWF)-cleaving protease based on decreased collagen binding affinity of degraded vWF: a tool for the diagnosis of thrombotic thrombocytopenic purpura (TTP). Thromb Haemost 82:1386–1389

33. Cruz MA, Whitelock J, Dong JF (2003) Evaluation of ADAMTS-13 activity in plasma using recombinant von Willebrand Factor A2 domain polypeptide as substrate. Thromb Haemost 90:1204–1209

34. Kokame K, Matsumoto M, Fujimura Y, et al (2004) VWF73, a region from D1596 to R1668 of von Willebrand factor, provides a minimal substrate for ADAMTS-13. Blood 103:607–612

35. Kokame K, Nobe Y, Kokubo Y, et al (2005) FRETS-VWF73, a first fluorogenic substrate for ADAMTS13 assay. Br J Haematol 129:93–100

36. Meyer SC, Sulzer I, Lämmle B, et al (2007) Hyperbilirubinemia interferes with ADAMTS-13 activity measurement by FRETS-VWF73 assay: diagnostic relevance in patients suffering from acute thrombotic microangiopathies. J Thromb Haemost 5:866–867

37. Kato S, Matsumoto M, Matsuyama T, et al (2006) Novel monoclonal antibody-based enzyme immunoassay for determining plasma levels of ADAMTS13 activity. Transfusion 46:1444–1452

38. Kasper CK, Aledort L, Aronson D, et al (1975) Proceedings: a more uniform measurement of factor VIII inhibitors. Thromb Diath Haemorrh 34:612

39. Ishizashi H, Yagi H, Matsumoto M, et al (2007) Quantitative Western blot analysis of plasma ADAMTS13 antigen in patients with Upshaw-Schulman syndrome. Thromb Res. 120:381–386

40. Rieger M, Ferrari S, Kremer Hovinga JA, et al (2006) Relation between ADAMTS13 activity and ADAMTS13 antigen levels in healthy donors and patients with thrombotic microangiopathies (TMA). Thromb Haemost 95:212–220

41. Soejima K, Nakamura H, Hirashima M, et al (2006) Analysis on the molecular species and concentration of circulating ADAMTS13 in Blood. J Biochem (Tokyo) 139:147–154

42. Yagi H, Ito S, Kato S, et al (2007) Plasma levels of ADAMTS13 antigen determined with an enzyme immunoassay using a neutralizing monoclonal antibody parallel to ADAMTS13 activity levels. Int J Hematol. 85:403–407

43. Moake JL, Rudy CK, Troll JH, et al (1982) Unusually large plasma factor VIII: von Willebrand factor multimers in chronic relapsing thrombotic thrombocytopenic purpura. N Engl J Med 307:1432–1435

44. Furlan M, Robles R, Solenthaler M, et al (1997) Deficient activity of von Willebrand factor-cleaving protease in chronic relapsing thrombotic thrombocytopenic purpura. Blood 89:3097–3103

45. Kinoshita S, Yoshioka A, Park YD, et al (2001) Upshaw-Schulman syndrome revisited: a concept of congenital thrombotic thrombocytopenic purpura. Int J Hematol 74:101–108

46. Kokame K, Matsumoto M, Soejima K., et al (2002) Mutations and common polymorphisms in ADAMTS13 gene responsible for von Willebrand factor-cleaving protease activity. Proc Natl Acad Sci U S A 99:11902–11907

47. Schneppenheim R, Budde U, Oyen F, et al (2003) von Willebrand factor cleaving protease and ADAMTS13 mutations in childhood TTP. Blood 101:1845–1850

48. Matsumoto M, Kokame K, Soejima K, et al (2004) Molecular characterization of ADAMTS13 gene mutations in Japanese patients with Upshaw-Schulman syndrome. Blood 103:1305–1310

49. Studt JD, Kremer-Hovinga JA, Radonic R, et al (2004) Familial acquired thrombotic thrombocytopenic purpura: ADAMTS13 inhibitory autoantibodies in identical twins. Blood 103:4195–4197

50. Motto DG, Chauhan AK, Zhu G, et al (2005) Shigatoxin triggers thrombotic thrombocytopenic purpura in genetically susceptible ADAMTS13-deficient mice. J Clin Invest 115:2752–2761

51. Banno F, Kokame K, Okuda T, et al (2006) Complete deficiency in ADAMTS13 is prothrombotic, but it alone is not sufficient to cause thrombotic thrombocytopenic purpura. Blood 107:3161–3166

52. Hara T, Kitano A, Kajiwara T, et al (1986) Factor VIII concentrate-responsive thrombocytopenia, hemolytic anemia, and nephropathy: evidence that factor VIII:von Willebrand factor is involved in its pathogenesis. Am J Pediatr Hematol Oncol 8:324–328

53. Kokame K, Aoyama Y, Matsumoto M, et al (2008) Inherited and de novo mutations of ADAMTS13 in a patient with Upahaw-Schulman syndrome. J Thromb Haemost 6:213–215

54. Kawahara M, Kanno M, Matsumoto M, et al (2004) Diffuse neurodeficits in intravascular lymphomatosis with ADAMTS13 inhibitor. Neurology 63:1731–1733

55. Van der Plas RM, Schiphorst ME, Huizinga EG, et al (1999) von Willebrand factor proteolysis is deficient in classic, but not in bone marrow transplantation-associated, thrombotic thrombocytopenic purpura. Blood 93:3798–3802

56. Bennett CL, Kim B, Zakarija A, et al (2007) Two mechanistic pathways for thienopyridine-associated thrombotic thrombocytopenic purpura: a report from the Surveillance, Epidemiology, and Risk Factors for Thrombotic Thrombocytopenic Purpura (SERF-TTP) Research Group and the Research on Adverse Drug Events and Reports (RADAR) Project. J Am Coll Cardiol (in press)

57. Moake JL (2002) Thrombotic microangiopathies. N Engl J Med 347:589–600

Keyword Index